Environmental Science

Science

Systems and Solutions

Third Edition

Michael L. McKinney
University of Tennessee, Knoxville

Robert M. Schoch
Boston University

JONES AND BARTLETT PUBLISHERS

Sudbury, Massachusetts

BOSTON   TORONTO   LONDON   SINGAPORE

Dedication
To Patricia Anne Harshbarger McKinney,
A great mother and friend

World Headquarters
Jones and Bartlett Publishers
40 Tall Pine Drive
Sudbury, MA 01776
978-443-5000
info@jbpub.com
www.jbpub.com

Jones and Bartlett Publishers Canada
2406 Nikanna Road
Mississauga, ON L5C 2W6
CANADA

Jones and Bartlett Publishers
International
Barb House, Barb Mews
London W6 7PA
UK

Production Credits
Chief Executive Officer: Clayton Jones
Chief Operating Officer: Don W. Jones, Jr.
Executive V.P. & Publisher: Robert W. Holland, Jr.
V.P., Design and Production: Anne Spencer
V.P., Manufacturing and Inventory Control: Therese Bräuer
Editor-in-Chief, College: J. Michael Stranz
Director, Sales and Marketing: William Kane
Executive Editor: Stephen L. Weaver
Associate Managing Editor, College Editorial: Dean W. DeChambeau
Senior Production Editor: Louis C. Bruno, Jr.
Senior Marketing Manager: Nathan Schultz
Cover and Text Design: Anne Spencer
Composition: Graphic World, Inc.
Printing and Binding: Courier Kendallville
Cover Printing: Lehigh Press

Library of Congress Cataloging-in-Publication Data
McKinney, Michael L.
   Environmental science: systems and solutions / Michael L. McKinney, Robert M. Schoch.— 3rd ed.
     p. cm.
   Includes index.
   ISBN 0-7637-0918-2 (alk. paper)
    1. Environmental sciences. 2. Pollution—Environmental aspects. 3. Environmentalism.
   I. Schoch, Robert M. II. Title.

GE105 . M39 2003
363.7—dc21   2002040627

Commitment to the Environment
As a book publisher, Jones and Bartlett is committed to reducing its impact on the environment. This and other Jones and Bartlett titles are printed using recycled post-consumer paper. We purchase our paper from manufacturers committed to sustainable, environmentally sensitive processes. We employ the Internet and office computer networks technology in our effort toward sustainable solutions. New communication technology provides opportunities to reduce our use of paper and other resources through on-line delivery of instructors' materials and educational information—and of course, we recycle in the office.

About the Cover
The John Heinz National Wildlife Refuge at Tinicum, with the Philadelphia skyline in the background, protects the last acreage of freshwater tidal marsh in Pennsylvania. Photo credit: John and Karen Hollingsworth/U.S. Fish and Wildlife Service.

Printed in the United States of America
07 06 05 04 03   10 9 8 7 6 5 4 3 2 1

MICHAEL L. McKINNEY is Director of the Environmental Studies Program at the University of Tennessee, Knoxville. He is also a Professor in the Geological Science Department and the Ecology & Evolutionary Biology Department. Since 1985, he has taught a variety of courses, focusing on environmental science and biodiversity issues at the undergraduate level.

Dr. McKinney has two Masters degrees, one from the University of Colorado at Boulder and one from the University of Florida. He received his Ph.D. from Yale University in 1985. Since that time he has published several books and dozens of technical articles. Most of his recent research has focused on conservation biology. Dr. McKinney has received several teaching awards and a prestigious University award for creative research. He is currently working on a book documenting the harmful impact of urban sprawl on native species.

In addition to his scholarly work, Dr. McKinney is very active in promoting environmental solutions where he lives, the Southern Appalachian bioregion. He is on the Board of Directors of the Foothills Land Conservancy, which is the major private land trust that creates wilderness preserves around the Smoky Mountain National Park. In 2001, Dr. McKinney received the Environmental Achievement award from the city's main newspaper, the Knoxville *News-Sentinel,* given to the individual who has done the most to promote a better environment. Dr. McKinney is also an active member of the Tennessee Citizens for Wilderness Planning, the East Tennessee Sierra Club (Harvey Broome Chapter), the Southern Alliance for Clean Energy, the Tennessee Clean Water Network, and Ijams Nature Center. He writes a bimonthly column called the "Suburban Ecologist" in the *Hellbender,* the environmental newspaper of East Tennessee.

Dr. McKinney lives in Knoxville, Tennessee where he greatly enjoys hiking and promoting sustainable living with his wife, Teressa, who provides him with invaluable help, encouragement, and understanding.

ROBERT M. SCHOCH is on the faculty of the College of General Studies, Boston University, where he has specialized since 1984 in teaching undergraduate science, including environmental science, biology (for many years he coordinated the college's biology program), physical science, geology, and geography courses. Dr. Schoch always includes a strong environmental component in all of the courses he teaches. He is a recipient of his college's Peyton Richter Award for interdisciplinary teaching.

Dr. Schoch received his Ph.D. in geology and geophysics from Yale University in 1983 and is the author of technical and popular books (including *Phylogeny Reconstruction in Paleontology, Stratigraphy: Principles and Methods,* and *Voices of the Rocks*) as well as numerous articles on various aspects of paleontology and geology. In honor of his paleontological contributions, the fossil mammal genus *Schochia* was named after him in 1993. In some circles, however, Dr. Schoch is better known for his work in Egypt. Correlating ancient environmental changes with surface and underground weathering features, he concluded that the base of the Great Sphinx of Giza was initially carved much earlier than had been generally believed. This research, corroborated by independent studies undertaken by other geologists, has been the subject of many articles in the professional and popular archaeological and Egyptological literature and has also been featured in several television documentaries aired around the world. Understanding past environmental changes is important as we face future challenges.

Besides his academic and scholarly studies, Dr. Schoch is an active environmental advocate who stresses a pragmatic, hands-on approach. In this connection, he helped to found, and serves on the board of directors of, a local community land trust devoted to protecting land from harmful development. Likewise, Dr. Schoch takes an active part in "green" politics (thus far he has served eight years as an elected member of the city council of Attleboro, Massachusetts, where he lives with his wife, Cynthia, and their two sons, Nicholas and Edward).

Brief Contents

Earth at night.

Contents

CONTENTS

SECTION
3

Resource Use and Management 140

SECTION 4 Dealing With
Environmental
Degradation   **334**

CONTENTS

Preface

*The future which we hold in trust for our own
children will be shaped by our fairness to other
people's children.*
> —Marian Wright Edelman

*Nothing is more honorable to any large mass of
people assembled for the purpose of a fair discussion, than that kind and respectful attention that
is yielded not only to your political friends, but to
those who are opposed to you in politics.*
> —Stephen Douglas,
> from the Lincoln-Douglas debates

The critical importance of environmental science and environmental studies cannot be disputed as virtually everyone is aware of the issues—be they global warming, the
depletion of the ozone layer, the controversy over nuclear power, or the continuing problems of water pollution and solid waste disposal. *Environmental Science:
Systems and Solutions, Third Edition,* offers the basic
principles necessary to understand and address current environmental concerns.

We wrote this book to provide a comprehensive
overview and synthesis of environmental science.
Environmental Science: Systems and Solutions provides
the reader with the basic factual data necessary to understand current environmental issues. But to know the
raw facts is not enough. A well-informed person must understand how various aspects of the natural environment
interconnect with each other and with human society.
We thus use a systems approach as a means of organizing complex information in a way that highlights connections for the reader. The systems approach allows the
reader to take in the information without feeling overwhelmed, as often happens when large amounts of information are presented in a disorganized fashion. With
a subject as diverse as environmental science, it is easy to
get lost in the details. We have always kept the "big picture" in mind.

All too often environmental discussions become
bogged down in partisan rhetoric or "gloom and doom"
tactics. Our intention is not to preach but to inform.
Accordingly, in approaching what is often an extremely
controversial subject, we have adopted an objective and
practical perspective that tries to highlight what is going
right in dealing with modern environmental problems.
Furthermore, we have consciously aimed at being both
fair and balanced (presenting differing opinions and information) in our approach to many controversial issues.

A key concept among modern environmentalists is sustainability. In this book we have adopted the sustainability paradigm: We focus on sustainable technologies
and economic systems and the ways that sustainable development can be implemented around the world. Our emphasis is on specific examples that can give concrete
meaning to the concept: Sustainable technological and social solutions to environmental problems are discussed
throughout the book. We hope to inspire the reader to
move beyond simple awareness of current environmental problems to become an active promoter of sustainable
solutions to those problems.

Organization and What's New in this Edition

While building on the framework of the two previous
editions, this edition has been entirely revised, updated,
and to a certain extent reorganized (although not radically departing from the basic organization of the last
edition). In particular, while shortening and combining
some chapters (for example, fossil fuels and nuclear energy are now both covered in Chapter 7, whereas hydroelectric power is now included with alternative energy
sources in Chapter 8), we have expanded our coverage
of other areas. For instance, we have greatly expanded our
discussions of terrestrial biomes, genetically modified
foods, and environmental hormones, and we have added
coverage of such topics as El Niño, light pollution, and
technowaste (also known as e-trash). Perhaps most significantly, this edition includes a new chapter devoted to
land resources and management (Chapter 12). Still, despite this expanded coverage in some areas, with judicious
rewriting and editing, we have decreased the length of this
edition by over 100 pages. Furthermore, for ease of use,
the book is now divided into five major sections rather
than four. We believe that all of these changes will make
the book both more timely and more accessible to the
reader.

PREFACE

The five sections of the book are:

Section 1, The Environment and Humans (Chapters 1 & 2), introduces the systems approach and gives an overview of environmental science in Chapter 1, while Chapter 2 focuses on the increasing impact that the growing human population has had on all natural systems.

Section 2, The Environment of Life on Planet Earth (Chapters 3 through 5), describes how natural systems work, including both biological systems and physical systems. Here we introduce such concepts as populations, communities, ecosystems, the distribution of life on Earth, biogeochemical cycles, weather patterns and climatic zones, the rock cycle and plate tectonics, and natural hazards.

Section 3, Resource Use and Management (Chapters 6 through 13), deals with issues surrounding the use of natural resources by human society. Chapter 6 introduces the broad principles of resource management. The following chapters address energy use, water use, mineral use, and the use of biological resources (including agriculture and soil resources). A major theme is that humans have been rapidly depleting many of these resources and that we must begin using them in a sustainable manner if we are to survive and flourish in the future.

Section 4, Dealing with Environmental Degradation (Chapters 14 through 18), concentrates on various forms of pollution and waste—the results of dumping large amounts of the by-products of human society into the environment. Chapter 14 introduces the principles of pollution control, toxicology, and risk, while subsequent chapters deal with such subjects as water pollution, air pollution, the destruction of the ozone layer, global climate change, municipal solid waste, and hazardous waste. Every chapter includes discussions of how we can limit or mitigate the effects of excessive pollution, especially by limiting the production of pollutants in the first place, as well as by increased efficiency, reuse, recycling, and substitutions.

Section 5, Environmental Issues: Social Aspects and Solutions (Chapters 19 & 20), includes explicit discussions of economic, social, historical, and legal aspects of environmental issues. A major emphasis of the book is on solutions to current environmental concerns. Woven throughout the text are discussions and examples of environmentally friendly technological, legal, and economic solutions. We firmly believe that sustainable and realistic solutions must be implemented and that the root causes of the environmental problems we now face must be addressed. Such problems cannot be solved using science and technology alone; the human aspect must also be taken into account.

Using This Book for a Course in Environmental Science or Environmental Studies

This book was designed to be accessible to introductory non-major students, but it has enough depth and breadth to be used in a majors course. It can be adapted to either an environmental science course or an environmental studies course, and it can be utilized for either one or two semesters. Also, the book was designed so that the chapters need not necessarily be used in the order in which they appear. In particular, depending on the nature and emphasis of a specific course, an instructor may choose to use the chapters of Section 5 (Social Aspects and Solutions) at either the beginning or end of the course, or these or other chapters may be omitted entirely.

Assuming a standard 15 full weeks for a semester (usually about a week is lost due to holidays, exams, and the like), the chapters of this text might be assigned according to one of the following schedules:

For a comprehensive environmental science and environmental studies course:

Week 1: Chapters 1 & 2, An Overview of Environmental Science and Human Population Growth

Week 2: Chapter 3, An Overview of the Biosphere (including Populations, Communities, and Ecosystems) and Biogeochemical Cycles

Week 3: Chapters 4 & 5, The Distribution of Life on Earth and the Physical Environment of Earth

Week 4: Chapter 6, Resource Management

Week 5: Chapter 7, Fossil Fuels and Nuclear Energy

Week 6: Chapter 8, Renewable (including Hydropower) and Alternative Energy Sources

Week 7: Chapters 9 & 10, Water and Mineral Resources

Week 8: Chapter 11, Conserving Biological Resources

Week 9: Chapter 12, Land Resources

Week 10: Chapter 13, Food and Agriculture

Week 11: Chapters 14 & 15, Principles of Pollution Control and Water Pollution

Week 12: Chapter 16, Local and Regional Air Pollution

Week 13: Chapter 17, Destruction of the Ozone Layer and Global Climate Change

Week 14: Chapter 18, Municipal Solid Waste and Hazardous Waste

Week 15: Chapters 19 & 20, Economic, Historical, Social, and Legal Aspects of Current Environmental Concerns

For a basic environmental science course:

Week 1: Chapters 1 & 2, An Overview of Environmental Science and Human Population Growth

Week 2: Chapter 3, An Overview of the Biosphere (including Populations, Communities, and Ecosystems) and Biogeochemical Cycles

Week 3: Chapter 4, The Distribution of Life on Earth

Week 4: Chapter 5, The Workings of Planet Earth and Natural Hazards

Week 5: Chapter 6, Resource Management

Week 6: Chapter 7, Fossil Fuels and Nuclear Energy

Week 7: Chapter 8, Renewable (including Hydropower) and Alternative Energy Sources

Week 8: Chapters 9 & 10, Water and Mineral Resources

Week 9: Chapter 11, Conserving Biological Resources

Week 10: Chapter 12, Land Resources

Week 11: Chapter 13, Food and Agriculture

Week 12: Chapters 14 & 15, Principles of Pollution Control and Water Pollution

Week 13: Chapter 16, Local and Regional Air Pollution

Week 14: Chapter 17, Destruction of the Ozone Layer and Global Climate Change

Week 15: Chapter 18, Municipal Solid Waste and Hazardous Waste

For a general environmental studies course (emphasizing social and historical aspects):

Week 1: Chapter 1, An Overview of Environmental Science

Week 2: Chapter 17, Destruction of the Ozone Layer and Global Climate Change—Examples of the impacts humans are having on the environment

Week 3: Chapter 20, Historical, Cultural, and Legal Aspects of Current Environmental Concerns

Week 4: Chapter 2, Human Population Growth

Week 5: Chapter 6, Resource Management

Week 6: Chapter 7, Fossil Fuels and Nuclear Energy

Week 7: Chapter 8, Renewable (including Hydropower) and Alternative Energy Sources

Week 8: Chapters 9 & 10, Water and Mineral Resources

Week 9: Chapters 9 & 10, Water and Mineral Resources

Week 10: Chapters 11 & 12, Conserving Biological Resources and Land Resource Management

Week 11: Chapter 13, Food and Agriculture

Week 12: Chapters 14 & 15, Principles of Pollution Control and Water Pollution

Week 13: Chapter 16, Local and Regional Air Pollution

Week 14: Chapter 18, Municipal Solid Waste and Hazardous Waste

Week 15: Chapter 20, Environmental Economics

If this book is used for a two-semester course, some of the chapters should be used over a longer period than one week. In particular, we recommend that the following chapters be split as indicated and extended over two weeks:

Chapter 3, Populations, Communities, and Ecosystems/Biogeochemical Cycles

Chapter 4, Evolution and Biodiversity/Biomes on Earth

Chapter 5, General Physical Environment/Natural Hazards

Chapter 7, Fundamentals of Energy & Fossil Fuels/Nuclear Energy

Chapter 8, Renewable Energy/Energy Conservation

Chapter 13, Food/Soil Resources

Chapter 14, Pollution Control/Toxicology

Chapter 17, Destruction of the Ozone Layer/Global Climate Change

Chapter 18, Municipal Solid Waste/Hazardous Waste

Chapter 20, Historical and Social Perspectives/Environmental Law and Decision Making

If these chapters are used as suggested, then chapter or subchapter readings from the text will easily fit into a two-semester schedule (approximately 30 full weeks).

Pedagogical Features

Each chapter uses the same basic organizational format. Following an opening photograph and learning objectives, the chapter begins with an introduction that offers an overview of the subject matter of the chapter and places it in context.

The text is written so as to be interesting and accessible to the average reader and is profusely illustrated with color diagrams demonstrating basic concepts and key ideas. Additionally, numerous charts, tables, drawings, and photographs enrich the basic text. Throughout the text key terms denoting important concepts are in boldface type. Each chapter contains a number of Case Studies, many with thought-provoking questions to promote critical thinking.

A Study Guide at the end of each chapter includes a bulleted summary, a list of the chapter's key terms, and several kinds of questions. Answers to the odd-numbered questions are on the text's Web site at www.environment.jbpub.com.

- **Study Questions.** These review questions generally test objective knowledge and require fairly short answers. Some require more analytical and critical thinking skills.

- **What's the Evidence?** New to this edition, these innovative questions ask the reader to review the authors' arguments and decide for her or himself whether the authors have successfully supported their conclusions. The questions then challenge the reader to support their own position if they disagree.

- **Calculations.** Calculation questions, written at a pre-calculus level or lower, are provided for courses that have a quantitative component.

- **Illustration and Table Review.** These questions are designed to help readers strengthen their data interpretation skills.

This book includes several special features. Inside the front cover is a map of North America showing the physical geography and political boundaries of all the states and provinces of the United States, Mexico, and Canada. Inside the back cover are physical and political maps of the world. These maps will serve as handy reference guides for the reader when various states, provinces, and countries are mentioned in the text. It is increasingly important that everyone be familiar with basic global political geography.

Section Opening Model. This text uses a global systems framework to reinforce the interconnectivity of natural and human systems and the importance of the input-throughput-output concept. The systems model, introduced in Chapter One (Figure 1-1), is repeated at the beginning of each later section to highlight the connections and the book's organization.

Case Studies. Thought-provoking case studies provide detailed examples of interesting environmental applications, experimental work, and controversies. Many of the Case Studies provide follow-up questions that ask the reader to examine the facts, the arguments, and the conclusions.

Fundamentals boxes. This new category of boxes reviews quantitative information covering human population equations and statistics, energy and thermodynamics, and common measures of energy and power.

The appendices that appeared in the previous edition—selected major pieces of U.S. environmental legislation, selected pieces of international environmental legislation, a list of selected environmental organizations and government agencies—are now available at the text's web site where they can be more readily updated. These appendices should be useful to students as they peruse the text or decide to delve deeper into environmental issues.

The book concludes with an English/Metric Conversion Table, a glossary of key terms, and a detailed index.

Web Enhancement

Environmental Science, Third Edition, is supported by an extensive web site at **www.environment.jbpub.com**.

Students will find a variety of study aids and resources, all designed to explore in more depth the basic science, concepts and controversies of environmental science.

eLearning. The central learning component of the site, eLearning, is an on-line interactive study guide with activities that will help students make full use of today's learning technology. Students will find Practice Quizzes, Virtual Flash Cards of key terms, answers to the text's odd-numbered questions, and Integrated Exercises. These exercises require students to synthesize material from several chapters, and are designed to enhance critical thinking and research skills. Students can research each issue through the links provided or use Internet search tools to find their own sources.

Links. EnviroNet Links directs students to useful independent Web sites that offer additional coverage or alternative perspectives on material discussed in the text.

Readings and Resources. This area offers bibliographies of the classic literature, additional research references, and the primary sources for the information in this text.

On-line Appendices: Legislation and Organizations. To save on paper and allow for updating, we've placed the following appendices on line. Links to Web sites offering information on careers in environmental science and environmental studies are also included in this area.

- Selected Major Pieces of U.S. Environmental Legislation

- Selected Pieces of International Environmental Legislation

- Selected Environmental Organizations

Ancillary Materials

To assist you in teaching this course and supplying your students with the best in teaching aids, Jones and Bartlett Publishers has prepared a complete ancillary package available to all adopters. Additional information and review copies of any of the following items are available through your Jones and Bartlett Sales Representative.

Instructor's ToolKit CD-ROM

The Instructor's ToolKit CD-ROM provides adopters with the following traditional ancillaries. All the materials are cross-platform for Windows and Macintosh systems.

The *Instructor's Manual,* provided as a text file, includes complete chapter lecture outlines, learning objectives, discussions of common student misconceptions, and answers to the even-numbered Study Questions in the text.

The *Test Bank* is available as straight text files and as part of the Diploma™ Test Generator software included on the CD. The test bank contains approximately 2,000 multiple-choice, true/false, fill-in-the-blank, matching, short-answer, analogy, and quantitative questions. The Test Generator software helps choose an appropriate variety of questions, create multiple versions of tests, even administer and grade tests on-line.

The *PowerPoint™ Lecture Outline Slides* presentation package provides lecture notes, graphs, and images for each chapter of Environmental Science. Instructors with the Microsoft PowerPoint™ software can customize the outlines, art, and order of presentation.

The *Kaleidoscope Media Viewer* provides a library of all the art, tables, and photographs in the text to which Jones and Bartlett Publishers holds the copyright or has digital reprint rights. The Kaleidoscope Media Viewer uses your browser—Internet Explorer or Netscape Navigator—to enable you to project images from the text in the classroom, insert images into PowerPoint presentations, or print your own acetates.

Transparency Acetates

A full-color transparency acetate set provides clear, effective illustrations of 100 of the most important pieces of artwork from the text.

For the Student

Outlooks: Readings in Environmental Literacy, edited by Michael McKinney and Parri S. Shariff, is a collection of 50 articles to supplement materials that students will encounter in their course work. The articles come from a variety of general interest and science magazines. Jones and Bartlett makes this ancillary available with the text, or it can be purchased separately.

Environmental Ethics: A Case Studies Approach is an introduction to environmental ethics followed by a collection of approximately 60 case studies, prepared by Robert Schoch, exploring controversial issues in environmental science. Each case study is followed by questions that emphasize critical examination of the issues and the impact poor ethical decisions can have on the environment. At the same time, the questions show that there are no easy answers to environmental issues.

Michael L. McKinney, Knoxville, TN
Robert M. Schoch, Attleboro, MA

ACKNOWLEDGMENTS

Acknowledgments

As authors, we are ultimately responsible for the content of this book, but dozens of people have provided help, encouragement, and advice. In particular, we are grateful for the advice of many teachers and practitioners of environmental science. Due to its depth and breadth, environmental science contains far more information than only two people can master, and we drew heavily on the expertise of people who have specialized in its many subfields. We therefore wish to express our deep appreciation to the reviewers of various editions of this book listed below.

We thank all of the wonderful people at Jones and Bartlett who made the revision and updating of this book possible; in particular, we must mention Louis Bruno, Anne Spencer, Stephen Weaver, Rebecca Seastrong, and Nathan Schultz. Kathy Smith copyedited the manuscript and helped to proofread the galleys, Elizabeth Morales prepared new artwork for this edition of the book, and John May of Graphic World was the compositor for the current edition of the book. Last, but certainly not least, we must single out Dean DeChambeau at Jones and Bartlett, who has been involved with this book since the first edition. Dean helped to develop this book at the very beginning. Through the years, he edited the text, helped update statistics and facts, and he researched photographs, coordinated artwork, and continually offered invaluable advice and encouragement. If it were not for Dean DeChambeau, this book would not exist.

Finally, we are deeply indebted to the patience and support of our families—and in particular to our wives, Teressa McKinney and Cynthia Schoch—while we worked on this book.

MLM
RMS

Clark E. Adams, Texas A & M University

David A. Adams, North Carolina University

John W. Adams, University of Texas, San Antonio

Michael Albert, University of Wisconsin, River Falls

Sara E. Alexander, Baylor University

Gary L. Anderson, Santa Rosa Junior College

Richard D. Bates, Rancho Santiago College

Mark C. Belk, Brigham Young University

Charles F. Bennett, University of California, Los Angeles

Keith Bildstein, Winthrop College

Gerald Collier, San Diego State University

Harold Cones, Christopher Newport University

Carl F. Chuey, Youngstown State University

Lorraine Doucet, University of New Hampshire, Manchester

Nicholas P. Dunning, University of Cincinnati

L. M. Ehrhart, University of Central Florida

Gina Gupta, Clemson University

George W. Hinman, Washington State University

Gary J. James, Orange Coast College

Robert L. Janiskee, University of South Carolina

James R. Karr, University of Washington

Michael G. King, College of the Redwoods

Clifford B. Knight, East Carolina University

Cindy M. Lee, Clemson University

Jack Lutz, University of New Hampshire

Timothy F. Lyon, Ball State University

Theodore L. Maguder, St. Petersburg Junior College

Kenneth E. Mantai, SUNY, Fredonia

Heidi Marcum, Baylor University

Priscilla Mattson, Middlesex Community College

W. D. McBryde, Central Texas College

Richard L. Meyer, University of Kansas

Henry R. Mushinsky, University of South Florida

Muthena Naseri, Moorpark College

Arnold L. O'Brien, University of Massachusetts, Lowell

Charles E. Olmsted, University of Northern Colorado

Nancy Ostiguy, California State University, Sacramento

Richard A. Paull, University of Wisconsin, Milwaukee

Adrienne Peacock, Douglas College

Charles R. Peebles, Michigan State University

R. H. Pemble, Moorhead State University

Chris E. Petersen, College of DuPage

Dennis M. Richter, University of Wisconsin, Whitewater

Gordon C. Robinson, University of Manitoba

C. Lee Rockett, Bowling Green State University

Paul Rowland, Northern Arizona University

David B. Scott, Dalhousie University

Ray Sumner, Los Angeles Valley College

R. Bruce Sundrud, Harrisburg Area Community College

Peter G. Sutterlin, Wichita State University

S. Carl Tobin, Utah Valley State College

Jerry Towle, California State University, Fresno

Lee B. Waian, Saddleback College

Linda Wallace, University of Oklahoma

Joel Weintraub, California State University, Fullerton

Frank Williams, Langara College

Richard J. Wright, Valencia Community College

Bruce Wyman, McNeese State University

Craig ZumBrunnen, University of Washington

The Environment and Humans

Science alone does not have and never will have solutions to the fundamental environmental problems of our time, which are religious in the largest sense of the word, dealing as they do with values and the human spirit. If we remember this at all times, our science will then be freed to play the part that is expected of it in the battle to save the life on this planet.

David Ehrenfeld, biologist and writer, The Arrogance of Humanism

Much of the world's land area has been disturbed by humans. In this photo, red indicates land that is greatly altered and yellow shows the partially altered land area. Only the green area remains in a relatively natural state, and it is disappearing rapidly. (*Source:* Lee Hannah, Conservation International.)

Global warming, ozone depletion, overpopulation, hazardous waste, extinctions, and many other problems are so vast that we may conclude that individual actions cannot make a difference. But this is not true. Global environmental problems are the result of the local actions of many individuals, so the problems can only be solved if we change these many local actions. One of the most common sayings in environmentalism is "think globally, act locally." The large scale of many environmental problems should not overwhelm you. Simple measures taken by each of us can save water, decrease pollution, reduce extinctions, and help solve many other environmental problems. In the photo, volunteers work to restore the Indian River Lagoon in Florida by removing the invasive Brazilian Pepper bush.

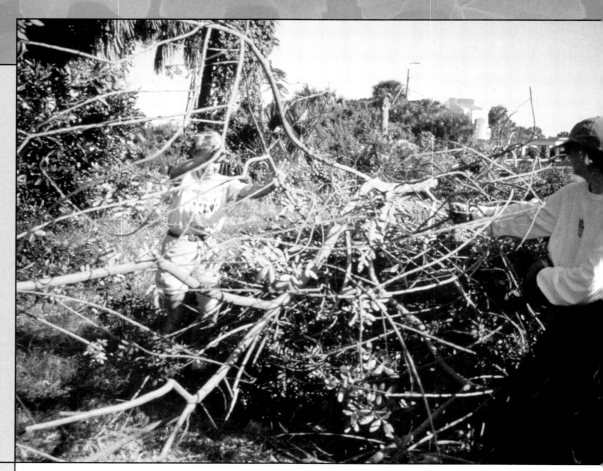

CHAPTER

1

Environmental Science: An Overview

The critical importance of environmental science and environmental studies cannot be disputed. Virtually everyone is aware of environmental issues such as global warming, the depletion of the ozone layer, the controversy over nuclear power, the global loss of species and biodiversity, the introduction of hormone-altering pollutants (also known as environmental hormones) into the natural environment, the use of and controversy surrounding genetically engineered crops, or the continuing problem of solid waste disposal. Politicians who once focused primarily on such issues as economic productivity and national security now increasingly acknowledge that environmental sustainability is a necessary prerequisite if we are to ensure our survival. The United Nations Conference on Environment and Development (popularly known as the "Earth Summit"), held in Rio de Janeiro in June 1992, was the largest international summit ever held up until then, attracting delegates from over 175 countries, including more than a hundred heads of state. In 1997, over 160 nations met in Kyoto, Japan, and signed a climate treaty protocol. The signatories of the Kyoto Protocol have met since then to work out agreements on methods to cut greenhouse gas emissions. In August-September 2002, the United Nations held a World Summit on Sustainable Development in Johannesburg, South Africa, where government officials and other interested parties from around the globe gathered to discuss many important, long-term environmental issues. No citizen of the Earth can afford to be ignorant of environmental issues.

Our planet has existed for more than four billion years, yet never before has one species dominated the Earth and other species so completely. Humanity now stands at a unique crossroads: Because of human technological capabilities and impacts, the world is not the same as it was in the past, and in the future it could be very different indeed. In all likelihood, the next few decades will witness drastic changes in the Earth and its inhabitants. It is up to all of us as individuals to try to influence the outcome of such changes so that the welfare of both humans and the environment is best served.

While this book will present many important facts, it also has two larger goals. One is to help you sort through the huge amount of environmental information available and focus on important issues. Today, especially with the widespread availability of the Internet and the World Wide Web, little bits of information on almost any issue related to the environment are readily available. Yet, even as we become overwhelmed with factoids, the bigger picture, a larger perspective, is often lost. It can often be difficult to distinguish the truly important from the merely interesting or even the downright trivial. As the old saying goes, it can be hard to see the forest because all the trees block the view.

The second goal of this book is to get you, the reader, seriously thinking about how you can put an understanding of environmental issues to work to effectively help society make the fundamental changes needed to build a world that can sustain many generations of people, with a decent standard of living, while minimizing human impact on the natural environment. We encourage you to become actively involved in environmental concerns.

What Is Environmental Science?

Environmental science involves all fields of natural science as they bear on the physical and biological environment around us. Aspects of biology, geology, chemistry, physics, meteorology, and many other disciplines must be considered when studying environmental science. Unfortunately, a major component of modern environmental science involves addressing current environmental problems that have been brought about by human activities. Therefore, in addition to presenting the scientific concepts underlying environmental issues, we also discuss social aspects of environmental problems. Not only technology and scientific understanding, but laws, ethics, economics, and other aspects of human behavior will play a key role in solving our current environmental problems.

About This Book—A Guide to the Environment

This book will provide you with an overview of the natural environment and the way humans are increasingly affecting it. Through an understanding of environmental science you will be in a position to make sound decisions that will affect the future of the entire planet. An important aspect of genuine understanding is the ability to relate seemingly diverse and disparate facts and phenomena to one another within a larger picture. A truism in environmental science is that everything is related to everything else. The environment is a finely tuned grand, and extremely complex, system in which if one component is modified or tampered with, there will be ramifications in other, often unexpected, portions of the system. Thus, it is extremely beneficial to use a **systems approach** in analyzing global environmental issues (discussed in more detail later in this chapter).

Figure 1-1 shows how the book is subdivided into the five major sections (each containing a number of chapters). These sections reflect the input-output, or throughput, processes that can be used to describe the global environment, how the environmental system is modified, and solutions to environmental problems (such as the depletion of resources or pollution of nature) brought on by human activities.

Section 1: The Environment and Humans.   The natural environment and the humans living on planet Earth form a complex, integrated system, as introduced in this chapter. Chapters 1 and 2 also describe the explosive growth of the human population, the development of modern technological and industrial society, and some of the ramifications this growth and development

Objectives

After reading this chapter, you should be able to explain or describe:

What is meant by *environmental science*

How the environment works as a system

The key traits of the environmental system

The concepts of consumption and exponential growth

A brief history of environmental movements

The concept of a *Commons*

Throughput

What is meant by *true costs* and why humans avoid them

Figure 1-1 A system can be technically defined as a "set of components functioning together as a whole." A systems view allows us to isolate a part of the world and focus on those aspects that interact more closely than others. For example, a cell in the system we call a human body generally interacts much more closely with other cells in the body than with the outside world. By focusing only on those cells that function in digestion, we confine our view further, to the digestive system. The key point here is that most systems are hierarchical: They are composed of smaller sets of systems made of smaller interacting parts.

Population   Consumption

1. The Environment and Humans (Chapters 1-2)

Lithosphere   Biosphere

Hydrosphere   Atmosphere

2. The Environment of Life on Planet Earth (Chapters 3-5)

Lithosphere   Biosphere

Hydrosphere   Atmosphere

3. Resource Use and Management (Chapters 6-13)

Lithosphere   Biosphere

Hydrosphere   Atmosphere

4. Dealing with Environmental Degradation (Chapters 14-18)

Lithosphere   Biosphere

Hydrosphere   Atmosphere

5. Environmental Issues: Social Aspects and Solutions (Chapters 19-20)

Figure 1-2 Even parts of remote Antarctica, shown here, have become dumping grounds for human waste.

have had for the natural environment. Many environmentalists consider human population growth, or, more precisely, the impact of humans on the environment, to be the single most important issue that the world currently faces. Studying Figure 1-1, you will see that conceptually, the raw elements of the natural environment pass through human society to produce the global environment in which we currently live. For all practical purposes, there is no pristine wilderness remaining anywhere on Earth (**Figure 1-2**).

Section 2: The Environment of Life on Planet Earth. Earth and life form an interconnected system based on energy flow and matter cycles. Study of these natural systems includes the disciplines of biology, chemistry, geology, climatology, and many other natural sciences. Learning the basic principles of these fields permits a better understanding of natural systems and how these systems can be utilized, protected, or disturbed. Humans utilize, modify, and are dependent on these natural systems.

Section 3: Resource Use and Management. As we utilize resources, we run the risk of overusing and depleting them. Depletion of environmental resources is one of the two basic types of environmental disturbance. It includes depletion of water, minerals (including fossil fuels), soils, wildlife, and many other resources. In order to avoid depletion problems, we must learn to manage our resources wisely.

Section 4: Dealing with Environmental Degradation. Discharging waste into the environment is the other

basic type of environmental disturbance. It includes pollution of air, water, land, and biological communities. There are ways to utilize resources productively without the levels of pollution and concomitant environmental damage that have been so widespread in the past. Environmentally friendly options to address these problems are explored.

Section 5: Environmental Issues: Social Aspects and Solutions. Technological solutions are insufficient unless society is willing to employ them. Social solutions are interwoven throughout this book, but these concluding chapters allow us to focus on specific social institutions, and their legal and economic aspects. Ultimately, through a combination of scientific analysis, technological advances, and the continuing development of social institutions, humanity must lessen its detrimental impact on the environment if we are to avoid a very bleak postindustrial age. The global environment of the future will not be the preindustrial environment of the past because today we humans have a tremendous impact on the environment. Our actions today will have a profound effect on the environment of the future. By our deeds we will determine whether our future environment will be more or less hospitable to both our species and numerous others.

Beyond "Information Overload": Environmental Wisdom

One of the challenges we face today is how to cope with information overload. Newspapers, radio, television, and the Internet/World Wide Web bombard us daily with data and statistics. Feeling overwhelmed, most people react by "tuning out."

Rather than quit trying to assimilate all this information, we might recall Henry David Thoreau's advice to "simplify! simplify!" Thoreau has become a hero to many environmentalists because he long ago predicted many of the problems that we now face. His advice to simplify was a response to what he saw as a tendency for civilization to become increasingly more complex and removed from the natural world that nourished it. This, he said, led to anxieties and spiritual impoverishment despite material wealth. On a pragmatic level, it also leads to a weakened and damaged natural environment, which is potentially less capable of sustaining human life.

We can heed Thoreau's advice by mentally stepping back and keeping our priorities set on what we think is important. The systems approach to environmental science greatly facilitates this process. In this way, we can focus on the information that we can use, instead of trying to learn it all (which no

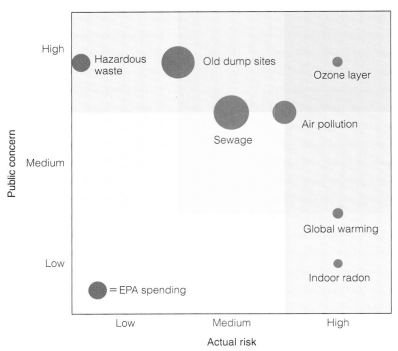

Figure 1-3   The actual risk of a hazard compared to the amount of public concern about the risk. There is often little correlation between the risk and the concern. As a result, large amounts of money (size of "dot") are spent on small risks while large risks, such as global warming, get less money. (*Source*: U.S. Environmental Protection Agency.)

education and practical experience, so environmental wisdom takes time to develop. Wisdom also means that we take a broad view in solving problems and weigh all kinds of information, social and economic as well as scientific and technical. Environmental science is often called **holistic**, meaning that it seeks connections among all aspects of a problem.

Lack of environmental wisdom is costly in many ways. It is costly to other species, to our quality of life, to future generations, and often to human happiness itself. But the most easily measured costs are economic. A good example is illustrated in **Figure 1-3**, which shows that money is often poorly spent on environmental problems in the United States. Problems that pose substantial risks such as ozone depletion may receive less money than hazardous wastes and other issues that are less threatening overall. Such spending inefficiencies occur because people often lack adequate information about the true risk of environmental problems. For example, a small group of people may panic when they discover that there is a toxic waste dump in their own neighborhood. Furthermore, even if they do have adequate information, many people fail to take a holistic approach and they ignore such aspects as long-term global impacts.

one can do). Ideally, we can strive to seek what might be called **environmental wisdom**. Wisdom is the ability to sort through all the facts and information to make reasonable decisions and plan sensible long-term strategies. Wisdom is gained through

Beyond the Quick Fix: Building a Sustainable World

Many environmentalists point out that past efforts have tended to focus on short-term, emergency actions rather than long-term solutions. Examples of this approach include cleanup of wastes and pollution after they are produced (**Figure 1-4**) and trying to save species only when they are nearly gone. Besides being less effective, such piecemeal, late-acting remedial solutions are almost always the most expensive way to solve environmental problems.

In the last decade, the rising costs and inefficiencies of cleaning up pollution after it is produced have led to a search for better approaches to solving environmental problems. Generally, holistic approaches have been able to solve problems more cheaply and efficiently. By examining society and the environment as an interconnected system, we can often solve many problems at once. The **input** of materials and energy through society is often called **throughput**. Environmental resources are referred to as **sources** of throughput, and environmental reservoirs that receive throughput are called **sinks** and are the ultimate repository of societal **output**. This throughput model is the basis for many systems approaches that attempt to link social systems such as the economy to natural systems. As illustrated in

Figure 1-4   Cleaning up toxic waste is an expensive after-the-fact approach. It is much cheaper to design methods that produce less waste.

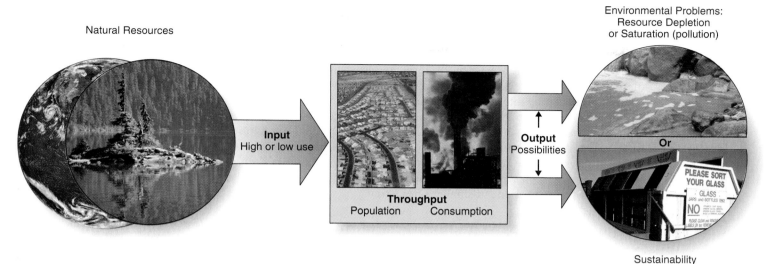

Natural Resources

Input
High or low use

Throughput
Population   Consumption

Output
Possibilities

Environmental Problems:
Resource Depletion
or Saturation (pollution)

Or

Sustainability

SOURCES

SINKS

Figure 1-5   Humans use resources from the natural environment and deposit wastes back into the natural environment. Materials and energy that humans draw from are known as *sources* and environmental reservoirs that receive the products of human society are known as *sinks*. Environmental problems can arise from resource depletion and pollution of sinks. Controlling consumption and reducing the flow of material through society can alleviate many environmental problems.

Figure 1-5, environmental problems arise from (1) resource depletion and (2) pollution. Past efforts at pollution control were largely "end-of-pipe" solutions, cleaning up waste after it was produced. Reducing the flow of material through society can also control pollution. Such **input reduction**, which conserves resources and reduces pollution at the same time, is now widely accepted by environmental economists and others as a better solution to most environmental problems. Input reduction illustrates the kind of fundamental change needed to build a society that can be maintained for many years without degrading the environment.

Figure 1-5 also identifies the two basic causes of most current environmental problems: human population pressures and consumption of material goods. Both need to be addressed if long-term solutions are to be achieved. Population and consumption have both led to increased resource depletion and pollution. Two key solutions proposed are to address population issues and develop environmentally "friendly" technologies. Many newer technologies, such as solar and wind power, are much less harmful to the environment than the traditional, nonrenewable-resource-intensive, highly polluting technologies of the industrial age, such as the use of fossil fuels.

Overview: The Environment as a System

The entire environment of planet Earth can be divided into four spheres (refer back to Figure 1-1): the biosphere (living organisms of Earth), atmosphere

(the gaseous envelope surrounding the planet), hydrosphere (the liquid water on the surface of Earth), and lithosphere (the stony or rocky matter composing the bulk of planet Earth). Matter cycles and energy flows through these spheres (**Figure 1-6**). A systems approach provides a convenient overview and manner of understanding how our environment works.

PLANET EARTH

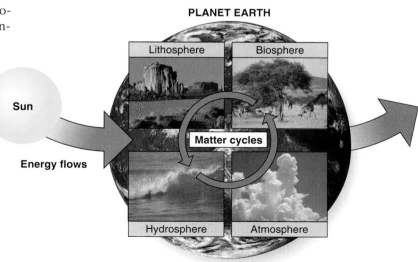

Lithosphere   Biosphere

Sun

Matter cycles

Energy flows

Hydrosphere   Atmosphere

Figure 1-6   The Earth is open to energy, which flows into and out of it. Essentially, however, the Earth is closed to matter, which cycles over and over through the four spheres.

What Is a System?

A *system* can be technically defined as a "set of components functioning together as a whole." A systems view allows us to isolate a part of the world and focus on those aspects that interact more closely than others. For example, a cell in the system we call a human body generally interacts much more closely with other cells in the body than with the outside world. By focusing on only those cells that function in digestion, we confine our view further, to the digestive system. The key point here is that most systems are *hierarchical*: they are composed of smaller sets of systems made of smaller interacting parts.

Three Key Traits of the Environmental System

We can analyze the global environment in terms of three system traits: openness, integration, and complexity.

Openness refers to whether a system is isolated from other systems. An **open system** is not isolated in that it exchanges matter or energy with other systems. A **closed system** is isolated and exchanges nothing.

The law of entropy means that energy cannot be recycled; rather, "high quality" energy (such as visible light) is degraded to "lower quality" energy (such as waste heat). Therefore, any system that does not have a renewing supply of energy from outside will eventually cease to exist. Not surprisingly then, the Earth is an open system in terms of energy. Figure 1-6 shows how energy flows from the Sun and is often radiated back into space. In contrast, the Earth, for all practical purposes, is a closed system in terms of matter. If we discount the relatively small amount of matter added from meteorites and other space debris, the Earth contains all the matter it will ever have. Driven by energy from the Sun, this matter cycles over and over among the four spheres, often moving back and forth among the gaseous, liquid, and solid states, and participating in the metabolism of living things.

Integration refers to the strength of the interactions among the parts of the system. For instance, the human body is a highly integrated system whose cells are interdependent and in close communication. The loss of certain cells, such as those composing the heart or brain, can result in the death of all the other cells in the system (the whole organism) because the cells are so interdependent. At the other extreme are systems with very weak integration, such as the cells in a colony of single-celled organisms (such as the green algae, *Volvox*). Removal of many cells will have little effect on the remaining cells because they are less dependent on each other.

The degree of integration of the global environmental system is under debate. At one extreme are scientists who argue that the global system is a superorganism: the lithosphere, hydrosphere, atmosphere, and biosphere are intimately interconnected by many complex pathways. According to this **Gaia hypothesis**, the Earth is similar to an organism, and its component parts are so integrated that they are like cells in a living body. Many scientists, however, believe that the global environment is less integrated than the Gaia hypothesis argues. This does not mean that the environment is "unconnected" or even as weakly connected as a colony of cells. As we will describe in future chapters, many kinds of matter cycles and energy flows interconnect the spheres and cycle within the spheres as well. The true level of integration in the global system is probably somewhat less than a "superorganism" but considerably more than a loose collection of independent parts, such as a sponge (a sponge can be considered a colony of semi-independent cells). Because the system is so vast and the interconnections so complex, the exact level of the Earth system's integration will probably be the topic of much debate for many years.

Complexity is often defined as how many kinds of parts a system has. This definition conforms to our intuition: A tiny insect seems more complex to us than a large rock because it has many more "parts." The insect has more complex molecules, as well as more different types of molecules, and these are used to construct cells and organs. This example also illustrates that complexity is often hierarchical, with smaller components being used to construct larger ones.

As you would expect, the environment is enormously complex. The four spheres, with their matter cycles and energy flows, have trillions and trillions of different components operating at many spatial and temporal scales. Organisms, soils, rainwater, air, and many other components interact in complicated ways. The individual spheres themselves are extremely complex. Even with advanced computers, no one has been able to predict the weather, or even climate, very far into the future because the atmosphere is so complex. Indeed, the many interactions make unpredictability a basic characteristic of complex systems. This inability to predict how the environment will respond to changing conditions is perhaps the major reason for so much controversy and inaction over environmental problems.

Systems and Chaos Theories: Ways to Study Complexity

Visit www.environment.jbpub.com for more information.

Studying complex systems can be difficult because they have many parts that often interact in different ways. Over the last few decades, researchers have developed several methods of studying complexity. Systems theory and chaos theory, for example, try to produce general "laws" of complexity. Such laws would not only make complex systems more understandable, but also allow us to predict more accurately how these systems will behave. Think how important such predictions could be in a complex system such as the stock market or the weather!

Unfortunately, none of these theories has been entirely successful in providing a complete understanding of complexity or producing accurate predictions of how any complex system will behave. For example, despite thousands of studies and computer models, no one knows what exactly will happen to either the stock market or the biosphere on a certain day in the future. Nevertheless, these theories have provided a better idea of how complex systems will generally behave under a given set of conditions. Thus, we have a general understanding of how a lake ecosystem will respond to excess nutrients even though we cannot specify every event.

Systems theory (or general systems theory) was one of the first widely used attempts to find "laws" of complex systems. It grew rapidly in the late 1940s during the boom in automation and information technology, so it has traditionally focused on how systems are regulated and become unregulated. Systems theory treats a complex system as a "black box" with inputs and outputs. Such a system is kept at equilibrium by negative feedback processes, defined as processes that counteract perturbations. An example is a thermostat that turns on a furnace to produce heat when a house is cold and turns on air conditioning when it is hot. In contrast, positive feedback processes amplify perturbations. For example, a cooling global climate can cause more snow to remain on the ground, which leads to more global cooling because the snow reflects sunlight back into space. This causes yet more snow, and so on (a snowball effect).

Systems theory has been widely used as a convenient scheme to classify processes such as positive feedback. However, it has often been criticized as too general or vague because by treating a system as a "black box," the theory omits many of the details of how the system operates. Therefore, more recent efforts to study complex systems have focused on more mathematical, rigorous descriptions of them. One theory that has received much attention is chaos theory. A chaotic system is one whose workings are extremely sensitive to even the slightest change: Just the slightest perturbation can become greatly amplified through positive feedback. The classic example is the weather, as first described by E. Lorenz, who helped discover chaos theory in the early 1960s. Lorenz created a set of equations that precisely described atmospheric conditions and showed how even tiny changes in one of the parameters could cause a massive alteration of the weather in a few days. This is often called the "butterfly effect" by analogy with the idea that a butterfly flapping its wings in South America could eventually affect the weather in North America (**Figure 1**). By creating tiny changes in atmospheric turbulence, which in turn create cascading effects on larger air flows, the butterfly could have a major impact. (Of course, the chances that it actually will produce such a major impact are extremely low.)

Chaos theory's most important finding so far is that even simple systems, such as several atoms, often have chaotic properties. Consequently, predicting their precise behavior very far into the future is nearly impossible. How can we make predictions when a minute unseen change, as in the butterfly effect, can have cascading effects? Many theorists think it will always be impossible to accurately predict precise future behaviors in complex systems, which have even more potential for chaos than simple systems. Nevertheless, by applying chaos theory, patterns of regularity can be discerned and studied.

Figure 1 Can a butterfly in the rainforest affect the weather in North America?

Major Obstacles: Delayed and Unpredictable Impacts. Unfortunately, social responses to environmental problems are greatly hindered by two of the key traits of the environmental system: its moderate integration and high complexity. Any system that is integrated, such as the environment, can transmit disturbances from one part of the system to another. Integration results from connectedness so that resource depletion or pollution of one part of the environment can have cascading, or domino, effects into other parts. For example, removal of one species from an ecosystem will often affect the abundance of many other species, such as those that prey on or compete with it. Burning sulfur-rich coal affects the atmosphere as air pollution, but it also affects the hydrosphere when it falls as acid rain to acidify lakes. The biosphere is also affected because aquatic organisms in the lake can die from the more acidic lake water. The burning coal can even affect the lithosphere when the acid rain dissolves limestone and other alkaline rocks to form caves and sinkholes. This example shows how just one activity, burning coal, can affect all four spheres of the environment. Such wide-ranging cascading effects are anything but rare, as we will see in later chapters. This connectedness of the environment means that virtually any action has a number of consequences, many of which are unforeseen and unintended. Such cascades are so important that the biologist Garrett Hardin has formulated the **first law of ecology:** "We can never do merely one thing." This is often called the "law of unintended consequences."

That the environment is only moderately integrated ("loosely connected") greatly hinders our ability to observe, predict, and ultimately correct the unintended consequences of our actions. The indirect connections and interactions in the environment create delays, or long lag times, before cascading effects become visible (**Figure. 1-7**). Returning to the acid rain example, it may take many years of coal burning before fish populations in lakes are affected. In other cases, the impacts can be delayed for centuries, millennia, or even longer. Environmental impacts on a global scale will usually take an especially long time to occur. Many decades and probably centuries will pass before the full effects of added atmospheric carbon dioxide on global temperature are observable. This lag time is one of the main reasons for the debate over global warming. To make matters more complicated, corrective measures may themselves have unanticipated consequences.

As we have seen, the complexity of the environment leads to unpredictability, which hinders social responses to environmental problems. As Figure 1-7 illustrates, the unintended cascades we cause not only take a long time, but also occur as unexpected, complicated chains of events. How many people associate global warming with disease? Perhaps very few, yet there has been strong evidence for many years now that many tropical diseases are spreading due to global warming. Back in 1993 a paper was published in the medical journal *The Lancet* concluding that a 1991 cholera outbreak in South America was related to localized warming of Pacific Ocean waters due to global climate changes. The paper argued that the warming had caused the rapid growth of plankton that harbor the cholera bacterium, leading to thousands of deaths (**Figure 1-8**). Since then, many other studies have linked global warming with the spread of diseases.

In addition to having many interactions, complex systems are unpredictable because some of the interactions exhibit positive feedback. **Positive feedback** occurs when part of a system responds to

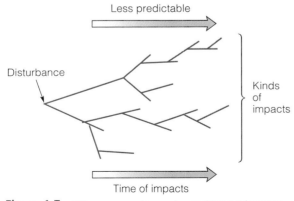

Figure 1-7 "We can never do merely one thing." Whenever we do something, we cause a cascade of impacts. These impacts become less predictable with time.

Figure 1-8 Burial of cholera victims in Peru.

Figure 1-9 Society accelerates the cycling of matter through the four spheres, which depletes resources (matter inputs to society) and causes pollution (matter outputs by society). This acceleration is due to increasing population (P) and consumption (C). Slowing population growth and reducing personal consumption will reduce this acceleration.

change in a way that magnifies the initial change. For example, evidence indicates that a slight increase in average global temperature can cause a further increase by melting some of the glaciers and snow that reflect sunlight back into space. Instead of reflecting light, more of the Earth's surface becomes available to absorb heat. Another example is poverty in developing countries that results from overpopulation; the poverty leads to high reproductive rates and thus still further overpopulation. In nontechnical terms, positive feedback is often referred to as a "snowball effect" or "vicious circle."

Society in the Environmental System
Figure 1-9 illustrates how modern society is embedded within the environment, being dependent on it for the materials and energy needed to maintain civilization. It also shows how industrialized society accelerates the cycling of matter and the flow of energy through itself and the four spheres.

As we have seen, fundamentally, all environmental problems involve either depletion (consumption) of sources or pollution (waste) of sinks. Thus, we can measure the net environmental impact of society by these two processes. Depletion occurs when the accelerated cycling and flow remove matter and energy faster than natural processes are renewing them. Conversely, pollution occurs when the accelerated cycling and flow are discharged into the environment, overwhelming the local natural purification processes (**Figure 1-10**).

Problems of environmental depletion and pollution both exhibit the delayed and unpredictable impacts we have discussed. Much of the remainder of this book is concerned with these issues (Figure 1-1).

Figure 1-10 Pollution occurs when natural purification processes are overwhelmed, such as by large amounts of nutrients or poisons. Shown here is acid mine damage.

What Is Environmental Impact?

Environmental impact refers to the alteration of the natural environment by human activity. We have already identified two basic types of environmental impact: resource depletion and pollution (Figure 1-5). In other words, there is either too much input (resource depletion) and/or too much output (pollution). Input reduction seeks to slow both depletion and pollution, while output reduction just slows pollution.

Population and consumption are two main forces accelerating resource depletion and pollution. The following equation is a simple way to remember this:

Impact = Population × Consumption, or
I = P × C

It is easy to see why the number of people affects impact (**Figure 1-11**). Consumption has historically tended to increase the effect of each person. For example, a baby born in the United States has many times the impact of a baby born in a developing country because over a lifetime, the U.S. resident consumes many more resources, such as fossil fuels, and produces much more pollution.

Both population and consumption per person have been increasing very rapidly worldwide. This has led to an extremely rapid increase in environmental impact. This section briefly examines both of these factors.

Exponential Growth

Exponential growth of any kind is caused by multiplicative processes. It occurs in population growth because biological reproduction is inherently multi-plicative: most organisms have the ability to produce more offspring than are necessary to simply replace the parental generation. If conditions allow, the population of each successive generation will be larger than the previous generation. Exponential growth usually has an initial "lag phase" of slow growth followed by a period of increasingly accelerated growth. This pattern occurs because, initially, small numbers are being multiplied together; as larger numbers are multiplied, the rate of growth increases "explosively." Consider, for example, a pond where the algae cover starts from a single algal cell and doubles in size each day. For a long time, you would see nothing happening. Once larger areas were covered, however, the algae would spread extremely fast: One day, the pond would be only half covered; the next day, it would be completely covered.

Exponential Growth of the Human Population

The current (2003) world population stands at approximately 6.3 billion people. For millions of years, relatively few humans were on Earth at any given time. The total human population on Earth remained quite low until the development of agriculture. Even then, population growth did not become explosive until the 1900s, when it was aided by the global spread of industry and advances in modern medicine, especially the great reduction in infant deaths. At present, each year the world population experiences a net gain of slightly over 77 million people, the rough equivalent of about one-quarter of the population of the United States.

How long will human population growth continue before it encounters environmental limitations? The answer depends on how many people the Earth can support, a question that is much debated, in part, because the answer depends on how high a standard of living one assumes. However, many estimates predict that the Earth can adequately and sustainably support between 6 and 8 billion people at a reasonable standard of living (though perhaps lower than that of the typical American or Western European). If this is true, population growth clearly must decline very soon if overshoot is to be avoided. Unfortunately, almost all population projections by the United Nations, the World Bank, and other organizations indicate that world population will probably not stabilize until it exceeds 9 billion, sometime in the middle to late twenty-first century.

Exponential Growth of Consumption

The second basic factor in our impact equation, I = P × C, is consumption. Here we use the term con-

Growth in Population (*P*) X Growth in Consumption (*C*) = Impact (*I*)

Figure 1-11   Personal consumption (C) has increased the throughput per person. This is multiplied by the number of people (population or P) to obtain the total resource depletion and pollution produced by society.

High input

Low input

High output

Low output

SOCIETY

sumption in the economic sense, to refer to the purchase and use of material goods and services in our daily lives. Consumption thus includes all the things we do and buy to meet our wants and needs. Consumption has increased the overall environmental impact by increasing the impact per person:

> Overall impact = Population ×
> Consumption
> = Number of Individuals ×
> (Impact/individual)

This impact per person occurs through an increase in both resource depletion and pollution.

Because natural resources are used to make the increasing amount of things we buy, consumption has tended to increase per capita resource use and pollution for many types of resources and pollutants. **Figure 1-12** illustrates the fact that solid waste output has grown dramatically in the United States in the last few decades.

Personal consumption in wealthy nations could easily be reduced with no reduction in the quality of life. For example, many new sustainable or "green" technologies are being developed that can greatly reduce individual impact. More efficient technologies, such as fuel-efficient cars, use fewer resources and produce less pollution when an individual uses them. Alternative technologies can eliminate many impacts altogether. Replacing coal-burning machines with solar-powered machines conserves nonrenewable fossil fuels and eliminates many air pollutants released by burning coal.

Exponential Growth of Environmental Impact

The exponential increase of population and traditional industrial technology (which correlated with increasing consumption) have caused an exponential increase in environmental impact, I. For example, Figure 1-12 illustrates how the overall impact of municipal solid waste of U.S. society has grown as a result of both increased population and increased waste per person.

The increase in total solid waste in Figure 1-12 is typical of the pattern seen in many other kinds of pollution. A society that produces such waste is likely to be consuming many resources that ultimately generate the waste. For example, global consumption of fossil fuel has risen exponentially due to the growing world population and the spread of technologies that use fossil fuels (increasing per capita use of fuels). As consumptive technology and population grow, more materials and energy move through society. This accelerates the depletion of

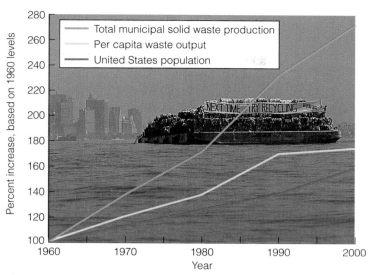

Figure 1-12   Trends in U.S. total solid waste production, output per person, and population. The background photo shows Greenpeace protestors on a New York garbage barge. (Chart is based on data from the U.S. Environmental Protection Agency and the U.S. Bureau of the Census).

environmental resources. In addition, the materials and energy that move through society must have somewhere to go when society is finished with them. Solid waste, air and water pollution, and other outputs usually end up in the environment. In the fossil fuel example, fossil fuels are the sources showing consumption impact, and the atmosphere is the sink showing pollution impacts of carbon, leading to global warming. Carbon is the material whose throughput is being accelerated by growing population and consumptive technology.

Acceleration of throughput resulting in increases of both depletion and pollution in industrialized societies is evident by almost any measure:

1. Developed nations contain less than a quarter of the world's population, but consume over four-fifths of the world's natural resources and nearly three-quarters of the world's energy.
2. The United States consumes far more meat, fossil fuels, and pesticides per person than the vast majority of other nations, while producing more solid waste per person than almost any other nation.
3. The United States has less than one-twentieth of the world's population but produces nearly one-fifth of the world's greenhouse gas pollution (and many other air and water pollutants) and between one-third and one-half of the world's solid waste (estimates vary).

In summary, there are two basic kinds of environmental impacts, depletion and pollution. Both have increased because growing consumption of resources increases pollution as throughput is accelerated. The main causes of increased throughput have been the growth of (1) population and (2) consumption.

History of Environmental Impact and Environmental Movements

Considering that the human species (*Homo sapiens*) has inhabited Earth for 100,000 or more years, the current massive global environmental impact of humans is a very recent development. Indeed, our relationship with the environment has dramatically changed as we and our technology have evolved.

Conceptually, we can distinguish five basic developmental stages in the relationship between humans and their environment (illustrated in **Figure 1-13**). These stages are largely determined by the economic activity in which humans engage using the technologies available to them at the time. This activity, in turn, affects how humans impact the environment. Notice that at any one time on Earth different human populations may be in different stages simultaneously, yet the general global trend has been to progress through these stages.

1. *Gathering and hunting*. Early humans were largely at the mercy of their environment, so

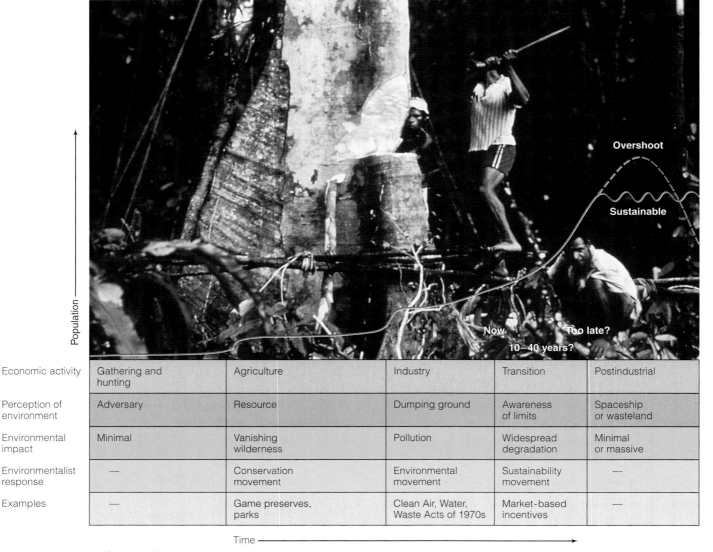

Economic activity	Gathering and hunting	Agriculture	Industry	Transition	Postindustrial
Perception of environment	Adversary	Resource	Dumping ground	Awareness of limits	Spaceship or wasteland
Environmental impact	Minimal	Vanishing wilderness	Pollution	Widespread degradation	Minimal or massive
Environmentalist response	—	Conservation movement	Environmental movement	Sustainability movement	—
Examples	—	Game preserves, parks	Clean Air, Water, Waste Acts of 1970s	Market-based incentives	—

Figure 1-13   Population and personal consumption have increased the human impact on the environment in exponential fashion. Environmentalists have responded with the conservation movement, the environmental movement, and the sustainability movement. Many people believe that we are currently in a transition stage and that humans have no more than 50 years to prevent "overshoot" and attain sustainability.

they might have viewed it in adversarial terms. Weather, predators, food shortages, and disease were constant threats. During this stage, however, humans also had a closeness and intimacy with the environment that has been generally lost since that time. Population levels typically remained low, and any human-induced effects on the environment (such as purposefully setting forest fires) were relatively localized.

2. *Agriculture and conservationism.* The shift from hunting and gathering to cultivating food is one of the most profound milestones in human evolution. It allowed a great increase in population size and permitted people to settle down in large towns and cities. But agriculture also had a major impact on the environment. People began to view land as a resource to be exploited wherever needed. As land was cleared and cultivated,

however, the wilderness vanished. Toward the end of the agricultural stage, the loss of wilderness became so great that alarmed citizens began conservation movements to set up preserves for the remaining wilderness. In the United States, this happened in the late nineteenth century and is often associated with the Sierra Club and President Theodore Roosevelt (**Figure 1-14**). Today many developing countries still have agricultural economies, and their vanishing wilderness, especially tropical rainforests, has stimulated the growth of conservationism.

3. *Industry and environmentalism.* As nations industrialize, population grows faster, and the environment is perceived more and more as both a source of raw materials and a place to dispose of the concentrated waste byproducts of industry. The result is a rapid increase in air and water pollution, as well as problems with solid and hazardous waste disposal. Toward the end of this stage, pollution becomes so widespread that antipollution social movements emerge. In the United States, these social movements began in the early 1960s and peaked in the 1970s. When people talk about "environmentalism," this antipollution movement is often what they mean. Several early books heralded this new awareness, including *Silent Spring* (1962) by Rachel Carson, which warned of pesticide pollution; *The Population Bomb* (1968) by Paul Ehrlich; and *The Limits to Growth* (1972) by Donella H. Meadows and others. The antipollution movement added litigation and citizen activism to the lobbying tactics used by the conservationists. These efforts and rising public concern led to the passage of landmark environmental legislation by the U.S. Congress:

1970	National Environmental Policy Act (NEPA) requires environmental impact studies before land development projects.
1970	Environmental Protection Agency (EPA) created.
1970	Clean Air Act Amendments (CAAA).
1972	Clean Water Act (CWA) reduces pollution of lakes and rivers.
1972	Coastal Zone Management Act (CZMA) begins cleanup of coastal ocean waters.
1973	Endangered Species Act (ESA) enacted to preserve endangered species.
1974	Safe Drinking Water Act (SDWA) requires EPA to set and enforce drinking water standards.

Figure 1-14　Theodore Roosevelt (left) and John Muir, founder of the Sierra Club, in Yosemite.

1976	Toxic Substances Control Act (TSCA) helps limit the amount of poisonous chemicals made and sold in the United States.
1980	Comprehensive Environmental Response, Compensation, and Liability Act (CERCLA), or "Superfund," began systematic cleanup of large waste sites.
1990	The Oil Pollution Act (OPA) strengthens EPA's ability to prevent and respond to catastrophic oil spills.
1990	The Pollution Prevention Act focuses industry, government, and public attention on reducing the amount of pollution through cost-effective changes in production, operation, and raw materials use.

Many more federal, state, and local laws and amendments have been passed since 1990. The efforts to clean up pollution have been highly successful in some areas, including lakes and rivers, toxic waste dumps, and ocean dumping. It's not only the United States and Canada that are making great strides in cleaning up wastes and preserving the environment, many other countries are as well, including those of Western Europe. At an international level a number of important pieces of environmental legislation have been established, such as the 1973–1975 Convention on International Trade in Endangered Species of Wild Fauna and Flora (CITES) and the 1987 Montreal Protocol on Substances that Deplete the Ozone Layer.

4. *Transition and sustainability.* Although some forms of pollution have been reduced, many other environmental problems have increased. In the United States, for example, species of wildlife are becoming imperiled at increasing rates as habitat is destroyed. Groundwater contamination has worsened, and there are many thousands of hazardous and radioactive waste sites that will likely not be cleaned up for centuries. Despite recycling and precycling (the reduction of packaging material by manufacturers) efforts, the amount of solid waste produced per person continues to climb. Globally, the EPA has cited global warming, ozone depletion, and increasing species extinction as the greatest environmental threats to future generations. These problems are caused by both the rapidly increasing populations in developing countries and the increasing demands for goods and services by the populations of developed countries. No single group or nation carries all of the blame.

We are currently in this fourth stage—the transition. Environmental problems have become so widespread that they demand large-scale regional and global solutions involving many aspects of society. Beginning in the early 1980s, a sustainability movement emerged to try to deal with these problems. Unlike the conservation and antipollution movements of the past, which emphasized specific problems rather than a global approach, the sustainability movement seeks long-term coexistence with the environment. **Sustainability** means meeting the needs of today without reducing the quality of life for future generations. This includes not reducing the quality of the future environment. Sustainability can be achieved in many ways, such as through the use of sustainable ("green") technologies that utilize renewable resources (for instance, solar power) and recycle many materials. These technologies allow a **sustainable economy** that produces wealth and provides jobs for many human generations without degrading the environment.

The sustainability movement uses three approaches not attempted by previous environmental movements. First, it focuses explicitly on trying to reduce society's use of all resources. Emphasis is thus on input reduction, as opposed to end-of-pipe solutions. *Waste is viewed as a symptom, not a cause, of the environmental crisis.*

Second, the sustainability movement is more holistic. It realizes the necessity of addressing the social, and especially economic, causes of environmental degradation. This has led to an increasing appreciation of the role of poverty and other economic factors that encourage people to deplete resources and pollute. Market-based solutions are becoming more popular, and less emphasis is placed on the legal solutions used in the past. For example, many experts now agree that it is often more effective and cheaper for society to tax coal, gasoline, and other polluting substances than to pass laws specifying how much pollution may be emitted. Higher gasoline prices encourage people to drive less or buy fuel-efficient cars.

Third, the sustainability movement has encouraged the growth of thousands of local community action groups, as opposed to the national groups that dominated the conservationism and environmentalism periods. Many of the national and international environmen-

Figure 1-15   Citizen protest is an increasingly popular and effective way to promote local environmental sustainability.

tal organizations arose well before 1980. For instance, the Sierra Club was founded in 1892, the National Audubon Society in 1905, the National Wildlife Federation in 1936, and the World Wildlife Fund in 1961. Since the exposure of the toxic dump at Love Canal, New York (1978) and the Three Mile Island nuclear accident in Pennsylvania (1979), residents of communities have become increasingly active in addressing local environmental problems. Vocal debates over incinerators, landfills, land development, and many other environmental issues now often dominate the local news (**Figure 1-15**). Such local participation is often called **grassroots activism**. Recently, grassroots groups with common interests have begun to network by establishing regional and national newsletters and computer networks. By the mid-1990s, many of the national organizations had begun to experience a decline in membership. Although some people suggest that this decline means environmental interest is waning in the United States, others argue that it simply reflects the transfer of environmental allegiance from national to grassroots organizations.

All three characteristics of the sustainability movement arose from the desire to find better ways to solve widespread environmental problems. Grassroots activism is often the best way to deal with local issues. The rise of input reduction and economic approaches reflects the need to reduce social costs. While litigation and lobbying are still used, the sustainability movement also uses direct action

and lifestyle changes. In keeping with its more holistic approach, this movement is also **ecocentric** ("environment-centered") instead of anthropocentric. In other words, the sustainability movement seeks to preserve nature for reasons beyond simply improving the quality of human life. Nature is viewed as having a high "intrinsic" value that is not related to human needs. Nature is worth protecting for its own sake.

5. *Postindustrial stage—sustainability or overshoot?* Today we live in a time of rapid change and technological advancement. Over the next few decades, the decisions we make as a global community will almost certainly determine the long-term fate of the environment for many future generations. For this reason, we can refer to the current stage of many developed countries as a "transition" stage. The next, future, stage of society has been called the "postindustrial" stage.

Just what form postindustrial society will take is open to speculation and debate, but two possible alternatives have been proposed. One is a sustainable future, where the human population stabilizes and technology becomes less environmentally harmful. The second alternative is overshoot, where population climbs so high and technology is so harmful that the environment is degraded to the point that relatively few people can be supported, at least with a decent standard of living. Which alternative comes into being could depend on the collective actions of humanity during this first half of the twenty-first century.

Why Do People Deplete and Pollute? The Environment as a Commons

In 1968, biologist Garrett Hardin wrote a famous essay called "The Tragedy of the Commons." He argued that property held in common by many people will be destroyed or at least overused until it deteriorates. He gave the example of a pasture where each herdsman in the village can keep his cattle. The herdsman who overgrazes the most will also benefit the most. Each cow added by a herdsman will benefit the owner, but the community as a whole will bear the cost of overgrazing. Because the benefit of adding another cow goes to the individual and the cost of overgrazing goes to the community, the "rational" choice of each individual is to add cows. The commons thus rewards behaviors that lead to deterioration, such as overgrazing, and punishes individuals who show restraint. Those who add fewer cows will simply obtain fewer benefits while the commons itself deteriorates anyway because of the individuals who continue to add cows. This problem with common property was known long before Hardin's eloquent essay. For instance, the ancient Greek philosopher Aristotle noted that "what is common to the greatest number has the least care bestowed upon it."

Hardin's pasture exemplifies the problems that arise when any part of our natural environment is treated as common property. Unless there is some kind of regulation, overexploitation will likely occur via both input and output impacts:

1. Commonly held resources, such as the pasture, will become depleted through excessive consumption.
2. Commonly held environmental sinks will be overwhelmed by pollution.

Many local, regional, and global environmental problems illustrate this view of the commons as a source or a sink:

	Problem
Atmosphere as global common sink	Global warming, ozone loss
Atmosphere as regional common sink	Acid rain
Atmosphere as local common sink	Urban smog
Ocean as global common sink	Ocean pollution
Ocean as global common resource	Many fish species overfished
Rainforest as common resource	Global warming promoted by deforestation Biodiversity reduced by deforestation

Notice that some of these commons are shared by many nations. The international nature of many environmental problems adds greatly to the complexity of solving them because international agreements are required. Consider the Mediterranean Sea. A confined shallow ocean basin surrounded by many nations, this sea is one of the most overfished and polluted large bodies of water on Earth (**Figure 1-16**).

Saving the Commons: Reducing Throughput by Paying True Costs

If we are to save the environmental commons, the throughput of matter and energy through all societies must be reduced. This will slow both depletion and pollution. Increased impact (I) has been driven by increasing population (P) and consumption (C), so their growth must be reduced.

The economic forces that promote population growth and consumption have arisen because the environment has been undervalued in the past. The true costs of using the environment as both a source and a sink have not been incorporated into global economic activity. Traditional economics has considered the environment as a "free" commodity, such as the atmospheric sink, or a source of very cheap materials, such as cheap timber or ores. This view has led to rapid consumption of resources with little incentive for conservation. It has also contributed greatly to the rapid population growth of developing nations. Poverty is strongly correlated with high population growth rates, and people in resource-rich tropical developing nations are often underpaid for their resources (as compared to prices in many developed nations).

Reducing Consumption: Defusing the Bomb of the North

Rapid population growth is often described as a "time bomb" that will greatly degrade the environment in coming years. While population growth is especially rapid in the Southern Hemisphere, the industrialized nations of the Northern Hemisphere are also contributing to environmental degradation through their habit of excessive and wasteful resource consumption developed during a time of abundant resources. Thus, consumption may be viewed as the environmental "time bomb" of the North. The solution is to reduce consumption or C in the impact equation $I = P \times C$.

There are three basic ways to reduce consumption. None of these ways needs to involve painful self-sacrifice or a lowered quality of life. To the contrary, reducing consumption can improve

Figure 1-16   A computer-enhanced image of pollution in the Mediterranean Sea. Red, yellow, and orange areas are concentrations of plankton growth promoted by discharge of raw sewage. (Blue indicates clear water.) Many coastal cities lack sewage treatment.

the quality of life in many ways by improving human health and the environment.

- *Reduce material needs.* Psychological studies have statistically documented what many religions have long taught: Increasing a person's material wealth will rarely produce an increase in happiness or contentment. The "treadmill of wealth" says that the more we accumulate, the more we want. When Americans are surveyed, the large majority consistently respond that they have far too much "stuff."
- *Use less technology to meet our needs.* Long before Thoreau, many people observed that some machines can have detrimental effects on humans. Riding bicycles to work (instead of cars), buying products with fewer artificial chemicals and less superfluous packaging, and using fewer unnecessary convenience appliances are but a few ways that people have reduced their reliance on technology.
- *Use sustainable technology to meet our needs.* Another way to reduce technological impact is to use technologies that are much more "environmentally friendly" than the fossil-fuel-based, industrial technologies of the past. **Sustainable technology** permits humans to meet their needs with minimum impact on the environment. It produces a sustainable economy that provides jobs for many generations without degrading the environment.

There are many kinds of sustainable technologies, ranging from direct solar and wind power to recycling. While advances in pollution cleanup technologies are often heralded, true sustainability results from input reduction: Slowing resource depletion also slows pollution. The three basic ways that sustainable technologies achieve input reduction, or conservation, are (1) efficiency improvements, (2) reuse and recycling, and (3) substitution.

Efficiency improvements reduce the flow of throughput by decreasing the per capita resource use. The United States, more than almost any other country, uses technologies developed during times of abundant resources. As a result, it generates much waste, providing enormous opportunities to save many resources by relatively simple changes in existing technologies. The amount of waste is so great that, in many cases, the large amount of resources saved will more than compensate for the cost of investment to make the change.

Let us use the example of energy, which drives all economies. Each year, the United States spends about 10%–11% of its national wealth to pay for the energy needed to produce that wealth. In contrast, Japan spends only about 5%–6% of its national wealth on energy because it uses energy about twice as efficiently. It is estimated that changes in U.S. technology, ranging from high-mileage cars to super-efficient heating and cooling systems to new lighting technologies could reduce overall energy consumption by up to 80%. For example, the incandescent light bulb converts 95% of the electricity it uses to waste heat instead of light. Switching to more efficient compact fluorescent bulbs would allow the United States to shut down all its nuclear power plants. Conservation can reduce environmental damage.

In addition to reducing environmental damage, efficiency improvements have two immediate economic advantages. First, as noted, the improvement usually pays for itself. The cost of a technological change that increases efficiency is usually recovered within a few years, and sometimes almost immediately. A study by a major beverage company found, for instance, that each reduction of one-thousandth (.001) of an inch (.0245 mm) in the thickness of their aluminum cans reduces aluminum costs by $1 million per year. (Of course, there is a limit to how thin aluminum cans can be and remain structurally sound.) Also, increased efficiency tends to produce many more jobs than wasteful technologies. The traditional energy industries, such as oil, coal, and nuclear power, are much less labor-intensive than the energy conservation industry that designs and maintains many kinds of energy conservation equipment.

Many other resources in the United States could also be utilized much more efficiently. To list just a couple of examples:

1. *Wood.* The United States converts only about 50% of raw timber directly into furniture and other refined timber products, compared to 70% in Japan.
2. *Water.* The United States uses over twice as much water per person as most other nations on Earth. The average U.S. farmer could typically cut water use by more than half by adopting water conservation measures such as microirrigation that pipes water directly to crops instead of using evaporation-prone irrigation ditches.

Reuse and recycling are the second best way to accomplish input reduction. **Reuse** refers to using the same resource over and over in the same form. An example would be soda bottles that are returned, sanitized, and refilled. **Recycling** refers to using the same resource over and over, but in modified form. The soda bottles, for instance, could be melted to produce new glass bottles. Wastewater, paper, plastics, and many other resources can also be recycled.

In general, reuse is less costly than recycling because the resource is not modified. Both measures are often less costly than extracting "virgin" resources such as aluminum ore or cutting trees for paper, because they usually consume less energy and fewer natural resources than is needed to make products from virgin materials. Recycling aluminum cans, for example, can save up to 95% of the energy cost of cans from newly mined aluminum. Recycling and reuse are thus less costly than using natural raw materials in both economic and environmental terms.

Substitution of one resource for another can benefit the environment in a number of ways. A renewable resource can be substituted for a nonrenewable one, or a less-polluting resource can substitute for a highly polluting resource. Often the newly substituted resource provides both benefits. Substituting renewable, cleaner alternative fuels such as solar and wind energy for fossil fuels is an example. Another example is making products from paper instead of plastics, which are generally made from fossil fuels and last longer in the environment (although there are now biodegradable plastics and plastics produced from plants are being developed). Like conservation, reuse, and recycling, substitution also often yields economic benefits. Studies routinely show that substitution of solar and wind power technologies for the fossil fuel energy now consumed in the United States would produce three to five times as many jobs as now exist in fossil fuel industries.

Promoting Sustainable Technology: Paying True Costs. Some people are surprised to learn that many sustainable technologies were invented centuries ago. Wind and water power are examples. The solar cell was invented in the 1950s. These technologies have failed to become widespread largely for social and economic reasons, not technical ones. When nonrenewable resources are cheap, they will be wasted because people have no incentive to increase efficiency, recycle/reuse, and substitute renewable resources. **Figure 1-17a** shows the situation when resources are cheap and sinks are free and are treated as "commons." Note that there is much throughput, which leads to high rates of resource depletion and pollution.

(a) Cheap natural resources

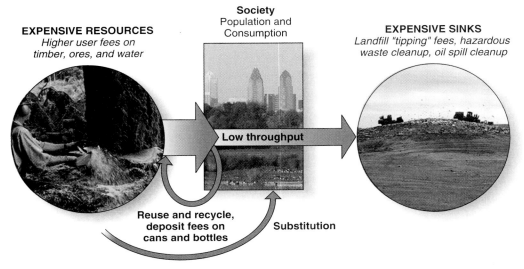

(b) Environmental costs included

Figure 1-17 (a) Excluding environmental costs by allowing cheap resources and free sinks promotes high throughput (and therefore much depletion and pollution). (b) Including environmental costs by imposing user fees and deposit fees promotes an increase in efficiency, recycling, and all other forms of input reduction.

For instance, many private and government studies show that the "true cost" of gasoline is at least $5.60 per gallon when all environmental costs are included. These environmental costs include smog and other urban air pollutants, global warming, contamination of groundwater by leaking underground oil tanks, oil spills, and many other well-known impacts of gasoline use. Yet the price of gasoline in the United States has remained far below $5.60, although most other industrial nations have higher prices.

When market prices do not reflect all the true costs of a product or service, economists call this a **market failure** (also called an "externality"). Many environmental problems can ultimately be traced to market failures. To list just a couple of examples:

1. Electricity from nuclear energy does not include the cost of disposing of nuclear waste; electricity produced by coal burning omits the cost of most air pollution.
2. Water used by many U.S. farmers does not reflect the fact that in many regions the groundwater is being depleted much faster than it is being replenished by rainfall.

Many economists suggest society can correct market failures by imposing **green fees** that adjust the costs of products and services to include environmental costs (**Figure 1-17b**). Fees that increase the price of a resource are particularly effective because they promote conservation of the resource and also reduce pollution by reducing throughput. For example, fossil fuels taxes, such as a "gas tax" or "carbon tax" covering all fossil fuels, would encourage reduced and more efficient use of these fuels.

Other green fees include deposits, such as on cans or bottles, to encourage recycling (Figure 1-17b). Charging for garbage by the bag and imposing effluent (pollution) fees on factories are other examples. These fees motivate people to reduce the waste they produce. Other fees can be levied for using sinks, such as imposing "tipping fees" for landfill use or making polluters pay for cleanup (Figure 1-17b). Another benefit of green fees is that they may improve the environment at lower monetary costs than simply passing regulations and arresting violators.

Although the United States is the world's largest resource consumer and polluter (by most measures), it unfortunately lags far behind most of the industrialized world in the use of green fees and other economic incentives to conserve resources and reduce pollution. The general trend has been the continuation of cheap resources and free sinks. In fact, in many cases the United States actively discourages conservation and encourages pollution by subsidies that reward these activities (as opposed to green taxes that discourage them).

Figure 1-18 shows an example of great importance, the cheap cost of oil, and thus the retail price of gasoline at the pump, in the United States since the early 1980s, which has led to a reversal of the previous trend toward more fuel-efficient cars. Nevertheless, many observers think that a "greener" economy is inevitable because of consumer demand for environmental quality and the economic benefits of sustainable technologies such as increased efficiency, reduced resource imports, and more jobs. There are signs that U.S. business is responding to "green" pressures. We are seeing more companies

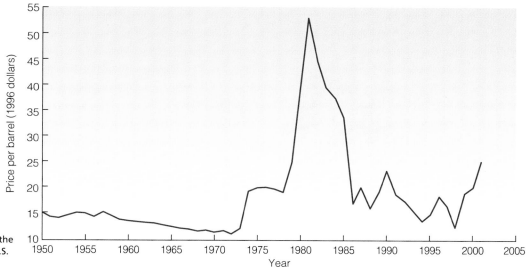

Figure 1-18 Price of oil per barrel in the United States (in 1996 dollars). (*Source*: U.S. Department of Energy.)

advertising that their products are made from reused or recycled materials, or are "environmentally friendly." This trend has been accompanied by increased discussion of green fees and the lowering or removal of government subsidies on timber, grazing lands, ore deposits, and oil companies.

Reducing Population: Defusing the Bomb of the South

Over 90% of world population growth occurs in developing nations. Most developing nations are in the Southern Hemisphere, so this cause of throughput is especially important there. To defuse this so-called **population bomb,** we must address its central causes, which are often economic. **Figure 1-19** illustrates how population growth is often driven by poverty. Poor people tend to have more offspring for a variety of reasons, including lack of education about birth control, lack of economic opportunities for women, and the need for children to perform chores and care for the aged. The result is a vicious circle in which population growth leads to poverty because the increasing number of people reduces the standard of living. The growing population reduces the amount of money per person available for investment in education, equipment, and other needs for economic growth. Impoverished peoples tend not to focus on environmental concerns, but rather to turn their attention to simply surviving from one day to the next.

Can this poverty-population-environment cycle of human misery and environmental destruction be broken? Many people still subscribe to the **fallacy of enlightenment,** or the idea that education will solve the problem. Education is extremely important,

but often it is not enough; realistic solutions that remove the root causes of the problem must be found. Attempting to protect the environment by legal means, such as creating game reserves, is often not effective either because desperate people will break laws. Decades of experience show that the economic causes of this cycle must be addressed. Eliminating poverty will not only reduce many causes of population growth, but will reduce environmental degradation as well.

Reducing Population Growth by Paying True Costs. It has long been known that the economic woes of the developing nations must be solved. In 1961, the United Nations General Assembly pledged that developed nations would donate 1% of their gross national product (GNP) to developing nations. However, instead of improving, the economies of many developing Southern Hemisphere nations have declined since 1961. Indeed, the gap in wealth between the developed and developing nations has widened at an increasing rate. Between 1961 and 1999, the gross economic product per capita increased $15,000 for industrial countries. Africa's gross economic product per capita has increased by approximately $600.

This gap continues to grow—because developing countries now pay a huge amount of interest on money that was lent to them many years ago. Developing countries now owe over $1.5 trillion. As a result of interest on this debt, the net flow of money has been from South to North since 1982. Rich countries now receive more money from poor countries (over $50 billion per year) than they transfer to them. This is sometimes called the **debt bomb.**

Some people argue that the solution to the debt bomb is increased foreign aid. They note that the 1% of GNP pledged in 1961 to developing nations has never been achieved. In recent years the United States spent less than 0.2% of its GNP on foreign aid annually.

Some economists argue that foreign aid to poor countries should not be viewed as charity. They note that these poor but resource-rich nations would not need charity if they were paid the true value for their resources. This view is becoming increasingly common among the people living in the Southern Hemisphere. They maintain that if people in the developed Northern Hemisphere wish to save rainforests, they should pay for the environmental goods and services the forests produce. This would mean paying much higher prices for tropical forest products (such as fruits) and services (such as the trees' absorption of carbon dioxide to slow global warm-

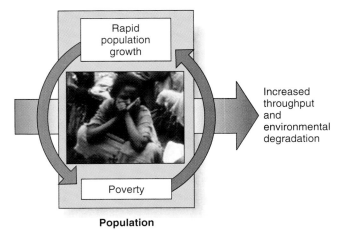

Figure 1-19   Poverty and population growth feed on each other, and both produce environmental degradation.

ing). Although it is very difficult to estimate the true environmental costs and values in most cases, nearly all estimates indicate that developing nations are greatly underpaid for their resources.

Appropriate payment for environmental goods and services could help eliminate poverty, probably much more effectively than foreign aid donations. These payments would break the vicious poverty-population growth cycle and reduce population growth as well. A key necessity is that this increased wealth be used to buy sustainable technologies that focus on efficiency, recycling/reuse, and renewable resources. This proposal contrasts with older views that promoted the "modernization" of developing countries following the same developmental pathway as Europe, North America, and Japan.

The Role of the Individual

Writer Wendell Berry said that the roots of all environmental problems ultimately lie in the values of the individuals who comprise society. Or, as writer Paul Hawken put it, the environment is not being degraded by corporate presidents; it is being degraded by popular demand. By this he means that companies only produce things that people buy, and they cause depletion and pollution only as long as society permits such behavior to be profitable. For example, green fees, such as a carbon tax, could be used to correct the market failures that cause pollution. Yet these strategies are rarely applied in practice. People have generally voted down proposals to add green taxes, even where income taxes and other taxes would be reduced to compensate.

A good example is the "gas tax." In the early 1990s, the Clinton administration initially conceived of a gasoline tax of up to a few dollars per gallon. As gasoline was only about a dollar per gallon, this would have raised gasoline costs closer to, but not fully up to, the estimated true environmental cost. The goal of the tax was to promote fuel conservation and the use of alternative fuels and reduce air pollution including carbon dioxide. After extensive congressional debate, a gas tax of only about four cents was imposed rather than the two or three dollars originally conceived.

Values on the Here and Now: Why We Avoid True Costs

Why do individuals have values that lead to environmental degradation by "popular demand"? Many writers have argued that most large-scale problems arise because individuals are not good at dealing with problems beyond their own immediate situation. We tend to focus on the here and now of current time and local space. We think most about things we can see around us and things that have just recently happened to us or will soon happen to us.

This focus on the here and now is why people have been reluctant, especially in the United States, to pay true costs for goods and services that deplete or pollute the environment. Economists call this discounting the future and discounting by distance. **Discounting the future** results from focusing on the *now*: Environmental costs of our actions on future generations are not fully paid. Cheap gasoline and minerals, for instance, lead to rapid depletion of these nonrenewable resources, making them unavailable to future generations. These resources are cheap because their current cost does not incorporate their value to future generations. Nor does the cost include the future effects of their use, such as global warming or other pollution hazards. **Discounting by distance** results from focusing on the *here*: Environmental costs of our actions on people living in another area are not fully paid. Cheap ivory, imported animals, and tropical timber, for instance, lead to rapid depletion of those tropical resources, thereby degrading the environment for the people who live there. These resources are so cheap because their cost does not incorporate their full value to their local environment.

From an individual viewpoint, being preoccupied with immediate concerns has been a good survival trait. In our evolutionary and historical past, individuals needed to be aware of activity in their immediate vicinity to avoid predators and other dangers. However, modern technologies have led to problems that occur on far larger scales of time and space; their effects are global and last many centuries. Such problems require long-range planning that will span many generations and involve international cooperation. In their book *New World, New Mind* (1989), Robert Ornstein and Paul Ehrlich state that they believe our minds have a "putting out fires" approach to problems; another term might be a "band-aid" approach. They indicate that only recently, with the sustainability movement, has there been widespread interest in addressing the systematic causes that underlie these problems.

A Solution: Values beyond the Self

It can be argued that the single greatest obstacle to building a sustainable society is this tendency of the human mind to focus on the here and now. It limits our ability to make the systemic social changes needed to solve regional and global problems. This

CASE STUDY 1-2

What Are Your Values? Would You Decline $600,000 to Save a Rainforest?

Although many of us say we are pro-environment, we never really know how strong our values are until we are asked to make personal sacrifices for them. For example, in this chapter, we have asserted that the environment of future generations is often discounted or devalued. Most of us would like to change that situation, but just how much are we willing to sacrifice?

Consider the case of Miguel Sanchez, a subsistence farmer in Costa Rica. He was offered $600,000 for a piece of land that he and his family have tended for many years. This land consists of spectacular old-growth rainforest and a beautiful black sand beach in an area where land is rapidly disappearing to development. The money was offered by a hotel developer. Sanchez declined, knowing that this was more money than he, his children, and all his grandchildren would ever accumulate. Sanchez explained his decision: "I have no desire to live anywhere else. Money can be evil. People will ask me for money."

Miguel knows that cleared forest means less rain, less water in the rivers, less escape from the blazing tropical Sun. But he is reluctant to talk about his decision. Hundreds of other farmers and fishermen in his area feel the same way and are fighting development.

(*Source*: Adapted from A. Carothers, "Letters from Costa Rica," *E Magazine*, September 1993.)

Questions

1. How does Miguel's sacrifice compare to sacrifices you have made to promote long-term environmental health? What would you do if you were in Miguel's position?

2. Comment on this statement from writer Andre Carothers: America has "a skewed moral universe, where people regularly swoon over each other's negligible acts of charity." The writer's point is that we make a big deal about recycling cans, buying organic foods, and giving tiny donations to environmental funds. Is this a fair criticism or unjustified cynicism?

---

is why David Ehrenfeld, in *The Arrogance of Humanism* (1981), warns that history shows that humans have never successfully managed anything for very long. How can we expect to solve global environmental problems with such limitations?

Figure 1-20 illustrates how people can extend their sphere of concern, beginning with close relatives and progressing through other social groupings to include all people. Ultimately, all living things can be included, as in the philosophy of "deep ecology." If we are to place a higher value on the environment of the future and in other parts of the world, such extended spheres of concern are needed. If many people adopt this view, society will be more willing to pay green taxes, vote for politicians who favor sustainable lifestyles, and, in general, promote solutions that require looking beyond the here and now.

Most sociologists believe that the trend has been in the opposite direction until now. The growth of modern civilization has promoted individualism that has progressively shrunk the sphere of concern. Large cities have fostered a loss of community and the fragmentation of social groups, including even families. Can this trend be reversed? No one knows, but some encouraging signs have appeared. For example, the grassroots activism of the current sustainability movement is based on community concerns. There are also many examples of exceptional unselfishness in individuals.

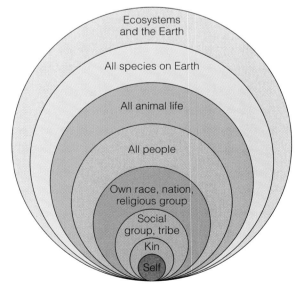

Figure 1-20 An ethical sequence in which the individual concerns extend outward beyond the self to progressively more inclusive levels. (*Source*: R. Noss, 1992).

Computer Networking: The Environmental Internet and You

An increasingly common way for individuals to enhance their role in solving environmental problems is to use the Internet and World Wide Web. While a few environmentalists have been critical of this technology, it has been difficult to avoid a tool that can be so useful to them.

In particular, the Web can help environmentalists in two important ways: networking and information access. Networking allows individuals and organizations to maintain a constant dialogue. This is extremely important, for example, for organizing political and other campaigns at local, national, and even international scales. Information access allows individuals and organizations to support their opinions and campaigns with "hard" data. Given the huge amount of environmental information that is now available on many issues, such easy and rapid access is of critical importance.

Toward a Sustainable World

While no one knows if sustainability will be attained, there is certainly no shortage of debate and speculation. At one extreme are **cornucopians**, who argue that human ingenuity has always overcome environmental limitations. They suggest that such inventions as genetic engineering could make agriculture more productive, for example, so that the planet will be able to support many more people at a high standard of living. At the other extreme are **cassandras** who argue that humans have always altered the environment and managed things poorly; they insist that exponential growth of populations and consumptive technology will finally degrade the environment so much that it will lead to overshoot (see Figure 1-13). Cornucopian and cassandran views have been common, in various forms, throughout history.

As is often the case, extreme views can be counterproductive. Too much optimism, such as a cornucopian view, can lull people into inactivity because they think the future will take care of itself. Too much pessimism, such as a cassandran view, can cause despair, which leads to inaction because people think that the future is bleak no matter what they do. Many writers therefore suggest a more moderate view that still acknowledges that urgent environmental problems exist. These problems need to be addressed within the next few years, or at least decades, or they may lead to large-scale environmental degradation and overshoot. This moderate view relies on the **precautionary principle**, which says that, in the face of uncertainty, the best course of action is to assume that a potential problem is real and should be addressed. In other words, we are "better safe than sorry." The worst that can happen with this approach is that society will become more efficient, less wasteful, and less polluting, even if environmental problems are not as bad as some people argue. On the other hand, the results could be disastrous if the precautionary principle is not used and problems are indeed as urgent as some people say.

Assuming that a person wishes to rely on the precautionary principle, which specific environmental problems should be of most concern? A good starting place is the following priority list based on material published by the U. S. Environmental Protection Agency and other sources:

Highest Priority. Species extinction, ecological disruption and habitat destruction, stratospheric ozone depletion, and global climate change.

Middle Priority. Pesticides and herbicides, lake and river pollution, acid rain, and airborne toxics.

Lowest Priority. Oil spills, groundwater pollution and depletion, airborne radiation, waste dumping, thermal pollution to water bodies, and acid runoff.

Note that the highest priority problems are global and very long-term in their scope. Lower priority problems tend to be geographically more localized and can (in theory) be solved within shorter time spans.

Visualizing a Sustainable World

Recall that sustainability means meeting the needs of today without degrading the environment for future generations. **Table 1-1** summarizes many of the changes that are necessary to create a sustainable society. Most of these changes will be discussed later in the book, but for now some major points can be noted. First, many aspects of society are affected: scientific paradigms, role of the human, values toward nature, land, and people. Second, many social institutions are affected: religion, education, political systems, and economic systems. Third, technology and agriculture are affected. Fourth, many of these changes are ultimately based on increasing the importance of the community, from encouraging community values among individuals to creating decentralized, community-based economies and political systems.

The need for such pervasive changes is reflected in academic disciplines such as the social sciences and humanities, which have traditionally studied society as an entity distinct from the natural environment, or at most only moderately influenced by the natural world. For example, economics once fo-

TABLE 1-1   Transition from the Industrial to the Ecological Age

Some of these may be considered speculative attempts to understand the personal and humanistic aspects of sustainable living.

	Industrial Age	Sustainability or Ecological Age
Scientific paradigms	Mechanistic Earth as inert matter Determinism Atomism	Organismic Gaia: Earth as superorganism Indeterminacy, probability Holism/systems theory
Role of the human	Conquest of nature Individual versus world Resource management	Living as part of nature Extended sense of self Ecological stewardship
Values in relation to nature	Nature as resource Exploit or conserve Anthropocentric/humanist Nature has instrumental value	Preserve biodiversity Protect ecosystem integrity Biocentric/ecocentric Nature has intrinsic value
Relation to land	Land use; farming, herding Competing for territory Owning "real estate"	Land ethic: think like mountain Dwelling in place Reinhabiting the bioregion
Human/social values	Sexism, patriarchy Racism, ethnocentrism Hierarchies of class and caste	Ecofeminism, partnership Respect and value differences Social ecology, egalitarianism
Theology and religion	Nature as background Nature as demonic/frightening Transcendent divinity Monotheism and atheism	Animism: everything lives Nature as sacred Immanent divinity Pantheism and panentheism
Education and research	Specialized disciplines	Multidisciplinary, integrative
Political systems	Nation-state sovereignty Centralized national authority Cultural homogeneity National security focus Militarism	Multinational federations Decentralized bioregions Pluralistic societies Humans and environment focus Commitment to nonviolence
Economic systems	Multinational corporations Competition Limitless progress "Economic development" No accounting of nature	Community-based economies Cooperation Limits to growth Steady state, sustainability Economics based on ecology
Technology	Addiction to fossil fuels Profit-driven technologies Waste overload Exploitation/consumerism	Reliance on renewables Appropriate technologies Recycling, reusing Protect and restore ecosystems
Agriculture	Monoculture farming Agribusiness, factory farms Chemical fertilizers and pesticides Vulnerable high-yield hybrids	Poly- and permaculture Community and family farms Biological pest control Preserve genetic diversity

Source: From Ralph Metzner, "The Emerging Ecological World View," in Mary Tucker and John Grim, eds., *World Views and Ecology* [Cranbury, New Jersey: Associated University Presses, 1993]: 170–71. Reprinted with the permission of Associated University Presses.

cused almost exclusively on the flow of goods through society, ignoring the costs of depletion and pollution that the flow imposed on the environment. Similarly, the environment was studied as a largely separate entity by other disciplines such as ecology. The new field of environmental economics is a more holistic discipline that includes both society and the environment. It studies not just the flow of goods within society but also the effects on the environment.

Other socially oriented disciplines are also rapidly developing an interest in studies that incorporate the environment. Findings from such fields are incorporated throughout this book, and are also emphasized in Section 5 (Environmental Issues: Social Aspects and Solutions).

SUMMARY

- Environmental science uses a holistic approach to view the interconnected aspects of environmental problems and apply solutions.
- Human-environmental interactions evolve through five basic stages: gathering and hunting; agriculture and conservationism; industry and environmentalism; transition and sustainability; and the postindustrial stage.
- We are in the transition phase, which has produced the sustainability movement.
- The sustainability movement:
  - Focuses on reducing societal consumption of resources;
  - Seeks to solve holistically the structural social causes of environmental problems such as poverty.
  - Relies on grassroots activism.

- The postindustrial stage will likely witness either sustainability or catastrophic overshoot of human civilization.
- Environmental impact—the alteration of the natural environment by human activity—involves either resource depletion or pollution.
- Impact is promoted by population and consumption: $I = P \times C$.
- The world's population continues to increase by about 77 million people a year.
- Traditionally, resource-intensive industrial technology and consumption have increased exponentially.
- Impact is often measured as throughput—the movement of materials and energy through society.

KEY TERMS

Cassandras
Closed system
Complexity
Consumption
Cornucopians
Debt bomb
Discounting by distance
Discounting the future
Ecocentric
Efficiency improvements
Environmental science
Environmental wisdom
Exponential growth

Fallacy of enlightenment
First law of ecology
Gaia hypothesis
Grassroots activism
Green fees
Holistic
Input
Input reduction
Integration
Market failure
Open system
Openness
Output

Population bomb
Positive feedback
Precautionary principle
Recycling
Reuse
Sinks
Sources
Substitution
Sustainability
Sustainable economy
Sustainable technology
Systems approach
Throughput

STUDY QUESTIONS

1. What are the two major goals of this book?
2. Define *environmental science*. What fields of study does it encompass? What relevance does it have to humans?
3. Name several costs of a lack of environmental wisdom.
4. What is often the most expensive way to attempt to solve environmental problems?
5. What is the throughput model and the systems approach as applied to environmental science?
6. Why is input reduction a good solution to many environmental problems?
7. Name the two basic causes of environmental problems.
8. What are the key traits of the environmental system?
9. State and explain the "first law of ecology."
10. What is environmental impact? Identify two basic types of environmental impact and explain their causes.
11. Why was the cultivation of food such an important development in human technological evolution?
12. What are the five basic developmental stages in the relationship between humans and the environment? Briefly describe each.
13. Name three pieces of landmark environmental legislation enacted by Congress between 1970 and 1990.
14. Define *sustainable technology* and a *sustainable economy*.
15. Explain the concept of "The Tragedy of the Commons" and how it applies to environmental issues.

STUDY GUIDE

STUDY QUESTIONS

16. What are some basic ways that have been suggested to reduce consumption? Does a reduction in consumption need to entail a decreased quality of life?

17. What is a green fee? What does "market failure" refer to?

18. What is "discounting the future"?

19. What is the "precautionary principle" and how can it be applied to environmental issues?

WHAT'S THE EVIDENCE?

1. The authors contend that many people suffer from information overload and thus "tune out." Do you really believe that this is true? Isn't it better to have more information rather than less? Cite examples that either support or contradict the authors' assertion relative to information overload and a lack of environmental wisdom on the part of some people.

2. The authors suggest that in the context of the history of human environmental impact, the United States and other developed nations are currently in a transition stage and the decisions we make over the next few decades (in your lifetime) will determine the long-term fate of the environment. Do you believe this assertion? Or is it simply alarmist rhetoric on the part of the authors? Cite specific and general examples to support your position.

ILLUSTRATION AND TABLE REVIEW

1. Study Figure 1-12. Which of the three statistics shown (total solid waste production, per capita waste output, population) has increased most quickly over the last few decades? Why? How are these three factors related to one another?

2. From Figure 1-18, determine the approximate price per barrel of oil in 1980.

3. From Figure 1-18, determine the approximate price per barrel of oil in 2000. How does this compare to the price twenty years earlier?

WWW.ENVIRONMENT.JBPUB.COM

www.environment.jbpub.com

Connect to this book's web site: http://www.environment.jbpub.com/ The site features eLearning, an online review area that provides quizzes, chapter outlines, and other tools to help you study for your class. We've posted the answers to the book's odd-numbered questions. You can also follow useful links for more in-depth information. The site provides sources for the information in this chapter and recommendations for further readings.

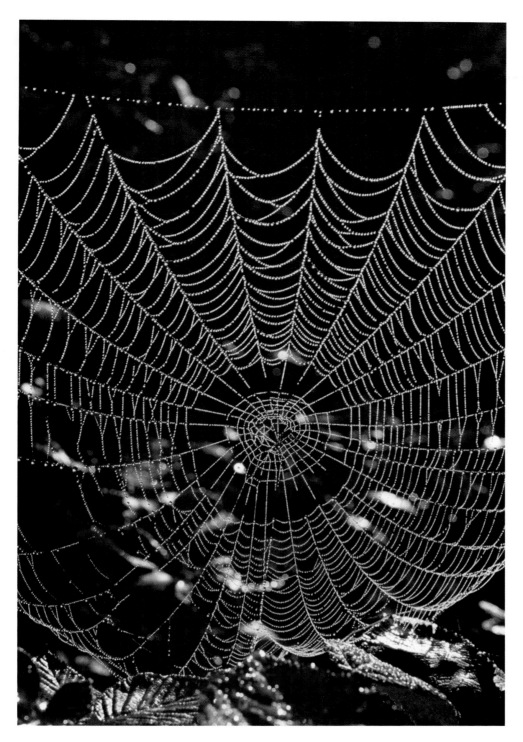

It really boils down to this: that all life is interrelated. We are all caught in an inescapable network of mutuality, tied into a single garment of destiny. Whatever affects one directly, affects all indirectly.

—Martin Luther King, Jr. (1929–1968), Trumpet of Conscience

The world's population growth rate is now declining. But back in the late 1960s the world population was growing at a record rate of approximately 2% a year, and some pessimists feared the worst—massive famine, disease, wars, civil unrest, and widespread global ecosystem collapse. These dire predictions have not come to pass on a worldwide scale, but certain regions are experiencing these problems. Millions of people in Africa have been displaced by drought, war, or political upheaval. Shown here is a relief camp in Ethiopia.

CHAPTER 2

Human Population Growth

Earth is now a very crowded planet. Currently (2003), approximately 6.3 billion people inhabit Earth. Nearly 9,000 people are added each hour, over 200,000 persons a day. The global population increased by approximately 77 million persons in 2001 (131.5 million babies were born and roughly 54.5 million individuals died), a rate of about 1.26% per year. The growth rate, however, appears to be decreasing.

Human overpopulation is one of the central issues in environmental science. As noted in Chapter 1, high population levels increase both major types of environmental problems: (1) resource use and (2) pollution and waste. For example, our species alone utilizes, either directly or by diverting it from other uses, an estimated 40% of the world's terrestrial green plant production. The other 60% is divided among the remaining 4 million to 50 million or more terrestrial species (each species being composed of untold numbers of individuals) with which we share the globe.

Demography is the study of the size, growth, density, distribution, and other characteristics of human populations. No longer can population, or overpopulation, be viewed simply as a cause of environmental destruction; rather, it must be viewed as a major symptom of underlying social, economic, and environmental issues. Poverty, high fertility, and environmental degradation go hand in hand, each reinforcing the other. In this chapter we will review the early development of human society, trace the initially slow and subsequently **exponential growth** of the human population, and begin to explore the ways in which human population pressure has progressively modified the environment. We will also briefly discuss what is being done to address the current "population crisis."

World Population Changes over Time

Natural animal populations often exhibit an "S-shaped" growth curve—slow growth (**lag phase**), then rapid growth (**exponential phase**), then slowing down as limits are reached, and finally a leveling out (**Figure 2-1**). The same curve can be applied to human populations. On a global scale, the human population is still in the rapid growth phase.

Starting Slow: The Lag Phase

About one million years ago, the total population of our ancestors was perhaps only some 125,000. From this small size, the population slowly increased until by about 8000 B.C. it numbered about 5 to 10 million (**Figure 2-2a**). This slow increase was due to (1) the invention and development of tools, which enabled food and other necessities to be procured more efficiently and thus allowed more people to be supported in the same area, and (2) migration to new areas, as from the Old World to the Americas. Although we tend to focus on chipped stone tools because they are most readily preserved in the archaeological record, many organic tools (woven nets, baskets, clothing, tents and other shelters) also were invented. Also "discovered" and used during this period was fire—to heat, cook food, and scare away potentially harmful predatory animals.

Before 10,000 years ago, humans had spread through most of the world, reaching all of the mainland landmasses except Antarctica and the islands of the Caribbean, Polynesia, Madagascar, and New Zealand, all of which were colonized subsequently. This simple spread of people to new areas allowed the world population to increase.

The extent to which early humans impacted and permanently modified their environments remains in question. Some researchers believe that through overhunting and competition for environmental resources, the Upper Paleolithic hunters of Europe, Africa, Asia, and the Americas were responsible for global patterns of extinction among large mammals, such as the woolly mammoth and the woolly rhinoceros, during the later Pleistocene (around 100,000 to 10,000 years ago).

Human modification of the environment is thus nothing new; many habitats that we think of as "natural" may in fact be the product of human intervention over tens of thousands of years. What is new at the beginning of the twenty-first century is the extent, degree, and rate at which humans are modifying their environment—much more quickly than we or the rest of nature can adapt to the new conditions.

The Agricultural (Neolithic) Revolution: Beginning Exponential Growth

The advent of domestication and agriculture about 10,000 years ago led to a sharp rise in the human **birth rate** for reasons discussed below. Though often referred to as the **agricultural (Neolithic) revolution**, this was actually a gradual process that extended over eight or more millennia and occurred independently at different rates in different parts of the world. Full domestication of many plants and animals seems to have been established in the Old World by 8000–5000 B.C. Once agriculture and domestication were entrenched, the delicate near balance between births and deaths, which apparently had held for hundreds of millennia, was broken, and the world's human population began to increase dramatically (see Figure 2-2). By A.D. 1, the world population had increased to roughly 300 million, and cities such as Rome had populations as high as one million. By contrast, the world population had taken a million or more years to reach a total of, at most, 10 million people.

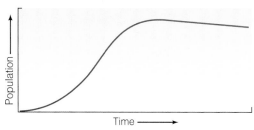

Figure 2-1 A typical organismal population growth curve.

Figure 2-2   World human population changes over time. (a) Estimated population growth in billions since the neolithic age. (b) World population projections until 2050 (medium-, high-, and low-fertility variants) in billions. Note that the time scales for the two graphs are different. The first graph is in one-hundred-year increments and appears to show a vertical and unending rise in population. The second graph is in ten-year increments, and shows the population beginning to level out at under 10 billion in approximately 50 years (using the medium or low variant). (*Sources:* (a) U.S. Census Bureau, International Programs Center, 2001; (b) United Nations Population Division, *World Population Prospects: The 2000 Revision, February, 2001.*)

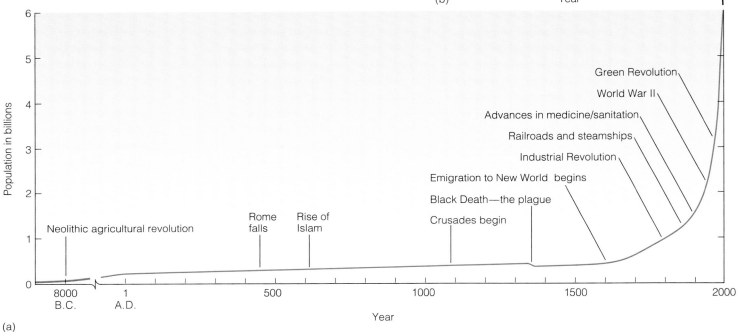

The dramatic increase in human population has been attributed to several factors:

1. Settlement on farms may have allowed women to bear and raise more children; freed from the nomadic lifestyle, they no longer had to carry young offspring for great distances. Previously, women may have chosen to raise fewer children, perhaps using natural contraceptives and practicing primitive forms of abortion or infanticide to get rid of unwanted offspring.

2. Children are more useful in agricultural communities than in gatherer and hunter cultures and therefore more highly valued. This would have increased the incentive to have more children.

3. Agriculture and domestication may have made softer foods available, which allowed mothers to wean their children earlier. Thus, women could bear additional children over the course of their lifetimes.

4. Agriculture and domestication, by their very nature, allowed and promoted higher densities of people. Indeed, with farming, one family or group of persons could raise more food than they personally needed. This surplus led directly to the rise of cities and civilization because it allowed people to develop and concentrate on manufacturing, trading, and other specializations. In a classic example of positive feedback, this in turn led to

rapid advances in technology, art, and other innovations.

After late ancient times, the world's population slowly but steadily increased, except for a slight decline in the fourteenth century due to the plague (see Figure 2-2a), until the mid-seventeenth century when it totaled approximately 500 million. Since about 1650, the human population has grown at an ever increasing rate, reaching 800 million around 1750, 1.2 billion in 1850, slightly over 2.5 billion in 1950, 4.44 billion in 1980, and 6.1 billion in 2000. Another way of looking at this is in terms of how long it has taken to add each additional billion people to the world's population (**Table 2-1**). While it took until approximately 1800 for the human population to reach the first billion, it took 130 years to add the second billion. More than four billion were added in the twentieth century, with the last billion added in just twelve years. Even as population growth slows in the twenty-first century, still another four billion could be added to the globe by 2100. Recent United Nations projections estimate that the world population will reach 6.8 billion in the year 2010 and 9.3 billion in 2050 (using a medium-fertility projection, generally considered as "most likely"; **Figure 2-2b**). However, these estimates are based on many uncertain variables, such as how quickly modern **birth control** methods will spread and **fertility rates** drop. Even assuming that continued advances are made in decreasing global fertility rates and increasing global use of contraceptives, the global population may not stabilize until it reaches around 10 billion toward the end of the twenty-first century!

TABLE 2-1	How Long Has It Taken to Add an Additional One Billion People to Earth's Human Population?
World Population reached:	
1 billion in ~1800	
2 billion in 1930 (130 years later)	
3 billion in 1960 (30 years later)	
4 billion in 1974 (14 years later)	
5 billion in 1987 (13 years later)	
6 billion in 1999 (12 years later)	
World Population may reach:	
7 billion in 2013 (14 years later)	
8 billion in 2025 (17 years later)	
9 billion in 2044 (19 years later)	

Source: United Nations Population Division, *World Population Prospects: The 2000 Revision*, February, 2001.

Carrying Capacity

Many natural populations of organisms increase very rapidly when invading a new area, but then population growth slows as the upper limit of the number of individuals that the area can support is approached. Eventually, an equilibrium population size is reached. This pattern can be conceptualized in terms of the **carrying capacity** for the particular population of organisms in a given area. The carrying capacity can be thought of as the number of individuals of a certain population that can be supported in a certain area for a prolonged period of time given the resources of that area.

Applying these generalizations to human world population growth, we can ask: What is the carrying capacity of Earth relative to humans? This question has yet to be satisfactorily answered. Some believe that humans have already exceeded the world's carrying capacity (see Case Study 2-1 for an early proponent of this idea). How could this be? Remember that many people alive today are unable to meet their basic nutritional requirements. When a species lives within its carrying capacity, it does not degrade the resources upon which it depends. While degrading our resources (for example, through topsoil erosion and the pumping of underground water for irrigation faster than it is renewed), we produce enough food globally to feed our population of over 6.3 billion, but an estimated 500 million to 1.1 billion of those people are undernourished (estimates vary depending on the definition of "undernourished"). Given current global food production, it has been suggested that our 6.3 billion could be fed adequately if everyone kept to a vegetarian diet.

On the other hand, some persons believe we have not even begun to approach the physical limits on human population size. With appropriate technology and distribution systems for foodstuffs and other necessities, some argue that Earth might be able to support 40 to 50 billion humans (although for how long is another issue). The human carrying capacity also depends on the standard of living that is acceptable for the average human. To a certain extent, we can conceptually increase the world's carrying capacity by lowering the standards at which we are willing to exist.

One can argue that the world's human carrying capacity has systematically changed over time. Although physical and biological factors, such as the withdrawal of the ice and general global warming at the end of the Ice Ages about 10,000 years ago and the evolution or introduction of new plant or animal species, may have allowed human densities to increase, the prime factors increasing the human carrying capacity have probably been our ability to

CASE STUDY 2-1

Thomas Malthus, the Original Population Pessimist

Visit www.environment.jbpub.com for more information.

The English political economist Thomas Robert Malthus (1766–1834) is generally credited with being the first modern pessimistic thinker concerning population growth rates and the overpopulation problem. Indeed, the term **neo-Malthusian** is often used to refer to those who believe that the modern rapid increase in human population is extremely detrimental. Neo-Malthusians generally believe that we will run out of resources and seriously damage or destroy our environment unless we can control our breeding.

In his work *An Essay on the Principle of Population, as It Affects the Future Improvement of Society* (1798, revised and enlarged in 1803), Malthus suggested that while the size of a population increases geometrically or exponentially, the means that support the population tend only to increase arithmetically; thus, increasing populations invariably outstrip their resource bases (**Figure 1**). As this occurs, the poor get poorer and more desperate, leading to misery and vice. If humans do not intervene of their own accord, such as through "moral restraint" (restraint in breeding), then the population increase will ultimately be checked by natural means, such as widespread famine, disease, and possible warfare. Although Malthus did not necessarily advocate direct birth control, he did suggest that early marriages should be avoided and self-restraint should be cultivated. Most neo-Malthusians are strong proponents of accessible birth control and family planning services.

Since his *Essay* first appeared, Malthus's ideas have been widely discussed—both admired and heavily criticized—and have greatly influenced subsequent generations. The effects of his writings have been profound in economic, political, social, and biological circles. It was from reading a version of Malthus's *Essay* that Charles Darwin hit upon the idea of natural selection and "survival of the fittest" as the major mechanism underlying biological evolution.

Figure 1 A "Malthusian view" of an overcrowded London of the future is shown in this 1851 George Cruikshank etching.

grow our own food and our increasingly advanced technology and culture.

The development of stone tools, the harnessing of fire, and especially agriculture and domestication allowed humans to increase in population size from the hundreds of thousands to the hundreds of millions. Essentially, humans increased the carrying capacity of Earth for themselves.

Consider the last several hundred years. Specific developments that have increased the human population since the mid-seventeenth century include advances in agriculture, such as growing legumes, which replace nitrogen in the soils. This allowed fields to be cultivated continuously, rather than having to lie fallow every third year; the result was in-

creased food production that could support a growing population. Also important was the development of modern theories of disease (the "germ theory") and sanitation, which decreased infant mortality and increased life expectancy. Vaccines for diseases were developed. Food preservation and storage were vastly improved. And, of course, the **Industrial Revolution** allowed the cheap and efficient mass production of necessary commodities. Since the mid-nineteenth century, we have seen continued improvements in mass production of foods and goods, increasing advances in medical fields, and the so-called **Green Revolution** of the mid-twentieth century, which allowed greatly increased crop yields (see Chapter 13). All of these developments have not

FUNDAMENTALS 2-1 Fundamentals: Human Population Equations and Statistics

The basic equation describing the growth of a population (the change in the size of a population, represented by ΔN, that takes place over some interval of time, represented by Δt) is $\Delta N/\Delta t = rN$, where N is the population size, t is time, and r is the intrinsic rate of increase of the population (birth and recruitment rate minus death and emigration rate). If we are discussing the global human population as a whole, there is no recruitment or emigration, so the intrinsic rate of increase is simply the birth rate minus the death rate.

In discussing human populations, the following basic statistics are commonly used:

Crude birth rate. The number of births per year per 1000 members of a population. The crude birth rate is commonly determined by dividing the total number of births in the given population by the midyear population size (which gives the number of births per individual per year) and then multiplying by 1000.

Crude death rate. The number of deaths per year per 1000 members of a population. The crude death rate is calculated in a manner comparable to the crude birth rate.

Rate of natural increase (*essentially r*). The crude birth rate minus the crude death rate. The rate of natural increase can be expressed in terms of number of additional humans per 1000 members of the population at midyear. Thus, a rate of natural increase of 20 per 1000 (= .02) means that the population is increasing overall by 20 individuals per 1000 members of the population each year. A population of 10,000 at the midpoint of one year would grow to a population of 10,200 by the midpoint of the following year. When the rate of natural increase is a negative number, the population is decreasing in size over time rather than increasing.

Percent annual growth (*or change*) *of a population.* The rate of natural increase expressed as a percentage of the given population. For example, if the rate of natural increase is 20 per 1000 (20/1000 = 0.02), the percent annual growth is 2%.

Doubling time of a population. The amount of time that a population of a given size at time zero, increasing at a fixed rate, will take to double in size. Essentially, the annual growth rate of a population is equivalent to the compounding of interest on money in the bank; that is, as dollars of interest added to the account will themselves earn interest, so too will persons added to a population give birth to more people. To demonstrate, if the annual growth rate is 1.0%, then the doubling time will be 70 years (rather than 100 years as it would be if "interest" were not compounded). Likewise, the doubling times of populations

with growth rates of 2.0, 3.0, or 4.0% per year are 35, 24, and 17 years, respectively. The doubling time of a population can be calculated roughly by dividing the percent annual growth into 70. For example, a population with a percent annual growth of 2.0 will double in size in about 35 years (70/2.0 = 35).

General fertility rate. The total number of births in a population in any given year as a function of the total number of women in their reproductive years, variously defined as from age 15 to age 44 or 49. Often the general fertility rate is expressed in terms of number of births per 1000 women in their reproductive years and thus is calculated similarly to the crude birth and death rate, except that the number of births for a given year is divided only by the midyear number of reproductive-aged women in the population.

Total fertility rate (*TFR*). Basically, the number of children a woman of a given population will have, on average, during her childbearing years. Replacement level is generally considered to be approximately 2.1 children per woman. (Women must replace not only themselves, but their male partners as well. All other factors being equal, more males tend to be born than females—about 105.5 male births for every 100 females born—and not all children make it to their reproductive years.)

Infant mortality rate. The number of babies that die before their first birthday, given that they are born alive. Babies that are born but die shortly thereafter are included in both the crude birth rate and the crude death rate.

Life expectancy at birth. The average number of years that a typical newborn baby can expect to live when it is born. Life expectancy can change over time; for instance, in certain populations the biggest hurdle to the newborn may be to survive for its first five years.

Carrying capacity. The equation $\Delta N/\Delta t = rN$ maps an exponential J-shaped curve of population growth. This equation predicts astronomical population sizes after relatively short periods of time, and it does not take into account any factors that may limit the growth of a population. The equation for the J-shaped curve can be modified as follows:

$$\Delta N/\Delta t = rN\,[(K - N)/K]$$

where K is the equilibrium limit of the population, or the carrying capacity. This gives an S-shaped curve. The population size increases, but eventually levels out as factors that limit population growth take effect—such as lack of food, space to live, and so forth.

only increased the world's theoretical human carrying capacity, but also promoted the actual growth of the world population.

In the last few decades, though, major environmental changes such as the buildup of greenhouse gases and the destruction of the ozone layer have been taking place on a global level. Some observers have interpreted these changes as indicating that we have finally reached, and perhaps begun to exceed, Earth's carrying capacity for humans.

Growth Rate and Doubling Time of the World's Human Population

While the human population **growth rate** has generally decreased in the last quarter century, over the broad span of human history, estimates show that the **doubling time** for the human population was getting smaller and smaller until just recently, meaning that the growth rate was increasing. Between 10,000 years ago and A.D. 1650, the **annual growth rate** was less than 0.1% a year for a doubling time of over 1000 years. Between 1650 and approximately 1800, however, the human population doubled again, reaching one billion at the start of the nineteenth century. This was a doubling time of only 150 years, or an average annual growth rate of just over 0.46%. In the next 130 years or so, the population doubled again, giving an average annual growth rate approaching 0.54%. In the late 1960s and into the 1970s, the annual growth rate hovered around 2.0%, and thus the doubling time was about 35 years. An annual growth rate of 2.0% may seem small, but it is enormous—especially in comparison with the annual growth rate during the vast majority of human history and prehistory. The human species was adding more and more individuals at a faster and faster pace.

Between 1970 and 2001, the annual rate of growth of the world's population began to decrease, from about 2.0% per year to between 1.2% and 1.3%. According to some UN projections, using a medium-fertility projection, the annual growth rate could continue to decline until it reaches about 0.47% in 2045–2050. In the last half of the twenty-first century, it may continue to decline until it reaches zero or possibly even a negative number (that is, the population would shrink) in future centuries. This steady decline in the annual growth rate of the world population is extremely significant—but any growth rate above zero means that the already huge global population is continuing to increase in size. If there are 9.3 billion people on Earth in 2050 and the growth rate at that time is 0.47%, that will mean that nearly 44 million people will be added to the world's population in 2050 (**Figure 2-3**). However, this is less than the number of people currently added to the world on an annual basis. An annual growth rate of 0.47% corresponds to a doubling time for the population of about 148 years.

Today (2003) the world population stands at 6.3 billion, and opinions differ as to where it will stand in the future. In 1988, the United Nations predicted a global human population of 6.25 billion in the year 2000. But that was when the growth rate was 1.7% and the global average fertility (children per woman) was approximately 3.6. It's possible the world's population may now finally be coming under control. Another billion people were added to Earth in the twelve years from 1987 to 1999, but the growth rate continues to drop dramatically and the end may be in sight. Today, the global average fertility rate is 2.68 and many countries, especially in Europe, have below-replacement fertility levels. Their populations are decreasing rather than increasing (see further discussion below). After the global population is stabilized at around 10 billion it could even begin to fall during the twenty-second century as populations age and fertility rates continue to decline.

In Chapter 1 we discussed the conceptual equation I = P × C (Impact = Population × Consumption). Once the population (P) aspect of this equation is relatively stabilized, we can focus on environmentally friendly technologies that can serve to feed, clothe, shelter, and otherwise provide for all the people on Earth while also lessening the negative human impact on the planet. When we began writing the first edition of this book in the late 1980s, we were much less optimistic about the long-term future of humans on Earth than we are now.

Distribution of the Earth's Human Population

The present human population is distributed somewhat unevenly over the Earth (**Figure 2-4; Table 2-2**)—

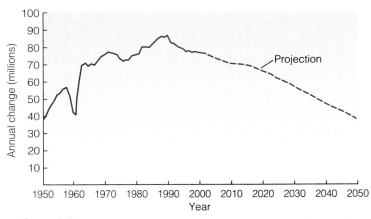

Figure 2-3   Annual increases in world population, 1950–2050. (*Source*: U.S. Census Bureau, International Programs Center, 2000.)

Figure 2-4 Map of world population densities. Note that one square kilometer equals approximately 0.386 square mile; thus a population density of 10 persons per square kilometer equals approximately 26 persons per square mile.

TABLE 2-2   **Population Distribution for World and Major Areas, 1750–2050 (millions)**

Major area	1750	1800	1850	1900	1950	2000	2050
World	791	1000	1262	1650	2520	6057	9322
Africa	106	109	111	133	221	794	2000
Asia	502	649	809	947	1399	3672	5428
Europe	163	208	276	408	548	727	603
Latin America and the Caribbean	16	25	38	74	167	519	806
Northern America	2	7	26	82	172	314	438
Oceania	2	2	2	6	13	31	47

Sources: United Nations, 1973. *The Determinants and Consequences of Population Trends*, Vol.1. (United Nations, New York); United Nations, *World Population Prospects: The 2000 Revision*, February 2001. (United Nations, New York).

even more so in terms of access to and use of resources. A select few tend to be moderately to very well off, while many lead mediocre, marginal, or substandard existences. Likewise, such basic statistics as infant mortality, crude birth and death rates, and longevity vary widely from country to country. In 2000, approximately 1.2 billion people (or 19%) lived in **more developed countries** or MDCs (basically the countries of North America, Europe, Japan, Australia, New Zealand, and the former Soviet

Union), whereas the remaining 4.9 billion people (81%) lived in **less developed countries** or LDCs. As **Figures 2-5** and **2-6** and Table 2-2 demonstrate, past and predicted future changes in population size are unevenly distributed among the continents and between the MDCs and LDCs. Whereas the population growth curve for Europe is essentially flat, it is sharply rising for Asia and Africa.

The population of Asia is increasing by approximately 49 million a year, while that of Africa

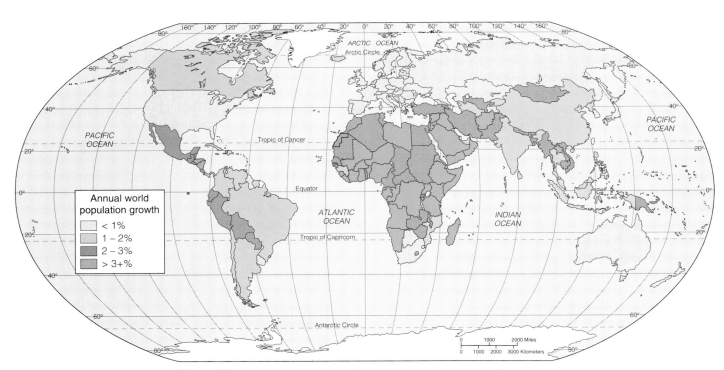

Figure 2-5   Expected annual population growth rates around the world in 2000–2005. (*Source*: United Nations Population Division, *World Population Prospects: The 2000 Revision*, February, 2001.)

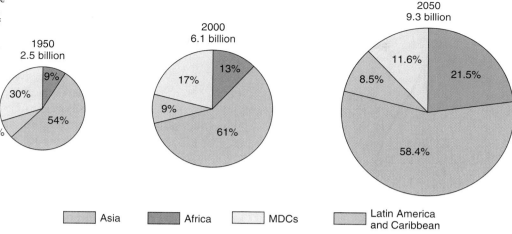

Figure 2-6   The shifting balance of the world's population by region, 1950–2050. Note that no growth in the population of the present MDCs is predicted between 2000 and 2050, yet the size of the world's population will continue to increase. Therefore, people of those industrial countries will account for a much smaller percentage of the world's population in 2050. In contrast, the populations of Asia and Africa will increase sharply, with the highest growth rates occurring in Africa. In 2050 the number of Asians could be nearly the number of all people on Earth in 1998. (*Source*: United Nations Population Division, *World Population Prospects: The 2000 Revision*, February, 2001.)

is increasing by 18 million a year, and the collective population of Latin America and the Caribbean is increasing by almost 7 million a year. Africa has the highest growth rate, standing currently at 2.4% a year. At the other end of the scale, Europe has the lowest growth rate of any continent (0.1%), and Eastern Europe actually has a negative growth rate (−0.5%) which means that the population is declining. At the beginning of the twenty-first century, only ten countries accounted for 60% of the world's population growth (**Table 2-3**). Of these, China and India accounted for about a third of the world's population growth, but the United States was in the top ten. Based on UN median variant projections, in 2050, the United States could be the third most populous country in the world (with an estimated population of 397 million), after India (2050 esti-

mated population of 1.572 billion) and China (2050 estimated population of 1.462 billion). By 2050, however, China may actually be experiencing negative population growth. It is predicted by the UN that the more developed regions of the world (including Japan and all of Europe) will reach a collective population peak of about 1.181 billion in 2050, and then decline after that. However, during the same time, the population of the less developed regions overall will continue to increase in size (with the fastest growth in Africa), reaching a projected 8.2 billion in the mid-twenty-first century.

Age Structures

When comparing populations between different countries and regions, one must consider the **age**

No.	Country	2001 Population	Net Addition	Percent	Cumulative Percent
TABLE 2-3		**Top Ten Contributors to World Population Growth, 2000–2005 (net annual additions in thousands)**			
1.	India	1.008 million	15,929	20.7	20.7
2.	China	1.275 billion	9,246	12.0	32.7
3.	Pakistan	141 million	3,818	5.0	37.7
4.	Nigeria	113 million	3,172	4.1	41.8
5.	Bangladesh	137 million	3,023	4.0	45.8
6.	Indonesia	212 million	2,679	3.4	49.2
7.	United States of America	283 million	2,567	3.3	52.5
8.	Brazil	170 million	2,136	2.8	55.3
9.	Dem. Rep. of the Congo	50 million	1,852	2.4	57.7
10.	Ethiopia	62 million	1,611	2.1	59.8
	Sub-total		46,003	59.8	59.8
	World total		76,857	100	100

Source: United Nations Population Division, *World Population Prospects: The 2000 Revision*, February, 2001.

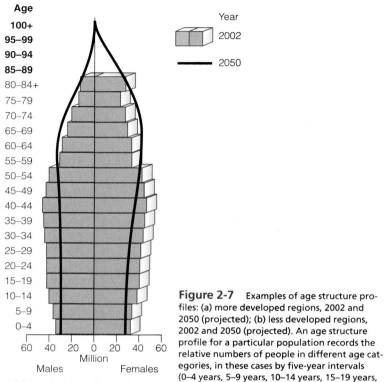

Age

100+
95–99
90–94
85–89
80–84+
75–79
70–74
65–69
60–64
55–59
50–54
45–49
40–44
35–39
30–34
25–29
20–24
15–19
10–14
5–9
0–4

Year

2002

2050

60   40   20   0   20   40   60
Million

Males   Females

(a) More developed countries

Figure 2-7 Examples of age structure profiles: (a) more developed regions, 2002 and 2050 (projected); (b) less developed regions, 2002 and 2050 (projected). An age structure profile for a particular population records the relative numbers of people in different age categories, in these cases by five-year intervals (0–4 years, 5–9 years, 10–14 years, 15–19 years, and so on). (*Source*: U.S. Bureau of the Census, International Database.)

structure profiles or the population age profiles. The age structure of a particular population is essentially a frozen profile of the population at any one instant (**Figure 2-7**); the age structure of a population will often change over time.

There are two common patterns in age structures today. A typical Western industrialized country, which is an example of an MDC, has a relatively flat, vertical, or uniform age structure profile (Figure 2-7a). That is, in each age category from about age 0 to 30 or older, there are approximately the same number of people or even slightly more people in older categories up until about age 45. Of course, as people get much older than 50, their numbers drop off dramatically as they die of old age. In comparison, a typical LDC (Figure 2-7b) has an age profile that is strongly skewed toward the younger categories, indicating that there are many more younger people than older people.

In absolute numbers, a population that is skewed toward the young will continue to grow even as the birth rate falls. In fact, the fertility rate of a population can drop to, or below, the replacement fertility rate (approximately 2.1 children per woman) and the population will still continue to increase in size for some time. As the population ages, more women will reach their reproductive years (generously regarded as 15 to 49 years in many UN statistics) and bear offspring. Thus, even as each woman, on the average, may have fewer children, many more women will be having children. In con-

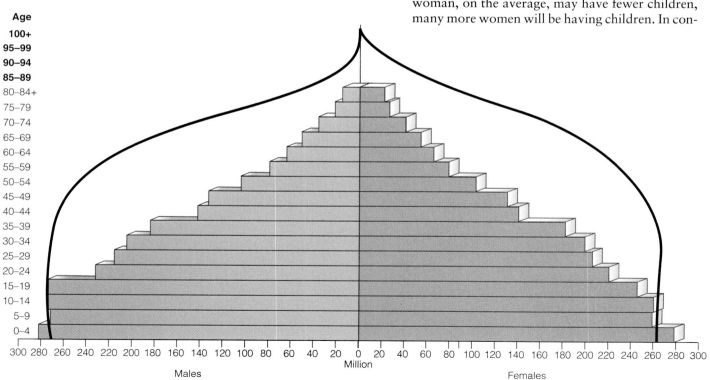

Age

100+
95–99
90–94
85–89
80–84+
75–79
70–74
65–69
60–64
55–59
50–54
45–49
40–44
35–39
30–34
25–29
20–24
15–19
10–14
5–9
0–4

300 280 260 240 220 200 180 160 140 120 100 80 60 40 20 0 20 40 60 80 100 120 140 160 180 200 220 240 260 280 300
Males   Million   Females

(b) Less developed countries

trast, a flat age structure profile combined with replacement fertility will result in stable population numbers. As individuals age and move from one age category to the next, and ultimately to death, they will be replaced by equal numbers of individuals being born.

Figure 2-8 shows the age structure profiles for the world as a whole in 2002 and the year 2050 (estimated projection). Currently, there are an estimated 1.61 billion people on Earth of prime reproductive age (15–29) and another 1.81 billion people under age 15 who will enter their reproductive phase in the next generation. This is a primary reason why the world's population will continue to grow in the twenty-first century even as fertility rates decline.

Life Expectancies and Infant Mortality Rates

Not all people have the same chance of living to a ripe old age, and in fact, exactly what age you can expect to live to is highly dependent on what region of the world you are born in. To dramatically demonstrate this point, we can refer to the following data compiled by the U.S. Bureau of the Census.

Life Expectancies at Birth	2002
Western Europe	78.5
North America	77.6
Latin America and the Caribbean	69
Asia	66.6
Sub-Saharan Africa	48

As can easily be seen, there is a thirty year discrepancy in **life expectancy** for a person from Sub-Saharan Africa versus a Western European. Intimately related to this is the large discrepancy in infant death rates around the world. Of the 55 million people who died in 2001, 6.9 million were infants who died before their first birthday. In the nations of Sub-Saharan Africa such **infant mortality** can represent 22% of all deaths, whereas in more developed countries (for example, those of Europe, North America, and Japan), infant deaths account for fewer than 1% of all deaths.

One factor that has recently had a major effect on life expectancies in many Sub-Saharan African countries is the spread HIV/AIDS (Human Immunodeficiency Virus/Acquired Immunodeficiency Syndrome) as well as the resurgence of various other parasitic and infectious (communicable) diseases. According to data compiled by the World Health Organization (WHO), worldwide in 1999, one-sixth of all deaths were attributable to infectious diseases including acute lower respiratory infections, tuberculosis, diarrhea, HIV/AIDS, and malaria.

Returning to the subject of HIV/AIDS, it is estimated that 40 million people worldwide were infected as of 2001; 94% of those infections occur in developing countries and 28.1 million live in

Figure 2-8   Age structure profiles of the world as a whole in (a) 2002 and (b) 2050 (estimated projection). (*Source*: U.S. Bureau of the Census, International Database.)

(a) 2002

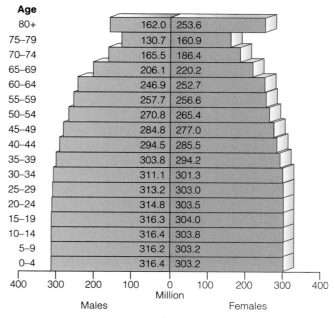

(b) 2050

Sub-Saharan Africa. AIDS has killed 22 million people since the epidemic began and is expected to kill another 15 million by 2005 in the most affected countries. HIV continues to spread rapidly; there were approximately 5 million new HIV infections in just 2001.

Overall, as of 2001, an estimated 8.4% of all adult Sub-Saharan Africans were infected with HIV, but some countries are harder hit than others. Currently the country most devastated by HIV/AIDS is Botswana, where approximately one out of every three adults is infected by HIV. As a result of this epidemic, the life expectancy of Botswanians has been dropping, from 61 years in 1995 to less than 40 years in 2001 and a mere 36 years in 2005. Furthermore, the annual population growth rate of the country has steadily declined from 3.5% in 1985 to 1.6% in 2000. By 2010, Botswana's growth rate is expected to be 0.5%.

Another country very hard hit by HIV/AIDS is Zimbabwe where an estimated 25% of all adults are infected. As in Botswana, both the life expectancy and annual population growth rate of Zimbabwe have been declining. Without AIDS, Zimbabwe could have expected an annual growth of 3.5% between 2000 and 2005; it's now expected to be 1.7%. In all, thirty-five Sub-Saharan countries are particularly hard hit by HIV/AIDS—a "hard hit" country is one containing either over a million infected individuals or an adult prevalence of infection over 2%. But it is not just in Africa that HIV/AIDS is a major problem. In Asia, the nations of Cambodia, India, Myanmar, and Thailand have been particularly hard hit, and in Latin America and the Caribbean area, the countries of Brazil and Haiti have major segments of their populations that are infected. AIDS is causing decreases in life expectancies and slowing of population growth in all of these countries (mentioned above); however, due to other demographic factors, their populations are projected to continue to increase in the twenty-first century.

Total Fertility Rates
The **total fertility rate** (TFR) is an important statistic to examine when looking at human populations around the world. The TFR is basically the number of children a woman of a given population will have, on average, during her childbearing years. **Replacement level** TFR is generally considered to be approximately 2.1 children per woman. (Women must replace not only themselves, but their male partners as well. All other factors being equal, more males tend to be born than females—about 105.5 male births for every 100 females born—and not all children make it to their reproductive years.)

Based on United Nations data, during the period of 1995–2000, the total fertility rate was at or below the replacement level (2.1) in sixty-four countries, or 44% of the world's population. This includes most of the countries of the more developed regions along with a number of developing countries, particularly in Eastern Asia (China, for instance, has a TFR of 1.1). This data is indicative of a dramatic decline in TFRs around the globe. During the period of 1965–1970, for instance, not a single major country had a TFR below replacement level. China had a TFR of 6.0 in 1965–1970. The global TFR in 1965–1970 was approximately 5.9, whereas in 2000, it is estimated at 2.8.

Of course the flip side of the above data is that even at present approximately 56% of the planet's population is still characterized by above-replacement fertility. Furthermore, just because a country has achieved a TFR below replacement level, that does not mean that the size of the population will decline in the near future. A large proportion of young persons in a population, as well as **immigration**, can cause an increase even when the TFR is below 2.1. For instance, the population of China, with a TFR of 1.1, is still expected to grow from 1.27 billion in 2000 to an estimated 1.46 billion in 2050 (based on medium variant UN projections), an addition of 190 million people or nearly 15% of its 2000 population. China's estimates have improved, however. Projections in 1998 predicted that China would have an additional 220 million people by 2050, an 18% increase. The population of the world as a whole will continue to increase until the global TFR is at or below replacement level *and* the age structure distribution of the planet is no longer skewed toward the young side.

Aging of the World Population
With the decline of fertility rates, the age structure of the world's population is changing. The distinct trend is toward older median ages and higher proportions of older people (60 years old and older) relative to children (less than 15 years old), as indicated by the data in Tables 2-4 and 2-5. However, notice that in general, the less developed regions are lagging behind the more developed regions in these trends. In approximately 2050, the less developed regions will be at the stage where the more developed regions were in about 2000. This, along with discrepancies in total fertility rates between more and less developed regions, is a primary reason why most of the world's population growth during the twenty-first century will take place in the less developed regions.

TABLE 2-4	Median Age by Major Area, 1950, 2000, and 2050 (medium variant projections)		
	Median Age (years)		
	1950	**2000**	**2050**
World Total	23.6	26.5	36.2
More developed regions	28.6	37.4	46.4
Less developed regions	21.4	24.3	35.0
Africa	19	18.4	27.4
Asia	22	26.2	38.3
Europe	29.2	37.7	49.5
Latin America and the Caribbean	20.1	24.4	37.8
Northern America	29.8	35.6	41.0
Oceania	27.9	30.9	38.1

Source: United Nations Population Division, *World Population Prospects: The 2000 Revision*, February, 2001.

TABLE 2-5	Percentages of the Population Less Than 15 Years Old and 60 and Over					
	Less than 15 Years			60 and Over		
	1950	**2000**	**2050**	**1950**	**2000**	**2050**
World	34%	29.8%	20.4%	8%	10%	21.9%
More Developed Regions	27%	18.2%	15.6%	12%	19.5%	32.3%
Less Developed Regions	38%	32.7%	21.2%	6%	7.6%	20.2%

Based on United States Census Bureau, International Programs Center, 2001.

The Consequences of Overpopulation

Rapid population growth and overpopulation have many far-reaching effects ecologically, economically, and societally. The increasing population is putting an increasingly greater burden on the Earth's natural resource base and environment. As Paul Ehrlich (1988) pointed out:

> One can think of our species as having inherited from Earth a one-time bonanza of nonrenewable resources. These include fossil fuels, high-grade ores, deep agricultural soils, abundant groundwater, and the plethora of plants, animals, and microorganisms. These accumulate on time-scales ranging from millennia (soils) to hundreds of millions of years (ores) but are being consumed and dispersed on time-scales of centuries (fuels, ores) or even decades (water, soils, species).

Most people readily acknowledge that fossil fuels, such as oil and coal, are nonrenewable. However, many fail to realize that, from the human perspec-

tive, soils and much fresh water that is pumped from underground aquifers are also nonrenewable resources (see Chapters 9, 12, and 13). Topsoils are being eroded at a tremendous rate, and in many regions the water table is being drawn down to alarmingly low levels. Another example is phosphorus, an extremely important element that is mined from nonrenewable rock deposits primarily in order to make modern fertilizers. A 1971 study concluded that known supplies of phosphorus would be exhausted by 2100, and that without phosphate fertilizers, the Earth can support only one to two billion people.

Such ecological damage is not solely a function of population growth. Given the discrepancies in affluence and consumption among the Earth's peoples, not everyone impacts equally on the environment (remember the equation $I = P \times C$ from Chapter 1). Persons in rich, industrialized countries (the MDCs) typically cause much more ecosystem damage per capita than persons in poor, nonindustrialized countries (LDCs). Based on such considerations, Paul Ehrlich has suggested that one new American baby

TABLE 2-6	Selected Examples of Rapid Population Growth in Cities Located in Developing Countries		
	Population in Millions		
City	1950	1995	Projection for 2015
Bombay, India	2.9	18.0	26.1
Lagos, Nigeria	0.28	13.4	23.2
Dhaka, Bangladesh	0.41	12.3	21.1
Sao Paulo, Brazil	2.4	17.7	20.4
Karachi, Pakistan	1.0	11.7	19.2

Source: United Nations Population Division, *World Urbanization Prospects: The 1999 Revision*.

and 250 new Bangladeshi babies pose an equal threat to the environment. The United States contains less than 5% of the world's population, yet consumes approximately 25% of the world's energy resources and produces about a quarter of the world's pollution. Likewise, U.S. citizens consume approximately 12 times as much energy per capita as the average citizen in a developing country. On the other hand, the United States also produces 28% of the world's gross domestic product.

Rapid population growth and overpopulation lead to increased urbanization, increased unemployment, and in many cases, spreading poverty. Projections indicate that 60% of the world's population will live in urban areas by 2030. In the developing world, many cities are growing at phenomenal rates and to phenomenal sizes (Table 2-6), but much of the growth is in the form of slums, shanty towns, and squatter settlements that lack such necessities as adequate housing, safe drinking water, and proper sanitation systems. It should be pointed out, however, that extreme poverty is not always directly caused by overpopulation. Economic shortsightedness, government mismanagement, or deliberately harmful political policies can also be the cause of the suffering.

Population growth leads invariably to competition for limited resources, often culminating in outright armed conflict. This may take the form of increasing local crime and violence, particularly among various ethnic and racial groups, some of whom may have voluntarily migrated or been forcefully displaced due to increasing population pressures. Sometimes the conflict escalates to actual warfare, primarily domestic civil wars. There have been more than 20 million war-related deaths in the world since 1945, and over 75% of these came in domestic civil wars. The majority of these deaths have occurred in highly populated countries. Bangladesh, with one of the highest population den-

sities in the world, accounts for at least 1.5 million of these deaths.

The Population of the United States and Canada

The 2001 population of the United States was approximately 283 million, or 4.6% of the world's population of 6.1 billion, making it the third most populous country on the planet (based on UN statistics). Although the total fertility rate is currently about 1.93 (2.1 is generally accepted as replacement level fertility), the population is still growing by approximately 2.5 million every year, in part due to immigration. It is projected to reach approximately 397 million in 2050. The rate of growth is slowing, however. By 2050, the rate is projected to level off at 0.47%. For comparison, the current growth rate is about 0.88%.

Another way to look at the growth is by percentage of increase. The population increased by approximately 13% during the 1990s. For comparison, between 1940 and 1990, the U.S. population increases have varied from just under 10% to over 18% per decade.

The continued growth of the U.S. population in the late twentieth century was primarily attributable to two factors: (1) The age structure of the population was such that a large number of women (many of whom were Baby Boomers, discussed below) were in or entering their reproductive years and bearing children; consequently, the birth rate was appreciably higher than the death rate. (2) Large numbers of immigrants, legal (about 600,000 a year) and illegal (numbers unknown), continued to enter the country—some 6 million immigrants entered the country legally during the 1980s, along with an estimated 2 million undocumented immigrants. The proportion of U.S. population growth attributable to immigration rose by more than 20% between World War II and the end of the century. Often foreign-born women in the United States had higher fertility rates than did American-born women. As the twentieth century closed, it was estimated that immigration accounted for over 25% of the growth of the U.S. population.

That the U.S. population is still growing faster than those of most of the industrialized nations of the world is worrisome, not only for the citizens of the United States but for the Earth. The United States, it can be argued, consumes much more than its fair share of world resources and produces an equivalent amount of pollution and waste; this inordinate consumption will only get worse as the U.S. population increases.

Even within the borders of the United States, the increasing population pressure will undoubtedly carry serious consequences. The infrastructure and social services will be further stressed. Dump sites for waste, both hazardous and nonhazardous, are in short supply. Freshwater supplies are already pressed; periodic water shortages have been reported around the country, and these will only increase. The agricultural capacity of the United States is decreasing, even as its population is increasing, due to the conversion of an estimated 3 million acres (1.2 million hectares) of agricultural land each year to other uses.

Another concern is that the age structure of the United States underwent major changes in the twentieth century, and continues to shift. After World War II (during the years of 1946–1964) birth rates soared in the U.S., giving rise to the "Baby Boom Generation." This large segment of the U.S. population demands appropriate resources relative to their age. In the 1950s, the Baby Boomers needed to be educated, so schools were built. Later, colleges and universities expanded to meet the needs of the Baby Boomers. In the 1970s, Baby Boomers entering the work force drove unemployment rates down and fueled the economy with increased consumption of goods and services. In the early twenty-first century, Baby Boomers are starting to retire and some fear this may put a strain on certain aspects of the American economy. In the late 1990s, the ratio of working to retired stood at about 5 to 1, but by 2020 it may be only 2.5 to 1. The exact consequences of this demographic shift are uncertain.

Although Canada is much less densely populated than the United States, in the past it has had an even higher population growth rate than the United States. However, Canada's birth rate is now projected to decline more quickly (total fertility rate for 1995–2000 was about 1.6) and the population has been aging; like the United States, the Canadian population underwent a 1950s baby boom. In 1950, the Canadian population stood at 13.74 million, in 1960 at 17.9 million, in 1990 at 26.64 million, and in 2000 at 30.75 million. It is projected to reach 34 million in 2015, 36.7 million around 2025, and perhaps 40 million about 2050. Geographically, Canada is a vast country, with an area slightly greater than the entire United States (including Alaska and Hawaii). Yet large portions of Canada, such as the Arctic regions, are inhospitable to dense human habitation. Consequently, over three-quarters of all Canadians are urban dwellers. The majority of Canadians live within about 200 miles (322 km) of the U.S. border, many being concentrated in the metropolitan areas of Toronto, Montreal, and Vancouver.

Solving the World's Population Problem

Industrialization, Economic Development, and the Demographic Transition

How do we solve the world's population problem? Many people have subscribed to the following proposition, as succinctly stated by Dr. Nathan Keyfitz (1991):

> The one thing that we know for sure about population is that people in the developed condition do not have many children; whatever problems the rich countries face, rapid population growth is not one of them. Hence, all that is needed is for the poor countries to develop and they too will be spared the troubles arising from rapid population growth.

This is the theory of **demographic transition**—essentially, that as a nation undergoes technological and economic development, its population growth rate will decrease. According to this theory, early in its history a nation will be characterized by both high birth rates and high death rates, and population growth will be relatively slow. With initial economic and technological development, the death rate will drop, while the birth rate remains high. Usually, this trend is ascribed to better nutrition, increased and advanced medical technology, more adequate shelter, and the like. Cultural norms are modified more slowly, however, and parents continue to produce large families, even though they do not "need" to because a higher percentage of their offspring survive and reach adulthood. Also, with initial economic and technological development, children will be in less demand as a labor force and will no longer be an economic asset. Still, because children continue to be produced at high rates, the nation will undergo rapid population growth, and the size of the population will increase dramatically.

Finally, industrial, technological, and economic development will mature and peak, and the culture will change to meet the new conditions. Parents will no longer produce as many children as did previous generations. Why this is the case is still under investigation. As nations industrialize, many of the attributes and nuances of the indigenous culture are progressively lost. In many westernized societies, children are no longer needed as a form of "social security" to support their parents in their old age. High-technology material goods are in demand, the dietary intake tends to include more meat, and having a large number of children is de-emphasized (**Figure 2-9**). In terms of a cost-benefit analysis, having more children is an economic burden. Until they are adults, children produce little, if anything. In one way or another, consumer goods appear to substitute for children. At any rate, values and cultural norms change, and the birth rate falls until it ap-

proximates the death rate; during this time, the nation's population growth rate slows down, perhaps even reaching zero or a negative number.

Problems with the Demographic Transition Model

The demographic transition model seems to describe well the population histories of some technologically advanced nations, especially in Europe. But does this model hold the key to the world's population problem? This is now being seriously questioned. Countries in Africa, Asia, and Latin America, for instance, have been heavily affected by westernization. Their death rates have been lowered, but their birth rates remain high. Consequently, their populations are growing at staggering rates. These nations are partly, but not fully, industrialized. With partial industrialization, their environments and resources are being destroyed, but they do not have the money to industrialize further. The longer a nation remains stuck halfway through the demographic transition, with high birth rates and low death rates, the larger its population will become. The larger the population becomes, the poorer the nation will be, the more poverty will manifest itself, and the harder it will be to end the cycle. A partial developmental program seems to be much worse than no developmental program at all; partial development has a very strong destabilizing effect. As Jodi Jacobson has pointed out, "slower economic growth in developing countries plagued by debt, dwindling exports, and environmental degradation means that governments can no longer rely on socioeconomic gains to help reduce births" (Jacobson, 1988).

The demographic transition model has also been criticized on another, related issue. The industrialized, highly developed nations of the North (such as the United States and the European nations) reached their current level of development and affluence not only by degrading their own immediate environments, but also by exploiting the territory of the developing nations in the Southern Hemisphere. For example, for every acre (or hectare) of land farmed in the United Kingdom, another two acres (or two hectares) of land are farmed elsewhere in the world in support of the United Kingdom. Perhaps the classic example is the Netherlands, with a very dense population of over 1100 people per square mile (or over 425 people per km²); in contrast, the United States has a density of about 75 people per square mile (29 per km²). The Netherlands can support such a dense population only by importing massive amounts of foodstuffs and other resources. Essentially, the argument is that it would

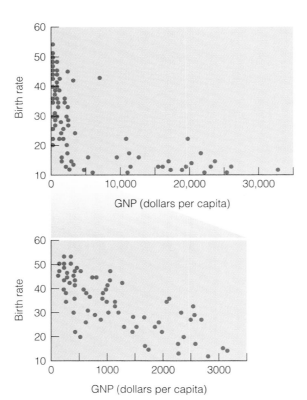

Figure 2-9   Increasing GNP (gross national product) per capita correlates with decreased birth rates (crude birth rates per 1000 members of the population), based on a study of 107 countries. (*Source*: From H. R. Pulliam and N. M. Haddad "Human Population Growth and the Carrying Capacity Concept." In *Bulletin of the Ecological Society of America* [September 1994, pp. 141–157].)

be impossible for the entire world to consist of only developed countries; without the developing countries, the developed countries could not exist at their present population densities and levels of affluence. In order for a would-be developing country to become a fully developed country, it needs suitable developing countries to exploit.

It is also questionable whether the correlation between development, industrialization, and lower fertility seen in Europe and the United States is a causal relationship. Not only are fertility rates lower in the industrialized nations, but so are rates of malnutrition, infant mortality, and illiteracy. This evidence suggests that fertility declines are caused primarily by rising levels of education, nutrition, and infant survivorship rather than by industrialization and development per se. A case in point would appear to be Costa Rica, which cut its fertility rate by 53% between 1960 and 1985 without either major industrial development or a major effort in family planning. Instead, Costa Rica promoted education and health issues and waged a war on poverty.

Another factor that must be considered is that as the growth rate of a particular population declines, the age structure profile of the population will change. For example, the typical LDC currently has an age structure profile that is skewed toward the younger age groups. But as population growth rates decline and birth rates approximate death rates, the population will "mature" and have many more persons in the older age categories than previously. Such a shift in age structure can seriously affect the culture and society. More adults in their prime will need jobs, and the elderly, who were once relatively rare, will demand a larger proportion of the resources with fewer younger people to support them.

Contraceptives, Abortion, and Reproductive Rights

An obvious way to address the population problem is to continue and expand family planning services, the distribution of contraceptives (birth control devices; see **Figure 2-10**), and the availability of safe, legal abortions, while promoting the "**reproductive rights**" of women (essentially the right to be able to choose not to bear children) throughout the world. Indeed, much progress has been made in these areas, but it is also a highly sensitive, emotionally charged subject (**Figure 2-11**). **Contraception, abortion,** and reproductive rights in general go against the grain of some major religions and cultural values. Some people in developing countries view family planning as genocide, for the developed, industrialized nations are encouraging the undeveloped and developing nations to curtail their pop-

Figure 2-10 (a) Contraceptive use by women strongly correlates with decreased fertility rates (based on a study of 50 countries). (b) Women attending a family planning class in India. (*Source*: (a) From H. R. Pulliam and N. M. Haddad, "Human Population Growth and the Carrying Capacity Concept." In *Bulletin of the Ecological Society of America* [September 1994], pp. 141–157.)

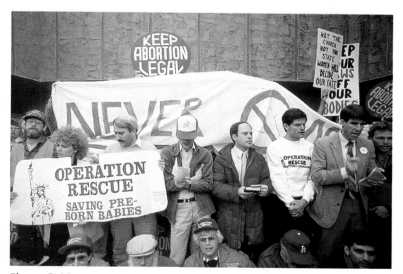

Figure 2-11 Birth control, family planning, reproductive rights, and especially abortion are highly sensitive, emotionally charged issues for many people. Even in the United States, where abortion is legal, many find it morally repugnant and wish to make it illegal.

ulations. Many African governments have traditionally opposed family planning in the belief that curtailing population growth would hurt them economically; without people, leaders of these countries felt they could not reach their full economic potential. But this view is changing.

Easy access to contraception is often taken for granted by many people in developed countries such as the United States and Canada. However, access to and use of various methods of contraception varies widely between developed and developing regions (**Figure 2-12**). Of all regions, Africa is the farthest behind in the use of contraception; over three-quarters of African couples do not use any form of contraception. In contrast, the rate of non-use in developed countries is less than a third. It is estimated that well over 100 million women in the developing world have an "unmet need" for contraception and family planning services (data from Population Action International). What this means is that these women have indicated that they do not want to have children, at least not at the present time, yet they are not using any form of contraception. In some cases this non-use of contraception may be due to deeply held religious or cultural practices, but in many cases it is simply due to a lack of informa-

tion about, and accessibility to, methods of contraception. Because of nonexistent, or inadequate, preventative measures to stop conceptions, on the order of 50 million abortions are carried out annually around the world. Many of these abortions occur in developing countries under unsafe (and sometimes illegal) conditions, resulting in an estimated 50,000 to 100,000 deaths to the women involved each year. Numerous studies from around the world have demonstrated that, in general, greater access to contraception methods and more effective forms of contraception (for instance, sterilization, pills, condoms, and IUDs (intrauterine devices) are all typically much more effective than using methods relying on withdrawal prior to ejaculation or "rhythm" methods), is positively correlated with declining numbers of abortions.

Female Education and Status
Empirically, it has been found that one of the best ways to decrease the growth rate of a particular population is to increase the average educational level and societal status of women in that population (**Figure 2-13**). Female education is apparently successful in decreasing fertility rates for several reasons. Women typically apply their knowledge to improving the home situation of their families—for instance, by assuring better nutrition, health care, sanitation, and so on. Rates of infant and child mortality decline as a result, causing the population to become more receptive to the notion of smaller families. Educated women are also more likely to be willing and able to use contraceptive methods effectively. Finally, through education, women can gain status and prestige in ways other than through bearing numerous offspring. Educated

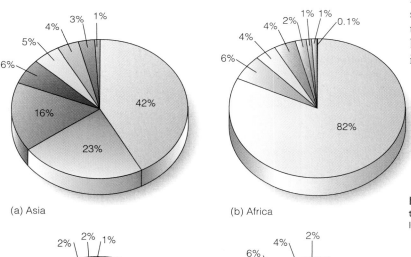

(a) Asia

(b) Africa

Figure 2-12   Contraceptive use by region in different part of the world, middle 1990s. (*Source*: Population Action International.)

(c) Latin America

(d) Developed countries

None

Female sterilization

IUD

Male sterilization

Non-supply method

Oral contraceptives

Condom

Other supply method

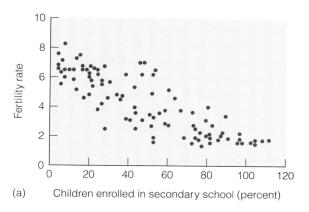

(a)   Children enrolled in secondary school (percent)

Figure 2-13   (a) Increased average education levels correlate with decreased fertility rates, based on a study of 105 countries. This correlation holds at all levels, among both males and females, and at all ages. Particularly important for decreasing the growth rate of a population, however, is the education of women. [Note that UNESCO (United Nations Education, Scientific and Cultural Organization) defines the school enrollment ratio as total enrollment, regardless of age, divided by the population of the age group that typically corresponds to a specific education level (thus if enough younger or older children or adults are enrolled in school the ratio can go over 100%).] (b) Women attending class, Jalalabad, Afghanistan. (*Source:* (a) From H. R. Pulliam and N. M. Haddad, "Human Population Growth and the Carrying Capacity Concept." In *Bulletin of the Ecological Society of America* [September 1994], pp. 141–157.)   (b)

women can choose to pursue career opportunities that do not involve staying home and having children. And educating the women of the world is relatively inexpensive. Estimates are that raising the education level of all women in developing nations to at least equality with men would only cost about $6.5 billion per year, less than is spent on lawn care annually in the United States, and much less than is spent on video games each year in the industrialized nations.

Unfortunately, the status of women is relatively low in many nonwesternized societies, including Islamic societies, traditional Latin America, and traditional African societies. In many traditional African societies, for example, women carry a large economic and labor burden, yet have almost no legal independence or rights. A groom buys his wife from the bride's family and pays a high price in the expectation that she will be fertile and produce many offspring to help with the chores. As long as a woman's social and economic status is tied to the number of children she bears, little headway can be expected in the area of decreasing fertility rates.

Another sign of the generally low status of women in developing countries is the low priority placed on women's health. It has been estimated that roughly a half million women of reproductive age die in the developing counties each year due to complications of pregnancy, childbirth, and abortion. A Worldwatch Institute report estimated that 60% of these deaths could be avoided by spending less than $2.00 a year on each woman in those countries.

Of course, educating women and improving their status will irrevocably alter these traditional societies. Some have questioned whether it is "right" for Westerners to interfere in the natural development and order of non-European cultures.

Economic Incentives, Disincentives, and Government Regulation of Childbearing

Private organizations and national governments have attempted to control population growth through voluntary programs, such as increasing the accessibility of modern birth control without mandating its use. But as the overpopulation problem has intensified, many governments have decided they must regulate childbearing.

To encourage their citizens to voluntarily limit the number of offspring, governments have offered a variety of economic incentives for using contraceptives, being sterilized, or limiting the number of children in a family (**Figure 2-14**). For example, in parts of Taiwan during the 1970s, families that had two children or fewer received annual payments,

Disease and Global Population: The Plague and AIDS

Visit www.environment.jbpub.com
for more information.

In the mid-fourteenth century, Europe, as well as much of Asia and Africa, was devastated by an outbreak of plague, which contemporaries called the Black Death (**Figure 1**). The disease actually took three forms: bubonic plague, pneumonic plague, and septicemic plague (all caused by the bacterium variously known as *Pasteurella pestis* or *Yersinia pestis*). It is estimated that at least 25 million, and perhaps as many as 75 million, of Europe's population of 100 million died between 1347 and 1351. As the plague raged, the social structure of Europe was destroyed. Contemporary accounts report that government and law enforcement, religious ceremonies, and medical practice disappeared in areas where the plague was worst. In an ecological sense, the plague can be viewed as a classic case of a density-dependent mechanism that served to limit the population. Six and a half centuries later, it should serve as a warning to us as we continue to overcrowd our world.

Modern research suggests that the fourteenth-century plague may have originated among the wild rodents that live in the Kirghiz Steppes of Asia, one of several apparently permanent reservoirs of plague (others are in China, India, the southern former Soviet Union, and the western United States). Under normal conditions, the plague does not spread beyond these reservoirs, but occasionally it does—with devastating effects. In the early fourteenth century, a major outbreak ravaged China, India, and other eastern and Middle Eastern populations, killing millions. The plague spread west, and rumors of the terror spread even more quickly.

In the mid-fourteenth century, the city of Genoa in Italy controlled the Crimean port of Caffa on the Black Sea, which was attacked by Mongolian Tartars (Tatars) who had traveled west across Asia from the Kirghiz Steppes. While laying siege to Caffa, the Tartars began to die of a strange disease, characterized by fever, delirium, pneumonia, and in some cases enlarged, pus-filled, lymph nodes (buboes) that opened to the skin and drained spontaneously. The Tartars decided to withdraw, but before leaving, they catapulted the bodies of their dead comrades into the city—an early and particularly gruesome instance of biological warfare. The inhabitants of Caffa subsequently contracted the plague, and when infected Genoese returned to Italy in 1347, they carried the disease to Europe.

Medieval Europeans had no idea what caused the plague or how to control it. It is now known to be caused by a bacterium that can be carried by rodents, such as rats and squirrels, and is transmitted from rodents to hu-

Figure 1 Victims of the Black Death appear in this fourteenth-century French fresco. A physician lances a plague-caused bubo (a swollen and inflamed lymph node) on a woman's neck; on the left is another victim with an enlarged bubo under his arm. (*Source:* The Granger Collection, New York.)

mans by a flea. For tens of thousands, or even millions, of years, populations of *Yersinia pestis* have been living in the guts of fleas that feed on rats and infect them with the plague. Once the rat dies, the fleas seek another host, carrying with them the plague bacilli. Eventually, the plague-carrying flea also dies, but often not before it has infected other mammalian hosts and thus indirectly other fleas that feed on the same host. In the case of Caffa, rats probably fed on the bodies of the dead Tartars and became infected, fleas fed on the rats, and the fleas subsequently bit and infected humans.

Even today bubonic plague is not well understood—and isolated cases and small outbreaks continue to occur among humans. Why does it remain contained among rodent colonies for centuries, then suddenly and unexpectedly spread at an alarming rate among the human population? And why does the outbreak subside? Do only individuals who have some type of natural resistance to the plague remain? Are reservoirs of plague perhaps always in waiting, ready to spread when conditions are right? Unfortunately, we may not know what those "right" conditions are.

We should remember that, even with our advanced medical knowledge and technology, we could conceivably find ourselves again facing an unknown or poorly understood, but rampant and devastating, disease. Some would even suggest that the current AIDS crisis is a roughly analogous situation (**Figure 2**). Globally, from 1980 through 2001, an estimated 62 million people were infected with HIV (the cause of AIDS) and 22 million succumbed to the disease. This means that roughly 0.5% or more of the world's population was infected with HIV. Over two-thirds of all HIV infected persons live in sub-Saharan Africa, where in some countries 20% or more of all adults are infected. AIDS takes not just a toll in human lives, but also has a devastating effect on local economies and social structures. AIDS reduces the number of healthy workers in their prime and increases the number of dependent people, including the sick, the young, and the old. Scarce resources are diverted to caring for AIDS victims, and as AIDS continues to spread, the productivity of the workforce continues to decrease. Children in particular are being hit hard; teachers are being lost to AIDS and furthermore many children must leave school to help support the family after a parent contracts AIDS. The full effect of the AIDS epidemic will not be seen for another generation or more.

Figure 2 The AIDS Memorial Quilt on display in Washington, D.C. There are over 44,000 panels with 84,000 names, representing approximately 19% of the AIDS deaths in the United States.

and in Thailand, some families who utilized contraceptives could rent a team of water buffalo at a reduced rate. South Korea and Pakistan allow income tax deductions for families with two children or fewer. The government of Singapore has withheld employment benefits, housing subsidies, and preferred school admissions from families with more than three children. In extreme cases, such as in China and India, sterilizations and abortions have allegedly been performed without the full consent of the individual concerned.

Many people question whether such policies are appropriate. For one thing, bribes and payoffs may be most attractive to the poor, and some people may not fully understand what sterilization entails. In addition, the case has been strongly argued

Figure 2-14 The People's Republic of China is currently the most populous nation on Earth, although by the middle of the twenty-first century this distinction will be held by India. Here a Beijing couple, with their only child, stand in front of a billboard encouraging only one child per family.

that childbearing is, and should remain, a private family matter. But given the nature of the world today, it can also be argued that childbearing is no longer a "private" matter; the number of children a couple bears affects more than that couple and their immediate family. Most governments of developing countries now acknowledge the need for family planning, and according to the United Nations, only Cambodia, Iraq, Laos, and Saudi Arabia actively restrict access to family planning services.

Population Control around the World

Government efforts to curb birth rates have met with notable successes, as well as some failures. Great strides have been made in contraceptive use and availability; 9% of the world had access to contraceptives during the period 1960–1965, but by the end of the twentieth century, this percentage had risen to an estimated 53% (married women between 15–49) and slowly continues to rise. Yet this is not enough. The United Nations estimates that if we are to achieve a stable world population, 75% of all couples worldwide will need to use contraception. Contraceptive use is also very unevenly distributed (see Figure 2-12); for instance, contraceptive use is slightly above 70% in the United States and about 81% in China, but only 1% in Chad. This is part of the reason why over 95% of global population growth is projected to take place in the developing nations of Africa, Asia, and Latin America during the next 35 years.

In the 1990s, the typical completed family in the developing nations (excluding China) had three to five children (and in many families, especially in East Asia, three or fewer). Although this number represents a drop from the typical six children of the early 1960s, a United Nations Population Fund report observed that cutting the fertility rate from four children to two will most likely prove more difficult than reducing it from six to four. Will contraceptive use continue to increase, and fertility rates and average completed family sizes decrease, at the rates predicted? It is difficult to know. However, it is just this cut in fertility rate and completed family size—from about four children to two—that is essential if we are to eventually stabilize the world's population. The world has made great strides along these lines in the past few decades. Fertility rates in more than 60 countries are already at or below replacement levels. Let us hope that this portends well for the future.

SUMMARY

- Earth's human population has grown explosively in the last few centuries, and particularly over the last few decades, reaching a current total of over 6 billion people.
- For hundreds of thousands or millions of years, our ancestors subsisted as gatherers and hunters at very low population levels.
- Major increases in human population came with the agricultural revolution (about 10,000 years ago) and the Industrial Revolution (in the last 250 years).
- Current population growth is not distributed evenly over the globe; the greatest increases are occurring in the developing countries.
- In some Western industrialized countries, such as in Europe, the population has stabilized or is even declining.
- The United States has a high population growth rate compared to other developed countries, but part of the increase in population is due to immigration.
- Population is one of the major factors in the $I = P \times C$ equation (impact equals population times consumption) and thus is a major concern of environmentalists.

- Many different solutions have been suggested as ways of dealing with overpopulation.
  - Increased industrialization and economic development, which will force a demographic transition and thus halt population growth in developing countries.
  - Increasing medical care, family planning services, and the use of contraceptives.
  - Increasing the educational level and social status of women in particular; studies indicate that there is a positive correlation between women's educational levels and decreased fertility rates.
- Some governments have used voluntary (such as economic incentives and disincentives) and mandatory (allegations of forced contraceptive use, sterilizations, and abortions) means to regulate population growth.
- The subject of human population is a complex, emotionally charged issue. The potential ramifications of increasing human numbers are a continuing theme that will recur throughout this book.

KEY TERMS

Abortion	Demography	Less developed country (LDC)
Age structure profile	Doubling time	Life expectancy
Agricultural (Neolithic) revolution	Exponential growth	More developed country (MDC)
Annual growth rate	Exponential phase	Neo-Malthusian
Birth control	Fertility rate	Percent annual growth
Birth rate	Green Revolution	Rate of natural increase
Carrying capacity	Growth rate	Replacement level
Contraception	Immigration	Reproductive rights
Crude birth rate	Industrial Revolution	Total fertility rate (TFR)
Crude death rate	Infant mortality rate	
Demographic transition	Lag phase	

STUDY QUESTIONS

1. Briefly summarize the history of human population growth over the last million years.
2. Explain how the agricultural revolution spurred human population growth.
3. List some factors involved in the exponential growth of the human population over the last four hundred years. What was the Industrial Revolution? What was the Green Revolution?
4. What parts of the world currently have high rates of population growth? Where are there low rates of population growth?
5. What are some of the potential consequences of global overpopulation?

STUDY QUESTIONS

6. Describe several alternative scenarios for global population growth through the year 2100.
7. Discuss the thesis that there is no such thing as human "overpopulation."
8. How have the growth rate and doubling time of the world's population changed over the millennia?
9. Compare and contrast the age structure profiles of populations from more developed and less developed countries.
10. Explain the concepts of life expectancy, infant mortality rates, and total fertility rates. What types of factors affect each of these in different parts of the world?

11. Briefly discuss how the equation $I = P \times C$ relates to the problem of global human overpopulation.
12. Describe the concept of the demographic transition model. What criticisms has this model received?
13. How does the education of women bear on the issue of global population?
14. Describe some of the controversies surrounding various methods of birth control and abortion.
15. Discuss the appropriateness of voluntary versus mandatory government regulation of childbearing.

WHAT'S THE EVIDENCE?

1. Throughout this chapter the authors present many statistics concerning populations on a global and national scale over time. How reliable do you think these statistics are? Can we be sure of how many people inhabited Earth in, for instance, 1800? How precisely can we know the number of people that are alive today? As a project, research population statistics from various sources (for instance, the U.S. Bureau of the Census, the Population Division of the United Nations, and the World Resources Institute) and com-

pare the various numbers and the data that are used to support them.
2. The authors suggest that the theory of the demographic transition—that increasing technological and economic development inevitably leads to a decrease in population growth rate—is not always true. What evidence is marshaled against the demographic transition model? Are you convinced that the demographic transition model, long held to be true by many scholars of demographic theory, is fundamentally flawed?

CALCULATIONS

1. If the population of Nigeria was 126.93 million at the midpoint of 1995 and had increased to 130.77 by the midpoint of 1996, what was Nigeria's percent annual growth over this period? At this annual growth rate, approximately how long will Nigeria's population take to double? (For information on population equations and statistics, see the Fundamentals Box on page 35.)

2. If the population of El Salvador was 5.77 million at the midpoint of 1995 and had increased to 5.88 million by the midpoint of 1996, what was El Salvador's percent annual growth over this period? At this annual growth rate, approximately how long will El Salvador's population take to double?

ILLUSTRATION AND TABLE REVIEW

1. Using the graph in Figure 2-2, estimate the global human population in A.D. 1000 and A.D. 1700. By what amount did the population change over this period? What percentage change is this?

2. Using the projections in Figure 2-2b, extrapolate each curve to the year 2150. Based on your extrapolations, what might the global world population be in 2150? Is there any validity to such an estimate based on extrapolating a projection? What assumptions is such an extrapolation based on?

3. Based on Figure 2-7, what percentage of the population of the less developed regions was nine years old or younger in 2002? What percentage of the population of the more developed regions was nine years old or younger in 2002?

4. Using the figures in Table 2-3, and assuming a total world population of 6.15 billion in 2001, what percentage of the human population lived in Pakistan in 2001? In Nigeria? In China? In India? In the United States?

5. Of the cities listed in Table 2-6, which had the fastest rate of growth between 1950 and 1995? By what percentage did Karachi grow between 1950 and 1995? By what percentage did Lagos grow between 1950 and 1995?

WWW.ENVIRONMENT.JBPUB.COM

www.environment.jbpub.com

Connect to this book's web site: http://www.environment.jbpub.com/ The site features eLearning, an online review area that provides quizzes, chapter outlines, and other tools to help you study for your class. We've posted the answers to the book's odd-numbered questions. You can also follow useful links for more in-depth information. The site provides sources for the information in this chapter and recommendations for further readings.

SECTION 2

The Environment of Life on Planet Earth

Tug on anything at all and you'll find it connected to everything else in the universe.

John Muir, Founder of the Sierra Club

Population

Consumption

1. The Environment and
Humans (Chapters 1-2)

Lithosphere

Biosphere

Hydrosphere

Atmosphere

2. The Environment of
Life on Planet Earth
(Chapters 3-5)

Lithosphere

Biosphere

Hydrosphere

Atmosphere

3. Resource Use and
Management
(Chapters 6-13)

Lithosphere

Biosphere

Hydrosphere

Atmosphere

4. Dealing with
Environmental
Degradation
(Chapters 14-18)

Lithosphere

Biosphere

Hydrosphere

Atmosphere

5. Environmental Issues: Social Aspects and Solutions
(Chapters 19-20)

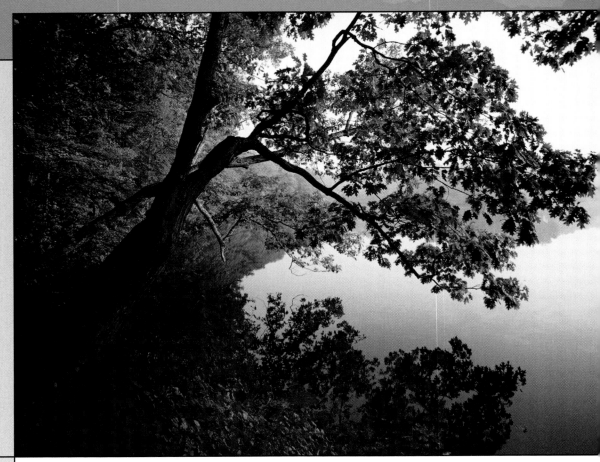

We live in a time of incredible biological diversity, with numerous individuals and species interacting to form populations, communities, and ecosystems that are essential to the maintenance of life as we know it. Charles Darwin, the founder of modern evolutionary theory, wrote in 1859 (*On the Origin of Species*), "It is interesting to contemplate an entangled bank, clothed with many plants of many kinds, with birds singing on the bushes, with various insects flitting about, and with worms crawling through the damp earth, and to reflect that these elaborately constructed forms, so different from each other, and dependent on each other in so complex a manner, have all been produced by laws acting around us [such as growth, reproduction, organismal variability, interactions among individuals and species, natural selection] . . . . There is grandeur in this view of life . . . from so simple a beginning endless forms most beautiful and most wonderful have been, and are being, evolved."

CHAPTER 3

The Biosphere: Populations, Communities, Ecosystems, and Biogeochemical Cycles

The environment is everything that surrounds you, including the air, the land, the oceans, and all living things. For convenience, the natural environment can be subdivided into two parts: (1) the physical environment, which includes nonliving things, and (2) the biological environment, which includes all life-forms. The physical environment can be further subdivided into the three basic states of physical matter: solid, liquid, and gas. This division creates four "spheres" that compose the natural environment. The three physical spheres are called the **lithosphere** (*lithos* = rock), **hydrosphere** (*hydro* = water), and **atmosphere** (*atmos* = vapor). The biological environment is called the **biosphere** (*bios* = life).

Figure 3-1 shows how the four spheres form the outermost layers of the planet Earth. They are underlain by a very thick molten layer of rock belonging to the mantle and a heavy metallic core.

Three major points about Figure 3-1 should be stressed. First, these four spheres make the Earth a dynamic planet. The lithosphere creates rocks as hot magma, generated by the "internal heat engine" of radioactive minerals. Sometimes this magma erupts onto the surface in the form of volcanoes. The atmosphere weathers and erodes (degrades) the rocks with chemically reactive gases. The hydrosphere also erodes rocks and transports them as sediments that are often ultimately deposited in ocean basins. The moving lithosphere then carries these sediments to great depth where they are reheated and remelted into magma, beginning the cycle anew. The life of the biosphere inhabits parts of the other spheres, relying on the cycling of chemicals through all of the spheres. Life also affects the other spheres by altering the cycling of chemicals, such as oxygen. Many rocks (for instance, most limestones) are the result of direct biological activity, and recent studies have discovered anaerobic bacteria living in rocks over 9000 feet (2740 m) below the Earth's surface.

Second, the figure illustrates that the spheres closely interact with one another. Matter is transported both within and between the spheres in various kinds of cycles. For instance, gases are expelled from the lithosphere (by volcanoes) and the biosphere (by plants and animals) into the atmosphere. Similarly, water vapor moves from the oceans to the atmosphere and back again. Moreover, there are also numerous cycles within each sphere. For example, new rocks are now being created in the lithosphere while simultaneously older rocks are being eroded away. Thus, the Earth is an active, integrated system, with many cycles that constantly transport matter both within and between the spheres at various rates.

Third, these dynamic and interactive spheres have evolved through time. Figure 3-1 depicts the environment as it exists today, but the Earth is about 4.6 billion years old. The current lithosphere, originally formed from a molten layer on the hot, young Earth, probably originated some 4 billion years ago. The oceans and atmosphere date back to at least 3.8 billion years ago, and the earliest known fossils (single-celled microbes) representing the biosphere are at least 3.5 billion years old.

Objectives

After reading this chapter, you should be able to explain or describe:

The four spheres of the environment and how they are interrelated

The major aspects of the physical environment

What ecology is

The basic principles of population dynamics

Human impact on natural populations

Biosphere interactions (communities and ecosystems)

Ecosystem productivity

Matter cycling

The major biogeochemical cycles and their features

Major energy flows on Earth

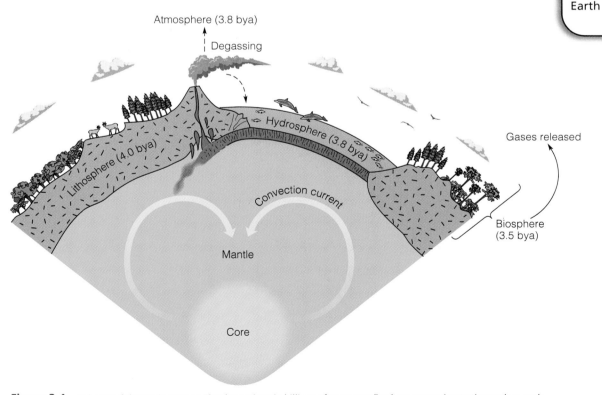

Figure 3-1 Spheres of the environment. The dates given in billions of years ago (bya) are approximate times when each of the major spheres of the environment originated: lithosphere 4.0 bya, atmosphere and hydrosphere 3.8 bya, and biosphere 3.5 bya. Note that the Earth is a dynamic planet and that all of the spheres have evolved and changed since its origin.

Water, Air, and Energy—Three Major Aspects of the Physical Environment

The Earth's surface is almost entirely covered with water—not just the seas and oceans, but numerous lakes, ponds, swamps, and even morning dew on the "dry" land. It is enveloped in an atmosphere composed predominantly of nitrogen, or N_2 (78.1%), and oxygen, or O_2 (20.9%). It is also bathed in radiation—energy in the form of heat (infrared radiation) and light—from the Sun. All of these components work together to make the surface conditions of our planet very active and constantly changing.

The Earth's atmospheric envelope is divided into layers, distinguished mainly by the changing temperature gradient encountered as one moves up into the atmosphere (**Figure 3-2**): the troposphere, stratosphere, mesosphere, and thermosphere. The changing temperature gradient is the result of the interplay of the energy received directly from the Sun, the gaseous constitution and density of the atmosphere at different levels, and the distance the energy (heat) reflected and emitted from the Earth's surface reaches into the atmosphere.

In terms of energy flow, the Earth's surface is an open system: It constantly receives energy from the Sun, and it continuously loses most of this energy back into space. Approximately 30% of the incoming solar energy that reaches the upper atmosphere is immediately reflected back to space, and another 19% is absorbed by water vapor in the atmosphere, so only about 51% reaches Earth's surface. Very little (less than 1%) of the Sun's energy that reaches Earth goes into **photosynthesis**, and thus the direct support of living organisms. The vast majority of the energy that reaches Earth goes into evaporating surface waters; it thus reenters the atmosphere and is eventually dissipated into space as heat (electromagnetic radiation in the infrared range).

The movement of water about the surface of Earth, driven by energy from the Sun, constitutes the **hydrologic cycle** (**Figure 3-3**). Approximately 97.4% of the water on or near the Earth's surface is found

Figure 3-2   The structure of the Earth's atmosphere. The tropopause, stratopause, and mesopause are the boundaries between the troposphere and stratosphere, stratosphere and mesosphere, and mesosphere and thermosphere respectively.

in the seas and oceans, the remaining water is bound up as ice (primarily in the polar ice caps), stored temporarily as groundwater, found in freshwater lakes and streams, or in the form of atmospheric water vapor at any one point in time. The shifting of water over the globe through the hydrologic cycle redistributes heat and generally makes for more equitable climates. The hydrologic cycle also plays a prominent role in the weathering and decomposition of rocks. Together with other atmospheric movements (influenced by many factors, including the rotation of the Earth around the Sun to give the seasons and the movement of predominant winds over the surface of Earth; see Chapter 5), the hydrologic cycle is also responsible for **weather**, or the short-term, daily perturbations in the atmospheric/hydrologic cycles. **Climate** is the average weather over time ranges of decades to millennia to hundreds of millions of years.

Figure 3-3
The hydrologic cycle.

Temporary storage in snow fields and glacial ice 1.98%

Condensation 0.001%

Temporary storage in lakes, rivers, and plants 0.014%

Regional water table

Evaporation from oceans

Precipitation

Transpiration

Ocean 97.4%

Streamflow to oceans

Groundwater flow 0.59%

Saline groundwater

Biosphere Interactions: Populations

The biosphere today, as in the past, is hierarchical: Organisms, composed of atoms, molecules, and cells, are grouped into populations. Populations form communities, which then form ecosystems. Finally, ecosystems, when considered together, form the biosphere, which subsumes all life on Earth. **Figure 3-4** illustrates the interconnectedness of life; each level is composed of the units in the level below. Furthermore, the hierarchy is not static. Each level is dynamic, with many interactions occurring among its units. For instance, the organisms that compose populations interact in many ways, as do the populations that compose communities.

Ecology is the study of how organisms interact with each other and their physical environment. Because interactions occur at many levels in the hierarchy of life, ecology is very complex and often yields generalities and predictions that are less precise than we might wish. Partly because of this complexity, ecology has come to have a somewhat different meaning in popular usage (see Case Study 3-1, What Is Ecology?).

The science of ecology, however, has advanced far since its formal origin over a century ago to provide us with a general understanding of how interactions in the hierarchy of life occur. Interactions within and among coexisting populations are among the most fundamental, so let us begin with them.

Population Dynamics

A **population** is a group of individuals of the same species living in the same area. There is no limit on the size of the area so a population may be all the bass in a lake or all the black bears in Alaska. Whatever the species or area, all populations undergo three distinct phases during their existence: (1) growth, (2) stability, and (3) decline (**Figure 3-5**).

Growth occurs when available resources exceed the number of individuals able to exploit them, such as when a population is first introduced into an area. Individuals tend to reproduce rapidly, and death rates are relatively low because of the relatively abundant resources. Stability occurs when population growth levels off, as the environment becomes saturated with individuals of that population. However, stability is usually preceded by a population "crash" because the rapidly growing population abruptly exceeds the resources, causing an "overshoot" (Figure 3-5). Even stable populations fluctuate, sometimes widely, so stability does not mean that population abundance is unchanging. Usually, stability is the longest phase by far. Decline refers to the inevitable decrease in abundance that leads to extinction of all populations in the long term. Each of these three phases is affected by many factors; the study of how these factors interact is called *population dynamics*.

Figure 3-4   The cosmos can be depicted as a hierarchy from subatomic particles up through galaxies. Life, or biological entities, is at an intermediate level, ranging from cells through the biosphere.

Growth of Populations

Nearly all populations have one key trait in common: They will grow exponentially if left unchecked. This exponential growth occurs because individuals usually produce many more than one offspring in their lifetime. Furthermore, each of those offspring can, in turn, have many more. As a result, most populations have the potential to increase very rapidly. The potential for increase in a given population is called the **intrinsic rate of increase**, symbolized by r.

The exact intrinsic rate varies among populations depending on several factors, many of them genetically determined. Two of the most basic are the birth rate and death rate:

Intrinsic Rate of Increase =
Birth Rate − Death Rate

This equation shows that the net rate of increase in a population is governed not only by the rate at which the population can multiply, but also by the rate at which individuals die. For example, bacterial populations can have birth rates of millions per day while a pair of elephants may produce only one offspring every few years. But bacteria also die by the millions every day, so the net intrinsic rate is dramatically lower than if it were determined by births alone.

Population Stability: Regulation of Growth

Common sense tells us that nothing grows forever, and this is certainly true of populations. Growth is limited by the complex interaction of many factors, including other species.

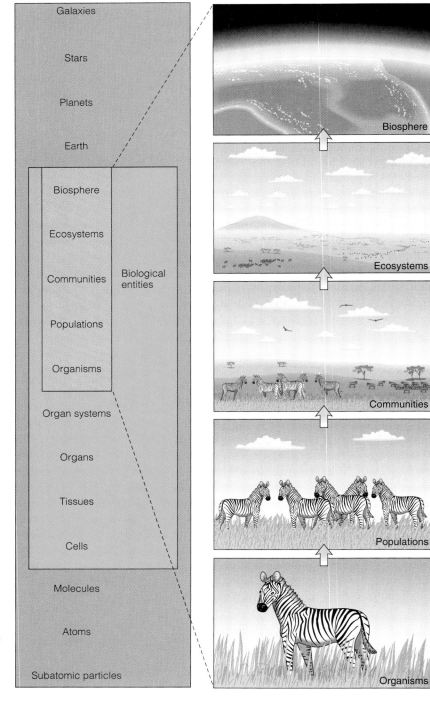

Figure 3-5   Initially, a population will sometimes undergo exponential growth until it exceeds the capacity of the environment to support it, causing "overshoot." This phase is sometimes followed by a period of relative stability and eventually by decline.

What Is Ecology?

The prefix *eco-* is derived from the Greek word *oikos*, which means home. Since *logia* means "study of" in Latin, *ecology* is the study of home. The word *economics* has the same derivation: *nomos* is Greek for "managing," so economics means managing the house. Ironically, although the words ecology and economics derive from the same word, they are often thought to represent opposing interests. Even more ironically, studying the home and managing the home seem to have resulted in very different sets of priorities about how the home (that is, our environment) should be treated.

Henry David Thoreau was apparently the first to use the word ecology in one of his letters in 1858, but he did not give it a specific definition (and possibly Thoreau's ecology was simply a misspelling for some other intended word). Instead, the German biologist Ernst Haeckel is generally credited with introducing, in 1866, the word (originally spelled *oecologie*) as it is now used in biology to mean the study of organisms and their interactions with each other and their physical environment.

Although scientists still use this technical definition, ecology has come to mean many other things, especially to nonscientists. As environmental awareness has risen over the last few decades and environmental issues have come to be discussed in social rather than scientific contexts, ecology has taken on new meanings. Thus, many people who express concern about the environment call themselves ecologists and are interested in ecology. This label has even been extended to political ideologies that reflect these concerns. Thus, the political scientist William Ophuls has said that ecology is a profoundly conservative doctrine in its social implications.

The confusion arises when ecology is expanded beyond its restricted meaning as a branch of biology that studies natural environments (ecosystems) to refer to the social and political ideas of people who are actively concerned with preserving those natural environments (ecosystems). For reasons of clarity, we might do better to use *environmentalist* as a general term for someone who actively wants to preserve the natural environment, while reserving *ecologist* for scientists who study it. (Of course, a person can be, and often is, both an ecologist and an environmentalist.) Ophuls should say that environmentalism is a conservative doctrine because the study of something is not a doctrine at all: a doctrine is a system of beliefs.

In addition to being expanded to mean social and political environmental action, ecology has also been generalized as an academic term. Many students, perhaps even yourself, have enrolled in ecology courses expecting to learn about water pollution, solar power, and many other aspects of environmental problems. They are often surprised to find that the course focuses on the study of natural communities, unaltered by humans. While the study of natural laws governing such communities is basic to understanding how humans affect them, many students desire a broader perspective that includes pollution and other ways that humans modify the natural world. Environmental science, such as the course you are enrolled in now, has arisen in recent years to fulfill this need. Because environmental science is such a broad area of study, it is taught in a variety of traditional science departments, including biology, geology, chemistry, and geography. College courses labeled ecology are often primarily for students majoring in biology or who have an interest in the specific study of natural communities.

Visit www.environment.jbpub.com for more information.

The simplest way to examine abundance limits is to consider a species of microbe in a glass container (but the same analysis can be applied to any biological population, such as sheep on an island). When a few individuals are first admitted to the container, they will likely experience rapid population growth for the reasons we have just discussed. At some point, however, the limited space and food in the container will cause the multiplicative growth phase to enter a slower growth phase as net reproduction slows. Growth continues to slow until the population size stabilizes; that is, the population remains roughly constant, fluctuating around some average size. This average abundance where the population levels off is the **carrying capacity**, or the maximum population size that can be sustained by an environment for a long time. As noted earlier, the carrying capacity is usually briefly exceeded by an overshoot phase before the population levels off in the stability phase.

Four Basic Abundance Controls. What causes population size to level off and stabilize? Clearly, the ultimate cause is to be found in the environment: Factors external to the population must be limiting the rate at which individuals can reproduce and survive. Ecologists typically divide these environmental factors into two basic categories, the (1) physical and (2) biological environment. Within the biological environment, we can identify three subcategories of biological interactions that can limit population growth:

1. Physical environment
    Physical limitations
2. Biological environment
    Competition
    Predation
    Symbiosis

In Chapter 1 we noted that a system can be defined as a "set of components functioning together as a whole." The abundance controls on populations that are discussed here are in fact natural systems that act together to control the sizes of populations.

Limitations of the physical environment include a vast number of constraints such as water supply, space availability, or soil and light in the case of plants. All constraints of the physical environment are determined by the population's **habitat**, which is the place where it lives. For example, the remaining Florida panther population currently lives in a wetland habitat including the Florida Everglades. California's red-legged frog, made famous in Mark Twain's short story, "The Celebrated Jumping Frog of Calaveras County," was once common in most of the state. This large frog can now be found only in isolated pockets along California's

central coast and in a few sites in the Sierra Nevada (**Figure 3-6**). It only takes one aspect of the physical environment to limit population growth. For instance, even if light, soil, temperature, and many other resources can support more plants, population abundance may be constrained far below that limit if a single resource, water, is scarce. This observation, which was made famous by the German botanist Justus Liebig in 1840, has come to be called the **law of the minimum**: Growth is limited by the resource in the shortest supply.

Competition occurs when organisms require the same limited resource. Competition can take many forms, but two of the most important are exploitative and interference competition. Exploitative, or "scramble," competition occurs when both competing populations have equal access to the resource: The population that exploits the resource the fastest is the "winner." Interference competition occurs when one of the competitors prevents the other from gaining access. For instance, many plants secrete toxins that wash into surrounding soil to prevent other plants from encroaching and taking sunlight and soil nutrients.

A basic determinant of any competition is the organism's **niche**, which is often defined as the organism's "occupation" or how it lives. The niche can be thought of as how the organism fits into the local "system," such as the local habitat, ecosystem, or biome. The niche includes what the organism eats, how it eats, and what eats the organism. The degree to which the niches of competing species overlap indicates the extent of similar resources required and therefore the strength of the competition between the species. **Competitive exclusion** occurs when niche overlap is very great and competition is so intense that one species eliminates the second species from an area. Although competitive exclusion has received much theoretical attention, some ecologists think it is relatively rare in nature. In contrast, species often compete for only one or a few resources, but each species can also utilize resources that the others will not. For example, foxes and owls in a forest may compete for mice, but each will also eat other prey that the other usually does not pursue. Such cases of minor niche overlap do not result in complete competitive exclusion. Instead, each competitor merely limits the abundance of the other. When the single-celled eukaryote *Paramecium aurelia* is raised alone, its population size is nearly twice as large as when it must share resources with a second closely related species.

An important result of competitive exclusion is that if one of two competing species is removed, the remaining species may increase in number. Thus

Figure 3-6   The red-legged frog.

the population of *P. aurelia* also nearly doubles after the competitor is removed. This is often called *ecological release* because the species is "released" from one of the factors limiting its abundance. In this case, some ecologists would call it "competitive release," specifying release from competition. Humans often cause ecological release by selectively exterminating unwanted species; as a result, abundance controls are upset, and competitor species experience population booms. When farmers spray pesticides on crops, for example, they often find that another species increases its population numbers.

Predation occurs when organisms consume other living organisms. Predators can be divided into (1) carnivores, which prey on animals, and (2) herbivores, which prey on plants. As with competition, the extent to which predation limits abundance of prey varies considerably. In the most extreme cases, a predator drives the prey species to extinction. Many of the known cases of extinction have occurred when humans introduced a predator into a new area, especially an island. Much more commonly, predators simply limit abundance and do not drive the prey to extinction. In the case of many herbivores, even as they prey on the plant (eating, for instance, some leaves or fruit), they do not even kill the individual prey organism.

But why don't predators keep reproducing until they have eaten all the prey? At least three major reasons have been identified. For one thing, prey species have often evolved protective traits, such as camouflage, poisons, spines, large size, and so on. In addition, prey species often flee to refuges such as burrows and tree-tops where predators cannot reach them. Experiments have shown that an insect or microbe predator population in a container will consume the prey population unto extinction unless the prey can escape to a refuge. Finally, predators tend to engage in prey switching. When one prey species becomes scarce, predators switch to another prey, thereby allowing the first prey species to rebound in numbers.

Ecological release is also seen in predation control of abundance. When predator populations are greatly reduced by disease or other causes, or when prey populations move to an area without major predators (such as islands), the prey species may exhibit "predatory release" and expand in the same way populations expand when experiencing competitive release. Today, humans are a major cause of predator release both by killing off predators, such as mountain lions, and by introducing species into areas without major predators.

Symbiosis (*sym* = "together"; *biosis* = "life") encompasses many kinds of interactions, sometimes

TABLE 3-1	Symbolic Classification of Symbioses[a]		
Form of Symbiosis		**Species a**	**Species b**
Mutualism		1	1
Predation and parasitism		1	2
Commensalism		1	0
Competition		2	2
Amensalism		0	2
Neutralism		0	0

[a] 1 = benefit, 2 = detriment, 0 = no effect

including competition and predation. As **Table 3-1** shows, these interactions can be classified according to whether they benefit or inhibit one or both populations that are interacting. **Mutualism** benefits both species. An example of mutualism is the coexistence of algae within the tissue of the tiny animals that build coral reefs. The algae are provided with a protected living space while the coral animals are supplied with nutrients from the algae's photosynthesis. Similarly, a lichen is actually a fungus and algae growing together symbiotically. Mutualism is thus a factor regulating the abundance of certain types of algae because their numbers are influenced by the abundance of coral or fungal hosts.

Parasitism is similar to predation in that one species benefits while harming the other species (**Figure 3-7**). The main difference is that parasites act more slowly than typical predators and do not always kill the prey (host). **Commensalism** occurs when one species benefits and the other is not affected. For example, Spanish moss (a lichen) hangs from trees for support, but causes the trees no great harm or benefit. Barnacles attach to crab shells or other organisms in a similar way, deriving food from

Figure 3-7 Parasites, such as this leech, often do not kill their hosts.

Figure 3-8 Commensal organisms such as these barnacles derive benefits from the turtle without harming it.

the surrounding water (**Figure 3-8**). **Amensalism** occurs when one population inhibits another while being unaffected itself. Harm to one species is simply an incidental byproduct of the actions of another. For example, when elephants crash through vegetation, they often have a detrimental effect on it, while gaining relatively few benefits (of course, the elephants get where they want to go).

The Real World: Complex Interaction of Abundance Controls
Our brief description of the various controls on abundance (physical fluctuations, competition, predation,

and symbioses) has greatly oversimplified by discussing each control separately. In reality, most natural populations are governed not by a single control, but by several controls often acting simultaneously and forming a system. When two populations of beetles interact in the laboratory, for example, competition leads to a high abundance of one species and a very low abundance of another. When a third species, a parasite, is introduced, the formerly abundant species may become the rarer one if it is more susceptible to the parasite (**Figure 3-9**).

Now consider that natural communities are often composed of hundreds to many thousands of species. Any single population will often have many competitors, predators, and symbionts such as commensals (including microbes). Abundance changes in any of these could have a major influence on the abundance of that population. At the same time, changes in physical conditions could also influence any of the populations, with each responding differently to a physical change. This complex interaction of abundance controls in natural ecosystems means that small changes in one control, such as the abundance of a prey species, can have a cascading domino effect throughout the ecosystem, causing other changes in the abundance of other species. The abundances of all species are therefore nearly always fluctuating in response to natural fluctuations in physical conditions and in other species.

Population Decline
While even stable populations rise and fall in abundance during fluctuations, given enough time, eventually most populations decline to zero and become extinct. Extinction is the elimination of all individuals in a group. In a local extinction, all individu-

(a) Parasite absent

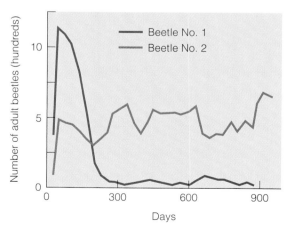

(b) Parasite added

Figure 3-9 (a) With two competing species of beetles, the superior competitor can drive the other to a very low abundance. (b) The addition of a parasite can drive the formerly abundant species to rarity and allow the formerly rare species to increase. (*Source*: Modified from T. Park, *Ecological Monographs* 18 [1948]: 265–308. Reprinted with permission of Ecological Society of America.)

als in a population are lost, but a new population can be reestablished from other populations of that species living elsewhere. Species extinction occurs when all populations of the species become extinct.

The inevitability of population decline is clear when we consider that over 99% of all species that have ever existed are now extinct. The vast majority of these became extinct before humans, so decline and extinction are natural processes. Ultimately, decline and extinction are caused by environmental change: One or more of the abundance controls becomes altered, leading to an abrupt or gradual decline in abundance. For instance, many extinctions in the fossil record resulted largely from physical changes such as cooling climates or asteroid impacts. Similarly, the fossils reveal cases where competition from new groups apparently caused decline, such as when placental mammals invaded South America and replaced most marsupials.

Human Impact on Population Growth and Decline

Increasing human populations and traditional industrial technology have caused progressively greater disruption in natural populations. Pollution, agriculture, and many other kinds of human alterations of the environment have destabilized populations by affecting the various abundance controls. Depending on the population, humans may cause (1) population growth by removing previous limitations or (2) population decline by imposing new limitations.

Human Impact on Population Growth.   Populations grow when there is an excess of resources relative to the number of individuals available to exploit them. Humans commonly contribute to population growth in the four ways listed in **Table 3-2**; each involves removing a control on abundance.

Increase of available resources can be planned or unplanned. Agriculture and animal domestication are obvious examples of how humans can deliber-

ately increase the populations of plants and animals that they favor. We do this by providing far more food and other resources than those organisms would find in their natural state. The effect of humans on cat numbers provides a striking example. In England alone, domestic cats occur in such numbers that more than 300,000 cats must be destroyed each year. Yet before humans began to domesticate them a few thousand years ago (to catch mice in grain storage areas), the small, wild ancestors of our pet cats were somewhat rare and were probably limited to the Middle East and Europe, a relatively small area compared to their present range. Similarly, corn, potatoes, and many other domesticated organisms had much smaller ancestral ranges.

Unplanned increases in available resources by humans also have major effects. Pollutants, for example, generally represent unplanned releases of substances into water or air. These substances are often nutrients that are in short supply. Recall from Liebig's law of the minimum that growth is limited by the resources in shortest supply. Phosphorus and nitrogen are very often the two most **limiting nutrients** for plants in water or on land. When fertilizers rich in these nutrients are carried into rivers and lakes, the result is often runaway plant growth. This enrichment of nutrients in waters is called **eutrophication**. Eutrophication is a classic example of "too much of a good thing" because the decomposition of the accumulating plants uses up increasing amounts of oxygen, causing fishes and other organisms to suffocate (**Figure 3-10**).

Competitive release is common when humans try to eliminate a population of one species, allowing its competitors to increase in numbers. The most

TABLE 3-2	Four Ways That Humans Cause Population Growth	
	Examples	
Increase available resources	Agriculture Nutrient pollution in lakes	
Competitive release	Poisoning of insect pests	
Predator release	Overhunting of large carnivores	
Introduce to new areas	Game releases	

Figure 3-10   An overabundance of phosphorous and nitrogen, often from fertilizers, can cause eutrophication in a lake.

Figure 3-11   (a) This map shows the rapid migration of the European starling across North America after its introduction in New York. (b) The European starling is now one of the most common birds in North America. (*Source:* (a) C. B. Cox and P. Moore, *Biogeography* [Cambridge, MA: Blackwell, 1993], p. 62. Reprinted by permission of Blackwell Science, Inc.)

(a)

(b)

economically important examples occur when farmers try to eradicate pests from crops with pesticides. Because poison tolerance varies among species, some pests will not be killed by the poisons and will actually increase in numbers when their competitors are gone. This is called a "secondary pest outbreak." A dramatic example occurred in Central America where cotton crops were sprayed in 1950 to kill boll weevils. The effort was highly successful until 1955, when populations of cotton aphids and cotton bollworms soared. When a new pesticide was used to remove them, five other secondary pests emerged. Such experiences have led to new methods of pest control (Chapters 13 and 14).

Predator release is common where humans hunt, trap, or otherwise reduce populations of predators, allowing the prey species's population to increase. For example, large mammalian predators such as wolves and panthers have long been the target of ranchers and farmers because they prey on domesticated animals. The result has been a rapid increase in the predators' natural prey. Deer have shown an especially spectacular rise. Indeed, most experts estimate that more deer are now living in the United States than were here before Europeans arrived. Unfortunately, the excess populations often lead to overgrazing and death by starvation. For this reason, game hunting by humans is a justifiable activity, at least based on ecological criteria. By prudently culling deer and other prey, humans are essentially carrying out the role of predators that no longer exist.

Introduction of nonnative (also called "exotic") species into new areas may be the single greatest alteration of nature carried out by humans so far. Few people realize the enormous scale on which humans have, either accidentally or purposely, transferred organisms from one area to another. For example, more than 1500 insect species and 24 families of fishes have been successfully introduced into North America. The extent of this introduction is even more impressive considering that the vast majority of populations die out when initially introduced. An analysis of ballast water in a tanker revealed that it contained the live larvae of 367 species of marine organisms carried from Japan to the Oregon coast. Such ballast waters are routinely discharged. Once an organism is introduced and established, expansion can be quite rapid. The European starling covered North America in just a few decades after it was introduced in New York for a Shakespearean play in the early 1900s (**Figure 3-11**).

TABLE 3-3	Four Ways That Humans Cause Population Decline and Extinction

The actions are listed in approximate order of importance in causing extinctions. In other words, habitat disruption causes the most extinctions today; secondary extinctions cause the least.

	Examples
Change physical environment:	
1. Habitat disruption	Draining a swamp, toxic pollution
Change biological environment:	
2. Introduce new species	New predator
3. Overkill	Big-game hunting
4. Secondary extinctions	Loss of food species

Human Impact on Population Decline   Humans are causing the decline and extinction of species at thousands of times the natural rate in nearly all parts of the biosphere. We will examine the causes of extinction in more detail in Chapter 11, but note here that humans alter the environment in four basic ways to cause population decline (**Table 3-3**). All are directly related to the abundance controls. The first affects the physical environment and the other three ways affect the biological environment.

- Habitat disruption occurs when humans disturb the physical environment in which a population lives. This disturbance can range from minor, such as mild chemical changes from air pollution, to major, such as total destruction of a forest by bulldozers or fire.
- The introduction of species who are competitors, predators, or symbionts (including diseases and parasites) in the native biological system. As we have seen, many new species have been introduced in various parts of the world. Island species are especially susceptible to introduced species. An example of competitive decline in island species is the introduction of rabbits to Australia; the voracious appetite and rapid reproduction of rabbits led to decreased abundances of many native marsupials.
- Human overkill is the shooting, trapping, or poisoning of certain populations, usually for sport or economic reasons. It is very difficult for humans to cause the extinction of "pest" species, such as roaches or mice, in this way because they are so abundant and reproduce so rapidly. However, overkill has been very effective in eliminating populations of large

animals because they are much less abundant and reproduce much more slowly. Leopards, elephants, rhinos, pandas, and many other large animals comprise a disproportionate number of threatened and endangered species in the world (Chapter 11).
- Secondary extinctions occur when a population is lost due to the extinction of another population on which it depends, such as a food species.

It is not necessary for these environmental changes to reduce a population to zero in order to cause extinction. Even if many individuals survive, the population may never recover if it becomes too small. Small populations are beset by many breeding problems, such as inbreeding and lack of mates, and are also more likely to become extinct from environmental fluctuations such as hurricanes, droughts, and so on. Ecologists often refer to the **minimum viable population** (MVP): If population abundance drops below the MVP, the population will probably never recover.

Population Range

In addition to varying through time, abundance may vary geographically. Populations tend to have a maximum abundance near the center of their geographic range, which is the total area occupied by the population (**Figure 3-12**). This central maximum occurs where the physical and biological factors that control abundance are the most favorable. As one moves away from this central optimum into the zone of physiological stress, abundance generally begins to decline. This decline is usually gradual because both physical and biological limiting factors tend to follow a gradient. Physical conditions such as temperature and salinity may change gradually (for instance, latitudinal temperature gradients) or abruptly (such as where the water in a lake meets the shore line) geographically, as can biological limits such as the abundance of competitors and predators. The zone of physiological stress grades into or meets the zone of intolerance, where the population is absent. The zone of intolerance occurs because some limiting factor has become so great that the species can no longer survive. Around deep-sea thermal vents, an extreme example, the zones of physiological stress and intolerance may be very narrow indeed. Water coming out of the vent may be up to 400°C, yet the water temperature only an inch or so (2 to 3 cm) from the vent can be 2°C, nearly freezing! Organisms living near the vent utilize the chemical nutrients pouring out of the vent, but may not be able to tolerate the high temperatures of the vent itself.

Figure 3-12   Organisms tolerate a range of conditions, but thrive in an optimum range.

Lower limit of tolerance

Upper limit of tolerance

| Organisms absent | Organisms infrequent | Greatest abundance | Organisms infrequent | Organisms absent |

| Zone of intolerance | Zone of physiological stress | Optimum range | Zone of physiological stress | Zone of intolerance |

Population

Low   Environmental Gradient   High

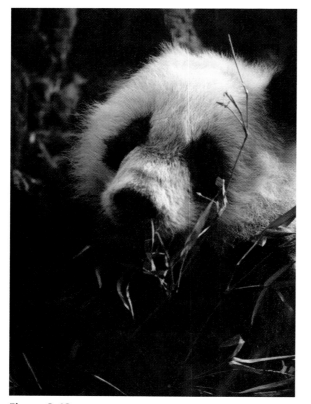

Figure 3-13   The giant panda, which generally relies on bamboo shoots and roots as a major part of its diet, is an example of a specialized species. Such species are easily pushed to extinction.

Abundance is almost never uniformly distributed throughout a population's geographic range because the environment is rarely uniform enough to follow perfectly smooth gradients. Instead, many irregularities occur; ecologists call this the *patchiness of the environment*. In populations that inhabit a very patchy environment, individuals are commonly clumped together with gaps in between.

The size of the geographic range of populations also varies among species. **Endemic species** are localized and may have just one population that inhabits only a small area. This pattern is especially common in tropical organisms and in organisms that are highly specialized to live on resources with limited distributions or that have narrow environmental tolerances (**Figure 3-13**).

Human Impact on Population Ranges.   Humans have both decreased and expanded the geographic ranges of populations. On the one hand, range decrease has commonly accompanied declining populations and extinction. Habitat destruction such as cutting down trees can also reduce a population's range. If the habitat disappears, the population's geographic range can be reduced to zero. This is a major reason why tropical extinctions are particularly destructive; with the high number of endemic populations, it does not take much habitat destruction to reduce the geographic range to zero. On the other hand, we saw how humans have often expanded geographic range by domesticating and introducing wild species into new areas (see Figure 3-11).

Biosphere Interactions: Communities and Ecosystems

Now that we understand how population dynamics work, we can carry this concept into an examination of communities and ecosystems. A community consists of all populations that inhabit a certain area. The size of this area can range from very small, such as a puddle of water, to large regions encompassing many hundreds of square miles (or thousands of km²). The extent of a community thus depends on the size of the area one wishes to denote. An ecosystem is the community plus its physical environment. Therefore, ecosystems can also be delineated at many spatial scales.

Because ecosystems include both organisms and their physical environment, the study of ecosystems often tends to focus on the movement of physical components, such as the flow of energy and the cycling of matter, through the "system." This approach is often called the "functional" view. In contrast, because communities consist only of organisms, one can focus on describing how organisms are distributed in communities through time and space: This approach is called the "structural" view. We turn first to the community, or structural, view.

Community Structure

Populations are not randomly distributed in communities. Indeed, discovering the various spatial patterns by which species are distributed has been one of the major accomplishments of ecology over the last century. Two of the most important patterns are the open structure of communities and the relative rarity of most species in communities.

Open Structure of Communities. A key question about community structure is whether the populations that comprise a community have about the same geographic ranges and abundance peaks. Proponents of closed community structure argued that populations in most communities had similar range boundaries and abundance peaks. A closed community is thus a discrete unit with sharp boundaries called **ecotones** (**Figure 3-14a**). In contrast, proponents of open communities argued that most communities had populations with highly varied abundance peaks and range boundaries. Open communities were said to have populations distributed more or less randomly relative to one another (**Figure 3-14b**).

Decades of data collection have finally resolved this debate, showing that most communities are indeed open. For example, populations of the many plants comprising a typical forest community have highly varied ranges that almost appear randomly distributed. In the case of forests, tolerance to moisture is a major determinant: Some species do best in wetter areas while others thrive in a wide range of moisture. As a result, population ranges very often overlap to various degrees depending on how similar the populations' tolerances to moisture are. Areas where drastic changes occur in the physical environment are the major exception to the generally open structure of communities. In such areas, boundaries between communities can be sharp because population tolerances are abruptly exceeded. A beach where land and sea come together, producing an ecotone, is an example. Because changes in the physical environment are usually more gradual, population abundance between communities also changes gradually.

(a) Closed communities

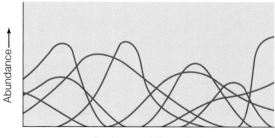

(b) Open communities

Figure 3-14   Each curve represents the abundance of a single species. (a) In closed communities, the species boundaries tend to coincide. (b) In open communities, the species boundaries are more randomly distributed. (*Source*: R. E. Ricklefs, *Ecology* [New York: W. H. Freeman, 1990], p. 659. Copyright © 1990 by W. H. Freeman and Company. Used with permission.)

The fact that communities are not closed makes their study more complicated because it is difficult to characterize, describe, and even name a community when it changes gradually in nearly every geographic direction. Ecologists often use advanced statistical methods, such as gradient analysis, to quantitatively describe the spatial changes in species composition. At any point along the gradient, the community is usually assigned a name on the basis of the most common species. One might refer to an oak-hickory forest community or a barnacle-blue mussel tidal zone community.

Another major implication of open communities is that communities are not tightly integrated assemblages of organisms that can be destroyed in an all-or-nothing fashion. Some ecologists once argued that communities were closed, highly integrated units forming a "superorganism." They maintained that if we destroyed just one or a few populations, the whole superorganism would die, just as an organism dies if a key organ is removed. The fact that species come and go in communities, however, means that communities are generally not as integrated as the superorganismic concept suggested. This point is significant because many nonscientists still accept the superorganismic concept of biological communities.

Most Species Are Rare in Communities.   A second basic population pattern is the relatively low abundance of most populations in communities. As we have seen, populations tend to have their peak abundance near the center of their range, where optimum conditions prevail. In the example of the forest community above, optimum moisture level seems to be a primary determinant of abundance. At any given point on the gradient, however, a large percentage of the individuals belong to just a few tree species. Even at their maximum abundance, most populations are much less abundant than a few dominant populations such as beech, maple, oak, and pine.

This high population abundance of only a few species is found in nearly all natural communities and among many kinds of organisms. From deep-ocean to mountain communities, populations of just a few species almost always dominate in abundance, with many species being represented by only a few individuals. The result is a logarithmic pattern in that when we plot the number of individuals per species, the number of species with few individuals rises rapidly.

What causes this abundance dominance by a few species? This question is currently being debated among ecologists, but most agree that the general cause is related to resource partitioning. In any environment, a limited amount of resources is available. Because of their evolutionary history, individuals of only a few species are best able to exploit a large part of the available resources. These species were the first to evolve the ability to obtain and eat a common food in the community. Other species must partition the remaining resources and are therefore less abundant. A species that is very rare in one community, however, may be very abundant in a nearby community if conditions are different enough that the species's particular adaptations are more effective in exploiting the resources there.

Community Diversity

Diversity refers to how many kinds of organisms occur in a community. Diversity is therefore often expressed in terms of species richness, or the number of species in a community. Many factors influence diversity, and the importance of any single factor varies with the particular community. Nevertheless, most of these factors can be summarized by two diversity trends. The **latitudinal diversity gradient** describes how species richness in most groups steadily decreases going away from the equator. Consequently, richer communities are found in tropical areas. For instance, 2.5 acres (1 hectare) of tropical forest typically contain from 40–100 tree species. In contrast, a typical temperate zone forest has about 10–30 tree species, while a taiga forest in northern Canada has only 1–5 species. Furthermore, the number of insect species living on those trees increases with the kinds of trees (resources) available to exploit, so tropical forests also have vastly more kinds of insects and other animal species per acre. As a result of this richness, habitat destruction in tropical countries generally leads to many more extinctions per acre than destruction elsewhere.

Ecologists generally agree that this gradient is largely due to three interrelated factors: (1) environmental stability, (2) community age, and (3) length of growing season. Greater environmental stability in equatorial areas means that communities are exposed to less environmental change on a daily, seasonal, and even hundred-year basis. This stability allows more kinds of species to thrive because high disturbance or stress generally reduces diversity. Equatorial communities are older because they have been less disturbed by advancing ice sheets and other climatic changes over the long span of geologic time. Hence evolution has had more time to create new species. The longer growing season in equatorial areas leads to more photosynthesis and plant growth, which forms the food base for all life. As we will see, higher plant productivity supports a greater diversity of organisms that depend on the plants.

The second important diversity trend is the **depth diversity gradient** found in aquatic communities. This gradient shows how species richness increases with water depth, down to about 6560 feet (2000 m) deep, and then begins to decline. This gradient is due to (1) environmental stability and (2) supply of nutrients. Environmental stability increases as one moves away from the higher water energies of the beach and shoreline. As we have seen, stability allows more species to thrive. Similarly, as one moves offshore, the amount of nutrients from land runoff begins to diminish. Thus, even though deep water is very stable, it contains insufficient nutrients to permit the high productivity seen in shallower waters. Marine life depends especially on land runoff to supply limiting nutrients such as phosphorus.

To summarize, four major factors may increase diversity in any community: increasing environmental stability, age, growing season, and supply of nutrients. Stability provides an accommodating environment for diversity to proliferate, age provides the time, and the last two factors provide the energy and nutrients to supply many types of organisms. In general, the more of these factors that a community has, the more species-rich it will be.

Community Change through Time

Biological communities change through time. This is not surprising since the physical environment that ultimately supports life is always changing. Whether we perceive change as "fast" or "slow" depends almost entirely on the time scale we are using. Two time scales are particularly useful for examining change in communities: ecological time and geological time. Ecological time focuses on community events that occur on the order of tens to hundreds of years. These events are most relevant to our own human time scale. Geological time focuses on community events that are longer, on the order of thousands of years or more, such as evolution.

Community Succession. Community succession is the sequential replacement of species in a community by immigration of new species and the local extinction of old ones. Community succession is initiated by a disturbance that creates unoccupied habitats for colonizing species. These colonizers usually have a hardy nature and are adapted for widespread dispersal and rapid growth, characteristics that enable them to become the first species to appear and thrive. This initial community of colonizing species is called the **pioneer community**. Eventually, other species migrate into the community. These new species are usually poorer dispersers and slower growing than the colonizers, but they are more efficient specialists and better competitors and therefore begin to replace the colonizers. This process continues, as still newer species migrate in, until the **climax community** is reached. The climax community continues to change, although at a much slower pace.

Succession has been most fully documented in forest communities (**Figure 3-15**). Pioneering plant

Figure 3-15 Ecological succession occurs when biological communities become established in a sequence, from pioneer through climax communities.

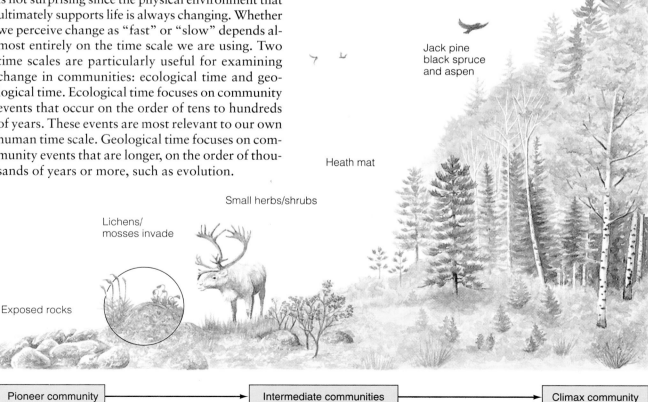

Balsam fir, paper birch and white spruce

Jack pine black spruce and aspen

Heath mat

Small herbs/shrubs

Lichens/ mosses invade

Exposed rocks

| Pioneer community | → | Intermediate communities | → | Climax community |

species include lichens, mosses, and herbs, which give way to shrubs, small trees, and finally large trees in the climax community. Various animals, such as the bird and mammal species shown, also appear in sequence. The animal sequence is determined largely by the appearance of the plants they rely on.

Succession is characterized by a number of trends (Table 3-4). One of the most basic is the decrease in productivity. Pioneering plants tend to be smaller in size and exhibit rapid growth that maximizes productivity. In later stages, as more specialized, slower-growing species begin to migrate in, productivity declines. As these later species immigrate, diversity increases because the more specialized species more finely subdivide the resources. In addition, the later species generally are characterized by larger sizes and longer life cycles. Together, these trends result in more biomass in later stages because living tissue accumulates. **Biomass** is the total weight of living tissue in a community. Finally, later stages tend to have populations controlled mainly by biological or density-dependent controls such as competition and predation. In contrast, early stages mainly show physical or density-independent controls, such as physical disturbance. Some communities, such as a beach, are generally in a constant state of physical disturbance so early-successional populations become permanent dwellers.

Succession occurs because each community stage, from pioneer to climax, prepares the way for the stage that follows. Each preceding community alters soil conditions, nutrient availability, temperature, and many other environmental traits. For example, the pioneering stage of a forest stabilizes the soil of a bare patch of land, begins to accumulate nutrients in the soil, attracts pollinating insects, retains water, and provides ground shade, among many other processes that make the environment more livable for later stages. The process of "preparation" is ironic in the sense that species in each community often bring about their own demise. Yet these early species are adapted by evolution to be colonists and can nearly always migrate to other, newly disturbed areas that permit them to persist. Thus, nearly all natural landscapes (including sea-bottom areas) consist of a mosaic of undisturbed patches intermixed with patches that are disturbed to varying degrees. Ecologists think that this "patchiness" of the natural environment is crucial for maintaining diversity because it allows species from different stages of succession to exist simultaneously.

Human Disturbance of Communities

Community structure is determined by species distributions, so whenever species distributions change, the structure is altered. Humans can change natural species distributions in many ways—from introducing new predators into pristine ecosystems to outright annihilation of large areas of habitat. However, the basic effect of nearly all human activity is community simplification: the reduction of overall species diversity (number of species).

In many cases, humans simplify communities on purpose. The farmer's agricultural and the suburbanite's horticultural communities of plants, insects, and other animals are common examples. In such cases, we seek to grow only certain species, creating a much lower diversity than normally is found in that area. The extreme case is called **monoculture**, meaning that only one particular species is grown. An example is a wheat field. Monocultures and other forms of extreme community simplification are very susceptible to diseases and other forms of destruction, such as the Irish Potato Famine. It is interesting to note that most of the plant species we cultivate for food and pleasure are species from pioneering communities. For example, corn, wheat, and many other plants are grasses that were originally adapted to colonizing disturbed areas. Humans favor them as foods because they are fast-growing, rapidly reproducing organisms.

In other cases, humans inadvertently simplify communities. Construction, road building, pollution, and many other aspects of "development" act as

TABLE 3-4	Trends in Ecological Succession	
	Stage in Ecosystem Development	
Attribute	**Early**	**Late**
Biomass	Small	Large
Productivity	High	Low
Food chains	Short	Long, complex
Species diversity	Low	High (?)
Niche specialization	Broad	Narrow
Feeding relations	General	Specialized
Size of individuals	Smaller	Larger (?)
Life cycles	Short, simple	Long, complex
Population control mechanisms	Physical	Biological
Fluctuations	More pronounced	Less pronounced
Mineral cycles	Open	More or less closed
Stability	Low	High
Potential yield to humans	High	Low

Source: Adapted from R. Smith, *Elements of Ecology and Field Biology* [New York: Harper & Row, 1977]. Table 8-9: *Trends in Ecological Succession from Elements of Ecology and Field Biology* by Robert Leo Smith. Copyright 1978 by Robert Leo Smith. Reprinted by permission of HarperCollins Publishers, Inc.

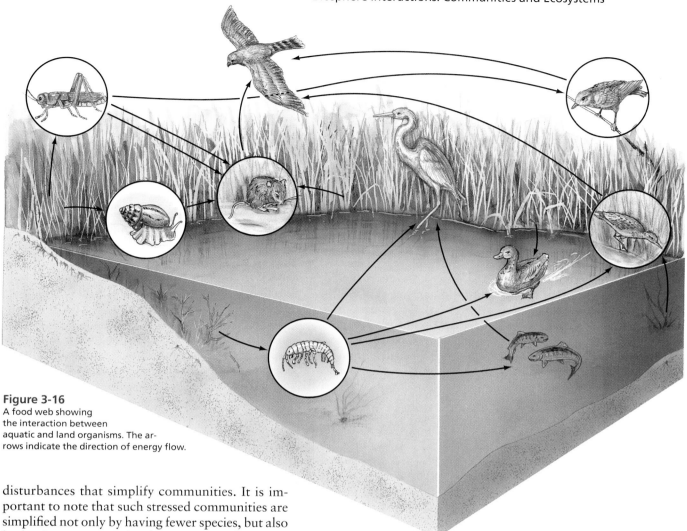

Figure 3-16
A food web showing the interaction between aquatic and land organisms. The arrows indicate the direction of energy flow.

disturbances that simplify communities. It is important to note that such stressed communities are simplified not only by having fewer species, but also by having some species that are superabundant. Although most species cannot tolerate the stressful conditions, some find the new conditions beneficial. For instance, some organisms thrive in highly polluted waters and even use the pollutants as food. Even in these inadvertent disturbances, we often favor early-successional species. Whether we are building roads, farms, cities, or lawns, one of our first actions is to bulldoze or otherwise remove the climax community. Because colonizing disturbed environments is what early-successional species are adapted to do, they have tended to thrive as we have expanded. Indeed, the term *weed* is virtually synonymous with early-successional species, which also include "weedy" animals, such as some mice and many insects.

Ecosystems and Community Function
Although communities vary in structure, certain basic processes, or functions, unite them all. The most basic processes are (1) energy flow and (2) matter cycling. All organisms must eat (take in energy and matter) to stay alive, causing energy and matter to move through the community. All energy and matter ultimately come from, and return to, the physi-

cal environment. We must therefore observe the ecosystem (community plus physical environment) to understand the complete process.

Energy flows and matter cycles through all four of the environmental spheres. The movement of energy and matter through ecosystems represents movement through one of those spheres, the biosphere, back into and through the other three physical spheres.

Energy Flow through Ecosystems. Energy flows are routed through feeding relationships in the ecosystem. An organism's niche ("occupation") in the ecosystem is closely associated with feeding. Energy flow through any ecosystem can be represented by the food web and the biomass pyramid. The **food web** describes the complex interrelationships by which organisms consume other organisms. The food web in **Figure 3-16** illustrates how even aquatic and land organisms prey on each other.

Although food webs are adequate for graphically depicting the feeding relationships in any given ecosystem, the **biomass pyramid** provides a more basic understanding of energy flow (**Figure 3-17**). Biomass is the weight of living matter. The first trophic (feeding) level consists of the producers in the ecosystem, which produce the food used by all other organisms. Usually, the producers are plants, producing the food by photosynthesis; in a few deep-sea ecosystems, organisms produce food by chemosynthesis based on heat energy and compounds from underwater hydrothermal vents instead of the Sun's energy. All levels above this first level contain consumers. First-order, or primary, consumers are generally herbivores that directly consume the producers, deriving energy from the chemical energy stored in the producers' bodies. As a marine example, first-order consumers include the crustaceans and other organisms that eat the phytoplankton. In a forest ecosystem, first-order consumers include deer and other plant eaters. Above the first-order consumers are the second-order (or secondary) consumers, which feed on the

first-order consumers. In a marine ecosystem, second-order consumers may consist of fishes, lobsters, and other species. In a forest ecosystem, second-order consumers include wolves, panthers, and other meat eaters (carnivores) that eat the deer and other first-order consumers (**Figure 3-18**). Third-, fourth-, and even higher-order consumers can occur in some ecosystems. Decomposers are a special type of consumer. Decomposers, such as many bacteria, consume the tissue of dead organisms from all levels of the food pyramid. Although they are inconspicuous, decomposers are extremely important in energy flow; in virtually all ecosystems, they consume the largest part of the energy flow.

Why does the biomass pyramid form? Biomass declines with each higher trophic level because progressively less food is available. Much of the food an animal consumes is not passed on to the animal that eats it. Instead, much of the food is (1) lost as undigested waste, or (2) "burned up" by the animal's

Figure 3-17   A biomass pyramid. In most land food webs, biomass decreases from one feeding (trophic) level to the next highest.

Figure 3-18   The gray wolf is a predator high on the biomass pyramid.

metabolism to produce heat (**Figure 3-19**). For example, a deer excretes about 25% of its ingested calories as undigested waste. Of the 75% that is digested, most is lost as metabolic waste products (such as urine) and, especially, body heat generated from movement and other kinds of maintenance. Thus, of all the calories eaten by the deer, less than 20% are converted into the deer's body tissue, which can be eaten by wolves or other animals that feed on the deer. Other organisms are more energy-efficient. Insects and other cold-blooded organisms can convert ingested calories into tissue much more efficiently than mammals because they have slower metabolisms. Even so, these organisms convert less than 50% of ingested calories into tissue. The result is a "leakage" of energy between each feeding level. This inefficiency is why feeding relationships form a pyramid. In general, about 80%–95% of the energy is lost in the transfer between each level, depending on the organisms involved. Because so little energy is left, very few ecosystems have food pyramids with more than five levels. This is also why large carnivores are rare: They are the organisms at the top.

Figure 3-19   Flow of biomass and energy through a food pyramid. Note that much of the biomass is consumed by decomposers.

Ecosystem Productivity

The amount of food generated by producers at the base of the food pyramid varies greatly among ecosystems. Productivity is the rate at which biomass is produced in a community. **Net primary productivity (NPP)** is the rate at which producer, usually plant, biomass is created. Among the most productive terrestrial ecosystems are tropical forests and swamps (Table 3-5), which produce plant biomass (NPP) at many times the rate of deserts. Temperate communities such as grasslands and temperate forests have intermediate productivities. The main reason for this pattern is that productivity on land increases where the growing season is longer. As **Figure 3-20** illustrates, productivity tends to increase toward the equator where winters are milder and shorter. Deserts are the exception to this trend because lack of water limits growth, even though the growing season is long.

In terms of productivity per unit area, among the most productive aquatic ecosystems are estuaries and reefs, which may be up to 10 times more productive than certain other freshwater or marine ecosystems (Table 3-5). By this measure, the open ocean is the least productive by far. This point is crucial because the open ocean constitutes about 90%

TABLE 3-5	**Ecosystems and Productivity**			
	Net Primary Productivity, Per unit area per year (g/m² or t/km² per year)[a]			
Ecosystem Type	Area (10^6 km²)[b]	Normal Range	Mean	World Net Primary production (10^9t per year)[c]
Tropical rainforest	17.0	1000-3500	2200	37.4
Tropical seasonal forest	7.5	1000-2500	1600	12.0
Temperate evergreen forest	5.0	600-2500	1300	6.5
Temperate deciduous forest	7.0	600-2500	1200	8.4
Boreal northern forest	12.0	400-2000	800	9.6
Woodland and shrubland	8.5	250-1200	700	6.0
Savanna	15.0	200-2000	900	13.5
Temperate grassland	9.0	200-1500	600	5.4
Tundra and alpine	8.0	10-400	140	1.1
Desert and semidesert shrub	18.0	10-250	90	1.6
Extreme desert, rock, sand, and ice	24.0	0-10	3	0.07
Cultivated land	14.0	100-3500	650	9.1
Swamp and marsh	2.0	800-3500	2000	4.0
Lake and stream	2.0	100-1500	250	0.5
Total continental	**149**		**773**	**115**
Open ocean	332.0	2-400	125	41.5
Upwelling zones	0.4	400-1000	500	0.2
Continental shelf	26.6	200-600	360	9.6
Reefs	0.6	500-4000	2500	1.6
Estuaries	1.4	200-3500	1500	2.1
Total marine	**361**		**152**	**55.0**
Full total	**510**		**333**	**170**

[a] t/km² = g/m² = metric tons/km² = approximately 2.85 tons per square mile.

[b] 10^6 km² = approximately 386,000 square miles.

[c] 10^9 t = 1 billion metric tons = approximately 1.102 billion tons.

Source: M. Begon, J. Harper, and C. Townsend, *Ecology*, 2nd ed. [Cambridge, MA: Blackwell, 1990]. Reprinted by permission of Blackwell Science, Inc.

of the ocean. Therefore, 90% of the ocean, which is over half of the Earth's surface, is essentially a "marine desert" in terms of productivity per unit area. Due to the size of the open ocean, however, it does make a significant contribution to the world's net primary productivity (Table 3-5). Reefs and estuaries are highly productive ecosystems despite their small areas. Unlike terrestrial ecosystems, the productivity of these nearshore areas is not strongly determined by the length of the growing season. Instead, nutrient availability tends to be the main limiting factor in marine ecosystems. The open ocean is relatively "starved" for some nutrients, especially phosphorus, because the source of the nutrients is runoff from land. Zones of upwelling, the upward transport of water to the nearshore surface from the ocean depths, can be highly productive, however. The upwelling currents often carry many nutrients that have settled and been swept up from the ocean bottom. The upwelling zones west of Peru, which support a great fishing industry, are an example.

Figure 3-20   Annual net primary productivity (NPP) on the Earth's land surface. In photosynthesis, plants extract carbon from the atmosphere and utilize the carbon to form biomass. Therefore, one way to measure NPP is in terms of the amount of carbon converted into biomass per unit area per year. In this figure NPP is measured using the units grams of carbon per square meter per year (g C m^{-2} yr^{-1}). [1 gram C per m^2 per year is approximately 0.029 ounce C per yd^2 per year.] Note that productivity is high in tropical forests and very low in arid regions. (*Source*: J. M. Melillo, et al., *Nature* 363 [1993]: p. 237. Reprinted with permission from *Nature*. Copyright © 1993 Macmillan Magazines Limited.)

Net secondary productivity (NSP) is the rate at which consumer and decomposer biomass is produced. In other words, NSP includes all biomass except plants. A general rule of ecology is that primary and secondary net productivity are correlated: Communities that have high primary productivity almost always have high secondary productivity. If the base of the food pyramid is producing much biomass, the organisms that consume and decompose plants will usually produce more biomass, too.

Human Disturbance of Energy Flow and Productivity. The extent to which humans have altered ecosystem energy flow is demonstrated by a startling statistic: Nearly 40% of the potential terrestrial NPP and about 2% of the oceanic ecosystem NPP is directly used, diverted, or lost (such as when forests are paved over to construct shopping malls) due to the activities of humans. This means that nearly half of the energy potentially converted by land plants is largely not available to species in natural ecosystems to trickle upward into the food pyramids. This energy loss to nature is particularly striking considering that up to 90% of all species on Earth live on land. If we add the NPP of the aquatic food pyramid to that of the land pyramid, humans redirect an estimated 25% of global NPP.

Another way humans disturb productivity is by causing extinctions. Case Study 3-2 discusses evidence that humans not only "usurp" natural productivity, but also reduce what ecosystems can produce.

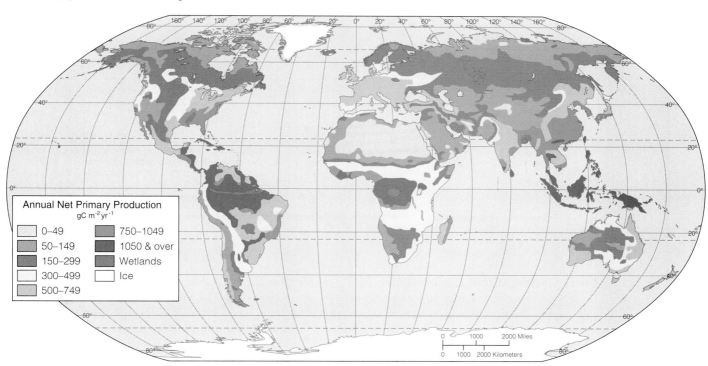

Annual Net Primary Production
gC m^{-2} yr^{-1}

- 0–49
- 50–149
- 150–299
- 300–499
- 500–749
- 750–1049
- 1050 & over
- Wetlands
- Ice

CASE STUDY 3-2

Does Extinction Reduce Ecosystem Productivity? How Experimental Ecology Answers Key Questions

Visit www.environment.jbpub.com
for more information.

If an ecosystem loses plant species, is its primary productivity reduced? After all, the ground could be completely covered with just one species. Some ecologists, however, have argued that the more plant species in an ecosystem, the more biomass it can produce because they provide buffers against seasonal and other environmental changes. If one species suffers from cold, for instance, another can take over the photosynthetic processes.

After years of debate, evidence is accumulating that plant diversity does indeed tend to increase primary productivity. John H. Lawton and his colleagues at the Imperial College in England have performed a series of experiments that measured productivity of ecosystems under environmentally controlled laboratory conditions. As **Figure 1** shows, their general finding was that plant productivity remained relatively high during the initial loss of species in very species-rich communities. But as the number of species continued to decline, productivity began to decrease until species-poor communities, with 1 to 5 plant species, showed significantly lower productivity. Much further work is needed to verify this finding, but it shows how experimental methods can answer crucial environmental questions that otherwise become embroiled in fruitless debates.

Another series of experiments conducted by David C. Tilman of the University of Minnesota produced another key finding: Increased diversity also increases ecosystem resistance to disturbance. **Figure 2** shows that in Minnesota grasslands, species-poor communities produce much less relative biomass during drought years than species-rich communities. Apparently, having more species helps buffer the ecosystem against disturbances because some species can tolerate the disturbance better than others. By having more species, an ecosystem is more likely to have at least some species that can tolerate a disturbance and continue to produce biomass.

The studies by Lawton and Tilman illustrate how ecological experiments can answer crucial questions about how extinction is impacting the environment. Species loss not only diminishes our world aesthetically and economically, it apparently impairs (1) ecosystem functioning such as biomass production and (2) ecosystem resistance to disturbance. A species-poor ecosystem has lower productivity during normal years, and this reduced productivity is even more drastically lowered during times of stress.

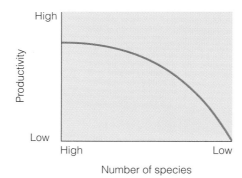

Figure 1   Declines in the number of species ultimately lead to declines in productivity.

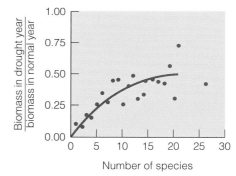

Figure 2   Communities with more species produce relatively more biomass in drought years than communities with fewer species. (*Source*: Redrawn from J. Lockwood and S. Pimm, Do Species Matter? *Current Biology* 4 [1994]: p. 456. Reprinted with permission from *Current Science*.)

Matter Cycling through Ecosystems

The second basic ecosystem function, matter cycling, occurs because, unlike energy, matter is not always converted into less useful forms when used. Dozens of elements are cycled through ecosystems in biogeochemical cycles, which carry the elements through living tissue and the physical environment such as water, air, and rocks. An example is the biogeochemical cycle of carbon (**Figure 3-21**), which has a major influence on global climate due to the fact that carbon dioxide is a potent "greenhouse gas" (Chapter 17).

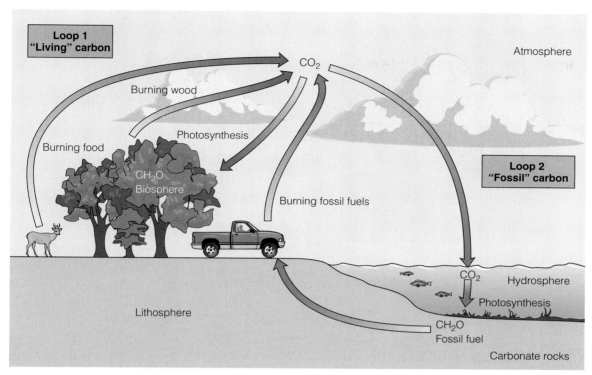

Figure 3-21   The carbon cycle. Loop 1 is "living" carbon that is still actively circulating among living organisms and their environment. Loop 2 is "fossil" carbon; it consists of carbon that is bound in molecules such as coal deposits that are deeply buried until released by burning or some other process. Combustion (and respiration) and photosynthesis ultimately cause carbon to move through both cycles.

Most of the elements that cycle through ecosystems are trace elements, used in small amounts by organisms. Living things use carbon, hydrogen, oxygen, nitrogen, sulfur, and phosphorus in large amounts, however. Because organisms both metabolize and store these elements, ecosystems exert great control over how fast elements cycle. Some elements cycle in a matter of days while others may be buried for millions of years. Carbon may spend millions of years underground stored in fossil fuels such as coal and oil or as limestone.

Ecosystems are generally very efficient in cycling matter, in that most matter is cycled over and over within the ecosystem itself (**Figure 3-22a**). For example, the carbon atoms in a plant will be incorporated into a deer. These, in turn, will be incorporated into the tissue of a wolf that eats the deer. When the wolf dies, decomposers will incorporate the same carbon atoms. All of these changes take place within the ecosystem. Nevertheless, a small amount of matter will be lost from the ecosystem over time. Leaching from rainfall will carry off carbon in the form of decaying organic matter, leaves, and so on. In undisturbed ecosystems, this output loss is roughly balanced by an equal input gain of the same matter. For instance, carbon enters the ecosystem via weathering of rocks and is carried into the ecosystem by rainwater. In undisturbed natural ecosystems, both the input and the output are small relative to the amount of matter "locked up" and recycled within the biomass of the ecosystem itself.

Both the rate and efficiency of matter cycling vary among ecosystems. The cycling of matter is generally faster in tropical ecosystems, such as tropical rainforests and coral reefs, because biochemical reaction rates tend to increase with temperature. Matter cycling is also especially efficient in tropical ecosystems, where high rainfall will leach elements from the soil unless plants incorporate them quickly and efficiently into their tissue. Similarly, coral reefs thrive mainly in nutrient-poor tropical waters, so the elements in the nutrients must be utilized quickly and recycled very efficiently into the tissues of the marine life.

Human Disturbance of Matter Cycling.   Matter cycling in the ecosystem is disturbed when humans alter the balance between the input and output of matter by creating (1) excess output, or (2) excess input (**Figure 3-22b**). Excess output occurs when humans suddenly release the large quantity of matter retained in the biomass of the ecosystem. For instance, in **slash**

Ecosystem

Matter input
(rock weathering, rain)

Matter output
(runoff, leaching)

Matter cycles

Excess output
(such as slash and burn
agriculture)

Excess input disturbance
(such as eutrophication)

Figure 3-22   (a) A healthy ecosystem cycles most of its matter over and over through the food web. Excess input and excess output are symptoms of an unhealthy ecosystem. An example of excess input, where too much matter enters into the system, is fertilizer or nutrient excess. An example of excess output is the rapid loss of matter caused by slash and burn agriculture. (b) Slash-and-burn farming of the rainforest releases nutrients normally stored in plant biomass.

(a)

(b)

and burn agriculture, trees are cut down and burned. The burning releases the nutrients into the soil for agriculture. Unfortunately, the nutrients are quickly leached out of the soil where rainfall is heavy, as in the tropics where slash and burn techniques are common. This massive output of matter from the ecosystem is not fully replaced by input for many hundreds or perhaps thousands of years. During this time, the area can sustain only a relatively barren ecosystem with a fraction of its former diversity. In the meantime, farmers must move on and burn another area of tropical forest to produce arable land. This practice is contributing to massive tropical deforestation worldwide and the release of billions of tons of carbon into the atmosphere annually. Another example of excess output is the massive burning of fossil fuels. These emissions are contributing to the likelihood of global climate change (Chapter 17).

Disturbance by excess input commonly occurs when runoff from agricultural activity carries large amounts of fertilizer, organic waste, and other nutrients into natural ecosystems. This also destroys diversity because the excess nutrients cause eutrophication, leading to unrestrained growth of some organisms, such as algae in a lake. When the algae die, the decay of their now-abundant bodies by bacteria uses up so much oxygen that fishes and many other organisms die.

Biogeochemical Cycles: An Introduction

When observing the cycles of matter such as water, rocks, nutrients, and other substances within and among the spheres, scientists often find it useful to focus on the cycles of chemical elements that compose those substances. This approach can be used to simplify our models of environmental cycles because just a few basic elements participate in many of the most important cycles on Earth. These cycles of chemical elements through the atmosphere, lithosphere, hydrosphere, and biosphere are called **biogeochemical cycles**.

Among the most important biogeochemical cycles are the six cycles that transport the six elements most important to life: carbon, hydrogen, oxygen, nitrogen, phosphorus, and sulfur. Of the approximately 90 elements that occur naturally on Earth, these six comprise the vast majority of atoms in the tissue of all living things. As **Table 3-6** shows, oxygen alone accounts for more than 62% of the weight

TABLE 3-6	Atomic Composition by Weight of Three Representative Organisms		
Element	**Human**	**Alfalfa**	**Bacterium**
Oxygen	62.81%	77.90%	73.68%
Carbon	19.37	11.34	12.14
Hydrogen	9.31	8.72	9.94
Nitrogen	5.14	0.83	3.04
Phosphorus	0.63	0.71	0.60
Sulfur	0.64	0.10	0.32
Total	97.90	99.60	99.72

TABLE 3-7	The Relative Abundance by Weight of Some Chemical Elements in the Earth's Crust	
Element (Chemical Symbol)	**Relative Abundance**	
Oxygen (O)	46.6%	
Silicon (Si)	27.7	
Aluminum (Al)	8.1	
Iron (Fe)	5.0	
Calcium (Ca)	3.6	
Sodium (Na)	2.8	
Potassium (K)	2.6	
Magnesium (Mg)	2.1	
Phosphorus (P)	0.07	
Carbon (C)	0.03	
Nitrogen (N)	Trace	

Each of the many biogeochemical cycles has different pathways of transport and temporary storage reservoirs. The **carbon cycle** in Figure 3-21 is a typical biogeochemical cycle. Like many cycles, the carbon cycle appears at first glance to be quite complex, with many pathways (arrows), but closer inspection shows that these pathways are based on just two processes: withdrawal from and addition to the atmosphere.

1. Withdrawal of carbon is largely driven by photosynthesis whereby plants take carbon out of the atmosphere where it resides as carbon dioxide. The CO_2 is combined with water (H_2O) to form biochemical molecules such as sugars ($(CH_2O)_n$) and oxygen. Photosynthesis is conveniently written as:

$$CO_2 + H_2O + energy \longrightarrow (CH_2O)_n + O_2$$

This reaction is called photosynthesis because it requires energy from the Sun (*photo* = light; *synthesis* = combine).

Figure 3-21 also shows two "loops" of photosynthesis. Loop 1 illustrates the pathway of "living" carbon in the ongoing photosynthesis of modern plants. Loop 2 shows how "fossil" carbon forms. Fossil carbon is stored carbon that has been temporarily withdrawn from use by living organisms. Carbon is stored in the lithosphere, when plants, such as tiny plankton in marine and fresh waters, die, sink to the bottom, and are buried. After millions of years of burial, these dead plants, and the carbon in them, can become fossil fuels such as petroleum. Similarly, clams and other ocean shellfishes withdraw carbon for use in constructing their shells. When the shellfishes die, their shells contribute to the vast amounts of limestone.

However, the large majority of the Earth's carbon now resides in the ocean. As **Figure 3-23** illustrates, the ocean stores much more carbon than is

of the human body and more than 77% of the weight of the alfalfa plant. Carbon and oxygen together account for more than 80% of the weight of a human.

Table 3-7 shows the relative abundances of the most common elements in the Earth's crust. Oxygen is the most abundant, just as it is most common in the human body. But the second most common human element, carbon, is hundreds of times rarer in the crust. Instead, silicon, which is virtually absent from the human body, is extremely abundant in the crust. You can see other major discrepancies by comparing Tables 3-6 and 3-7. These discrepancies illustrate how life is chemically distinct from its environment. Without biogeochemical cycles to transport and store temporary concentrations of matter for food and other uses, life could not survive.

Billion Metric Tons

Atmospheric 748

Terrestrial 2,000

Geological 4,000

Oceanic 38,000

(a)

(b)

Figure 3-23
(a) Major reservoirs of the carbon cycle, in billion tons of carbon. The oceans are the largest reservoir by far. (b) Limestone, such as the white cliffs of Dover, is formed from shells of clams and other marine life. (*Source:* (a) Adapted from W. M. Post, et al., "The Global Carbon Cycle," *American Scientist* 78 [1990]: 315.)

found in the other three sinks—the atmosphere, lithosphere (geological), and living organisms on land (terrestrial biosphere).

2. Addition of carbon to the atmosphere often occurs from combustion. Combustion or "burning" is essentially the reverse of photosynthesis; oxygen (O_2) is combined with plant matter (($CH_2O)_n$) to release CO_2 and H_2O:

$$(CH_2O)_n + O_2 \longrightarrow CO_2 + H_2O + energy$$

Setting fire to either living matter (loop 1) or fossil fuel (loop 2) will therefore release carbon dioxide. Similarly, when we digest food, we are carrying out combustion: Our bodies take the oxygen we inhale and use it to break down the biochemical molecules we eat, such as plant foods. We then use the energy given off to move around, grow, and maintain our bodies. The carbon dioxide produced is exhaled into the atmosphere (loop 1). This process of biological combustion is called **respiration**.

Another important biogeochemical cycle is the **nitrogen cycle** (**Figure 3-24**). Nitrogen is of fundamental importance to all organisms, being a necessary component of DNA (the genetic material), RNA, amino acids (of which proteins are formed), and many other organic molecules. Nitrogen is also quite abundant, or so it would seem, since the atmosphere is 78% nitrogen gas (N_2) by volume. However, this atmospheric nitrogen is not directly available to organisms; rather it must be turned into a more chemically reactive form, such as NH_4^+ (ammonium) or NO_3^- (nitrate). A small amount of atmospheric N_2 is converted to such usable forms through the action of rain and various chemically reactions, but most of the usable nitrogen is converted from atmospheric N_2 by certain **nitrogen-fixing** prokaryotes ("bacteria"). Nitrogen-fixing bacteria in the soil and nitrogen-fixing bacteria in root nodules of legumes and certain other types of plants produce ammonia (NH_3). Most soils are slightly acidic, and the ammonia can gain a hydrogen ion (H^+) to become NH_4^+ that is then utilized by plants. Ammonia may also evaporate into the atmosphere where it can form NH_4^+ (ammonium) that is then brought back to the surface of Earth in rainfall.

In the soil, aerobic bacteria can also transform NH_4^+ to nitrite (NO_2^-) and then nitrate (NO_3^-) in the process known as *nitrification*. The nitrate can also be used by various plants. Animals gain necessary nitrogen by eating plants or other animals. There are also bacteria that convert nitrate to N_2, which is released to the atmosphere (the process known as *denitrification*). Still other bacteria, as well as various funguses, decompose dead organic matter and in the process turn nitrogen that is bound up in organic molecules back into ammonium (this is known as *ammonification*), which can then be reassimilated into other organisms. Decomposition and reassimilation of nitrogen accounts for the vast majority of nitrogen cycled through most natural ecosystems. In modern times nitrogen fixation is also carried out artificially during the manufacturing of fertilizers, and this has significantly increased the amount of nitrogen chemically available to organisms.

Biogeochemical Cycles: Major Features

The biogeochemical cycles as a group can be analyzed in terms of a number of important features. These features in-

Figure 3-24   The nitrogen cycle.

Figure 3-25 Recycling rates of water, oxygen, and carbon dioxide through the four spheres. (*Source*: L. Laporte, *Encounter with the Earth* [San Francisco: Canfield, 1975], p. 22. Modified by permission of Leo F. Laporte.)

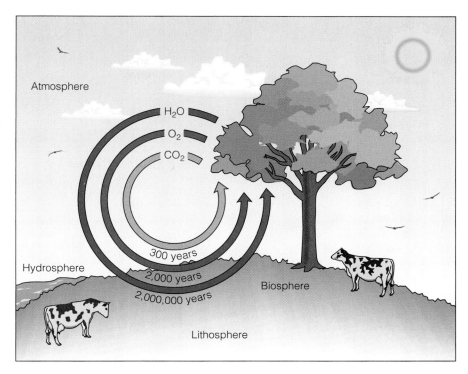

clude the cycles' pathways, their rates of cycling, and the degree to which they are being disturbed by human activities.

A Variety of Pathways. Each biogeochemical cycle has many different pathways; many chemical and physical processes help to cycle each atom. For instance, the carbon cycle transports carbon through all four spheres. From the atmosphere, the carbon dioxide (CO_2 molecule) dissolves in water in the hydrosphere, where plankton use the carbon to build body tissue (CH_2O molecule), thereby moving the carbon into the biosphere. When the plankton are buried and converted to fossil fuel, the carbon is converted to complex hydrocarbon molecules such as oil and becomes part of the lithosphere. When the oil is burned, the carbon is released back into the atmosphere, where it may recycle through a different set of pathways. The CO_2 gas may be absorbed by a tropical tree the next time instead of by plankton.

Of course, each element has a different set of potential biogeochemical pathways. For example, unlike carbon, phosphorus generally does not cycle through the atmosphere because it does not easily form a gas. During its cycle, phosphorus often combines with different atoms than carbon does and therefore forms different molecules and undergoes different chemical reactions.

Variable Rates of Cycling. Biogeochemical cycles vary in their rate of cycling. **Figure 3-25** shows the average amount of time water, oxygen, and carbon dioxide molecules take to make a complete cycle through the four spheres. Clearly, cycling rates can vary drastically over many orders of magnitude. Carbon takes only hundreds of years to cycle, whereas water takes about two million years. These times are only approximate because specific molecules may cycle much more rapidly or slowly depending on the pathway they follow. A small number of carbon atoms (fewer than 1 in 10,000) in the active, living loop, for instance, may be diverted and stored as oil deposits for over 200 million years and cycle through the much slower fossil loop.

Why do substances cycle at such different rates? Two major determinants are (1) the chemical reactivity of the substance and (2) whether it has a gaseous phase (occurs in the atmosphere) somewhere in the cycle. The high chemical reactivity of carbon causes it to participate in many chemical pathways and is a major reason why it cycles so quickly. In addition, carbon is abundant as the gas carbon dioxide. Because gas molecules move much more quickly than more tightly bonded molecules in liquids or solids, the existence of a gaseous phase allows the substance to be transported more rapidly.

Although oxygen and water cycle more slowly than carbon dioxide, they actually cycle at relatively fast rates compared to many other substances. Like carbon dioxide, oxygen and water are chemically reactive and have a major gaseous phase. For instance, the average water molecule has a **residence time** of 10 days in the atmosphere, during which time it may move thousands of miles (thousands of kilometers) before falling back to the surface of Earth as a liquid.

Thus, to find a substance that has an extremely slow cycling time, we should look for one that has no gaseous phase and is also relatively unreactive in natural systems. Phosphorus is a good example because it not only has a very slow cycling rate but is one of the six most important elements of life and is therefore of much interest and very well studied. Because of its chemical and physical properties, phosphorus does not form a gas and does not read-

Figure 3-26
(a) The global phosphorus cycle.
The amount that flows on Earth is much smaller
than the amount stored in rocks and sediment. (b) Phosphate is mined in
Florida, and other areas, where ocean waters deposited phosphorus-rich sediments.

ily combine with other substances. Its main mode of transport is water, which moves much more slowly than air, and even in water phosphorus is relatively insoluble. As **Figure 3-26** illustrates, large amounts of phosphorus become "locked up" in storage for long periods of time as sediments in the deep ocean and the Earth's crust. Only relatively slow and rare events, such as upwelling ocean currents from the deep sea or weathering of phosphorus-rich rocks, recycle the phosphorus.

Instead of the few hundreds to few millions of years typical of cycles with a gaseous phase, phosphorus requires many tens of millions of years to complete its biogeochemical cycle. This slow cycling rate drastically reduces the availability of this critical nutrient with profound effects for the biosphere. Phosphorus is usually the nutrient in shortest supply in most ecosystems and is therefore labeled the *limiting nutrient*. As we saw already, the sudden availability of limiting nutrients in natural systems causes rapid growth.

The Effects of Human Activity.   Biogeochemical cycles are crucial to all life, but are being greatly disturbed by human activity. As human population and technological ability rapidly increase, huge quantities of materials are extracted through mining and other means and redistributed through all the spheres. The net result has been the disturbance of nearly all biogeochemical cycles. The most com-

mon type of disturbance is acceleration of the cycles: materials are being rapidly mined and otherwise extracted from storage reservoirs (sources) and, after use, are rapidly deposited back into the environment (sinks). This increased rate of cycling from source to sink leads to the two basic environmental problems: depletion and pollution (Chapter 1). Indeed, a basic definition of pollution is a temporary concentration of a chemical above levels that normally occur in its biogeochemical cycle.

No one really knows just how drastic or dangerous this acceleration of natural cycles will ultimately prove to be. No one doubts that major consequences will occur, however, and because biogeochemical cycles are global in nature, many of these consequences will occur on a global scale. Carbon provides a prominent example of how humans are actively disturbing a major cycle, with potentially drastic consequences. The burning of fossil fuels, such as coal, petroleum, and natural gas, has released increasing amounts of carbon dioxide into the atmosphere. Global release of carbon has in-

Figure 3-27 The two laws of thermodynamics. The first law is illustrated by the conversion of Sun energy (a) to food (chemical) energy (c). The second law dictates that heat loss (b) during conversion causes the amount of usable food (chemical) energy to be less than the Sun energy. In this case, it is much less. (Source: E. P. Odum, *Ecology and Our Endangered Life Support Systems* [Sunderland, Mass.: Sinauer, 1989], p. 70. Reprinted by permission of Sinauer Associates, Inc.)

(a) Sun's rays, 100 units
Dilute energy form

(b) Heat, 98 units
Very dilute (dispersed) energy form

Maple leaf
Energy conversion system

(c) Sugars, 2 units
Concentrated energy form

creased exponentially from about 1 billion metric tons per year in 1940 to an estimated 6.5 billion metric tons per year at the end of the century. It is estimated that about half of this 6.5 billion metric tons per year is absorbed by plant life and the oceans, but the remainder is accumulating in the atmosphere in the form of carbon dioxide.

This accumulation of atmospheric carbon dioxide has many potential global consequences. For example, many studies have shown that the rate of plant growth will generally increase from increasing photosynthesis. The most publicized consequence is global warming (see Chapter 17). Carbon dioxide is a "greenhouse gas," meaning that it increases the ability of the atmosphere to trap heat. How much carbon dioxide can be added to the atmosphere before significant global warming will occur is much debated.

Although the consequences of disturbing the carbon cycle may be especially dramatic—rising temperatures could cause dramatic changes in weather and climate, shift rainfall patterns, melt glaciers, and raise sea levels—all of the many dozens of biogeochemical cycles are increasingly being disturbed by humans. Some examples include excess phosphorus, which causes runaway plant growth in natural waters; nitrogen emissions, which contribute to smog; and sulfur emissions, which cause acid rain. These and other disturbances are discussed in detail in later chapters.

Energy Flows

The **first law of thermodynamics** says that energy cannot be created or destroyed, but can be transformed. The **second law of thermodynamics** says that when energy is transformed from one kind to another, it is degraded, meaning that the energy becomes less capable of doing work. For example, only about 30% of the chemical energy in gasoline is converted to the energy of motion in a typical gasoline-powered automobile. Similarly, photosynthesis uses solar energy to create food from carbon and other atoms. Food represents chemical energy, which is stored in the bonds between atoms, so photosynthesis is a transformation from solar to chemical energy. As **Figure 3-27** illustrates, this trans-

formation is far from 100% efficient. Most of the incoming energy is "lost" as heat. Heat is considered to be low-quality energy and is capable of doing less work than high-quality energy.

Entropy refers to the amount of low-quality energy in a system. If entropy is very high, matter will tend to disorganize to simpler states. The second law of thermodynamics is sometimes called the *law of entropy* because all energy transformations will increase the entropy of a system unless new high-quality energy, such as sunlight, enters the system to replenish it.

Some energy transformations are more efficient than others. By carefully refining our technology, humans have managed to achieve much greater efficiencies. For example, the best solar (photovoltaic) cells convert as much as 30% of sunlight to electricity. Other examples of energy conversion are the use of nuclear energy to generate electrical energy and the conversion of gasoline (chemical energy) into mechanical energy ("energy of motion"). In each of these and all other energy transformations, however, engineers long ago accepted the fact that no matter how advanced our technology, some loss of usable energy will always occur, if only a few percent.

Because of this loss, energy cannot be recycled like matter. Ultimately, all the energy in a system will become relatively useless (transformed to heat) unless new energy flows into the system. Thus, we say that matter cycles, but energy must flow from one source to another. On Earth, the vast majority of new energy flows from the Sun. This energy originates with nuclear reactions at the Sun's core and travels as light energy across 93 million miles (150 million km) to strike the Earth. Upon striking the Earth, the energy is transformed in many ways, depending on where it strikes. Collectively, the various flow pathways of all energy on Earth are called the Earth's **energy budget**. Over 80% of the incoming light is either directly reflected back into space or absorbed and reradiated back into space as heat. The remaining energy powers the hydrologic cycle by evaporation, generates wind, powers photosynthesis, and in general drives many of the cycles within and between the spheres that we discussed earlier. Photosynthesis, for all its importance in sustaining most life on Earth, including humans, uses a tiny fraction, just 0.06%, of solar radiation received by Earth. And given the inefficiency of photosynthesis, much of this is wasted.

Actually, not all of the Earth's energy comes from the Sun. A small fraction comes from two other sources. One is the Moon's gravitational pull, which causes tides in the ocean. This tidal energy is thousands of times less than the amount of energy provided by the Sun. The second source of energy is the Earth's own internal geothermal energy, which is generated by radioactive minerals deep within the Earth. The heat diffuses outward to the Earth's surface and melts rocks to drive the tectonic cycle of the lithosphere and the emission of volcanic gases into the atmosphere. It also produces the heat that creates deep-sea vents in the ocean floor, where hot magma provides nutrients and energy to rich marine communities.

Human Use of Energy Flows.   Humans are using ever-greater amounts of the energy flows on Earth.

Modern civilization is built upon fossil fuels, which are fossilized plant, animal, or microorganism materials that store solar energy from millions of years ago as chemical energy. Because they take so long to form, fossil fuels are called **nonrenewable resources**. If rates of use continue to rise, some estimates suggest that the known world supply of oil will be used up within the next 100 years and world coal supplies within 300–400 years. Besides depletion, another and perhaps more immediate problem with fossil fuels is the release of pollutants. Almost all major air pollution, including acid rain, smog, carbon monoxide, and greenhouse gases, is caused by burning fossil fuels. Although many of these pollutants can be controlled by smokestack devices, fuel cleansing, and other technical solutions, there is no economical way of removing the carbon because so much is produced. Burning just a gallon (3.785 liters) of gasoline produces over 20 pounds (9 kg) of carbon dioxide, and coal produces even more. Fossil fuels also cause many other pollution problems, such as seepage from storage tanks into groundwater and ocean spills.

Instead of using "fossilized" solar energy, it would be less damaging to use the solar energy flow as it strikes the Earth today. Because the Sun will last about five billion more years, solar energy will not soon be depleted, and the potential supply is vast. In just one month, the Earth intercepts more energy from the Sun than is contained in all the fossil fuels on the planet. Solar energy, along with tidal energy, is a form of **renewable energy**.

SUMMARY

- Earth's natural environment can be divided into the lithosphere, hydrosphere, atmosphere, and biosphere.
- Water, air, and energy are three major aspects of the physical environment.
- The field of ecology examines how organisms interact with each other and the physical environment.
- Populations of organisms typically go through three distinct phases: growth, stability, and decline.
- Human populations can impact upon, and cause disruptions of, natural populations.
- Populations interact with one another and their physical environment in communities and ecosystems.
- Communities can change through the sequential replacement of certain species by

others in a process known as community succession.
- Energy flow through an ecosystem can be studied using food webs and biomass pyramids.
- Productivity is the rate at which biomass is produced by a community.
- Energy flows and matter cycles through ecosystems.
- Biogeochemical cycles are cycles of elements and substances (for example, carbon and nitrogen) that move through the atmosphere, lithosphere, hydrosphere, and biosphere.
- According to the laws of thermodynamics, energy cannot be created or destroyed, but it is degraded (less capable of doing work) when transformed from one form of energy to another.

KEY TERMS

Amensalism
Atmosphere
Biogeochemical cycle
Biomass
Biomass pyramid
Biosphere
Carbon cycle
Carrying capacity
Climate
Climax community
Commensalism
Competition
Competitive exclusion
Depth diversity gradient
Ecology
Ecotone
Endemic species
Energy budget

Entropy
Eutrophication
First law of thermodynamics
Food web
Habitat
Hydrologic cycle
Hydrosphere
Intrinsic rate of increase
Latitudinal diversity gradient
Law of the minimum
Limiting nutrient
Lithosphere
Minimum viable population
   (MVP)
Monoculture
Mutualism
Net primary productivity (NPP)

Net secondary productivity (NSP)
Niche
Nitrogen cycle
Nitrogen fixing
Nonrenewable resource
Parasitism
Photosynthesis
Pioneer community
Population
Predation
Renewable energy
Residence time
Respiration
Second law of thermodynamics
Slash and burn agriculture
Symbiosis
Weather

STUDY GUIDE

STUDY QUESTIONS

1. Briefly describe the four spheres that compose the natural environment. What are a few ways that humans have affected each?
2. Describe the composition of Earth's atmosphere, including the major constituent gases and the layers of the atmospheric envelope.
3. Briefly describe the energy flow on the surface of Earth. What is the hydrologic cycle?
4. Define *ecology*. Need an ecologist necessarily be an environmentalist?
5. What are the typical phases that a population of organisms will pass through over time?
6. List the basic categories of abundance controls that regulate the sizes of populations in nature. Give an example of how humans have, perhaps inadvertently, affected the size of a specific population of organisms.
7. What is the "law of the minimum"?
8. Discuss the difference between closed and open structure communities. Why is it thought that most communities are open in structure? What is the exception to this?
9. How can a community change over time? What are some of the basic trends in community succession?
10. How do humans affect the biosphere? Include the following in your discussion: population growth, population decline, pop-

ulation ranges, communities, energy flow and productivity, and matter cycling.
11. Distinguish between food webs and biomass pyramids. Why are both important to consider when discussing ecosystems? How does each relate to energy flow through an ecosystem?
12. How does net primary productivity generally correlate with latitude?
13. Name the six elements most important to life.
14. Review the carbon cycle, including the two "loops" of the carbon cycle. How does the carbon cycle illustrate the major features of a biogeochemical cycle?
15. Where is the majority of carbon stored?
16. Why do different substances cycle at different rates?
17. Why do large amounts of phosphorous become locked up?
18. What is the limiting nutrient in many ecosystems?
19. Describe an example of a biogeochemical cycle that has been disturbed by humans.
20. How is energy different from matter? How is it similar?
21. Why can't energy be recycled like matter?
22. What are the two laws of thermodynamics? What is entropy?

STUDY GUIDE

WHAT'S THE EVIDENCE?

1. The authors assert that humans, in many cases inadvertently, have had major effects on some populations of organisms. Do the authors present enough evidence to convince you that this is the case? In some cases might a natural population have increased or decreased even if humans had not altered its environment? Are human-induced effects necessarily detrimental?

2. The authors state that humans redirect an estimated 25% of global net primary productivity. What, exactly, does this mean? Does this sound reasonable to you? That is, is it believable that humans could redirect such a large percentage of NPP? Or would you expect it to be larger? Cite some examples of how humans redirect NPP.

CALCULATIONS

1. If a population doubles each month and begins with 10 individuals, what will the size of the population be after one year?

2. What is the intrinsic rate of increase of a population with a birth rate of 100 per day and a death rate of 98 per day?

ILLUSTRATION AND TABLE REVIEW

1. Based on Table 3-5, which ecosystem type covers the largest area of Earth? What percentage is this of the total area of Earth?

2. Based on on Table 3-5, which ecosystem type is the most productive? What percentage of the total area of Earth does it cover?

WWW.ENVIRONMENT.JBPUB.COM

www.environment.jbpub.com

Connect to this book's web site: http://www.environment.jbpub. com/ The site features eLearning, an online review area that provides quizzes, chapter outlines, and other tools to help you study for your class. We've posted the answers to the book's odd-numbered questions. You can also follow useful links for more in-depth information. The site provides sources for the information in this chapter and recommendations for further readings.

A herd of springbok, southern Africa.

Life has co-evolved with the Earth's atmosphere, hydrosphere, and lithosphere to create an integrated "living" planet. Living things help cycle the elements and sustain an oxygen atmosphere while the Earth provides a habitable environment for life.

CHAPTER 4

The Distribution of Life on Earth

This is a special time in Earth's long history. Continental drift (plate tectonics) has caused the continents to be dispersed and farther apart from one another than during most of their past. The global climate is also relatively cool compared to what it has been for most of Earth's history. Life must adapt to its physical environment, so our biosphere is, not surprisingly, also unique. For example, the Earth holds an unusually high number of species, in part because the wide separation of continents has promoted evolution from geographic isolation.

In this chapter, we examine how species originated and discuss estimates of the number of species on Earth today. We will see that life arose rather early in Earth's history and has diversified into many species through the process of evolution via natural selection. Indeed, there are now so many species that biologists have scarcely begun the task of classifying and describing them all. Sadly, as species continue to disappear from human activities, some species will clearly be gone before they are found and described. There is a lack of professional biologists who are trained to study insects and other groups that are extremely abundant. This information is very important in debates about how fast species are going extinct and what percentage of Earth's species are disappearing.

We will also see how species are distributed on Earth. Species are neither randomly nor evenly distributed because their distribution is closely related to the physical conditions to which they are adapted. Rainfall and temperature are especially crucial in determining which suites of species can survive in any given region. Some species are narrowly adapted to very limited conditions in one area whereas other species, such as the wolf, were once found across many continents. Human impacts, like species distributions, are not randomly or evenly distributed. Some regions, such as deserts, are more sensitive than others to human disturbance. Other regions such as grasslands, with rich soils for crops, are impacted because they are resource-rich and are rapidly exploited by humans.

Evolution of the Biosphere

Primitive fossilized bacteria at least 3.5 billion years old have been found. This indicates that life arose not long after the molten Earth first became cool enough to support life. This very early appearance of life implies that natural processes readily produce life under appropriate conditions. Beginning in the early 1950s with the work of Stanley Miller and Harold Urey, scientists have shown that complex molecules possessed by all living things are readily produced under laboratory conditions that duplicate early environments on Earth. As **Figure 4-1** shows, the early atmosphere is thought to have been composed of ammonia (NH_3), methane (CH_4), water vapor (H_2O), and other gases. When these are subjected to electricity, which simulates lightning and sunlight, chemical reactions occur that produce **amino acids**. Amino acids are complex molecules that are the building blocks of proteins. Proteins make up enzymes and many other components of life such as muscles, hair, and skin.

Of course, protein molecules alone are not living things. Organisms are composed of molecules organized in very complex ways. The basic organizational unit of life is the cell. Remarkably, in the late 1950s, Sidney Fox found that heated amino acids and other simple organic molecules can form cell-like structures sometimes called **protocells**. These structures are not true cells but have many cell-like properties such as being semipermeable to certain materials.

Producing amino acids and protocells in the lab is far from creating life in the lab. Even the simplest bacteria are vastly more complex than these protein and protocellular building blocks. Nevertheless, the readiness with which these first steps toward life occur, combined with the fossil record, supports the idea that life readily arose through natural processes.

Diversification of the Biosphere

Evolution through Natural Selection.   Once life originated, it began to diversify into different kinds of organisms through biological evolution. As Charles Darwin documented in 1859, biological evolution occurs due to natural selection of individual variation:

1. Nearly all populations exhibit variation among individuals.
2. Individuals with advantageous traits will tend to have a reproductive advantage (they will be "naturally selected"), producing more offspring that themselves survive and reproduce, passing on the advantageous traits.
3. Advantageous traits will therefore become widespread in populations.

Figure 4-1   Experimental apparatus that Miller and Urey used to show that organic molecules could be produced from the chemical components of the Earth's early atmosphere.

Variation in neck length in giraffes is an example. We know from fossils that early giraffes had relatively short necks. A few individuals, however, had genes that produced slightly longer necks. In some localities, these giraffes had a feeding advantage because they could browse on leaves in taller trees. Consequently, in populations living where tall trees were common, longer-necked giraffes tended to have more offspring and passed on genes for longer neck growth so that longer necks became more common.

If this process occurs with many traits over a long period of time, the population will eventually become very different from other populations and will create a new species. **Figure 4-2** illustrates how isolation of populations promotes speciation. The different populations are exposed to different environments, which favor different traits, causing the populations to diverge through time. But when do

two different populations become two different species? Biologists generally define a species as a group of individuals that can interbreed to produce fertile offspring (this definition applies only to sexual organisms; the issue of defining nonsexual species is a matter of debate in biological circles). A new species is therefore formed when members of a diverging population can no longer successfully mate with populations of the ancestral species. Closely related species, which have often diverged relatively recently, are grouped together within the same genus. Similar genera are then grouped together within the same family. This method of classifying species according to hierarchically nested categories is a form of **taxonomy**. Applying this traditional system to humans, we have the following:

Kingdom Animalia
  Phylum Chordata
    Class Mammalia
      Order Primates
        Family Hominidae
          Genus *Homo*
            Species *H. sapiens**

*Note that a species name must always be associated with a genic name or abbreviation (in this case *H.* for *Homo*). Thus humans are the species *Homo sapiens*; it is incorrect to call humans simply *sapiens*.

(a)

(b)

(c)

(d)

(e)

Figure 4-2 (a) Reduction of habitat may leave small isolated populations, called peripheral isolates. (b) Barriers may form to isolate small populations. (c) Out-migration to an isolated area may form small separated populations. In all cases (a–c), the small populations can evolve into new species. Isolated by thousands of miles of ocean, the Hawaiian honeycreeper has evolved into many forms, including species with short (d) and long (e) beaks.

Where does individual variation come from? This question is crucial because without the initial variation in the population, natural selection would have nothing to act upon. This question troubled Darwin himself because he was not aware of the work of Gregor Mendel who is credited with discovering the laws of inherited variation in 1865. Indeed, Mendel's work was not well publicized and was only rediscovered by scientists in the early 1900s. Mendel discovered that traits are passed on by **genes**, which are the basic units of heredity. According to the Human Genome Project, humans have an estimated 30,000 to 40,000 genes that determine our traits. Variation in a population, or gene pool, occurs because individuals possess different sets of gene forms (alleles) that produce different traits. But how do different sets of alleles arise? Reproduction is one way. Alleles are shuffled when sperm and egg cells are formed, causing offspring from the same parents to have different combinations of alleles. A second cause of genetic variation is **mutation**, which is a spontaneous change in a gene or allele. Genes are composed of **DNA** molecules, and mutations occur when DNA molecules are altered. Mutation is the ultimate source of all genetic variation.

Patterns of Diversification. Evolution through natural selection has produced an increasingly diverse biosphere, with the total number of species becoming greater through time. Initially, evolution was relatively slow. The right side of **Figure 4-3** summarizes the evolution of life on Earth. For about 2 billion years after the first appearance of fossils, relatively few species of simple single-celled organisms, such as various bacteria and cyanobacteria (formerly known as blue-green algae), appear in the fossil record. A major change occurred about 1.5 billion years ago when more complex cells, called eukaryotes evolved. These cells had a true nucleus, chromosomes, and specialized cellular organelles such as mitochondria. Apparently, such complex cells were a primary impetus of increasing rates of evolution. Multicellular organisms, including sponge-like and jellyfish-like creatures, appeared in the oceans by at least 1 billion years ago. These

Figure 4-3 The left side of this diagram summarizes the changes in the Earth's continents, oceans, and atmosphere, while the right side summarizes the evolution of life on Earth. (mya = millions of years ago)

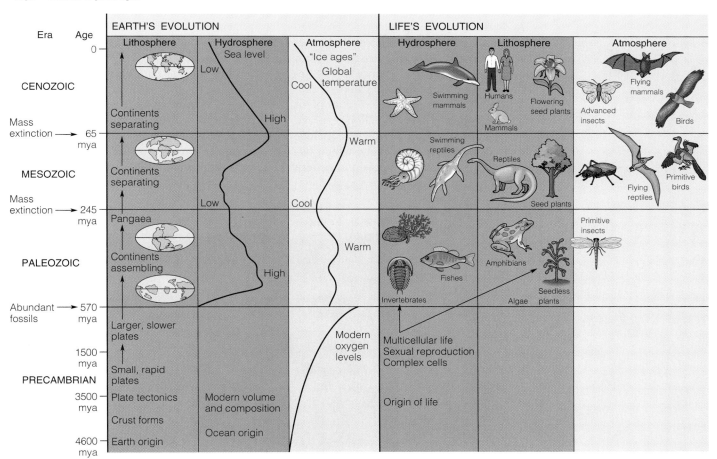

CASE STUDY 4-1

Hellbenders and Alligators: The Asia–U.S. Connection

Visit www.environment.jbpub.com for more information.

The geographic distribution of species is partly dependent on their biological adaptations, such as how well a species can tolerate cold winters. In many cases, however, the history of a group also plays an important role in its current distribution. A good example of this is the presence of some plants and animals in eastern North America.

As early as 1750, the famous biologist Linnaeus noticed a curious similarity between many species in eastern North America with species in Eastern Asia. A century later, Charles Darwin discussed this relationship as well. For example, a giant salamander called the Hellbender, found only in the southern Appalachian region, belongs to a salamander family with all of its other species in Asia (**Figure 1**). Similarly, there are only two species of alligators in the world; the American alligator, which is found throughout the southeastern United States, and the Chinese alligator. Primitive fishes show this pattern too. The paddlefish is an archaic fish with a long snout able to detect magnetic fields of its prey in murky waters. The only species of this family outside of the United States occurs in the Yangtze River system of China.

An even more striking relationship is seen among plants where the broadleaf deciduous forests of eastern Asia and eastern North America have a great resemblance to one another. Most notable are similarities among primitive flowering plants, descended from plant families that appeared over 70 million years ago when dinosaurs were still alive. These include members of the magnolia family, which has many species in both eastern Asia and eastern North America. In some cases, the species themselves are very similar, such as the tulip poplar that occurs in both regions. Even some herbs have striking similarities. The medicinal herb, ginseng, has but two closely related species, with one occurring in eastern North America and the other in eastern Asia.

What is the cause of this biological connection between Asia and eastern North America? Many people have suggested the ancestors of the North American plants and animals may have colonized North America in the same way as the ancestors of the American Indians. These early arrivals migrated across the Bering Sea land-bridge from Asia through Alaska several thousand years ago. However, the very ancient geological age of nearly all of these groups suggests that they became separated much longer than a few thousand years ago. Paleobiologists have found that about 40 million years ago there was a land-bridge connection across the North Atlantic. It ranged across northern Asia through northern Europe into Greenland and northern Canada. The climate of the Earth was much warmer at this time, so it permitted the migration of species adapted to warm climates. By 30 million years ago, the Earth had cooled to the point where such migration was no longer possible. These groups have been isolated from Asia ever since.

This example illustrates how understanding the history of life helps us value its preservation. We see that many of the species that constitute the eastern North American forests are part of a very ancient ecosystem. These species belong to groups with a very long evolutionary heritage, and they contain very ancient genetic diversity.

Questions

1. Name three species that might be called "living fossils" because they have a very long evolutionary heritage and are similar to their distant ancestors.
2. Botanists have long noted that ornamental plants from eastern Asia are among the easiest plants for landscapers to cultivate in eastern North America. Why is this?

Figure 1 Hellbender Salamander

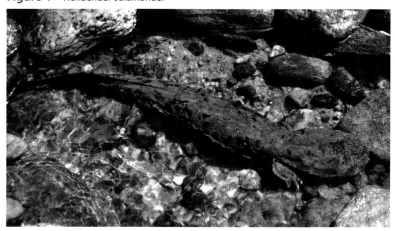

probably evolved from colonies of single celled **eukaryotes,** such as protozoa, that became progressively more specialized and integrated.

About 570 million years ago, the fossil record shows a rapid diversification sometimes called the **explosion of life,** when most of the major groups of animals first appear. Note in Figure 4-3 that this explosion corresponds to the time when modern oxygen levels were attained in the atmosphere. This permitted the evolution of more complex animals, which have a greater metabolic need for oxygen.

Following the explosion of life, living things diversified into new environments. As shown in Figure 4-3, life colonized the land (lithosphere) and the air (atmosphere) during the Paleozoic Era, which began with the explosion of life and ended with a global mass extinction. The Mesozoic Era, sometimes called the age of dinosaurs, was the second major era, and it also ended with a mass extinction. The third and last era is the Cenozoic Era, sometimes called the age of mammals. Although catastrophes, especially from asteroid and comet impacts and global climate change, have temporarily caused species numbers to decrease at very rare intervals through mass extinctions, the overall trend throughout these eras is toward increasing numbers of species as life has adapted to new environments and found new ways of doing things through mutation and natural selection. Ironically, we now live in a biosphere that is one of the most diverse in life's long history, although this will not be true in a few decades if current rates of species extinction continue (Chapter 11).

What Is Biodiversity?

Biological diversity, or **biodiversity,** has many definitions, but they all involve the variety of living things in a given area. The U.S. Office of Technology Assessment, for instance, defines biodiversity as "the variety and variability among living organisms and the ecological complexes in which they occur."

One reason for the lack of a precise definition is that life is hierarchical. For example, genes occur in cells, cells occur in organisms, and organisms occur in ecosystems (Chapter 3). At what level do we measure the diversity of life in an area? The number of genes? Organisms? Ecosystems? All can be used.

Measuring Biodiversity

Although one can measure biodiversity by counting the variety of genes or ecosystems in an area, the most common method is to count species. Species diversity, or species richness, is the number of species that occur in an area. Using species richness is largely

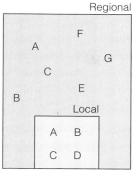

(a) Low diversity　　　　(b) High diversity

Figure 4-4 Letters (A, B, C, and so on) represent species living in an area: (a) low diversity at both the local and regional scales; (b) high diversity at both scales.

a matter of convenience: It is easier to tabulate the number of species in an area than to count genes or ecosystems. Fortunately, species diversity is generally a good indicator of genetic and ecosystem diversity as well. Nevertheless, even species richness omits important information about biodiversity such as the abundance of each species.

Biodiversity can be measured at all geographic scales, from local to regional to global. **Figure 4-4a** illustrates low diversity at both the local and regional scales. Regions with low diversity often have fewer species at the local level as well. As **Figure 4-4b** illustrates, regions with high diversity tend to be composed of local biological communities with high diversity. The tropics, for instance, tend to have very high species richness at both the local and regional level, compared to areas of similar size in temperate zones.

Global Biodiversity: How Many Species?

No one knows how many species live on Earth. **Taxonomists,** the biologists who classify and describe organisms, have described about 1.8 million species. Of these, 56% are insects, and 14% are plants; vertebrates such as birds, mammals, and fishes are just 3%. Only about 15% of described species occur in the oceans.

But these 1.8 million species are a highly biased sample and may not reflect the true species richness of the biosphere. For one thing, vertebrates are much more widely studied and therefore much better known than nonvertebrates: Only one or two new species of birds are described each year, whereas hundreds of new invertebrates are found. A recent study reported over 4000 bacteria species in a single gram of Norwegian soil. Most of these species were previously unknown.

Another problem is that most biologists have been concentrated in North America and Europe, but most species live in the tropics. The areas with most species are thus the least known! Similarly, most biologists study life on land, but the oceans cover 70% of the Earth and contain unknown numbers of species. Hence, the described species likely considerably underestimate invertebrates and species that exist in the tropics and the oceans.

How Estimates Are Made.   Describing all species on Earth could take centuries, so biologists have devised a number of ways to develop immediate biodiversity estimates of species numbers from limited information. The following are just three examples of the many methods of estimating global diversity from small samples:

1. **Rainforest insect samples.** Terry Erwin of the Smithsonian Institution has become well known for his studies of the very diverse tropical insect communities. Since most species are insects, and most insects are tropical, these studies are important for global estimates (**Figure 4-5**). Erwin (and others) use insecticides to kill and collect all the insects on the canopy (upper branches) of a tropical tree. In Panama, for example, 1200 beetle species were found on the canopy of a single tree. Since about 40% of all insects are beetles, we can estimate that perhaps 3000 insect species occur there. But only about two thirds of the species occur on the canopy (the rest occur on roots, bark, and other places). The total number of insects on this one tree is thus estimated at 4500 species! Since many of these insects occur only on one tree species, and since there

are about 50,000 species of tropical trees, it seems that many tropical insect species must exist. Erwin estimates the insect species on Earth at more than 30 million.

2. **Ecological ratios.** Another method is to use well-studied groups to predict the diversity of less-studied groups that are associated with them. In Europe, for example, there are about six fungus species for each plant species. Plant species have been relatively well described, and it is estimated that 270,000 plant species exist worldwide. Thus, there may be as many as 1.6 million fungus species if the 6 : 1 ratio is applicable throughout the world. Only 69,000 fungus species have been described.

3. **Species-area curves**. The species-area curve has been very influential since the 1960s. Whenever the number of species is counted in a gradually enlarged area of sampling, the result is a curve as shown in **Figure 4-6**. The number of species rises rapidly at first, but it slows as the area of sampling increases because the same species are encountered again and again. Repeated surveys in the tropics and temperate areas have shown that small areas of rainforest contain many more species, often over 100 times more species, than areas of temperate forest. By using the known shape of the species-area curve, one can predict how many more species will be found in larger unsampled areas of tropical and temperate regions.

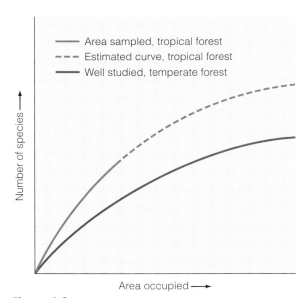

Figure 4-5   Each of the tropical trees in this expanse can be home to hundreds of insect species.

Figure 4-6   A species-area curve plots the number of species found in increasingly larger areas. The temperate forests are well studied compared to tropical forests. Only small areas have been examined in most tropical forests so many species diversity estimates for large areas are derived from extrapolated projections as indicated by the dashed line.

Conservation and the Geography of People

Visit www.environment.jbpub.com
for more information.

We see in this chapter how species are not randomly distributed on Earth. As you might suspect, people are not randomly distributed either. Far more people live in some areas, such as near coastlines, than in other areas, such as deserts. Do you suppose that large concentrations of humans tend to occupy the same areas that are favorable for high species diversity and density? If true, this has enormous implications for conservation.

This hypothesis seems likely at first glance. We have seen, for example, that biomes such as the desert, tundra, and deep ocean have a low productivity caused by reduced light and photosynthesis. This not only lowers the species diversity and density of these areas but it also makes them less favorable places for agriculture. In addition, many people prefer to avoid the cold climate of the tundra or simply cannot live near the deep ocean. But this is only a hypothesis based on the initial evidence presented in this chapter. What does more detailed evidence indicate?

A paper published in the journal *Science* by Andrew Balmford and others[1] analyzed a map of sub-Saharan Africa at a very high degree of resolution (one degree). They compared the number of bird, mammal, snake, and amphibian species in each square on a grid with the human population density of that square. They found that, at least for most of Africa, the scenario seems, unfortunately for most nonhuman species, to be true. Humans do tend to occupy the same areas that are richest in species diversity. This is not entirely because of plant productivity. Other reasons are that humans prefer to live in areas of fertile soils and low elevations, and near water bodies. All of these geographic factors also tend to support relatively high numbers of species.

Another important aspect of species geography is how it relates to the nature reserves that are set aside to protect species. Are these nature preserves geographically selected so that they protect areas of high species diversity? A paper published in the journal *Ecological Applications* by J. Michael Scott and others[2] examined the distribution of over 2500 protected areas in the lower 48 states of the United States. They found that less than 6% of the total area is protected and that the reserves were disproportionately located in areas of higher elevations and poor soils. This pattern occurs because land that is set aside for preserves has historically been mountainous land that was not productive for agriculture and was thus of little economic value. As you might suspect, this is bad for the preservation of biodiversity because many areas of high species diversity, such as wetland and riparian habitats only occur at low elevations. A more effective reserve system must add land that is more representative of all habitats, even if it requires setting aside land that is desirable for agriculture and other human uses.

Questions

1. Name two state parks in your state. Do you know anything about their history? Why were they located where they are?
2. California has the highest population for any state, and also the most native plant species. What reasons can you give for its high rank for both population and biodiversity?

[1]A. Balmford, J. Moore, T. Brooks, N. Burgess, L. Hansen, P. Williams, and C. Rahbek. 2001. Conservation conflicts across Africa. *Science* 291: 2616–2619.

[2]J. Michael Scott, F.W. Davis, G. McGhie, R. Wright, C. Groves, and J. Estes, 2001. Nature reserves: do they capture the full range of America's biological diversity? *Ecological Applications* 11: 999–1007.

Biodiversity Today: A Rare Wealth in Time

All methods of estimating global biodiversity involve a certain amount of extrapolation, so understandably there is still wide disagreement over how many species exist. Estimates range from as low as 5 million species to as high as 100 million. Biologists generally agree, though, that fewer species live in the ocean than on land. Estimates for the ocean range from 1 million to 10 million species. Despite having fewer species, the oceans have more fundamental biodiversity: 32 phyla are found in the oceans compared to just 12 phyla on land. A phylum, such as echinoderms or mollusks, is a much more distinct taxonomic unit of biodiversity than a species.

Whatever the exact number of species today, the fossil record indicates that we live in a special time of Earth's history. Detailed compilations of fossil data show that the number of families in the oceans has generally increased through time (**Figure 4-7**). Evolution has produced new families that have added to overall biodiversity. Estimates based on species produce similar results. We apparently live

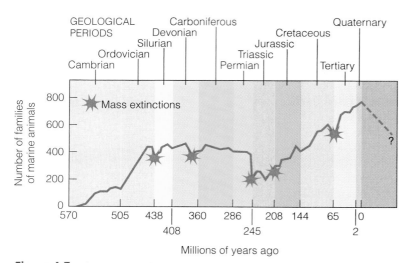

GEOLOGICAL PERIODS

Cambrian Ordovician Silurian Devonian Carboniferous Permian Triassic Jurassic Cretaceous Tertiary Quaternary

★ Mass extinctions

Number of families of marine animals: 0, 200, 400, 600, 800

Millions of years ago: 570 505 438 408 360 286 245 208 144 65 2 0

Figure 4-7   The number of families in the ocean has generally increased until the present. Five mass extinctions have temporarily lowered biodiversity. Will a sixth mass extinction occur in the near future? (*Source:* Modified from Raup and Sepkoski (1982), University of Chicago, Department of Geophysical Sciences.)

at a time when global biodiversity is near its all-time peak. As we discussed in Chapter 3, the individual organisms composing this diversity are organized into populations, and all the populations of a certain geographic area form a community. In the remainder of this chapter we will briefly describe some of the broad types of natural communities that occur on Earth today.

Biomes and Communities

A **biome** is a large-scale category that includes many communities of a similar nature. Many thousands of communities exist on the Earth. Rather than de-

scribe each in detail, we will examine the basic categories into which communities can be grouped. The most basic distinction is between terrestrial (land) and aquatic (water) dwelling communities. There are seven major types of terrestrial biomes and two major types of aquatic biomes:

1. *Terrestrial.* Tundra, grassland, savanna, desert, taiga, temperate forest, tropical forest (including tropical rainforests).
2. *Aquatic.* Marine, freshwater.

Both terrestrial and aquatic biomes (and thus the communities within them) are largely determined by climate, especially temperature. Climate is so important because it affects many aspects of the physical environment: rainfall, air and water temperature, soil conditions, and so on. Many secondary factors, such as local nutrient availability, are also important, however. In all cases, biomes illustrate the key point that species will often adapt to physical conditions in similar ways, no matter what their evolutionary heritage. For instance, a desert biome in the western United States looks superficially similar to a desert biome in North Africa even though the plants have different ancestries.

Terrestrial Biomes

Of the seven basic land biome types commonly distinguished (tropical forests, savanna, grasslands, deserts, temperate forests, taiga, and tundra—each of these is described and illustrated below), the tundra and desert biomes represent adaptations to the extreme conditions of very low temperature and low water, respectively. Not surprisingly, communities in these biomes tend to have the least number of species because organisms have difficulty adapting to the extreme physical conditions. In contrast, the tropical rainforests tend to be richest in species, in part because the tropics have the most moderate conditions.

Figure 4-8 diagrammatically illustrates the distribution of some of the major land biomes by precipitation, altitude, and latitude. This figure demonstrates the importance of temperature, which decreases with both increasing altitude and increasing latitude; similar changes result in both cases.

Also note the importance of precipitation, ranging from the very moist equatorial regions to the desert biome. A more detailed global view in **Figure 4-9** shows some of the true complexities of the lati-

Figure 4-8   Vegetation changes with altitude in the same way as with latitude because both changes are associated with decreasing temperatures.

Feet: 20,000 15,000 10,000 5,000 0
Meters: 6,000 5,000 4,000 3,000 2,000 1,000 0

Elevation

Polar

Tundra

Coniferous and deciduous forest

Humid subtropical

Humid continental

Subarctic

Dry

Latitude in degrees: 0 30 60 90

Tropical wet   Tropical wet & dry

tudinal pattern. For example, tropical forests are not always neatly confined to equatorial areas.

We will now describe seven major terrestrial biomes (listed in approximate order going from the equator to the poles).

Tropical Forests.   While rainforests are the best-known kind of tropical forest, there are actually several types. Tropical seasonal forests occur in dry areas of the tropics, where pronounced dry periods cause the trees to be deciduous (drop their leaves). Tropical seasonal forests are common in parts of Southeast Asia, especially India, where they are often called "monsoon forests" because the rain-bearing monsoon winds carry the moisture that restarts forest growth after the dry season. In contrast, the tropical rainforest is an evergreen forest where plants show no major seasonal changes. Another type of

tropical forest is the tropical woodland. This occurs in parts of the tropics where rainfall is relatively rare throughout the year, such as areas of Central America and central Africa. Because of the moisture limitations, photosynthesis and plant productivity is only about one-third that of the tropical rainforest.

Tropical rainforests are found on most major continents, mainly in equatorial regions (Figure 4-9). These represent the most complex and diverse biome, containing perhaps over 50% of the world's species while occupying only 7% of the land area. Unfortunately, about half of this area has been deforested by humans, so many rainforest species are on the brink of extinction. The high biodiversity of tropical rainforests is partly due to the relatively constant temperatures and high water availability: Daily and seasonal changes (fluctuations) are usually less

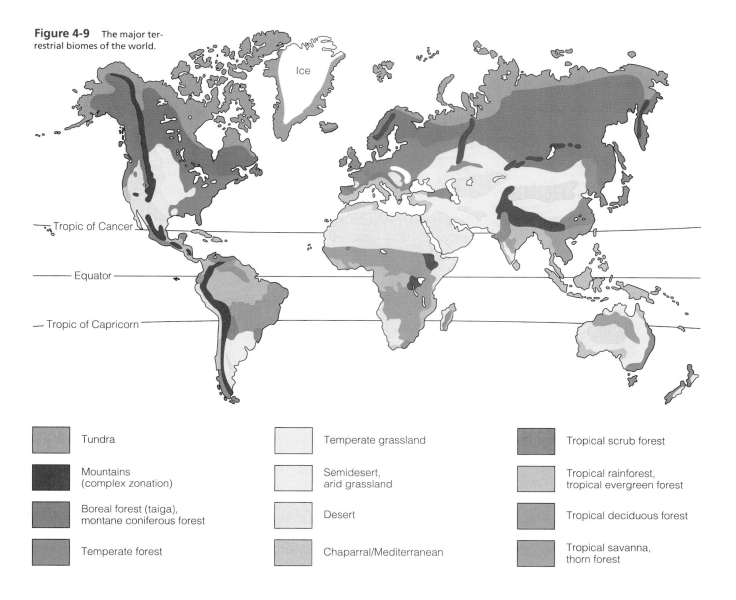

Figure 4-9   The major terrestrial biomes of the world.

Tundra

Mountains (complex zonation)

Boreal forest (taiga), montane coniferous forest

Temperate forest

Temperate grassland

Semidesert, arid grassland

Desert

Chaparral/Mediterranean

Tropical scrub forest

Tropical rainforest, tropical evergreen forest

Tropical deciduous forest

Tropical savanna, thorn forest

than a total of 9°F (5°C). Rainfall is very heavy, ranging from 80 to 175 inches (200 to 450 centimeters) per year. In comparison, New Orleans, Louisiana receives an average of 57 inches (145 centimeters) of rainfall annually. The average temperature of the tropical rainforest biome is 77°F (25°C). In comparison, New Orleans, Louisiana has an average temperature of just 68°F (20°C).

This lack of pronounced seasonal temperatures, including longer days and more direct sunlight, provides more opportunity for plant growth and productivity in this biome. The result is an ecosystem of very high species diversity and structural complexity (**Figure 4-10**). This structural complexity is seen in the many layers of trees forming a multistoried canopy. The canopy blocks out most of the sunlight so that, contrary to popular conception, the forest floor is relatively dark and does not support dense vegetation. Indeed, in order to photosynthesize, plants living at ground level must have special adaptations for coping with the lack of sunlight. Many plants, for example, have very broad leaves (such as the common "elephant ear" plant) to maximize surface area and very dark green pig-

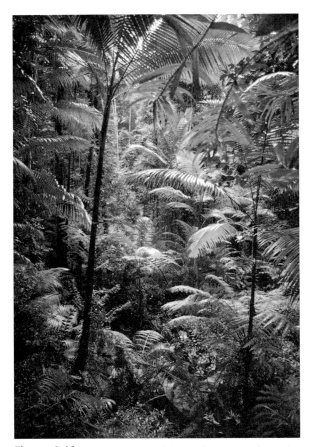

Figure 4-10   The complex (multilayered) structure of a tropical rainforest.

mentation (chlorophyll) to maximize light absorption. Another common adaptation is found among epiphytes, which are plants that grow on the branches of trees, and vines that grow upward along the trunks of trees to capture light in gaps high in the canopy.

The high species diversity of tropical rainforests has also led to the misconception that they must have very fertile soils. In fact, the soils are quite poor (low in phosphorus, nitrogen, and other nutrients) because they are depleted by both the intense competition for nutrients by the abundant plantlife and the high rainfall that quickly leaches away most free nutrients. The result is that nearly all nutrients are present as biomass incorporated into the tissue of plants and animals living in the rainforest. The only way to release these nutrients for farming is to cut and burn the vegetation by "slash and burn" agriculture, which is a major cause of rainforest destruction (Chapter 11).

Unlike the temperate forest biome (below), tropical forests on different continents vary widely in the genera and families of plants that compose them. This is probably a result of the relative isolation and rapid evolution of species in tropical ecosystems. Also, most tropical forest plants (especially trees) belong to botanical families that are rarely, if ever, found in temperate latitudes. This gives the rainforest its exotic appearance to visitors from North America and other temperate latitudes.

In most ecosystems, there is a strong correlation between the diversity of plant species and the diversity of animal species. Thus it is not surprising that the diverse tropical plant life supports a proportionately rich diversity of animal life. It has been estimated, for example, that over 90% of earth's insect species are found here. This high diversity is true for many other animal groups such as mammals and especially birds, which are much more diverse in the tropics than other biomes (**Figure 4-11**).

Savannas.   Savannas occur in tropical and subtropical areas that are not wet enough to support forests. Probably the best known of these are the savannas of Africa and South America, but they also occur in tropical regions of Australia and Asia (Figure 4-9). These regions are warm and have prolonged dry seasons with annual rainfall averaging 35–60 inches (90–150 cm). Such climatic conditions support vegetation of open grasslands with scattered shrubs and trees (**Figure 4-12**). The grasses grow rapidly during the rainy season and dry up during the dry season while the few trees survive by having deep root systems to access groundwater supplies. In areas of increasing rainfall, woodland

Figure 4-12   Savanna in Kenya.

Figure 4-11   Macaws range from southern Mexico down into central South America. This is a Blue and Gold Macaw.

begins to replace the savanna as tree density increases. Some savanna-like environments with dense scrubby evergreen vegetation are called **chaparral**. Examples occur in California and the Mediterranean region (Figure 4-9).

Many people are familiar with the big-game animals of the African savanna including such large herbivores as wildebeests, zebras, rhinoceroses, and buffalos. The most notorious predators of these creatures are the prides of lions that roam the savanna. Savannas also support many species of insects including ants, beetles, grasshoppers, and termites. The termites may be familiar for the huge mounds they build. These mounds are extremely important to the savanna ecosystem because they are made of digested plant material. Their construction accelerates the breakdown of these nutrients thereby enhancing soil quality. The deep and complex passageways in these mounds help carry badly needed rainwater into the soil instead of it being carried away or evaporating.

During the dry season, the dead grass material frequently catches fire so that savannas periodically experience large fires. Vegetation has thus evolved

ways to be fire-resistant such as rapid regeneration from unburned roots. Humans have often taken advantage of this process by purposely setting fires to promote new growth that is favorable for grazing animals.

Deserts.   Deserts occur where rainfall is very scarce, less than 10 inches (25 cm) per year. In warm regions, the problem of scarce water is aggravated by high evaporation and water loss. As a result, deserts have among the lowest rates of plant growth and productivity of any area on Earth (Chapter 3). An important consequence is that desert ecosystems will take a longer time than nearly all other ecosystems to recover from human disturbances—often centuries. The most extreme deserts, including the Sahara of Africa and Gobi of Mongolia, average less than 4 inches (10 cm) of rain per year, sometimes going for years with no rain at all. As a result, extreme deserts support very little life of any kind. Most extreme deserts occur in a belt that is about 20 to 35 degrees latitude North and South of the equator (Figure 4-9). Rising hot equatorial air produces heavy tropical rains so that when the cooler air sinks to the Earth at about 20 to 35 degrees North and South latitude, it is quite arid.

Semideserts, such as those of the southwestern United States, are less extreme and receive between 4–10 inches (10–25 cm) of rainfall per year. As a result, semideserts can support a fairly diverse array of species, usually typified by widely spaced, thorny "desert scrub" vegetation. Desert plants, often called **xerophytes**, have many interesting adaptations that

Figure 4-13   A variety of cacti and other semidesert plants.

allow them to survive the scarce and unpredictable water supply. The wide spacing of plants allows maximum moisture per plant. The hard exterior of the plants maximizes water retention and the presence of thorns reduces the consumption of the plants by animals seeking to extract the stored water. Such water-storing plants are called succulents; cacti are the most familiar example to many people (**Figure 4-13**). Deciduous plants shed leaves during dry periods to conserve water. An adaptation of nearly all perennial plant species is dormancy, when photosynthesis is greatly reduced, during periods of very high temperature. Night-time gas-exchange (transpiration) is also favored. Conversely, these plants are capable of

rapid growth during brief periods of rainfall. Annual plants are common in deserts. In these plants, the adult dies and the species survives as seeds which germinate when enough rainfall occurs. Some plants, especially trees, have very long tap roots that can reach into deep underground water supplies. Small trees, such as the Joshua tree and mesquite, can occur where water supplies are adequate, but large trees such as the desert cottonwood occur only at oases (springs) or dry riverbeds where deep groundwater supplies are present.

Animal adaptations are often similar to plants. Nearly all animals have mechanisms that increase the efficiency by which water is retained. The ability of the camel to go for weeks with very little water is legendary. Also very common is the habit of restricting activity to night time, with day time spent in deep burrows as a means of escaping the heat. Desert rattlesnakes and small rodents illustrate this. Many animals, such as the desert toad, take this one step farther and enter a dormant stage for long periods, becoming revived during the brief periods of rain. It is during this period that reproduction usually occurs.

Temperate Grasslands.   Temperate regions with scarce rainfall, about 10–30 inches (25–75 cm) per year, tend to have grasses and other herbaceous plants as the most prominent plants. Examples include such vast expanses as the Great Plains of North America, the pampas of South America, and the steppes of Russia (Figure 4-9). "Tall-grass prairies" have grasses, often standing over a meter tall, that grow in the moister areas. "Short-grass prairies" consist of clumps of short grasses, such as bunch-grass, that grow in drier regions. Because of the aridity, grasslands regularly experience fires, especially during the dry season. Trees are not common, but occur locally along stream banks and steep slopes where shade from the sun provides more moisture and protection from fires.

Humans have drastically altered much of this biome for agriculture because it is a very favorable environment for growing wheat and other grain crops. Economically, this is the most important biome, providing grazing land for sheep, cattle, and other food animals, as well as the richest cropland in the world (**Figure 4-14**). The rich soils are formed by the relative lack of rain and held in place by grass roots. It is estimated, for example, that over 95% of the original Great Plains biome has been drastically modified in the last 200 years by farming and grazing. In its natural state, grasslands are populated by burrowing animals such as prairie dogs and the area is extensively grazed by native herbivorous mammals, for example, bison in North America. While grasses have evolved to tolerate significant

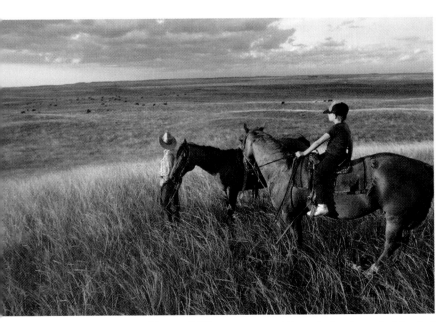

Figure 4-14   A rancher and his son look out over their herd of cattle grazing on temperate grasslands.

amounts of grazing, the introduction of non-native grazers such as cattle has led to overgrazing because, unlike native grazers, they tend to eat the plant faster than it grows. It may not always be obvious that non-native grazers are more destructive than native species. For instance, why didn't the millions of bison that once roamed the American Plains cause just as much impact as some herded cattle do now? In fact, bison could cause localized, temporary damage in the sense of short-term "overgrazing," but the bison were migratory and moved on to other areas, allowing the plants to grow back. Also, bison under natural conditions could be more selective than domesticated cattle in what they ate, focusing for instance on just certain tender or tasty parts of a plant, and some bison would also selectively feed on a variety of leaves, twigs, and bark. The overall effect was that bison could live sustainably in their natural environment. Another problem is the invasion of non-native grasses, including cheatgrass in the United States, which replace the native grasses after such disturbances as fires and overgrazing. Because of this extensive loss of native grassland plants, substantial efforts have been made toward the ecological restoration of the prairies in North America.

Temperate Forests. Temperate regions with adequate rainfall, 30–60 inches (75–150 cm) per year, support forests of broad-leaved deciduous trees that show colorful seasonal changes before dropping their leaves (**Figure 4-15a**). The seasonal leaf dropping is an adaptation to the very cold winters when water becomes frozen and photosynthesis is difficult or impossible. Examples are found in Europe, parts of Asia, and the eastern United States (Figure 4-9). These regions share many closely related tree species including the oaks, chestnuts, beeches, and birches that indicate a shared common evolutionary heritage dating back to geologic periods when the continents were interconnected (Chapter 5). Within the United States, oak and hickory dominate in the Southeast, and beech and maple dominate in the Northeast. However, substantial numbers of conifers (evergreen cone-bearing trees such as pines) occur in many of these forests mainly as early colonizers of disturbed areas or on poorer soils to which deciduous trees are not well adapted.

These forests lack the spectacular diversity of tropical forests, but are still more diverse than coniferous forests. Compared to the tropical rainforests to the south, temperate forests have a well-developed understory vegetation of many shrubs, herbs, ferns, mosses and small trees because so much light reaches the ground when the tall trees seasonally lose their leaves. Also unlike the rainforest where much animal life is found in the trees, there is a diverse array of species that dwell on the ground. Large herbivores include elk and deer and carnivores include wolves, panthers, and foxes (**Figure 4-15b**).

(b)

Figure 4-15   (a) Colorful Fall foliage in an eastern US deciduous forest. (b) Fox.

(a)

In the eastern United States, this biome has been drastically modified throughout nearly all of its range. These forests were cleared for farming by Europeans during the early years of colonization. In some areas, especially the northeastern United States, the land is no longer farmed so that ecological succession is beginning to return the land to its former state and young temperate forests are now common.

Taiga.   Also called coniferous or boreal forests, these occur in a broad belt in northern North America and Asia (Figure 4-9). Very little taiga occurs in the Southern Hemisphere as these latitudes are mainly occupied by ocean waters. Diversity and plant productivity are relatively low because of the stress imposed by long cold winters with little precipitation. Trees are dominated by conifers (evergreens), which are tolerant of dry, cold conditions (**Figure 4-16**). Prominent types of evergreens include spruce, firs, and pines whose needles conserve water and withstand freezing better than leaves. During the long summer days, plants grow rapidly. Marshes, ponds, and lakes are common, often forming in depressions carved out by the glaciers present here just a few thousand years ago.

Common animals of the taiga include herbivores such as moose, elk, deer, and the snowshoe hare and carnivores such as timber wolves, brown bears, lynx, and wolverines. The animals have adapted to the cold in various ways, including exceptionally thick fur and fat storage abilities that help sustain them through the long winters.

Compared to other biomes, the taiga has been less modified by agriculture. One reason is that cleared taiga makes poor cropland. In addition to the cold climate and short growing season, the thin soils are very acidic (pH of 4.5–5.0) compared to the soils of grasslands or temperate forests (pH of 6–7). The acidic soil is produced by the microbial decay of pine needles and other coniferous ground litter. However, extensive tracts of the lower-latitude taiga have been drastically modified by clear-cutting for timber.

Tundra.   Tundra occurs in the far north of Canada and Eurasia, between the taiga and shores of the polar seas (Figure 4-9). As with taiga, little tundra occurs in the Southern Hemisphere because these latitudes are occupied by ocean waters. Tundra is the low-lying vegetation that occupies an extensive treeless plain whose topsoil is frozen all year except for about 8 weeks during summer (**Figure 4-17**). Below this is permafrost soil, which is frozen all year long. This vast region is not only very cold but also very dry. It receives less than 10 inches (25 cm) of precipitation per year, or about the same as the desert biome. Few species are able to adapt to such harsh conditions so the tundra has relatively low plant species diversity. Not surprisingly, this region also has relatively low plant productivity because of the limited opportunity for photosynthesis.

During the brief summer thaw, the tundra becomes boggy and is pocketed with many lakes and streams. Lichens (algae and fungi symbionts), grasses, mosses, and small shrubs are dominant plants. These species grow and reproduce very rapidly as the growing season is often less than 8 weeks. During this period the tundra teems with life as animals seek to utilize these rich but short-lived plant resources. Insects hatch and many animals migrate into the area for feeding. Prominent animals include caribou (reindeer), which migrate into the area to feed on the fast-growing mosses and lichens (**Figure 4-17b**). More permanent residents include the arctic hare, arctic fox, and snowy owl. Some small rodents such as lemmings survive the winters underground and emerge in summer. They often show drastic population fluctuations. Many migratory birds arrive for the rich summer growing season, characterized by billions of insects, bright flowers, and marshy conditions. Like the taiga, the harsh nature of this biome has kept it from being extensively occupied by humans and thus it is among the most unmodified of the biomes.

Aquatic Biomes

In general, conditions in water are much less harsh than those on land. Water has many fewer temperature fluctuations and provides more buoyancy for organisms as support against gravity. All of these dif-

Figure 4-16   Taiga.

Figure 4-17   (a) The Arctic tundra. (b) Caribou.

(a)

(b)

ferences occur because water, as a liquid medium, is much denser than the gaseous air. Most important perhaps, living in water eliminates the danger of drying out, whereas water is often scarce on land. Not surprisingly, life originated in the oceans and took many millions of years to adapt to land. Yet even though water covers about 71% of the Earth's surface, most of it contains relatively little life; most of the open ocean is a vast aquatic desert with few nutrients.

Aquatic communities do not divide into distinctive biomes like those found on land because the liquid state of water makes local conditions very important in determining what can live in a particular environment. Water is a powerful solvent and also readily carries many substances in suspension; these substances, which range from toxins to nutrients, influence life locally. Furthermore, water readily transports heat so warm currents, such as the Gulf Stream, can warm large areas even near the poles. This prevents a simple latitudinal gradient from forming. Therefore, ecologists often designate only two major aquatic biomes: the marine biome and the freshwater biome.

Marine Biome.   Marine waters differ from fresh waters in containing more dissolved minerals (salts) of various kinds. On average, marine waters have about 3.5% salt, mainly sodium chloride, but with many hundreds of other materials as well. The marine biome is the largest biome by far (over 70% of the Earth's surface), but can be subdivided relatively easily into (1) **benthic** (bottom-dwelling), and (2) **pelagic** (water column) zones. Benthic communities (**Figure 4-18**) are further subdivided by depth: littoral (shore), continental shelf, and abyssal (deepsea). Benthic organisms (**Figure 4-19a**) include burrowers (such as worms), crawlers (such as snails), and stationary filterfeeders (such as barnacles). Littoral (shoreline) species have adaptations to avoid drying out. Barnacles, for example, can tightly seal their hard exterior to retain water when the tide goes out. Many other crustaceans dig deeply into the wet sand. At the other extreme, deep-water benthic species are adapted for withstanding the huge water pressures found in the deep sea (**Figure 4-19b**).

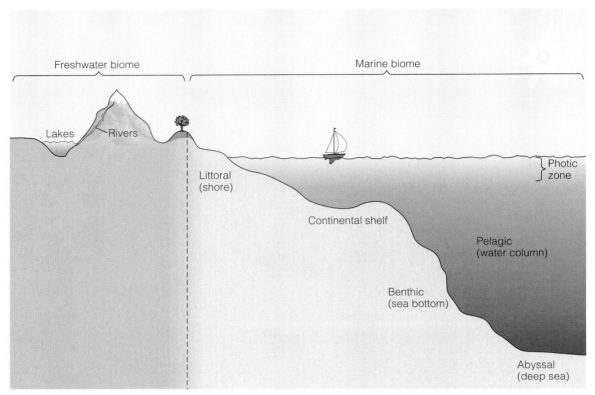

Lakes — Rivers

Freshwater biome

Marine biome

Littoral
(shore)

Continental shelf

Photic
zone

Pelagic
(water column)

Benthic
(sea bottom)

Abyssal
(deep sea)

Figure 4-18   The two aquatic biomes are freshwater and
marine. The marine biome can be divided into benthic and
pelagic zones.

Pelagic organisms (**Figure 4-20**) include (1) planktonic organisms (floaters) such as diatoms and
(2) nektonic organisms (swimmers) such as fish and
squids. The **photic zone** is the upper part of the biome
where light penetrates; usually, this zone extends to
about 150 feet (46 m) below the water surface. The
photic zone is the main zone of photosynthesis and
therefore is crucial to life in the biome. The photo-
synthesizing plants, which form the base of the marine food pyramid, are mainly tiny planktonic
organisms including diatoms. Much of the ocean is

too deep for light to penetrate and drive the photosynthetic food base, though, so life is largely absent.
Another reason why much of the ocean is a "biological desert" is that many nutrients in the ocean flow
in from land and consequently are found only close
to shore. Finally, the deep ocean is poorly oxygenated,
which also restricts the ability of species to live there.

(a)

(b)

Figure 4-19   (a) This bloodworm lives in salt marsh sediments. Some of these worms are free living while others are
tube builders. (b) Spider crabs around a deep sea vent.

(a)

Figure 4-20 (a) Diatoms are microscopic phytoplankton responsible for much of the ocean's primary production. (b) French angelfish.

(b)

Estuaries are transitional ecosystems between the ocean and freshwater biome. Estuaries, such as the Chesapeake Bay of Maryland, occur where rivers and streams flow into the ocean (**Figure 4-21**). This produces wide fluctuations in the salinity of these waters to which organisms must adapt. On the other hand, the flowing rivers provide a rich supply of nutrients rarely found in such abundance anywhere else on Earth. Typical animals that live in estuaries include oysters which form vast reefs that support many other species such as sponges and barnacles that attach to the reef. The reef traps nutrients and this attracts many fishes and other species. A single reef has been found to support over 300 species of organisms.

Freshwater Biome. Only 2% of the Earth is covered by liquid freshwater. The freshwater biome includes both running waters, such as rivers and streams, and standing waters, such as lakes and ponds. In general, rivers and lakes are larger and more permanent than streams and ponds. The faster motion of running waters tends to keep them more highly oxygenated and easier to clean up after being polluted. A river or stream is an open ecosystem in that it derives most of its organic material from upstream or runoff from the land. Many species feed on this detritus. Of special ecological importance is the **riparian zone**. This is the area along the river or stream bank, consisting of cattails, reeds, willows, river birches, and other plants that tolerate high moisture. These plants are very important buffers that reduce flooding impacts and water pollution by absorbing water. They are also critical habitat for many animal species including numerous highly endangered wetland species. Riparian plants are also important for aquatic life: they help protect the water from direct sunlight and reduce heat stress and provide many nutrients for aquatic life in the form of leaves and other detritus.

The slower motion of water in lakes and ponds leads to stratification of the water: The uppermost layer of water has plenty of oxygen while the oxygen decreases with depth. The uppermost layer is also much warmer during the summer and cooler during the winter than the lower layers. There is a sharp

Figure 4-21 A satellite image of Chesapeake Bay.

boundary, the **thermocline**, between the warm surface water and the colder deeper waters. Summertime swimmers often notice this abrupt transition to cold water when diving. Most of the mixing between the uppermost and deeper layers occurs during seasonal changes known as spring and fall overturn. As with rivers and streams, lakes also have a transitional zone with land, sometimes called the **shore zone**. It functions in the same way as the riparian zone, reducing flooding and heat stress and providing nutrients and critical habitat for many wetland and aquatic species.

As in the ocean, freshwaters have two basic life zones, the bottom-dwelling benthic organisms and the swimming pelagic organisms. Familiar benthic organisms include freshwater mussels that filter nutrients from the water and a variety of insects and other crawling or burrowing invertebrates such as crayfishes. Of course, fishes are the most familiar swimming organisms, and range from catfish that prefer bottom waters to perch and trout that prefer shallower depths. Amphibians are abundant in many freshwaters as they require water to reproduce. Other vertebrates include turtles and a few water-adapted mammals such as otters, beavers

Figure 4-22   Beavers are highly dependent on water and can be considered freshwater pelagic mammals.

(**Figure 4-22**), and muskrats. Algae are usually an especially important type of plant in freshwater ecosystems although water lilies, various "grasses" and other plants occur as well.

STUDY GUIDE

SUMMARY

- Primitive bacteria arose at least 3.5 billion years ago.
- New species originate through natural selection, which passes on beneficial genes.
- Biodiversity is often measured as number of species.
- Biodiversity is estimated by: rainforest sampling, ecological ratios, and species-area curves.

- Biomes are suites of communities that share many basic adaptations.
- Biomes are largely determined by climate, especially temperature and rainfall.
- The seven terrestrial biomes are: tropical forests, savanna, grasslands, deserts, temperate forests, taiga, and tundra.
- The two major aquatic biomes are: marine and freshwater.

KEY TERMS

Amino acids
Benthic
Biodiversity
Biome
Chaparral
DNA
Estuaries
Eukaryotes

Explosion of life
Genes
Mutation
Natural selection
Pelagic
Photic zone
Protocells
Riparian zone

Shore zone
Species richness
Species-area curve
Taxonomists
Taxonomy
Thermocline
Xerophytes

STUDY QUESTIONS

1. What are amino acids and what is the significance of Miller and Urey's, and later Fox's, experiments with amino acids?
2. Describe the process of evolution by natural selection. How has the physical environment affected the history of life?
3. If a population consisted of a small number of individuals, what effect might natural selection have on its survival?
4. Describe how Mendel's discovery helped in the understanding of variation.
5. Explain two basic causes of genetic variation.
6. Approximately how many genes do humans have? Why is this number significant?
7. What factors may have caused the "explosion of life"?
8. Define *biodiversity*. Is there only one way to define this concept? How can biodiversity be measured or estimated?
9. Explain why tropical rainforests have such high biodiversity.
10. Do tropical rainforests have fertile soils? Why or why not?
11. Describe four adaptations that plants can have for survival in the desert.
12. How are the two fundamental aquatic biomes recognized?
13. Describe why riparian zones are so important.
14. What is the sharp boundary between warm and cold lake waters?
15. Explain why much of the ocean can be considered a "biological desert."
16. Name the major biomes on Earth. Which biome or biomes do you live in? Would you consider it one of the more favorable biomes for human habitation? Why or why not?

WHAT'S THE EVIDENCE?

1. The authors state that new species may arise as a result of natural selection of individual variation and the isolation of populations. What evidence is given to support this? Can you think of evidence for other causes of evolution? If so, list them.
2. The authors state that the number of species on Earth could be anywhere from 5 to 100 million species. What is the evidence given to support this estimate?

CALCULATIONS

1. If taxonomists have described approximately 1.8 million different species, and 56% of these are insects, how many of the described species are insects?
2. If taxonomists have described approximately 1.8 million species, but there are actually 30 million different species on Earth, what percentage of the species have taxonomists described?

ILLUSTRATION AND TABLE REVIEW

1. Using Figure 4-8, explain how there can be a polar climate at the Equator.
2. In which biomes would you expect to find the type of diversity shown in Figure 4-4(a)?

WWW.ENVIRONMENT.JBPUB.COM

www.environment.jbpub.com

Connect to this book's web site: http://www.environment.jbpub.com/ The site features eLearning, an online review area that provides quizzes, chapter outlines, and other tools to help you study for your class. We've posted the answers to the book's odd-numbered questions. You can also follow useful links for more in-depth information. The site provides sources for the information in this chapter and recommendations for further readings.

Our physical Earth is still a young and evolving planet. We, as small creatures crawling on the surface of the globe, are still subject to the mercies of nature in many respects even as we affect the global ecological balance. We must never forget that we cannot totally isolate ourselves from our physical surroundings on this small world. (*Source*: Frederic Edwin Church, *Cotopaxi* (1862). Oil on canvas, 48 in × 7 ft. 1 in. Copyright © The Detroit Institute of Arts, Founders Society Purchase with funds from Mr. and Mrs. Richard A. Monoogian, Robert H. Tannahill Foundation Fund, Gibbs-Williams Fund, Dexter M. Ferry, Jr. Fund, Merrill Fund, and Beatrice W. Rogers Fund.)

CHAPTER 5

The Dynamic Earth and Natural Hazards

The Earth has evolved from a relatively homogeneous collection of particles (cosmic dust) to form a well-differentiated, complex, dynamic, recycling machine. Our planet may seem eternal and unchanging, yet it was very different 4 billion years ago, and it will continue to evolve and change. From a human perspective, such changes are slow and gradual, taking thousands to hundreds of millions of years. We must have a clear understanding of the delicate, and in many ways fortuitous, development of the present state of the Earth to fully comprehend just how fragile our environment really is. Furthermore, despite our sometimes seeming hubris, we humans and our civilization pale in comparison to the forces of nature. Certain phenomena that can be tragic to us, such as the eruption of a major volcano that sends plumes of ash into the atmosphere, with resultant worldwide weather changes that may destroy crops and bring about mass starvation and epidemics (and indeed this has happened, for instance, in the early nineteenth century following the 1815 eruption of the volcano Tambora in Indonesia), are in reality just the normal workings of planet Earth. "Natural hazards" are an integral part of the natural world. We, as small creatures crawling on the surface of the globe, are still subject to the mercies of nature in many respects even as we tamper with the global ecological balance. We must never forget that we cannot totally isolate ourselves from our physical surroundings on this small world.

Objectives

After reading this chapter, you should be able to explain or describe

The basic internal structure of Earth

Plate tectonics

The composition of matter

The rock cycle

Climate, weather, and large-scale atmospheric and seasonal cycles

The origin of the solar system and Earth

The basic types of natural hazards

Workings of Planet Earth Today

Earth and Its Neighboring Planets

Earth is the third planet out from the center of the solar system (moving out from the Sun, the planets are: Mercury, Venus, Earth, Mars, Jupiter, Saturn, Uranus, Neptune, and Pluto), but it is very different from the other planets. Unlike its two immediate neighbors, Venus and Mars, which both have atmospheres that are 95% to 97% carbon dioxide, Earth has an atmosphere composed of about 78.1% nitrogen, 20.9% oxygen, and only 0.03% to 0.04% carbon dioxide. As a result, much of the sunlight and heat reaching the Earth is scattered back into space, and Earth has an average surface temperature of 57° to 59°F (14° to 15°C). At this temperature, water can exist in a liquid state (water quickly evaporates on the hot surface of Venus and remains frozen on Mars), making it possible for life to be maintained on Earth. As far as we are aware, no other planet is inhabited by lifeforms (although it has been suggested that structures in a meteorite that originated from Mars may represent microscopic organisms, and it is certainly conceivable that life could exist elsewhere in the solar system and universe).

The gases surrounding the Earth (**Figure 5-1**) and the liquids (primarily water) on its surface are continually swirling and moving, causing the degradation, erosion, and destruction of topographic highs such as mountains. The surface of our Moon is pockmarked with craters due to the lack of an atmosphere to weather them away. There are very few craters on Earth because they quickly erode away. Why then does the Earth's surface have any relief at all? The reason is that the Earth has an "internal engine."

The Earth's interior is seething and churning. Rocks slide and move plastically around one another. The interior is very hot, and at breaks and cracks, molten rock erupts at the surface as volcanoes. In other areas ancient rock is dragged back toward the center of Earth only to be reheated, melted, and rejuvenated. The vast majority of the Earth's surface is relatively young rock. In comparison, some of Earth's close neighbors—the Moon, Mercury, and Mars—are relatively inactive. The Moon and Mars show little or no evidence of plate tectonic activity, and their surfaces are covered with very ancient rocks, generally dating back more than 3 billion years. Mars has large volcanoes, including Olympus Mons, which rises 17 miles (27 km) high and is 340 miles (550 km) in diameter at the base, but apparently shows no evidence of true plate tectonic activity except perhaps very early in its history. Of the planets close to Earth, only Venus shows some definitive evidence of limited crustal movements comparable to the plate tectonics seen on Earth. But any tectonic movement on Venus appears to have been relatively limited; most of that planet's geological activity seems to be in the form of active volcanoes situated over stationary "hot spots" where molten rock wells up from the interior.

Present-Day Deep Structure of the Earth

In order to understand natural processes on the surface of the Earth, we must have a clear understanding of the structure of the Earth. The internal make-up of the Earth accounts for volcanoes and earthquakes. It even affects the composition of the atmosphere, oceans, and ultimately the nature of life on the Earth.

If we could cut a slice through the Earth, we would see that it is formed of concentric rings of differing constitutions—similar to the layers of an onion (**Figure 5-2**). From the outside, the first major layer is the Earth's gaseous envelope or atmosphere. It extends in rarefied form for hundreds of miles above the Earth's surface, although almost the total mass is contained within

Figure 5-1   The Earth seen from the Moon. The Earth is a "living planet" with an active atmosphere and an internal heat engine driving many geological processes. In contrast, the Moon is a "dead planet" (our Moon is larger than Pluto and nearly as large as Mercury) with virtually no atmosphere or active geological processes on its surface. As a result, the Moon's surface is pockmarked with craters that have not eroded away.

Figure 5-2   Schematic section through the Earth.

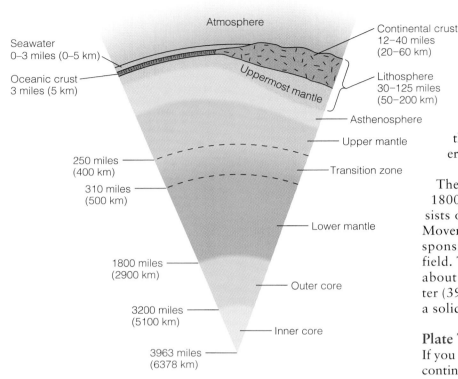

the bottom 12 to 13 miles (20 km). The Earth's surface is covered with seawater or dry land (including land that contains freshwater lakes or ice patches) in a ratio of about 7 to 3. If all water were removed from the surface, we would observe two basic terranes: ocean basins, which average about 3 miles (5 km) depth below sea level and are floored with basaltic-type rocks, and continents and continental islands, which on average rise several hundred feet (a few hundred meters) above sea level, and are founded on granitic-type rocks. Actual ocean basins cover only about two-thirds of Earth's surface; in some places, the oceans lap onto the shallow continental margins.

The thin **oceanic crust** averages only about 3 miles (5 km) thick. It is composed of about 50% (by weight) silicon dioxide (perhaps most familiar as the mineral quartz) and contains higher amounts of iron, magnesium, and calcium than are generally seen in continental crust. The **continental crust**, ranging from about 12 to 40 miles (20 to 60 km) thick, is about two-thirds silicon dioxide.

Below the crust, extending to a depth of about 1800 miles (2900 km), is the **mantle**, which is composed of about 45% silicon dioxide and 38% mag-

nesium oxide. The relatively rigid upper layer of the mantle with its attached crust is known as the lithosphere. The **lithosphere** is the major unit of movement in plate tectonics (discussed below). Below the lithosphere lies another layer of the mantle, the relatively weak and soft **asthenosphere** (literally, weak or glassy sphere; see Figure 5-2).

Beneath the mantle is the Earth's **core**. The outer core extends from a depth of about 1800 to 3200 miles (2900 to 5100 km) and consists of a liquid iron-nickel (mostly iron) alloy. Movements within the outer core probably are responsible for generating the Earth's magnetic field. The inner core, extending from a depth of about 3200 miles (5100 km) to the Earth's center (3963 miles [6378 km] deep), is composed of a solid iron-nickel alloy.

Plate Tectonics

If you look carefully at a map and imagine that the continents are pieces of a jigsaw puzzle, you will notice that the east coasts of North and South America appear to fit into the west coasts of Europe and Africa. This simple observation, and a wealth of other data, led to the hypothesis of drifting continents, proposed by the German scientist Alfred Wegener in about 1912. Today a variation of the theory of continental drift, known as **plate tectonics**, is accepted by virtually all earth scientists. Indeed, plate tectonics forms the unifying theory for most of the geological structures observed on the surface of the planet. It is the active process of plate tectonics that not only moves continents, but also raises mountains, creates new sea floor, destroys and recycles old sea floor, and causes volcanoes to erupt and earthquakes to occur. Plate tectonics plays a major role in natural biogeochemical cycles on Earth—cycles which are now being disrupted by human activities (Chapter 3).

Today the Earth's lithosphere is divided into about eight major tectonic plates and numerous smaller ones (**Figure 5-3**). These plates are in continuous motion, sliding past one another, colliding, or pulling apart from each other. Essentially, the continents are carried on the tops of the plates. The plates do not move very fast from a human perspective, about 0.8 to 12 inches (2 to 30 cm) a year, but this is fast enough to have caused major changes in the positions of the continents over the last few hundred million years (**Figure 5-4**).

Figure 5-3   The Earth's tectonic plates. The plates are in continuous motion relative to one another. Notice the relationship between plate boundaries, intense earthquakes, and major volcanic eruptions.

Ridge axis
------- Transform
------- Convergent plate boundary
------- Uncertain plate boundary
------> Motion of plate
········ Zones of extensions within continents

••• Earthquakes
▲▲ Volcanoes

Figure 5-4   Continental movements over the last 240 million years. (a) The Earth about 240 million years ago. (b) The Earth about 70 million years ago. (c) The modern Earth.

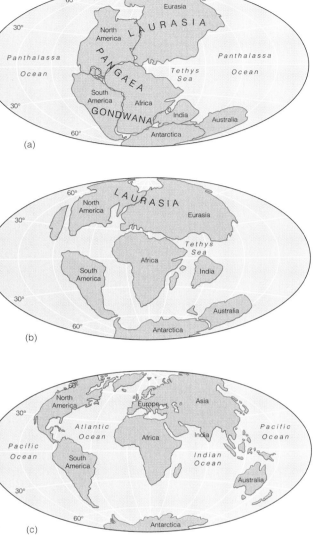

Prior to about 2.5 billion years ago, the Earth's surface was probably covered by small, rapidly moving "platelets," which slowly coalesced into larger, thicker plates. Only in the last 800 million years has the crust been characterized by modern-style, large, thick plates. About 240 million years ago, all of the continental landmass formed a single supercontinent, Pangaea, which has since split up into the present-day continents.

What causes the plates to move? This has been a topic of heated discussion over a number of years. Today most scientists believe that convection currents in the molten mantle move the lithospheric plates (**Figure 5-5**). Heat continually flows out from the hot center of the Earth. As in a pot of cooking soup, the hot liquid rises to the surface and flows there for some distance; as the liquid cools, it sinks below the surface, only to be heated up and start the cycle once again. This is known as a **convection cell**. Giant convection cells are believed to exist in the liquid mantle, and as they cycle, the flowing mantle drags the overlying, rigid lithosphere plates along. As the Earth cools over time, the rates of motion of the plates are decreasing, and they are continuing to enlarge and thicken.

At divergent plate boundaries, places where two lithospheric plates are moving away from each other, a gap or void is left in the solid crust

Figure 5-5   Convection cells in the Earth's mantle.

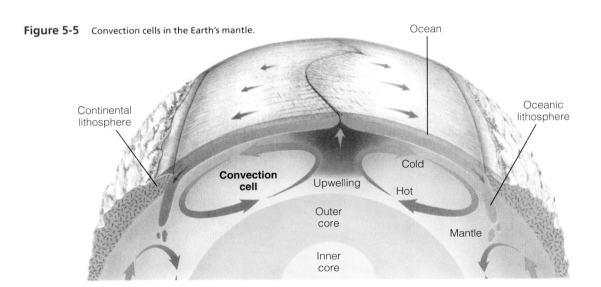

Figure 5-6 Principal types of plate boundaries.

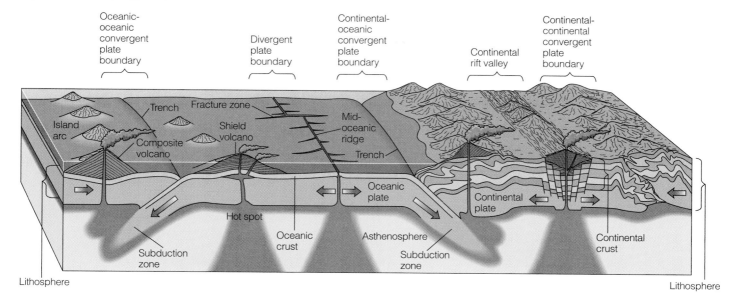

Oceanic-oceanic convergent plate boundary

Divergent plate boundary

Continental-oceanic convergent plate boundary

Continental rift valley

Continental-continental convergent plate boundary

Island arc

Trench   Fracture zone

Composite volcano

Shield volcano

Mid-oceanic ridge

Trench

Oceanic plate

Continental plate

Hot spot

Oceanic crust

Asthenosphere

Continental crust

Subduction zone

Subduction zone

Lithosphere

Lithosphere

(**Figure 5-6**). Initially, the rock in this area may start to collapse, forming a rift valley, but as the plates move apart, hot molten rock from the mantle wells up to fill the void and form new oceanic crust. This process is often referred to as seafloor spreading. The classic example of a divergent plate boundary is the Mid-Atlantic Ridge, which runs roughly north-south through the Atlantic Ocean.

Along deep-sea rift zones, hot water (about 750°F, or 400°C) emanates from hydrothermal vents. These vents often support unique ecological communities that are based on geothermal energy rather than the solar energy most ecosystems require (**Figure 5-7**). Bacteria use the energy found in hydrogen sulfide (H$_2$S) and similar chemicals emanating from the vents to produce organic molecules from inorganic molecules. Crabs, clams, tube worms, and fishes then feed on the bacteria. Thus these vents dramatically illustrate the interrelationships between geological processes and living organisms.

In some places, convergent plate boundaries occur where two lithospheric plates converge or collide (see Figure 5-6). If the colliding edges of the two plates both bear oceanic crust, either plate may be forced, or subducted, under the other plate. A subduction zone is formed under the leading edge of the plate that remains on top. On the surface, where the plates are downwarped, a deep trench forms. As the subducted plate moves down into the hot mantle, it melts, and the lighter rock components rise to the surface, forming volcanoes. New continental crustal material, formed from the recycled oceanic crust, collects behind the trench.

The northern and western edges of the Pacific Ocean (known as the "ring of fire" for their numerous volcanoes) are lined with convergent plate boundaries (see Figure 5-3).

Figure 5-7 A deep-sea hydrothermal vent supports a unique biotic community, which is based on geothermal energy rather than the solar energy most systems require. Bacteria use the energy found in hydrogen sulfide and similar chemicals to produce organic molecules. Crabs, clams, giant tube worms, and fishes then feed on the bacteria.

If the leading edge of one colliding plate contains continental crust and the leading edge of the other plate contains heavier oceanic crust, the plate with oceanic crust will always be subducted under the plate bearing lighter continental crust, which "floats" on top. If leading edges of colliding lithospheric plates both contain continental crust, neither can be subducted because they are both relatively light. Instead, the continents crash into one another, crumple, and deform, often raising imposing mountain ranges. The mighty Himalayas are the result of two continental landmasses crashing into one another. Eventually, the plates lock and relative motion between them stops.

Finally, two plates may slide past one another; their edges grind and slip against each other along what are commonly termed *transform faults*. This is the least dramatic type of relative plate motion in that it usually does not involve either volcanoes or mountain building. Earthquakes are common along transform faults, however, as is well known by residents of California, where the Pacific plate slides north past the North American plate. The surficial rocks do not flow past one another slowly and smoothly, but tend to "hang up." Pressure builds, and then is suddenly released—the rock "jumps" and an earthquake occurs.

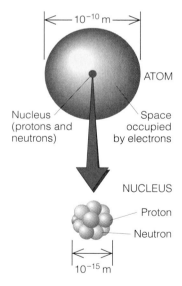

Properties of Atomic Particles

Particle	Charge	Mass	Location	Symbol
Proton	Positive	1.673×10^{-24}g	Nucleus	p
Neutron	Neutral	1.675×10^{-24}g	Nucleus	n
Electron	Negative	9.110×10^{-28}g	Orbitals	e

Figure 5-8   The basic structure of an atom.

The Composition of Matter on the Earth

Rocks are by far the most common substances on the Earth's surface. Indeed, everything is rock or made primarily of materials that were once components of rocks. A clear understanding of rocks and the rock cycle are necessary in order to fully understand the cycling of matter through the environment (Chapter 3). Rocks themselves are composed of minerals, which in turn are composed of elements.

Minerals are naturally occurring, inorganic solids that have a regular internal structure and composition—they are said to be crystalline. Some well known minerals include quartz, diamond, garnet, and pyrite. In all, over two thousand different minerals are known. Minerals are composed of combinations of atoms of elements.

The **elements** such as iron, hydrogen, oxygen, carbon, mercury, and gold are the fundamental substances of our world. An element cannot be broken down chemically into other elements. There are close to 90 naturally occurring elements found in measurable amounts on Earth today, and additional elements have been artificially synthesized under laboratory conditions.

Atoms are the smallest units of an element that retain the physical and chemical properties of the element. When combined together, substances composed of different elements can take on new properties. Water, for instance, is a **compound** composed of atoms of the elements hydrogen and oxygen. Atoms themselves are divisible into even smaller particles. In the center of each atom is a nucleus composed of protons and neutrons (**Figure 5-8**). Protons carry positive charges and neutrons are electrically neutral. Most of an atom's mass is contained in its nucleus, where the number of protons determines to what element the atom belongs. For example, all atoms of hydrogen have only one proton in their nucleus, whereas all atoms of carbon have six protons, and those of gold have 79 protons in their nucleus. The number of protons in the nucleus is referred to as the atom's atomic number and is characteristic of the atoms of a particular element.

The number of protons and neutrons in the nucleus is the atom's atomic mass number. Atoms of the same element can have varying numbers of neutrons in their nucleus. In nature one can find carbon atoms with atomic mass numbers of 12, 13, or 14. These variants are known as *isotopes* of the element carbon. Some isotopes are unstable and undergo radioactive decay by emitting particles spontaneously so as to change into a more stable form of atom (see Chapter 7).

"Orbiting" (they are not literally orbiting in an everyday sense) around the protons and neutrons of the atom's nucleus are electrons (Figure 5-8), small, negatively charged particles that electrically balance the protons in an electrically neutral atom. If a particular atom has more electrons than protons, it will be a negatively charged ion. If it has fewer electrons than protons, it will be a positively charged ion.

Single atoms are much too small for us to perceive in isolation. Substances that we are familiar with are composed of tremendous numbers of atoms, which are almost always bonded together in various arrangements. Depending on how the atoms are arranged, we refer to states of matter composed of atoms as solids, liquids, or gases.

Rocks: Their Origin and the Rock Cycle

Most rocks are composed primarily of minerals and are classified into three basic categories: igneous, sedimentary, and metamorphic (**Figure 5-9**).

Igneous rocks formed, or crystallized, from extremely hot molten rock known as magma ("ig-neous" = fire-rock). The lava that flows out of a volcano and hardens into rock is a well-known type of igneous rock. Igneous rocks that formed from lava flows or volcanic ash ejected from a volcano are known as *extrusive igneous rocks*. Most igneous rocks, however, form deep underground as liquid magma slowly cools and crystallizes; such rocks are known as *intrusive igneous rocks*. Typical intrusive igneous rocks include basalt, gabbro, and granite. Igneous rocks are the most common rocks, forming the vast bulk of the Earth's crust and all of its rocky inside (where it is not molten). On the surface of the Earth, however, igneous rocks are often covered by a blanket of sediments or sedimentary rocks.

Having formed deep inside the Earth, most igneous rocks are inherently unstable at the surface. Over time, the rocks weather, broken down by both mechanical (for example, running water, ice, wind, or the roots of trees) and chemical agents (such as acids and solvents found in nature—for instance, the ubiquitous carbonic acid of rainwater), and disintegrate into fragments. These fragments are known as sedi-

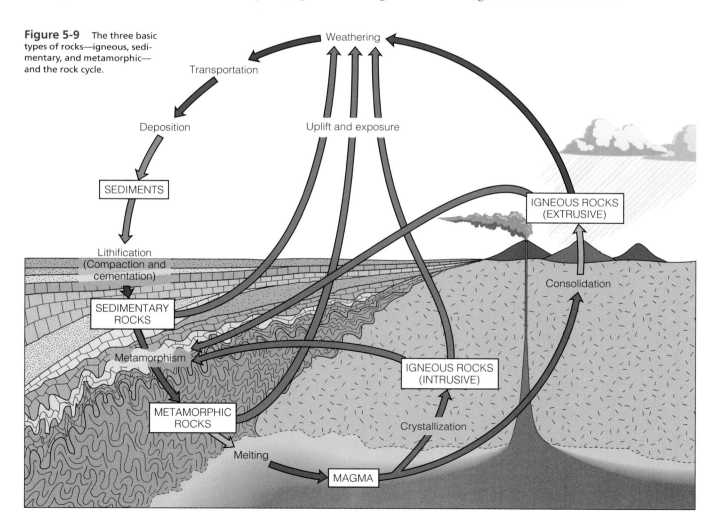

Figure 5-9 The three basic types of rocks—igneous, sedimentary, and metamorphic—and the rock cycle.

ments; they generally consist of either small pieces of rock or single mineral grains, and come in various sizes—such as gravel, sand, silt, or mud. The sediments are physically transported from higher to lower elevations due to gravity and by running water, wind, or ice flows (glaciers), until they are deposited in a resting place. As piles of sediments accumulate, they become compacted and cemented together, or even crystallized. Minerals may precipitate between the sediment grains, binding them tightly together to form a **sedimentary rock** (Figure 5-9). Sedimentary rocks cover approximately 75% of Earth's surface. Among the most common sedimentary rocks are sandstones and shales.

Not all sedimentary rocks are formed from the weathered remains of other rocks. Many limestones, for example, are formed from sediment that consists entirely of the old calcium carbonate shells or skeletons of sea organisms. Fossils, the remains of ancient organisms, can be found encapsulated in sedimentary rocks such as limestones, sandstones, and shales. Peats, coals, oils, and other fossil fuels, formed of the fragments of many plants and other organisms, are components of some sedimentary rock layers (and in such cases, a "rock" like coal may be composed of non-mineral material). Other types of sedimentary rocks are formed from the precipitation, or "sedimentation" of substances out of an aqueous solution, perhaps as it evaporates. For instance, large deposits of rock salt (halite) often form in this way.

Like igneous rocks, sedimentary rocks weather and break down. The sediments formed from sedimentary rocks can then be recycled to form other sedimentary rocks. The next generation of rocks may weather and the resulting fragments be recycled again, with the pattern occurring over and over.

The remaining class is **metamorphic rocks** ("metamorphic" = change of form). Metamorphic rocks are made when rock—igneous, sedimentary, or in some cases another metamorphic rock—is subjected to great temperature (but not hot enough to completely melt the rock) or pressure, or both (Figure 5-9). The rock undergoes some combination of mineralogical changes (certain minerals are turned into other minerals) and/or textural changes (mineral grains may grow in size). In some cases metamorphic rocks may be produced by hot fluids released at the edge of a molten rock body that penetrate preexisting rock. Some common metamorphic rocks include slate (produced from the metamorphism of shale or mudstone), marble (produced from limestone), quartzite (produced from quartz sandstone), and various schists and gneisses (produced by the extreme metamorphism of various

other types of igneous, sedimentary, and metamorphic rocks).

Geologists often refer to the **rock cycle** (Figure 5-9). Effectively, all rock/earth material on the Earth's surface begins as igneous rock. Initial crystallization may take place at a rift zone where two lithospheric plates diverge or by extrusion from a volcano. This new rock may weather directly into sediments or become metamorphosed. Ultimately, it may be carried to a subduction zone where it is remelted, rises toward the surface, and eventually is added to the continental crust. From there it may be metamorphosed or remelted again, or it may be weathered at the surface, with the fragments eventually forming sedimentary rocks. The sedimentary rock may be weathered, broken into fragments, and made into other sedimentary rocks, or more sediments may be piled on top of it, until it is buried deep below the surface. With increasing temperatures and pressures, the sediments may be metamorphosed and eventually find themselves in a setting where they actually melt. If so, they will form genuine igneous rocks upon cooling and crystallization, thus continuing the rock cycle.

By human standards, the rock cycle is very slow, taking on the order of millions to hundreds of millions of years. Furthermore, it is not a complete or closed cycle. Once continental (granitic type) material is formed, it remains continental material, even if it cycles from igneous to sedimentary, to metamorphic, back to igneous, and so on. As described in the section on plate tectonics, continental material is too light to be subducted back into the mantle; thus, it is not recycled in that sense. More continental crustal material is in existence today than ever before in Earth's history.

The Surface of the Earth: Climate and Weather

In Chapter 3 we discussed the Earth's atmosphere, energy flow from the Sun to the Earth, and the hydrologic cycle driven by the Sun's energy. As we mentioned there, the **hydrologic cycle** combined with various atmospheric cycles and the rotation of the Earth around the Sun (discussed below) is responsible for both **weather** (short-term perturbations in the atmospheric/hydrologic cycles) and **climate** (average weather conditions over time ranges of decades to millennia to hundreds of millions of years).

The Earth's surface is divided into several major climatic belts that roughly correspond to latitudinal changes (north and south); these climate belts are based on large-scale **atmospheric cycles**.

Working out from the equator, one finds the rain-forest belt (centered about the equator), the desert belts (centered about 30 degrees latitude north and south), the temperate regions, and the polar regions.

In the atmosphere, convection cells form; these are similar to those believed to occur in the mantle. The rays of the Sun fall most directly on the equator, making the equatorial regions hotter than the rest of the Earth's surface. The hot air rises, carrying with it large quantities of water vapor (warm air can hold more moisture than cool air). The rising air leaves a relative void, and cooler, drier air closer to the surface flows in from the north and south to fill the void (**Figure 5-10**). This new cool air heats up, picks up moisture, and then itself rises from the equator. The rising air quickly cools, however, and loses its capacity to hold moisture. Consequently, vast quantities of water are released in the vicinity of the equator, producing the tropical rainforests. As the air continues to rise and cool, it moves either north or south, continues to lose moisture, and eventually begins to sink as dry air in the region of 20 to 35 degrees latitude North and South, giving rise to the subtropical high-pressure zones that produce some of the world's great deserts. Once the cool air approaches the surface, it flows south or north toward the equator. On reaching the equator, the air again heats and absorbs moisture, and the cycle continues. Not all of the air sinks at the subtropical high pressure zones, however. Some of the air continues toward the poles, where it eventually sinks and takes part in polar atmospheric cycles. Cold surface winds tend to blow from the poles toward the equator.

Of course, the atmospheric cycle is not quite that simple. Two effects in particular modify it. First, the Earth spins on its axis, causing the air in the convection cells to cycle at an angle rather than directly north and south. This is known as the *Coriolis effect*. To understand the Coriolis effect, imagine yourself standing on the geographical North Pole. The Earth is rotating toward the east. If you throw a ball, it will land to the west of where you aimed it—the Earth turns under it as it travels through the air. This effect produces the trade-winds that once carried sailing vessels across the oceans.

Second, the Earth's topography—the positions of its mountains, valleys, lakes, and oceans—greatly modify both weather and climate. The oceans that

Figure 5-10   General circulation in the Earth's atmosphere.

cover much of the surface contribute greatly to the overall climate (**Figure 5-11**). Moving air masses create ocean currents, which are then modified by the Coriolis effect and the continental landmasses that the currents bump into. Due to the great capacity of water to absorb and retain heat, ocean currents play an important role in distributing heat over the Earth's surface.

On land the topography greatly influences weather patterns. For example, mountains may create rainshadows, causing deserts to form on their leeward sides. When a warm moist air mass encounters a mountain, the air begins to sweep up the mountain's windward side. In so doing, the air rises, cools, and loses much of its moisture as precipitation. Once across the mountain, the now dry air flows down the other side, becoming warmer as it reaches lower elevations. Because the air has left its moisture on the other side of the mountain, a desert is created. In North America, a rainshadow effect produces deserts east of the Sierra Nevada, and in South America, the Andes Mountains are responsible for the extremely dry deserts found along the western coasts of Peru and Chile.

Local and global temperature and climate are also influenced by the albedo, or the proportion of the incoming solar radiation that is reflected back to space by various surfaces. Large, vegetation-free

(a)

(b)

Figure 5-11   (a) Global surface ocean currents. (b) "Heat" photo of the Gulf Stream flowing northward past the U.S. coastline.

Dust in the atmosphere can also have a high albedo and reflect significant amounts of radiation including heat back into space. The dust and ash emitted by the volcano Tambora in 1815 had worldwide ramifications for the weather, including a "year without a summer" for New England in 1816. The impact of a huge meteorite could throw up blankets of dust and particulate matter that would have the same effect. Conversely, other additions to the atmosphere can lead to higher temperatures. Various greenhouse gases, most notably carbon dioxide, when added to the atmosphere (either artificially or naturally), create a blanket that allows radiation to enter, but blocks the escape of heat (see Chapter 17), thus leading to global warming.

Rotation, Orbits, and Seasons

In addition to its short-term, day-to-day fluctuations, weather is seasonal, changing on an annual basis. These cyclical variations in the weather are due to the orientation of the Earth's axis. As you know, the Earth rotates on its axis once a day as it orbits around the Sun once a year. The Earth's axis of rotation is not perpendicular to the plane of its orbit, but rather is inclined at an angle of about 23.5 degrees. Consequently, as the Earth follows its orbit, different parts of the surface are exposed to more direct rays from the Sun (**Figure 5-12**). In the Northern Hemisphere during the summer, the axis is tilted toward the Sun, causing sunlight to fall more directly on this part of the world. During the

deserts and blankets of ice or snow have a large reflective capacity—a high albedo. Oceans and other large bodies of water, which have the ability to absorb heat, as well as areas covered with thick vegetation, have a low albedo.

Figure 5-12   Rotation of the Earth around the Sun.

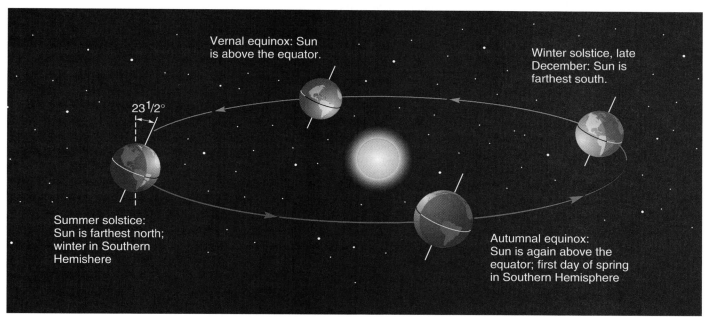

winter months, when the axis is tilted away from the Sun, the sunlight reaches the surface at an oblique angle. The seasons create fluctuating surface temperatures that stay within a moderate range. Without the seasons, large amounts of heat might be permanently trapped at the equator, and huge regions of permanently frozen wasteland would exist at moderate and high latitudes.

Origin and Physical Development of the Earth

The Earth is not a stagnant, unchanging planet. Since its origin around 4.6 billion years ago, our planet has been progressively developing and evolving. Atmospheric composition, climate and weather, continental and ocean positions, surface topography, sea level heights, and many other factors have varied in the past.

Therefore, if we are to evaluate the modern environment in a holistic context (taking into account the functional relationships between the parts and the whole), we need to take an historical perspective. For instance, to evaluate the current concentration of greenhouse gases and the predictions of global warming, we need to consider the changing composition of the atmosphere through geologic time and the oscillations in climate that have occurred over the past few million years (see Chapter 17). Similarly, we must compare present extinction rates of species to extinction rates over the last 500 million years (see Chapter 11).

Origin of the Solar System and Earth

The present universe is believed to be between 8 and 20 billion years old. Our little bit of the universe—our Sun and solar system—is much younger, though. The Sun and solar system originated about 5 billion years ago when a gas and dust cloud in our region of the galaxy began to collapse and coalesce. The material rotated more and more swiftly around its center, flattening the cloud into a disc shape and causing the accretion and condensation of particles of matter.

In the middle of the disc, the proto-Sun (the material that would become our Sun) condensed, and at various distances from the center, eddies concentrated particles in clusters that formed planetesimals (small planet-like bodies). Through gravitational attraction and accretion, these bodies eventually became our familiar planets. By about 4.6 or 4.5 billion years ago, the Sun ignited, giving off tremendous heat and light. The influence of the Sun may have produced the basic arrangement of the planets we observe today, with the innermost planets being the "rocky" planets (Mercury, Venus, Earth and its Moon, and Mars) that can withstand the higher temperatures closer to the Sun, and the outer planets (Jupiter, Saturn, Uranus, Neptune, and Pluto) being frozen, gassy, or icy planets.

On the Earth, a period of intense heating caused the various components to melt and to separate into layers of heavier and lighter components. The dense iron-nickel components sank to the core while the lighter silicates rose and formed the mantle and

El Niño and Integrated Earth Systems

The El Niño Southern Oscillation (ENSO) is a periodic weather phenomenon that occurs approximately every three to seven years (as measured during the last century) over the tropical Pacific Ocean and influences rainfall and temperatures around the globe. An analysis of El Niño demonstrates just how integrated and finely balanced various global systems can be, with changes in one region having worldwide repercussions. This should be kept in mind as we consider potential global climate change (see Chapter 17) and other possible human-produced assaults on the environment. Indeed, some researchers question whether some weather phenomena that in recent years have been attributed to global warming are in reality part of the El Niño phenomenon, or if El Niño events may have increased in intensity and frequency over the last hundred years (and especially over the last few decades), perhaps driven by increasing global warming.

Under normal conditions, the western portion of the tropical Pacific Ocean, east of Australia, has the highest surface water temperatures (refer to **Figure 1a**) and is a relatively low-pressure area (meaning the air is heating and

rising). In contrast, over the eastern portion of the tropical Pacific, air pressures are higher, air is descending to the surface of the ocean, and the air flows from the high pressure to low pressure zones, creating westward blowing winds (the trade winds). The winds blowing against the surface of the ocean water from east to west set up surface currents in the same direction and actually cause warm water to "pile up" along the edge of the western tropical Pacific basin and the movement of the warm water to the west reinforces the pressure gradient which in turn reinforces the strength of the trade winds. As a result of the water piling up, the ocean's surface can actually be about twenty inches (a half meter) higher in Indonesia than Ecuador. This piling up of warm water in the western Pacific also causes relatively cold and deep water to well up along the west coast of South America, bringing with it rich nutrients important to the fishes and other ocean life of the region. (In the oceans, deep waters tend to be relatively rich in nutrients because it is in the shallow, warm, upper layers that much life lives, depleting the local nutrients, and as organisms die some of their remains tend to descend into the deep waters and dissolve, adding nutrients to the cold depths. In places of deep-water upwelling, these nutrients are brought to the surface and support an abundance of life.)

During El Niño episodes, however, the relative air pressure differences over the western and eastern regions of the tropical Pacific decrease or even reverse, with higher pressures and descending air in the west and lower pressures and ascending air in the east (see **Figure 1b**). This is known as the "Southern Oscillation" in air pressures, and causes the warm water that is normally piled up on the

(a) Normal conditions

Figure 1 (a) Schematic diagram of normal (non-El Niño) conditions in the Pacific. North and South America are shown schematically on the right and a portion of Australia is shown schematically on the left. Warmest surface waters are shown in red and coolest surface waters in light blue. Thick white arrows show the direction of surface ocean currents. The thermocline marks the transition between relatively warm surface waters and cooler deep waters. (b) El Niño conditions in the Pacific. Notice the reversal of direction of surface currents, the change in surface water temperatures, the lowering of the thermocline along the western coast of South America (meaning that deep, nutrient-rich waters are not coming to the surface there as they do during non-El Niño conditions), and the change in atmospheric convection patterns. (*Source*: National Oceanic and Atmospheric Administration.)

(b) El Niño conditions

western end of the tropical Pacific to "slosh" or flow back toward the east, and if the pressure is high enough in the west and low enough in the east, winds and surface ocean currents may actually reverse directions. Also, the upwelling of cold, deep waters along the west coast of South America is decreased.

An El Niño episode not only affects the tropical Pacific, but also has ramifications for weather patterns worldwide, in part by modifying and displacing the jet streams (high altitude wind currents). Although they can vary greatly depending on the magnitude and duration of a typical El Niño event (which may last a number of months or over a year, from beginning to end), some of the effects of an El Niño include droughts in the western Pacific, such as in the region of Australia, Indonesia, and southeast Asia; warm weather in parts of Alaska, northwestern and northeastern Canada, and India; cooler weather in the southeastern United States; and increased rainfall (with possible flooding) in Peru as well as parts of the southeastern coast of South America, the southern United States (especially the southeast), and parts of Europe and central Asia. Outbreaks of infectious diseases have been correlated with extreme weather conditions caused by El Niño events, as have major fires in Australia, Indonesia, and elsewhere.

Over the last two decades, El Niño events have occurred during 1982/1983, 1986/1987, 1991/1992, 1993, 1994, 1997/1998, and 2002/2003. Of these, 1982/1983 and 1997/1998 were particularly strong events, with 1997/1998 being considered the strongest El Niño ever documented (although records have only been kept for about a century). Sometimes, but not always, El Niño events are preceded or followed by what are known as La Niña events (also known as El Veijo or anti-El Niño). Unusually cold and widespread temperatures in the tropical Pacific characterize La Niña events, and the worldwide weather effects of a La Niña can be more or less the opposite of an El Niño. Many researchers consider El Niño to be the warm phase of the ENSO cycle and La Niña to be the cold phase.

Peruvian fishermen working off the coast of South America have recognized El Niño events for over a hundred years. The Spanish name itself refers to "The Little One," "Little Boy," or "Christ child" since the phenomenon often begins around the season of Christmas and the New Year. (Originally "El Niño" referred to warm ocean currents that appeared around Christmastime every year.) La Niña refers to "Little Girl" (as opposed to "Little Boy") and "El Viejo," an alternative name for the cold phase of ENSO, means "The Old One."

No one really knows what causes the initiation of an El Niño event. It appears that in many cases, relatively short-lived winds might blow against the normal trade winds, pushing warm water to the east. This eastward movement of warm water could lower the air pressure in the central and eastern tropical Pacific (since the warm water will heat the air above it and cause it to rise), which in turn will weaken the westward moving trade winds that depend on a strong air pressure gradient from east to west, and in a self-perpetuating cycle, the warm water could continue to flow east (remember that the warm water is normally piled to the west and held there by the trade winds) even as the trade winds continue to weaken; ultimately a full-fledged El Niño might occur.

As far as humans are able to determine, essentially stochastic, or random, processes may cause the initiating wind described in the last paragraph. It has been suggested that large storms originating in the Indian Ocean and moving east into the Pacific may be the triggering factor for an El Niño event. Another possibility that has been suggested is the influence of Rossby waves (also known as *planetary waves*), which are very low frequency, long, slow-moving waves found in both the atmosphere and oceans. Rossby waves are generated by the rotation of Earth and travel only from the east to the west. The reflection of the Rossby waves in the ocean off of the rim of the western Pacific Ocean basin may, it has been speculated, somehow slowly build up over several years and initiate a chain of events leading to an El Niño. It is also quite conceivable that not all El Niño events are caused by the same triggering factors, or are perhaps initiated by a complex set of factors.

The El Niño phenomenon is just one example that illustrates how complex and deeply integrated the various systems of Earth are, and also points to the fact that our understanding of the workings of Earth is far from complete. We should keep this in mind as we ponder the possible outcomes of human-induced environmental change, whether on a local level or a global level.

crust. Many sources for the heat that led to this differentiation have been suggested, such as the impact energy of meteorites and asteroids, but in all likelihood, the main source was the radioactivity of elements trapped within the planet.

Origin of the Oceans

By about 4 billion years ago, the Earth probably had an ocean. We can assign an approximate date because among the oldest known rocks, formed some 3.8 billion years ago, are sedimentary rocks that show evidence of water weathering and erosion. But where did the water come from? Various hypotheses have been proposed. Many scientists believe that during the early accretion of the Earth, much water (initially in the form of water vapor) and large amounts of other gases were released by volcanoes. Modern volcanoes also spew out water, although much of their output may actually be recycled seawater. A significant amount of water may also have been deposited by icy comets that collided with the primitive Earth.

Over time, the oceans have accumulated salts as weathered material from the rocks of the continents has washed into them. The oceans contain ions, or salts, of such elements as sodium, magnesium, and chlorine along with smaller amounts of calcium, silica, potassium, and many other elements, and various dissolved gases including oxygen, carbon dioxide, and sulfur dioxide.

Origin of the Atmosphere

The earliest atmosphere of which we have any direct evidence, that of about 3.5 to 4 billion years ago, was very different than the atmosphere of today. It almost certainly originated from the Earth's interior as the planet underwent melting and segregation into a core, mantle, and crust. This early atmosphere probably contained large amounts of nitrogen gas, appreciable amounts of carbon dioxide, some methane and ammonia, and virtually no molecular oxygen.

So where did the present atmosphere's oxygen come from? Two basic processes are known to produce oxygen gas in some abundance in the natural world: (1) the breakdown of water into hydrogen gas and oxygen gas by ultraviolet radiation and (2) plant photosynthesis, the process of using the energy of sunlight to combine carbon dioxide and water into carbohydrates, giving off free oxygen gas as a byproduct. The latter process is much more efficient at producing oxygen.

Currently, photosynthesis produces an estimated 20 billion metric tons of oxygen per year. Many scientists believe that the high oxygen level of our current atmosphere is largely the result of photosynthetic organisms dumping this once lethal toxic waste into the environment over the last 3 billion years. As the levels of oxygen in the atmosphere increased, organisms evolved the ability to utilize oxygen in an efficient form of metabolism known as aerobic respiration; ultimately these changes allowed the evolution of very complex, multicellular life-forms.

Natural Hazards

Humans have always had to deal with "unpredictable" **natural hazards**—earthquakes, volcanic eruptions, floods, avalanches, droughts, fires, tornadoes, hurricanes, and so forth (**Table 5-1**). For most of human history such phenomena were beyond human control: They were neither caused by humans, nor was there much humans could do to predict their occurrences or mitigate (make less severe) their consequences.

Although many acts of nature still cannot be controlled, we have learned to better predict their occurrences and mitigate their effects. Furthermore, we have come to understand that some types of phenomena have a large anthropogenic (produced or caused by humans) component. Flash floods and avalanches may be caused, or at least exacerbated, by human deforestation in hilly regions. Abnormal droughts, storms, and other unusual weather phenomena may be caused or magnified by human interference with the atmosphere, such as the emission of high levels of greenhouse gases. Earthquakes, at least

TABLE 5-1	Estimated Deaths Due to Major Natural Hazards 1960–2000		
Hazard Type	**Estimated Total Deaths**	**Examples of Major Events**	**Deaths**
Tropical cyclones	850,000	East Pakistan (Bangladesh) 1970	500,000
Earthquakes	650,000	Tangshan, China 1976	250,000
Floods	60,000	Vietnam 1964	8,000
Avalanches, mudslides	50,000	Peru 1970	25,000
Volcanic eruptions	36,000	Colombia 1985	23,000

on a local level, may be caused by the pumping of fluid wastes into rocks lying deep below the surface.

As the Earth's human population increases, the damage done by naturally occurring, periodic "disasters" is magnified. A coastal area may be hit by a major storm once every century, a river may swell over its "normal" banks covering the floodplain only once every few centuries, or a naturally set forest fire may periodically burn off dead underbrush and litter on the forest floor. Viewed on a temporal scale of millennia (thousands of years), such events form an integral part of the natural ecosystem cycle; from a natural, holistic perspective, they are not disasters at all. But when dense human populations inhabit a coastal area or a river's floodplain and depend on a forest for lumber products or for scenic beauty, these events are perceived as terrible disasters—and so they are from a human perspective. But they are disasters that, given a little knowledge of the natural world, should have been easily anticipated. In this section, we briefly introduce some of the basic types of natural hazards that civilization must face as we continue to live in a natural environment.

Earthquakes and Volcanoes

Earthquakes and volcanic eruptions are geological phenomena that humans may be able to predict (though not always), but can virtually never control. With few exceptions, earthquakes and volcanoes are caused by processes that take place deep within the crust and mantle. Volcanic eruptions are usually accompanied by at least minor to moderate earthquakes, but very large earthquakes can occur in the absence of volcanic activity. As we have already discussed, earthquakes and volcanic activity are associated with the boundaries of the lithospheric plates (refer to Figure 5-3), and can be explained in terms of moving and/or subducting plates. But earthquakes (and less commonly volcanoes) can also occur in the middle of plates; indeed, no place on the surface of Earth is immune to earthquake activity.

Earthquakes. Earthquakes are essentially shock waves that originate when large masses of rocks suddenly move relative to each other below the Earth's surface. For instance, along a plate boundary where two lithospheric plates are sliding past each other, the plates may "hang up," allowing strain (frictional drag) to accumulate until it is finally relieved by rock movement—causing an earthquake. Earthquakes can originate up to 450 miles (700 km) below the surface. When the sudden rock movement occurs, various types of shock, or seismic, waves are transmitted through the Earth, causing the shaking or trembling

felt by humans on the surface. A large earthquake may be detectable around the world.

Earthquakes are a constant threat to human life and well-being in some areas. Perhaps the most devastating set of earthquakes occurred near Shensi, China, in 1556, killing an estimated 830,000 people. A dozen earthquakes that caused the death of 100,000 people or more have been recorded. Estimates are that, on average, earthquakes kill at least 10,000 people a year and cause about $500 million in property damage. Japan, a nation always at risk for earthquakes, suffered devastating consequences in 1995 when a quake hit the Kobe area killing over 5000 people, injuring another 25,000, and initially leaving over 300,000 homeless. More than 17,000 people were killed in the Izmit, Turkey, earthquake in 1999 (**Figure 5-13**). In 2001, a quake struck India, killing over 18,000, injuring over 166,000, and leaving 600,000 homeless. The damage estimates were placed at $1.3 billion U.S. dollars.

The intensity, magnitude, or strength of an earthquake is commonly measured either on the Richter scale (in North America) or the Mercalli scale (in Europe). The Richter scale, devised by Charles Richter in 1935, is based on the amplitude of the seismic waves recorded by seismographs (instruments that record the motions of the Earth's surface) coming from a particular earthquake. The Richter and related scales measure the energy released by an earthquake, or in crude terms, how much "shaking" of the ground occurs during an earthquake. The Richter scale is a logarithmic scale so that every unit corresponds to a 10-fold increase in the amplitude of the seismic waves. Thus a 7.0 earthquake is characterized by seismic waves with an amplitude of 100 times that seen in a

Figure 5-13   The 1999, Izmit, Turkey earthquake (magnitude 7.4) killed more than 17,000 and additional thousands were missing and presumed dead. An estimated 500,000 people were left homeless. Before the earthquake reduced them to rubble, there were probably four or five buildings here, each five or six stories high.

5.0 earthquake. Theoretically, the Richter scale has no upper limit, but some of the largest recorded earthquakes (such as the Lisbon earthquake of 1755 immortalized in Voltaire's *Candide*) have been ranked at about 8.9-9.0 (**Table 5-2**).

The Mercalli scale (**Table 5-3**) is more qualitative, based primarily on observations of the earthquake's effects close to its origin. This scale classifies earthquakes into a dozen basic categories, from instrumental (earthquakes so small that they can be detected only on seismographs) to catastrophic (earthquakes where local destruction is virtually total). The Mercalli scale is easily related to the Richter scale in an approximate way (Table 5-3).

Much research has recently gone into devising ways to predict earthquakes in both the long term and the short term. Some work has attempted to discover long-term earthquake cycles, based on the movements of the lithospheric plates and also perhaps correlated with such phenomena as meteoritic impacts, variations in the Earth's geomagnetic field, sunspot activity, and variations in the length of the day. Such work is very difficult to pursue, and no method has been able to predict long-term earthquake events with any degree of accuracy.

As for the short term, researchers have identified several precursors that often signal the probability of a major earthquake in earthquake-prone areas. The most commonly utilized precursors are unusual land deformation, seismic activity, geomagnetic and geo-electric activity, groundwater fluctuations, and natural phenomena such as animal behavior and unusual weather conditions (see Case Study 5-2).

Although we usually do not think of earthquakes as being caused by human activity, some earthquakes are indeed anthropogenically induced. Earthquakes have been induced by nuclear blasts, conventional blasting, mining activities, fluid injection and extraction from rocks deep underground, and the building of dams and reservoirs. A now-classic example of human-induced faulting and seismic activity occurred near Denver in the early 1960s, when nerve-gas waste was disposed of by pumping it down a well to great depths where it would be below groundwater supplies. Pumping the waste down the well at high pressures triggered a series of earthquakes. Since then, it has been verified experimentally in oil fields that pumping fluids into the ground (and in some cases extracting them, such as the pumping out of oil) can induce seismic activity.

The most important cause of human-induced earthquakes seems to be the construction of large dams and reservoirs. In the case of at least six major dams around the world, including Hoover Dam on the Colorado River, earthquakes of a magnitude greater than 5 on the Richter scale (moderately strong earthquakes, capable of minor damage) have apparently been induced by the impounding of water in large reservoirs. Over 1000 earthquakes of various magnitudes have been felt since Hoover Dam was constructed in 1935; before 1935 the area was not known for earthquake activity. Worldwide, dozens of dams have been associated with seismic phenomena. It is thought that water in the reservoir may penetrate the underlying bedrock and cause rock slippage that generates earthquakes. The huge mass of water in the reservoir also exerts tremendous pressures on the underlying rocks, and this can cause downwarping and subsidence of the land's surface.

TABLE 5-2	Some of the Largest Earthquakes of the Last Century (plus the famous Lisbon earthquake), by Estimated Magnitude on the Richter Scale, Along with Some Relatively Recent Major Earthquakes	
Year	**Location**	**Magnitude**
1755	Lisbon, Portugal	9.0
1905	Kangra, India	8.6
1906	Valparaiso, Chile	8.6
1906	San Francisco, United States	8.3
1911	Tienshan, China	8.4
1920	Kansu, China	8.5
1923	Yokohama, Japan	8.3
1927	Nan-Shan, China	8.3
1933	Japanese trench	8.9
1950	North Assam, India	8.7
1960	Chile	9.5
1964	Anchorage, Alaska	9.2
1976	Tangshan, China	8.0
1977	Sumba, Indonesia	8.0
1977	Argentina	8.2
1979	Indonesia	8.1
1985	Michoacan, Mexico	8.1
1989	San Francisco Bay area	7.1
1990	Northwest Iran	7.7
1994	San Fernando Valley, Calif.	6.6
1995	Osaka, Kyoto, Kobe, Japan	7.2
1997	Northeast Iran	7.1
1998	Northern Afghanistan	7.1
1999	Izmit, Turkey	7.4
1999	Taiwan	7.6
2001	El Salvador	7.7
2001	Bhuj, India	7.7
2002	Central Alaska	7.9

TABLE 5-3	The Mercalli Scale of Earthquake Intensity			
Scale	Intensity	Description of Effect	Maximum Acceleration (mm sec^{-2})	Corresponding Richter Scale
I	Instrumental	Not felt except by a very few under especially favorable circumstances.	<10	
II	Feeble	Felt only by a few persons at rest, especially on upper floors of buildings. Delicately suspended objects may swing.	<25	
III	Slight	Felt quite noticeably indoors, especially on upper floors of buildings, but many people do not recognize it as an earthquake. Standing automobiles may rock slightly. Vibration like a passing truck.	<50	<4.2
IV	Moderate	During the day felt indoors by many, outdoors by few. At night some awakened. Dishes, windows, doors disturbed; walls make cracking sound. Sensation like heavy truck striking building. Standing automobiles rock noticeably.	<100	
V	Slightly strong	Felt by nearly everyone, many awakened. Some dishes, windows, etc., broken; a few instances of cracked plaster; unstable objects overturned. Disturbances of trees, poles, and other tall objects sometimes noticed. Pendulum clocks may stop.	<250	<4.8
VI	Strong	Felt by all, many frightened and run outdoors. Some heavy furniture moved; a few instances of fallen plaster or damaged chimneys. Damage slight.	<500	<5.4
VII	Very strong	Everybody runs outdoors. Damage negligible in buildings of good design and construction; slight to moderate in well-built ordinary structures; considerable in poorly built or badly designed structures; some chimneys broken. Noticed by persons driving automobiles.	<1000	<6.1
VIII	Destructive	Damage slight in specially designed structures; considerable in ordinary substantial buildings, with partial collapse; great in poorly built structures. Panel walls thrown out of frame structures. Fall of chimneys, factory stacks, columns, monuments, walls. Heavy furniture overturned. Sand and mud ejected in small amounts. Changes in well water. Persons driving automobiles disturbed.	<2500	
IX	Ruinous	Damage considerable in specially designed structures; well-designed frame structures thrown out of plumb; great in substantial buildings, with partial collapse. Buildings shifted off foundations. Ground cracked conspicuously. Underground pipes broken.	<5000	<6.9
X	Disastrous	Some well-built wooden structures destroyed; most masonry and frame structures destroyed with foundations destroyed; ground badly cracked. Rails bent. Landslides considerable from river banks and steep slopes. Shifted sand and mud. Water splashed (slopped) over river banks.	<7500	<7.3
XI	Very Disastrous	Few, if any (masonry) structures remain standing. Bridges destroyed. Broad fissures in ground. Underground pipelines completely out of service. Earth slumps and land slips in soft ground. Rails bent greatly.	<9800	<8.1
XII	Catastrophic	Damage total. Practically all works of construction are damaged greatly or destroyed. Waves seen on ground surface. Objects are thrown upwards into the air.	>9800	>8.1

Source: United States Geological Survey

Predicting Earthquakes

The careful periodic surveying of known benchmarks on the Earth's surface can lead to the detection of movement within the crust. Benchmarks may have moved relative to each other, suggesting possible faulting. In some areas where an active fault zone has been identified, accurate surveys across the fault zone can detect very slight movements (on the order of millimeters) that may be precursors of larger earthquake activity to come. In known earthquake-prone areas, instruments called *tiltmeters* can be used to monitor the surface of the Earth; if the surface starts tilting rapidly, an earthquake is usually imminent.

Daily monitoring of seismic activity is useful in an earthquake-prone area. Changes in the background seismic activity, especially increases in activity (sometimes referred to as foreshocks), can herald a major earthquake. The Earth's surface is characterized locally by magnetic fields and electric currents flowing through the rocks. Anomalous geomagnetic and geoelectric activity may be detectable months, days, or hours before an earthquake strikes. Another technique that has been used very successfully in China is to monitor groundwater levels in wells. Fluctuations in normal water levels may precede an earthquake by hours to over a week. Radon levels in groundwater may increase for years before a major earthquake; this is probably due to the movement of water through new cracks that open up in the rocks before the onset of an earthquake.

Some of the most interesting and useful predictors are also extremely low-tech and poorly explained from a scientific perspective (indeed, they almost verge on superstition), namely, the behaviors of various animals in the days and hours before an area experiences an earthquake. The Japanese have observed that catfish exhibit unusual behavior before an earthquake, becoming very active and even jumping out of the water (**Figure 1**). All kinds of domestic animals become restless and exhibit odd behaviors prior to an earthquake: dogs bark, pigs become very aggressive, horses refuse to go into their stables, and so forth. Wild animals also show unusual behavior: Burrowing animals leave the ground, rats run around randomly, snakes attempt to leave the area, and worms crawl from the ground to the surface in large numbers. In the San Francisco area, the behavior of zoo and marine animals is monitored daily as part of the local earthquake prediction system.

Possibly, certain animals are sensitive to vibrations (or ultrasound) generated by small earthquakes undetectable to humans that precede a major earthquake. Perhaps some animals are extremely sensitive to the smell of methane that may leak from the ground prior to an earthquake. Another possibility is that electrostatic particles may be coming from the ground; furry and feathered animals, such as mammals and birds, are generally very sensitive to electrostatic charges.

Reports of unusual weather before major earthquakes have often been dismissed as nonsense. However, some researchers have suggested that degassing of methane and other gases from below the Earth's surface may precede a major earthquake and could account for unusual weather phenomena such as strange mists, glowing skies, and flashes of light. If electrostatic particles are released prior to an earthquake, this also might help explain unusual weather phenomena.

All in all, predicting earthquakes is still a tricky business, perhaps more art than science. But China in particular has a very good track record; the Chinese have been able to predict a number of major earthquakes in time to evacuate cities before they were hit. Although substantial financial losses may still be unavoidable, by removing people from the area, the death toll can be kept at a minimum.

Figure 1 The Japanese have used catfish behavior as a predictor of earthquakes, and according to Japanese legend, earthquakes are caused by the movement of a giant catfish. This antique illustration shows people attempting to subdue the giant catfish that causes earthquakes.

Given that humans have inadvertently caused earthquakes, some people suggest that in the future we might be able to intervene to control natural earthquake activity. Perhaps, in an active earthquake zone, we could selectively induce a series of small earthquakes by either injecting or extracting fluids from the rocks. The small, relatively harmless earthquakes might relieve the strain on the rocks and thus allow us to avoid a single large, very destructive earthquake.

Volcanoes.   **Volcanoes** are basically spots in the Earth's crust where hot, molten rock (magma) wells up to the surface (**Figure 5-14**). Active volcanoes are found almost exclusively in three geologic settings: at convergent plate margins where one lithospheric plate subducts under another, melting the rock, which rises to the surface as volcanoes (for instance, in Indonesia and elsewhere along the Pacific rim); at divergent plate margins where magma wells up to the surface as two lithospheric plates pull apart from each other, forming a rift (as in the middle of the Atlantic Ocean); and over mantle hot spots, areas that lie over a hot mantle plume that breaks through the crust and spews molten rock onto the Earth's surface (as in the Hawaiian Islands).

Volcanoes can be classified according to whether they extrude predominantly basaltic or andesitic magma. Volcanoes at divergent plate boundaries and hot spots tend to produce basaltic magma that is relatively silica-poor and originates from the mantle. Andesitic magma contains a higher percentage of silica and is generally formed from the remelting, differentiation, and recrystallization of previously existing crustal or mantle material. Thus andesitic volcanoes are commonly found in subduction zones where rock is remelted.

Basaltic magmas are hotter and much less viscous than andesitic magmas, which tend to contain a much higher percentage of gases (often predominantly water vapor). Consequently, whereas lava may flow smoothly out of the crater of a basaltic volcano, andesitic volcanoes tend to be much more explosive, shooting out steam and other gases, rock fragments of various sizes, and volcanic ash.

Volcanic eruptions can affect humans on both a local and a global scale. On a local level, volcanic eruptions can destroy local towns and cities. In A.D. 79, ash falls and mudflows triggered by the eruption of Mount Vesuvius in Italy buried the towns of Pompeii and Herculaneum. The eruption of Mount Pelée on Martinique in 1902 destroyed the city of St. Pierre, killing some 30,000 people in a span of two minutes (**Figure 5-15**). Large volcanic eruptions can have worldwide effects by

Figure 5-14   Volcanoes, evidence that the Earth is still geologically young and active, inspire fear and awe in humans. Illustrated here is the 1980 eruption of Mount St. Helens, Washington.

spewing dust, ash, and gases (including material that can form acid rain) into the atmosphere, affecting global weather patterns. For example, the 1883 eruption of Krakatoa, a volcano in the Sunda Straits between Sumatra and Java, spewed 4 cubic miles (18 km^3) of rock, ash, and other debris into the atmosphere to heights of 50 miles (80 km). The materials initially reduced the amount of incoming solar radiation reaching the Earth's surface by an estimated 13%. Even two years later, the amount of incoming solar radiation over France was still 10% below normal. In such examples we can see that natural phenomena (volcanic eruptions) may dramatically alter regional and global

Figure 5-15   Seen here are the remains of St. Pierre, a once thriving city of 30,000 on the island of Martinique, that was destroyed by the 8 May 1902 eruption of Mount Pelée.

weather patterns. This is an important consideration to take into account as one debates whether recorded patterns of apparent climate change over the last century are partly or wholly induced by human activities.

The eruption of Krakatoa also set off tsunamis that caused damage throughout the Pacific basin. Tsunamis (Japanese for "harbor wave") are huge waves, sometimes caused by explosive volcanic activity in or near the oceans or by large undersea earthquakes (usually registering higher than 6.5 on the Richter scale) or landslides. Fortunately, tsunamis are relatively rare; however, they can be extremely destructive. The waves can range in height from less than 3 feet (1 m) to nearly 200 feet (60 m).

In 1692, an earthquake destroyed Port Royal, Jamaica, and the resulting tsunami threw harbored ships inland over two-story buildings. The Lisbon earthquake of 1755 sent a wave across the Atlantic Ocean that temporarily raised the sea by 10 to 13 feet (3 to 4 m) in Barbados.

Even with modern technology and knowledge, predicting future volcanic activity is extremely difficult. Although we know where to expect active volcanoes relative to lithospheric plates and can usually determine if a volcano is active or dormant (based on the time since its last eruption), we cannot predict exactly the timing and intensity of future eruptions. Researchers use many of the same techniques employed for predicting earthquakes. Seismic activity, land deformation, geomagnetic and geoelectric parameters can be monitored. Any anomalous behavior, like a change in seismic activity or

active deformation of the sides of a volcano, can herald an imminent eruption. But estimating the strength and type of eruption that will occur is difficult. The analysis of gases given off by volcanoes holds some promise for predicting volcanic activity, at least in the short term. However, sampling gases being vented from an active volcano that may be near eruption can be very difficult and dangerous.

Land Instability

A very widespread form of natural geologic hazard, usually brought on by human ignorance of the principles of geology and soil mechanics, is the collapse of soil or weathered rock material. Landslides, rockfalls, and avalanches may bury roads, buildings, and other human structures. Soils may fail to support buildings, thus causing them to collapse. Such phenomena may occur simply because humans erect structures in areas that are geologically unstable and thus unsuitable for building. Human activity, such as clear-cutting a mountainside or pumping out groundwater, may induce land instability. Of course, natural phenomena such as earthquakes and volcanic eruptions may also induce landslides, rockfalls, avalanches, surface subsidence, and so forth.

Weather Hazards

Hurricanes, typhoons, tornadoes, droughts, floods, heat waves, wind storms, dust storms, and other "irregular" weather patterns can wreak havoc on human settlements. Full discussion of this large and complex topic is beyond the scope of this book, but here we will briefly mention some major types of storm hazards. Due to possible global warming brought on by the greenhouse effect (see Chapter 17), some researchers expect a dramatic increase in irregular weather patterns, particularly the number of violent storms, in the relatively near future (see further discussion in Chapter 17).

Tropical Cyclones. Tropical **cyclones** are intense storms that develop over warm tropical seas (**Figure 5-16**). When they occur in North America, tropical cyclones are generally referred to as *hurricanes*; when they occur in Southeast Asia, they are known as *typhoons*. Only storms in the Indian Ocean and around Australia are commonly or colloquially called *cyclones*. Approximately 100 to 120 tropical cyclones develop worldwide every year, and many of them travel far inland as well as posing hazards to coastal areas. Every five years or so, a hurricane along the east coast of North America crosses the Appalachian Mountains and enters the Great Lakes region.

Figure 5-16   Tropical cyclones, more commonly known as hurricanes and typhoons, are among the most destructive of natural hazards. Here we see just some of the damage caused by Hurricane Andrew in 1992.

In terms of deaths, tropical cyclones are among the worst natural hazards that humans currently face. Between 1960 and 2000, approximately 850,000 people died as a result of tropical cyclones compared with 650,000 people killed by earthquakes. Of the people who died in tropical cyclones, however, 500,000 lost their lives in a single event—when a cyclone hit Bangladesh (then East Pakistan) in 1970. The death rate was so high because Bangladesh is immensely overcrowded and people live in a low-lying deltaic area that, given the regularity of storms, is virtually unsuitable for human habitation. In addition, the islands of the delta were swamped with salt water that damaged or destroyed over 990,000 acres (400,000 hectares) of rice paddies; about a million head of livestock were killed; and over 50 million people were significantly affected by the storm (the total population of Bangladesh is over 115 million). Bangladesh was hit by another tropical cyclone that caused 100,000 deaths in 1985 and by yet another that killed 140,000 people in 1991.

Realistically, humans can do very little about tropical cyclones or most natural climatic hazards, except to take preventive measures long before the disasters hit. For instance, areas prone to tropical cyclones should not be heavily settled. Early predictions of impending storms can help the public take proper precautions, such as evacuating low-lying areas that will be hardest hit. Unfortunately, in countries like Bangladesh many people have no place to live except areas that are prone to tropical cyclones, and even when an imminent cyclone is detected, evacuating hundreds of thousands to millions of people at once is virtually impossible.

Tornadoes and Other Strong Wind Phenomena. Tornadoes are a classic American phenomenon; about 80% of all tornadoes occur in the United States, mostly in the Great Plains (about 600 a year in this area). The typical tornado consists of a rapidly rotating vortex of air that forms a funnel (**Figure 5-17**); tornadoes may be associated with hurricanes, but not necessarily. A tornado may be relatively harmless as long as the funnel does not touch the surface of the Earth. But when they do touch down, tornadoes are among the most intense and destructive phenomena found in nature. As a result of winds of 250 to 300 miles per hour (400 to 500 km/hr) or greater and changes in air pressure found within the funnel, tornadoes can lift objects weighing hundreds of tons. A tornado that moves over water can suck in millions of tons of water, perhaps temporarily draining a river or lake and even carrying the fishes or other organisms in

Figure 5-17   A vortex of rotating air, with wind speeds as high as 300 miles an hour (500 km/hr), a single tornado can cause incredible destruction in a small area. The tornado shown here is in its early stage of formation near Union City, Oklahoma.

the lake to a new area (where it may literally "rain" fish once the energy of the tornado dissipates).

Despite their destructive force, tornadoes that touch the surface of the Earth are fairly localized phenomena. Their contact with the ground generally lasts only half an hour or less; at most the ground path of a single tornado might be 1100 yards (1 km) wide and 10 or 20 miles (a few tens of kilometers) long.

Although most humans associate tornadoes with trouble, in actuality tornadoes serve a number of important geological and ecological functions. They transport large quantities of clay and silt particles from one area to another. They also serve as an important biotic dispersal mechanism; tornadoes have been proposed as the primary mechanism by which various species are dispersed extremely rapidly over fairly large areas. Seeds, pollen, microorganisms, insects, and even such animals as fishes, frogs, toads, and turtles can be transported alive for a few hundred miles (several hundred kilometers) in a tornado. Tornadoes may explain how aquatic species could migrate between unconnected lakes in parts of North America.

Tornadoes are not the only weather phenomenon associated with strong winds that pose a hazard for humans. Much more important, in terms of their frequency and worldwide extent, are dust storms. There have been some very major dust storms indeed. In 1901, a dust storm carried approximately 150 million metric tons of dust from the Sahara in Africa to western Europe and the Ural Mountains. In 1928, a dust storm removed an es-

Figure 5-18   A dust storm approaching a Kansas town in 1935.

timated 15 million metric tons of soil from Ukraine and deposited it in Romania and Poland. The dust storms that occurred in the midwestern United States during the 1930s are legendary (**Figure 5-18**). Over 2.5 million acres (1 million hectares) of agricultural land were severely damaged by the resulting loss of topsoil.

In recent years, satellite tracking of dust storms and dust movements around the planet has demonstrated just how pervasive dust movements are on Earth. By one estimate, 11.8 million metric tons a year are blown or drift (sometimes at altitudes of

Figure 5-19   A satellite image of sand and dust blowing out to sea from the Sahara.

over 10,000 feet [3048 meters]) from Africa to the Northeastern Amazon Basin in Brazil (**Figure 5-19**). In 2001, a huge Asian dust cloud blew east across the Pacific Ocean, across the entire continental United States, and some remnants of it made their way to the mid-Atlantic Ocean. Some of the dust below your feet or on your windowsill could literally be from anywhere in the world. And the dust that is transported around the world serves a vital function—it brings fresh nutrients from one part of the globe to another. Iron and various minerals in desert dust that lands in the oceans provide essential nutrients upon which the marine organisms depend. According to some research, the canopies of Central and South American rain forests receive valuable nutrients from the dust that is transported across the Atlantic from the Sahara Desert. Furthermore, it is not only inorganic or nonliving dust per se (finally ground minerals and nonliving organic matter) that can be transported around the globe, but also airborne microbes and pollens. Dust carried by global atmospheric winds and currents demonstrates once again that our planet is ultimately a single, integrated system.

Floods.   A **flood**, quite simply, is a high flow of water that overruns its normal confinement area and covers land that is usually dry. Natural floods are a normal part of the physical environment, but because they often seem "unusual" from a human perspective—occurring irregularly over periods of decades, centuries, or millennia—they may take us by surprise. Cumulatively, floods are among the most destructive of all natural hazards. In China, flooding along the Yangtze and Yellow Rivers has taken the lives of millions of people over the last century and a half (many were drowned or died of flood-associated diseases or flood-caused starvation when crops were destroyed); in the United States, floods have caused billions of dollars worth of damage and killed hundreds of people (**Figure 5-20**). The human losses due to floods are greatly exacerbated by the fact that riverfront properties are often considered prime real estate, many towns and farms are located in natural floodplains, and large cities are often situated along rivers.

Other than storm surges, where the sea level rises locally relative to a low coastal area (such as during a hurricane), two basic types of floods can be distinguished: riverine floods and flash floods. Riverine floods occur along major streams or rivers when particularly heavy rainfalls or the rapid melting of snow causes a large amount of water to flow through the drainage basin. Riverine floods may

Figure 5-20 Salvage begins amidst flooding in the Midwest during June of 1994.

also occur when dams, either natural or human-made, break from the strain imposed by high water levels and flood conditions or other causes. Flash floods generally occur in drainage basins containing small, shallow, or ephemeral (periodically dry) streams when heavy rainfall overloads the system with water.

In the natural environment, rivers and streams are generally short-term phenomena. Rivers continually change their course and flood their banks; the flooding serves to transport and redistribute sediments and nutrients carried by the water. The annual floods of the Nile, bringing fresh nutrients to restore the fertility of the land, were the lifeblood of the ancient Egyptian civilization. Flooding can thus be viewed as a small part of the larger hydrologic and rock cycles.

When humans develop a drainage basin, however, they tend to build permanent structures as if the rivers and streams are permanent, sedentary features of the landscape. All too often humans do not properly consider the natural periodic occurrence of floods. Consequently, when the occasional major

flood does occur (perhaps only once every hundred or thousand years), the toll in human life and property loss can be immense. To make matters worse, many human activities inadvertently promote flooding. Farming, overgrazing, deforestation, paving large expanses, and other aspects of development limit the ability of water to seep into the ground (infiltration). As a result, there will be more runoff—more water will travel overland when it rains—which can promote flooding. Mining, construction, and other human activities can cause stream channels to become filled with sediment, hindering their ability to carry water quickly, and this too promotes flooding.

Even human activities that are specifically directed at controlling or avoiding flood conditions can actually promote flooding. Dams built to control flood surges may burst and cause even more flooding. Furthermore, even a properly maintained dam will eventually collect silt, ending its usefulness. Water reservoirs behind dams may flood large tracts of forest or agricultural land, and as these artificial pools sit stagnant, they may become a breeding ground for disease vectors and pests. Dams also disrupt stream flow, causing areas downstream to undergo abnormal erosion due to sediment starvation (the sediments once carried by the stream are trapped behind the dam). Of course, some forms of aquatic wildlife can be severely disrupted by artificial dams.

Channelization, the artificial straightening of a stream or river to increase its capacity to carry large amounts of water quickly downstream, has been used in some areas as a flood control measure. Such channelization often damages the natural aquatic life in the stream, and it can further exacerbate flooding conditions downstream by carrying even more water at a faster rate. Likewise, for thousands of years, humans have built dikes, levees, and flood walls to keep a stream in its channel and avoid flooding. But if, or when, the levees or dikes are breached, the resulting flood can be devastating. All the evidence points to the conclusion that artificial containment structures can only be a temporary measure at best; ultimately, they are bound to fail. In 1927, a breach in an Army Corps of Engineers' levee allowed the Mississippi River to flood over 2.3 million acres (nearly a million hectares) of land despite the efforts of 5000 workers to repair the levee. In China, the Yellow River has been artificially contained by levees and dikes for hundreds of years; periodically, they break (for instance in 1887 and 1939), flooding up to 10,000 square miles (26,000 km²) of land at a time.

Today, rather than emphasizing engineering works such as the construction of dams and levees

Figure 5-21   The Sahel region of Africa has experienced extreme droughts in recent years, yet the people continue to attempt to farm the land in an effort to survive.

to control flooding, many authorities are advocating stricter zoning laws and better floodplain management policies. Ultimately, development may need to be strictly limited in flood-prone areas.

Figure 5-22   Elk take refuge in the East Fork of Bitteroot River during an August, 2000, fire in Montana.

Droughts.   Droughts—periods of abnormally low rainfall over an extended period of time in a particular area—can have devastating effects. Because of drought, agricultural productivity may drop dramatically, and even drinkable water for humans may be in short supply. Droughts, and likewise periods of abnormally high rainfall that may cause flooding, are a typical aspect of the climate in the long run. But in the short run, humans may settle in an area based on abnormally good (perhaps moister than usual) weather conditions, which may last for decades, and then be hard hit when drier weather conditions set in again. Changing human land-use patterns, especially the destruction of tropical forests and human-induced desertification, may affect circulation patterns globally, increasing local droughts. Additionally, it is possible that global warming from the human-induced greenhouse effect may shift precipitation patterns, causing once well-watered areas to become drier.

Major droughts have been experienced in recent years in a number of places around the world, in particular, Ethiopia, Sudan, and especially the Sahel region (Chad, Niger, Mali, Burkina Faso, Mauritania, Senegal, and Gambia) south of the Sahara in Africa (**Figure 5-21**). This area is instructive, for prolonged dry conditions have been made worse by human practices. The declining rainfall in the area has led to reduced plant growth, which reduces the amount of evapotranspiration (water given off from the soil by evaporation and water given off from plants by transpiration), which decreases the local moisture content of the atmosphere, which further decreases the amount of rainfall locally. Over time the soil dries out and heats up. The lack of vegetation allows the wind to carry more dust into the lower atmosphere, which can add to the heating of the air higher up in the troposphere, contributing to atmospheric instability and reducing the amount of dew that forms at night. All of these factors enhance and accelerate drought conditions in the Sahel region. Furthermore, human overpopulation has exacerbated the situation. Shrubs and trees are cut down and burned as fuel or fed to animals. Virtually all of the arable land (land fit to be cultivated or tilled) is under cultivation, often utilizing inappropriate Western plowing, which leads to progressive destruction of the soil structure and erosion of the topsoil. A hard crust remains on the surface of the land, preventing water from infiltrating the soil on the rare occasions when rain occurs. Together these factors are causing severe desertification in the once semiarid Sahel region.

Figure 5-23 The unconsolidated sediments found along many coasts can spell disaster for the local homeowner when a major storm strikes. This house has started to tip after a Northeaster storm hit the Outer Banks of North Carolina.

Fires

Fires are a natural part of many ecosystems; particularly severe natural fires may be the result of drought conditions and lightning (**Figure 5-22**). Since prehistoric times, humans have set fires—some of which invariably get out of control. Forest fires, grassland fires, and bushfires are particularly feared in the United States and Australia, but they pose a threat to forests around the world. Major fires in the past include the 1871 forest fires that destroyed 4.2 million acres (1.7 million hectares) of forest in Wisconsin and Michigan and killed 2200 people, and the 12 million acres (4.8 million hectares) of forest that burned during 1980 in Canada. The Ash Wednesday fires of 16 February 1983 in South Australia and Victoria (southern Australia) destroyed over 1.2 million acres (over half a million hectares) of land, including urban areas (nearly two dozen towns were destroyed), forest, and pasture; killed over 300,000 sheep, 18,000 cattle, and 76 people; and injured another 3500 people. By far the greatest known fire incident was the Great Siberian fire of 1915, which burned 390,000 square miles (1 million km^2) of land in Siberia following a severe drought. Huge amounts of smoke were injected into the atmosphere blocking incoming solar radiation and suppressing ground temperatures.

For the past century, a basic principle of U.S. wilderness management has been to fight fires, whether they are set by humans or occur naturally. Timber interests saw fighting fires as preserving a resource; people interested in recreation or preservation saw the prevention of fires as a way to protect the wilderness. But by the 1940s, some foresters had begun to question the wisdom of suppressing all natural forest fires. Slowly, they came to realize that fires play a critical role in nature. Fires promote the decomposition of some forest litter (dead leaves, branches, and so on) and are an essential component of the biogeochemical nutrient cycle. Periodic fires increase the biotic diversity (the range of different organisms) of an area because they maintain more open habitats necessary for some species, especially certain birds and larger mammals. Certain plants need periodic fires as part of their life cycle; for example, the seeds of the giant sequoia (California redwood) and the jack pine must be exposed to the heat of a fire before they will germinate.

Coastal Hazards

Coastal regions along the oceans and larger lakes are among the most sought-after places for human occupation; in the United States, over half of the population lives on or near such coasts, and the same holds true around the world. Yet geologically, coastal areas can be very unstable and prone to change. Natural erosional processes are ongoing along many coasts, and human habitation may accelerate the process, for instance, by destroying wild plants that help stabilize soils and sands. In the United States approximately 25% of the coastline experiences severe erosion (**Figure 5-23**). The shifting sand dunes and unconsolidated sediments found on many coasts provide a poor foundation for homes or other buildings. Wave action and moderate storms can cause a shore to recede (in some cases at a rate of 60 feet [20 m] a year), leaving structures dangling on the edge of cliffs or eventually collapsing into the water.

Storm surges—sudden local rises in sea level caused by hurricanes and typhoons—pose particular threats to coastal regions. Likewise, tsunamis can wreak havoc on coastal areas. Yet despite the risks, coastal areas remain popular places of residence and recreation.

SUMMARY

- Major components of the Earth's internal structure include the crust, lithosphere, asthenosphere, mantle, and core.
- The lithosphere (upper crust and mantle) is divided into plates that move relative to one another.
- Plate tectonics, the unifying theory of geology, can account for the distribution of major features (such as mountain ranges, earthquake zones, and volcanoes) on the surface of Earth.
- Rocks are formed primarily of minerals which are in turn formed of atoms.
- Different types of rocks—igneous, metamorphic, and sedimentary—can be transformed from one to another during the rock cycle.
- Weather is short-term perturbations in the atmospheric/hydrologic cycles and climate is average weather over many years.
- The rotation and orbit of Earth accounts for seasonal changes in weather.
- The Earth originated about 4.5 to 4.6 billion years ago and has been undergoing changes ever since.
- Natural hazards from a human perspective include earthquakes, volcanic eruptions, floods, land instability, cyclones and other storms, droughts, and fires.

KEY TERMS

Asthenosphere
Atmospheric cycles
Atom
Climate
Compound
Continental crust
Convection cell
Core
Cyclone

Drought
Earthquake
Element
Flood
Hydrologic cycle
Igneous rocks
Lithosphere
Mantle
Metamorphic rocks

Mineral
Natural hazards
Oceanic crust
Plate tectonics
Rock cycle
Sedimentary rocks
Tornado
Volcano
Weather

STUDY QUESTIONS

1. How does planet Earth compare to its nearest neighbors in the solar system?
2. If one could cut a slice through the Earth, what would be encountered?
3. Describe modern plate tectonic theory.
4. What is the composition of matter on Earth?
5. What are the differences and similarities among sedimentary, igneous, and metamorphic rocks? Where does each form and occur?
6. What is the rock cycle?
7. What is the hydrologic cycle?
8. What are some major factors contributing to large-scale atmospheric cycles and climatic belts?
9. What happens when warm, moisture-laden air rises?
10. Explain the causes of the seasons.
11. How, and when, did our solar system and Earth originate?
12. What are some of the most dangerous natural hazards?
13. What approaches have scientists used in their attempts to predict future earthquake activity?
14. Describe some of the more important weather hazards. What, if anything, can humans do to prepare for such disasters?
15. After a major flood, should people whose homes and businesses were destroyed by the flood be allowed to rebuild on the floodplain? If so, under what conditions and who should pay for the rebuilding (insurance companies or the federal government)? If you feel people should not be allowed to rebuild, why not? Justify your answer.

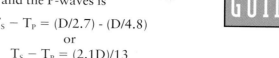

STUDY GUIDE

CALCULATIONS

1. When an earthquake occurs in the Earth's crust, caused by a rupture of the rock along a fault, seismic (shock) waves are released. Various types of seismic waves are generally recognized, including primary P-waves (compressional waves) and secondary S-waves (shear waves). P-waves travel faster than S-waves; in a certain context, P-waves may travel at a speed of 4.8 miles per second, and S-waves at a speed of 2.7 miles per second. Given this information, then the time, T_S (in seconds), that S-waves require to travel a certain distance (D) in miles from the epicenter (the point on the Earth's surface at, or immediately above, the rupturing fault that caused the earthquake; the epicenter is generally most damaged by the shock waves) to a seismograph can be expressed as follows:

$$T_S = D/2.7$$

Likewise, the time P-waves require to arrive can be expressed as

$$T_P = D/4.8$$

The difference in arrival time between the S-waves and the P-waves is

$$T_S - T_P = (D/2.7) - (D/4.8)$$
or
$$T_S - T_P = (2.1D)/13$$

Therefore, the distance (in miles) from the earthquake's epicenter to the seismograph is

$$D = 13(T_S - T_P)/2.1$$

Using this equation, if the difference in arrival time between S-waves and P-waves for a certain earthquake is 6 minutes and 23 seconds at a particular seismograph station, how far away is the epicenter of the earthquake from this seismograph station?

2. Using the equation in Question 1 (above), if the difference in arrival time between S-waves and P-waves for a certain earthquake is 4 minutes and 16 seconds at a particular seismograph station, how far away is the epicenter of the earthquake from this seismograph station?

ILLUSTRATION AND TABLE REVIEW

1. Study Table 5-3, The Mercalli Scale of Earthquake Intensity. Have you ever experienced an earthquake? If so, where would you judge it fell on the Mercalli Scale? If not, either talk to someone firsthand about an earthquake experience and try to place it on the Mercalli Scale, or research an earthquake and place it on the Mercalli Scale.

2. Referring to Figure 5-10, at which latitudes do high-pressure zones occur? At which latitudes do low-pressure zones occur?

WWW.ENVIRONMENT.JBPUB.COM

www.environment.jbpub.com

Connect to this book's web site: http://www.environment.jbpub.com/ The site features eLearning, an online review area that provides quizzes, chapter outlines, and other tools to help you study for your class. We've posted the answers to the book's odd-numbered questions. You can also follow useful links for more in-depth information. The site provides sources for the information in this chapter and recommendations for further readings.

Resource Use and Management

Once we see our place, our part of the world, as surrounding us, we have already made a profound division between it and ourselves. We have given up the understanding...that we and our country create one another, depend on one another, are literally part of one another; that our land passes in and out of our bodies just as our bodies pass in and out of our land...It is for this reason that none of our basic problems is ever solved.

Wendell Berry, poet, essayist, and social commentator

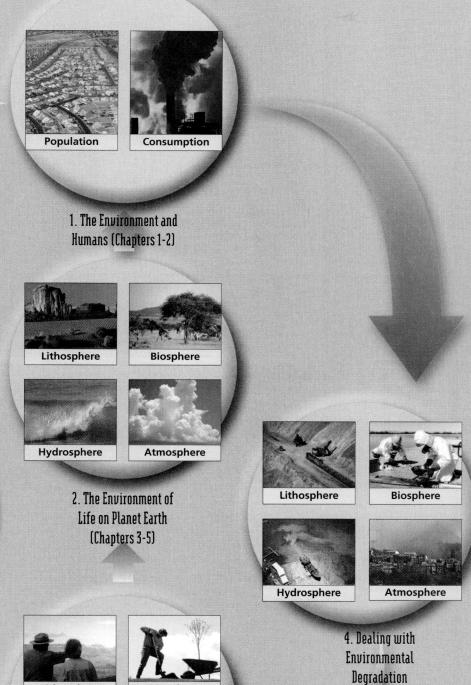

Population

Consumption

1. The Environment and
Humans (Chapters 1-2)

Lithosphere

Biosphere

Hydrosphere

Atmosphere

2. The Environment of
Life on Planet Earth
(Chapters 3-5)

Lithosphere

Biosphere

Hydrosphere

Atmosphere

3. Resource Use and
Management
(Chapters 6-13)

Lithosphere

Biosphere

Hydrosphere

Atmosphere

4. Dealing with
Environmental
Degradation
(Chapters 14-18)

Lithosphere

Biosphere

Hydrosphere

Atmosphere

5. Environmental Issues: Social Aspects and Solutions
(Chapters 19-20)

Since there are now well over six billion people on Earth, it is imperative that we learn to manage our resources in a sustainable manner, not utilizing resources faster than they can be replaced. As an example, instead of destroying old growth, mature forest for wood and pulp products—forest that, even under ideal circumstances, will not grow back for many human generations—we can raise fast-growing trees on tree farms (as shown in this photograph).

CHAPTER 6

Principles of Resource Management

In this chapter, we take an overview of the nature of resources and how to better manage them. As we illustrate in the diagram for Section 3 (page 141), natural resources from all four spheres may suffer from depletion. Depletion occurs when a resource is utilized faster than it is replaced by natural processes. An oil deposit that took millions of years to form, for example, may be extracted and burned in just a few years, or a species that took many thousands of years to evolve may be driven to extinction in a short period of time.

The ideal goal of much resource management is sustainable resource use, which seeks to conserve or slow down the rate of resource exploitation to the point where the resource can be replaced by nature. Tree harvesting and ocean fishing are examples. In other cases, if the resource is in immediate threat of disappearing, sustainable management must go farther and actually "preserve" the resource by stopping its current exploitation. Endangered species and ecosystems, for instance, must often be preserved immediately, or they will disappear in a few years.

Recall (Chapter 1) that stopping or slowing down the rate of resource use not only slows depletion but also usually has the added benefit of reducing pollution. A sustainable society would have "throughput" reduced to the point where exploited resources (inputs) are being renewed by natural processes, and pollution (outputs) can be safely absorbed by natural processes. Historically, resource management has not achieved this goal of sustainable use because of social, economic, and political pressures that emphasize rapid exploitation of resources.

The Need to Manage Resources

To some people, the concept of resource management reflects human arrogance. They argue that viewing the natural environment as a "resource" is a very narrow, anthropocentric (human-centered) approach to nature. Also, some question whether we are able to manage resources effectively. Many debates in environmental ethics revolve around whether humans have a right to "tamper" with nature, and if so, how much tampering is justified. Ethics aside, the assumption that humans are able, as a practical reality, to effectively manage nature is not demonstrated by past human history. David Ehrenfeld (in his book *The Arrogance of Humanism*) and many others have argued that humans have never effectively managed anything for very long.

In spite of these valid concerns, the need for resource management is inescapable. As human populations and technologies grow, inevitably pressures to exploit the environment will increase. Proper management can help minimize environmental damage. Careful planning of water use, for instance, could spare water for native ecosystems that would have been used for agriculture. Furthermore, management can help undo past damage. Elimination of alien (introduced) species, for example, is a common management strategy for some biological communities. Thus, although resource management is not an attractive concept in some ways, it is preferable to the alternative, which is uncontrolled resource exploitation. Global society will be facing many difficult environmental challenges in the future, and making informed decisions about how to use resources is essential for success.

What Is Resource Management?

Increasing resource use tends to be "bought" with increasing environmental costs. Mining, for example, tends to degrade the land more than tourism. But increasing resource use also tends to provide high short-term economic benefits. The history of the United States demonstrates that the economy has generally rewarded those entrepreneurs who most rapidly exploited natural resources. This process seemed justified to most people because modern society has traditionally ignored most environmental costs. For instance, **benefit-cost analysis (BCA)** is a method of comparing the benefits of an activity to its cost. When the benefits (calculated in a dollar amount) are greater than the costs (calculated in a dollar amount), there is said to be a net benefit to society. If you ignore most environmental costs, any short-term economic benefits of resource use will seem worthwhile: When environmental costs are very low (artificially), the benefit will be greater than the costs. For example, clear-cutting a virgin forest could yield enormous profits (benefits) in the short term.

A more realistic way to analyze resource use is to include the long-term economic benefits of not using them. When this is done, less resource use often translates to greater economic benefits. The total economic value of a rainforest, for instance, is usually greater if the forest is utilized over a long time span for tourism, pharmaceuticals, native foods, and other uses than if it is cut down for a one-time, short-term gain in lumber that leaves the forest unusable for decades or centuries. The total value of the rainforest is enhanced even more if extremely long-term environmental benefits are included, such as the value of the forest to future generations.

Resource Management: Preservation, Conservation, and Restoration

Proper resource management is based on the recognition that less resource use can lead to long-term economic benefits and reduced environmental costs. Such management, while recognizing that some resource use is unavoidable, thus seeks to minimize use where possible. There are three basic options that resource management can apply to minimize resource use: preservation, conservation, and restoration.

Preservation refers to nonuse. A "preserve," national park, or wilderness area is an ecosystem that is set aside and (in theory at least) protected in its pristine, natural state. **Conservation**

(input reduction) attempts to minimize the use of a natural resource. Use can be minimized through efficiency improvements, recycling or reuse, and substitution of other resources. Finally, **restoration** seeks to return a degraded resource to its original state. For example, attempts are being made to redirect the Kissimmee River of Florida into the original path that it followed before it was altered by humans. The goal of the rapidly growing field of restoration ecology is to attempt to return many ecosystems, such as tall-grass prairies and wetlands, to their original state.

A Brief History of Preservation, Conservation, and Restoration

When the national parks and national forests were being established in the early 1900s under Theodore Roosevelt, there was a lively debate over how much public land should be allotted to preservation and how much to conservation. By prohibiting most forms of resource use except tourism, national parks are an example of preservation (**Figure 6-1**). In contrast, national forests (and most other federal lands) permit timber cutting, mining, grazing, and other uses. The promoters of conservation won the debate, and most federal land has permitted these resource uses. In theory, such uses represent "conservation" because the resources are supposed to be closely managed in a way that minimizes damage to the land. This is rarely realized in practice, however, and many federal and state public lands have suffered extensive damage from overuse.

The abuse of public lands is often called the **tragedy of the commons.** A "commons" refers to public lands that are utilized by many people.

History shows that such public property is almost invariably abused unless there are strict controls or regulations implemented to protect it. Early American settlers, for example, found that shared lands were much more frequently overgrazed by farm animals because no single landowner had a vested interest in protecting the land. In addition to federal and state lands, many other environmental resources are "commons" and suffer from this problem; the oceans, atmosphere, and rivers, are a few examples. In all cases, they are polluted and abused because few people see them as "their" responsibility. In the case of public lands, many environmental groups have argued for setting aside more land as wilderness areas to be regulated by government agencies. Some groups, especially the Nature Conservancy, buy such land and set it aside as private preserves.

Restoration, the newest type of resource management, has become much more common in the last fifteen years. Aquifers, ecosystems, lakes, soils, and many other environments are being restored by a growing number of restoration specialists. Preservation and conservation are more cost-effective than restoration, which can be extremely expensive. Nevertheless, because so many environments are highly degraded from past abuses, restoration will undoubtedly become increasingly common. Restoration is most effective at the landscape level. It does little good to restore a lake acidified from mining runoff to its normal chemistry unless the surrounding land is treated to help reduce acidic runoff. Similarly, reintroducing wolves into a small forest is unlikely to be successful unless the natural ecosystem for the entire region is prepared to support them. A small forest is not large enough for a self-sustaining wolf population (Chapter 11).

Who Cares? The Many Values of Natural Resources

Up to this point, we have talked about the value of resources in economic terms. But as **Figure 6-2** shows, people can place at least five values, sometimes called the **five e's,** on natural resources. One of the e's, **ethical value,** is what philosophers call an **intrinsic value.** This is the value of a resource unto itself, regardless of its value to humans. Does a mountain have as much right to exist as you do? Does a worm? If you say "yes," then you place a high intrinsic value on these natural resources. Intrinsic values are **ecocentric** (environment-centered).

The other four e's are what philosophers call **extrinsic values** (Figure 6-2). These are values that are external to a resource's own right to exist, referring

Figure 6-1. The U.S. national park system represents an attempt to preserve nature. This scene shows the pristine beauty of Grand Teton National Park.

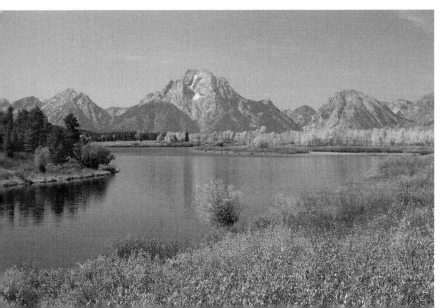

instead to its ability to provide something that is of value to humans. Such values are **anthropocentric** (human-centered). Extrinsic values are more utilitarian, or practical, than intrinsic values and therefore tend to be more widely discussed in political and economic debates on resource management. **Esthetic** (aesthetic) value is the value of a resource in making the world more beautiful, more appealing to the senses, and generally more pleasant. The value you place on a mountain hike in the cool morning air might be an example. Some people place no value on this and would pay nothing for it. Others find it indispensable. **Emotional** values include the value of a resource beyond sensory enjoyment. Some people, for example, develop very strong emotional bonds to certain natural areas or certain plant or animal species. This is sometimes called a "sense of place." Many psychologists consider natural habitats to be important for mental health, especially in children.

Economic values are directly involved with tangible products that can be bought or sold: food, timber, energy, and so on. As we previously discussed, society needs to focus more on long-term economic values, which actually provide more income over the long run. The value of resources for tourism, native fruits, or other sustainable products is ultimately much greater than the value of their destructive uses. **Environmental service** values are the value of resources in providing intangible services that allow humans (and other life) to exist on Earth. Plants help purify air and produce oxygen, and plant roots and soil microbes purify water; ultimately, all food relies on a variety of environmental services.

Some people place all five values on all environmental resources. How many values would you place on a forest? How many on a beach? Many people would place only economic values on such resources. Logging, mining, and other types of har-

vesting that destroy the resources are called **direct values**. Most environmental problems arise when resources are appreciated for only their direct value. Placing only direct values on natural resources artificially "discounts" their true value to society and to future generations. For example, four of the five e's—esthetic, emotional, environmental services, and ethical values—are not direct values. They represent what economists call **indirect values**, meaning that they are valued in ways that do not involve direct mining, harvesting, or other destruction of the resources. If resource prices incorporated both indirect values and long-term direct values, the prices would reflect the resources' true environmental cost.

While the direct value of resources provides immediate financial rewards, harvesting the direct value often destroys (1) the long-term economic value and (2) many or all of the indirect values. If society included all the indirect values of existing natural resources in the pricing of consumer products to reflect their **true environmental costs**, it would encourage less destructive harvesting of natural resources. A major problem is that indirect values such as ethical and emotional values are often subjective and very difficult to calculate (Chapter 19). Including the true environmental costs in the prices of products would encourage the marketplace to preserve and conserve resources. More sustainable uses of resources, such as recycling, organic farming, and ecotourism would become more profitable businesses. As long as only direct (short-term) values are considered, overuse and exploitation will be encouraged and rewarded.

Kinds of Resources

A **resource** is a source of raw materials used by society. These materials include all types of matter and energy that are used to build and run society. Minerals, trees, soil, water, coal, and all other naturally occurring materials are resources. **Reserves** are the subset of resources that have been located and can be profitably extracted at the current market price. Raw materials that have been located but cannot be profitably extracted at the present time are simply called resources, as are those raw materials that have not been discovered.

Renewable resources can be replaced within a few human generations. Examples include timber, food, and many alternative fuels such as solar power, biomass, and hydropower. **Nonrenewable** resources cannot be replaced within a few human generations. Examples include fossil fuels such as oil and coal, and ore deposits of metals. The phrase "a few human generations" is necessary because some re-

Figure 6-2 The five e's representing five potential values of environmental resources.

sources are replaceable on very long, geologic time scales. Oil, coal, soils, and some metallic mineral deposits may form again if we wait for thousands to hundreds of millions of years. However, these rates of renewal are so many thousands of times slower than the rates of use that, realistically, they are nonrenewable on a human scale. In contrast, solar energy is actually supplied faster than we can use it.

The definition of renewability is sometimes blurred. It may take centuries or even many thousands of years to replace very old groundwater in deserts, while groundwater in rainy tropical areas may be replaced in a few days. Thus, deep groundwater in deserts, sometimes called "fossil groundwater," is essentially a nonrenewable resource. Although nonrenewable resources cannot be replaced through natural processes on a human time scale, some (such as certain metals) can be recycled many times.

Patterns of Resource Depletion

There are two basic inputs from the environment: matter and energy. Matter constantly cycles through society and the environment, whereas energy primarily has a one-way flow. Because of this difference, matter and energy are depleted differently.

How Matter Resources Are Depleted

Matter resources are depleted by being dispersed. Ore deposits are unusually concentrated deposits of minerals that are normally found in more dilute form in the Earth's crust. When we mine and process the ore into metals to build cars and other refined products, the atoms may be eventually dispersed (such as when gears wear down from grinding against one another) or lost to further human use when we dispose of the products in landfills and elsewhere. Of course, the **urban ore** of a landfill may later be mined for its metal content. Similarly, rapid erosion depletes soil not because the nutrients and minerals in the soil are destroyed, but because the soil is dispersed, ultimately into the oceans.

These examples are nonrenewable matter; when dispersed, molecules of metals and soils will stay dispersed unless we spend much energy and money to reconcentrate them. In the case of renewable matter resources, dispersal still occurs, such as when we build houses from timber, but we can regenerate the timber relatively quickly. Renewable matter resources are often biological resources that can be regrown.

How Energy Resources Are Depleted

Energy has a one-way flow through society because it is transformed to an unusable form, "waste heat," when we use it. Energy resources are therefore depleted when they are transformed this way. This is a key difference from matter, which can often be recollected and reconcentrated if society has enough cheap energy. In contrast, once energy is transformed, it is lost forever; waste heat can never be reconcentrated. If we burn oil or coal to release their chemical energy to drive an engine, that energy can never be reused.

We do have a major source of renewable (renewable in the sense of continuously entering the Earth system) energy, however: the Sun. This source of renewable energy could potentially keep society running for many millions of years. Examples of the Sun's energy include direct solar power, biomass, hydropower, and wind (Chapter 8).

Bubble Pattern of Depletion

Unsustainable use of many resources exhibits a bubble pattern of depletion. The best-known example is the so-called **Hubbert's bubble** of oil depletion. This was named after M. King Hubbert, who accurately predicted the bubble pattern of oil depletion in the United States. Since the 1950s when his predictions were first made, they have proven to be strikingly accurate. U.S. oil production peaked around 1970 and has been declining since, as the richest reserves are steadily depleted (**Figure 6-3**).

The bubble pattern has two main causes: exponential exploitation and exponential depletion. Because both use and exhaustion are exponential, they tend to form a mirror image. The left, or ex-

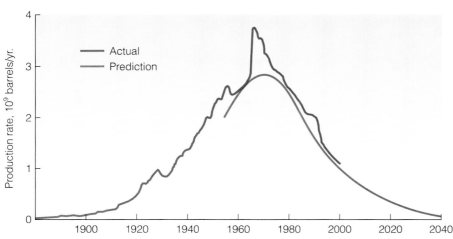

Figure 6-3   Predicted U.S. oil production by M. King Hubbert and actual U.S. oil production until 1999. Both curves peak around 1970. (*Source*: U.S. Department of Energy.)

ploitation, side of the bubble is exponential because resources are exploited very quickly once society discovers their utility. But because all resources on Earth are finite, limits to growth eventually occur, and demand begins to exceed supply. During this time, society usually intensifies its efforts to obtain more of the resource through further exploration and increased technological applications. However, these efforts usually soon encounter what economists call the **law of diminishing returns**, meaning that increasing efforts to extract the resource produce progressively smaller amounts. The result is shown in the right, or depletion, side of the bubble. Production declines exponentially because the most easily extracted concentrations of the resource become exhausted.

As supplies of the resource decline, prices rise, sometimes leading to unemployment and other unpleasant changes. Historically, society has responded to the increase in resource prices by switching to another resource. In the 1800s, for example, England switched from wood to coal as an energy source when the forests were decimated. Then, in the 1900s, the English economy switched to oil as its primary energy source because it became cheap and abundant. This pattern is common, whereby societies have tended to switch from one unsustainable resource to another. The way to break this "cycle of unsustainable use" is to switch to sustainable uses.

Problems with Past Resource Management

Society can respond to a diminishing supply of a resource in two ways: (1) intensify efforts to extract more of the resource or (2) reduce the need for the resource. The first response has typically been, and still is, much more common. The usual result was a bubble pattern as depletion accelerated when supplies ran low. The mining of lower-grade, high-volume materials also led to pollution problems. Two influential concepts were advanced to justify these past intensification efforts to extract more of a diminishing resource.

Net Yield of Nonrenewable Resources
Intensified extraction of nonrenewable resources is based on the concept of **net yield**, which holds that a resource can continue to be extracted as long as the resources used in extraction do not exceed the resources gained. Because energy prices have been relatively low over the last 90 years, miners have simply switched to lower-grade deposits when high-grade deposits were depleted (**Figure 6-4**). For example, at the current market price of copper, copper ore that is as low as 0.5% copper content in the rock can profitably be mined. Switching to lower-

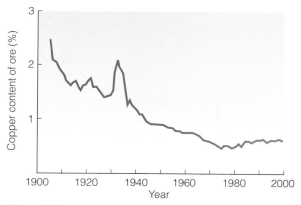

Figure 6-4   Since 1905, the quality of copper ore mined in the United States has declined from 2.5% to about 0.5%. (*Source*: U.S. Bureau of Mines.)

grade deposits greatly increases the available supplies because nearly all nonrenewable resources are characterized by an inverse quality curve: Higher-grade deposits of coal, oil, minerals and other resources are much rarer than lower-grade deposits.

The pursuit of net yield has incurred great environmental costs by accelerating both depletion and pollution. Depletion accelerates because increasingly lower grade ores must be mined and more energy and other resources must be expended to extract the ore. Pollution accelerates because increasing amounts of waste rock are produced when lower-grade ores are mined (**Figure 6-5**). For exam-

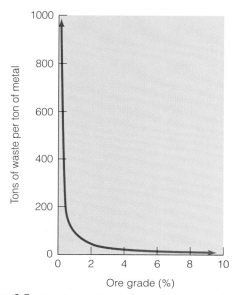

Figure 6-5   For all ores, the amount of waste rock and other waste materials increases exponentially as ore grade decreases. (*Source*: D. H. Meadows, D. L. Meadows, and J. Randers, *Beyond the Limits* [Post Mills, VT: Chelsea Green, 1992], p. 87. Reprinted from *Beyond the Limits*. Copyright ©1992 by Meadows, Meadows, and Randers. With permission from Chelsea Green Publishing Company, White River Junction, Vermont.)

CASE STUDY 6-1

How to Exceed Maximum Sustainable Yield: Overfishing the Oceans

Visit www.environment.jbpub.com
for more information.

In 1981, the United States declared a 200-mile (322 km) boundary around its shores, banning most foreign trawlers that had fished U.S. waters. The ban was an attempt to save the beleaguered New England fishing industry. The measure worked until Americans themselves began overfishing. Encouraged in part by a federal loan program, they began building

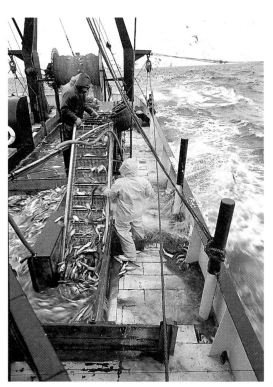

Figure 1   New England fishermen are struggling to make a living because of the decline of many species and government restrictions.

more boats, equipping them with powerful fish-finding electronics. They lobbied to remove quotas on catches. And, for a while, they caught a lot more fish. But the trawler catch in New England peaked in 1983 and has since generally continued to fall (**Figure 1**).

The story is the same throughout the world. Stocks of flounder and haddock are near record lows. The cod population is down. Bluefin tuna and swordfish have been depleted. There have been booms and busts before, but scientists say that this time it is different. The fleets are so big and the technology so advanced that fish no longer have anywhere to hide. The global fishing industry has more than doubled in the last three decades. Dozens of species have been fished to commercial extinction and the rate of growth in marine harvests has plummeted to nearly zero. In 2002, the United Nations Food and Agriculture Organization warned that about one-quarter of the world's marine resources were either overexploited or already depleted. About half of the marine fishing grounds were classified as "fully exploited," meaning that increased fish production from these regions is unattainable.

In addition, the composition of global catches has changed. Fishers respond to reduced population size by catching huge numbers of juvenile fish. Many of the fish have not had a chance to spawn, which accelerates the long-term decline. There is also a trend toward catching fish lower on the "food chain." As fishers deplete large, long-lived predatory species, such as cod, tuna, shark, and snapper,

ple, the Goldstrike mine of Nevada, the largest gold mine in the United States, moves 295,00 metric tons of rock and waste to produce just over 100 pounds (45 kg) of gold. In 1991, globally 898 million metric tons of copper ore was mined to produce about 8 million metric tons of copper. In the past, some geologists worried that eventually humans might run out of ores to mine. Many environmentalists now feel that the bigger issue is mining-related environmental degradation. Mining may have to be curtailed due to the pollution generated by metal extraction rather than because the ores are actually depleted.

Maximum Sustainable Yield of Renewable Resources

Another very influential concept in resource management has been the **maximum sustainable yield (MSY)**, which holds that the optimum way to exploit a renewable resource is to harvest as much as possible up to the point where the harvest rate equals the renewal rate. As an example, a person could withdraw as much groundwater as needed up to the point where the withdrawal rate equaled the recharge rate from rainfall. Taking less could be considered underutilization and taking more would lead to depletion.

they move down to the next level—to species that tend to be smaller, shorter-lived, and less valuable. As a result, fishers worldwide now fill their nets with plankton-eating species, such as mackerel, jacks, and sardines, and invertebrates like squid, oysters, mussels, and shrimp. Initially, this transition brings new bounty: At higher trophic levels, fishes are larger, but there are fewer of them. At lower levels, the species are smaller but more plentiful. Relieved of predatory pressures and competition for food, the smaller species are free to fill in the empty niche once occupied by their predators. This is one reason why global fish catches have not declined even more than they have in the past 25 years, despite severe overfishing.

Some people in the fishing industry say they can't back off. They have to keep the boats going nonstop to make enough to survive. But others and conservationists insist that current measures, such as minimum net mesh sizes and occasional closures of overfished waters, simply are not enough. Strict catch quotas, trip limits, even moratoria on new boats are needed, or more people will be forced out of the business in the long run. Legislation has been introduced to reduce the size of the New England fleet by buying out vessels with money from a tax on the diesel fuel the fishing boats use. Some have suggested that the government should subsidize the fleets by meeting mortgages for those who do not fish a certain number of days. And severe restrictions are being placed on fishing in some areas, such as the once-fertile Georges Bank ecosystem off the coast of Massachusetts. Some way of rationally allocating fishery resources is needed to guarantee that the industry remains intact.

This is a classic example of the "tragedy of the commons." The oceans are used by many individuals and nations. Thus, many nations must agree on laws that govern access to ocean resources, and, perhaps more difficult, these laws must be enforced in ways that prevent overfishing.

One solution is to set aside marine "reserves": strategically located areas where fishing is prohibited to rebuild fish populations and restore the marine ecosystem. But 95% of the world's fishers are from developing countries and work in traditional fisheries where implementation of no-fishing zones has met with tremendous resistance among people economically dependent on fishing.

The successful establishment of marine protection programs requires that the immediate and long-term economic benefits for the fisheries be secured through incentives. Co-management of marine resources by governments, private fisheries, and communities should encourage and reward fisheries in their effort to sustainably manage marine resources.

Questions

1. How has the composition of catches changed through time as overfishing occurred?
2. How does overfishing relate to the tragedy of the commons? What are some possible solutions?

MSY has been especially influential in the management of renewable biological resources such as commercially important fishes and wildlife. The basic principle was that fishes, game, and other populations could be harvested up to the point where the population's ability to reproduce itself was impaired. For instance, individuals could have trouble finding mates if the population was overharvested. MSY actively encourages limited harvesting because it holds that if population abundance becomes too high, further population growth is inhibited by crowding and competition (**Figure 6-6**). Thus, MSY aims at a balance between too much and too little

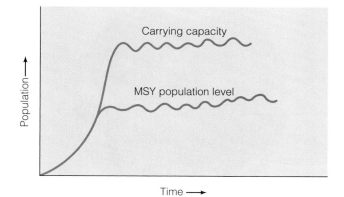

Figure 6-6 Maximum sustainable yield (MSY) occurs at population levels around one-half of the carrying capacity.

harvesting to keep the population at some intermediate abundance.

Although it is widely practiced by state and federal government agencies regulating wildlife, forests, and fishing, MSY has come under heavy criticism by ecologists and others for both theoretical and practical reasons. An important theoretical shortcoming is that MSY does not take large environmental fluctuations into account: A bad winter or some other natural catastrophe could reduce the population unpredictably. More important perhaps are the many practical problems with MSY. Calculating the point at which population growth begins to slow from competition is very difficult. Indeed, in many cases such as marine fishes, just estimating the population size can be difficult. Consequently, estimating MSY with certainty is virtually impossible, although more sophisticated methods, such as refined statistical sampling models for estimating population size, are improving the calculations. Still, the decline in commercial fishing in many areas indicates that humans are harvesting far more than the MSY.

Despite these and other problems, many agencies continue to rely on MSY. A variation on the MSY concept known as the **optimum sustainable yield (OSY)** is gaining popularity with some resource managers. OSY states that the optimum harvestable rate for a renewable resource must consider many factors, not just the maximum yield. How will harvesting affect other species in the ecosystem? How will it affect other human uses of the ecosystem, such as recreation? When such additional benefit-cost questions are included, managers usually find that the harvests recommended by OSY are less than the MSY.

Conservation: Reducing the Need for Resources

The basic problem with net yield and maximum sustainable yield is that their main goal has been to maximize resource use. Their focus was on short-term economic gain. But long-term economic and other gains (such as esthetic, emotional, and environmental services—for instance the service that a forest provides in producing oxygen and removing carbon dioxide from the atmosphere) are greater if the emphasis is shifted to reducing resource use. This can be done in three ways: preservation, conservation, or restoration of what has been degraded. Instead of maximizing resource use, the focus should be on accomplishing more with the resources that are used.

Conservation, or input reduction by lowering resource use, will thus be a fundamental part of a sustainable society. Recall from Chapter 1 that conservation (1) slows depletion of resources, (2) reduces pollution by slowing the flow of matter and energy (throughput) through society, and (3) saves money. For example, burning less coal by increasing a power plant's efficiency not only saves coal but produces less acid rain and other pollution. It is also often cheaper to design power plants to burn less coal than to pay for all the pollution-control devices needed to trap the air pollution in the smokestack and then dispose of the trapped pollutants. Until recently, supplies of most resources were so abundant, especially in the United States, that little attention was paid to input reduction; as a result, depletion rates were high, and pollution was controlled by scrubbers and other forms of output, or "end-of-pipe," reduction.

Ways to Conserve: Efficiency, Recycling, and Substitution

Figure 6-7 shows three basic ways to conserve or reduce the need for a resource: efficiency improvements, reuse/recycling, and substitution.

Efficiency improvement is generally most effective and economically cheapest because many technologies and activities are wasteful and inefficient. Different resources will require different methods of input reduction, however. It is very difficult, for example, to find affordable substitutes for water in many of its uses, such as agriculture and drinking, so increased efficiency and reuse/recycling of wastewater are common.

Efficiency Improvements.   Efficiency improvements occur when the same task is accomplished with fewer resources. An example would be lighter, more fuel-efficient cars to conserve fuel and

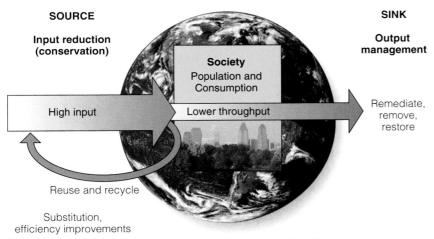

Figure 6-7   Input reduction (= conservation) can be achieved by efficiency improvements, reuse/recycling, and substitution. All three methods reduce the flow of materials and energy (throughput) into and out of society, slowing both depletion and pollution.

building materials. Between 1973 and 1992, the average efficiency of cars made in the United States doubled from 14 to 28 miles per gallon (6 to 12 km per liter). Such cars perform the same tasks as less fuel-efficient cars but use fewer resources in doing so. As another example, about two-thirds of the water used in irrigation is lost to evaporation. Using "micro-irrigation," where water is carried by pipes and sprayed through small holes, reduces water loss to less than 20%.

Japan in particular and the European countries are leaders in devising technologies that improve efficiency. The **energy intensity index**, which equals energy consumption/gross domestic product, measures how much energy is used to produce the same unit of wealth. The index for the United States in the late 1990s was about 11.0 billion BTUs consumed per $1 million GDP. In contrast, the index for Japan was about 3.8 to 6.7 billion BTUs per $1 million GDP (see Table 8-2 in chapter 8). Japan uses much less energy to produce a given amount of wealth, from manufacturing cars to heating hotel rooms. Material efficiency indexes, which indicate usage of other resources, show a similar pattern of waste for the United States. For example, Japanese technologies convert about 70% of raw timber into finished wood products, while in the United States only about 50% is converted.

The relative inefficiency of resource use in the United States is a direct result of its historical endowment of abundant resources. Costs for energy and other resources were low, so efficiency was not encouraged. Ultimately, however, inefficiency is costly to any economy. According to many experts, energy inefficiencies account for at least 50% of U.S. energy use. With such inefficiency, money invested in energy-saving technologies would more than pay for itself by reduced energy costs. The Worldwatch Institute estimates that U.S. electricity could be reduced by 70% through efficiency gains, at no net cost because energy savings would equal the investment needed to make the changes. Increased efficiency leading to conservation of minerals and other resources would achieve similar savings.

Besides the economic savings, environmental savings from efficiency increases are also enormous. Less depletion saves more resources for future generations. Less resources extracted means less degradation of the land. Pollution control from reduced processing and usage can also be significant. To continue with the energy example, the pollutants released by the burning of fossil fuels are reduced when efficiency is increased. It has been estimated that national investments in energy conservation would not only save from $10 billion to $110 billion per year, but would reduce greenhouse gas (global warming) emissions up to 40%.

Reuse and Recycling. Reuse occurs when the same resource is used again in the same form, such as refilling soda bottles. Recycling is similar, but the resource is not reused in the same form. The soda bottles, for instance, may be remelted to make new bottles or other glass containers. Like improved efficiency, reuse/recycling reduces depletion of resources and pollution from resource extraction and use (Figure 6-7). Reuse/recycling is useful in reducing solid waste, which is important because landfills are being permanently closed at the rate of about two per day. Many areas of the United States, especially in the Northeast, are running out of space and few new landfills are opening up. Because about 33% of U.S. garbage consists of packaging, **precycling**, which is the reduction of packaging material by manufacturers, is becoming more important. Concentrated foods, for example, can often be packaged in smaller containers. Precycling is not recycling, but is conservation by increased efficiency; the same task is accomplished, but fewer resources are used. Despite such efforts, the amount of solid waste in the United States continues to grow (Chapter 18).

The technological and economic aspects of recycling can be very complex. A simplified recycling scheme begins with **virgin resources,** which are the original natural resources being extracted. Both extraction such as mining and processing during product manufacture usually create pollution. The **recycling loop** begins before the purchased product is discarded; the discard is reprocessed into the same or perhaps another product. Note that, unlike the reuse loop, the reprocessed product does not have to be the same as the original. For example, while paper is often recycled into more paper, car tires are often recycled into road asphalt, shoes, playground structures, and many other products. As long as the discard is being used in place of a virgin resource, recycling occurs. Unfortunately, some manufacturers claim their products are recycled even though they do not meet this criterion of being used instead of a virgin resource (see Case Study 6-2, Is This Recycled or Not?).

The recycling loop is closed when someone buys a product containing recycled material (**Figure 6-8**). This slows depletion of virgin resources and reduces pol-

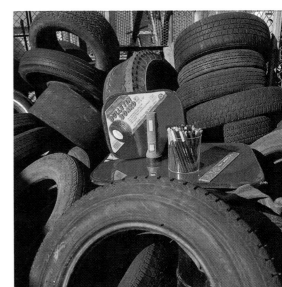

Figure 6-8 Pens and many other products can be made from recycled tires.

CASE STUDY 6-2

Is This Recycled or Not?

Visit www.environment.jbpub.com
for more information.

Since Earth Day 1990, many products touted as "environmentally friendly" have entered the market. Sometimes called green marketing or "greenwashing," promoting products as green can be very successful because of increasing concern over the environment. Unfortunately, green marketing has been plagued by misunderstandings and even false claims. A major reason is the lack of clear definitions. For example, what exactly does it mean when a product is advertised as being biodegradable or recycled (**Figure 1**)?

The definition of biodegradability came to widespread public attention when certain "biodegradable" trash bags were found to degrade only under laboratory conditions, and not under those found in landfills. As a result, the company had to remove this claim, and the term biodegradable is now reserved for products that decompose under natural conditions. Another problem is that decomposition must involve true chemical destruction of

the product into harmless components and not just reduction into smaller particles, such as smaller plastic pieces. But even if we define biodegradable as chemical decomposition in natural conditions, there is still plenty of room for vagueness. What are natural conditions? What if decomposition takes five years? Ten years? One hundred years?

The definition of a "recycled" product is perhaps even more difficult to specify. Many kinds of products are recycled in many ways, and the remanufactured products are often only partly composed of recycled material. Yet buying recycled material is essential to close the recycling loop. Paper recycling is a key illustration of these difficulties because it is so important: Paper is the largest component of municipal solid waste, comprising over 40% of landfill volume. But the United States has no national legal standards defining recycled paper. As a result, some paper that is sold as recycled actually contains only a tiny fraction of recycled paper. Furthermore, even this recycled paper can be preconsumer, meaning that it is wastepaper generated by papermills, such as cuttings, scraps, and flawed batches. Such preconsumer waste is often not as desirable as recycling postconsumer waste, which is discarded by consumers; recycling postconsumer waste is more helpful to the environment because that is the source of so much landfill waste.

The lack of clear meaning for product environmental claims is harmful to nearly everyone. Consumers suffer when they pay extra for a product that is not what they think; companies that really try to produce environmentally friendly products suffer when they are undersold by companies that market more harmful products as environ-

Figure 1 Three types of recycling symbols commonly used in the United States. (a) This symbol simply means that the object is potentially recyclable, not that it has been or will be recycled. (b) This symbol indicates that a product contains recycled material, but it does not indicate how much recycled material is in the product (it could be only a very small amount). (c) This symbol states explicitly the percentage of recycled content found in the product.

lution in two basic ways. Most obviously, it reduces the amount of solid waste that would have been discarded into landfills, incinerators, and other means of disposal. In addition, recycling reduces the pollution that would have been produced from the extraction of the virgin resources. Unfortunately, the recycling loop is often not closed because, even though people are willing to sort and return recyclable waste, there is often little consumer demand to purchase the recycled product. The reasons for

this are usually economic: When products made from virgin resources are cheaper, the loop goes unclosed. If the costs of products made from virgin resources were increased to reflect their true cost to the environment, the price differential would disappear, and the loop could be closed.

Substitution. Substitution occurs when one resource is used instead of another. Substitution can also help reduce both depletion and pollution probmentally friendly; and the environment suffers when we use harmful products when we think we are not.

The solution to such problems is to create clear definitions for environmental marketing claims. Those who falsely make such claims can then be legally prosecuted, or at least conscientious consumers can avoid the products. Currently, some standards have been established by state and local laws and nonbinding agreements between consumer groups and companies. Unfortunately, these vary from place to place. Therefore, support is growing for a single set of federal guidelines that would apply all over the United States. For example, the Environmental Protection Agency (EPA) currently has a paper guideline specifying that recycled paper should contain at least 50% recycled fiber. But companies are not required to use this definition, which also omits how much of the recycled fiber must be postconsumer. In 1992, the EPA and the U.S. Office of Consumer Affairs and the Federal Trade Commission joined forces in developing voluntary guidelines for the use of environmental terms on product labels; this work is ongoing.

Until enforceable federal regulations are established, consumers and private organizations must educate themselves about what is being marketed. Two private nonprofit organizations, Green Cross Certification Company and the Green Seal, test products and affix "seal of approval" logos if the product passes the tests. Similar organizations also exist in other countries, such as the West German Blue Angel program, Canada's Environmental Choice, and Japan's Eco-mark. Unfortunately, certification by the Green Cross and Green Seal is largely voluntary: Many companies never submit products for certification.

Consumers should also be aware that many recycling logos on boxes and cans don't mean very much. For example, consider the familiar logo with arrows in a triangle that appears on many food and paper products. The arrows alone indicate that the product is recyclable; arrows within a circle mean "made from recycled materials." But this logo alone does not indicate the amount of recycled material, how easily recycled the product is, or other key information. For example, many products are recyclable but are never recycled because the manufacturer does not repurchase its own recycled material. The recycling loop remains "unclosed."

If you want to be a green consumer, here are a few basic rules: avoid needless shopping (conservation is always the "greenest" activity you can do); avoid vague environmental claims such as "environmentally safe"; where possible, try to buy products certified by the Green Cross or Green Seal; seek out minimal packaging and reusable containers; and look for specific information that a company may provide such as "made from 30% postconsumer recycled fiber" (this is preferable to a product made of, say, 10% postconsumer recycled fiber).

Questions

1. What does "biodegradable" really mean? Do you think that paper is more biodegradable than standard (petroleum-based) plastic? Why?
2. What are some basic rules for being a green consumer? How often do you really try to follow these rules when shopping?

lems. It helps with depletion because when one resource is being depleted, a more common substitute can be used at a cheaper price. For instance, aluminum, a very common metal in the Earth's crust, can sometimes be substituted for the much rarer and more expensive copper in making alloys, equipment, and other uses. Substitution helps with pollution problems when the extraction, processing, or disposal of the substituted resources produces less pollution. For example, many plastics last for 50 to 100 years in the environment before they significantly decompose. Also, plastics are made of a nonrenewable resource, petroleum, whereas trees are renewable. Such considerations have led to the substitution of paper for plastic in many items, such as drinking cups and food containers.

Although it can be useful in reducing resource depletion, substitution is often less desirable than efficiency improvements or reuse/recycling. Instead of reducing overall resource depletion, substitution often simply switches depletion from one resource to another. This can be satisfactory if the new resource is renewable, as in paper, or very abundant, as with glass made from sand. Also, substitution often does not reduce pollution, solid waste, or other output problems. For instance, the use of paper products offers many environmental advantages over plastic, but it may do little to solve landfill space problems. Indeed, paper is already the largest component of city garbage.

Resource Economics

All environmental problems are closely intertwined with economic causes. This is especially true of resources because their extraction and use are directly determined by their profitability.

Resource Overuse: Ignoring Environmental Costs

A basic reason why resources are overused is that they are too cheap. The price paid for metals, timber, petroleum, and many other virgin natural resources does not reflect their true environmental costs. These true costs would include esthetic, emotional, environmental services, ethical values, and long-term economic values. The price paid by consumers for natural resources now usually omits many of these. The price of metals, for instance, rarely incorporates many of the environmental costs of mining: costs of water pollution on nearby aquatic ecosystems, esthetic loss from a cratered landscape, and so on. In Chapter 1, we saw how many environmental economists suggest that green fees, such as taxes, can incorporate environmental costs into natural resource use. Mining companies, manufacturers, and consumers all respond to higher prices by increasing efficiency, reusing/recycling more, and substituting more.

Another way to incorporate long-term and indirect values is to use basic market principles: Consumers can demand sustainable products such as those made with recycled materials. When the profit made from such sustainable activities exceeds the profit made from the destruction of resources, sustainable activities will be rewarded. At present, this is rarely the case, and unsustainable activities are more profitable.

Economics of Recycling.   Recycling is an excellent example of the key role that economics plays in promoting resource conservation. Although about 80% of U.S. household garbage can be recycled as compost, aluminum, paper, and many other components, only about 13% of municipal solid waste

was being recycled in 1992 because the economic incentives were lacking; by the end of the twentieth century approximately a quarter was being recycled. Returning used items for collection is not enough. The recycling loop is closed only if collected items are used again to make a product that is repurchased. The loop will often go unclosed if products made from recycled materials cost more than products made only from virgin resources. Consumers will often choose to buy the cheaper product, and the waste people thought they returned for recycling will be taken to landfills because no one will buy it.

Widespread recycling is unlikely to occur until it is economically feasible, even though the number of U.S. communities requiring recycling has risen dramatically such that currently nearly half of all U.S. communities now have recycling programs in place. As many communities have learned, simply passing laws requiring recycling is not enough. Seattle, Washington, which resolved to recycle 60% of its garbage by 1998, is a good example. By 1993, Seattle had the highest rate of recycling of any major U.S. city, with 42% of its garbage being recycled and 90% of all single-family homes participating. By charging an extra fee for each barrel of garbage left out for pickup, the city encouraged people to sort their waste and deposit it in recycling containers. Unfortunately, the city has had difficulty closing the loop on some materials, such as glass and paper, which were sometimes sent to landfills. Even though the average price of used newspapers dropped from $20 to $5 per ton (0.907 metric tons) in 1991, it was still often cheaper to make paper from trees, so no recycler would buy them. More recently, however, the price of paper has risen and the demand for recycled paper has increased especially since the federal government now requires all its offices to use recycled paper.

Why is recycled material sometimes more expensive than virgin material? A major reason is the cost of energy, including the energy of human labor. The cost of **embodied energy**, or the energy used in producing a product, is often the main factor determining the product's retail cost. The energy consumed in recycling some materials is much less than the energy consumed in processing virgin material. For example, mining and processing aluminum and many other metals is very energy-intensive. Because recycling aluminum uses much less energy, there is a strong cost incentive to reprocess it and close the loop. In everyday terms, recycling just one aluminum can saves enough energy to run a television set for three hours. The aluminum recycling industry is one of the fastest growing industries in the United

States; it earns approximately $1 billion annually and recycles an astounding 50 billion aluminum cans per year, equal to 200 cans per U.S. citizen. This example shows how both the environment and jobs can benefit from recycling.

Recycling other materials, such as newspaper and glass, saves considerably less energy than recycling aluminum, however. When the costs of human energy, such as collecting and sorting the newspaper and glass, are added on, the slim profit margin may disappear, and the loop may sometimes be unclosed as in the Seattle example.

The problem of closing the recycling loop has many possible solutions. One option is to appeal to consumers to purchase recycled products, even if they cost more. For instance, some students buy recycled notebook paper even if it is more expensive because they want to encourage recycling. To encourage such buying on a wide scale, society can levy higher "green taxes" on virgin resources to boost the price of products made from them. If the taxes are high enough, the recycled product will be cheaper, and the loop will be closed through consumer demand. Lastly, human ingenuity and innovation may devise new recycling processes that lower costs or new products that appeal to consumers. For example, the growing problem of recycling used tires has led to dozens of products that incorporate them, from sandals and tennis shoes to asphalt. When products that people want are created, there is an economic incentive to close the loop. Many businesses are finding new uses for plastics, such as stuffing for ski jackets and even "lumber" to build benches, fences, and playground equipment.

Jobs and Life in a Sustainable World

A sustainable world meets today's needs without reducing the quality of life of future generations. Modern resource exploitation is not even close to being sustainable. **Figure 6-9** illustrates how the consumption of virgin raw materials such as minerals has increased much faster in the United States than the population. Since 1900, materials consumption increased about six times faster than population growth. The average American today consumes six times more resources than the average American in 1900.

Figure 6-10 shows that many jobs in the U.S. high-throughput society are concentrated at both ends of the flow of ma-

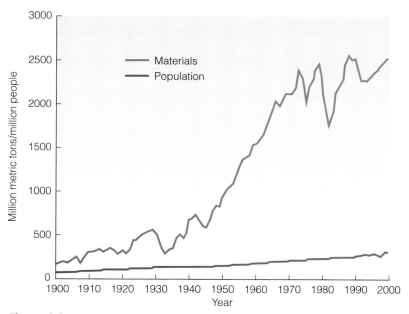

Figure 6-9   Consumption of materials in the United States has grown much faster than population. (Note that one metric ton = approximately 1.102 tons.) (*Source*: U.S. Bureau of Mines.)

terials through society: in resource extraction and waste cleanup. This leads to unsustainability because resources are being used at a vastly higher rate than they are being replaced by natural processes.

Figure 6-10 also shows what an economy could look like in a low-throughput, sustainable world.

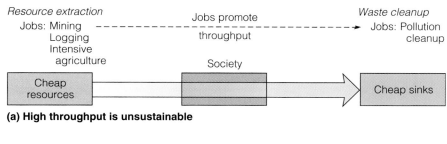

(a) High throughput is unsustainable

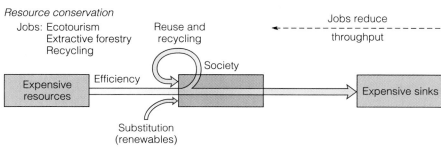

(b) Low throughput is sustainable

Figure 6-10   (a) Cheap resources and environmental sinks promote an unsustainable, high-throughput economy with jobs concentrated in resource extraction and waste cleanup. (b) Expensive resources and environmental sinks promote a low-throughput, sustainable economy with jobs concentrated in resource conservation (input reduction) activities.

CASE STUDY 6-3

Measuring Consumption: What is Your Ecological Footprint?

We all depend on the natural environment for resources such as food, energy, the absorption of waste products and other life-support services. Sustainability requires that nature's productivity not be used more quickly than it can be renewed, and that waste is not discharged more quickly than nature can absorb it. The ecological footprint, developed by Mathis Wackernagel and William Rees, is an increasingly popular way of measuring how sustainable our lifestyles are.

The ecological footprint measures how much land is required for people to support their lifestyle. How much land, for example, is required to grow the crops for the food you eat, to cover the housing you occupy, the roads you use, and the waste that you produce? As you might expect, the footprint varies among nations, and generally increases with rising wealth and especially consumption. The ecological footprint of the average Canadian, for instance, adds up to 4.8 hectares (11.9 acres). This is the total amount of land required for food, housing, transport, consumer goods and services. Energy is a large component of the footprint: Some 2.9 hectares (7.2 acres) are necessary for the long-term provision of a biological substitute for fossil fuels. The second largest component at 1.1 hectare (2.7 acres) is agriculture for food supply and consumer goods. Forestry takes up 0.6 hectare (1.5 acres) to supply the plant resources for housing and consumer goods. Finally, the built environment takes up 0.2 hectares (.5 acre) for housing and transport.

The average U.S. citizen has the largest footprint in the world, exceeding 5 hectares (over 12.4 acres). In fact, if everyone on earth lived like the average U.S. citizen, it would require at least three earths to provide all the material and energy she or he currently uses. Estimates show that the ecological footprint of today's global consumption in food, forestry products, and fossil fuels alone might already exceed global carrying capacity by roughly 30%. About 3/4 of the current consumption goes to the 1.2 billion people who live in affluence, while 1/4 of the consumption remains for the other 5.1 billion people.

Of course, even within nations, personal lifestyle choices have a strong influence on a person's ecological footprint. Do you drive a large car? How far do you commute? For example, compare how much ecologically productive land is necessary to commute by bicycle, bus, or car. Most of the car's land is required to absorb CO_2. Most the biker's land is required to provide the extra food for quenching the biker's hunger. Obviously, the biker has a much smaller footprint for travel needs. The same is true for many other choices such as housing, food, energy and water consumption. Someone who takes longer showers, for example, is increasing his or her ecological footprint compared to someone who does not. In general, as personal consumption increases, as it has been in wealthy nations over the past decades, so does one's ecological footprint.

Questions
1. What, exactly, does the ecological footprint measure? How can it be reduced?
2. Must wealth always produce a larger ecological footprint? Explain.

Figure 6-11 "Solar farms," such as this one in California, can potentially provide many high-tech careers.

Virgin resources would be more costly, reflecting their true long-term value and promoting highly efficient use of resources, reuse/recycling, and substitution of more abundant and more renewable resources. Jobs would be concentrated at the input end of the throughput flow and would involve sustainable resource management activities such as ecotourism, recycling, and the renewable energy industry (**Figure 6-11**). These, and similar activities, reduce throughput and slow nonrenewable resource depletion and pollution. In contrast, in a high-throughput society today, jobs actively promote nonrenewable resource depletion and pollution.

In many cases, sustainable activities produce more jobs than unsustainable ones. Extracting re-

TABLE 6-1	Some Suggestions for Creating a Sustainable Society

Environmentally harmful (unsustainable) activities can be solved in many ways. All of these solutions use efficiency, recycling, and substitution promoted by economic incentives to reduce throughput.

Harmful Activity	Example	Solutions
Extraction	Mining high-grade ore and then moving on to a new site because the land is artificially cheap, while ignoring lower-grade ore on the already-disrupted site	Higher land prices (through elimination of subsidies and addition of full costs of environmental disruption) would increase incentives to use more efficient extraction technologies, reducing the area of land disrupted.
Manufacture	Making paper from 90%–100% virgin wood fiber	Most paperboard, paper packaging, and office paper can be made with less than 50% virgin input with no loss of quality, potentially saving millions of trees each year.
Product design	Designing discount products—from umbrellas to televisions to houses—that compete for low retail prices but do not last	Design emphasizing durability and reparability would reduce the number of times the consumer has to replace the product, and would thus reduce materials consumption.
Community development	Planning communities in which residences are far from workplaces and services	Planning that puts people closer to what they need and that makes efficient use of already developed land would reduce the use of cars and thus the need for materials-intensive construction projects such as roads and bridges.
Direct consumption	Stressing immediate convenience of consumption and disposal as the ultimate good, without considering the prospects for sustainable consumption	Making changes in our consumption patterns to promote a culture of conservation: Copying on both sides of the page, using canvas shopping bags, reading books from the library instead of buying new copies, and taking public transportation could ultimately save both money and materials.

Source: Modified from Worldwatch Paper #121, *The Next Efficiency Revolution* by John E. Young and Aaron Sachs, with permission of Worldwatch Institute, Washington, D.C. Copyright © 1994.

sources often produces few jobs. For each person in the United States employed in logging timber, more than 25 people are employed in making furniture from it. Collecting used material and making furniture, or any item, from recycled or reused material would employ many more people than logging. Similarly, mining metals in the American West, for example, employs just 0.1% of all workers there. Tourism, a much cleaner industry, employs a much larger proportion of workers. Employment in the U.S. automobile manufacturing industry increased only 3% between 1980 and 1990. But employment in the auto repair industry increased by 50%, growing nearly 17 times faster. Auto repair is a form of efficiency improvement because it lets society accomplish the same task (car driving) with fewer resources (fewer new cars). Indeed, designing more durable and repairable products is an important step toward reducing throughput and curing the "throwaway" mentality. Table 6-1 lists many other specific opportunities for jobs to make society more sustainable.

SUMMARY

- Resources are raw materials used by society.
- The need for resource management is inescapable.
- Resources should be managed for both living and future generations.
- Preservation, conservation, and restoration manage resources for future generations.
- The "five e's" describe five key values of resources.
- Matter resources are depleted by dispersion.
- Energy resources are depleted when energy is changed to a less usable form.
- As a resource is depleted, we should conserve, not intensify its use.
- Net yield and maximum sustainable yield represent intensification.
- Three ways to conserve resources are: raise efficiency, recycle/reuse, and substitute.
- Conservation is typically discouraged because natural resources are undervalued.
- Sustainable industries not only conserve resources but often produce more jobs.

STUDY GUIDE

KEY TERMS

Anthropocentric
Benefit-cost analysis (BCA)
Conservation
Direct values
Ecocentric
Economic value
Embodied energy
Emotional value
Energy intensity index
Environmental service
Esthetic value
Ethical value

Extrinsic value
Five e's (Ethical, Esthetic,
    Emotional, Economic,
    Environmental service)
Hubbert's bubble
Indirect values
Intrinsic value
Law of diminishing returns
Maximum sustainable yield
    (MSY)
Net yield
Nonrenewable resources

Optimum sustainable yield (OSY)
Precycling
Preservation
Recycling loop
Renewable resources
Reserves
Resource
Restoration
Tragedy of the commons
True environmental costs
Urban ore
Virgin resources

STUDY QUESTIONS

1. How are "matter" resources depleted?
2. How are benefit-cost analyses sometimes used to analyze environmental problems?
3. What does conservation seek to minimize?
4. What is the "tragedy of the commons"?
5. List the five e's.
6. Distinguish between extrinsic and intrinsic values. Give examples of both.
7. Distinguish between indirect and direct values. Give examples of both.
8. What are "true environmental costs"?
9. Distinguish between reserves, resources, renewable resources, and nonrenewable resources.
10. How is depletion of matter resources different from depletion of energy resources? Give examples.
11. How has the past pursuit of net yield incurred environmental costs?
12. What are the two main causes of the bubble pattern?

13. Distinguish between optimum sustainable yield and maximum sustainable yield.
14. Maximum sustainable yield (MSY) seeks to keep the harvest rate equal to what? How is MSY related to carrying capacity?
15. What are the criticisms of maximum sustainable yield?
16. List and briefly describe some basic ways to conserve and reduce the need for resources.
17. What is the energy intensity index? What does it mean? What is the historical reason why the United States has a much higher index than Japan?
18. Give an example of a commonly recycled material in the United States. What materials are rarely recycled?
19. What happens to throughput in a sustainable world? What kinds of jobs are promoted?

WHAT'S THE EVIDENCE?

1. The authors assert that many resources, when used unsustainably, exhibit a bubble pattern of depletion due to exponential exploitation and exponential depletion. Cite some examples of the bubble pattern. Can you think of a resource that is currently being exploited, or could potentially be exploited, that might not fit the bubble pattern? If a resource does not fit the bubble pattern, does that necessarily mean that it is being used sustainably?
2. The authors suggest that in many cases, long-term economic and other gains (for instance, environmental services and esthetic values) can be increased by reducing resource use, at least in the short term. Can you cite examples either supporting or countering the authors' suggestion along these lines? Is this just a matter of how one defines "long-term economic and other gains" (perhaps one doesn't put any value on esthetic aspects of nature)?
3. The authors state that environmental sustainability often produces more jobs than unsustainability. What evidence is given to justify this? Can you think of evidence that contradicts this?

CALCULATIONS

1. Assuming an energy intensity index of 11.0 for the United States, how would this index change if the U.S. gross domestic product remained constant while energy consumption was reduced by 50%? How would it change if U.S. energy consumption increased by 10% while the gross domestic product remained constant?

2. Packaging accounts for about 33% of U.S. municipal solid waste. In 1999, the average American discarded 4.6 pounds (2.1 kg) of garbage per day. How many pounds of this 4.6 pounds was packaging? How much of the 4.6 pounds would be packaging if packaging were reduced to 20% of U.S. municipal solid waste?

ILLUSTRATION AND TABLE REVIEW

1. Based on Figure 6-3, what was the peak U.S. oil production (around 1970)? How much did this differ from Hubbert's predictions of the 1950s? For the decade from 1980 to 1990, were Hubbert's predictions more accurate than for the previous decade?

2. Based on Figure 6-9, what was the U.S. annual consumption of materials in 1950? In 1960? By what percentage did U.S. annual consumption of materials increase between 1950 and 1960?

WWW.ENVIRONMENT.JBPUB.COM

www.environment.jbpub.com

Connect to this book's web site: http://www.environment.jbpub.com/ The site features eLearning, an online review area that provides quizzes, chapter outlines, and other tools to help you study for your class. We've posted the answers to the book's odd-numbered questions. You can also follow useful links for more in-depth information. The site provides sources for the information in this chapter and recommendations for further readings.

Traditionally energy problems have been addressed simply by harnessing more energy—increasing the usable energy supply. This strategy involves both increasing the rate at which current technologies can produce energy and developing new energy-producing technologies. Another way to address the world's energy concerns is from the demand side. Attacking the demand side of energy use need not lead to a reduction in living standards or a change in lifestyle for the worse.

CHAPTER 7

Fundamentals of Energy, Fossil Fuels, and Nuclear Energy

Human society is dependent upon a continuous flow of energy (see Figure 1-9, page 11); indeed, all living organisms require energy. (For a discussion of what energy is, see Fundamentals 7-1, page 162.) Without energy, life could not exist. One of the prime concerns of any nation is to ensure that its citizens have ready access to the energy they need, whether that energy takes the form of food, heat for a home, power to drive machinery, electricity to run appliances, or gasoline to fuel an automobile. As technology has advanced, the amount of energy utilized by humans has increased dramatically. Energy consumption per capita in a technologically developed society today is at least a hundred times what it was when humans first evolved and over six times what it was two hundred years ago. Of course, more people are living on the Earth today than at any one time in the past, so total energy use has increased even more dramatically than per capita energy use: It has been estimated that today the world as a whole uses over 70 times as much energy as it did in 1865.

Energy is necessary for industrial development; thus, most of the world's energy flow historically has been through industrialized countries. Crude estimates indicate that the typical European uses 4 times as much energy as the typical Asian resident (excluding the Middle East), and, to consider a specific example between two countries, a citizen of the Netherlands uses 39 times as much energy as someone living in Bangladesh. The average North American uses 4.9 times as much energy as the global per capita average, and 41 times as much energy as a Bangladeshi. Certainly, this massive use of energy has brought material prosperity to the industrialized world, but at a substantial cost. The more energy humans use, the greater their detrimental impact on the environment. One could argue that virtually all our environmental problems can be traced back to energy use (or overuse). (Remember the equation $I = P \times C$ [Impact = Population $\times$ Consumption] introduced in Chapter 1.) The use of **fossil fuels** (oil, coal, and natural gas) is the major source of air pollution, acid rain, and greenhouse gases. Mining for coal or pumping oil can destroy vast tracts of land. Dams necessary for **hydroelectric power** production can wreak havoc on natural ecosystems and spread disease. Mining and processing of uranium to fuel nuclear power plants is fraught with hazards. Finally, conventional fossil fuel sources—upon which we rely most heavily—are nonrenewable and are in the long run a limited resource.

Thus, energy consumption by humans raises two basic concerns: (1) Where will the energy come from in the future as we deplete the most convenient, high-quality supplies of fossil fuels upon which much of modern industrial society is based? (2) How can we avoid the environmental degradation so often associated with the levels of energy consumption that characterize modern civilization?

We can take two basic approaches to these energy problems: We can address the supply side or the demand side. We can view energy as a commodity that we produce, and then attempt to maximize production of useful energy (such as refined oil or usable electricity); furthermore, some people suggest that we use more energy to solve any environmental problems that energy use may initially cause. Or we can consider energy to be a valuable resource and attempt to minimize its consumption, such as by increasing the efficiency of our energy-consuming devices, and thus minimize any detrimental environmental effects that energy use may entail.

An Overview of the Current Global Energy Situation

Worldwide consumption of energy has increased dramatically since the middle of the twentieth century (**Figure 7-1**), and it continues to increase, although at different rates in different parts of the world. During the decade from 1990 to 1999, for instance, global energy consumption increased by over 10%, but due to major decreases in energy use in Eastern Europe and the former Soviet Union during the 1990s, overall energy use actually decreased slightly among all developed nations, whereas it continued to increase steadily among developing nations through most of the decade (with a very slight decline at the end of the century as the global economic situation deteriorated). In North America (Canada and the United States), energy consumption increased by approximately the global average during the 1990s. No one can predict the future with certainty and politics and economics strongly influence energy production and consumption; thus it is unclear what the long-term future trend of global energy use will be. Based on the past, it will probably continue to increase (as shown in the projects presented in Figure 7-1). In 1999, world energy consumption (based on data compiled by the United States Department of Energy, Energy Information Agency) was approximately 382 quadrillion BTUs (382 quads; see **Fundamentals 7-2** on page 190). This includes the consumption of **petroleum** products (liquid petroleum products generally, often referred to as "oil," 152 quads), natural gas (87 quads), coal (85 quads), hydroelectric power (27 quads), nuclear power (25 quads), geothermal power, wind power, and so on (see Figure 7-1c and d). Besides commercial energy consumption, traditional fuels, such as fuelwood, charcoal, and animal and plant wastes, are consumed by many peoples. However, such traditional fuels are estimated to supply less than 10% of the energy supplied by the combined sources listed above. The largest per capita consumptions of traditional fuels are in South America and Africa.

Locally and regionally, the energy mix may differ significantly from that of the world as a whole, and even on a global scale there have been changes in the mix in recent years. As an example, until 1999 (or, according to some statistical compilations, until 2000 or 2001), more coal was used on a global scale than natural gas. In 1986, the ratio was 86 quads of coal to

(a)

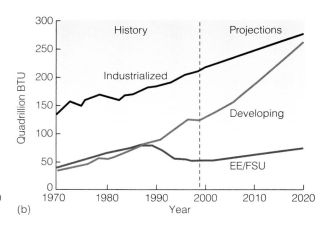

(b)

Figure 7-1 World energy use, 1900 to 1999 with projections to 2020. (a) Overall world energy consumption, 1900 to 1999 with projections to 2020. (b) World energy consumption by region, 1970 to 1999 with projections to 2020. Note that in this graph "Industrialized Countries" refers basically to North America, Western Europe, Japan, Australia, and New Zealand; "EE/FSU" is Eastern Europe and the former Soviet Union. (c) World energy consumption by fuel type, 1970 to 1999 with projections to 2020. (d) World energy consumption shares by fuel type for 1970 and 1999, and projections for 2010 and 2020. (1054 petajoules equals approximately one quadrillion BTU) (*Source*: United States Department of Energy, Energy Information Administration, *International Energy Outlook* 2002.)

(c)

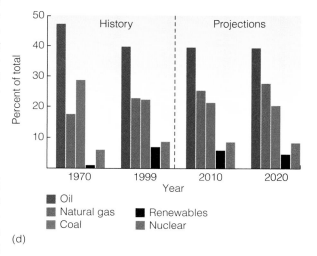

(d)

62 quads of natural gas, but usage of natural gas increased even as the usage of coal fluctuated, until natural gas overtook coal globally (see statistics for 1999 in the previous paragraph). In many developing countries, however, coal is still the dominant source of energy. Thus, in the Far East (including China, which relies heavily on coal), several times as much coal is burned as natural gas (in terms of energy equivalents), and China in particular also burns more coal than oil.

Power Plants—Supplying the People with Electricity

In modern, industrialized societies such as the United States, a major form of energy consumption is the burning of fossil fuels to power vehicles and machinery directly or to heat buildings. Another large portion of the energy sector consists of producing **electricity**, which is then distributed to consumers for running electrical appliances, lighting, heating (to some extent), air conditioning, and so forth. Although at present electricity provides only about 12% of the world's energy, it is often viewed as indispensable to modern society. Electricity is most often generated at electrical **power plants**, usually operated by private or semipublic utility companies, and then distributed to consumers.

Commercial power plants can take many forms, some of which are described in more detail later in this chapter and in the next chapter. At present, electricity is most commonly generated by using a **turbine** to drive a **generator** that transforms mechanical energy into electrical energy. A turbine is simply a machine, something like an old-fashioned

FUNDAMENTALS 7-1 Energy, Work, Power, and Thermodynamics

Energy, Work, and Power

What is energy? To a physicist, *energy* is simply the ability to do work. But what is work? Work (W) is defined as a force (F) applied to a material object times the distance (d) that the material object is moved. This is expressed in the equation:

$W = Fd$ (Work equals force times distance)

A commonly used unit of force is the newton (named after Sir Isaac Newton), which is defined as the force that will accelerate a mass of one kilogram one meter per second per second (in a vacuum with no frictional resistance). Distance can be expressed in meters. A force of one newton times one meter is the unit of work known as the **joule** (**J**). Energy has the same units as work; thus, the energy that makes it possible to do a joule of work is a joule of energy.

Energy, the capacity to do work, can take many different forms and be stored in many different ways. Mechanical energy involves objects, their motion, and position. One form of mechanical energy is known as potential energy—energy stored by virtue of the position of an object. A brick raised over your head before you drop it has potential energy, as does the wound spring of a watch or the water behind a hydroelectric dam built across a river. When you drop the brick or the water is released from the dam, the potential energy is turned into kinetic energy, or the energy of motion. The water dropping from the dam can be used to turn the blades of a turbine. When the turbine is connected to a generator, it will convert the mechanical energy to electrical energy. Electrical energy involves the forces of charged particles, such as electrons, acting upon one another.

Energy can take the form of heat, which is the random motion of atoms and molecules in a substance. As the brick you drop hits the ground, or the water of a waterfall hits bottom, the brick and the ground, or water, will heat up—the macroscopic kinetic energy will be converted to energy at an atomic and molecular level. You can demonstrate this convincingly by pounding hard on a nail with a steel hammer and then immediately feeling the heat radiating from the nail's head.

Often energy of any form is expressed in units based upon heat energy. One **calorie** (also known as a gram-calorie, and commonly abbreviated cal) is the quantity of energy that, when converted completely to heat, will warm one gram of water by one degree Celsius (or centigrade, commonly abbreviated ° C). A kilocalorie (or kilogram-calorie, commonly abbreviated kcal) is equal to 1000 calories; sometimes the term Calorie (with a capital C) represents a kilocalorie, as in the rating of foods by energy content. One kilocalorie is equal to 4184 joules of energy. Another commonly used energy unit is the British thermal unit (**BTU**). One BTU is the amount of energy that, when converted to heat, will raise the temperature of one pound of water one degree Fahrenheit (° F). One kilocalorie is equal to 3.968 BTU.

When discussing energy consumption or production on a national or global scale, very large-scale units are used. Some of the more common units are listed in Fundamentals 7-2, page 190. Two frequently used units are the "**quad,**" or quadrillion BTU (Q), and the **petajoule** (1×10^{15} joules, abbreviated PJ). One quad is equivalent to approximately 1054 petajoules. Another common unit is the metric ton of oil equivalent (toe), equal to 41.868 gigajoules. Thus, a thousand metric tons of oil equivalent (1000 toe) is equal to 41,868 gigajoules or 0.041868 petajoules, and a million metric tons of oil (1,000,000 toe) is equal to 41.868 petajoules. As an example, in 1999, the United States consumed nearly 38 quads worth of petroleum, or the equivalent of about 40,000 petajoules of oil (this represents about 950 million metric tons of oil equivalent, or 6.5 to 7 billion barrels of oil—the numbers are staggering).

Other forms that energy takes include the energy stored in the chemical bonds of substances (such as the energy in the food you eat or the gasoline that fills the tank of an automobile), the energy of light (electromagnetic radiation), and the energy that binds the particles composing the nucleus of an atom. Energy is the capacity to do work, but this is not the complete story. How fast work is done—how fast energy is converted from one form to another—is the concept of power. **Power** is defined as work (requiring energy) divided by the time period over which the work is done:

Power = work/time (Power equals work divided by time)

A common unit of power is the **watt**, defined as one joule of work (or energy) per second. A kilowatt (kW) is equal to 1000 watts; a **megawatt** (**MW**) is equal to 1000 kilowatts, or 1,000,000 watts; a **terawatt** (**TW**) is equal to one billion (10^9) kilowatts, or 10^{12} watts. The unit horsepower is equal to three-quarters of a kilowatt, or 750 watts. Thus, a one-horsepower electric motor can convert 750 joules of electrical energy into mechanical energy per second. If we multiply a unit of power by time, we arrive at another unit for energy. For example, a one-kilowatt motor running for one hour will convert 1000 joules of electrical energy per second into mechanical energy; since there are 3600 seconds in an hour, it will convert 3,600,000 joules of energy in a kilowatt-hour (kWh). In other words, one kilowatt-hour is equal to 3,600,000 joules, or approximately 860.4 kilocalories (since 1 kilocalorie is equal to 4184 joules).

The Laws of Thermodynamics

In order to study energy and energy conversions, one must know the laws of **thermodynamics** (the term *thermodynamic* means "heat movement").

The first law of thermodynamics is the law of the conservation of energy. Energy can be neither created nor destroyed—it is simply changed from one form to another (of course, it has been demonstrated that matter can be converted into energy, and vice versa, in nuclear reactions). Within an isolated (closed) system, the total quantity of energy will always be the same. However, not all energy is equivalent from a human perspective. One must take into consideration not only the quantity of energy, but also its form or quality. As energy is transformed from one type to another—for instance, from mechanical to electrical energy—the transformation will not be 100% efficient (some of the energy will be lost as waste heat). This leads to the second law of thermodynamics, which states that heat or energy cannot be transformed into work with 100% efficiency, and heat will always flow spontaneously from an object of higher temperature to an object of lower temperature. Another way of considering the second law of thermodynamics is to note that highly organized energy (such as electricity or mechanical energy) always ultimately degenerates into disorganized energy (such as heat).

Entropy, the concept of the amount of disorder or randomness in a system, always increases. For example, a liter of gasoline represents a relatively ordered, low-entropy system. The large, complex organic molecules in the gasoline store much high-quality energy. When this gasoline is burned in a car, some of the stored energy is converted into high-quality, usable mechanical energy. But much of the original energy is transformed into low-quality, high-entropy heat, and the gasoline is reordered into simple combustion products such as carbon dioxide, carbon monoxide, and water—randomness and entropy are increased in the system.

According to the laws of thermodynamics, entropy in the universe as a whole is continually increasing. In certain situations, entropy may decrease locally, but only at the expense of increasing entropy somewhere else. Living systems create and maintain order where there was previously disorder, but only by utilizing vast quantities of energy. Plants convert sunlight—a form of energy—into complex biological molecules that contain high-quality energy. These organic molecules are the high-quality energy reservoir upon which virtually all other life depends. Humans, in their quest for energy, are currently highly dependent upon the high-quality energy stored in the fossil fuels (oil, coal, natural gas) formed from the organic molecules of countless organisms over millions of years.

As described by the second law of thermodynamics, no energy transformations are 100% efficient. But from a pragmatic point of view, just how efficient (or inefficient) are our machines that transform energy from one form to another?

Efficiency is the useful work that is performed relative to the total energy input of a system. Thus, an engine that is 25% efficient will convert one-quarter of the energy contained in its fuel to mechanical work. In any engine, some energy is lost as heat because of friction. Furthermore, a power plant or any other heat engine usually achieves the most efficient transformations of energy at a slower rate than humans desire. We prefer to have our energy converted quickly, so that we can drive faster or heat more houses within a specific timeframe, even if this takes a heavy toll on the efficiency of the energy exchange, meaning that we use considerably more fuel.

Figure 7-2 A coal-fired power plant, Shawville, Pennsylvania.

water wheel, that can convert the lateral motion of a flowing fluid (such as water, steam, or some other gas) into the rotational or turning motion of a shaft or axle. This turning motion can then be used to turn the generator that transforms mechanical energy into electricity.

The more common types of commercial power plants are mainly distinguished by the source of the energy that drives the turbine. In hydroelectric power plants, water falling from a dam drives the turbine. In wind-powered plants, the turbines are driven by wind—indeed, they take the form of huge windmills. In geothermal power plants, hot steam and other gases originating from the interior of the Earth may be used to drive the turbine. Most commonly, however, large power plants use artificially created steam. Water is heated to extremely high temperatures (600° F or higher/over 300° C), and the resulting pressure of the expanding steam is used to turn the turbines. The fuel that the power plant consumes provides the heat for generating the steam. In the United States, the most common fuel is coal. A coal-fired power plant (**Figure 7-2**) is, in the simplest sense, a huge furnace used to turn water into steam. Another common way to turn water into

steam is by harnessing the tremendous amounts of heat given off by nuclear reactions—this method of generating electricity is commonly referred to as nuclear power. Other methods of heating water to produce steam, and ultimately electricity, include burning wood or other biomass, burning oil, burning natural gas, or using the focused energy of the Sun to heat water to high temperatures.

An internal combustion engine or turbine can also be connected to a generator. An internal combustion turbine is essentially an engine that burns oil or natural gas; rather than heating water first, it uses the hot gases from the burned fossil fuel directly to turn the turbine. Conventional steam-generating power plants, whether coal-burning or nuclear-fired, are typically large. They are extremely expensive, costing billions of dollars each, and take many years or even decades to build and put on line. Once built, these large plants are relatively efficient in converting fuel into electrical energy. Conversely, standard internal combustion turbines are smaller, less expensive to purchase, and can be placed into operation more quickly. The drawback is that they are much less fuel-efficient than larger plants, and the fuel they burn—oil or natural gas—

is usually more expensive than the fuel (such as coal or uranium) of larger, steam-based power plants. However, newly developed gas turbines and combined-cycle (gas and steam) plants are greatly increasing realized efficiencies.

Utilities that produce electrical power must have a considerably larger generating capacity than they actually need at most times. In fact, to ensure reliable production, the local grid of power plants usually must be able to generate approximately 20% more electricity than is normally used at any one time. This extra generating capacity is known as the *reserve capacity*. Electricity is not used at an equal rate throughout the day or the year. Depending on the area, more electricity may be used during the day when businesses are operating and, in hot climates, when air conditioners are running at maximum capacity. The amount of electricity needed at the time of highest demand is known as the **peak load**. With current technology, it is very difficult to store electricity, and it is also not possible to turn large electrical power plants off when they are temporarily not needed. As a result, utilities must be able to satisfy peak load requirements, even if this means generating more electricity than is needed at other times of the day or year.

Another reason that utilities need to maintain a substantial reserve capacity is that at any one time, more than 10% of the generating capacity may be shut down for repairs, routine maintenance, or changing the fuel (in nuclear plants). When a plant shuts down temporarily, power is supplied to customers from the reserve capacity of other plants owned by the utility and connected to the same power grid (the set of power lines that connect power plants to their customers and to each other). Different electrical utility companies may purchase power from each other if their power lines interconnect. Even with interconnecting lines, however, transporting electricity over long distances is extremely inefficient and expensive because much of the electrical energy is lost in transit (although superconducting technology may help to alleviate this problem).

When the capacity of a utility's power plants is exceeded, **brownouts** and **blackouts** can occur. In a brownout, demand is only a few percent above the generating capacity of the utility; the voltage the consumers receive is inadequate, with the result that the lights often dim. Blackouts are often associated with major breakdowns, either of a power plant or in the grid that distributes power. In a classic blackout, a region served by the utility is left without electricity for an extended period of time. Localized blackouts can also occur if demand significantly exceeds the utility's capacity to generate electricity. In such an instance, the utility may have to institute a rolling blackout—the utility stops and later restores electrical service to a sequence of neighborhoods in turn, each losing electricity for an hour or two during the day. Such a rolling blackout forcibly reduces the consumption of electricity. Needless to say, brownouts and blackouts are very disruptive to modern societies that are extremely dependent upon electricity.

Hard versus Soft Energy Technologies

Large, modern, electricity-generating power plants have been characterized by physicist and energy consultant Amory B. Lovins as **"hard" technologies.** Hard energy technologies depend on large-scale plants that are complex, expensive, and centralized. The energy they generate, such as electricity, is of a very high-grade form and must be distributed to consumers through an elaborate system of hardware, such as power lines. Large coal-burning or oil-burning electrical power plants, major hydroelectric plants, and nuclear power plants are all classic examples of hard technologies. Another example would be petroleum production and refinement facilities connected to consumers by an elaborate system of pipes. Vast infusions of money are necessary to establish a hard technology and its accompanying infrastructure. Because of the substantial investments required, it can be very difficult to modify or abandon a hard technology once it is in place.

As Lovins points out, however, the end uses of the high-quality energy delivered by a hard technology may be extremely low-grade applications, for instance, heating water or open space in buildings. Rather than burning oil or coal at a power plant to generate electricity and then transmitting the electricity to the consumer who uses it to generate heat, it is much more efficient to burn the oil or coal on site to generate heat directly. Or better yet, passive solar systems might be used to heat (and even cool) a building.

The alternative to hard energy technologies are **"soft" technologies,** such as using passive solar energy, burning wood or coal in a stove, using an onsite windmill or small water wheel, and even increasing energy efficiencies and promoting conservation measures. Soft technologies are small-scale, local (thus they do not require elaborate distribution systems), and usually much more environmentally friendly than hard technologies. Soft energy technologies are often relatively inexpensive to install and operate, and they deliver energy that

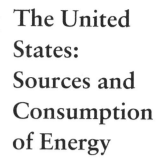

The United States: Sources and Consumption of Energy

Even today, much of the world, especially in developing countries, depends on the traditional forms of energy used for centuries. But modern industrialized Western society seems inconceivable without the intensive energy use to which North Americans and Europeans have grown accustomed. As an example of energy sources and consumption in a modern industrialized nation, let us take a look at the United States.

The United States is both the leading producer and consumer of primary energy in the world. Currently, the United States produces approximately 17% to 18% of the world's energy, but it consumes approximately 22% to 23% of the world's energy; thus, the United States consumes over 25% more energy than it produces. The next two largest producers and consumers of energy are China (producing about 11% to 12% of the world total and consuming about the same amount as it produces), and the Russian Federation (producing 9% to 10% of the world's energy, but accounting for only 6% to 7% of world consumption). Saudi Arabia is the world's fourth leading energy producer (accounting for approximately 5% of world production), yet its energy consumption is just 1% of the world total. Japan produces slightly over 1% of the world's energy, yet accounts for over 5% of world consumption.

Energy production and consumption in the United States have generally increased over the last 40 plus years (**Figure 1**), but consumption has grown faster than production. This means that energy imports (principally, oil and other petroleum products and natural gas) have substantially increased over the decades.

For example, consider 1999, a typical year and the latest for which a reasonably accurate and complete analysis is available. In that year, the United States consumed some 97 quads of energy and energy materials, including relatively small amounts of energy exports and non-fuel uses of petroleum products such as oils for lubrication and hydrocarbons used as raw materials for the petrochemical industry (see **Figure 2**). A quad is one quadrillion BTUs (approximately 1054 petajoules) or the equivalent of about 171–172 million barrels of oil. So in 1999, the United States consumed the equivalent of 16.6 billion barrels of oil (the equiv-

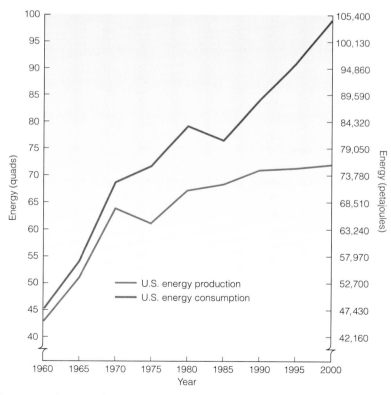

Figure 1 U.S. energy production and consumption, 1960–2000. Notice that in 1960, the United States produced almost as much energy as it consumed, but since then, the trend has been for consumption to outpace production. Today, more than 25% of the energy consumed in the United States is imported. (*Source*: Data from Energy Information Administration, U.S. Department of Energy, as published in *The World Almanac and Book of Facts, 2002* [New York: World Almanac Books, 2002], page 157.)

alent of over 55 barrels of oil for every man, woman, and child in the United States).

Of the nearly 97 quads of energy consumed in 1999, 23.4 quads, or over 23%, were lost during electrical transmission. Transmitting electrical power through standard power lines is very inefficient. Over twice as much electrical power is lost during transmission from large, centralized power plants as is delivered in useful form to consumers. Looking at the data in Figure 2, we see that over 25% of our energy goes toward transportation. Inspecting Figure 2 further, we notice that our modes of transportation are relatively inefficient. We waste about four times as much energy as we actually use. This is due in large part to the notoriously inefficient motor vehicles that we drive; with their low gas mileages, compared to the efficiencies that could be achieved even with today's technology, these vehicles waste most of the energy value of the gasoline. Most environmentalists advocate a major effort to increase the fuel efficiencies of cars and other vehicles. After energy lost during transmission through power lines, the inefficiency of our motor vehicle fleet is the largest single drain on the nation's energy budget.

Twenty-five percent of the energy used to heat, cool, and light our commercial and residential buildings, run our televisions and small appliances, and so forth is lost. Better insulation, more efficient refrigerators, and other energy saving devices could help decrease the energy losses in this sector. In addition, as we shall discuss further in Chapter 8, energy needs could be considerably reduced through the increased use of old and new technologies, such as passive and active solar heating, compact fluorescent light bulbs, and many other energy conservation methods. The industrial sector uses energy with the greatest efficiency in the United States, but even so, about one-fifth of the energy is wasted.

Looking at the supply side of Figure 2, about 80 quads of our energy comes from fos-

sil fuels; currently, then, the United States is truly dependent on fossil fuels. The remainder comes from hydroelectric power, nuclear power, biomass, and to a very small extent, alternative energy sources (geothermal power, wind power, solar power, and so on). For its single largest energy source, oil, the United States is heavily dependent on imports (over half the oil we use is imported).

Questions

1. Some people feel that the United States has a "moral obligation" to the rest of the world to cut down on its energy consumption, especially since the United States consumes substantially more energy than it produces. What is your opinion on this matter? If the United States suddenly slashed its energy consumption by 30% and stopped the importation of all fossil fuels, how would global trade relations be affected? How would a 30% cut in U.S. energy consumption benefit the environment?

2. If you were a cabinet member and the president charged you with promoting the reduction of U.S. energy consumption, which user sector in Figure 2 would you address first? Where do you think the most gains in energy savings can be made in the short term? In the long term?

3. Japan is the world's fourth largest consumer of primary energy, using over 5% of the world supply. Yet Japan produces just over 1% of the world's primary energy. Thus, in global energy matters, Japan could be seen as the inverse of a country like Saudi Arabia. Why does the world need both "Japans" and "Saudi Arabias," or does it? How does this either support or counter the argument that every nation should be energy self-sufficient?

is of the appropriate grade—such as high-grade electricity from photovoltaic cells to run computers and other electronic devices, and low-grade heat for space or water heating. Examples of both hard and soft technologies are discussed in this chapter and the next.

The Fossil Fuels

The fossil fuels are currently the primary sources of commercial energy used worldwide. This is due to many factors. Historically, fossil fuels have been available in a plentiful supply that was easy to mine or otherwise obtain. Fossil fuels provide a form of high-

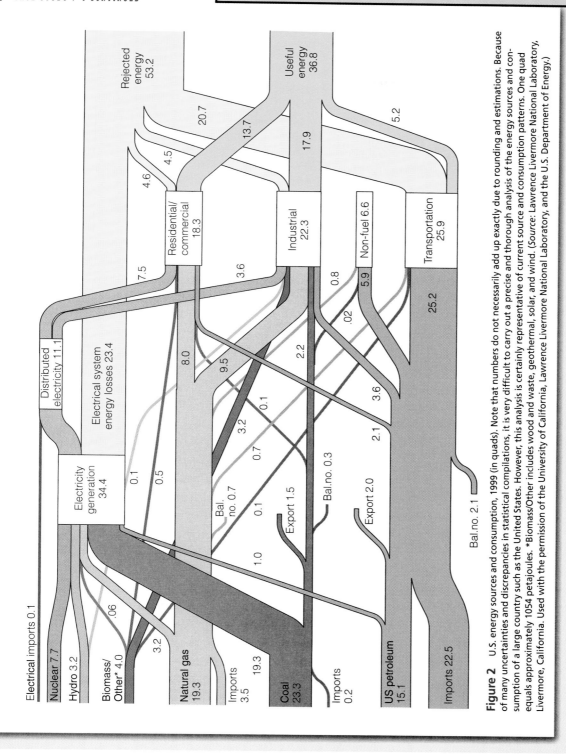

Figure 2 U.S. energy sources and consumption, 1999 (in quads). Note that numbers do not necessarily add up exactly due to rounding and estimations. Because of many uncertainties and discrepancies in statistical compilations, it is very difficult to carry out a precise and thorough analysis of the energy sources and consumption of a large country such as the United States. However, this analysis is certainly representative of current source and consumption patterns. One quad equals approximately 1054 petajoules. *Biomass/Other includes wood and waste, geothermal, solar, and wind. (*Source:* Lawrence Livermore National Laboratory, Livermore, California. Used with the permission of the University of California, Lawrence Livermore National Laboratory, and the U.S. Department of Energy.)

grade, concentrated energy that is easily transported and can be put to many different uses. Generating electricity using fossil fuels (specifically, natural gas and coal) has typically been much less expensive than using competing technologies such as nuclear power plants, solar thermal power, or wind power.

Oil

Oil (and its liquid petroleum products) is the most widely used form of commercial energy. It remains relatively cheap, is easily transported, and can be put to a variety of uses—from heating buildings and running machinery to (in modified form) fueling

The Hidden Costs of Foreign Oil Dependence

A secure supply of energy is critical for any nation, for without energy the country runs the risk of collapse. When a nation is heavily dependent upon foreign oil (or foreign natural gas, coal, or uranium), it may need to spend vast sums of money, and even enter into armed conflict, to guard its energy supplies. We saw this happen in the Persian Gulf. In January 1991, the United States attacked Iraqi forces that had taken over the country of Kuwait, a small, oil-rich nation on the northwestern corner of the Gulf, and threatened to disrupt the free flow of oil. Six weeks later, the Gulf War was over, but beyond the loss of life, it cost approximately $40 billion outright and caused considerable environmental damage to the area. Even before the Gulf War broke out, the hidden costs of Gulf oil meant that this supply was far from a bargain. P. R. Ehrlich and A. H. Ehrlich (1991, *Healing the Planet*) have estimated that prior to January 1991, the costs of the routine U.S. military maneuvers that took place in the Gulf to guard the oil added approximately $25 to the real cost of a barrel of oil. Between 1974 and 1989, the price of oil on the world market fluctuated between $12 and $48 a barrel, so this hidden cost effectively doubled the cost of oil. Of course, this extra $25 per barrel was not added to the price

of the oil, but was paid by American taxpayers as part of the country's outlays for defense. Today (2002), some analysts estimate that when all of the costs are added up (military training, equipment, transportation of troops and supplies, maintaining a ready state of alert, and so forth), the United States effectively spends $50 billion a year protecting oil supplies in the Middle East.

In addition to the cost in lives and money spent to wage the war and then rebuild the damaged infrastructure of Kuwait and Iraq, the war extracted a very high price in terms of environmental damage. During the war, Iraq resorted to ecological sabotage. Many Kuwaiti oil wells were deliberately damaged and set on fire (**Figure 1**), while oil reservoirs were allowed to leak onto the land and into the water. Millions of barrels of Kuwaiti oil leaked into the Gulf, forming an oil slick tens of miles long that damaged coasts and wildlife. Approximately 700 wells were set on fire, burning about 3 million barrels of oil a day. Even after the war ended in late February 1991, many of these wells continued to burn—the last one was not capped until November 1991. As they burned, they released millions of tons of smoke, carbon dioxide, carbon monoxide, sulfur dioxide, and other pollu-

vehicles. Indeed, worldwide, more than half of all oil is used for transportation.

Crude oil is an assortment of many different organic compounds. In the refining process, crude oil is turned into automobile gasoline, heating oil, jet fuel, tar for roads and roofs, motor oil, and innumerable other substances. Oil forms the raw materials to manufacture various plastics, fertilizers, medicines, and many products indispensable to the industrialized world. However, oil is a nonrenewable resource. The vast majority of the remaining reserves are found in the Middle East; by most estimates, two-thirds of crude oil reserves are located in the Middle East, and a quarter of global oil reserves may occur in just Saudi Arabia, while Iraq contains about 10% of global oil reserves, and Kuwait almost as much. The Gulf War of 1991 was effectively a battle over oil supplies and it is quite possible that hostilities of the future could revolve around limited supplies of petroleum. Some proponents of the use of fossil fuels, such as oil, contend that current known reserves are a poor measure

of how much oil actually still exists. They note that estimates of proven oil reserves, as well as coal reserves and natural gas reserves, have often been revised upward as more oil is found. This is true to a certain extent, but it is also true that in general over the past two decades the new sources of fossil fuels that have been discovered have been smaller and of lower quality than past discoveries. Also, because of their inaccessibility, some proven reserves are much more expensive (in terms of dollars and energy) to produce than others. For instance, it can cost five to ten times as much to produce a barrel of oil from Alaska or the North Sea as it does to produce a barrel of oil in the Middle East.

Even if we can stretch our use of fossil fuels by a few decades, either by finding new sources (such as undiscovered oil fields) or by utilizing lower-quality sources (such as oil shale or forcing the remaining oil out of "depleted" wells), the fact remains that fossil fuels are a finite, nonrenewable resource. They will run out eventually. Looking well into the future, fossil fuel resources may become so tants into the atmosphere. Millions of tons of acid rain came down in the Gulf region. During the summer of 1991, the smoke and air pollution caused the temperatures in Kuwait to be 10° to 27° F (5.6° to 15° C) lower than normal. Smoke and acid rain from the burning oil were detected over a thousand miles (over 1600 km) away. Another hundred or so oil wells spilled oil all over the Kuwaiti desert—reportedly, the lakes of oil were 6 feet (2 meters) deep in some areas. Scarce fresh water supplies throughout the region were contaminated.

Although the burning oil wells were dramatic and horrible, in hindsight, many experts suggest that the worst long-term damage to Iraq and Kuwait occurred when the vegetation and soils of the fragile desert ecosystems were subjected to massive bombings, trenches dug by bulldozers, and the military maneuvers of soldiers and motorized armored vehicles. As a result, an estimated 25% of the land surface of Kuwait has been damaged. In both countries, wind erosion has increased significantly, causing an increase in sandstorms and an intensification of desertification processes. In an area already heavily dependent upon imports of food, the loss of what little soils there were can only worsen the situation.

Figure 1 The Bergan oil field fire in Kuwait is reflected in a pool of oil. The fire was deliberately set by the Iraqis during the Gulf War, 1991.

valuable as a source of the hydrocarbons used in manufacturing plastics, fertilizers, and so forth that the price will rise to the point where they will no longer be economically feasible as fuels. Furthermore, even before our last supplies of oil, coal, and gas are depleted, we may be forced to switch to non-carbon-emitting, non-fossil fuels because of the deleterious effects of greenhouse gas accumulation, acid precipitation, and ground-level air pollution (see Chapters 16 and 17). Both the sources of fossil fuels (the absolute amount of fossil fuels that occur on Earth) and the sinks for fossil fuels (the ability of the Earth to successfully absorb the waste products of the burning of fossil fuels) are already limiting their use. Currently, the burning of oil and related liquid fuels releases about 9.5 billion metric tons of carbon dioxide per year into the atmosphere, over 30% of all carbon emissions. Additionally, burning oil releases such atmospheric pollutants as nitrogen oxides, sulfur dioxide, and various hydrocarbons.

Besides its contribution to atmospheric pollution, a particular concern with oil is that its use rou-

tinely results in huge amounts of raw oil being dumped into the environment (**Figure 7-3**). Some of this happens accidentally, as during an oil spill (a tanker may collide with shallow rocks, or a pipeline may burst), and sometimes oil is released intentionally as a terror tactic or act of war. For instance, during the Persian Gulf War in 1991, Iraq deliberately released an estimated 11 million barrels of oil from tankers off the coast of Kuwait. The *Exxon Valdez* disaster off Alaska on March 24, 1989, which dumped a quarter million barrels of oil into Prince William Sound, was well publicized, but it was certainly not the largest recent spill. Later in the same year, on December 19, 1989, an explosion on the Iranian supertanker *Kharg-5* caused 450,000 barrels of crude oil to be released into the Atlantic 400 miles north of the Canary Islands; a 100-square-mile slick formed. In 1996, the supertanker *Sea Express* ran aground at the port of Milford Haven, Wales, spilling approximately half a million barrels of oil. While such spills are usually avoidable, much more crude oil is released intentionally during

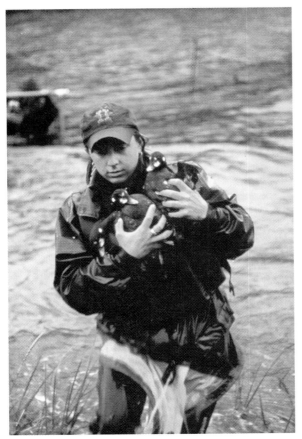

Figure 7-3   An oil spill response worker takes three seabirds to be cleaned after the *Exxon Valdez* accident in Prince William Sound, Alaska.

normal operations; the oil-carrying tanks of many ships are routinely flooded with seawater to clean them out, and the contaminated water is pumped back into the sea. An estimated 3 to 6 million metric tons, or roughly 22 to 44 million barrels of oil are released into the world's oceans every year. Even more of the oil found in the ocean comes from rivers that carry untreated domestic and industrial hydrocarbon wastes out to sea. However, natural seepage also accounts for an uncertain portion of the oil in the ocean.

World oil consumption increased steadily from 1950 (that year, 470 million metric tons of oil equivalent were consumed) until 1979, when an estimated 3103 million metric tons of oil equivalent were consumed. From 1979 until 1994, oil consumption dipped, then stabilized, and finally only exceeded 1979 consumption in 1990 (3136 million metric tons consumed). Since the early 1990s, global oil consumption has slowly but steadily increased, reaching an estimated 3511 million metric tons in 2001.

The reason for the general decline and stabilizing of oil consumption between 1979 and 1990 seems to have stemmed from the energy crisis of the 1970s and the growing realization of the environmental damage incurred by oil use. More emphasis was placed on energy efficiency and the use of alternative energy sources. Most recently, however, oil production has again been rising due to several factors. Various developing countries are encouraging the exploration and development of their oil reserves (in part to meet the increasing demand for oil in these countries and in part to raise much needed revenues). Central Asia in particular is experiencing an oil boom, and advanced technologies are allowing the location and exploitation of oil reserves that were previously economically inaccessible.

At present there is no imminent shortage of oil. Some analysts believe that oil prices and production will be driven by demand well into the future, and that the theoretical maximum level of production will never be reached simply because the demand will not be there. Hopefully, the world will break its heavy oil dependence during this century. Still, until that time comes, oil will continue to supply approximately 40% of the globe's commercial energy.

Coal

Like oil, **coal** is a fossil fuel, but in solid rather than liquid form. It is much more abundant than oil, and significant coal deposits are much more evenly distributed throughout the world than are major oil fields. Given these factors, some analysts predict that worldwide use of coal will increase in the decades to come. At current consumption rates, the world's supply of coal could probably last for about two hundred years or more. Of course, this would be considerably shortened if coal consumption increased dramatically, perhaps as a replacement for oil once that resource is exhausted. A major concern over the burning of coal is that, like oil, it emits large quantities of carbon dioxide (coal emits 25% more carbon dioxide than an equivalent amount of oil and 80% more carbon dioxide than an equivalent amount of natural gas) into the atmosphere, promoting the greenhouse effect. Also, coal burning releases such substances as sulfur dioxide and nitrogen dioxide, major contributors to acid rain. New techniques are being developed to ensure the more complete burning of coal with less pollution. In **fluidized bed combustion,** for example, very small coal particles are burned at very high temperatures in the presence of limestone particles (the limestone helps capture sulfur and other pollutants) while air is blown through them.

At present, coal is used directly in only a handful of major contexts, such as to produce electricity in coal-fired power plants, to heat large plants and buildings, and in the production of iron and steel. In the modern world, coal use on a smaller scale has been generally phased out—few people burn coal directly to heat their homes, for instance—because it is bulky, cumbersome to handle and transport on a small scale, and extremely dirty and polluting. To overcome some of these disadvantages and increase the use of coal, various organizations are working on coal gasification and liquefaction techniques that can produce natural gas, alcohol, and oil replacements from solid coal. The main drawback with such technologies is that at the moment they are not economically competitive with natural gas and petroleum products. Methane or gasoline produced from coal can cost over 50% more than the equivalent substances refined from natural gas or petroleum taken from the ground. Furthermore, while transforming coal into liquid and gaseous hydrocarbons may help mitigate pollution and the greenhouse effect, these new products are certainly not pollution-free. At best, such techniques can only serve as short-term, stop-gap measures to fend off the fossil fuel crisis.

Since the middle of the twentieth century, global consumption of coal has doubled, climbing from 1043 million metric tons of oil equivalent per year in 1950 to 2298 million metric tons of oil equivalent in 1996. Since 1996, global coal consumption has seen a slight decline overall, coming in at an estimated 2034 million metric tons of oil equivalent in 2000. The worldwide decline in coal use, it has been suggested, may be a symptom of global recession, but it is also probably a function of more efficient technologies being put into place, and the substitution of oil and especially natural gas in place of coal for some purposes. Analysts are divided when it comes to predicting the future of worldwide coal use. On the one hand, it is an abundant, widely distributed fuel utilized by both developed and developing countries. United States coal consumption accounts for around a quarter of the world total. Many developing countries are dependent on coal, especially if they cannot afford oil, natural gas, or other sources of energy. China, with a population of over 1.3 billion, extracts about 60% of its total energy requirements from coal. India is also a leading user of coal. If the developing world continues to expand its use of coal, worldwide consumption could be driven up.

On the other hand, some analysts predict that worldwide coal use has already reached a plateau. General improvements in energy efficiency, as have been instituted in Russia and other republics of the former Soviet Union, may be eliminating the need for major increases in coal use. Increasingly, governments are instituting and enforcing environmental regulations that discourage the use of coal. When it comes to meeting environmental standards, coal is a poor match for other forms of energy. Additionally, there is a trend toward eliminating the subsidies for coal production and use that historically have been provided by many governments, including Germany and the United Kingdom. Often coal workers lose their jobs as a result, however, and major strikes by miners (for instance, in England in 1992) have slowed progress toward eliminating coal subsidies. Only time will tell whether worldwide coal use will increase or decrease.

Natural Gas

A fossil fuel that takes a gaseous form, **natural gas** is currently the second most commonly used fuel. Its use is growing rapidly due to its versatility and relative abundance; in addition, it burns more cleanly than other fossil fuels (producing less carbon dioxide and other pollutants than coal or oil).

Nevertheless, there is some fear that economies based on natural gas will not be able to last much longer than economies that depend on oil. Based on current consumption rates, the world's known natural gas reserves will be depleted by the middle or end of the twenty-first century. However, it is now widely believed that natural gas resources are much more abundant than once assumed. In the past, natural gas was often found as a byproduct of the search for oil, and in some cases, the natural gas was simply released or burned off the top of oil pockets. Today natural gas reserves are being discovered in areas not associated with oil deposits, and various research groups have suggested that our lack of understanding of natural gas geology may have led to overly conservative estimates of the extent of natural gas resources. It is now known that whereas oil is primarily limited to relatively shallow deposits, methane (natural gas) can be found at much more extreme depths. At such depths, large amounts of natural gas may be highly compressed into small volumes. New exploration and drilling techniques are making it easier to locate and tap such natural gas resources. Natural gas has also been discovered in association with coal seams. Another advantage of natural gas is that deposits are somewhat more evenly distributed around the globe than are major petroleum fields. While the majority of proven reserves of oil occur in the Middle East, significant proven reserves of natural gas are found in the former Soviet Union and the Middle

East, and smaller reserves in North America, Latin America, and Western Europe.

Worldwide production (and likewise consumption) of natural gas has grown approximately twelve-fold since the middle of the twentieth century, from 187 million metric tons of oil equivalent in 1950 to 2277 million metric tons of oil equivalent a year in 2000. Some analysts suggest that natural gas use may nearly double over the next two to three decades and then continue to be used at a level of about 3700–4000 million metric tons of oil equivalent a year for several decades. Ultimately, some predict, natural gas will be the fossil fuel to usher in the final transition to a solar-hydrogen and alternative energy global economy (see Case Study 8-1, page 202 in Chapter 8).

Can Fossil Fuel Supplies Increase on a Human Timescale?

Of course, the simple answer to this question is no. The geological processes involved in the formation of fossil fuel resources are imperceptibly slow from a human perspective. Yet a superficial reading of certain statistics may suggest that our fossil fuel supplies have actually increased over time. For example, in 1970, the ratio of oil reserves to production was 31 years, but by 1989, the ratio had increased to 41 years. Even more dramatically, over the same period the ratio of reserves to production for natural gas increased from 38 to 60 years! In the year 2000, over a decade later, the ratio of oil reserves to production was 38 years and for natural gas it was still 60 years.

Superficially, it may appear that there was more oil and natural gas available in 2000 than thirty years earlier. Of course, this is not the case—there were about 600 billion fewer barrels of oil (about 88,388 million metric tons of oil equivalent) and 2000 trillion fewer cubic feet (51,228 million metric tons of oil equivalent) of natural gas in the world (the amount consumed between 1970 and 2000). What changed was that more oil deposits and natural gas deposits were discovered between 1970 and 2000, thus increasing the known reserves of these fossil fuels. Some might argue that the successful history of oil and gas exploration between 1970 and 2000 demonstrates that there are still plenty of deposits to be discovered, but this too is a false argument. There is an absolute amount of fossil fuels still buried in the crust of the Earth. With exploration and discovery, some of this fixed amount is shifted from the category of "undiscovered reserves" to "known reserves." Every time we do this, the amount of undiscovered reserves is decreased. By definition, we can never know precisely the quan-

tity of unknown reserves that remain (for they are "unknown"), but we can estimate their quantity by various means. One way to gauge the extent of undiscovered reserves is by how easy it is to find them—the more abundant the undiscovered reserves are, the easier they are to locate.

In fact, in many parts of the world in the last thirty years, and since about 1940 in the continental United States, it has become progressively more difficult, with progressively more sophisticated instrumentation and deeper drilling rigs in more obscure places, to locate previously undiscovered reserves of oil. This clearly indicates that we are quickly locating the last major fields of undiscovered reserves. It would be unwise to anticipate that the rate of past oil discoveries will continue into the future.

Why We Must Stop Burning Fossil Fuels

Whether we are in danger of running out of fossil fuels or not, most people who have seriously considered the matter agree that we must greatly curb our use of these fuels. One consideration is the degree to which the United States and many other industrialized societies are dependent on foreign oil. Even more importantly, the burning of coal, and to a lesser extent the burning of petroleum products and natural gas, has created environmental havoc on a global scale—the potential for global warming due to the "greenhouse effect," acid rain, air pollution, damage to the land's surface as a result of mining and drilling activities (especially harmful is stripmining for coal), water pollution, and so forth (**Figure 7-4**).

The fossil fuels are composed principally of carbon and hydrogen derived from once living organisms. When a fossil fuel is burned, the carbon in the fuel combines with oxygen (O_2) from the air to form carbon dioxide (CO_2), and the hydrogen combines with atmospheric oxygen to form water (H_2O). During these reactions, energy in the form of heat is released, and the released heat also causes the reaction to proceed further.

A high-grade coal is almost pure carbon tainted by small quantities of other elements (such as sulfur and various metals). As the coal is burned, it gives off mostly carbon dioxide, plus small amounts of sulfur dioxide and other pollutants. Carbon dioxide is the dominant greenhouse gas. Natural gas, in contrast, contains a large amount of hydrogen (the formula for methane is CH_4); thus, when it is burned, less carbon dioxide is emitted per unit of heat given off. Some of the gas given off by methane takes the form of water vapor, which does not appear to pose any environmental problems. For these reasons, natural gas is considered to burn much

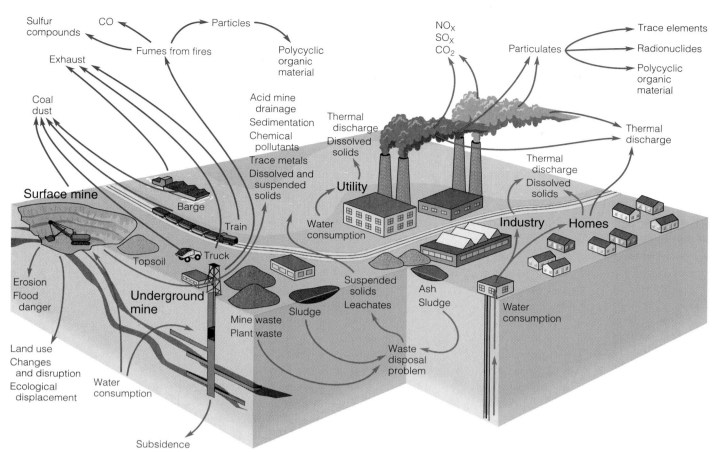

Figure 7-4 Environmental impacts of the coal fuel cycle. Similar impacts are caused by the oil and natural gas cycles. (*Source*: C. Hall, C. Cleveland, and R. Kaufman, *Energy and Resource Quality* [Niwot, Colo.: University Press of Colorado, 1986], p. 366; by permission of the publisher.)

Figure 7-5 Without any attempt at reclamation, a strip mine is very damaging to the land. The rows of rubble are called tailings and nothing will grow on them for years to come.

more cleanly than coal, yet even natural gas combustion produces significant amounts of carbon dioxide. Furthermore, unburned natural gas, if released into the atmosphere, is itself a powerful greenhouse gas. Chapter 17 discusses the greenhouse gases and their effects in detail.

Fossil fuel procurement is also extremely damaging to the environment. Oil and natural gas prospecting and drilling can involve the construction of access roads, the movement of heavy equipment, the erection of large drilling rigs or platforms (both on land and at sea), and so forth—all of which degrade the natural environment. Oil spills, either at wells or during transportation, have gained notoriety, but perhaps more damaging are "routine" leaks and emissions of petroleum compounds from wells, pipes, and tankers. Coal mining is extremely damaging to the environment and dangerous to the miners (**Figure 7-5**). Strip-mining may involve the removal of over 100 feet (30 meters) of soil to get to the coal, and in the United

CASE STUDY 7-3

The Arctic National Wildlife Refuge

The Arctic National Wildlife Refuge, often abbreviated as ANWR, covers about 20 million acres (8 million hectares) in the northeastern corner of Alaska (**Figure 1**). This amount of area is approximately the size of South Carolina, and is about 5% of the area of the state of Alaska. ANWR is at the center of a major national environmental controversy: Should oil drilling be allowed? This question has been debated in the U.S. Congress for many years.

The amount of oil underground in ANWR that is economically feasible to extract is estimated to range between 3.2 billion barrels to 7.8 billion barrels. People who favor drilling in the Refuge argue that this large supply of oil could be used to reduce U.S. energy dependence on oil imports. They

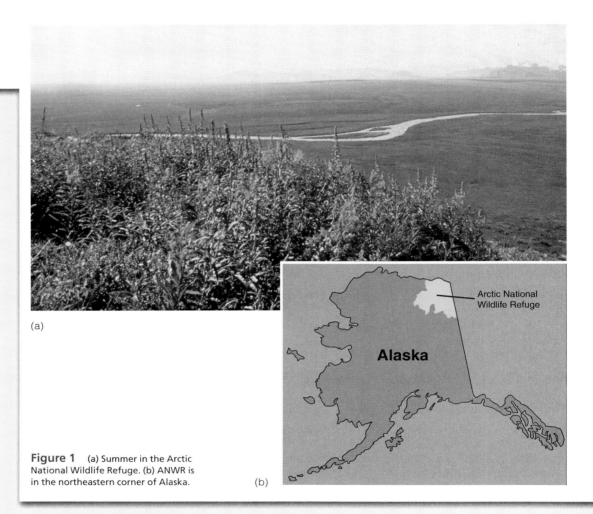

Figure 1 (a) Summer in the Arctic National Wildlife Refuge. (b) ANWR is in the northeastern corner of Alaska.

States alone, it is estimated that over 1.3 million acres (500,000 hectares) have been strip-mined and not yet reclaimed, while another 65,000 acres (26,000 hectares) continue to be strip-mined annually. Underground coal mines are little better. They can result in land subsidence, while water seeping into them can form acids (from the sulfur and other compounds that commonly occur in coal) that then leak out and contaminate land, streams, and lakes. Underground coal fires are often virtually impossible to extinguish and may smolder for decades; hundreds of such uncontrolled fires are reported in the United States each year. Coal dust can cause explosions. Coal washing, in which the usable coal is separated from impurities, results in unsightly and dangerous waste banks that can catch fire.

Coal mining is an extremely dangerous and physically demanding profession. Black lung disease plagues many miners, and miners are regularly also argue that the environmental impacts are minimal because the land in ANWR is tundra, with much less biodiversity than is found in rainforests and other highly productive ecosystems. Furthermore, they note, total area affected by drilling is much smaller than the entire Refuge and considerable money will be spent to reduce the impact of both the drilling and the roads and large pipelines needed to distribute the oil.

Opponents to the drilling argue that the amount of oil from ANWR is not that large compared to the U.S. demand. The United States consumes about 7.3 billion barrels per year, so the Refuge contains enough oil to supply the nation's needs for only approximately between 6 months to one year. These opponents argue that if the money was spent promoting energy conservation and alternative energy instead, this would not only improve the environment but also lead to a more permanent reduction in oil imports.

Drilling opponents also contend that, although the land within ANWR is tundra, it is one of America's richest areas for unique wildlife. It is a calving ground for an immense caribou herd and a place for half of the polar bears in Alaska to den. It is also where the eastern, western, and Rocky Mountain flyways converge for 130 species of migrating birds that head to the Arctic. Thousands of musk ox, artic foxes, arctic hares, wolves, and other wildlife make the Refuge their permanent home.

To what extent this wildlife will be harmed by oil drilling is a topic of much scientific interest. A study released by biologists in the Department of Interior in 2002 concluded that the caribou herd, which uses the coastal plain for calving each summer, "may be particularly sensitive to development" as there is a distinct lack of quality habitat for the herd elsewhere. Oil development will most likely result in restricting the location of concentrated calving areas and lead to fewer calves being able to survive and, in turn, possibly a decline in the herd. Similarly, snow geese, among the millions of migratory birds on the coastal plain, may be displaced because of increased activity, including air traffic. It cannot be assumed the geese will find adequate feeding areas elsewhere, the study says.

There is also concern over the impact of drilling on Native Americans that live in the region. The only permanent human settlement in ANWR is the town of Kaktovik, with a population of just 260. The Inupiat Eskimo in the town own shares, granted to them from native-rights settlements, in the Arctic Slope Regional Corporation. The corporation would benefit from oil drilling in the area. The mayor of Kaktovik, Lon Sonsalla, thinks that most residents support drilling, but he says of its prospects, "We're on a wait-and-see basis." The Inupiat's neighbors to the south, the Neets'aii Gwichin, live in Arctic Village, which sits on the southern boundary of the reserve. The Gwichin people strongly oppose any oil development in the coastal plain, as their cultural tradition is linked to the caribou, which they believe will be endangered by development.

Questions

1. List both the advantages and disadvantages of drilling in the Arctic refuge. Do you think the advantages outweigh the disadvantages? Why or why not?
2. Name three other areas of the world where oil drilling is common. Do you think that drilling in these areas is less harmful to the environment? Explain.

killed in accidents. The average coal miner dies several years earlier than other members of the population at a similar socioeconomic level.

Nuclear Power

Nuclear power supplies roughly 6.5% of the world's commercial energy overall, but accounts for a much larger percentage of the electrical generating capacity in certain industrial countries. More than 430 commercial nuclear reactors are currently in operation in some 30 countries, and approximately 25 nuclear plants are under active construction. The United States, with just over 100 nuclear power plants, generates approximately 20% of its electricity from nuclear power, while many other countries depend on nuclear power for more than 25% of their electricity (**Table 7-1**). The leading user of nuclear power in this respect is France, which generates over 75% of its electricity from nuclear power.

TABLE 7-1	Some of the Leading Countries in the Use of Nuclear Power		
Country	Number of Reactors	Electricity from Nuclear Generators (% of total)	Nuclear Energy Supplied, 2000 (billions of kilowatt-hours)
Belgium	7	56.8	45.4
Bulgaria	6	45.0	18.2
Canada	14	11.8	68.7
Czech Republic	5	20.1	13.6
Finland	4	32.2	21.6
France	59	76.4	395.0
Germany	19	30.6	159.6
Hungary	4	40.6	14.2
India	14	3.1	14.2
Japan	53	33.9	304.9
Korea, S.	16	40.7	103.5
Lithuania	2	73.7	8.4
Russia	29	15.0	119.7
Slovakia	6	53.4	16.5
Slovenia	1	37.4	4.5
Spain	9	27.6	59.3
Sweden	11	39.0	54.8
Switzerland	5	38.2	25.0
Taiwan	6	23.6	37.0
Ukraine	13	47.3	72.4
United Kingdom	35	21.9	78.3
United States	104	19.8	753.9

Source: Data compiled from *The World Almanac and Book of Facts, 2002* (New York: World Almanac Books, 2002, p. 162).

Japan continues to invest in nuclear power, and India and China, among other countries, are developing their nuclear capacity. In contrast, Austria is nuclear-free and it, along with the Philippines, does not plan to develop nuclear power. Due to public pressure, Sweden has been dismantling its nuclear power plants.

Nuclear power was first developed during World War II by the United States, but initially research focused on military applications such as the use of nuclear technology to build bombs and power submarines. The U.S. Atomic Energy Act of 1954 paved the way for private industry to develop nuclear energy, and by 1983, 80 nuclear power plants were operating in the United States. As nuclear plants proliferated in the 1960s and 1970s, however, the public became increasingly concerned about their safety. It was also discovered that nuclear power was not as inexpensive and efficient as had been predicted. In fact, in the United States electricity generated by nuclear power plants is currently more expensive than electricity generated by any other major technology. Nuclear-generated electricity can cost two to four times as much as electricity generated at a plant powered by natural gas, coal, or wind. Nuclear power plants are very expensive to build; mining and refining uranium ore is a considerable undertaking; accidents and malfunctions incur significant monetary losses; and the ultimate decommissioning of plants will require considerable sums of money (which could ultimately raise the cost of nuclear power to even higher levels).

Since 1978, there have been no new orders for nuclear power plants in the United States. Many other developed countries have overtaken the United States in the research, development, and implementation of nuclear power capability. Still, with more than a hundred operating nuclear power plants, the United States has the largest total monetary investment in nuclear power. Some observers

Figure 7-6   Diagram illustrating a nuclear chain reaction.

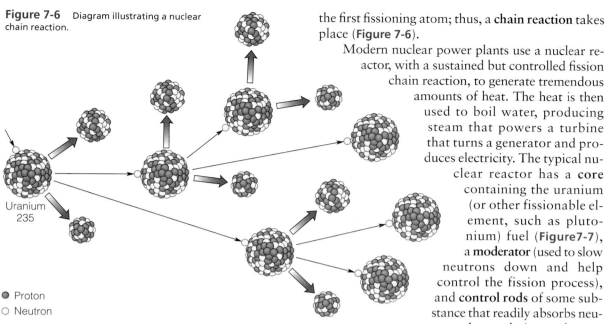

Uranium
235

● Proton
○ Neutron

the first fissioning atom; thus, a **chain reaction** takes place (**Figure 7-6**).

Modern nuclear power plants use a nuclear reactor, with a sustained but controlled fission chain reaction, to generate tremendous amounts of heat. The heat is then used to boil water, producing steam that powers a turbine that turns a generator and produces electricity. The typical nuclear reactor has a **core** containing the uranium (or other fissionable element, such as plutonium) fuel (**Figure 7-7**), a **moderator** (used to slow neutrons down and help control the fission process), and **control rods** of some substance that readily absorbs neutrons, such as cadmium or boron.

When the control rods are fully inserted, they absorb enough of the neutrons to halt the nuclear chain reaction. By selectively inserting the control rods, the number, or flux, of neutrons can be regulated and thus the rate of fission and energy release controlled. In some reactors, if the control rods are removed completely and excess heat is not removed quickly from the reactor core, the core may get so hot that it will begin to melt.

Water typically circulates around the core to cool it and is heated in the process. Because the wa-

worry that the United States is no longer a leader in advanced peacetime nuclear technology. Others feel that we are still too dependent on nuclear power.

Principles of Nuclear Power

Nuclear energy potentially encompasses two different, but related, types of reactions: fission and fusion. During nuclear **fission**, a radioactive isotope of a heavy element, such as a **uranium** or **plutonium** atom, is split into daughter products, and simultaneously energy is released. During **fusion**, isotopes of some light element (or elements) are fused together to make a heavier element (for instance, heavy isotopes of hydrogen bonding to form helium); in the process, energy is also released. All commercially operating nuclear reactors are run by fission. The Sun generates light and energy through fusion reactions, but on Earth controlled fusion is still only in the developmental stage.

Fission.   In fission, a neutron penetrates the nucleus of a fissionable atom (easily split by neutron penetration, such as uranium-235 [$^{235}_{92}$U]) and causes the nucleus to split into two or more smaller nuclei (these smaller nuclei, although nonfissionable, are highly radioactive). In the process, a tremendous amount of energy is released. The newly formed nuclei are positively charged and repel each other; as a result they travel apart at high speeds and correspondingly high kinetic energies. In addition, if other fissionable atoms are nearby, they may be induced to fission by the extra neutrons released by

Figure 7-7   The interior of the containment structure of the Trojan nuclear power plant, Rainier, Oregon. In the center of the photograph is the reactor core positioned under 35 feet (10.7 meters) of water.

ter exposed to the core potentially becomes radioactive, it is recycled and recooled in the primary water loop. The primary water loop, in turn, heats a secondary water loop that produces steam to operate a turbine. The turbine is connected to a generator that produces electricity (**Figure 7-8**).

The reactor core and primary water loop are housed within a thick steel tank, or pressure vessel, that is designed to contain all radioactive traces. In most reactors, the reactor vessel is housed in a **containment structure** composed of thick steel-reinforced concrete. The containment structure is designed to protect the outside environment from major radioactive contamination if the nuclear reactor should fail.

The principal type of reactor used in the United States is the **light water reactor (LWR)**. Currently, all commercial nuclear power plants in the United States, and about 75% of the nuclear plants worldwide, are LWRs (some U.S. research and military reactors are not LWRs). Light water reactors are named after their moderator, which is ordinary or light water (as opposed to heavy water, which contains an abundance of the hydrogen isotope deuterium). Chernobyl-type reactors use a graphite moderator and have other basic design differences from American-type LWRs.

The fuel used in typical LWRs is uranium that has been slightly **enriched** in fissionable U-235. The fuel has to be so enriched because ordinary water, while slowing the neutrons down, actually captures too many of the neutrons to sustain a chain reaction with uranium composed of its natural isotopic abundances. Typically, an LWR's fuel contains about 3% U-235 and 97% U-238.

U-238 is not normally fissionable, but it is a fertile isotope that, by absorbing a neutron, can give rise to a fissionable element. As U-235 fissions, it gives off neutrons. Some of these neutrons go into fissioning other U-235 atoms, but others will be absorbed by U-238 atoms. A U-238 atom that captures a neutron will ultimately be transformed to plutonium-239 ($^{239}_{94}$Pu). Pu-239 is a fissionable isotope, and once formed, it can undergo fission with a release of energy just as U-235 does.

Breeder reactors take advantage of the ability of nonfissionable U-238 to be converted to fissionable Pu-239. With a breeder reactor, the 99.3% of natural uranium that is nonfissionable can be put to use, thus greatly extending the amount of energy that can be generated from uranium ore deposits. Furthermore, **thorium**-232 ($^{232}_{90}$Th), an element that is several times more abundant in the Earth's crust than uranium, can be used in a breeder reactor to produce fissionable U-233.

To recover the fissionable plutonium that is produced in a breeder reactor, the used fuel and uranium/plutonium blanket around the core must be reprocessed. In reprocessing, the fissionable materials are isolated and concentrated to be used as fuel for a nuclear reactor. But this same fissionable material could be used to make an atomic bomb. A terrorist group or hostile country would need to steal only a small amount of weapons-grade, concentrated fissionable material to produce an atomic bomb. In addition, given the high concentrations of fissionable material that could accumulate in a breeder reactor and the fact that Pu-239 is fissionable by fast neutrons, some people fear that a breeder reactor might even have the potential to explode. In contrast, light water nonbreeder reactors theoretically never have high enough concentrations of fissionable materials to produce an actual atomic explosion, although an uncontrolled chain reaction could result in a meltdown, inducing nonatomic explosions (such as huge quantities of steam breaking containing

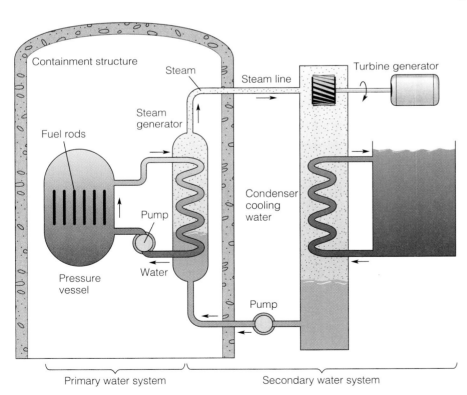

Figure 7-8   A schematic diagram of a typical pressurized light water reactor power plant.

pipes and vessels, or water being converted to hydrogen and oxygen, which could then explode).

Even if we could ensure that no fissionable material would be lost or stolen from a **reprocessing facility**, reprocessing spent nuclear fuel presents other problems. In particular, it is extremely dangerous and expensive. The spent fuel is highly radioactive and must be dissolved in strong toxic acids to extract the uranium and plutonium. Because of the high levels of radiation involved, much of the reprocessing must be done by remote control. The plants must be meticulously designed to avoid radiation leakage. For these reasons, along with concern over nuclear proliferation (the proliferation of atomic bombs), reprocessing is not being actively pursued in the United States.

Fusion.   During fusion, the nuclei of light elements are combined, or fused, together to form a heavier element, releasing a large amount of energy in the process. Extremely high temperatures are necessary to initiate fusion reactions, and on Earth the first artificial fusion reaction was attained in 1954 in the form of a successfully detonated hydrogen bomb. Controlled and sustained fusion reactions are still in the developmental stages and are not yet economically feasible as a source of commercial energy.

Methods currently being developed are based on the fusing of either two deuterium atoms (hydrogen atoms having a neutron as well as a proton in the nucleus), or a deuterium and a tritium atom (a hydrogen atom with two neutrons and a proton in the nucleus), to form a single helium atom plus a neutron. These reactions require temperatures in excess of 72 million degrees F (40 million degrees C). Two basic approaches are being developed to contain controlled fusion reactions: (1) magnetic confinement and (2) high-energy lasers and particle beams. (Experiments are also being pursued to produce "cold" fusion reactions; for instance, attracting deuterium [heavy isotopes of hydrogen] nuclei to an electrode immersed in heavy water [deuterium and oxygen] and forcing them to fuse and emit energy. Such research, however, has failed to produce unambiguous and universally accepted evidence of true nuclear fusion.)

In magnetic confinement, powerful electromagnets surround the fusion chamber and repel the fusion materials as they approach the walls of the chamber. Various types of magnet confinement reactors have been designed and tested experimentally; the most common and successful design is known as a tokamak. A tokamak is a large (the size

of a house) machine with a doughnut-shaped vacuum reactor vessel surrounded by magnetic coils.

Fusion reactions can also be initiated by bombarding a fuel pellet, containing deuterium or a deuterium-tritium mixture, with laser or particle beams from all directions. For a split second, the fuel pellet will remain suspended within the reaction chamber and implode with tremendous force and pressure, attaining enormous temperatures. If the pressure and temperature are great enough, fusion will occur. In many ways, fusion would be an ideal energy source especially if deuterium-deuterium fusion reactions can be harnessed. Deuterium makes up only a small percentage of all hydrogen atoms (about 1 out of every 6700 hydrogen atoms), but hydrogen is the most abundant element in the universe. The deuterium contained in 30 gallons (113.5 liters) of seawater, if totally fused, would release as much energy as 10,000 gallons (37,850 liters) of gasoline. Some researchers have estimated that if the deuterium in the top 10 feet (3 m) of water in Earth's oceans were used as a fuel in fusion reactors, it could satisfy the energy needs of all humans for about 50 million years.

Another advantage is that fusion does not produce fissionable materials that could be incorporated into the production of bombs. Fusion also produces much less radiation than fission, although some undesirable radioactivity is associated with fusion reactions. A sustained fusion reaction produces a dangerous number of neutrons. Every time two hydrogen atoms fuse to form a helium atom, a neutron is produced. These neutrons would cause the internal parts of the reactor to become highly radioactive. The neutron bombardment would weaken the fusion reactor over time, as would the intense heat generated by the fusion reaction.

Some people predict that fusion power could solve many of the world's fundamental material problems. The extremely high temperatures generated during the fusion process could be used to vaporize any and all waste materials; wastes would be broken down to their constituent atoms. The atoms could then be sorted out according to weight or other physical properties and reused—the ultimate recycling. Piles of waste—garbage, sewage, refuse—could be vaporized and sorted into piles of carbon atoms, oxygen atoms, iron atoms, gold atoms, copper atoms, and so on. The raw atoms could then be used to produce whatever is needed. Furthermore, with a virtually unlimited source of energy, humans could synthesize their own elements by fusing hydrogen nuclei together. We are unlikely to exhaust the supply of hydrogen since it is estimated that

Figure 7-9   The fuel cycle for electricity generation using a fission reactor.

over 90% of the atoms in the universe are hydrogen atoms.

The Nuclear Fuel Production Process

Conventional, nonbreeder fission reactors utilize uranium as their primary fuel source. The uranium must be extracted from uranium ore (**Figure 7-9**). Most uranium ore contains a relatively small percentage of actual uranium—typically, only 0.1%

Figure 7-10   An overhead view of the uranium-enrichment "converters" at the Paducah Gaseous Diffusion Plant in Paducah, Kentucky. Solid yellowcake is converted to the gas uranium hexafluoride (UF_6). The uranium hexafluoride gas is then run through a series of diffusion barriers that filter out some of the slightly heavier U-238 molecules and thus leave the mixture enriched in U-235. The plant has been in serious trouble in the last few years for releasing contaminated wastes, exposing its workers to excessive radiation levels, and possibly causing illness in nearby residents.

uranium metal by weight. Once mined, ore is concentrated during the milling process. There, the ore is crushed and leached using various chemicals (such as sulfuric acid or sodium carbonate); then, the uranium is precipitated from the solution. The result of the **milling process** is the formation of the substance U_3O_8, referred to as **yellowcake** (due to its color) or natural uranium.

Yellowcake is composed of 85% pure uranium by weight, but the uranium is still in its natural isotopic mixture of 99.3% nonfissionable U-238 and only 0.7% fissionable U-235. To be used in a typical American nuclear reactor, it must be enriched in the fissionable U-235 (**Figure 7-10**). The enriched uranium is made into fuel pellets, which are placed into fuel rods and used to power a reactor. The remaining depleted uranium (high in nonfissionable U-238, but low in U-235) is a radioactive waste.

Mining and processing uranium can have a major impact on the environment. A typical light water fission reactor with a 1000-megawatt capacity may require more than 140,000 metric tons of uranium ore to produce the fuel it consumes in a year. The mining operation to obtain this raw ore may extend over 18 acres (over 7 hectares) and displace as much as 2.5 million metric tons of rock and earth.

Uranium Resources

Uranium resources, like fossil fuels, are a nonrenewable source of energy. Assessing potential uranium resources is very difficult, but clearly some high-grade deposits of uranium ore (as in the western United States) are already being depleted. Like other mineral resources, concentrated uranium deposits are not spread evenly over the globe. Significant concentrations occur in Algeria, Australia, Brazil, Canada, the Central African Republic, Gabon, France, India, Niger, South Africa, Spain, and the United States. Based on current consumption, a world dependent on fission power, and not utilizing breeder reactors, might deplete these high-grade, currently economically minable uranium resources in only a couple of centuries.

Other sources of uranium exist. Seawater, for example, contains uranium, but it is so diffuse, large amounts of energy would be required to extract it. On a more positive note, uranium that would be uneconomical to mine for its own sake is recovered as a byproduct during the mining and refining of certain other minerals. Byproduct uranium has been produced from phosphate plants and from the mining and refining of copper. The end of the Cold War also saw the recycling and processing of weapons-grade uranium from nuclear warheads into fuel for power plants. Some lower-grade deposits of uranium also have the potential to be utilized. An example is the Chattanooga shale, which covers large areas in the east-central United States. Its uranium content is so low, however, that a 1000-megawatt fission reactor would require 30 times more shale to power it than if the uranium were obtained from higher-grade ore.

Advantages and Disadvantages of Nuclear Power

The Safety Record of Nuclear Power

Advocates of nuclear power rightfully claim that, in practice, nuclear power generation has thus far proven to be the safest form of large-scale commercial power generation. Nuclear power plants, so far, are much safer (in terms of human lives lost) and less damaging to the environment than fossil fuel-burning power plants, and they are also safer than hydroelectric power plants. Nuclear plants have a solid safety record despite several widely publicized accidents and the fact that minor mishaps occur routinely at nuclear facilities. Other sources of electricity generation also have many accidents, although they do not seem to attract as much attention from the media and the public as accidents at nuclear power plants. Furthermore, coal-fired power plants kill people on a routine basis—the pollution they give off literally kills innocent members of the public. Additionally, the air pollution, acid rain, and greenhouse gases they release are causing untold property damage (estimated at billions of dollars a year in the United States alone), disrupting ecosystems, and potentially changing the global climate (see Chapter 17). Hydroelectric power plants, although apparently clean and natural, cause untold environmental havoc by flooding upstream areas and reducing water flow downstream. The disruption they cause to natural water flow can encourage the proliferation of disease-bearing organisms, and a large dam failure could conceivably kill hundreds of thousands of people and cause billions of dollars worth of property damage.

Drawbacks of Nuclear Power

Two major categories of negative environmental impacts are associated with the use of conventional nuclear fission reactors. The first are the types of impacts that occur with any large power plant that uses a bulky fuel. Enormous amounts of energy, land, and materials must be utilized to build the plant; mining the enormous amounts of uranium ore needed to feed the plant involves substantial energy

and potential environmental degradation; and during routine operations, large quantities of water are used for cooling, and disruptive amounts of waste heat are dumped into the environment (such thermal pollution can be extremely damaging to the natural flora and fauna). Eventually, after 30 or 40 years or less, an aging nuclear power plant must be decommissioned; this too requires enormous amounts of energy and materials.

One of the chief advantages of nuclear power is that it does not produce harmful, carbon-based greenhouse gases such as carbon dioxide, nor does it spew particulates, sulfur dioxide, and similar harmful substances into the environment. Looking at the entire process, however, and not just at the actual operation of a nuclear power plant, reveals that greenhouse gases are produced. The building and later decommissioning of nuclear power plants involve the use of energy derived from fossil fuels, as do the mining and processing of uranium ore and the transportation and storage of the uranium fuel and spent fuel. It is sometimes suggested that nuclear power helps to decrease the U.S. dependence on foreign oil, but in fact many processes associated with nuclear power are driven by oil. Furthermore, nuclear power is used almost exclusively to generate electricity, and only about 6% to 7% of the United States domestic production of oil goes toward generating electricity. The real competitor with nuclear power for electricity generation is coal. Coal burning is extremely dirty and is known to be causing severe environmental degradation. Nuclear advocates rightfully point out that the large-scale substitution of nuclear power plants for coal-burning plants would avoid many of the problems inherent in coal-burning technology. However, nuclear power generation contributes its own set of wastes and attendant problems.

The second category of disadvantages of nuclear power is inherent and specific to this technology: the dangers of **radioactivity**. The very basis of nuclear power production involves radioactivity. A typical modern nuclear power plant contains within its walls radiation equivalent to that of a thousand Hiroshima bombs, and this leads to fears that radioactive wastes may inadvertently leak into the environment, accidents may occur (perhaps resulting in a reactor meltdown or explosion), or fissionable isotopes may fall into the wrong hands. Compounding these concerns are a number of widely publicized mishaps and the fact that there is still no long-term, satisfactory method of disposing of the radioactive wastes generated by nuclear power plants. Many people are concerned that a single "worst-case" nuclear accident could nullify all the potential benefits of nuclear power. A 1982 study by the Sandia National Laboratory in New Mexico predicted that a major accident in the United States might cause 50,000 to 100,000 immediate deaths and up to another 40,000 subsequent deaths due to radiation-induced cancer; monetary damages would amount to at least $100 billion. Even the Chernobyl accident was not a "worst-case" scenario—indeed, some argue it was relatively minor compared to what could happen.

Chernobyl and Three Mile Island

The Chernobyl Disaster.   In April 1986, a major nuclear reactor blew up, creating the single worst nuclear power plant disaster so far (**Figure 7-11**). The Chernobyl nuclear reactor site, located on the Pripet River about 80 miles (130 km) from Kiev in Ukraine, consisted of four reactors. The Number 4 reactor exploded, releasing an estimated 185 to 250 million curies of radioactivity into the environment (official Soviet estimates initially stated that 50 million curies were released, but it is now known that these numbers are too low). Within the first few months, according to official reports, 31 people died as a direct result of the explosion and the release of radiation (this includes a dozen firefighters who died of radiation poisoning), but the subsequent reports of mortalities are much higher. More than 4000 Ukranians have died as a result of being exposed to radiation during the cleanup operations, according to Ukraine government figures. The number who will ultimately die prematurely because of the incident could be in the tens of thousands. Immediately following the accident, hundreds of people were diagnosed as having radiation sickness, and over 100,000 people living within 18.5 miles (30 km) of Chernobyl had to be evacuated. It has been estimated that as many as 40,000 to 70,000 additional cancer deaths (many of them occurring outside the former Soviet Union) may result from the effects of Chernobyl over the next several decades. Dr. Vladimir Chernousenko, a nuclear physicist who was the scientific supervisor of the emergency damage control team sent into Chernobyl after the accident has suggested that perhaps 35 million people have been damaged by the high levels of radiation released from Chernobyl.

When the Chernobyl reactor exploded, a thick cloud of radioactive gases rose about a mile (1.6 km) into the atmosphere. These gases very quickly covered much of the Northern Hemisphere, affecting areas over a thousand miles (over 1600 km) away. In Scandinavia, Germany, and Great Britain, plants and animals were contaminated by radioactivity for years and declared unsafe for human consumption. In Lapland, the reindeer were declared unfit for human consumption due to radioactive contamination. In Italy, thousands of tons of vegetables were unusable. In Corsica, as a result of eating local contaminated cheeses and other milk products, children were found to have large accumulations of radioactive iodine-131 in their thyroid glands (the gland in the neck that regulates growth). In all, over a hundred million Europeans were subjected to either voluntary or mandatory food restrictions over the next few years.

Three years after the Chernobyl accident, the official reports stated that the severely contaminated area around the nuclear power plant was much larger than first estimated. Over an area of approximately 3900 square miles (10,000 km^2), contamination was at a level of 15 curies or higher per square kilometer—a very unsafe level. According to some estimates, a quarter million people still lived in badly contaminated areas. Four years after the accident, 627,000 Soviets were under permanent observation for symptoms and effects of radiation poisoning.

The financial costs of the Chernobyl accident totaled about $13 billion four years after the incident, and no one knows what the ultimate costs may be. The direct costs are expected to be greater than the Soviet government's total investment in nuclear power before the accident.

The Chernobyl accident was initiated by operator error, but various design defects (such as inadequate control systems and the lack of a containment vessel) made matters worse. The accident occurred during a test of the backup electrical system. The operators were supposed to slow down the reactor so that it would run at a low level during the test, but

Figure 7-11   The damaged Chernobyl reactor as it appeared on May 28, 1986.

they found they had slowed it too much. To speed up the reaction again, they pulled out too many control rods and reduced the amount of water cooling the reactor. The reaction sped up, but quickly went out of control. Intense heat caused an explosion that blew the roof, weighing about 900 metric tons, off the reactor and shot radioactive elements into the atmosphere, and a major fire started. Studies indicate that the actual reactor core at Chernobyl underwent a meltdown, burning through a steel and gravel barrier 6 feet (1.8 m) thick; it remained exposed for 10 days, all the while releasing radiation.

Only the gallant but suicidal work of firefighters and other emergency workers kept the fire from spreading to an adjacent nuclear reactor. Thousands of tons of boron, lead, and other radiation-absorbing materials had to be dropped onto the reactor, but unfortunately much or all of this material may have missed the actual core. Nitrogen was pumped under the reactor vessel to cool it, and finally the damaged reactor was entombed in reinforced concrete to try to contain the remaining radiation (**Figure 7-12**). No one is sure how long this tomb will hold; the radiation may begin to destroy it and leak through. Already there are reports that this concrete tomb is riddled with holes, structurally unstable, and could collapse at any time. Approximately 1000 square miles (2600 km²) of land around Chernobyl will remain contaminated with high levels of radioactivity into the indefinite future.

Over a dozen reactors of the same design type are still operating in Russia, Ukraine, and Lithuania. The other three reactors at Chernobyl were finally shut down in 2000, fourteen years after the Number 4 reactor melted down. Despite the early predictions of nuclear advocates that the probability of a major catastrophe was negligible, we now know that catastrophes do occur. Chernobyl was not even the first major nuclear accident.

Three Mile Island. In March 1979, a nuclear meltdown nearly occurred in the United States. At the Three Mile Island nuclear power plant (**Figure 7-13**) near Harrisburg, Pennsylvania, a minor problem developed in the plumbing of TMI Unit 2, with the end result that cooling water drained away from the reactor and the core began to partially melt—the worst commercial nuclear power plant accident in the United States to date. Operator errors, a stuck valve, faulty sensors, and design errors are all partially to blame for what happened at Three Mile Island. Although some radioactive gas was released due to the accident (only one thousandth as much as at Chernobyl), fortunately the containment structure around the reactor held most of the radioactivity in—about 18 billion curies of radioactivity that could have been released had the structure failed (compare this to the mere 185–250 million curies released at Chernobyl). Nuclear advocates have hailed the Three Mile Island accident as "proof" that a more serious accident would not occur in the United States—for the containment structure did hold. Many experts, however, have concluded that luck

Figure 7-12 An aerial view of the entombed Number 4 reactor at Chernobyl.

Figure 7-13 The Three Mile Island nuclear power plant near Harrisburg, Pennsylvania.

more than anything else kept the Three Mile Island accident from being worse. Some scientists have calculated that the core, which reached temperatures in the range of 4800° to 5000° F (2650°–2750° C) or higher, was just short of becoming hot enough to totally melt down. If emergency measures had not been initiated when they were, given another 20 to 30 minutes, the core might have completely melted through the steel reactor vessel and containment unit, releasing all 18 billion curies of radioactivity.

No one knows for certain what the health effects of Three Mile Island are, in large part because no one is certain how much radiation was actually released. A few days after the accident, the then-governor of Pennsylvania, Richard Thornburgh, evacuated all young children and pregnant women from within a 5-mile (8 km) radius of Three Mile Island as a safety precaution. Over 150,000 other people within an approximately 15-mile (24 km) radius of the plant voluntarily left the area for several days or more. At the least, there is evidence that due to radioactive-induced damage to immune systems, Three Mile Island contributed to the premature deaths of some elderly people in the affected area. Dairy farmers reported that many animals died shortly after the accident, local residents have come down with leukemia and other cancers, and one study suggested that an increase in infant mortality and severe thyroid disorders in babies born after the accident was due to the effects of radiation exposure.

Safer Nuclear Reactors

The typical nuclear reactor is very large and complex. In the United States, the average nuclear reactor has about 40,000 valves, compared to about 4000 valves for a coal-burning power plant that generates approximately the same amount of electricity. Ten times as many valves means 10 times as many chances for a valve to go bad or be turned the wrong way, perhaps leading to an accident. The complexity of nuclear reactors makes them difficult to build, especially where the designs are not standardized and each reactor is essentially designed and built from scratch. During the building of nuclear reactors, such errors as improper weldings, air bubbles in cement, unstable foundations, and even a reactor vessel that was initially installed backward (fortunately, someone noticed the mistake some months later before the plant was completed) have been documented.

This complexity makes nuclear reactors very expensive to build. Because of this great expense, commercial utilities tend to build large reactors that have the capacity to generate large amounts of energy that can be sold to cover the expense of the re-

actor and still generate a profit. Thus, commercial nuclear reactors in the United States typically have a capacity of 1000 megawatts or greater. Of course, the larger the reactor, the larger the potential catastrophe if anything malfunctions.

Addressing these sorts of concerns, a new series of nuclear fission reactors are currently being designed. These reactors will probably be smaller and will include such features as passive safety systems and modular designs.

Passive safety systems depend on natural forces such as gravity to cool and control the reactor, especially in the event that an accident occurs. Rather than relying on pumps and complex mazes of pipes to bring water to an overheating core, a passive design might have cooling water located in storage tanks directly above the reactor vessel. Under normal conditions, the water would not be released, but in an emergency, valves would open, and the water would be dumped directly on the core to cool it. Valves could even be installed that would open automatically if the temperature rose above a certain point. The core itself might be buried in a cavity underground to protect it from such dangers as an airplane crashing into it (perhaps an unlikely event, unless done purposefully by a terrorist group, and potentially disastrous if it should occur). Pilot tests of small experimental reactors incorporating such passive safety systems have met with success.

Building standardized reactors that are made of modular components (series of identical subunits) could reduce the risks of errors occurring during construction. The modules might even be prefabricated in a factory and then brought to the construction site. Assembly procedures could be streamlined, and the risk of potential errors reduced. Standardization and modular construction would also reduce costs and make it more practical to build smaller nuclear power plants. Such smaller reactors might also run at lower temperatures than the reactors currently in use; lower-temperature reactors would probably mean overall safer reactors. A reactor design currently under development in Great Britain has a planned output of only 320 megawatts.

Disposal of Nuclear Wastes

Although nuclear reactors have been on line since the 1950s, and massive amounts of radioactive wastes have been generated by nuclear reactors and nuclear weapons-manufacturing facilities, no satisfactory plan has been implemented to dispose of these wastes. Nuclear wastes include spent fuel, radioactive products generated in the core of a reactor during operation, contaminated materials and clothing, and radioactive mining wastes and tailings. (Wastes from the use of radioisotopes in medicine, smoke alarms, and the like are also radioactive, but pose a relatively minor problem.) Some of this material is highly radioactive and will remain so for hundreds of thousands or millions of years; other nuclear waste materials are characterized by low, but still very dangerous, levels of radioactivity. The amount of irradiated uranium fuel waste has more than doubled since 1989 and is continuing to accumulate at a fast pace (**Figure 7-14**). At present, it amounts to more than 190,000 metric tons.

This irradiated uranium fuel includes some of the most hazardous types of radioactive wastes. In the United States, irradiated uranium is less than 1% by volume of all radioactive wastes, but accounts for 95% of the radioactivity of radioactive wastes. The typical American commercial reactor discharges about 30 metric tons of irradiated fuel each year. Initially, this fuel is extremely radioactive and hot and decays relatively quickly—but it will continue to emit significant quantities of radiation for thousands of years.

Currently, more than 30 countries are operating nuclear power plants, yet nowhere in the world has a permanent disposal system for high-level nuclear wastes been fully developed, much less put into place. Instead, the spent fuel rods and other high-level wastes are generally stored "temporarily" on-site in pools of water (**Figure 7-15**). So much high-level, extremely toxic nuclear waste is now stored in this manner that many reactors are running out of temporary storage space. Low-level nuclear wastes, which occur in much larger volume, have generally been placed in steel drums and the like and buried in shallow landfills. At numerous such sites, the radioactive wastes have leached into the ground, contaminating soils and water.

Many suggestions for dealing with nuclear wastes have been put forth. Some people have suggested that we get them off the planet completely, perhaps by loading them onto rockets and shooting them into the Sun. The costs of such an undertaking would be almost incalculable given the volume of wastes that must be disposed of. Depending on the rocket engines used, massive amounts of energy would be required, and massive amounts of greenhouse gases would be added to the atmosphere during the process, effectively nullifying one of the primary advantages of nuclear power in the first place. Furthermore, the valuable raw materials that would go into the manufacture of such rockets would be lost forever, putting even more pressure on our dwindling mineral supplies. Finally, think of the disastrous consequences if a rocket ship laden with high-level nuclear waste were to explode in the atmosphere.

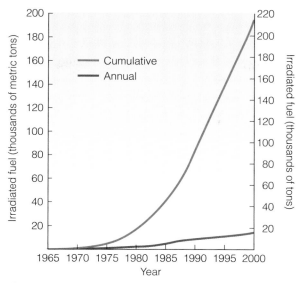

Figure 7-14   World generation of irradiated fuel from commercial nuclear power plants, 1965–2000. (*Source*: Based on data and projections from Worldwatch Institute, *Vital Signs 1995* [W.W. Norton, New York, 1995, page 89].)

Schemes to bury nuclear wastes at the bottom of the oceans or inside glaciers in Antarctica have also been proposed. But both the oceans and the ice are very delicate ecosystems, and contamination by nuclear wastes could have extremely damaging effects. The heat given off by radioactive wastes might even begin to melt the relatively unstable Antarctic ice sheets. Furthermore, both the oceans and ice have already been contaminated by nuclear wastes. Nuclear fallout from weapons testing is de-

Figure 7-15   A fuel rod storage pool at a nuclear power plant.

tectable even in the polar regions. From the 1940s through the 1960s, barrels of radioactive wastes were often dumped directly into the oceans. Now a number of these barrels are leaking and contaminating the seas.

Some persons have suggested that radioactive wastes could be inserted into subduction zones at colliding plate boundaries and thus forced deep into the Earth. This scheme appears to be extremely unrealistic, however. The initial insertion of the radioactive wastes would probably require deep excavation or drilling in a very active geologic terrane. Rates of subduction are so slow (several inches or tens of centimeters a year), that the wastes would take an inordinately long time to be subducted into the Earth. Finally, since subduction zones are very active geologically (ringed by areas of intense earthquake and volcanic activity), the radioactive materials could be spewed back to the surface before they are safely subducted into the lower layers of the Earth's crust.

Another suggestion is to store radioactive wastes on otherwise deserted islands, at least temporarily until more permanent disposal facilities are established. Again, such a scheme poses grave risks to the biosphere, and the islands would have to be heavily guarded on a permanent basis.

Another idea that is being actively explored by scientists in the United States, France, Japan, and Russia is to convert radioactive wastes to less dangerous, or shorter-lived, isotopes by bombarding the waste with neutrons. Such transmutation of elements would be technically difficult, however, and probably very energy-intensive and expensive.

The suggestion currently favored by many governmental and nongovernmental authorities in the United States and other countries that use nuclear power is to develop disposal facilities buried deep underground in geologically stable regions, perhaps in old mines. The U.S. National Research Council has concluded that geological burial is the "best, safest long-term option" for disposing of high-level radioactive waste. The problem is to find an underground formation that is isolated from the general environment (for instance, groundwater cannot be flowing through the site) and is not in an area of geological activity such as earthquakes or volcanic eruptions. One must be able to guarantee that the site will remain stable for tens of thousands to millions of years—a virtual impossibility given the active nature of the Earth's crust. Even if a geologically appropriate site is located, social and political considerations can be a problem. When a nuclear waste disposal facility is proposed, local residents tend to give a NIMBY ("not in my back yard") response.

Old salt mines have been proposed as storage facilities, but some mines have water percolating into them that forms a very corrosive brine. This solution will damage the steel drums containing the radioactive wastes and allow the contents to leak out; something other than steel drums may need to be used. In February, 2002, Secretary of Energy Spencer Abraham formally recommended to President Bush (a recommendation subsequently approved by the President and Congress) that Yucca Mountain be developed as a long-term national repository for high-level radioactive waste (**Figure 7-16**). If all goes smoothly, including still necessary approvals by the Nuclear Regulatory Commission, nuclear waste from 131 temporary storage locations in 39 states could begin being shipped to Yucca Mountain in 2010. Yucca Mountain is located about 85 miles (136 km) northwest of Las Vegas, Nevada, on land owned by the Shoshone Native Americans. From a political perspective, Yucca Mountain has advantages as a nuclear repository site: The population of Nevada is fairly small, which limits potential opposition, and the site is part of the former Nevada Test Site where nuclear weapons were tested, thus setting a precedent for nuclear-related activities in the area. Yet it is unclear if Yucca Mountain is appropriate from a geological perspective. Various U.S. Geological Survey geologists and engineers who have evaluated the site have raised questions about it. Yucca Mountain is surrounded by geological faults and volcanoes. Since 1857, eight major earthquakes have occurred within a 250 mile (400 km) radius of Yucca Mountain. Although none of the volcanoes in the immediate vicinity has erupted in historical times, estimates are that one volcano erupted a mere 5000 years ago.

Figure 7-16   Interior view of the storage facilities at the Yucca Mountain proposed national nuclear waste repository.

Thus, Yucca Mountain's geological stability for even the next 10,000 or 20,000 years seems questionable.

Another major problem with Yucca Mountain, or virtually any proposed nuclear waste repository on land, is that predicting the climate of the area for the next ten thousand years or so is impossible. At present, Yucca Mountain is located in a desert, and the water table is very deep below the surface. This is good because any water seeping into the repository could corrode canisters and mobilize radioactive waste. But given the potential for climatic change (see Chapter 17), there are no guarantees that Yucca Mountain will not experience significantly increased levels of precipitation in the future. Rainwater could not only seep into and leach the radioactive waste, resulting in radioactive runoff, but it could also cause the water table to rise and flood the nuclear waste. The net result could be massive leaking of high-level radioactive wastes into the groundwater system.

Decommissioning Nuclear Reactors

A nuclear reactor cannot last forever. The high temperatures and radiation bombardment, especially the neutron bombardment of the reactor vessel, cause it to weaken and become brittle over time. If allowed to run for a long enough period of time, eventually the whole plant will become contaminated and weakened by radiation. Pipes may begin to break, instruments may fail, and structural components collapse. The reactor vessel could even rupture. For these reasons, any nuclear power plant, even if it has a flawless record of operation, must eventually be taken out of service—it must be decommissioned.

Many researchers consider the appropriate life span of a good nuclear reactor to be in the range of 30 to 40 years. Until recently, nuclear power plants in the United States were only licensed for 40 years, after which they would have to be decommissioned. Since more than 60 nuclear power plants in the country began operating in the 1960s, we are currently approaching a time when many of these plants should begin planning for decommissioning. In June 1991, however, the Nuclear Regulatory Commission approved a plan that allows plants to apply for a renewal of their operating license for up to 20 more years. This ruling may push the costs and problems of massive decommissionings into the future, but many people worry that it may also increase the risks of major accidents as the aging plants continue to operate.

A fully operational large-scale commercial nuclear power plant has never been fully decommissioned in the United States, and scientists are not in complete agreement on what decommissioning will involve or how much it will cost. The spent fuel will have to be removed from the reactor site and eventually stored in some permanent storage facility. As far as the reactor complex itself, several options have been suggested. The plant could simply be closed up, fenced off, and put under guard. Such mothballing is perhaps the cheapest solution in the short run and also allows the plant to be reopened in the future. Or the plant could be further dismantled in the future, after the levels of radiation have died down. In some ways however, mothballing is also the most dangerous option because the radioactivity is still on site; if not guarded carefully, it could be disturbed purposefully or inadvertently, perhaps resulting in radiation leaking into the surrounding environment. A second option is to entomb the plant—seal it off by encasing it in concrete and steel so that nothing, including radioactivity, can get in or out. Entombment is currently being used to deal with the damaged Chernobyl reactor. Finally, a decommissioned power plant could be totally dismantled. All components, structures, and so on would be taken apart or cut up. All materials contaminated with radioactivity, including soil and water, would be removed to a safe facility for the permanent storage of radioactive waste.

Estimated costs to completely decommission and dismantle a nuclear power plant are $3 billion or more. For comparison, building a nuclear plant costs approximately $4 to $9 billion, and takes about 15 years. If the plant runs at full capacity for 40 years, it will generate about $12 billion worth of electricity. In reality, however, typical U.S. power plants operate at only about 60% capacity.

Global Nuclear Power Today and in the Future

The world's electrical generating capacity of nuclear power plants increased from 1 gigawatt in 1960 to 351 gigawatts in 2001 (**Figure 7-17**) (see Fundamentals 7-2 for an explanation of gigawatts and related units).

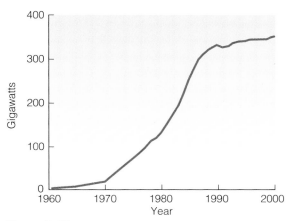

Figure 7-17   World electrical generating capacity of nuclear power plants, 1960-2001. (*Source*: Worldwatch Institute, *Vital Signs 2002* [W.W. Norton, New York, 2002, page 41].)

FUNDAMENTALS 7-2 Common Measures of Energy and Power

Joule (J). The basic unit of energy, heat, or work as measured using the metric system; equal to a force of one newton times one meter. From a macroscopic human perspective, this is a very small amount of energy. The burning tip of a wooden match gives off an estimated 1000 joules of energy.

Calorie (cal). A basic unit of heat energy in the metric system. A calorie is the quantity of energy that will warm one gram of water one degree Celsius. One calorie equals 4.184 joules.

Kilocalorie (kcal or Cal). 1000 calories or 4184 joules.

British thermal unit (BTU). The basic unit of energy in the English system. The energy that, when converted to heat, will raise the temperature of one pound of water one degree Fahrenheit. One BTU is equal to 0.252 kilocalories.

Quad (Q). One quadrillion (10^{15} or 1,000,000,000,000,000) BTUs of energy. One quad is equivalent to the amount of energy contained in approximately 171–172 million barrels of oil, 36 million metric tons of coal, or a trillion cubic feet of natural gas.

Petajoule (PJ). One quadrillion joules. There are approximately 1054 petajoules to a quad. One petajoule is approximately equivalent to the energy contained in 163,400 "U.N. standard" barrels of oil (see barrel) or 34,140 "U.N. standard" metric tons of coal (see metric ton).

Terajoule (TJ). 10^{12} joules. One thousand terajoules is equal to one petajoule.

Exajoule (EJ). 10^{18} joules. An exajoule is equal to 1000 petajoules or 1 million terajoules.

Barrel. A measure of volume of petroleum oil, containing about 42 gallons or about 159 liters. Depending on the quality of the particular oil, a typical barrel of petroleum may produce between about 5 and 6 million BTUs of energy when completely burned (see quad and petajoule).

Cubic feet (cu. ft). A measure of volume that is used to measure natural gas. A cubic foot of dry natural gas can produce approximately 1031 BTUs of energy.

Metric ton. 1000 kilograms (2204.6 pounds or 1.10231 English "short" tons). This unit is commonly used to measure amounts of coal (see quad and petajoule). A standard metric ton of coal may produce 27–28 million BTUs when completely burned, but the energy produced by an actual ton of coal will vary considerably depending on its grade and quality. Oil is also often measured in terms of metric tons. Depending on the type of oil, the exact number of barrels per metric ton will vary, but it is generally about 7.33 barrels of oil per metric ton.

Metric ton of oil equivalent (toe). Equal to 41.868 gigajoules. Thus, a thousand metric tons of oil equivalent (1000 toe) is equal to 41,868 gigajoules or 0.041868 petajoules, and a million metric tons of oil (1,000,000 toe) is equal to 41.868 petajoules.

Therm. A measure of natural gas, equal to 100,000 BTUs. A therm of gas is equivalent to approximately 97 cubic feet of natural gas.

Watt. A measure of power, energy consumption, or conversion. It is commonly used in connection with electricity. One watt is equal to one joule of work or energy per second. A 100-watt lightbulb consumes 100 joules of energy per second, or 360,000 joules per hour.

Kilowatt (kW). Energy conversion at the rate of 1000 watts (1000 joules per second).

Megawatt (MW). Energy conversion of 1000 kilowatts (1 million joules per second), or 1 million watts. Many conventional nuclear power plants have generating capacities of 1000 megawatts or more; that is, they can produce 1000 million (1 billion) joules of energy per second (enough energy to power 10 million 100-watt lightbulbs).

Gigawatt (GW). Energy conversion of 1 million (1,000,000) kilowatts.

Terawatt (TW). Energy conversion of 1 billion (1,000,000,000) kilowatts.

Horsepower (hp). A somewhat obsolete (except for cars and mechanical engines in some countries) term for power that is approximately equivalent to 750 watts or three-quarters of a kilowatt.

Kilowatt-hour (kWh). A measure of energy based on the concept of the kilowatt; the energy converted or consumed during an hour if the energy is being converted or consumed continuously at the rate of 1 kilowatt. A kilowatt-hour equals 3.6 million joules, 860.4 kcalories, and 3413 BTUs.

Kilowatt-year (kW-yr). The energy converted or consumed during a year if the energy is being converted or consumed continuously at the rate of 1 kilowatt. A kilowatt-year is approximately equivalent to the amount of energy given off when 1050 kilograms (approximately one metric ton) of coal are burned. If people consume 1 kilowatt-year of energy per year (1 kW-yr/yr), their average rate of energy consumption is 1 kilowatt.

Megawatt-year (MW-yr). The energy converted or consumed during a year at the rate of 1 megawatt. A megawatt-year equals 3.156×10^{13} joules, 7.542×10^9 kcalories, and 2.993×10^{10} BTUs.

Terawatt-year (TW-yr). The energy converted or consumed during a year at the rate of 1 terawatt. A terawatt-year is approximately equivalent to the energy given off by burning a billion metric tons of coal. The world as a whole consumed approximately 10 terawatt-years worth of energy in the single year 1980; thus, for 1980, we can express the total world energy consumption rate as approximately 10 TW (10 TW-yr/yr equals 10 terawatts). One terawatt-year is equal to approximately 31.6×10^{18} joules, 31,600 petajoules, or 31.6 million terajoules.

The heyday of nuclear reactor construction was in the 1970s and 1980s. In 1970, global nuclear electrical generating capacity was only 16 gigawatts, but twenty years later (1990), it was 328 gigawatts. In the last decade, increases in nuclear generating capacity have been small. The role that nuclear power will play in the global energy mix of the future is uncertain. In many areas there is strong opposition to nuclear power; thus in North America and Western Europe, no new reactors are currently being built.

Germany, for instance, plans to phase out the use of nuclear power and slowly shut down its 19 reactors; by 2025, there may be no commercial reactors operating in Germany. Still, in the year 2000, there were 25 nuclear reactors under construction (their combined capacity when brought on line would be 22 gigawatts). Most of this construction is located in Asia: China, India, Japan, South Korea, and Taiwan. Thus, it appears that the overall nuclear electrical generating capacity is shifting from the west to the east.

SUMMARY

STUDY GUIDE

- Global energy consumption has increased dramatically over the last half century.
- There are major discrepancies in energy use around the world; among the nations, the United States is both the largest producer and the largest consumer of energy, although the U.S. consumes considerably more energy than it produces.
- In order of importance (most to least) globally, the big five energy sources are: petroleum (oil), natural gas, coal, hydroelectric power, and nuclear power.
- Power plants convert various sources of energy into electrical energy.
- Hard energy technologies rely on large, centralized plants and complex distribution systems.
- Soft energy technologies are on-site, localized systems that are usually more environmentally friendly.

- Major advantages of fossil fuels are their historic abundance and relative ease of distribution and use; disadvantages are their growing scarcity and contributions to environmental problems such as local and regional pollution and global climate change as they emit large quantities of carbon dioxide.
- Nuclear power, based on fission reactions, was developed commercially after World War II (ended in 1945) as a means of generating electricity.
- A chief advantage of nuclear power is the lack of greenhouse gas emissions; drawbacks include massive amounts of radioactive wastes that must be disposed of and the fear of a major nuclear accident that could far exceed the Chernobyl disaster.

KEY TERMS

Blackout	Fossil fuel	Petroleum
Breeder reactor	Fusion	Plutonium
Brownout	Generator	Power plant
Chain reaction	"Hard" technology	Radioactivity
Coal	Hydroelectric power	Reprocessing facility
Containment structure	Light water reactor	"Soft" technology
Control rod	Milling process	Strip mining
Core	Moderator	Thorium
Electricity	Natural gas	Turbine
Enriched	Nuclear power	Uranium
Fission	Oil	Yellowcake
Fluidized bed combustion	Peak load	

STUDY QUESTIONS

1. Why is energy consumption important to society, and what concerns does it raise?
2. Name the basic laws of thermodynamics. Why are they important when considering energy transformations?
3. Summarize the current global energy situation. Which countries use the most energy? The least?
4. What is a power plant? How do most power plants operate?
5. Distinguish between "hard" and "soft" energy technologies, citing examples of each.
6. Why is modern industrial society so dependent on fossil fuels? Can we maintain this dependence in the future?
7. What does it mean to say that fossil fuels are a nonrenewable resource?
8. Discuss the pros and cons of each of the major types of fossil fuels: oil, coal, and natural gas.
9. Describe the environmental degradation attributed to the use of fossil fuels.
10. Discuss the potential consequences of a nation being overly dependent on foreign oil supplies.
11. Describe the current role of nuclear power in the global commercial energy budget.
12. Is nuclear power generation a "hard" or "soft" technology? Why?
13. Explain the principles behind modern nuclear fission power plants.
14. What is fusion? In what context have human-induced fusion reactions taken place on Earth?
15. Describe the nuclear fuel production process.
16. What are some of the advantages of nuclear power? Some of the disadvantages?
17. Briefly describe the Chernobyl and Three Mile Island accidents. How did these incidents affect the public's perception of nuclear power?
18. What methods have been suggested to dispose of the high-level and low-level radioactive wastes that result from nuclear power generation?
19. What steps are being taken to produce safer nuclear reactors?
20. Why is nuclear power so controversial? Is this controversy justified or simply the result of some people's ignorance of the technology involved?

WHAT'S THE EVIDENCE?

1. The authors assert that the burning of fossil fuels has created environmental havoc on a global scale. What are they referring to? Do you agree with the authors' assertion? Or are they simply being environmental alarmists who too quickly blame modern western technology for any ills, real or imagined? Cite evidence to defend your position.
2. The authors suggest that there is still no long-term, totally satisfactory method of disposing of radioactive wastes from nuclear power plants. However, the authors do discuss a number of options that have been suggested to deal with radioactive waste. What evidence do the authors present to bolster their claim that radioactive waste from nuclear power plants is a major problem? Taking the waste generated into consideration, what is your position on nuclear power? Cite evidence to defend your position.

CALCULATIONS

1. Plutonium-239 has a half-life of approximately 24,000 years. If we start with 7 ounces avoirdupois (198.44666 g) of plutonium-239, how much plutonium will remain after 120,000 years? (Express your answer in both ounces and grams.)
2. Radon-222 has a half-life of approximately 3.8 days. If we start with 9 ounces avoirdupois (255.14569 g) of radon-222, how much radon will remain after 19 days? (Express your answer in both ounces and grams.)

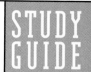

ILLUSTRATION AND TABLE REVIEW

1. Referring to Figure 7-1, world energy consumption increased by what percentage between 1970 and 1999? Based on the projections given in Figure 7-1, it is estimated that world energy consumption could increase by what percentage between 1999 and 2020?

2. Referring to Figure 1 of Case Study 7-1, how much more energy did the United States consume than it produced in 1965? Express U.S. energy consumption as a percentage of production in 1965. How much more energy did the U.S. consume than it produced in 2000? Express U.S. energy consumption as a percentage of production in 2000.

3. Referring to Table 7-1, list the top three countries in terms of their percentage of electricity supplied by nuclear generation. List the top three countries in terms of total amount of nuclear energy supplied.

WWW.ENVIRONMENT.JBPUB.COM

www.environment.jbpub.com

Connect to this book's web site: http://www.environment.jbpub.com/ The site features eLearning, an online review area that provides quizzes, chapter outlines, and other tools to help you study for your class. We've posted the answers to the book's odd-numbered questions. You can also follow useful links for more in-depth information. The site provides sources for the information in this chapter and recommendations for further readings.

Can you spot the solar-powered home? The townhouse on the right has an integrated photovoltaic (PV) roof that looks and performs like the standard metal roofing on the other units, but it also produces electricity. The solar roofing system was developed by United Solar Systems and Energy Conversion Devices. Decentralized, off-the-grid, alternative energy sources (alternatives to fossil fuels, nuclear power, and large-scale hydroelectric plants) such as solar, biogas, and wind power may someday supply electricity to the majority of the world's people.

CHAPTER 8

Renewable and Alternative Energy Sources

Solar power, geothermal power, wind power, and the burning of biomass (such as wood, dung, urban waste, methane produced from garbage, or alcohol fermented from plants) all provide significant amounts of energy in local settings, although they account for only a small percentage of the global energy budget. These, and even more minor sources of energy (such as ocean tidal power), present alternatives to the **"big five" energy sources**: coal, oil, natural gas, nuclear power (see Chapter 7) and hydropower (discussed below).

Along with hydropower, the alternative forms of energy generation are usually thought of as renewable, sustainable (or at least potentially sustainable), and environmentally benign. As we shall see, however, some potential environmental problems are associated with all of these sources of energy, and geothermal energy in particular is not strictly a renewable or sustainable energy source. Of the big five energy sources, hydroelectric power is most closely associated with the **alternative energy** sources. Hydroelectric power is renewable and sustainable (although overly large and mismanaged hydroelectric dams can lead to environmental abuse and long-term damage). Some futurists envision smaller-scale hydroelectric power plants playing an increasingly important role in the decades to come.

In the long run, it may be inevitable that hydroelectric power and alternative energy sources will have to make up a greater percentage of the energy produced once nonrenewable fossil fuel supplies are exhausted, especially if public confidence in nuclear power continues to falter. Alternative energy forms, such as electricity generated by wind or solar thermal power, stand to become increasingly popular as they become cost competitive with traditional methods of generating electricity.

Alternative energy sources have received a major boost with the rise of **independent power producers (IPPs)**. IPPs construct electricity-generating plants and then sell the electricity to the large utilities. In many cases, the IPP plants use alternative energy technologies and are relatively small, innovative, and environmentally friendly. Such independent power generation is supported in the United States by the Public Utility Regulatory Policies Act of 1978, which allows smaller, independent, relatively unregulated power companies to build generating plants that use renewable fuels (such as wind or solar power). But IPPs are not just an American phenomenon; they are appearing around the world and herald a new emphasis on alternative energy sources.

Along with the development of alternative energy sources, increased attention is being paid to energy conservation and efficiency—doing more with less energy. Already, great strides are being made in this area, and it is likely to attract greater interest in the future.

Hydropower

Hydroelectricity (**hydropower**) is the fourth largest source of commercial energy production and consumption globally, trailing far behind oil, coal, and natural gas; nuclear power comes in slightly behind hydroelectric power by most current estimates, although in recent years nuclear power has been catching up quickly. Unlike the fossil fuels and nuclear power, hydropower is a renewable resource. The basic principle behind this source of energy is the damming of rivers to create artificial waterfalls (**Figure 8-1**), although natural waterfalls are occasionally used. The falling water is used to turn turbines that drive electrical generators. The great advantage of hydropower is that once the dam is built and the turbines are in operation, it is a relatively cheap and very clean source of electrical energy. An operating hydropower plant produces no waste; it does not spew carbon dioxide or other pollutants into the atmosphere, as do fossil fuel-powered plants, and no toxic, hazardous, or radioactive waste is generated. An operating hydropower plant can be clean, safe, and efficient. For this reason, some people strongly support the construction of more hydropower plants to displace the use of fossil fuels and nuclear power. Nevertheless, the United States seems unlikely to build any more large-scale hydroelectric power plants in the near

Objectives

After reading this chapter, you should be able to explain or describe:

The concept of renewable and alternative energy sources

The principles of hydroelectricity

The pros and cons of hydropower

Advantages and disadvantages of biomass energy

Ways solar energy can be utilized by humans

Advantages and disadvantages of solar energy

Wind power

Geothermal energy

Ocean energy

Types of energy storage systems

Principles of energy conservation and efficiency

(a)

(b)

Figure 8-1 (a) An aerial view of Chief Joseph Dam, a very large hydroelectric power plant built on the Columbia River. It produces enough electricity to supply the entire Seattle metropolitan area. (b) The interior of the powerhouse, showing the turbines.

future, although they continue to be constructed in other parts of the world.

Despite its advantages, hydropower also has drawbacks. The construction of a major hydroelectric power plant is extremely expensive—it is a capital-intensive proposition—and in the process, large quantities of fossil fuels (to power machinery) and materials are consumed. Large-scale hydroelectric operations purposefully flood huge tracts of land and seriously disturb natural ecosystems. Forests and other natural habitats are destroyed, along with local villages (often inhabited by indigenous peoples); farmland is lost, people are displaced, erosion rates are increased, water may become polluted and choked with particulate matter, water temperature and flow are affected downstream (often harming plants and animals), and severe unanticipated flooding can occur.

Once a hydroelectric plant is built across a river, the mineral and dissolved oxygen contents of the water on both sides of the dam are often affected. Evaporation of water from the relatively still reservoir tends to increase the concentration of salts and minerals in the water, and water leaving the reservoir usually has a higher temperature and lower dissolved oxygen content than it did before the plant was built. These changes can radically affect the flora and fauna that live in the river and along its banks. The dams can also create conditions that favor the spread of disease. The larvae of malaria-carrying mosquitoes, the larval stages of flies that transmit the disease known as river blindness, freshwater snails that are the hosts for the parasites (schistosomes) that cause bilharzia, and other disease vectors can be favored by water changes brought about by artificial dams.

Large dams can also affect coastlines and coastal fisheries by obstructing the passage of nutrients and particulate matter to the sea. The construction of the Egyptian Aswan Dam resulted in the trapping of nutrients that previously flowed into the Mediterranean, resulting in damage to the Mediterranean sardine fishery. Similarly, by trapping sand and silt, dams can cause coastlines to flood. Normally, the new sand and silt would replace material that is eroded from the coastal beaches; when this supply of new material is cut off by the dam, the beaches continue to erode and the ocean creeps landward. After the Volta Dam was built in West Africa, some 10,000 people on the coast of Togo reportedly lost their homes to the Atlantic Ocean in this manner.

In many ways, giant hydroelectric power plants share the problems of other large power plants (fueled by coal, nuclear energy, or some other energy source). The setup expenses of a major hydropower plant are enormous. The plant must be carefully maintained and operated. The consequences of a major accident or failure, such as the dam cracking and giving way, could be catastrophic. Also, contrary to popular belief, most large hydropower projects have only a limited life span. The critical water reservoirs behind the dam often fill up with the sand and silt that are not allowed to pass downstream. Such silting can occur very rapidly, especially if the upstream watersheds are deforested and/or subjected to poor agricultural practices that lead to accelerated erosion. Once the reservoir is silted up, the usefulness of the hydropower plant is greatly diminished. The deep reservoir is necessary for storing energy during periods of high rainfall and runoff and for regulating production, so that a steady stream of electricity can be produced when needed. Without the reservoir, the plant becomes an intermittent source of electricity at best.

In the United States, traditionally hydropower plants have been very large. The first major hydropower plant was completed at Niagara Falls in 1895, and in the early and mid-twentieth century, numerous large plants were constructed. Hoover Dam, built in 1935, is typical of the dams of this era; it has an electrical energy output capacity of about 1300–1400 megawatts. But many operating hydroelectric plants in the United States are much smaller. At present, the country has more than 2000 hydropower facilities of all sizes, with an estimated total capacity of 73,000 megawatts. A recent Federal Energy Regulatory Commission study identified another 5000 sites that have the potential to sustain hydropower facilities. If all of these sites were developed, they would contribute another 74,000 megawatts of capacity. Thus, one can conclude that the United States could double its use of hydropower.

Such statistics can be misleading, however. Some of the best sites for further development of hydropower capacity are found in scenic areas, parks, and wildernesses. Federal, state, and local legislation and regulations (especially the federal 1968 National Wild and Scenic Rivers Act) will hamper or stop the building of hydropower plants on many prime sites. Studies of the remaining sites indicate that most would be uneconomical to develop given the current relatively inexpensive cost of energy in this country. But the primary hindrance to further development of hydropower seems to be the environmental and social disruption caused by large dams. By one estimate, 30 million to 60 million people worldwide over the last century have been forcibly displaced from their homes to make way for large hydroelectric and reservoir projects, and many of these persons belong to indigenous groups or are ethnic minorities with little political clout. Currently, work is underway on what is planned to be the world's largest hydroelectric dam, located in the Three Gorges region of the Yangtze River in China. The dam will stretch over a mile across the river and stand 575 feet high. When completed, in order to make room for the over 350-mile long reservoir, an estimated 1.2 to 1.9 million people will have to be relocated, with enormous social disruptions and costs. When completed, however, the generators on the dam will have a capacity of 18,200 megawatts, the equivalent of eighteen typical 1000-megawatt power plants. This could eliminate the need to burn annually tens of millions of tons of coal, a fuel that is dirty, adds enormous amounts of carbon dioxide to the atmosphere, and upon which China relies heavily. No radioactive wastes will be produced, as would be the case if nuclear power plants were built instead. On the other side, construction of any large dam and reservoir cannot help but alter the local environment and ecosystem, as vast tracts of land are flooded. In the case of the Yangtze River in particular, critics of the dam project predict that the current pollution problems of the river will increase as more sewage is dumped into the water by factories and industries that will locate on the edge of the reservoir, and the natural flow of the river will be disrupted, allowing pollutants to collect and stagnate. Currently, China lacks adequate treatment plants to handle the sewage already being dumped into the Yangtze. There is also the fear that the Three Gorges Dam may silt up and become useless in only a matter of decades.

Small-scale hydropower plants may be more compatible with the natural environment and entail fewer negative social impacts. Some small-scale plants do not even use dams. Located on fast streams and rivers, such damless facilities direct water flow directly through the turbines located in the middle of the channel or off to one side. Damless facilities can avoid many of the undesirable environmental impacts associated with dams (**Figure 8-2**). Unfortunately, damless plants require a fast rate of water flow and are suitable for only a very small number of sites.

Most types of small-scale hydropower facilities utilize dams, but many take advantage of smaller dams with lower heads (the distance the water drops from the top of the reservoir to the bottom of the dam). It has been suggested repeatedly that many already existing small dams could be fitted with electrical generators. In the United States alone, there are approximately 75,000 to 80,000 dams and reservoirs on rivers and streams, mostly small dams, that are not currently used to generate electricity. These dams are used for such purposes as navigation, flood control, irrigation, and other uses. Potentially, it has been argued, many of these dams could be fitted with the equipment to generate electricity, and some proponents of hydroelectric power advocate the further development of this potential resource. But such small-scale hydroelectric facilities can provide power only to people in the immediate vicinity of the plant or dam. Furthermore, such small operations may prove to be relatively unreliable, intermittent energy sources. Also, given the continued presence of cheap fossil fuels in the United States, the costs of converting these dams to energy-producing facilities may be prohibitive. Furthermore, the recent trend has been to decommission and remove dams so as to restore natural waterways and ecosystems. In the last 30 years, hundreds of small dams have been demolished in the United States. Another way of increasing the hydropower capacity of the country is to upgrade and refurbish existing hydropower plants to increase their capacities. This generally costs much less per kilowatt of capacity gained than building new plants. An added advantage of refurbishing and upgrading is that additional environmental deterioration can often be minimized. Replacing old turbines with new, more efficient versions, for instance, may place no additional pressure on the environment. Reconstructing and enlarging a dam, however, might. Of course, upgrading can only squeeze so much energy from existing hydroelectric facilities; at best, a 10% increase in overall capacity may be gained.

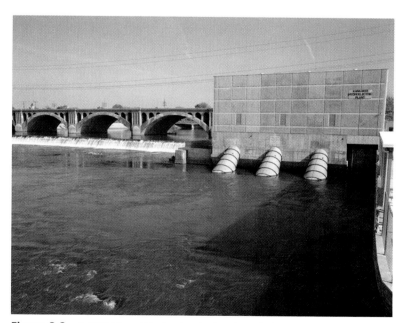

Figure 8-2   A 1200-kilowatt hydroelectric plant in Kankakee, Illinois. The hydroelectric plant provides all of the electricity for the city's wastewater treatment plant. Excess electricity is sold to a local power company at a profit to the city.

Globally, the prospects for further hydropower development look more promising than in the United States. Currently, the United States and Canada each contain about 13% of the world's hydroelectric capacity, the Russian Federation maintains about 6%, Brazil about 11%, and China currently 8% (but this will increase as China completes some major hydroelectric construction projects currently underway). The developing world in particular continues to increase its hydropower capacity, utilizing smaller facilities as well as huge centralized hydropower plants. Localized, small-scale energy production may be more feasible and compatible with local economies and energy use patterns in some developing countries than it is in many heavily industrialized nations that are wedded to centralized grid systems of power production and transmission. Thus, China, for instance, has more than 86,000 small hydroelectric power plants in operation. There are hundreds of thousands of dams already in existence around the world that are not currently used to produce hydropower. Converting these dams to small-scale hydroelectric power plants may be relatively efficient and inexpensive (not requiring inordinate up-front costs per conversion, even though smaller plants tend to be somewhat more expensive per kilowatt of capacity than large plants) and may also have minimal detrimental impacts on the environment.

Since the middle of the last century, worldwide hydroelectric generating capacity has increased by a factor of fifteen or more, and in the decade from 1987 to 1997, consumption of hydroelectric energy increased by 25%. Experts estimate that the current global capacity represents about 15% to 25% of the world's theoretical hydroelectric potential, but certainly 100% of the global potential will never be reached due to environmental and social concerns over hydropower. Although energy generated by hydropower will continue to increase globally over the next few decades, it is doubtful if in the foreseeable future it will ever contribute as much power as oil, coal, or gas currently do.

Biomass

The burning of biomass (**Figure 8-3**), especially fuelwood, has served as a major source of energy for most of recorded history. Even today, billions of people, perhaps more than half of the world's population, burn wood or other plant or animal products (such as animal waste) as their principal source of energy. Biomass provides as much as 10% to 15% of all energy consumed worldwide; in some developing countries, it provides up to 90% of the energy consumed. In the United States, about 3% of energy demands are supplied by biomass. Much of this is accounted for by the paper and pulp industry, which burns large quantities of wood and paper milling wastes to supply energy for its needs. Other substantial consumers of biomass in the United States include households that burn wood as a primary source of heat (about 5% fall into this category, and another 20% occasionally burn wood in a stove or fireplace), commercial industries and establishments that burn wood as a source of energy (in some cases, simply for space-heating purposes), and **waste-to-energy** facilities that burn municipal solid waste (see Chapter 18). Biomass-derived fuels, such as ethanol, currently account for a very small percentage of the U.S. energy budget, although about 8% of the gasoline sold in the United States contains ethanol.

Raw Sources of Biomass Energy
Sources of **biomass energy** can be classified into three major categories: wastes, standing forests, and energy crops. Wastes include such things as wood scraps, unusable parts of trees, pulp residue, paper scraps generated by the wood and paper industry, and municipal solid waste. During lumbering operations in timber forests, many sticks, branches, leaves, stumps, and roots are left over. Potentially, these could be collected and burned for energy (however, they are important sources of nutrients for regrowth). Likewise, agricultural wastes from food crops, such as leaves, stems, stalks, and even surplus or damaged food items, can serve as a source of biomass energy. In the United States, surplus corn is the principal raw material for the ethanol indus-

Figure 8-3   The Tracy Biomass Plant in the San Francisco Bay area provides renewable electricity from wood residues discarded from agricultural and industrial operations. The plant is rated at 21 megawatts.

try. A final large source of biomass energy is the organic material found in municipal solid waste (discussed in Chapter 18).

Natural standing forests afford a vast source of biomass energy, but unless harvested in a low-impact sustainable manner, they are a nonrenewable resource. Furthermore, diverting wood from forests for the purpose of energy use may compete directly with interests in other forest products, such as lumber for building and the furniture industry. Of course, clear-cutting forests for any purpose presents major environmental concerns (see Chapter 12). All too often, timbering results in soil erosion and ecosystem destruction.

Many advocates of increased use of biomass energy envision crops grown specifically for fuel purposes. Fast-growing varieties of trees, as well as grasses and other crops, could be commercially raised on large monoculture **energy farms**. The biomass would be harvested and the fuel burned directly or converted into other types of fuels (such as biogas, methanol, or ethanol). Coppicing trees (ones that will regenerate from stumps left in the ground) and perennial grasses may be particularly well-suited to energy farming. The crop could be clear-cut, but no further planting would be necessary because a new crop would grow from the stumps and roots already in the ground.

Such dedicated biomass plantations would require vast amounts of land—land that is fertile and could potentially be used for other purposes such as food crops. Indeed, growing crops for energy generally requires more land than any other energy generation technique (**Table 8-1**).

Environmental Advantages and Disadvantages of Biomass Energy

Biomass energy has become an extremely controversial subject among environmentalists, in part because of the diverse sources and technologies that fall into this category. The environmental appeal of biomass energy is that, in theory, fuelwood or other biomass fuel is a renewable energy source; in practice, however, it often is not. To be truly renewable and cause no deleterious environmental impact, every time a tree (or other plant) was cut down and burned, a new tree would have to be planted and assured of surviving. During growth, the new tree would theoretically absorb the carbon dioxide given off by the burning of the old tree. In point of fact, however, forests are often harvested for fuelwood in an unsustainable manner, leading to their de-

TABLE 8-1	Land Requirements for Various Electric Generation Technologies

Note that a square mile-year is a square mile used for one year, and a square kilometer-year is a square kilometer used for one year. One exajoule is equal to 1000 petajoules, which is equal to approximately 0.949 quadrillion BTU (quads); in 1999, the total world commercial energy consumption was approximately 403 exajoules or 382 quads (see Chapter 7). Also note that many alternative energy sources, such as wind and solar power, actually require less land than some conventional energy sources, such as coal and hydroelectric power. For instance, large hydroelectric power plants can require the flooding of vast tracts of land, and strip-mining for coal can devour huge land areas (often leaving the land devastated and virtually unusable once the coal has been mined).

Technology	Square Mile-Years per Exajoule[a]	Square Kilometer-Years per Exajoule[a]
Dedicated biomass plantation	48,250–96,500	125,000–250,000
Large hydro	3204–96,500	8300–250,000
Small hydro	66–6562	170–17,000
Wind[b]	116–6562	300–17,000
Photovoltaic central station	656–1274	1700–3300
Solar thermal trough	270–1158	700–3000
Bituminous coal	259–1274	670–3300
Lignite coal	2586	6700
Natural gas-fired turbine	77–259	200–670

[a]End-use energy figure averaged over assumed 30-year life cycles for power plants, mines, and so on.

[b]The lower range for wind includes only land occupied by turbines and service roads, while the higher number includes total area for a project.

Source: Data from L. R. Brown et al., *State of the World 1995* (New York: W.W. Norton, 1995), p. 73. Reprinted with permission of Worldwatch Institute, Washington, D.C., Copyright © 1995.

struction and ultimately to desertification. Burning forest or agricultural residues, dung, or other animal wastes can deprive the soil of nutrients that should be recycled back to the earth, reducing the soil's fertility and making it much more difficult for the land to support either natural vegetation or human-planted crops. Fuelwood gathering, biomass burning, and their attendant problems are causing an environmental nightmare in many developing countries.

Traditional biomass use, especially in developing countries, is typically very inefficient: From 3 to 10 times more energy may be used than is actually needed to accomplish the task at hand. For example, when food is cooked over an open fire, most of the heat is lost. Using a simple but efficient stove can cut the wood used by one-third to one-half. In some cases, wood use could be cut even further by introducing solar box cookers that can cook food by using the rays of the Sun instead of burning fuelwood or animal waste. Biogas digesters can be used to convert manure and other organic wastes to methane for use in cooking, heating, lighting, and other purposes.

In industrial countries, some people are strong advocates of the waste-to-energy approach, essentially burning garbage to generate electricity in specially designed incinerators. The major problem is that this approach discourages reuse and recycling, and the facilities can generate large quantities of air pollutants and hazardous waste.

Energy farms where crops such as fast-growing trees or grasses are raised specifically for use as fuel also have their advocates, but critics argue that these farms will suffer from the same problems as traditional food farms including soil erosion and massive pesticide, herbicide, and fertilizer use. Although soil damage and losses could be reduced by improving management techniques, raising energy crops would create another problem by displacing food crops in a world where there is already a shortage of food in many areas. Nor is raising energy crops on marginal lands that are not currently cultivated for food a solution, for those lands may be very prone to environmental degradation and perhaps should not be planted.

Furthermore, it is unlikely that energy crops, even under the best of circumstances, could ever supply the majority of the energy requirements of a major industrial nation. For instance, even if the entire annual U.S. corn crop were converted to ethanol, it would supply only a fifth of the country's motor-vehicle fuel needs. To make matters worse, in some cases, the biomass conversion processes are extremely inefficient. In some instances, converting corn to ethanol actually requires more energy than the energy value of the final ethanol produced, but this inefficiency occurs largely because corn is not really a good energy crop (it is used because there is a surplus left over from food production). Ethanol made from certain fast-growing grasses may contain five times as much energy as was used to grow and process the plants.

A major disadvantage of any biomass burning is that it is a major contributor to air pollution and the greenhouse effect when done unsustainably (that is, when replacements are not planted for all the materials burned). Biomass burning or combustion of biomass-derived fuels can emit carbon monoxide, nitrogen oxides, and particulate matter (such as ash and soot) into the air. The burning of biomass found in municipal solid waste can be extremely dangerous from this perspective, potentially releasing known carcinogens and heavy metals into the environment. Some have suggested that alcohol-based biofuels may release significant quantities of formaldehyde and other aldehydes (known carcinogens) when burned, but some studies indicate that such emissions from biofuels are less than those produced from comparable fossil fuels.

In summary, the long-term role of biomass in the energy mix of modern developed countries is uncertain. Biomass energy, in the form of biofuels such as ethanol, may be most important as a transitional measure between our virtually total dependence on fossil fuels and the full-scale introduction of a truly clean and environmentally benign energy system, such as a solar-hydrogen economy (see Case Study 8-1).

Solar Energy

Much of the energy we use currently is indirectly **solar energy**. Fossil fuels, wood, and other biomass combustibles are the result of organisms that trapped the Sun's energy into a form that we can conveniently employ. Hydropower and wind power also derive from the Sun, in that the Sun differentially heats the atmosphere, causing the winds, and evaporates water and recycles it as rain, which lets our rivers flow. Potentially, the Sun provides the Earth with more energy every day than humans would ever be able to use. While humans currently harness energy at a rate of about 13 terawatts (TW), the light and heat of the Sun deliver energy to the Earth's surface at a rate of about 80,000 TW. In just 20 days, the Earth receives energy from the Sun equal to all of the energy stored on Earth as fossil fuels.

Direct Use of Solar Energy and Passive Solar Designs

Humans have long been aware that they can harness solar energy directly. For thousands of years, people have understood how to build their homes so that they face the Sun and take maximum advantage of incoming solar energy that can be used to passively heat the living quarters. Similarly, gardeners have long known how to trap the Sun's energy in greenhouses, even simulating relatively tropical conditions in temperate climates. Modern solar technology encompasses two major themes: (1) simply trapping solar heat and light and redirecting it to purposes useful for people, and (2) directly or indirectly converting solar energy into electricity (**Figure 8-4**).

Sunlight striking the surface of the Earth at any one spot may be either direct or diffuse, although there is a continuous gradation between the extremes. Direct sunlight has traveled through the atmosphere with very little scattering or diffraction; generally, direct sunlight is bright and leaves sharp shadows. Direct sunlight can be concentrated with mirrors and lenses to form very energetic, focused beams. Diffuse sunlight has been scattered and diffracted by the atmosphere, clouds, and haze before reaching the surface of Earth. To the naked eye, diffuse sunlight does not appear as bright as direct sunlight, and it leaves soft shadows.

The distinction between direct and diffuse sunlight is important because some technologies require direct sunlight, whereas others can utilize either direct or diffuse sunlight. The appropriate solar technology will depend on the area because the amount of cloud cover and atmospheric turbulence experienced through the course of a year vary from place to place. In the United States, for instance, Albuquerque receives about 70% of its sunlight as direct radiation, whereas Boston receives over half of its sunlight in a diffuse form.

The time-honored way to utilize solar energy is through **passive solar design** of houses and other structures—using the inherent characteristics of the building to capture heat and light from the Sun. Such architecture has been practiced for thousands of years, but has to a certain extent been lost, or at least greatly de-emphasized, with the coming of the industrial age. With cheap, convenient fossil fuels readily available, architects and builders no longer worried about orienting houses toward the south (in the Northern Hemisphere) to capture the heat and light during the winter months. Likewise, windows were no longer required as a primary source of light once electric lights were widely introduced; in many buildings today, electric lights are used for illumination during daylight hours.

Figure 8-4   Rooftop photovoltaic modules provide electricity for a village health center in Satyanarayanpur, West Bengal, India.

With passive solar design, the entire building becomes a solar collector. Passive solar strategies include the installation of large, south-facing windows and few or no north-facing windows. The windows allow incoming sunlight, but trap heat. New types of glass with better thermal and light properties continue to be developed, including superinsulated glasses and electrochromic windows that can change their optical properties in response to small electrical currents and either block or admit sunlight as necessary to maintain a comfortable interior temperature. Windows and skylights also provide natural lighting that can displace artificial, energy-burning electrical lighting. Lightpipes, mirrors, and reflecting surfaces can be used to illuminate even the deep recesses of a building during the day.

Shade trees situated around a small building can keep it remarkably cool during even the hottest summers. Inside the building, windows and skylights can be adjusted to admit or block solar light and heat, natural ventilation will contribute greatly to efficient cooling, and water can be allowed to evaporate to cool the interior further.

The key to passive solar techniques is to design the building with such considerations in mind from the beginning. Adding passive solar systems to the initial design of a new house often raises the cost of construction by only 3% to 5% (usually less than 10%). Unfortunately, retrofitting an existing building so that it becomes an efficient passive solar system is usually very difficult or impossible, although many buildings can be improved by the installation of new windows, skylights, insulation, or thermal masses. Accordingly, some advocates of passive so-

CASE STUDY 8-1

The Natural Gas Transition to a Solar-Hydrogen Economy

Visit www.environment.jbpub.com
for more information.

It is very clear that in the future the world will have to stop burning fossil fuels at the high rate of the past few decades. Yet the world will continue to require large amounts of energy. The size of the human population continues to grow, and the developing nations rightfully desire to improve the living conditions of their peoples. A post-fossil fuel energy economy will be necessary.

The energy to be used in this post-fossil fuel world has been the subject of much discussion, but two major routes have been advocated: (1) a strong reliance on nuclear power supplemented by a mix of other energy sources, such as hydroelectric, solar, wind, and geothermal, or (2) a strong reliance on solar and other "clean" and renewable energy forms, combined with hydrogen gas as a way to store and transport energy conveniently—a so-called solar-hydrogen economy.

The idea behind a solar-hydrogen economy is that it would be based on renewable energy in the form of solar power plants (using photovoltaics or thermal solar technologies) and wind- and water-driven power plants, supplemented by such technologies as geothermal power, tidal power, and combustion of biomass to a limited (and sustainable) extent. The primary form of power generation would be electricity, but it would be produced intermittently when the Sun was shining brightly or the winds were blowing hard. Thus, peak electricity production and peak demand would not necessarily correspond in many areas, creating a problem, for electricity is notoriously difficult to store. Surplus electricity (not used directly by private consumers or industry)

would be used to subject water to electrolysis to form hydrogen gas. Essentially, an electrical current passed through water will break the water (H_2O) into hydrogen (H_2) and oxygen (O_2). The hydrogen can then be stored and burned when needed to generate electricity or heat, in a manner very similar to natural gas. When hydrogen gas is burned, it releases energy as it recombines with oxygen in the atmosphere to produce water.

Burning hydrogen gas is extremely clean (small amounts of nitrogen oxides [NO_x] may form, but hopefully with proper technology this hazard can be minimized). Furthermore, water to form hydrogen gas is extremely abundant, and the water can be recycled as the hydrogen combines with oxygen to form water once again.

Advocates of a solar-hydrogen economy point out that with the proper technology, much of which already exists, hydrogen gas can be used in place of virtually all fossil fuels. Hydrogen could be used to heat buildings, run appliances, power motor vehicles, and fuel airplanes. Alternatively, hydrogen-burning fuel cells where the hydrogen is once again chemically combined with oxygen to produce an electric current could be used to run electrical appliances or motors (**Figure 1**). Automobiles powered by electric motors and hydrogen-burning fuel cells could become standard in the next thirty years.

Hydrogen is bulky, however, relative to the amount of energy to be derived from a cubic unit, and it is dangerously explosive. Still, these drawbacks seem relatively minor. After all, fossil fuels, especially natural gas, can be

lar design believe that local building codes and ordinances should be modified to encourage passive solar design in new construction.

A major drawback of passive solar design is that it is not easily applied to larger, more compact buildings, such as large office buildings, skyscrapers, and major apartment complexes. But even in these cases, foresight can lead to the successful implementation of passive solar techniques. Buildings can be oriented relative to the Sun and windows placed appropriately. Interior atriums can reduce the need for artificial lighting and cooling. Natural ventilation can be used; in many large buildings today, the windows cannot be opened, and a mechanical

system must pump cool air throughout the building year-round. The bottom line is that even large structures can benefit from passive solar techniques.

Active Solar Techniques

A building's energy consumption can also be reduced by utilizing **active solar techniques**. One type of basic active solar system uses a **flat-plate collector**. In the simplest case, this device consists of a black metal plate that absorbs heat from the Sun when it is struck by sunlight (either direct or diffuse). The heat is transferred to a liquid (such as water or alcohol) carried in pipes in contact with the metal plate. The heat of the hot liquid can then be used as very dangerous, explosive substances, too. As for the bulk of hydrogen gas, this consideration may be negligible if, as some suggest, many hydrogen-powered vehicles may be up to twice as energy-efficient as their fossil fuel-powered equivalents.

Even strong proponents of a solar-hydrogen economy realize that the world cannot switch instantaneously to this type of energy mixture. They envision a transitional stage in which the world switches first to a relatively abundant and relatively clean fuel—natural gas. Natural gas use and technology are already familiar, and natural gas burns much more cleanly than oil or coal (substituting natural gas in many situations decreases carbon dioxide emissions by 30% to 65% and decreases other air pollutants by 90% to 99%). Natural gas is also relatively abundant. As natural gas technologies and infrastructures are developed and put into place, they will form the basis of future hydrogen use. Much of the hardware (piping systems, furnaces, engines, and so on) that is used for natural gas can easily be converted to hydrogen use when the time comes.

In summary, the steps to a sustainable solar-hydrogen economy would include the following: substitution of natural gas for other fossil fuels and the development of a natural gas-based technology and infrastructure; the concerted development of solar-powered electrical generating facilities (supplemented by other renewable energy sources) to replace other types of electric power plants; continued research and development in hydrogen production, storage, transportation, and use; and finally, the complete substitution of hydrogen for natural gas.

(a)

(b)

Figure 1 (a) A hydrogen fuel cell produces electricity from hydrogen and oxygen. (b) An example of a fuel cell, produced by Ballard Power Systems Inc. of North Vancouver, British Columbia, Canada.

desired (**Figure 8-5**). Water run through the pipes might be heated and used directly as hot water for washing or cooking; the hot liquid might be run through pipes to heat the interior of the building; or the heat could be stored in a thermal reservoir (such as a large tank of hot water) and used as needed.

Active solar collectors are usually placed on the tops of buildings, facing the predominant direction of the Sun in that part of the world. Putting solar collectors on roofs keeps them from taking up space that could be used for other purposes. Some systems collect solar heat during daylight hours and store a portion to heat the building at night or on a rainy day when the Sun is not at full intensity. Large

Figure 8-5 In an active solar heating system, flat plate collectors absorb heat. The heat is transferred to the fluid in the pipes, and the fluid is pumped to a storage tank, where the heat can be transferred again.

solar collectors can be built to serve a community; for instance, special ponds filled with a combination of salt and water can trap and store solar energy.

The major use of active solar collectors is to produce hot water and secondarily to provide space heating for building interiors. (Nearly 25% of all fossil fuel energy consumed in the United States goes toward space heating and cooling of buildings and heating hot water.) In the United States, active solar collectors are not used as much as they could be. With low energy prices and the large up-front costs of installing even a simple solar water heater, most Americans are unwilling (or unable) to make the necessary capital investment. Flat-plate collectors are relatively simple to manufacture and install, but their prices will remain high as long as there is little market for them. Some people predict that demand would increase if the systems were less expensive, but until demand actually does increase, manufacturers cannot justify investing in the equipment that will allow mass production and bring prices down. It seems to be a vicious circle that will be broken only by sharp increases in conventional fuel prices or government intervention (such as tax incentives to install solar collectors).

Heat collected by active solar collectors can potentially be used to cool buildings as well as to heat them. To actively cool a building using solar energy, a system similar to that used in a gas-powered refrigerator might be used. Currently, such systems are still in the developmental stage, but they could become widespread in the future.

Electricity from the Sun

Active **solar thermal technology** (that is, using the Sun to heat substances) is not limited to flat-plate collectors. In areas with a high incidence of direct sunlight, mirrors, reflectors, or lenses can be used to concentrate the Sun's rays to superheat a liquid; then the hot liquid can be used to generate electricity. For example, between 1984 and 1990, LUZ International Limited built nine solar thermal electric power plants in the Mojave Desert and sold electricity to the Southern California Edison Company. (Unfortunately, LUZ went bankrupt in 1991, putting a halt to its plans to build additional solar power plants. The plants it built are now owned by private investors and continue to operate. One reason that LUZ went bankrupt was a tax structure that requires solar power plants to pay much heavier taxes than conventional power plants.) In the solar thermal trough system used by LUZ, trough-shaped mirrors focus the sunlight on steel pipes encased in glass tubes (**Figure 8-6**). The Sun's rays heat a synthetic oil inside the pipes to

Figure 8-6   Kramer Junction Power operates the world's largest solar power facility in the Mohave Desert near Kramer Junction, California, where it covers well over 1000 acres (well over 400 hectares) and has a capacity of 354 megawatts.

more than 750° F (400° C). The hot oil is then used to turn liquid water into steam, which drives a turbine connected to an electric generator. Through advanced computer technology, light-sensitive instrumentation, and microprocessors, the mirrors track the Sun throughout the day to maintain the high temperatures necessary to produce the steam.

Other forms of solar thermal technologies include parabolic dish collectors that focus sunlight onto a single point; the heat can then be used to generate electricity. In another method, mirrors focus sunlight on a solar receiver mounted on top of a tower, which may be up to 660 feet (200 m) tall (**Figure 8-7**). The sunlight heats fluids, such as molten salts, in the receiver. The fluids can then be used to generate steam and power a turbine connected to an electric generator.

Sunlight can also be used to generate electricity through **photovoltaics**. Photovoltaics utilize semiconductor technology to generate electricity directly from sunlight. At present, various silicon substances are most commonly used for this purpose. A variation on simple photovoltaic systems, which produce electricity and send it to consumers directly, are photovoltaic systems that use some of the electricity they generate to split water into its component

Figure 8-7 Solar Two, a solar thermal demonstration project in the Mojave Desert near Barstow, California, operated between 1996 and 1999. Solar Two showed that an advanced molten salt system (a mixture of sodium and potassium nitrate) instead of a water-steam system, could more efficiently collect solar energy and dispatch electricity as needed. The mirrors focused sunlight on a solar receiver mounted on top of a tower. Once the salts were heated to approximately 1050° F (565° C), they were used to produce steam that drove a turbine connected to an electric generator, or they were stored and used to produce steam even after sunset.

gases, hydrogen and oxygen. The hydrogen can then be used as fuel.

Many photovoltaic cells are used in small consumer products, such as portable calculators, but the real potential lies in the application of photovoltaics to power homes, factories, and other commercial ventures. Photovoltaic systems are convenient because they can operate alone in remote areas with either direct or diffuse sunlight. They are already used for telecommunication and signaling devices, in space applications (such as satellites and manned space missions), and to provide power to small villages that are far from other sources of electricity (see Figure 8-4). In the United States and other developed countries, photovoltaic systems are already being used by some utility companies to supply extra electricity to grids during peak hours. The use of photovoltaics is expected to

increase in the coming decades, especially in developing countries.

The main constraint on the use of photovoltaics at present is their cost. Currently, generating electricity using certain photovoltaic systems is several times as expensive as using fossil fuels (of course, such comparisons are misleading because fossil fuels may not be available in a certain area, fossil fuels cause environmental damage that is not accounted for in their cost, and the cost of photovoltaic systems is primarily a capital expense). Since the mid-1970s, however, the manufacturing cost of photovoltaic cells has been steadily decreasing, from over $70 per watt in 1975 to $3.50 per watt in 2001, while their efficiency at converting sunlight into electricity has been steadily increasing. As the price has dropped, production and sales of photovoltaic systems have increased sharply from only 0.1 megawatts worldwide in 1971 to about 288 megawatts in 2000. Leaders in the production of photovoltaic units are Japan (where the government subsidizes rooftop solar systems), the United States, and Germany. As these trends continue, photovoltaic systems will become increasingly competitive, and expectations are that they will play an ever-larger role in our global energy budget. Although most existing photovoltaic systems are currently installed in developed countries, they are also increasingly being used in off-grid applications in developing countries of Asia, Latin America, and Africa. In South Africa, for instance, 350,000 solar home photovoltaic systems are planned for installation in the next few years.

Environmental Advantages and Disadvantages of Solar Energy

In many ways, sunlight is the ultimate and perfect renewable, and virtually limitless, source of energy for humankind. Of course, this energy from the Sun is often dilute, diffuse, and intermittent at any one point on the globe (the Sun only shines during the day), but if even a small fraction of this energy were harnessed, it could potentially solve many of our current energy problems. The main obstacle to utilizing solar energy appears to be its dispersed nature. Because of the diffuse nature of sunlight, solar systems will have to cover large areas (see Figure 8-6). A major solar-generating facility might cover a circular area 5 miles (8 km) in diameter with photovoltaic cells or mirrors and reflectors to capture sunlight. Many of these systems may be set up in deserts where the ecosystem is extremely fragile; solar power plants and the human activity they entail could be very disruptive to the local biota. Many types of solar systems will also need large amounts of water, a substance already in short supply in desert environments. Furthermore, while sunlight (the "fuel" of a solar energy plant) is free and clean, any large power plant is a capital-intensive investment initially. It has been estimated that the production and deployment of solar cell arrays (and wind turbines) can use about 3% of the fossil fuel that would be burned in a coal-burning power plant to generate approximately the same amount of electricity. This is clearly a drastic improvement, but it means that solar energy is not absolutely pollution-free. Furthermore, various solar technologies also generate toxic waste. Producing solar cells on a large scale requires large quantities of highly poisonous chemicals, such as hydrofluoric acid, boron trifluoride, arsenic, cadmium, tellurium, and selenium compounds. Photovoltaic cells and other solar energy systems have a life expectancy of 30 years or so. Thus, large portions of systems must be replaced or refitted every few decades. Doing so will entail manufacturing new systems, generating poisonous wastes and pollution in the process, and disposing of the used components.

Another major problem with using solar technology for electricity generation is that the power source is intermittent. Sunshine is not available at night or on rainy or overcast days; the flux of sunshine also varies seasonally (two or three times more solar energy is received during the summer than in the winter in some areas). Yet we use electricity 24 hours a day throughout the year. One solution to this problem is to collect energy when the Sun is out and put it into storage. But with current technology, the direct storage of electricity is notoriously difficult. Batteries are large, bulky, expensive, and dangerous (they contain many poisonous chemicals). Any battery system that is regularly subjected to charging and discharging degrades and must be replaced periodically.

An alternative is to pump water (using Sun-generated electricity) uphill into a reservoir during the daylight hours and subsequently release the water downhill as needed, using it to drive a generator that produces electricity. Essentially, this is a form of "artificial" hydroelectric power, and all of the problems associated with hydroelectric dams (flooding large areas, disrupting natural water flow, high cost of construction and maintenance, the chance of the dam breaking) are applicable to such a system. Another possibility is to compress gas into tanks and then release it again to drive generators when the Sun is not available. However, as described in Case Study 8-1, perhaps the most popular concept at the moment is to use electricity to produce hydrogen gas, which is clean burning and easily stored and transported.

Another way to deal with the intermittent availability of solar energy is to supplement solar systems with backup systems for generating electric power. But backup systems need to be able to start and stop on a regular, short-term basis, so they would almost inevitably have to burn fossil fuel. Except on a very small scale, backup systems do not make sense environmentally or economically. Essentially, the whole country would have to be covered with both a solar and a nonsolar electrical grid. A complete complement of fully operating nonsolar (that is, fossil fuel or nuclear) power plants would be needed to supply electricity at night, at considerable cost. A large power plant cannot simply be turned off during the day and turned on at night. Indeed, many people believe that at least in the foreseeable future, solar energy, and not conventional power, will be the backup. Solar energy can serve as a supplement to conventional power sources, especially during the summer when electricity to run air conditioners is in high demand.

If it comes to pass, the large-scale implementation of solar systems will entail a whole new set of environmental concerns. If past experience with other new technologies is any indication, many of the problems associated with solar technologies will not be anticipated—they will manifest themselves only after years of large-scale operation.

While large, centralized solar power plants are feasible in some areas such as desert regions, solar energy also lends itself to small-scale, decentralized, dispersed applications, such as passive and active lighting and heating of buildings, localized small-scale electrical generation, use of individual solar-powered ovens for cooking, solar hot water heaters, and so on. Widespread use of solar power in such situations could entail enormous energy savings.

Wind Power

Humans have been utilizing **wind power** for thousands of years for such applications as windmills and sailing ships. In the first quarter of the twentieth century, wind-powered irrigation pumps were very common in the United States, but their use gradually declined as inexpensive oil and gas became more readily available.

In the last 20 years, fluctuations in oil prices and the realization that our fossil fuel reserves are limited have helped rekindle interest in wind power. Today, wind generates approximately 1% of the world's electricity, but its use is growing quickly

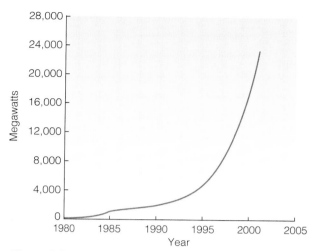

Figure 8-8   World wind energy generating capacity, 1980–2001. (*Source*: Worldwatch Institute, *Vital Signs 2002* [W.W. Norton, New York, 2002, page 43].)

year. Modern wind turbines are relatively simple machines that consist of blades and a rotor connected to an electrical generator, along with a control system, all mounted on top of a tall tower. In intermediate-sized and large wind turbines, a transmission and gears connect the rotor shaft to the generator; in some smaller wind turbines, the rotor is connected to the generator directly. Most wind machines have rotors with two or three blades mounted on the end of a horizontal shaft and look like huge fans or airplane propellers, but a few wind machines have blades that look like the end of an egg-beater mounted on a vertical shaft. Conventional propeller-like wind machines are controlled from the ground so that they face into the wind.

Currently, in the United States, the emphasis has been on intermediate-size machines with capacities of 50 to 500 kilowatts. The machines stand on towers about 100 to 165 feet (30–50 m) tall, and a single wind farm may contain thousands of machines. Such wind farms can feed electricity into the local utility grid.

Small wind machines with capacities of 0.25 to 50 kilowatts are also becoming more popular in the United States and other countries as well. Small wind turbines are well-suited to supplying power in rural locations that are not on a major grid, especially in developing countries. In some cases, small wind turbines can be combined with a photovoltaic system to increase the reliability of the power supply. Wind power offers many advantages. It is very safe and generally environmentally benign. Wind machines are typically erected along coasts (some can even stand in water offshore) or in mountain passes, where a constant and steady wind supply is assured; these areas are often not in high demand for other purposes. On farms, cattle can graze and crops can be planted around wind turbines. In the Great Plains, agricultural farms and wind farms can coexist peacefully on the same land (**Figure 8-9**).

Nevertheless, wind power has faced some heated local opposition. Wind farms have been subjected to the NIMBY (not in my back yard) syndrome. Some people complain that wind machines are noisy and say they are ugly and destroy the aesthetic qualities of the landscape. Perhaps the most serious concrete environmental problem with wind machines, however, has been bird deaths. Birds may collide with the blades or be electrocuted if they land on certain parts of the machine. In Altamont Pass in California, where a number of wind machines are located, several dozen rare golden eagles and scores of other large birds of prey have been

(**Figure 8-8**). Germany, the United States, Denmark, and Spain are leaders in the large-scale commercial application of wind power. Other countries, including China, are implementing the large-scale use of wind power. **Wind farms**—vast tracts of land covered with wind-powered turbines, some with 300-foot (91 m) blades—currently generate billions of kilowatt-hours of electricity per year at competitive prices. Indeed, like the costs of solar thermal and photovoltaic power, the costs of wind-derived electricity have decreased dramatically in the last 10 years. Conservatively, some researchers estimate that wind power could supply about 12% of the world's electricity needs. Other analysts have optimistically (but perhaps unrealistically) suggested that up to 100% of electricity needs could be met by wind power in some areas, such as Europe.

In the United States, as much as 20% of the demand for electricity could be met if wind power were fully developed. It cannot be developed in some areas because sufficient winds do not blow consistently. In general, the most suitable areas for wind turbines in the United States are along the west coast, along the northeast coast, along ridges in the Appalachian and Rocky Mountains, and in a large area of the Great Plains. Already, 1% of California's electricity demands are supplied by wind power, and wind farms with a capacity of 10 to 250 megawatts are being planned in several states including Iowa, Illinois, Minnesota, and Maine. Currently, the United States consumes wind-generated energy that is the equivalent of approximately 300,000 metric tons of oil each

Figure 8-9   Wind turbines can operate effectively on agricultural land.

killed by wind turbines. Researchers are currently trying to devise systems that will solve this problem.

Bird deaths aside, the major practical drawback of wind power currently is the intermittent nature of the energy source. Although prevailing wind patterns for any given location can be deter-

mined and predicted overall (as mariners did for centuries), on a seasonal and even daily basis, wind speeds and directions can be subject to rapid and wide fluctuations. Many large utilities are understandably wary of relying too heavily on the unknown "whims" of the wind. Furthermore, on a seasonal and daily basis, the strongest and steadiest winds may not blow during the time of peak demand for electricity. For these reasons, wind power has little chance of becoming a major or dominant source of electrical power until large-scale electrical storage facilities become readily available.

Despite its drawbacks, wind power capacity on a global scale continues to expand. In 1980, the global wind energy-generating capacity was a mere 10 megawatts; by 2001, it had grown to 24,800 megawatts—an increase of 37% over the previous year—and the growth shows no signs of abating. Indeed, even though wind power supplies only 1% of the world's electricity, it is one of the fastest growing sources of commercial energy.

Geothermal Energy

The interior of the Earth is very hot. This heat, which is primarily generated by radioactive decay within the Earth, reaches the surface by such means as molten rock, erupting volcanoes, and hot geysers and springs. The principle behind geothermal power is to tap and harness this natural heat from the Earth's interior. **Geothermal energy** can be used either directly for such purposes as heating buildings or indirectly to produce electricity (generally, either by utilizing naturally vented steam directly or by heating water to produce steam to drive a turbine generator). Geothermal energy has already found widespread use in some areas such as California, which produces about 6% of its electricity geothermally (**Figure 8-10**). In the United States as a whole, geothermal energy constitutes about 1.3% of the primary energy consumed (coming in after the fossil fuels, hydroelectric power, nuclear power, and biomass energy). Many researchers predict that geothermal energy use will continue to expand in the future.

Geothermal power is not practical everywhere; it is most easily utilized along plate margins and at other points where hot magma comes close to the surface (**Figure 8-11**). Thus, some of the most likely areas for developing geothermal power lie along the Pacific Rim from New Zealand to New Guinea, the Philippines, Japan, the western coast of the United States, Mexico, Central America, and the Pacific coast of South America. Iceland, which lies on the Mid-Atlantic Ridge, already heats some 80% of its

Figure 8-10   The Geysers is a steam-dominated geothermal field in Sonoma County, California.

Figure 8-11   Global locations of geothermal sites.

houses and many of its other buildings with geo-thermally heated hot water. Hawaii is also well-suited to the utilization of geothermal power.

There are four basic types of geothermal deposits: **hydrothermal fluid reservoirs** (the only deposits currently used commercially), geopressured brines, magma, and hot dry rock. Hydrothermal reservoirs are basically areas of the crust where hot rock occurs at relatively shallow depths and natural groundwater is heated, sometimes to extremely high temperatures. Such reservoirs may naturally manifest themselves on the surface as hot springs or geysers. The most useful, but also by far the rarest, are the steam-dominated hydrothermal reservoirs. At The Geysers in northern California, over two dozen power plants are fueled by the naturally produced and vented steam. Hot water-dominated hydrothermal deposits can also be used to run electricity-producing power plants; with lower-temperature deposits, the hot water is often used to heat homes and other buildings in the vicinity directly.

Geopressured brines are naturally occurring deposits of hot, salty water (brine) under pressure at depths of 10,000 to 20,000 feet (3000–6000 m). Such brines may contain significant quantities of dissolved gases, including methane. Potentially, these brines could be tapped for their heat content, dissolved gases, and pressure (which could be converted to useful forms of energy). Like fossil fuels, however, once a geopressured brine is exploited, it cannot be expected to renew itself on a human time frame.

Magmas, or hot molten rock, are also a potential source of geothermal energy. Where magmas approach the surface, such as around active or not long dormant volcanoes (for instance, on the Hawaiian Islands), pipes could be inserted into the magma and water circulated through the pipes. The heated water or steam could then be used to drive turbines. Although progress has been made in finding ways to prevent the pipes from melting—the magma is typically at temperatures of 1100° to 2400° F (600°–1300° C)—this power-generating technique is still a long way from commercial application.

The long-term future of geothermal power may lie in the developing field of hot dry rock technology. Although the temperature of the Earth increases with depth virtually everywhere, this geothermal gradient is much more pronounced in some areas, reaching 200° F per mile of depth (70° C/km) in some places. In hot dry rock technology, a hole is drilled some 2 to 6 miles (3–10 km) deep into the subsurface rock, until a sufficiently hot and thick layer of hard, dry rock is found. Once such a layer is located, two wells placed relatively close together are established in the rock. Next, the rock between the two wells is fractured. Then, cold water is pumped down one well, the hot rock heats the water, and it is withdrawn from the second well for use in generating electricity or for direct heating applications. Successful tests of hot dry rock techniques have been carried out in the United States, Great Britain, and Japan, but no commercial plants based on this principle have yet been constructed.

Although geothermal energy production is often considered relatively clean compared to traditional fossil fuels, it is not without environmental hazards. By its nature, it tends to expel excessive quantities of heat into the environment that may cause thermal pollution, killing plants and animals and disrupting natural ecosystems. Some of the hot underground water contains dissolved salts, other minerals, and heavy metals that are toxic pollutants on the surface. Hydrogen sulfide and other dangerous gases may be released from vents. Tons of hydrogen sulfide are released daily at The Geysers, despite the efficient scrubbers used at the plants to minimize air pollution. Furthermore, sludge is produced by the scrubbers and by the condensation of vented steam. Some sludge contains such high concentrations of heavy metals that it must be treated as a hazardous waste. Such waste could potentially be reinjected deep underground, but care must be taken to avoid contaminating freshwater aquifers.

Geothermal operations also require large quantities of water. Geothermal power plants currently in operation use water as a cooling and condensing agent, and hot dry rock techniques will also utilize considerable quantities of water.

In a sense, geothermal energy is an undepletable energy source, at least on a human timescale, because the Earth is not going to cool anytime soon. On a small scale, however, this resource can be depleted— and in some cases, quite rapidly. Naturally circulating hot water flow can be inadvertently diverted over time, thereby disrupting easy access to the energy. Hydrothermal fluid reservoirs often contain "fossil water" under pressure that is withdrawn by a power plant faster than it is replenished. The Geysers, for instance, has experienced significant drops in steam pressure due to overproduction since 1987, resulting in a 25% reduction in power output. And ulti-

mately, any hot rocks will cool down if heat is withdrawn from them quickly enough. Like any natural resource, geothermal energy must be managed carefully to maximize its potential.

Globally, the use of geothermal energy has increased dramatically over the last 50 years and accounts for over 40 million metric tons of oil equivalent each year. Approximately a third of world geothermal use is found in the United States. Other large users are the Philippines, Indonesia, Japan, Mexico, Italy, and Iceland.

Ocean Energy

The oceans are a vast, if sometimes diffuse and seemingly uncontrollable, source of energy. Various ways of exploiting this **ocean energy** are currently being researched. The waves, tides, and currents all have potential as energy sources. In 1995, a small electrical power station that runs on wave energy was installed off the coast of Scotland. The regularity of the tides can be used to drive mechanical systems, and this mechanical power can then be transformed into electricity (**Figure 8-12**). A **tidal power** station operates by allowing water at the high tide to flow into a reservoir or bay through sluice gates; then the gates are shut, forming a dam. At low tide, the water is released through a turbine and used to generate electricity. Ultimately, this energy comes from

(a)

Figure 8-12   The La Rance River Tidal Power Plant in St. Malo and Dinard, France. (a) Water from the Atlantic Ocean (upper part of photo) is trapped at high tide and then the water is released back into the sea at low tide. (b) The water passes through 24 turbines, like the one shown. These turbines are connected to generators that produce enough electricity for a city of 300,000.

(b)

the Moon, which generates the tides as it orbits the Earth.

A 240-megawatt tidal plant has been operating in France since 1966 and a 20-megawatt facility is currently operating in the Bay of Fundy, Nova Scotia. In most areas, the difference in height between high and low tide is not enough to efficiently drive a turbine, but due to local topographic conditions, the Bay of Fundy is an exception: There the difference between high and low tide can be as much as 53 feet (16 m). Another problem with tidal power is that it is very intermittent; power is generated only after high tides, and the times of the tides change daily. A final potential drawback is that the artificial damming and releasing of water may be extremely disruptive to local marine life.

Energy Storage

A major obstacle to the widespread use of many alternative fuels is the problem of **energy storage** and transportation. One of the reasons that the fossil fuels in general, and oil in particular, came to dominate the world energy mix is because of their ease of storage and transportation. Petroleum products pack a high quantity of energy in a small amount of space, and the energy is easily obtainable by combustion. Clean-burning hydrogen gas (see Case Study 8-1) offers some of the same advantages, but being a bulky and light gas, it is much more difficult to store. Electricity is one of the most versatile energy sources, but it is difficult to store in large quantities. Most electricity is shipped through wires as soon as it is produced and it is used almost immediately by a consumer.

Of the big five energy sources powering the modern world (oil, coal, natural gas, hydroelectric power, and nuclear power), the only one that actively stores newly transformed energy and then releases it as needed is hydropower. The artificial lake created by the dam at the power plant effectively stores energy when it stores water that was raised to higher elevations by evaporation and subsequent precipitation during the hydrologic cycle. When humans need energy, water is released through the dam to power turbines connected to electrical generators.

Electrical Energy Storage

The most common way to store electrical energy is with batteries. The principle of the **battery** was first developed by Alessandro Volta around 1800. All batteries work on basically the same principle. In the battery are substances that act as electron donors or receptors. When a battery is charged, electrical energy (essentially, a stream of electrons) is converted into chemical energy in the form of atomic arrangements. During discharge, the chemical process is reversed. Today, the most common batteries are of the lead-acid variety, utilizing lead plates and dilute sulfuric acid. Various types of miniature batteries and dry cells are commonly used in electronics equipment.

The problem with current battery technology is that the batteries tend to be heavy and bulky relative to the amount of energy they store. This is not a major problem for small applications, but for larger-scale, longer-term applications, it is a serious limitation. Also, batteries do not last forever; typical lead-acid batteries, for instance, can only be run through 500 to 2000 charge/discharge cycles.

Some futurists anticipate an increasing demand for medium-sized batteries such as those that can be used to power electric cars. Electric cars in existence today, using conventional battery technology, tend to have driving ranges of only about 70 to 100 miles (112–160 km) before recharging is necessary, and recharging can take three hours. Such a car might be appropriate for short commutes, but would be unsuited to long-distance travel. Also, while an electric car in isolation may give off zero emissions, it must be recharged from some energy source. If it is recharged by electricity produced by photovoltaic cells or a hydroelectric power plant, then it may indeed truly be a zero-emission vehicle. If it is recharged by electricity generated by a coal-burning electrical power plant, however, then it is ultimately still responsible for the emission of greenhouse gases and pollutants, although perhaps considerably less than a standard gasoline-powered car.

Currently, a new generation of lighter batteries that can store more energy per unit weight for longer periods of time is under development. These newer batteries utilize nonconventional combinations of elements as electrodes and electrolytes, such as zinc-bromine, zinc-chlorine, or hydrogen-nickel oxide. The next generation of batteries also promises to be longer lasting than current conventional batteries, but so far, they are also much more expensive.

Somewhat similar to batteries are **fuel cells**. In a typical fuel cell, hydrogen gas (H_2) is passed by a catalyst that separates the hydrogen atoms into hydrogen ions and electrons. The electrons, carrying energy, can be used to power an electric motor or virtually any other electric device. The hydrogen ions and electrons are ultimately combined with oxygen (O_2) to form water. Potentially, fuel cells could be used to power electric cars; the cars would carry refillable tanks full of hydrogen and would emit only water. Also, some fuel cells can utilize various hydrocarbon compounds, such as gasoline, natural gas, or methanol, as a source of hydrogen.

In such cases, there may be undesirable emissions, but they are much less than is the case with conventional fuel combustion systems.

Fuel cell technology is being aggressively pursued by a number of automobile makers, including such names as Ford, DaimlerChrysler, BMW, Volkswagen, Volvo, Toyota, Hyundai, and many more. For instance, Ford has introduced its P2000 Prodigy, a fuel cell-powered sedan that uses hydrogen and is designed to meet the same performance standards as the popular Ford Taurus. Many people in the industry believe that the future of motor vehicles, both private cars and public transportation vehicles such as buses, as well as trucks and other commercial vehicles, lies in hydrogen-powered fuel cell vehicles. Until the infrastructures are put into place to produce, transport, and deliver pure hydrogen gas per se, some futurists believe that service stations may utilize on-the-spot machinery to generate hydrogen gas from natural gas (which is already being used widely, although not primarily for motor vehicles). Such natural gas to hydrogen equipment will pave the way for a pure system of hydrogen storage and distribution. Perhaps by the year 2020, a large percentage of cars on the road will be running off of electric drives powered by hydrogen-consuming fuel cells.

Energy Conservation

Most authorities agree that the simplest and cheapest way of stretching our energy resources and mitigating energy-related problems is through improving **energy conservation** (decreasing the demand for energy) and increasing **energy efficiency** (increasing the usable output per unit of energy). Energy conservation and efficiency go hand-in-hand in that increasing efficiency is often the best way to conserve energy. A simple anecdote can help clarify the relationship of conservation to efficiency. Suppose that a room is illuminated at night using four standard 100-watt incandescent bulbs. One way to conserve energy would be to turn off three of the bulbs. But that strategy would sacrifice illumination. Another way to conserve energy, through increased efficiency, would be to replace the incandescent bulbs with four compact fluorescent lamp (CFL) bulbs. Each CFL bulb draws only 25 watts of electricity, yet produces the light equivalent of a 100-watt incandescent bulb. Thus, the energy consumption has been cut from 400 watts to 100 watts without sacrificing quality. But perhaps we should also rethink our standards when considering energy conservation. Perhaps the room is over-lighted with four bulbs so even more energy might be conserved by using only two 25-watt CFL bulbs and turning them off when no one is in the room.

A relatively small investment in energy efficiency can pay off handsomely in the long run. A car that gets better gas mileage may cost slightly more, but the difference in cost will be more than made up by the gas that is saved and the environmental degradation prevented. Likewise, even though CFL bulbs may cost $15 to $25 apiece initially, over their lifetime they may save $35 to $55 per bulb in electricity costs. Due to their advantages, in the last decade the global sales of CFLs has skyrocketed, rising from 45 million in annual sales in 1988 to over half a billion annually by the end of the century (and over 600 million in 2001). Just as investing in CFLs can save money in the long run, many other energy efficiency investments—from more efficient appliances and cars to utilities—can yield financial benefits. One study of Brazil concluded that investing $10 billion in making the country electrically more efficient could save approximately $44 billion in projected electrical needs.

Reductions in energy consumption need not result in a decrease in living standards—in many cases living standards are actually increased. But increasing energy efficiency may ultimately entail a qualitatively different lifestyle, as energy-intensive activities are replaced by more labor-intensive activities. Some analysts predict a **decentralization** of energy supplies and sources, especially as alternative energy sources continue to make inroads in the global energy budget. Residences and business may become partially or wholly energy self-sufficient as they consume less energy and use local energy-generating options such as photovoltaic cells mounted on the roof.

To many people, "energy conservation and efficiency" means more than reducing the energy used per se; it also means reducing the amount of energy obtained from nonrenewable fossil fuel resources. The aim is to reduce the environmental degradation that the big three energy sources (oil, coal, and natural gas) engender, especially the release of atmospheric carbon, which may be responsible for global warming (see Chapter 17). Studies of the **carbon efficiency** of the world economy show that from an economic point of view, the world has managed to raise its output per unit of carbon released. In 1950, one pound of carbon dioxide was released for every $0.29 worth of world economic output as measured by Gross Domestic Product (or a kilogram of carbon dioxide for every $0.64); currently, about $0.58 of output is generated for every pound of carbon dioxide released ($1.28 per kilogram). This increase in carbon efficiency over the last half-century is the result of many factors working together: improved energy efficiency, spurred on by the oil crisis of the 1970s; increased emphasis on energy sources that produce virtually no carbon (the only carbon emitted is during construction and maintenance), such as solar, wind, hydroelectric, geothermal, and nuclear power; shifting from the use of oil and coal to natural gas (which emits less carbon per unit of energy); and increasing emphasis on communications and various high-tech industries that use relatively little energy per economic contribution versus traditional heavy, energy-intensive industries.

Interestingly, the carbon efficiency of developing countries used to be higher than that of industrialized countries, but in the last few decades, this situation has generally been reversed. Formerly, the developing countries relied primarily on manual labor, but as they have begun to industrialize, many have become very energy-intensive. In many cases, these countries invest in less energy-efficient technologies simply because the up-front costs are lower. In contrast, many of the developed countries have been expanding their economic output while keeping carbon emissions stable. Japan's economic output is about $2.09 per pound ($4.61 per kilogram) of carbon dioxide; in contrast, the United States has an economic output of $0.66 per pound ($1.46 per kilogram) of carbon dioxide, and China has an economic output of only $0.11 per pound ($0.25 per kilogram). Theoretically, there is no limit to how high carbon efficiencies could climb. The exclusive use of virtually carbon-free energy sources would cause the carbon efficiency index to approach infinity.

Energy Efficiency at a National Level

At a national level, some countries are much more energy efficient than others. Energy efficiency can be measured in various ways: total energy consumption per year for a particular country; energy consumption per person per year in a particular country (per capita energy consumption); or primary energy consumption per unit of gross domestic product (that is, how energy efficient a country is in producing goods and services) (Table 8-2). For example, the People's Republic of China uses about half the

TABLE 8-2	Primary Energy Consumption, Energy Consumption per Million Persons, and Energy Intensity for Selected Countries, Late 1990s

Note: Primary energy consumption is in quads per year; energy consumption per million persons is in quads per year per million persons; GDP/EC is gross domestic product in millions of dollars per quad of primary energy consumed; energy intensity is billions of BTUs consumed per $1 million of gross domestic product = thousands of BTUs per $1 gross domestic product. For GDP/EC and energy intensity, two numbers are given: The first number was calculated using an estimate of purchasing power parity, which can be rather controversial and is subject to interpretation, while the second number (in parentheses) was calculated using the 1995 US$ exchange rate; it is the latter number that is usually referred to by many people when they discuss these issues. The numbers in this table are for comparative purposes among countries; the figures do not necessarily agree precisely with energy data given elsewhere in this book due to rounding and standardization among various countries in terms of minor energy sources (such as various types of domestic, traditional, or home use of energy) that are included or excluded from this table.

Country	Energy Consumption	Energy Consumption per Million Persons	GDP/EC	Energy Intensity
United States	85.89	.31	90,409 (90,657)	11.1 (11.0)
Russia	23.52	.23	27,390 (14,378)	36.5 (69.6)
China	43.65	.034	87,922 (19,129)	11.4 (52.3)
Japan	20.45	.16	148,434 (263,284)	6.7 (3.8)
Germany	13.79	.17	126,536 (180,347)	7.9 (5.5)
Canada	9.45	.30	72,055 (64,764)	13.9 (15.4)
France	9.83	.17	131,319 (162,237)	7.6 (6.2)
United Kingdom	9.05	.15	135,199 (129,539)	7.4 (7.7)
Ukraine	5.96	.16	18,645 (12,544)	53.6 (79.7)
India	18.31	.02	87,911 (20,634)	11.4 (48.5)
Italy	6.49	.11	179,870 (171,285)	5.6 (5.8)
Brazil	6.83	.04	155,193 (109,338)	6.44 (9.1)

Source: Figures calculated from data published in World Resources 2000–2001 (Washington, D.C.: World Resources Institute, 2000).

amount of energy used by the United States, even though China has more than four and a half times as many people. The per capita annual consumption of energy in China is just over a tenth of that in the United States. Yet, when we compare China's economic output to that of the United States, we find that China uses over four and a half times as much energy to produce the same amount of gross domestic product (calculated in terms of the 1995 US dollars exchange rate). In other words, in this respect, China is much less energy efficient than the United States.

The ratio of a country's energy consumption to its economic output is the country's **energy intensity** (see **Figure 8-13**). In general, developing countries have higher energy intensities than the developed countries—that is, they are less energy efficient even though they may use considerably less energy overall and per capita than the developed countries do. By becoming increasingly energy efficient, over the last several decades, various industrialized nations (such as Japan, France, Italy, Germany, and other western European nations) have been able to decrease their energy intensities while maintaining

strong economic growth and high standards of living. Thus, a country can increase its economic output while simultaneously decreasing its relative energy consumption—a concept virtually unimaginable before the 1970s.

Improving Energy Efficiency

Energy efficiency can be improved at every level. Electricity, for instance, tends to be very inefficient. It is estimated that the average American electric power company loses about two and a half times as much energy as it delivers. In each step of the conversion process, a sizable portion of the energy available is lost. More energy is lost as electricity is transmitted through power lines. Rather than turning the energy stored in coal into electric power and using it to run heaters that warm a building, it would be more efficient from an energy perspective—it would take much less coal—to heat the same building by burning the coal directly in a stove or furnace within the building.

Much energy could be saved, which is the same as gaining energy, by simply (1) not converting energy from one form to another unnecessarily, (2) increasing the efficiency at which energy is converted from one form to another, (3) not transporting energy unnecessarily, and (4) increasing the efficiency by which energy is transported. Unnecessary energy conversions and transport can be decreased by planning and decentralizing energy reserves and networks. Sometimes, several widely scattered, smaller, self-sufficient power plants may be more efficient than one large, centralized plant. Energy conversion and transmission can sometimes be made more efficient by simply readjusting and fine-tuning old equipment. More important in this area, though, is the development of new technologies. Superconductor technology, for instance, may dramatically increase the efficiency of all aspects of electrical energy use.

An old, but newly resurrected, technology, which can sometimes more than double the usable energy produced by a power plant, is **cogeneration**.

(a)

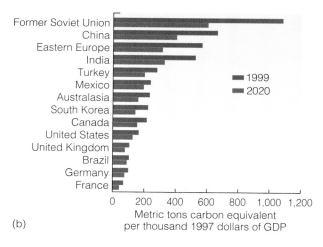

(b)

Figure 8-13   (a) World energy intensity by region, 1970 to 1999 with projections to 2020, as estimated by the Energy Information Administration of the United States Department of Energy. Note that in this graph "Industrialized Countries" refers basically to North America, Western Europe, Japan, Australia, and New Zealand; "EE/FSU" is Eastern Europe and the former Soviet Union. (b) Since much energy used in the world today is from fossil fuels and other carbon-emitting sources, energy intensity is related to the concept of "Carbon Intensity" (the amount of carbon released, much as carbon dioxide into the atmosphere, per unit of economic output). This graph is the Energy Information Administration's estimate of carbon intensity for selected countries in the years 1999 and 2020 (projection). (*Source*: United States Department of Energy, Energy Information Administration, *International Energy Outlook* 2002.)

The basic principle behind cogeneration is that a power plant simultaneously produces several types of energy, such as electricity and heat that can be used locally. Thus, an electrical power plant may heat water to produce steam; the steam in turn is used to power a turbine, which is connected to a generator that produces electricity. In conventional plants, the spent steam is treated as waste, and its heat content is discarded into the environment, often with deleterious results for the local wildlife. In cogeneration, the spent steam or hot water is used for some other purpose, such as to heat buildings. The main disadvantage to cogeneration plants is that the farther they are from where the heat is to be used, the less successful they tend to be; ideally, the power plant should be located right within the community it serves. For this reason, relatively small-scale cogeneration plants appear to have more potential for community acceptance and energy efficiency than large-scale versions.

Heavy and light industry can save much energy and increase energy efficiency in many ways. Tremendous energy savings have been achieved in this sector by using recycled materials. In general, recycling saves energy. Recycling steel, for example, uses only 14% of the energy that it would take to produce a ton of steel from the raw ores, and recycling aluminum saves about 95% of the energy that it would take to produce the same amount of aluminum from the raw ore bauxite.

In the United States, over a third of all energy consumption is used in residential and commercial buildings, primarily for lighting, heating, or cooling the interior space, and running small appliances. Energy can be saved by using more efficient heating, cooling, and lighting systems. Many existing buildings can be insulated much better than they are, and new construction materials and techniques enable new buildings to be superinsulated. In a well-designed, well-insulated, and sealed modern building, the interior will retain the heat given off by lights, human bodies, and various appliances; ideally, little additional heat should be needed. During the warm months, the insulation will keep heat out and cool air in the structure. Of course, a superinsulated building should also be well ventilated to prevent the buildup of harmful indoor pollutants. Such buildings should be designed and situated to take maximum advantage of passive solar heating during the cold months and to minimize such effects during the warm months. Using such approaches, a home can use 50% to 90% less energy than the average American house.

In western industrialized countries such as the United States, the transportation sector consumes about a third or more of all energy used, and worldwide, almost 25% of all energy used goes toward transportation. Automobiles, trucks, and other road vehicles account for about 80% of energy consumption in transportation. Fossil fuel, especially oil, products are used to fuel the vast majority of all transportation needs.

Currently, some 700 million vehicles are on the roads worldwide, and more than 75% of these are cars. Of these cars, slightly over a third are in the United States (although we have less than one-twentieth of the world's population), slightly over a third are in Europe, and the remainder are spread over the rest of the globe. The vehicle growth rate is very high, and based on present rates, there may be 1.1 billion vehicles on the roads by about 2020. Given the number of automobiles and other vehicles, it is imperative that something be done to control the amount of fossil fuel energy they consume and the amount of pollution they produce (see Case Study 8-2).

Encouraging Energy Savings—Voluntary versus Mandatory Measures

How do we encourage energy conservation? Will individuals, companies, and nations do it voluntarily? Or must conservation and efficiency measures be actively encouraged with subsidies and tax incentives and disincentives (for instance, high taxes on gasoline) or even mandated by law? These are difficult questions, and they elicit diverse reactions from various special interest groups. The health of our environment should not be seen as a "special interest," however; it is a necessity that should be everyone's concern.

Perhaps the most effective way to encourage voluntary energy conservation is through education. Sometimes a very practical approach that appeals to the individual's basic instincts can be effective. Improving the insulation in one's home or buying a more fuel-efficient car will not only repay the investment, and then begin to line the consumer's pocketbook, but will be good for the global environment as well. Also, advertising and popular culture can change our perceptions and instincts radically. In the past, wasting energy was a status symbol or considered "sexy"—for instance, we can point to the 1960s generation with their "muscle cars" that guzzled gas and generated tremendous amounts of horsepower for no practical reason. Today, there is a trend among some people to desire bigger and bigger SUVs (sport utility vehicles), which have poor gas mileages and appear to serve as status symbols the same way that other gas guzzlers did in earlier decades. With proper education (starting very early), perhaps being energy efficient could become a mark of status.

Improving the Fuel Economy of Gasoline-Powered Vehicles

Visit www.environment.jbpub.com
for more information.

Despite continued advances in the development of cars and trucks powered by electricity, natural gas, or biomass fuels, automobiles and light trucks that burn some form of conventional gasoline or diesel are likely to be the mainstay of the private transportation sector for many years to come. Thus, it is imperative that the fuel economy of the vehicle fleet be increased. The more miles per gallon (at an international level, often expressed in terms of liters per 100 kilometers, where one desires to decrease the number of liters used) we can squeeze out of our gasoline, the greater the reduction in environmental deterioration caused by petroleum burning and the longer our fossil fuel supplies will last. We know that we can improve the fuel efficiency of automobiles. The U.S. government first imposed fuel efficiency standards during the oil crisis of the mid-1970s, when the average fuel economy was 18 miles per gallon (13 L/100 km); by the first half of the 1990s, it had risen to a standard of 27.5 miles per gallon (8.5 L/100 km) for new cars sold in the United States. Using existing technologies, prototype automobiles have been built that attain average fuel economies of better than 78 miles per gallon (3.0 L/100 km).

What are some of the technical options that can be utilized to increase fuel efficiency? To begin with, up to 70% to 80% of the energy in gasoline can be lost because of inefficient engines. Newly developed engines, which burn "leaner" because of delicate electronic oxygen sensors, can increase fuel efficiencies by 20%. The development of efficient two-stroke engines, as compared to the standard four-stroke engines used in cars today, may help increase engine efficiencies. In addition, work is proceeding on ceramic materials that could be used in the manufacture of engines, particularly advanced diesel engines. The new ceramics can withstand much higher temperatures, thus reducing the need for cooling and decreasing the amount of energy lost.

Besides improvements in engine design, fuel efficiencies can be improved in many other ways. When power is transmitted from the engine to the wheels, some usable energy is lost. Some automobile manufacturers are designing more efficient transmission systems, including continuously variable transmissions that allow the engine to operate at its maximum efficiency even as the driver of the car speeds up or slows down.

Another area of promise is the increased streamlining of automobiles and the reduction of excess weight. Well-designed streamlined profiles not only may be aesthetically pleasing but also reduce aerodynamic drag and increase fuel efficiency. Likewise, excess weight adds to fuel consumption. Increasingly, automobile manufacturers are substituting composite plastics and lightweight metals, such as aluminum and magnesium, for the traditional steel in cars.

Perhaps the most exciting development is the prospect of storing excess energy from the engine and then applying it as needed. Think about driving a car: As you brake, enormous amounts of energy are lost because the engine continues to run, yet serves very little function (it may drive a generator or alternator to power the lights and other accessories). Likewise, when you sit at a red light or in heavy traffic, the idling engine consumes gasoline and wastes energy. Now some auto manufacturers, including Toyota and Volkswagen, are working on energy storage systems that will capture this excess energy, store it temporarily, and then apply it to the powering of the vehicle when needed. Some of the systems under development consist of a flywheel that is charged with the excess energy coming from the engine. In Volkswagen's system, the engine will actually shut down once the flywheel contains enough stored energy. The car then runs off the power of the flywheel until the energy level drops to a certain point; then the engine starts back up again. In another system, being developed at the University of Wisconsin with the support of Toyota, not only is excess energy from the engine saved and stored, but an advanced transmission system actually captures kinetic energy from the vehicle's wheels when it brakes and stores the energy in a flywheel.

Currently the most practical development along the lines of increasing fuel efficiency, decreasing emissions, and storing energy that would otherwise be lost, is the marketing of new hybrid gasoline-electric cars. Toyota's Prius is the world's first mass-produced hybrid car; the Prius was made available to consumers in Japan in December 1997, and in 2000, it was introduced to the American market (**Figure 1**). Honda was not far behind Toyota in the race to produce the first commercial mass-produced hybrid cars, and actually introduced the Honda Insight hybrid vehicle to America in 1999. Ford, GM, and

DaimlerChrysler reportedly plan to market hybrid gasoline-electric vehicles by 2003 or 2004. A typical hybrid car like the Prius has a gasoline engine and gas tank, an electric motor(s) and batteries, a generator (in some cases the electric motor can also serve as an electric generator when not being used as a motor), and a power-split device or transmission that will send engine power to where it is needed at a certain time—to the wheels driving the car, to a generator powering the electric motor, to the generator to recharge the batteries, or to some combination of the various options. Hybrid cars such as the Prius also typically are outfitted with a regenerative braking and coasting system such that when the car is breaking or coasting, an electric motor acts as a generator capturing and storing energy that would otherwise be lost and wasted.

From the point of view of the consumer, a hybrid gasoline-electric car is fueled only with gasoline. The battery for the electric motor is charged either by the gasoline engine driving a generator, or by otherwise "wasted" energy (such as kinetic energy during coasting or breaking) generating electricity; the batteries never need to be charged from an external source. The vehicle itself is propelled either by the gasoline engine driving the wheels, by the electric motor, or by both, depending on the most efficient method in a particular situation (automatically monitored and regulated by an on-board computer). In some cases, such as during braking and stop-and-go traffic, the gasoline engine automatically shuts down completely, resulting in zero exhaust emissions and no usage of fuel. When the gasoline engine is running, temperature and other parameters are constantly regulated to keep it and the catalytic converter running at the lowest level of emissions. As a result, hybrid cars such as the Prius can qualify as "Super Ultra Low Emissions Vehicles" under California's strict guidelines. Hybrid cars also attain remarkable gas mileages. The American version of the Prius is listed as 52 miles per gallon in the city and 45 miles per gallon on the highway; the Japanese version of the Prius, developed primarily for heavily congested, slow driving, is rated at an extraordinary 66 miles per gallon. These efficiencies are not only remarkable, but the reverse of what is expected of most standard gasoline or diesel vehicles where higher mileages are attained on the highway and lower mileages in stop-and-go city

traffic. But a hybrid car thrives in heavy, slow, congested traffic because the gasoline engine can completely shut down for periods of time, leaving the car powered by the electric motor, and during breaking, the generator recaptures energy. Thus, such hybrid cars may be ideal for city situations, both in terms of fuel economy and extremely low emissions.

It should be noted, though, that many environmentalists advocate not just increasing the efficiency of automobiles, but eventually eliminating them altogether. More and more emphasis is being placed on efficient public transportation and environmentally friendly modes of transportation, such as walking and bicycling. For the past decade, production and sales of automobiles worldwide have hovered around 34 to 41 million a year, and since 1993, it has been slowly creeping up from 34.2 million (1993) to 41.1 million in 2000 and then 40 million in 2001. Although the automobile fleet continues to increase, it is not growing as fast as it might. Even as tens of millions of new cars are produced each year, tens of millions of others are taken out of service. Currently (2002) there are an estimated 700 million passenger cars, trucks, and buses in service worldwide, and this may grow to 1.1 billion in 2020. As the environmental degradation caused by automobiles is recognized, conscientious citizens are restricting their auto use, and many cities, especially in Europe and Japan, are placing restrictions on the use and ownership of

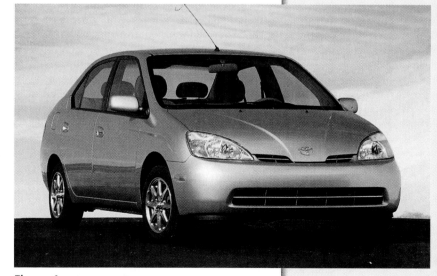

Figure 1 The world's first mass-produced gasoline/electric hybrid car, the Toyota Prius. In addition to getting an average of 48 miles to the gallon, the car qualifies as a super ultra low emissions vehicle in California.

cars. For instance, Amsterdam is phasing out the use of motor vehicles except for public transportation and delivery purposes.

Another way to decrease the number and use of automobiles is through car sharing. As of 2001, approximately 126,000 people worldwide, with 113,000 in Europe, participated in car-sharing organizations—and car sharing is expected to increase dramatically in the next decade. Members do not own private vehicles, but pay by the hour and kilometer (plus in most cases a refundable deposit, and perhaps an annual membership fee) to use a car as necessary from the car-sharing fleet. The fees paid by the car-sharing subscribers are used to cover the costs of maintenance, gasoline, insurance, and so forth. Among car-sharing participants in 2001, there were approximately 23 people per each shared car. Car-sharing works best in situations where participants do not need a car on a regular basis; for normal commuting, grocery shopping, and so forth, other modes of transportation—walking, bicycling, mass transport—are utilized.

There are alternatives to passenger cars and other motor vehicles. In many countries, such as China and India, bicycles are very important as a major mode of transportation. Currently, bicycles are being produced at a rate of about 90 to 100 million per year globally. This means that environmentally sound bicycles are being added to the transportation fleet much more quickly than automobiles. Indeed, in many parts of the world, bicycles greatly outnumber cars. In China, for instance, there are about 250 bikes for every automobile (**Figure 2**).

Figure 2   An example of a "bicycle culture," Nanjing, China.

Across the United States, electric utilities are adopting strategies that are both good for the environment, in that they save energy, and good for business, in that they reduce demand that cannot be readily met by the utility. This new approach is commonly referred to as "demand-side management" (DSM, also known as "negawatts"). DSM aims to improve the efficiency with which electricity is used, rather than stressing the production of ever-increasing amounts of electricity. At present, U.S. utilities are investing about $2 billion a year in DSM; it is an investment that is well worthwhile. DSM saves the company both short-term operating expenses and the considerable long-term expenses of building new power plants to increase production.

Much to the surprise of consumers, some utilities have found it cost effective to actively promote energy savings, sometimes even offering to pay for home improvements. Utilities have helped customers improve the insulation of their homes, distributed high-efficiency compact fluorescent lightbulbs, and promoted the installation of high-efficiency air conditioners and heaters. In the long run, the utility's savings can be split between stockholders and the consumers, who benefit from lower rates than would be the case otherwise.

Energy conservation can be promoted at every governmental level—local, state, and national. A first step is for government to set a positive example by using the most up-to-date energy-efficient technologies. The single largest consumer of energy in the United States is the federal government. The government can help fund the research and development of new products and then be their first consumer. In this way, the government can not only save energy but also help create a market for new, energy-efficient technologies. For instance, if all government vehicles had to meet very stringent fuel-efficiency requirements, automobile manufacturers would have an automatic incentive to build cars to these specifications. Once the cars were under production, they could be marketed to the general public relatively easily.

Governments could simply mandate energy efficiency measures within their jurisdiction, but such heavy-handedness is often unpopular. Through direct funding, the federal government in particular can encourage the research and development necessary to produce energy-efficient alternatives. Tax breaks and subsidies can be given to industries and individuals that are more energy responsible, and traditional energy consumption can be taxed heavily so that the price consumers pay reflects the real cost of the product. For example, many people contend that gasoline should be much more heavily taxed in the United States. In many European countries and Japan, gasoline costs several times the average price in the United States due to the heavy taxes imposed. These taxes are not just a way to discourage gasoline consumption and raise funds for the government, but reflect and help to cover the real costs of burning a gallon of gasoline. The cost of gasoline in the United States is not fully reflected in the price. The real cost involves much more than the cost of pumping the gas out of the ground and transporting it to the gas station; the real cost also includes such items as road building and maintenance, traffic regulation, health care needed due to traffic injuries, and the environmental degradation caused by burning that gallon of gas and releasing pollutants into the atmosphere. In the United States today, these additional costs, not covered by the price paid at the service station, are covered by government subsidies using money derived from taxes. Furthermore, the gallon of gas that the average American burns to drive less than 30 miles (or one liter to drive less than 12.7 kilometers) costs nature untold numbers of organisms and millions of years to produce.

Wouldn't it be fairer if the direct consumer, the person driving the automobile, paid a larger share of the true cost of a gallon of gas? Some people think so, but others argue that stiff gasoline taxes are regressive—they unfairly tax those who can least afford it because gasoline is a "necessity" (one has to drive to work) and not a luxury. A system could be devised that would exempt or otherwise subsidize low-income persons or impose different rates so that recreational driving would be more costly than driving to work. But such a system would be complicated and lend itself to cheating and abuse. Many people argue that what we need instead is more mass transportation systems, such as buses and trains.

An alternative is to tax cars according to their fuel efficiencies. If gas guzzler taxes were set high enough to pay for all the environmental damage a gas guzzler causes over its operating life, the average fuel efficiency of new automobiles would almost certainly increase quickly. Another suggestion is to tax the energy content of all fuels (not just gasoline), but then fuels that we might want to promote, such as clean-burning hydrogen gas, would be taxed too. Also, various "carbon taxes" have been suggested that could be imposed on all fuel sources. The basic idea is that a particular fuel, no matter how it is consumed, would be taxed in proportion to the amount of carbon dioxide and other greenhouse gases and pollutants it emits into the atmosphere per unit of energy it produces. Thus, if in a given context coal produces about twice the amount of greenhouse gases as natural gas does, then coal would be taxed twice as heavily per unit of energy generated. A very clean fuel, such as pure hydrogen, might not be taxed at all. The idea is to discourage the burning of some fuels while simultaneously encouraging the use of cleaner fuels that are less damaging to the environment.

SUMMARY

- Alternative energy sources, such as solar power, geothermal power, wind power, and biomass power, are alternatives to the "big five" energy sources: oil, natural gas, coal, hydropower, and nuclear power.
- Along with hydropower, most alternative energy sources are renewable (an exception in some cases is geothermal power), sustainable, and relatively environmentally friendly.
- Hydropower uses dammed water directed through generators to produce electricity.
- Although "clean" and renewable, the construction of dams can cause havoc to natural ecosystems and also have negative social impacts, such as when large numbers of people are exposed to water-borne disease vectors or must be displaced to make room for a reservoir.
- Biomass burning (wastes, wood, crops) produces greenhouse gases but can be renewable and sustainable; however, such use of biomass may work against recycling efforts and cause environmental problems.
- Solar energy takes many different forms—electrical solar power plants, active and passive solar techniques, photovoltaics; solar energy systems have only a limited lifespan and may

SUMMARY—cont'd

involve hazardous materials that are difficult to dispose of.

- Wind power has been utilized for thousands of years; modern wind farms are considered "ugly" by some people and have also been implicated in bird deaths.
- Geothermal energy draws heat from hot rocks and brines far beneath the surface of Earth, but may also bring toxic pollutants to the surface. Locally, high-quality geothermal reservoirs can be depleted.
- A major problem with many alternative energy sources is that the energy is intermittent (the Sun is only directly useful during daylight hours) and energy storage becomes a problem.
- Using electrical energy to produce hydrogen gas (from water) may be the way to conveniently store energy in the future—the hydrogen can then be burned or used in a fuel cell to produce electricity when needed.
- Energy conservation and increasing energy efficiencies are important ways of stretching energy resources and mitigating energy-related environmental problems.

KEY TERMS

Active solar technique
Alternative energy
Battery
"Big five" energy sources
Biomass energy
Carbon efficiency
Cogeneration
Decentralization
Energy conservation
Energy efficiency

Energy farm
Energy intensity
Energy storage
Flat-plate collector
Fuel cell
Geothermal energy
Hydropower
Hydrothermal fluid reservoir
Independent power producer
    (IPP)

Ocean energy
Passive solar design
Photovoltaic
Solar energy
Solar-hydrogen economy
Solar thermal technology
Tidal power
Waste-to-energy
Wind farm
Wind power

STUDY QUESTIONS

1. Define *alternative energy*. How do alternative energy sources differ from traditional energy sources?
2. List some of the important advantages and disadvantages common to many alternative energy sources.
3. Briefly describe the principles of hydroelectric power generation.
4. Discuss the pros and cons of hydroelectric power.
5. Describe the various types of biomass fuels and technologies.
6. What are the pros and cons of "waste-to-energy" techniques?
7. Describe some of the means currently being used to harness solar energy. Distinguish between active and passive solar techniques. How can electricity be generated from solar energy?
8. Describe the basic concept behind wind power.
9. What is *geothermal energy*? Is geothermal energy a renewable energy source?
10. What attempts are being made to harness ocean energy?
11. List the pros and cons of various renewable and alternative energy sources. Do you personally favor one over the other? Why? Are all such sources equally feasible in all areas? Explain.
12. List some of the major ways of storing electrical energy. Which are already in use today?
13. What is the distinction, if any, between energy conservation and energy efficiency?
14. Describe some ways that energy efficiencies can be improved.
15. What is a "solar-hydrogen economy"? Do you think it is feasible or just a pipe dream? Justify your answer.

STUDY
GUIDE

WHAT'S THE EVIDENCE?

1. The authors suggest that it may be inevitable that hydroelectric power and alternative energy sources will make up a greater percentage of the energy mix in the future. Why do they suggest this? What evidence do they base their suggestion on? Do you find their evidence persuasive? What energy mix nationally and globally do you envision in fifty years? Support your answer with appropriate evidence.

2. The authors seem to imply that decreasing the energy intensity of a country is a positive move. Discuss the concept of *energy intensity* (refer to Table 8-2). How does energy intensity relate to a nation's overall energy consumption? How does it relate to a nation's overall economic production? How does energy intensity relate to the per capita energy consumption of a nation's citizens? How does it relate to the per capita economic production? Some thinkers believe that all nations should strive to decrease their energy intensities, even if this means using more energy overall. What are some of the economic assumptions that underlie such a belief? Do you agree or disagree? Justify your answer.

CALCULATIONS

1. Referring to Table 8-1, calculate the amount of land required to generate 36.25 exajoules of electricity (the approximate amount of electricity generated in the United States during 1999) using solar thermal trough technology. (Express your answer in both square kilometer-years and square mile-years [one square kilometer equals 0.386 square miles]. Note that your answer will consist of a lower and upper range.)

2. Referring to Table 8-1, calculate the amount of land required to generate 36.25 exajoules of electricity using large hydroelectric power plants. (Express your answer in both square kilometer-years and square mile-years. Note that your answer will consist of a lower and upper range.)

ILLUSTRATION AND TABLE REVIEW

1. Referring to Figure 8-8, world wind generating capacity increased by approximately how much between 1990 and 2000? Express this in terms of an approximate percentage increase during this period.

2. Referring to Table 8-1, which electric generation technologies have the largest land requirements? The smallest?

3. Referring to Table 8-2, in terms of energy intensity, where does the United States fall relative to other nations listed in the table? Name a couple of countries that have lower energy intensities than the United States. Name a couple of countries that have higher energy intensities than the United States.

WWW.ENVIRONMENT.JBPUB.COM

www.environment.jbpub.com

Connect to this book's web site: http://www.environment.jbpub.com/ The site features eLearning, an online review area that provides quizzes, chapter outlines, and other tools to help you study for your class. We've posted the answers to the book's odd-numbered questions. You can also follow useful links for more in-depth information. The site provides sources for the information in this chapter and recommendations for further readings.

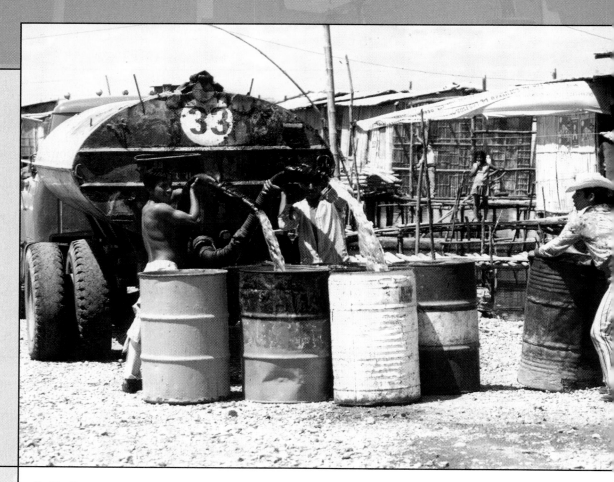

Approximately 1.5 billion people do not have access to clean drinking water, and almost half of the world's population (nearly 3 billion) lack access to adequate sanitation services. Even water from tanker trucks such as this one delivering water to a village in Ecuador may be of questionable quality. The lack of improved water supplies and sanitation have high health and economic costs.

CHAPTER 9

Water Resources

Water is one of the most remarkable materials on Earth. Despite its chemical simplicity—H_2O—water's properties make it absolutely essential for all life. The Earth is very rich in water, which covers over 70% of the planet's surface. If this water were evenly distributed, it would cover the entire Earth to a depth of 2 miles (3.2 km). More than 97% of this is saltwater, however, and is not usable by land life. About 30% of the world's renewable freshwater supplies are already being used; this is about eight times the yearly flow of the Mississippi River. Even so, enough freshwater is available to support over 20 billion people. Unfortunately, because of variable climatic and geologic conditions, freshwater is not uniformly distributed, so many areas of the world suffer from severe water shortages.

Such natural shortages are greatly aggravated by mismanagement of local water supplies, especially (1) lack of water conservation and (2) water pollution. Lack of water conservation is most evident in agriculture, which accounts for over 70% of water consumed worldwide and 80% consumed in the United States. For example, farmers in the western United States waste vast amounts of water because they do not pay its true cost. The government heavily subsidizes the water cost, so farmers do not conserve during irrigation, where much water evaporates. This wasted water could be used by such water-short urban areas as Los Angeles. In addition, the farmers in the drier western states often grow water-intensive crops, such as rice, tomatoes, and many fruits, that could easily be grown elsewhere where water is more plentiful.

After reading this chapter,
you should be able to ex-
plain or describe:

Properties of water and
the hydrologic cycle on
Earth

Water usage in the United
States

Global water usage and
supply

Types of water resources

Major groundwater
problems

Ways to increase effective
water resources

Benefits and drawbacks of
large dams

Ways to deal with water
scarcity

Mismanagement from water pollution is an especially large problem in developing countries. Indeed, a main cause of human misery in the world is not thirst, which very few people die from, but death and disease from polluted drinking water. Approximately 1.5 billion people around the globe do not have access to safe drinking water, and nearly 3 billion people (almost half the world's population) lack access to adequate sanitation services.

Water and the Hydrologic Cycle

Water: A Most Unusual Substance

Water is so common that we often assume it is a typical liquid. In fact, nearly all of its chemical and physical properties are unusual when it is compared to other liquids. These unique properties account for why water is the major component of most cells and living tissue. Some of these properties and the reasons they are so important to the Earth and its life include the following:

- *Density*. Water is the only common liquid that expands when it freezes. This causes ice to float so that it remains on the top of frozen lakes. If water were like most liquids, ice forming on lake tops would sink to the bottom, allowing the lakes to freeze solid each winter, killing fishes, plants, and other familiar life-forms. Also, much of the activity that shapes our evolving landscapes derives from this property: Much weathering of rocks occurs when water freezes in the cracks and expands, breaking the rock apart.
- *Boiling point*. If water were similar to most other liquids on Earth, it would boil at normal surface temperatures and thus exist only as a gas. This would render it useless to life, which needs water in the liquid state, such as in the bloodstream.
- *Specific heat*. Specific heat is the amount of energy required to raise the temperature of a substance. The specific heat of water is higher than any commonly known liquid except ammonia. It is five times higher than most common heavy solids. As a result, water takes much longer to heat up than most substances in our environment and also longer to cool down. In short, water temperature fluctuates less than land temperature so that large bodies of water, such as the oceans, serve as moderating influences on climate. Without the northward-moving Gulf Stream, for example, Great Britain would suffer much colder winters.
- *Solvent*. Water is a better solvent than any other common liquid, which means that it dissolves more substances. This property makes water the most effective liquid for transporting dissolved nutrients, as in the bloodstream, and eliminating wastes from living tissue. Indeed, water is crucial for transporting nutrients such as phosphorus throughout the biosphere. This dissolving ability also explains why water is the most important agent in eroding the landscape and why tropical areas are much more eroded than deserts. Unfortunately, this ability is also the reason water is so easily polluted and often stays polluted for such a long time.

Most of water's unusual properties are easily explained by the water molecule itself. Most important is that although the H_2O molecule as a whole is electrically neutral, it contains electric charges (**Figure 9-1**). These charges are distributed in a "bipolar" manner, meaning that the water molecule has a positive charge at one end and a negative charge at the other end. These charges create a relatively strong attraction between water molecules, which explains why water boils at such high temperatures. Without such bonding, water would boil at $-328°$ F ($-200°$ C). Strong electric charges are also important in increasing the chemical reactivity of a substance, which explains why water is such a good solvent.

The Hydrologic Cycle

While the "blue planet" Earth is rich in water, 97.4% of this water is saltwater, which contains about 35 parts per thousand (equal to 3.5%) dissolved substances (**Table 9-1**). Sodium chloride

Figure 9-1   The water molecule, consisting of two hydrogen (H) atoms and one oxygen (O) atom.

TABLE 9-1	Major Dissolved Substances in Seawater	
Ion	**Parts per Thousand**	**Percent**
Chloride	19	55
Sodium	11	31
Sulfate	3	8
Magnesium	1	3
Calcium	½	1
Potassium	½	1
Total	**35**	**99**

(NaCl), or "table salt," is the most abundant of these substances, with sodium and chlorine atoms making up about 86%. Magnesium, calcium, and other atoms in lesser amounts make up the rest. Indeed, saltwater in the sea contains dozens of elements, including gold, in trace amounts. Unfortunately, the vast majority of them occur in such small quantities that they are not economically recoverable because so much energy would have to be expended to obtain them. Nor is saltwater generally usable for drinking, agricultural, or industrial purposes. Drinking water requires dissolved substances of no more than 1 part per thousand. Industry requires even less because salts destroy the machinery.

We must therefore turn to the 2.6% of Earth's water that is freshwater to satisfy most of our needs. About three-fourths of this occurs as ice and is inaccessible (**Figure 9-2**). Ninety percent of this ice occurs in Antarctica. Only about 0.6% of Earth's water is freshwater in the readily available liquid state. The vast majority of this occurs as groundwater. Only a very tiny fraction (less than 0.014%) of Earth's water occurs as the freshwater lakes and rivers that we usually associate with water.

All of Earth's waters, whether freshwater or saltwater, are connected through the hydrologic cycle. The hydrologic cycle is the "great pump" that circulates water through the atmosphere, land, and oceans. The circulation is powered by energy from the Sun. This global water cycle involves two main processes: (1) evapotranspiration and (2) precipitation. **Evapotranspiration** is the transfer of water into the atmosphere (as the gas, water vapor) by evaporation and transpiration. Evaporation occurs from heating of liquid water by the Sun, and transpiration is the release of water vapor by plants. **Precipitation** occurs when water falls to the ground as rain, snow, sleet, or hail. In short, evapotranspiration removes water from the liquid state, and precipitation puts it back.

Let us follow this cycle by starting with the evaporation of ocean water, which amounts to about 102,000 cubic miles (425,000 km³) of water per year. Most of this precipitates back into the oceans, but about 9600 cubic miles (40,000 km³) falls on land. The cycle is completed when about 9600 cubic miles (40,000 km³) of water returns to the oceans as runoff via rivers or groundwater flow. The cycle is therefore in equilibrium in that the same amount of water taken from the sea by evaporation and precipitated onto the land is returned to the sea by runoff from the land. The most important part of the hydrologic cycle to humans is this 9600 cubic

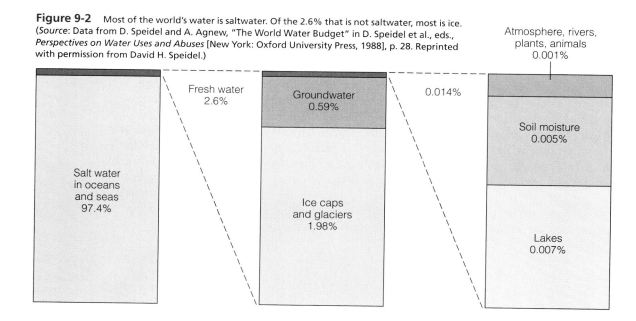

Figure 9-2   Most of the world's water is saltwater. Of the 2.6% that is not saltwater, most is ice. (*Source*: Data from D. Speidel and A. Agnew, "The World Water Budget" in D. Speidel et al., eds., *Perspectives on Water Uses and Abuses* [New York: Oxford University Press, 1988], p. 28. Reprinted with permission from David H. Speidel.)

TABLE 9-2	Water Used to Make Various Agricultural and Industrial Products					
Agricultural Products	**Gallons**	**Liters**	**Industrial Products**	**Gallons**	**Liters**
Egg, 1	40	151	Refine 1 gallon of crude oil	10	38
Milk, 1 glass	100	380	Sunday paper	280	1060
Flour, 1 pound	75	285	Aluminum, 1 pound	1000	3800
Rice, 1 pound	560	2120	Automobile, 1	100,000	380,000
Beef, 1 pound	800	3030			

Source: Based on U.S. Geological Survey data, 1992.

miles (40,000 km³) of runoff because it provides most of our water supply, as discussed below.

How fast does water move through this system? Each year, evapotranspiration removes an amount of water equivalent to a 39-inch (1 m) thick layer around the globe. It takes about 40,000 years to recycle all the water in the oceans. In contrast, the much smaller water reservoir of the atmosphere recycles in 9 to 10 days (depending on seasonal variation). Stream and river water is fully renewed about every two weeks.

Water Demand

The United States uses about 350 billion gallons (1.3 trillion liters [L]) of freshwater per day, or about 1250 gallons (4700 L) per person. This is more than any other nation and more than twice the average usage of water in Europe. Yet humans require only about 1 gallon (3.785 L) of water per day for our biological needs. Why do we use so much? In the United States, 41% of the water is used by agriculture, 38% to cool power generators, and 11% for industrial manufacturing.

Table 9-2 shows some examples of how much water is required to produce various agricultural and industrial products. Only about 10% of the water, or 140 gallons (530 L) per day per person, is directly used by the public. Some of this is for fire hydrants and other municipal uses. Table 9-3 shows the daily water use of an average American family of four; each person uses about 61 gallons (230 L).

All of these figures largely reflect the amount of water withdrawn from the water supply. **Withdrawn water** is water that is taken from its source (such as a river, lake, or aquifer), but it may be returned to its source after use. For example, a power plant may withdraw water from a river to cool generators, but return the water to the river when done.

Consumed water is water that is withdrawn but not returned to the original source: It is "lost" to the *local* part of the hydrologic cycle, usually by evaporation. Approximately a quarter of all water withdrawn in the United States is consumed, and agriculture accounts for 80% of the water consumed in the United States (**Figure 9-3**). California

Figure 9-3   Agriculture consumes more water than any other activity, largely because so much water is lost through evaporation.

TABLE 9-3	Daily Water Use for an American Family of Four People	
	Gallons	**Liters**
Toilet flushing	100	380
Showers and baths	80	303
Laundry	35	132
Dishwashing	15	57
Bathroom sink	8	30
Utility sink	5	19
Total	**243**	**921**

Source: U.S. Environmental Protection Agency, 1993.

agriculture alone accounts for about one-third of total U.S. water consumption. Of all the water consumed in agriculture, 85% is used for irrigation. Naturally, crops can't be grown without water, but most irrigation systems are extremely inefficient, in that on average only about 37% of the source water is actually absorbed by the plants being irrigated. The rest is mostly evaporated from the drainage ditches as the water travels to the crops. When irrigation pipes spray water upwards, most of that water can also be lost to evaporation.

In summary: Industry is the greatest withdrawer of water, but agriculture is the greatest consumer. Industry (including power plants) withdraws more water, but it returns a vastly higher proportion than agriculture does. Unfortunately, this returned water is not always clean and must be treated. Thus, even water that is not technically "consumed" is often rendered unusable for other purposes such as drinking.

Since 1950, total world water withdrawal has tripled (**Figure 9-4**) as the global population has increased. Because water conservation has led to a relative slowing of demand growth in industrialized countries in recent years, the rapid growth of global demand is caused mainly by increasing withdrawal in developing nations. Developing countries are using more water for the same basic reasons that their use of all resources is increasing: (1) increasing population growth (more people) and (2) increasing per capita demand (more water per person). Per capita demand rises with industrialization because water is needed to produce water-intensive consumer items.

As in the United States, agriculture is the main user of water worldwide, accounting for about 67% to 75%. Industry uses about 19% to 21%, and domestic use is about 6% to 9% (estimates vary). Use varies from country to country, however. In Nepal, for instance, 99% of all water withdrawal goes toward agriculture, whereas in Germany, 86% goes toward industry, and in Uganda, 62% goes toward domestic use.

Water Supply

Water shortages occur when supply does not meet demands. For now at least, such shortages are mainly regional problems that occur because the hydrologic cycle distributes water very unevenly.

Regional Water Shortages: Inequalities in the Hydrologic Cycle

Figure 9-5 illustrates the inequalities in the hydrologic cycle. **Surplus areas** receive more precipitation than is needed by well-established vegetation, including crops, and the local biota (including humans). In contrast, **deficit areas** receive less precipitation than is ideally needed by well-established vegetation and the local biota (including human occupants).

Figure 9-5 exhibits two main patterns:

1. Severe deficit areas are deserts found mainly in Australia, the western United States, and especially Africa and the Middle East. These correspond strongly to rainfall patterns: Most deserts tend to occur at about 30 degrees north and south latitude.

2. Major surplus areas are especially prominent in South America, Asia, and eastern North America. About two-thirds of stream flow in the United States is east of the Mississippi River. Most of the one-third found west of the Mississippi is confined to the Northwest.

Perhaps the most surprising aspect of these patterns is that so much of Africa is water deficit-ridden. Although we often think of Africa as a continent rich in lush, tropical forests, this is true of only the central part. North Africa and the Middle East, for instance, withdraw a huge portion of available water.

A Global Water Shortage?

Regional and local water shortages have always existed because of the inequalities of the hydrologic cycle. Even small bands of hunters and gatherers long ago experienced droughts and other serious shortages. The mysterious Anasazi "cliff-dwellers" of the American Southwest, for example, are thought to

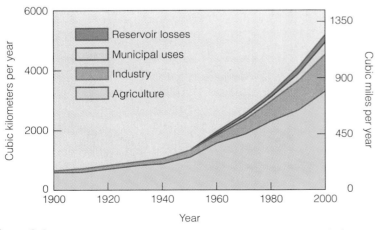

Figure 9-4   Estimated annual world water use, 1900 to 2000. Agriculture is the largest user by far. (*Source*: Adapted from *State of the World 1993*, edited by Lester R. Brown, et al., with the permission of W. W. Norton & Company, Inc. Copyright ©1993 by Worldwatch Institute.)

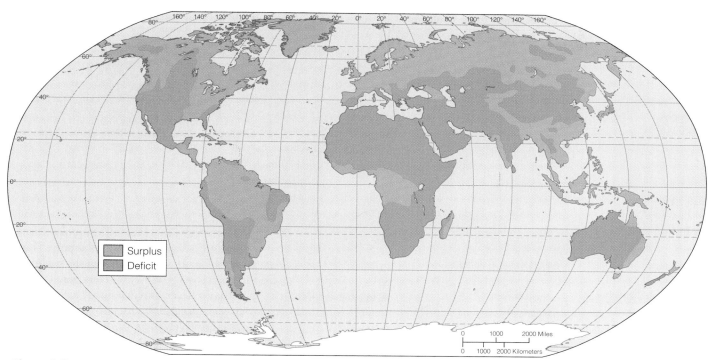

Figure 9-5   Regions of water surplus (green) and deficiency (brown) in the world. Deficit areas receive less precipitation than is needed by well-established vegetation and organisms; surplus areas receive more. (*Source*: M. Fal-kenmark, "Water and Mankind," *Ambio 6* [1977]: 5. Reprinted with permission of Ambio.)

Figure 9-6   Native Americans, often called "cliff dwellers," lived in these cliffs in the southwestern United States before a drought apparently forced them to leave.

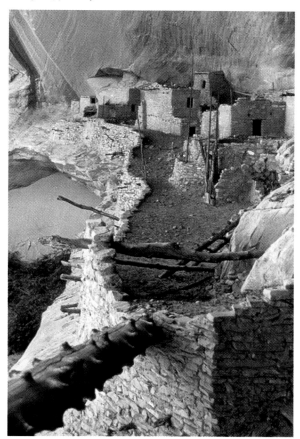

have abandoned their dwellings because of a long-term drought hundreds of years ago (**Figure 9-6**). With the rapid growth in population and industrialization (and consequent per capita water use), increasing demand has, of course, increased the extent of regional and local shortages.

Does this mean that countries with surplus water from the hydrologic cycle will not suffer water shortages? Not necessarily! A global view indicates that the entire hydrologic cycle is nearing the limits of use. If this occurs, even surplus areas will experience shortages. The hydrologic cycle annually creates about 9600 cubic miles, or 40,000 cubic kilometers, of freshwater runoff from the land to the sea. It is estimated that about 9000 cubic kilometers of this runoff are accessible in the form of a stable, predictable supply, such as rivers, that can be captured by dams, canals, and other methods (**Figure 9-7**).

Currently, humans use about half of the 9000 cubic kilometers (see Figure 9-4), but crude projections based on Figure 9-7 suggest that we are nearing the limits of the accessible water. Countries can be categorized according to the amount of water they have available for their populations. Nations with less than a thousand cubic meters of renewable water supplies per year per person are said to be "water-

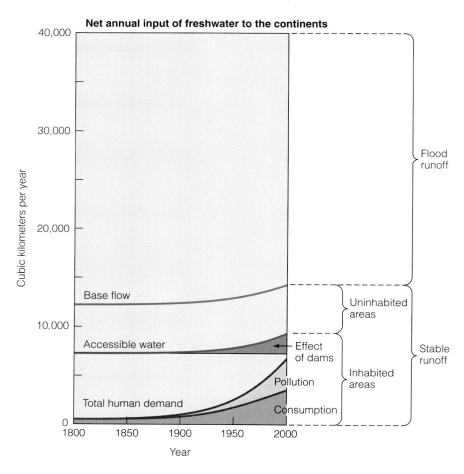

Net annual input of freshwater to the continents

Cubic kilometers per year

40,000

30,000

20,000

Base flow

10,000

Accessible water

Effect of dams

Total human demand

Pollution

Consumption

0

1800   1850   1900   1950   2000

Year

Flood runoff

Uninhabited areas

Stable runoff

Inhabited areas

Figure 9-7   Estimated total human demand for freshwater and amount of accessible freshwater, 1800 to 2000. Of the total input of freshwater to the continents, only about 9000 cubic kilometers is accessible in inhabited areas as of the year 2000. This barely exceeds the total amount of water used by humans either through direct consumption or by rendering it unusable through pollution. Note: 4.1655 km^3 equals approximately 1 mile^3. (*Source:* Reprinted from *Beyond the Limits.* Copyright ©1992 by Meadows, Meadows, and Randers. With permission from Chelsea Green Publishing Company, White River Junction, Vermont.)

ally have these two types of water resources, although, of course, in shorter supply. Thus, even the arid areas of the western United States have a few major rivers, such as the Colorado River, and groundwater supplies. Because precipitation is lower in those areas, however, these resources are replenished much more slowly when used.

Surface Waters.   Humans have traditionally used surface waters where possible because they are more accessible than groundwater. Surface waters include both (1) flowing waters, such as streams and rivers, and (2) basinal waters, such as ponds and lakes.

About 20% of the precipitation that falls on land flows over the surface, first as sheet wash, then as riverlets, and then as streams. The streams (often called "tributaries") eventually merge together into a large river. As the running water flows from the riverlets through the rivers, it carves out progressively larger channels, creating a tree-like pattern (**Figure 9-8**).

Stream and river valleys, from the Nile to the Mississippi, have always been among the most heavily populated areas. The reasons are many:

1. Soil on the floodplain is rich with nutrients deposited when the river overflows.
2. The water provides a ready means of transportation and shipping.
3. Moving water carries wastes away more readily than lakes and other standing supplies of water.
4. The river provides a source of clean drinking water.

These various uses often conflict, however, and the conflicts have intensified as growing numbers of humans have placed ever more pressures on stream and river resources. Rivers used for extensive waste disposal, for example, are not useful for drinking wa-

scarce," and nations with 1000 to 1700 cubic meters per person per year are "water-stressed." Various studies have demonstrated that as countries fall into the water stress and water scarce categories, their imports of food in particular must increase dramatically and the standard of living may decline. Today (2002), an estimated 505 million people live in water-scarce or water-stressed countries, and the group Population Action International has estimated that this number may rise to between 2.4 billion and 3.4 billion by 2025. Particularly affected are peoples in the Middle East, north, south, and east Africa, and southern and western Asia. Today, an estimated 4 million deaths each year (mostly of infants and young children) are the result of a lack of safe freshwater, and in many regions social and political tensions are increasing due to the scarcity of freshwater. The twenty-first century may see a series of armed conflicts over the control of river systems and freshwater supplies.

Types of Water Resources

Surplus areas are rich in water resources. These resources are usually categorized as either (1) surface waters or (2) groundwater. Deficit areas also usu-

The Global Safe Drinking Water Crisis

Many of the cities in developing countries do not have an adequate supply of safe drinking water. Limited water resources, inadequate facilities for treating and distributing water, and the absence of proper sewerage all contribute to the shortage (**Figure 1**). Despite efforts over the last half-century by industrialized countries and international agencies, the situation in these cities is actually worsening.

The poor in developing countries have long been perceived as being unable to pay for household water service. In reality, however, because the urban poor of these countries are generally not provided with public water service, they have to buy water, often of questionable quality, from private, unregulated vendors. Buying small amounts of water from vendors is more expensive by far than obtaining water from municipal sources.

A study of the water-vending system in Onitsha, Nigeria, found that these private vendors were responsible for more than 95% of water sales to the city's residents. The poor were annually paying water vendors twice the operational and maintenance costs and 70% of the annual capital costs of the new municipal water system. The new municipal water system in Onitsha was planned without any participation from the city's residents. As a result, they were unsure of its reliability and quality and were reluctant to connect to the system. Instead, they continued to pay high prices for a much inferior service.

Visitors to these cities may see an impressive skyline with modern hotels, offices, and apartment buildings. What they do not see is the absence of physical infrastructure, such as sewers and water lines, to serve these buildings. In cities where the public water supply is inadequate and developers sink wells, which lower the water table, land subsidence often results, and, in coastal areas, saltwater intrusion ultimately fouls the wells. The absence of sewerage is even more serious. The wastewater is discharged with little if any treatment into drainage channels and urban streams so that the surficial groundwater and the soil become badly contaminated. In Bangkok, for example, household wastewater is discharged into the klongs, or canals. Pipes containing water destined for homes run through the klongs, and when the water pressure in the pipes is low, wastewater in the klongs seeps into the pipes.

The lack of basic water and sanitation services leads to a number of problems. For example, although water-borne diseases are manageable in the industrialized world, they exact a heavy toll in developing countries. Also, many hours each day are spent fetching water—time that might be put to more socially and economically productive uses.

The problem is worsening. Eighteen out of a total of 22 "giant cities" (urban areas with more than 10 million inhabitants) are in developing countries, compared to only one out of a total of four such cities in 1960.

Lack of technology or funds cannot alone explain the failure of some developing countries to provide an adequate water supply and sanitation services. Even in cases where funds are available and the required technology involves only well-established practices, projects have not been sustained. As this suggests, solving the water supply problems of these countries will not be a simple matter and will require more fundamental changes than merely providing funding.

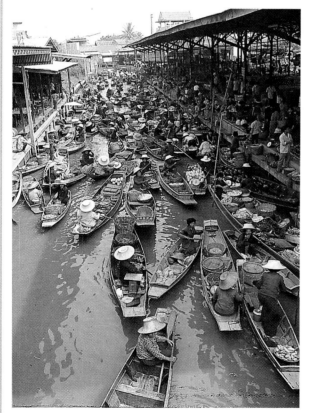

Figure 1 A "floating market" in Thailand illustrates the demands placed on water in urbanized areas of developing nations.

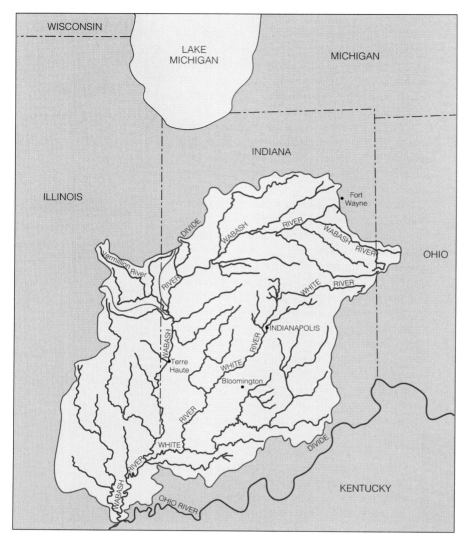

Figure 9-8   Drainage basin of the Wabash River showing tree-like pattern of tributaries that drain into larger channels.

Indeed, groundwater supplies the drinking water for about half the United States population and will likely provide much more in the future. **Groundwater** is a general term referring to water beneath the Earth's surface. It occurs because about 10% of the water falling as precipitation ultimately enters the ground, infiltrating down through the soil and rock (**Figure 9-9**). As the water is pulled downward by gravity, it fills the rock pores, called *voids*, until it is stopped by a layer of impermeable rock. As the infiltrating water "backs up," it forms the **zone of saturation**, which consists of voids that are fully saturated with water. Above this is the **zone of aeration**, which consists of voids that may be moist but are not saturated. The **water table** is the boundary between the zones of saturation and aeration. The rock that contains the zone of saturation is known as an **aquifer** (*aqua* = water; *fero* = containing). An aquifer must be a relatively permeable rock, such as sandstone, gravel, fractured limestone, or fractured granite.

Impermeable rocks that obstruct water flow are called **aquicludes**. Shales and other clay-related rocks are common aquicludes. Aquicludes can separate the zone of saturation into one or more aquifers, often called **confined aquifers** (**Figure 9-10**). These confined aquifers are supplied with water from one or more **recharge areas**, from which rainfall infiltrates into them. If the aquifer is sufficiently confined by aquicludes so that pressure is built up from the recharge, an **artesian well** can result. The pressure causes the water to flow freely, without the pumping required by most aquifers. Waste-well injection into deep confined aquifers was once a common method of waste disposal because it was erroneously thought that wastes would not leak into the higher aquifers that often provide drinking water. But aquicludes often have fissures or other openings that allow water to move between aquifers; "confined" aquifers are rarely completely confined. Waste-well injection has consequently been banned in many areas.

In areas of high rainfall, the water table tends to be near the surface. This is especially true in low-lying places such as Florida, which is near sea level.

ter or even recreational activities. In 1969, such large amounts of industrial waste were dumped into the Cuyahoga River near Cleveland, Ohio, that the river caught on fire. Such incidents led to the passage of the U.S. Clean Water Act and other legislation in the early 1970s, which has succeeded in keeping flowing waters available for many uses (Chapter 15).

Basinal freshwaters, such as ponds and lakes, are more readily polluted than flowing waters. Ponds and lakes form where water accumulates in basins (local depressions) in the land. Wastes are not carried away, but tend to accumulate in the basin. The water supply in even very large basins such as the Aral Sea in central Asia can be depleted if enough demands are made on it.

Groundwater.   The vast majority of accessible freshwater occurs as groundwater (see Figure 9-2).

In such areas, lakes and rivers are especially abundant because they occur where the land surface dips below the water table (**Figure 9-11**). Where the ocean is nearby, groundwater closest to the ocean will be composed of saltwater. The saltwater groundwater will migrate inland if humans withdraw too much water, as we will see shortly.

Groundwater Problems.   Nearly all areas of the Earth have groundwater supplies. Despite this wealth of water, two groups of problems have arisen that reduce groundwater's utility:

- Discharge problems
  1. Groundwater pollution
- Withdrawal problems
  2. Depletion
  3. Land subsidence
  4. Saltwater intrusion

Discharge problems arise from the discharge of toxins, metals, organics, and many other materials into the groundwater supply. Groundwater pollution is generally considered to be the greatest water pollution problem of the future. Until a few years ago, most research and legislation generally neglected groundwater pollution, concentrating instead on surface waters. However, groundwater's slow movement (an average of about 50 feet or 15.2 m per year) causes pollutants to stay in the water for long periods of time. Thus, seepage from surface wastes, well-injected wastes, septic tanks, and many other sources of pollution have created many long-term groundwater problems that will take much time and money to clean up.

Figure 9-9   Rainfall infiltrates the soil, moving downward to form the zone of saturation. The top of this zone is the water table.

Withdrawal problems arise from the removal of groundwater from aquifers. Depletion occurs when groundwater is withdrawn faster than the aquifer can be recharged. Aquifers can be depleted even in areas of very high rainfall because human-made pumps can remove groundwater at a much greater rate than the aquifers recharge. Recharge water must infiltrate through the sediment and rock, which usually takes years. Geologists estimate that the average aquifer takes at least 200 years to fully recharge. In such cases, groundwater becomes a nonrenewable resource that is "mined," in terms of a human lifetime. Much of central and south Florida, for example, is already suffering severe groundwater

Figure 9-10   Water from a recharge area flows into sandstone aquifers enclosed by aquicludes.

Figure 9-11   Florida's many lakes are caused by a combination of high rainfall and low-lying areas, including some created by sinkholes. This satellite image of central Florida shows hundreds of lakes and thousands of sinkholes. The area is underlain by limestone.

Figure 9-12   Areas of groundwater over-draft where groundwater is being pumped out faster than it recharges. Note the Ogallala (or "High Plains") aquifer overdraft in the central United States. (*Source*: U.S. Geological Survey.)

shortages despite its very high rainfall. Depletion is often greatly aggravated by paving over recharge areas, preventing water from infiltrating. Another problem is the destruction of swamps and other wetlands. These are major recharge areas because they retain water for long periods, purifying the water and allowing it to infiltrate the aquifer. For this reason, wetlands are often called the "kidneys" of the hydrologic cycle. When wetlands are paved over or filled in, the aquifer's recharge ability is reduced. The United States has already lost about half its wetlands to urban and agricultural development.

Recharge time is longest in areas with lower rainfall. Deserts and other water-deficit areas tend to have aquifers at great depth so that deep wells must be drilled to reach them. This groundwater is often called "fossil" water because it was deposited thousands of years ago, when the climate was wetter in the area. As a result of the water's age, its quality is often poor, and it may be highly saline. Even more important, when water is withdrawn in such arid areas, it often takes thousands of years to recharge. When high demand occurs in areas of low rainfall such as southern California with its dense population or the midwestern United States with its intensive agriculture, groundwater supply problems are especially acute (**Figure 9-12**). The great Ogallala

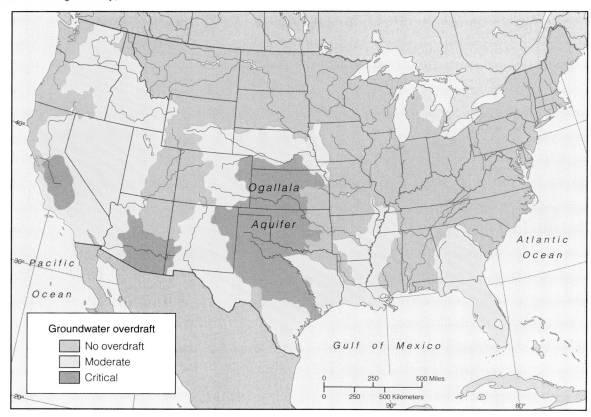

Figure 9-13   A cone of depression is created where over-pumping causes the water table to drop below its former level.

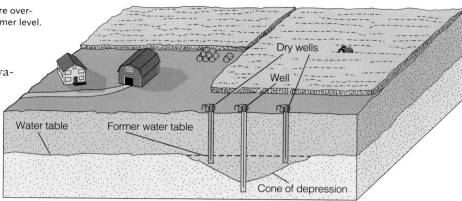

aquifer of the Great Plains provides the water for the "breadbasket" of the world. Yet this aquifer is perhaps half depleted, with recharge rates that are vastly slower in today's drier climate: Most of the Ogallala aquifer formed many thousands of years ago during the wetter Ice Age climate (see Case Study 9-2).

The rate of groundwater depletion can be estimated by the rate that the water table falls in an area. The global extent of groundwater depletion is apparent in this small sampling of global rates:

Water Table Fall Rate

Tucson, Arizona	7 feet (2.1 m) per year
Beijing, China	3 feet (0.91 m) per year
Manila, Philippines	30 feet (9.1 m) per year
Tamil Nadu, India	90 feet (27.4 m) per year

The second withdrawal problem, land subsidence, is more localized than depletion. Nevertheless, subsidence is very costly to those who live in areas where it occurs. Land subsidence means that the land "sinks," causing buildings, roads, and other surficial structures to sink with it. Subsidence occurs where depletion of groundwater causes the water table to fall. At first, this lowering is localized around the well, where water table drawdown forms a **cone of depression** (**Figure 9-13**). As more wells are drilled to draw water from the aquifer, the entire water table will lower. In many cases, a lower water table does not lead to significant land subsidence. This is very fortunate because water tables have been lowered in most parts of the United States and many areas of the world from rapid withdrawal, as we just have seen. In some cases, however, the grains of sediment composing the aquifer have very large voids or are otherwise supported by the water pressure of the groundwater. Removal of that water causes the voids to close up as the sediment or rock grains crowd closer together. This loss of rock volume underground causes the overlying land to subside.

Subsidence tends to be especially bad in urbanized areas (high withdrawal) with aquifers of sediment that are highly saturated with water. A notable example is Mexico City. It was built on an ancient lake bed so that massive groundwater withdrawal has caused the ground to sink by many feet. Some buildings must now be entered by the second floor. Houston, Texas; Venice, Italy; and other

coastal cities are suffering significant subsidence. Efforts have been made to reverse subsidence by pumping water back into some aquifers. Aside from the high cost, however, the ability of the aquifer to hold water is often permanently reduced because the void space, once diminished, cannot be recreated.

Sinkholes are a special type of land subsidence caused by water withdrawal. These are depressions in the ground that occur when the thin layer of rock overlying a previously existing underground cavern collapses into the cavern (**Figure 9-14**). Sinkholes are especially common in areas underlain by limestone, such as much of the southeastern United States and the Yucatan of Mexico. This limestone is often dissolved by groundwater to form underground caverns. Water pressure from the groundwater helps support the overlying land surface, so a drop in the water table from heavy withdrawal removes that support, causing the ground to collapse.

The third withdrawal problem is saltwater intrusion. Normally, the groundwater underlying coastal regions has an upper layer of freshwater,

Figure 9-14   Sinkhole in Winter Park, Florida, caused by collapse of the underlying limestone.

CASE STUDY 9-2

The Ogallala Aquifer

Visit www.environment.jbpub.com
for more information.

The Ogallala (or "High Plains") aquifer underlies most of Nebraska and sizable portions of Colorado, Kansas, and the Texas and Oklahoma panhandles (**Figure 1**). This area is one of the largest and most important agricultural regions in the United States. It accounts for about 25% of U.S. feed-grain exports and 40% of wheat, flour, and cotton exports. More than 14 million acres (5.7 million hectares) of land are irrigated with water pumped from the Ogallala. Yields on irrigated land may be triple the yields on similar land cultivated by dry farming (no irrigation).

The Ogallala's water was, for the most part, stored during the retreat of the Pleistocene continental ice sheets (the last Ice Age, which ended about 10,000 years ago). Present recharge is insufficient over most of the region (see Figure 1). Each year, farmers draw from the Ogallala more water than the entire flow of the Colorado River. In 1930, the average thickness of the saturated zone of the Ogallala aquifer was nearly 66 feet (20 m); currently, it is less than 10 feet (3 m), with the water table locally in various areas dropping by amounts ranging from 6 inches (15 cm) to 3.3 feet (1 m) per year. Overall, it is believed that the Ogallala will be effectively depleted within four decades. In areas of especially rapid drawdown, it could be locally drained in less than a decade.

Reversion to dry farming, where possible at all, will greatly diminish yields. Less vigorous vegetation may lead to a partial return to pre-irrigation, Dust Bowl-type conditions. Alternative local sources of municipal water are not at all apparent in many places. Planners in Texas and Oklahoma have advanced ambitious water-transport schemes as solutions. Texas is considering alternatives that would involve transferring water now draining into the Mississippi River from northeastern Texas across the state to the panhandle. Oklahoma's Comprehensive Water Plan would draw on the Red River and Arkansas River basins. Such schemes would cost billions of dollars, perhaps tens of billions, and could take a decade or longer to complete. Before the projects are completed, however, acute water shortages can be expected in the areas of most urgent need. Even if and when the transport networks are finished, the cost of the water may be 10 times what farmers in the region can comfortably afford to pay if their products are to remain fully competitive in the marketplace. There seem to be no easy solutions. Perhaps because of this, progress toward any solution has been slow. Meanwhile, the draining of the Ogallala aquifer continues.

It is noteworthy that in the short term, farmers have little incentive to exercise restraint, due in large part to federal policies. Price supports encourage the growing of crops such as cotton that require irrigation in the southern part of the region. The government shares the cost of soil-conservation programs and provides for crop-disaster payments, thus giving compensation for the natural consequences of water depletion. And, in fact, federal tax policy provides for groundwater depletion allowances (tax breaks) for High Plains farmers using pumped groundwater, with larger breaks for heavier groundwater use.

Figure 1 The Ogallala aquifer.

No saltwater intrusion

Saltwater intrusion

Further saltwater intrusion

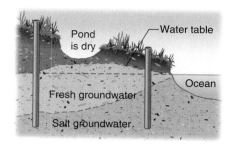

Figure 9-15   Saltwater intrusion into groundwater supplies is caused when rapid removal of freshwater allows saltwater to migrate farther inland, where it can reach freshwater wells.

underlain by saltwater. This layering occurs because rain, falling as freshwater, is less dense than saltwater; it therefore tends to form a lens, floating above the saltwater. This is also why oceanic islands are habitable by humans: Islands the size of several acres or more have their own underground freshwater supply from the rains. Unfortunately, rapid pumping of either coastal or island fresh groundwaters can cause the underlying saltwater to migrate upward, "intruding" into the freshwater (**Figure 9-15**). This degrades the quality of the freshwater for drinking, agriculture, and other uses. Saltwater intrusion is a significant problem in many coastal cities. Sections of the Gulf Coast and some parts of California are especially affected. Globally, such cities as Manila in the Philippines and Lima, Peru, have major problems with it.

These four groundwater problems are not exclusive: Some areas are plagued by all four. For example, some urban areas of the Gulf Coast, such as Tampa, Florida, have problems with groundwater pollution, depletion, land subsidence (and sinkholes), and saltwater intrusion.

Increasing Our Water Resources

Although much of the United States and many other parts of the world are facing critical shortages of water, these shortages are not insurmountable. Indeed, the main reason most shortages exist is simply that, until recently, water resources were taken for granted. On the positive side, this means that water resources can often be extended with relatively little sacrifice, simply by cutting the "fat" from past wasteful uses.

This section discusses five ways that water resources can be extended. The first three are familiar as methods of reducing the input of any resource: increased efficiency, recycling, and substitution. In the case of water, recycling refers to recycling waste-water. One of the many ways of substituting other liquids for freshwater is substituting plentiful seawater for naturally occurring freshwater by using desalination. The last two methods of extending water supplies are dams and canals, which do not reduce total demand. They shift water supplies from one region to another where they are needed more, extending water supplies locally but reducing them elsewhere.

Increased Efficiency

A major theme in environmental science is that input reduction is the best all-around solution to most environmental problems. The reduction of resource inputs by conservation not only prolongs the time that the resource will last, but reduces the pollution and other disturbances generated when the resource is extracted and used. Water resources provide many opportunities for conservation because they have been used so wastefully and inefficiently in the past. Water-rich areas have had no incentive to conserve.

Agriculture in particular and industry account for the large majority of water withdrawn and consumed in nearly all modern societies, so more efficient use in these areas is especially important. This will involve both technical and social changes. A classic example of technical change is **microirrigation** (sometimes called "drip" irrigation), whereby water is transported to crops via pipes instead of open ditches that promote evaporation. The water is then dripped onto the plants from tiny holes in the pipes installed on or below the soil. This increases the water that reaches the plant from 37% to as high as 95%. By saving nearly 300%, vast amounts of water are conserved. This method has been very effective in water-poor countries such as Israel. Only 3% of U.S. irrigation is by microirrigation, largely because of its expense; in contrast, Israel has about half of irrigated lands under microirrigation, and in Cyprus, better than 70% of its irrigated lands use microirrigation. The initial cost is about $1500 to $3000 per 2.5 acres (about 1 hectare), so it is used mainly on highly valued fruit and vegetable crops. But the use of microirrigation is rapidly growing in the United States, especially in the west, where incorporation of the true environmental costs of food production is raising prices (and some states are subsidizing the cost of irrigation). Similarly, industrial engineers have found it relatively easy to design production technologies that use much less water to produce everything from newspapers to aluminum. Recycling the same water during production is just one example of a design change. While such technical changes may temporarily increase costs, they often ultimately lower costs compared to what would be paid for a steadily decreasing water supply.

Conserving water often requires social changes in addition to technical change. Changes in diet can reduce agricultural water loss. Much more water is required to raise domestic animals for food than to produce the same amount of food as crops; in general, we use fewer resources when eating "lower on the food chain." The average person can usually do much to conserve water. Far from being painful, these changes require little personal sacrifice and may even save money. Simply shaving with a basin instead of running water reduces water used by 2000% (**Figure 9-16**). Installing low-flush toilets and taking shorter showers are especially important because flushing and showers are, by far, the two greatest personal uses of water (Table 9-3). Home lawns and gardens often use much water, so switching to plants with low water needs and other landscaping changes can save water. **Xeriscaping** (pronounced "zeriscaping") is landscaping designed to save water. It includes drought-tolerant plants and many other changes such as weed control, watering times, and height of mower settings.

Desalination

Since 97.4% of Earth's water is saltwater, removal of dissolved salts, called *desalination*, would obviously create a huge supply of freshwater. Unfortunately, desalination is expensive, even with the latest, most efficient technology. It is therefore economical only where other sources of water are even more expensive or are not available. Over 50% of the world's desalination capacity is located in the Middle East and North Africa, where over 1.5 billion cubic meters of freshwater are produced each year. Worldwide desalination plants produced about 3.0 billion cubic meters per year.

The primary reason for the high cost of desalination is the cost of energy. On average, seawater has about 35 parts per thousand of dissolved solids, which must be reduced to less than 1 part per thousand for drinking and industrial uses. About 200 pounds (91 kg) of salt are produced for every 1000 gallons (3800 L) processed. As a desalination plant often processes many millions of gallons per day, millions of pounds of salt must be removed daily. Disposing of all this salt is another major cost of desalination. Landfill costs are rising fast, and the concentrated salt cannot be dumped into the ocean without damage to ecosystems. Some economic uses, such as de-icing roads, have been found for this salt, but without costly treatment, it is generally not clean enough for many purposes. In some cases, costs of desalination may be reduced by producing water that is still too salty for drinking and industry, but is nevertheless useful for such purposes as watering salt-tolerant plants.

The membrane and the distillation desalination methods have proven to be the most effective in minimizing energy costs. The **membrane method** (also called the *filter and reverse-osmosis method*) removes dissolved ions by passing the liquid through a membrane at high pressure (**Figure 9-17a**). This method is

WATER SAVINGS GUIDE

CONSERVATIVE USE WILL SAVE WATER		NORMAL USE WILL WASTE WATER
Wet down, soap-up, rinse off 4 gallons	SHOWER	Regular shower 25 gallons
May we suggest a shower?	TUB BATH	Full tub 36 gallons
Minimize flushing Each use consumes 5–7 gallons	TOILET	Frequent flushing is very wasteful
Fill basin 1 gallon	WASHING HANDS	Tap running 2 gallons
Fill basin 1 gallon	SHAVING	Tap running 20 gallons
Wet brush Rinse briefly ½ gallon	BRUSHING TEETH	Tap running 10 gallons
Take only as much as you require	ICE	Unused ice goes down the drain
Please report immediately	LEAKS	A small drip wastes 25 gallons a day
Turn off lights, TV, heaters, and air conditioning when not in room	ENERGY	Wasting energy also wastes water

THANK YOU FOR USING THIS COLUMN . . . . AND NOT THIS ONE

Figure 9-16 Ways to conserve water. Note: 1 gallon equals approximately 3.785 liters. (*Source*: Reprinted by permission of the San Francisco Convention & Visitors Bureau.)

generally cheaper and faster than distillation. A large municipal plant can process several billion gallons per day. The membrane method is generally unable to desalinate normal saltwater on a large scale because it rapidly clogs the tiny pores. The method has proved highly effective for areas that need to desalinate brackish (moderately salty) water including brackish groundwater. Many cities near the ocean therefore pump brackish water from underground when using this method. An example is Santa Catalina Island off the coast of southern California.

The second desalination method, **distillation**, relies on heat: When saltwater evaporates, the dissolved solids are left behind. Such a device can be designed in many ways, but the most common is the **multistage flash distillation (MSF)** method. This pipes cold seawater through a series of coils in chambers that become progressively hotter; the heated seawater is then routed back, underneath the coils (**Figure 9-17b**). Freshwater condenses on the outside of the coils from the boiling seawater at the bottom. This method is over six times more effi-

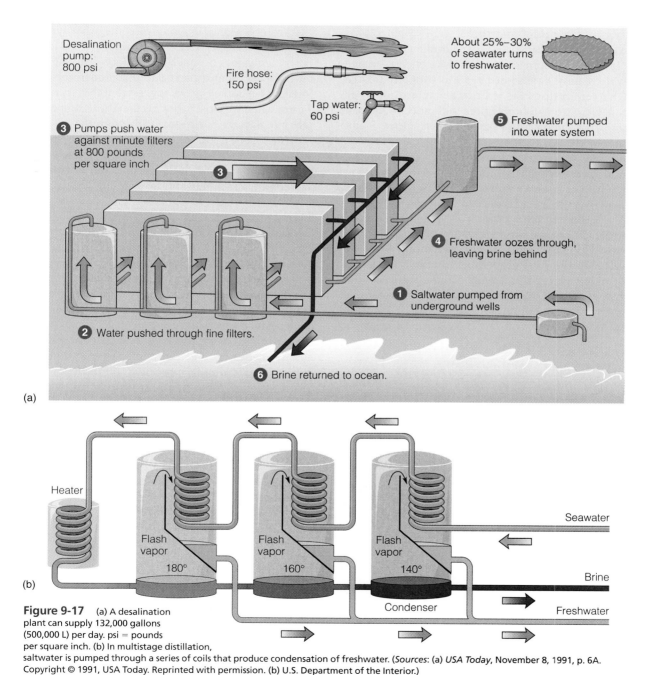

Figure 9-17 (a) A desalination plant can supply 132,000 gallons (500,000 L) per day. psi = pounds per square inch. (b) In multistage distillation, saltwater is pumped through a series of coils that produce condensation of freshwater. (*Sources:* (a) *USA Today*, November 8, 1991, p. 6A. Copyright © 1991, USA Today. Reprinted with permission. (b) U.S. Department of the Interior.)

cient than simply heating a single chamber, but still costly in terms of energy and therefore money. The high costs of MSF are often partially offset by locating MSF plants alongside electric power plants and using the heat they produce as a byproduct. Solar energy is used for desalination on a small scale, but sunlight is too dilute to make this method practical at large scales: Even under ideal conditions, only about 0.2 gallons per square foot (8.12 L per m³) can be collected per day. Many square miles of land would be required to rival the output of an average MSF plant.

Wastewater Reclamation

Most of us do not like the idea of drinking wastewater, especially reclaimed sewage. Nevertheless, wastewater can be safe to drink if it is properly treated. Chapter 15 discusses wastewater treatment in detail. Here we will simply note that most municipalities treat their wastewater and then release it into nearby rivers, lakes, or oceans. People downstream or elsewhere often drink and use this water. Many natural processes such as dilution help reduce pollutant levels, even if the water has traveled only a few miles, but constant testing is needed to guarantee safety.

Recently, many cities have begun to consider reusing their own wastewater directly. This so-called **closed loop reclamation** involves treating the wastewater to the level needed before direct reuse. For drinking water, this means using advanced—and very expensive—treatment to render the water safe. Even if this is done, the water must be stored in tanks or ponds for a few days before it is safe to

drink. A number of cities, especially in arid parts of South Africa, have successfully produced drinking water in this way. Denver, Colorado, has a well-known plant. However, it is often much more economical to use closed loop reclamation to produce water for nondrinking purposes and save natural sources for drinking. **Gray water** is untreated or partially treated wastewater that is used for watering golf courses and lawns instead of using cleaner water of drinkable quality.

Despite our emotional reactions to wastewater, it is generally much cheaper to treat wastewater than to desalinate seawater, especially if we only need to produce gray water or some other type of partially treated water. For instance, it typically costs eight times as much to produce drinking water from desalination as reclaiming moderately polluted water. Some experts suggest that the city of the future will have closed loop reclamation engineered into the natural environment through the "3 Rs": return, repurify, and reuse. In this system, treated wastewater is returned to irrigate crops; the water is renovated (repurified) by percolation as groundwater and is then reusable as drinking water.

Dams and Reservoirs

All of the ways of extending water resources discussed so far have a major advantage: They use water more efficiently and produce more for society in general. In contrast, dams and reservoirs, and canals simply redistribute water from water-surplus areas with low populations to water-deficit areas with high populations. This results in no net gain, and often a net loss for society in general because water is lost during storage and transport by evaporation, soil infiltration, and other processes. Nevertheless, redistribution is often the most economical way to meet society's water needs and can have minimal environmental impact if properly designed.

Dams: The Benefits and Costs. **Dams** are structures that obstruct river or stream flow to make lakes (**Figure 9-18**). By keeping water from flowing downstream, dams allow people upstream to have preferential use. There are more than 36,000 major dams in the world (and hundreds of thousands of smaller dams). Dams are built for one or more of three reasons: (1) provide better control of water flow to minimize flood damage; (2) create a reservoir, the lake, that can serve as a water source for nearby populations; and (3) provide hydroelectric power. Even well-designed dams have some environmental impacts. These include the following:

1. *Sediment accumulation.* Occurs in the lake as sediment is trapped by the dam. Unless the

Figure 9-18   A flood control dam on the Shenango River, Pennsylvania.

sediment is removed, the reservoir will fill up in a few decades to a few hundred years.

2. *Scouring downstream.* Occurs because the water that escapes the dam has little sediment. Such "sediment-starved" stream flow will scour out the river bottom on the other side of the dam.

3. *Water loss.* Occurs from evaporation of water from the lake. The scale of this loss is vast: In Texas, about three times as much water is lost from evaporation of reservoirs as is used by industry and cities. Attempts to reduce the loss by spreading a "skin" of chemicals have not proven effective.

4. *Salination.* Occurs because evaporation of water leaves dissolved minerals behind, causing the lake to become progressively saltier.

5. *Dam breaks.* Occasionally occur from heavy rainfalls, leading to disastrous flooding downriver.

6. *Biological disturbances.* Occurs because dams can disturb both land and aquatic life. One of the most obvious effects is when dams block migration patterns of fishes, such as spawning salmon in the northwestern United States. Fish ladders built to allow migration around the dam are often ineffective. A second cause of biological disturbance is loss of both land and aquatic habitats. Land habitats can be flooded upstream; downstream aquatic habitats can suffer from decreased water flow. One of the most infamous examples is the Aswan Dam on the Nile River, which led to massive disease outbreaks and loss of crop production along the Nile, among many other biological impacts. In recent years, there has been increasing pressure to dismantle unnecessary dams, and in the United States, hundreds of dams (mostly small dams) have been removed over the last few decades.

7. *Social disturbances.* Large dams in particular can impose heavy social damage, displacing people, forming reservoirs where stagnant water can breed disease vectors such as insects that inflict the human population, and hindering the natural ability of a river to break down and process human-dumped sewage. An example of the massive displacement of people by a single dam is the Three Gorges dam in China. By the time it is completed, 1.2 million or more people will have been relocated.

One often-proposed solution to the many environmental and social problems caused by dams is to make more use of natural methods of flood control, such as wetlands and local geological conditions. This suggestion runs counter to the traditional policy, often used by the U.S. Army Corps of Engineers, which is to alter natural conditions with dams, channelization, and other means according to economic and engineering criteria, without much consideration to environmental impacts. This alternative has received renewed attention since the great flood of 1993 in the Mississippi River system. Many environmental groups noted that dam breaks and many kinds of damage will occur again unless more consideration is given to conforming to natural conditions instead of trying to alter them.

Canals

Canals, or aqueducts, are artificial channels built to transfer water over long distances. As urban areas grow, they often exceed local water supplies, and canals become necessary for further growth. For example, New York City has imported water for over 100 years, moving to sources progressively farther away, such as upstate New York. Many long canals are even more important for large cities in dry areas. In California, two-thirds of the water runoff is north of San Francisco, while two-thirds of the water use is south of San Francisco. Since the early 1900s, the growth of cities in southern California, especially Los Angeles, has led to the construction of many canals to import water from the north and from the Colorado River. Population growth in southern California has continued to increase in recent years, causing ever greater water demands. This has led to increasing legal and social confrontations as people in other areas, especially farmers, object to the diversion of their water to urban areas. Such disputes illustrate the disadvantage of redistribution. There is no net gain for society, whereas conservation, desalination, and wastewater recycling result in greater amounts of available water. Cities such as Santa Barbara and San Diego are turning to all of these solutions and moving away from water importation.

In Florida, a vast artificial system created in the 1960s and earlier containing 1400 miles (2250 km) of canals drained away much of the water feeding the Everglades for use in urban centers and especially agriculture (**Figure 9-19**). Whereas the canals led to shortages and disputes over water rights in California, those in Florida illustrate another potential hazard of canals: habitat destruction and pollution. The Everglades ecosystem is one of the largest and most unique wetland habitats in North America. But it was drying up and becoming polluted as water was diverted away by canals. At the same time, straightening or **channelization** of the Kissimmee

Figure 9-19 (a) Computer models of the Everglades drainage system show that the number of days that the land is inundated (covered by water) has been greatly reduced since 1900. (b) Canals now drain rainwater away before the land can be saturated for long periods, especially the "permanent" wetlands shown in deep blue. (*Source*: Modified from South Florida Water Management District data.)

River was causing more pollution to flow into the Everglades. This pollution not only caused species extinction but affected water quality because the Everglades serves as a major recharge area for water supplies in South Florida. To help correct this situation, in 1992, the United States Congress authorized the Kissimmee River Restoration Project to restore, at a cost of tens of millions of dollars, 40 square miles (100 square kilometers) of riverine and floodplain ecosystems, including the restoration of 43 miles (69 kilometers) of meandering river channel that had been "straightened." The first phases of this project have now been completed, and ultimately it is planned that some 3000 square miles (7800 square kilometers) of floodplains, wetlands, and river channels will be restored at a cost of over $500 million.

Despite enormous economic and environmental costs, there will probably be increasing social pressures for water importation. Such pressures will grow not only because of increasing population growth in California and other parts of the West, but also because of agricultural needs. Recall that agriculture is the greatest consumer of water, and the West and Midwest are vast dry agricultural areas. Added pressure to import will come from the depletion of the Ogallala aquifer underlying the Midwest later this century.

Social Solutions to Water Scarcity

We have discussed many ways that society can extend water supplies, from conservation to redistribution by canals. Like all technical solutions to environmental problems, however, these solutions have important social aspects. For instance, conservation is strongly influenced by economic considerations, such as the price of water. Similarly, laws governing water use, especially in the western United States, have enormous influence on where dams and canals can be built.

Economics of Water Use

When a resource is cheap, it will be wasted. Only when the resource is costly do consumers have an incentive to conserve or reduce resource input by more efficient use, recycling, or substitution (**Figure 9-20**). In the case of water resources, wastewater reclamation is an example of recycling. Substitution could include switching to desalinated saltwater instead of depleting groundwater or using liquids other than water.

The low cost of water in the United States has offered little incentive to conserve. Most water supply systems charge only a fraction of the total cost of obtaining, storing, treating, and delivering water. The average family pays less than 1% of its income for water. As a result, many states and local areas are increasing water prices to provide more incentive to reduce water use: for each 1% increase in cost, water use in homes initially decreases by about half a percent. As costs rise, cities find it economical to monitor water supplies more closely, reducing waste.

Subsidizing Water for Agriculture. As the largest consumer of water, agriculture is an especially important target for conservation. California provides a good example. Farmers use 85% of California's water, much of it for the irrigation of water-intensive crops. Because of California's agricultural history, farmers not only have "first rights" to water but also pay far less (often less than 10% of the price to urban users) because their water is subsidized by the government. In the early 1990s, urban areas began to rebel against this practice and urged their

state legislators to reduce subsidies to farmers. As expected, this has led to less water use by farmers. One response has been to grow high-priced crops such as certain fruits, nuts, and vegetables instead of water-intensive crops such as rice. A similar situation occurs in many other parts of the western and southwestern United States, where federal taxes have historically funded projects to supply cheap water to farmers. Society as a whole does not necessarily suffer from refusal to subsidize cheap water because water-intensive crops can still be grown in other parts of the nation where water is more common, such as the eastern United States. They can also be imported from other countries, or perhaps most desirably, the higher costs will promote water conservation techniques, such as microirrigation, which are now rarely used in the United States.

Taxing Water Use in Industry and Households. Although water for industry and household use is less heavily subsidized, it is still very cheap. As a result, state and local governments have found that increasing the price of water via taxes or other means can greatly reduce the amount of water used. Often these taxes are levied as effluent charges, which is the cost of disposing of industrial wastewater. This illustrates again how input management not only conserves resources but also reduces pollution. For example, water withdrawals by U.S. manufacturing industries in the year 2000 were estimated to have been 60% less than in 1977 because of increased

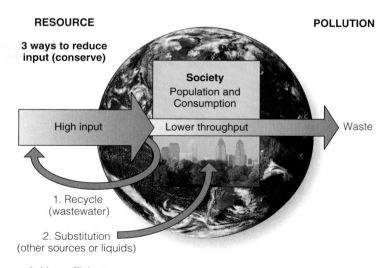

Figure 9-20   Three ways to conserve water (reduce water input to society), with examples: more efficient use, recycling, and substitution. If water resources become more costly to the water users, they will have an incentive to use these three methods of conservation. Conservation also tends to reduce water pollution because less wastewater is produced and water flow into natural waters is reduced.

taxes on wastewater disposal. Similarly, increased prices on domestic water encourage people to conserve by using low-flush toilets, taking shorter showers, and employing other means of conservation. In some countries, higher prices also encourage domestic water recycling and reclamation. Japan and Singapore, for example, have long used reclaimed wastewater in flush toilets.

Increasing the cost of a resource to its true or "environmental" cost will discourage both rapid depletion and pollution that occurs when a resource is very cheap. The improved efficiency of resource use often reduces not only overall environmental costs, but also economic costs to society. Industrial recycling of water is a good example. Cities can also gain from this. Beijing, China, has recently invested in more modern plumbing, leakage reduction, and water recycling. This will save billions of gallons of water per year, at a much lower dollar cost than the water development project originally planned to import water from farther away.

Legal Control of Water Use

As one of the most important components of living things, water probably has more laws governing its use than any other resource. In the United States, a hierarchy of laws extends from the federal to the lo-

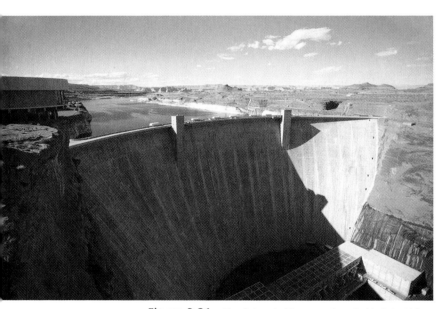

Figure 9-21   The Colorado River seen here behind the U.S. Bureau of Reclamation's Glen Canyon Dam is one of the few large sources of water in the southwestern United States.

cal level; as with all resources, federal law takes precedence when there is conflict between levels. Most direct regulation is found at the state level, however, for reasons that date back to homesteading and the use of the land when it was originally settled. This also explains why legal control of water is best developed for surface water, which has been intensively used through the years. In contrast, the rapidly growing use and pollution of underground waters have led to a confusion of legal activity in recent years. These difficulties are magnified because groundwater is much more difficult to monitor than surface water and it moves so slowly that depletion and pollution often take years to detect and remedy (Chapter 15).

Surface Waters.   Laws regulating surface waters can be divided into two basic types: riparian law and appropriation law. **Riparian law** dictates that the owner of land has the right to withdraw water that is adjacent to the land, such as from a river or lake. (*Riparian* means "bank," as in riverbank, in Latin.) A key provision is that water returned must be returned in a relatively unpolluted condition, so that owners of adjacent land, such as downstream, do not have their water rights violated. Riparian law is a common-law idea, meaning that it evolved through practical use, as opposed to being dictated from theory or administrative decree.

Riparian law is practiced mainly east of the Mississippi River. It worked well in those states, where water is relatively plentiful and areas were settled gradually. In contrast, all the western states have at least some degree of appropriation law. **Appropriation law** dictates that owners of land may be denied the right to withdraw water if a more beneficial use is found. In short, government, usually the state, can "appropriate" the water. The reason for this type of law in the western United States is obvious: Water has often been scarce so it had to be closely managed. Water from the Colorado River, for instance, is tightly regulated by state and federal agencies (**Figure 9-21**). Under riparian law someone could, in theory, own all rights to water in a river running through their land and restrict access by other people.

The greater state control of water under appropriation law makes it better suited for implementing environmentally optimum management. For example, if a study shows that an industry is being exceptionally wasteful, the state can threaten to appropriate the water if the industry does not take conservation measures. In contrast, under riparian law, if the industry owns the land along the river, the state may have to file suit and undergo expensive, prolonged litigation to make the industry change its ways.

Groundwater. Until recently, groundwater was virtually unregulated. Owners of land above the water were considered to have rights to the water, much as in riparian law. Even today, people in many rural areas are generally free to drill holes and withdraw water. In areas with rapid groundwater recharge or low population, this practice created few problems. Now, however, the rapidly falling water tables of many areas have required much stricter laws on drilling and withdrawal. Most of these are local, municipal ordinances that prohibit withdrawal inside city or county limits without a permit. Similarly, many cities regulate water usage, such as prohibiting lawn sprinkling during the afternoon when evaporation is high. Groundwater depletion is easier to regulate than groundwater pollution (discussed in Chapter 15). There is increasing pressure on the U.S. Congress to enact a single coherent set of laws regulating groundwater depletion and pollution to replace the piecemeal

regulation now occurring under many local and federal statutes.

Which Are Better: Economic or Legal Solutions?

The type of water resource problem determines whether economic or legal solutions are better. Most economists agree that market conditions provide the optimal allocation of resources when many parties want access to them. For example, in California, where many farmers and city dwellers want access to water, one effective solution is to make all parties pay a price for the water that incorporates the environmental costs. This could include taxing water, removing subsidies, or otherwise raising the costs to all parties. Such an economic solution not only tends to distribute costs more equitably, but is often the only practical way of controlling water use: Regulating a resource that is being used by many parties is often very costly because of the equipment and personnel needed to monitor use.

Legal solutions are therefore often most applicable where only a few parties use the resource, making monitoring possible. Also, legal regulation is often preferable where a resource is in very short supply, so that waste is minimized through stricter control than a market approach allows.

SUMMARY

- Water is a remarkable substance that is essential to life as we know it—unusual properties include its density, boiling point, specific heat, and solvency, which are all affected by the bipolar structure of the water molecule.
- Despite the abundance of water on Earth (mostly in the form of the oceans), there is only a limited supply of freshwater.
- The hydrologic cycle acts as a pump, cycling water through evapotranspiration and precipitation.
- From a human perspective, water is very unevenly distributed across the continents and nations; many areas are currently water-stressed.
- Withdrawn water is water removed from its source (such as a lake, river, or aquifer); con-

sumed water is withdrawn, but not returned to its original source.
- Worldwide, and also in the United States, the vast majority of withdrawn water goes toward agriculture; the next highest amount of use is by industry, followed by domestic use.
- Types of water sources include surface waters and groundwater; in many areas groundwaters are being polluted and depleted.
- Water resources can be stored and redirected using dams and canals, or effectively stretched and increased through efficiency measures (for instance, microirrigation), desalination, and wastewater reclamation.
- Taxes, subsidies, and various legal controls have been utilized to address issues of water scarcity.

STUDY GUIDE

KEY TERMS

Appropriation law
Aquicludes
Aquifer
Artesian well
Channelization
Closed loop reclamation
Cone of depression
Confined aquifers
Consumed water
Dams

Deficit areas
Distillation
Effluent charges
Evapotranspiration
Gray water
Groundwater
Membrane method
Microirrigation
Multistage flash distillation
  (MSF)

Precipitation
Recharge areas
Riparian law
Sinkholes
Surplus areas
Water table
Withdrawn water
Xeriscaping
Zone of aeration
Zone of saturation

STUDY QUESTIONS

1. Why is water an unusual substance? List and explain its properties. How does the structure of the water molecule explain water's unusual properties?

2. What percentage of the Earth's water is readily available in the liquid state as groundwater? As surface water?

3. The global water cycle involves what two main processes? Define these two processes. How much water returns to the oceans as runoff?

4. What are the major deficit areas of the world?

5. Why do hundreds of millions of people suffer from water shortages if we are not using all readily available fresh water?

6. Why have streams and river valleys, and areas near them, usually been among the most populated regions?

7. What are three withdrawal problems associated with groundwater? How does wetland destruction promote withdrawal and pollution problems in groundwater?

8. Why is agriculture a greater consumer of water than industry? How can microirrigation reduce this consumption?

9. What are the two main desalination methods? Explain how they function.

10. Why are dams and reservoirs less efficient and more harmful than other ways of extending water resources?

11. What happens when the cost of a water resource is increased? Name some ways to do this.

12. Discuss the economics of water use, focusing on ways to pay the true cost of water. Why is the true cost often underpaid? Give examples.

13. Name and define the two types of laws regulating water resources in the United States. Where are they used?

14. Discuss various ways of reducing water inputs to society, including increased efficiency of use, economic aspects, and legal aspects.

WHAT'S THE EVIDENCE?

1. The authors assert that natural water shortages can be greatly aggravated by human mismanagement of water supplies. What evidence supports this contention? Do you agree that people often mismanage water supplies? If so, how can we better manage water resources? Give specific examples. If you disagree with the idea that humans have historically mismanaged many supplies of water, cite evidence to support your position.

2. The authors appear to be critical of the large amounts of water used by the United States, including more water per person in the United States than among most other nations. Is such criticism well founded? Is the United States, in general, excessive in its water use? Or, being arguably the only superpower, is such water usage justified? Is it necessary? State your position and support your views with relevant evidence and persuasive arguments.

CALCULATIONS

1. Suppose a river carries 10 cubic miles of water into the ocean over 10 years. Convert this amount of water to cubic kilometers, given that 1 cubic mile = 4.168 cubic km.

2. Given the river in number 1, how much water would be carried to the ocean on average each day during the decade under consideration? Give your answer in both cubic feet and cubic meters.

ILLUSTRATION AND TABLE REVIEW

1. Referring to Figure 9-4, calculate the estimated world water use in 1960. In 2000? (Express your answers in units of cubic kilometers and cubic miles.) Estimate the percentage of total world water use that was used by industry in the year 2000.

2. Referring to Figure 9-7, calculate by approximately what percentage the estimated total human demand for freshwater increased between 1900 and 2000. During this same time period, by approximately what percentage did the amount of accessible freshwater increase?

WWW.ENVIRONMENT.JBPUB.COM

www.environment.jbpub.com

Connect to this book's web site: http://www.environment.jbpub.com/ The site features eLearning, an online review area that provides quizzes, chapter outlines, and other tools to help you study for your class. We've posted the answers to the book's odd-numbered questions. You can also follow useful links for more in-depth information. The site provides sources for the information in this chapter and recommendations for further readings.

In the 1960s and 1970s, some environmentalists predicted that by the year 2000, our finite supplies of many minerals and metals would be nearly exhausted and the prices for any remaining reserves would be astronomical. Yet today, many minerals appear to be more "common" than they were four decades ago. Mineral and metal resources really are finite, however. The open pit reserves of the Bingham Canyon copper mine in Utah, the single largest human-made excavation in the world (2.4 miles [3.9 km] across), are expected to last until 2015.

CHAPTER 10

Mineral Resources

Mineral resources are a classic form of resource that must be carefully managed. Mineral supplies are finite and nonrenewable. The extraction and processing of minerals can be extremely damaging to the environment. Yet modern technological society is heavily dependent upon minerals (see the diagram for Section III, page 141). As the world population continues to grow, and more nations become increasingly "developed," the need to extract and utilize mineral resources will increase.

So why are mineral and metal prices dropping even as the world's population is increasing? Many factors affect prices of resources. New deposits may be discovered, and new technologies may make it possible to economically recover minerals and metals from previously known, but lower-grade deposits. Recycling has decreased the need for new sources of metals and minerals, as has the continued substitution of relatively abundant materials for traditional, but scarcer materials (such as the use of glass optical fibers in place of traditional copper wire).

Even so, the trend of increasing mineral and metal consumption over time cannot be ignored. Remember the equation, $I = P \times C$ (Impact = Population $\times$ Consumption) from Chapter 1. Many researchers contend that our immediate concerns should not be the eventual depletion of our mineral and metal resources, but rather the environmental damage caused by mineral exploitation. Such thinkers emphasize the development of ever more efficient and environmentally friendly ways of mining, processing, using, and disposing of mineral and metal resources—a strategy that ultimately should benefit everyone. For example, the owner of the Bingham Canyon copper mine also boasts about having the cleanest smelter in the world.

Mineral resource refers not only to **minerals** in a strict sense (such as quartz, diamond, mica, or graphite), but also to material substances composed of minerals or extracted from minerals. Thus, mineral resources include ores, metals, gravel, sand, marble, granite, phosphate rock, and so on, including elements and compounds that can be extracted from earth materials. These are usually inorganic compounds, although some deposits may be concentrated or accumulated by life-forms; limestone, when composed of the calcium carbonate of the shells of sea creatures, is a classic example. As used in some contexts, the term mineral resource includes fossil fuels (which are discussed in Chapter 7).

Types of Mineral Resources

Minerals can be divided into different categories. **Metallic minerals** include iron, aluminum, copper, zinc, lead, gold, and so on. Metallic minerals are often referred to as **ferrous** (iron and related metals such as chromium, manganese, nickel, and molybdenum, which are commonly alloyed with iron) or **nonferrous** (gold, copper, silver, and so forth). **Nonmetallic minerals**, a second major category, can be divided into two groups: **structural materials**, such as building stone, sand, gravel, components of cement and concrete, etc., and **industrial materials**, such as fertilizer components (phosphorus and potassium), salts, sulfur, other nonmetals used in manufacturing, asbestos, abrasive minerals (such as emery, or corundum, and industrial diamonds), and so on. The gemstones and semiprecious minerals that serve no pragmatic function constitute a third category; they are purely of ornamental and aesthetic value. Finally, uranium and the fossil fuels (coal, oil, and natural gas—see Chapter 7) are sometimes regarded as a fourth category of **energy minerals**.

Relative Demands and Values of Mineral Resources— An Overview

Of all nonrenewable geological resources, including the fossil fuels (**Figure 10-1** and **Table 10-1**), the greatest demand in terms of total bulk quantity is for basic sand, gravel, crushed stone (road aggregates), and building stone. These are the raw bulk

TABLE 10-1	Representative Annual World Production of Selected Minerals
Mineral	**Production (Thousands of Metric Tons)[a, b]**
Metals	
Pig iron	552,000
Aluminum	18,100
Copper	8,920
Manganese	8,600
Zinc	7,300
Chromium	3,784
Lead	3,350
Nickel	949
Tin	216
Molybdenum	114
Titanium	102
Silver	15
Mercury	6
Platinum-group metals	0.3
Gold	0.2
Nonmetals	
Stone	11,000,000
Sand and gravel	9,000,000
Clays	500,000
Salt	191,000
Phosphate rock	166,350
Lime	135,300
Gypsum	99,000
Soda ash	32,000
Potash	28,125

[a]All data exclude recycling.

[b]One metric ton equals approximately 1.102 English tons.

Source: J. E. Young, *Mining the Earth*. In L. R. Brown et al., *State of the World 1992* (New York: W. W. Norton, 1992), p. 102. Reprinted with permission of Worldwatch Institute, Washington, D.C., Copyright © 1992.

Figure 10-1   Raw materials snakes: (a) in order of world production by quantity, and (b) in order of economic value. Note that while the relative ordering of the various materials is accurate, the relative sizes of the circles are merely diagrammatic. (*Source*: A. A. Archer et al., ed., *Man's Dependence on the Earth* [Paris: UNESCO, 1987], p. 68. © UNESCO 1987. Reproduced by permission of UNESCO.)

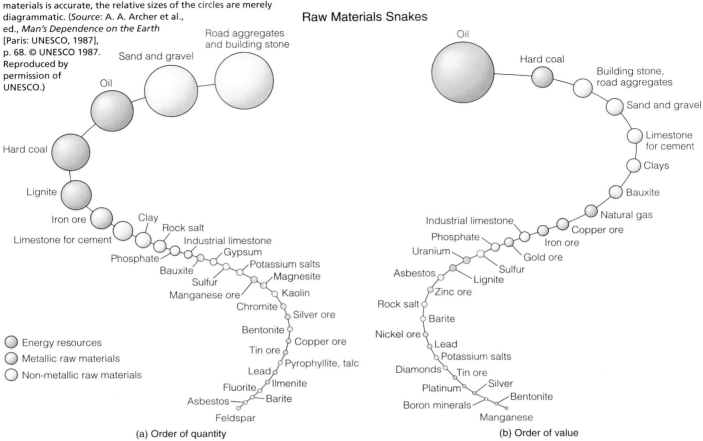

Raw Materials Snakes

(a) Order of quantity

(b) Order of value

Figure 10-2   Approximate annual per capita consumption of selected mineral resources by Americans.

Nonmetallics

Crushed stone
10,800 lbs
4900 kg

Sand and gravel
8313 lbs
3770 kg

Cement
845 lbs
383 kg

Clays
430 lbs
195 kg

Salt
375 lbs
170 kg

Phosphate
320 lbs
145 kg

(drawn $\frac{1}{10}$ scale)

Metals

Iron and steel
1240 lbs
562 kg

Aluminum
48.5 lbs
22 kg

Copper
19.8 lbs
9 kg

Zinc
13.2 lbs
6 kg

Lead
11 lbs
5 kg

materials used in constructing roads, bridges, buildings, and so forth. Despite the demand, however, they make up only a small percentage (about 4% to 10% of the total value of nonrenewable mineral materials, including the fossil fuels) of the total value of raw materials consumed each year due to their worldwide abundance. Virtually all of the cost involved with these materials pays for their extraction and shipment—the material itself may be almost "free." No one is concerned that we will run out of sand and gravel in the near future; indeed, it is doubtful that we could ever exhaust the supply of these materials, although certain fine building stones, such as high-grade marble from a particular quarry, may be exhausted and clean sand to fill holes (natural or human-made) may become scarce locally.

In terms of bulk, fossil fuels are the class of nonrenewable, mined resources that are next highest in demand. Because of their increasing scarcity and modern society's dependence on them, the fossil fuels are currently the most valuable sector of our mined materials (fossil fuels account for over 50% of the total value of nonrenewable mineral materials). Fossil fuels are discussed in Chapter 7.

The remaining mineral resources that are significant on a global scale are metals (iron, aluminum, copper, zinc, and so on) plus certain other elements and compounds (such as sulfur, phosphorus, salt, lime, soda ash, and potash). These substances form the basis of discussion for the remainder of this chapter.

Figure 10-2 shows per capita consumption of several mineral resources in the United States. United States mineral consumption has increased dramatically since World War II, nearly quintupling during that period to more than 2.8 billion metric tons (3.1 billion tons) annually in the late 1990s (**Figure 10-3**). Worldwide, an estimated 10 billion metric tons of minerals were being used each year in the late 1990s. Our industrialized economy is built on the consumption of ever larger amounts of materials. Some of this material goes into constructing **infrastructure**, such as roads, bridges, and hospitals, or producing other durable objects. A large part of it, however, consists of materials that are quickly discarded as wastes—such as the **tailings** from mining (the residue after the high-grade ore is extracted), inexpensive "disposable" electronic equipment, batteries, and cars (which are typically discarded after several years). Although some of this material can be recycled (see discussion later in the chapter), mineral resources are inherently nonrenewable. Once a particular copper or gold deposit is mined to exhaustion, it is gone forever; it does not "grow" back.

Mineral Deposits, Ores, and Reserves

Mineral resources are finite and exhaustible, just like fossil fuel resources. Furthermore, these resources are not distributed evenly over the face of the globe. Various types of **mineral deposits**, like fossil fuel deposits, formed only under special geological conditions.

On a global scale, a few elements (namely oxygen, silicon, aluminum, iron, calcium, sodium, magnesium, and potassium) are extremely abundant,

(a)

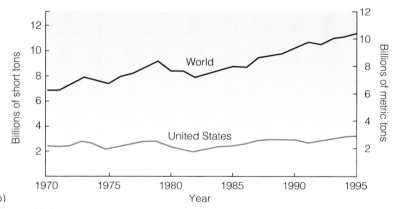
(b)

Figure 10-3 (a) Nonfood and nonfuel raw materials consumed in the United States, 1900–1995. (b) Nonfood and nonfuel raw materials use for the world and United States, 1900–1995. Note that the data in these graphs do not correspond exactly to the data in Table 10-1 due to discrepancies in reporting and estimating around the world; the information in the table and these graphs are presented to give a general picture of the magnitude of materials use and trends through the twentieth century. (*Source*: United States Geological Survey.)

making up over 99% of the Earth's crust by weight. But even for these abundant elements, the practically obtainable supply is not inexhaustible. To economically obtain most mineral products, we must locate mineral deposits formed by natural geological processes. But not all mineral deposits are usable. An **ore** deposit is a mineral deposit that can be economically mined at a certain time and place with a certain technology. Depending on supply, demand, and other factors, a mineral deposit that is not considered an ore deposit one year may be considered an ore deposit the next year (perhaps because the price of the mineral has risen enough to offset extraction costs).

Reserves are identified ore deposits that have yet to be exploited. Reserve estimates for a particular mineral will vary over time, increasing as new mineral deposits are located, as improved technology and more efficient methods of extraction decrease the costs of exploiting deposits, or as economic factors drive prices up, causing lower-grade deposits to become ore deposits. Reserve estimates will decrease as known deposits of a particular mineral are exhausted or if market prices fall, making it economically unfeasible to mine lower-grade deposits.

Iron and aluminum illustrate the concept of reserves. Given the abundance of these minerals in the Earth's crust, the supplies seem virtually inexhaustible. According to estimates, given current consumption rates, the iron and aluminum in the Earth's crust could last more than a million years and a hundred million years, respectively—without recycling. But when only proven, economically viable world reserves (in the form of ore deposits) of iron and aluminum are considered, estimates are that at present-day production and consumption rates (again, without recycling), there will be a severe shortage of these metals within a few hundred to a couple of thousand years. Which view is more realistic will be determined by the amount of energy and the kinds of sources that will be available in the future. With unlimited amounts of energy, we could extract virtually unlimited amounts of aluminum from very low-grade, diffuse deposits. But extracting from low-grade ores, even with unlimited energy, will surely cause environmental disruption and degradation on a scale even more massive than that of current mining.

Many substances are much rarer than aluminum or iron. For instance, copper makes up on average only 55 to 63 parts per million (ppm) of the Earth's crust, and tin only 2 ppm. To have a minable ore of most metals, the metal must occur in the rock at a concentration of tens to thousands of times higher than in ordinary rock. Even with unlimited amounts of energy, there is an upper limit to the quantity of minerals that can be produced. In the real world where energy is limited and minerals must be produced from ore deposits, taking world production, consumption, and reserves into account, it has been estimated that our known **virgin supplies of ores** for copper, tin, lead, zinc, and various other metals will be exhausted within a century.

Environmental Degradation Due to Mineral Exploitation

Today, vast amounts of land are devastated due to the direct effects of mining activities. Around the globe many millions of acres (millions of hectares) of land have been laid waste, trees destroyed, earth and rock churned up, and billions of tons of air and water pollutants and solid waste in the form of tailings have been generated (**Figure 10-4**).

The largest human excavations in the world are open-pit mines. Metal smelters typically produce enormous quantities of air pollution, including sulfur dioxide (causing acid deposition) and many toxic heavy metals—lead, cadmium, and arsenic among them. An estimated 8% of worldwide sulfur emissions into the atmosphere come from smelters. **Dead zones**, in which no vegetation or animal life can survive, cover thousands of acres (thousands of hectares) around many large smelters. The mining industry routinely uses hundreds of tons of mercury, cyanide compounds, and other very toxic substances to remove metals from ores. Acid drainage flows from abandoned mines and tailing piles. Removed overburden (rock and soil removed during mining), tailings, and other wastes pile up in massive amounts. On a global scale, mining operations have become a force as powerful as natural erosional processes. It has been estimated that more than 23 billion metric tons (25 billion tons) of nonfuel minerals are removed from the Earth annually (less than half of this is actually "used" or "consumed"), and including the overburden, the total amount of material artificially moved by humans in mining each year may be twice the amount of sediment carried annually by all of the world's rivers.

The mineral industry is one of the largest consumers of energy worldwide. Approximately 1% of all the world's energy goes toward aluminum production each year, and over 5% of the world's energy goes toward steelmaking annually. Much of

Figure 10-4   Tailings from an open-pit copper mine, Morenci, Arizona.

Trading a Mountain for a Hole in the Ground

Massive mining has the ability to obliterate natural landmarks. Goldsmith and colleagues tell the story of the obliteration of Mt. Fubilan in New Guinea (**Figure 1**):

For the Ok people of the highlands of central New Guinea, Mt. Fubilan was a sacred mountain, sitting on top of the land of the dead. In the late 1960s they were persuaded to lease their mountain to a mining company. To the utter astonishment of the Ok, the company began systematically to scoop away the peak of Fubilan. Within the next two decades, the 2,000-m (6,560-foot) peak will have ceased to exist. In order to exploit Mt. Fubilan's reserves of copper and gold, the mining company intends to remove the sacred mountain altogether. When the mine is finally exhausted, all that will be left is a hole in the ground, 1,200 m (3,900 feet) deep.

Questions

1. Is it "right" for the mining company to do what it is doing? On the one hand, everyone seems to benefit—the mining company makes a profit, the Ok profit monetarily from their mountain, and the world can use the minerals derived from Mt. Fubilan. On the other hand, it can be argued that we have a moral and aesthetic obligation to preserve natural wonders such as Mt. Fubilan. The mining company can be viewed as simply exploiting the Ok and their land. What do you think?

2. When the Ok leased the mountain to the mining company, do you believe they realized that it would be entirely destroyed? What responsibility, or moral obligation, does a mining company have to explain all of the "fine print" to people who may not be familiar with legal contracts?

3. Considering that the mountain was sacred to the Ok, what does its destruction say about the mining company's regard for the Ok religion and culture? Should primitive beliefs be allowed to stand in the way of modern technological progress?

(*Source*: E. Goldsmith, N. Hildyard, P. McCully, and P. Bunyard, *Imperiled Planet: Restoring Our Endangered Ecosystems* [Cambridge, MA: MIT Press, 1990], p. 201.)

Visit www.environment.jbpub.com for more information.

Figure 1 Mt. Fubilan in the process of being mined for gold and copper, 1984.

the energy used in the mining industry is acquired from burning fossil fuels, charcoal, or fuelwood (causing all of the problems associated with massive energy use).

We can get a glimpse of the magnitude of the total destruction by focusing on the United States. As of the early 1990s, abandoned and operating coal and metal mines covered an estimated 22 million acres (9 million hectares) of land; for comparison, 40 million acres (16 million hectares) of land in the United States are covered by pavement (roads, parking lots, and so forth). Each year, nonfuel mineral mining in the United States produces an estimated 1.0 to 1.3 billion metric tons (1.1 to 1.4 billion tons) of waste. This is about five to six times the amount of municipal solid waste produced by the United States annually.

Much of the volume of the waste from mining is overburden—the rock and dirt above the ore deposit that must be removed in order to obtain the ore. Overburden can be relatively safe and may be used in reclaiming land after mining, so it need not pose a major environmental problem. If the overburden is allowed to erode away and thus be dispersed, however, it may have a major negative impact on the environment: Once flourishing land loses its topsoil and associated flora, streams become clogged with silt, and wildlife is destroyed.

Generally more hazardous are the tailings (the leftovers from processing the ore). These may leach and produce acids, and the leachate may contain heavy metals or other toxic substances. An estimated 10,000 miles (16,000 km) worth of streams in the western United States have been damaged by leaching and drainage from mines and piles of mining wastes. Smelting and refining metals from minerals also releases pollutants into the atmosphere, such as heavy metals (like arsenic and lead) and sulfur oxides.

Mining can contribute directly to deforestation. Charcoal, a necessary fuel in many mining and refining operations, is created by partially burning wood. In Brazil, for example, 123,500 acres (193 square miles, or 50,000 hectares) of forest will be destroyed each year to produce charcoal to fuel the Grande Carajas iron ore mining and smelting project. If a proposed bauxite mine and smelter are also constructed, the devastation will be even worse. In the northwestern portion of the Brazilian Amazon, diamonds, uranium, and gold have been discovered in territory where the Yanomami Indians live. Miners are now invading the area and threatening both the forest and the native populations. Case Study 10-1 describes how mining has affected another native population.

Many mineral products have no ready substitutes. Petrochemical products, such as plastic compounds, can be substituted for some mineral uses, but these products have their own environmental drawbacks. The use of metals and other minerals is not in itself generally harmful to the environment; the vast majority of harm is done during the initial mining and processing of the minerals. In the past, the easiest, most convenient way to obtain more mineral products was usually considered to be the mining of more virgin ore. With reuse and recycling of already extracted mineral products, the mining of virgin ore could become a rare event.

Trends in Mineral Use

In global terms, mineral use has tended to increase over time. Before the nineteenth century, human use of minerals was relatively insignificant compared to the abundance of minerals in geological deposits. Since the Industrial Revolution and its associated technological developments and burgeoning population, however, the use of mineral resources has increased at very high rates. Between roughly 1750 and 1900, the world's population doubled, but mineral use increased by a factor of 10. Between 1900 and the present, mineral use increased

by 13-fold or more. The use of certain metals has increased even more dramatically. Current annual world production of crude pig iron is approximately 550 million metric tons (600 million tons), tens of thousands of times the production of three centuries ago, and steel production totals some 800 million metric tons a year (900 million tons). Copper and zinc production have increased by factors of 560 and 7300, respectively, in the last two hundred years. As noted earlier, mineral consumption in the industrialized countries is much higher than in the developing nations. Americans use about 25 times as much nickel per person as do citizens of India and four times as much steel and 23 times as much aluminum as the average Mexican. According to one estimate, the United States consumed more minerals between 1940 and 1976 than did all of humanity up to the year 1940.

Yet since the 1970s, the use of raw materials in the United States and Western Europe has leveled off (see Figure 10-3) or even experienced temporary declines. One reason for this trend is that Western industrial economies are placing more emphasis on high-technology goods and consumer services that utilize fewer raw materials; in addition, the infrastructure of most industrialized nations is firmly in place—major construction and public works projects are less common than they once were.

Nevertheless, in absolute terms, the demand for minerals is still extremely high, and the developing countries will need ever more minerals as they industrialize. Furthermore, within the next few decades, some industrialized countries may need to replace major segments of their aging infrastructures, thus once again sharply increasing the demand for mineral resources. Also, as new technologies are developed, such as high-speed railway systems, new infrastructure components may have to be built.

The bottom line is that on a global scale, humans are currently depleting their mineral resources very quickly. It is unclear, though, whether mineral scarcity will ever reach the point that we actually run out of most minerals. The effects of environmental degradation and attendant pollution caused by mineral extraction and refining may force us to curtail the exploitation of mineral deposits before they are exhausted.

What Accounts for the Prices of Minerals?

Minerals, including the precious metals, are artificially cheap. Since the middle of this century, the overall trend in the prices (using stable [noninflated] dollars) of various metals and other minerals has generally been downward. Prices have tended to decline even though mineral deposits are a nonrenewable resource, mineral consumption has been generally rising worldwide, and the environmental costs of mineral extraction and production are rapidly mounting. Given these factors, common sense might suggest that mineral prices should be increasing rapidly. Why then are they falling?

One reason is that known reserves for many metals and other substances have grown at least as fast as production over the last few decades. Given this record, most experts do not foresee any imminent shortages even when known reserves seem to be sufficient for only a few decades. Consequently, in the near future, the absolute limits to, and relative scarcity of, mineral resources will have little impact on the price of most minerals.

A second reason prices have declined is that the current price of most raw mineral products reflects only the immediate costs of their extraction. In many cases, as when the minerals are on public land obtained from the government, the source of the minerals, the ore body, is virtually free. Nor does the price the consumer pays include **externalities**. Externalities are environmental, social, and other costs that are not included in the prices of products that cause the costs. In this sense, mineral production is heavily subsidized by the public, for the public must ultimately bear the burden of the environmental devastation caused by mineral production. But even more immediately and directly, the mining and mineral industry is heavily subsidized. In the United States and many other countries, mining companies receive major tax exemptions and deductions from their gross income; the lost taxes are equivalent to subsidizing the industry to the tune of billions of dollars. Mining companies are often de facto exempt from many environmental regulations, including laws governing pollution emissions and land reclamation. Low-interest loans, loan guarantees, and direct investments by governments and financial institutions such as the World Bank contribute to maintaining mineral prices at artificially low levels.

Governments have traditionally supported mineral production for several reasons. Mineral production is historically associated with national security concerns—metals are needed to build weapons. Minerals are also important for the economies of many countries. The developed nations tend to think a healthy economy is synonymous with economic growth, and traditionally economic growth has been closely linked to manufacturing and heavy industry, which require large supplies of mineral products. Consequently, governments have tried to ensure that the manufacturing sector received a constant flow of inexpensive raw materials. Historically, opening up virgin mineral deposits was a way to attract people to frontiers that "needed" to be developed, or so politicians felt. For many present-day developing countries, mineral exports are the primary means of earning currency to pay off their international debts. Such nations have an incentive to open, subsidize, and maintain large mining operations, no matter what the environmental costs.

Dealing with Mineral Scarcity

There are two basic strategies for coping with mineral scarcities: increasing the supply of the substance (by locating new ore deposits or recycling old materials), or decreasing the demand for the substance (by finding alternatives or substitutes for the substance or simply eliminating the need for it through technological developments and changes in lifestyle).

Expanding the Resource Base

In the past, humans relieved the demand for more raw minerals by simply going out and discovering new mineral deposits, but the days of finding significant new ore deposits are almost surely over. The major geological provinces on land where deposits are found have for the most part already been heavily explored. The number of new discoveries of large and/or high-grade deposits has been declining for years. The high-grade deposits that remain to be discovered will be smaller, less accessible, and more expensive to exploit. Significant new mineral deposits may be found under the sea, but the technical difficulties and expense of mining in deep water have yet to be worked out. In general, drilling deeper into the crust is not a solution either, for most rich mineral deposits are found in the upper regions of the crust, and the cost of mining rises exponentially as one penetrates extreme depths.

Utilizing progressively lower-grade ores as high-grade deposits are depleted may sound feasible in theory, but in practice, it does not work for all minerals. For a few of the geologically common minerals (namely, iron, aluminum, manganese, magnesium, chromium, and titanium), higher-grade ores are rare, but the lower the grade of the ores, the more abundant they are. Of course, as lower grades of ore are worked, the extraction costs increase dramatically, as do the energy requirements, concomitant pollution, and other environmental costs associated with mining, processing, and disposing of huge volumes of rock. Also, the amount of waste produced for every ton of metal extracted rises exponentially, as progressively lower grades of ore are mined. In the

United States, for instance, the amount of waste generated by copper mining has increased dramatically as the average grade of copper ore has steadily dropped from about 2.5% in 1905 to about 0.5% today. The increase in waste simply reflects the fact that more raw rock must be mined, crushed, sorted, and processed in order to extract the metal. As more rock is mined and processed, more energy is consumed, and more of the landscape is devastated. Most of the energy consumed in mining is derived from fossil fuels; thus, mining ever lower grades of mineral ores contributes to the quickening pace of fossil fuel depletion and the resulting pollution generated when the fuel is burned.

Most metals are geologically scarce and are not found in a continuous gradation from high-grade ores to lower-grade deposits. Metals such as copper, tin, and zinc have been geologically concentrated into the high-grade, minable deposits that humans are currently exploiting. Once these deposits are depleted, the next lower grade of "ore" is essentially ordinary rock that may contain only approximately 0.1% to 0.001% of the amount of metals found in the high-grade ores. Enormous amounts of rock "ore" would have to be processed to obtain relatively small amounts of metal. Likewise, some people have argued that an abundant supply of metals is dissolved in seawater, but huge amounts of seawater would have to be processed to extract these metals because their concentrations are so low. For example, to extract a year's supply of zinc for the world, one would have to process approximately 75 billion metric tons (or 7.5×10^{13} kg) of ordinary

rock or 1.33×10^{17} (133,000,000 billion) gallons (503,400,000 billion liters) of seawater—assuming that the extraction processes were 100% efficient.

"Mining" ordinary rock or seawater for scarce minerals and elements not only would be energy and cost prohibitive, but would cause untold environmental devastation on a much larger scale than the considerable damage already caused by the mining of high-grade deposits.

Recycling

As we have seen, in an absolute sense, the stock of any mineral on and within the Earth's crust is finite and can only decrease. Yet, in a fundamental way, we never appreciably decrease the stock of any element on Earth (fissionable elements, such as uranium, used in nuclear processes would be the exception to this rule). As we use materials, the atoms of which they are composed physically still exist; they are simply in altered arrangements. Metals and many other substances can be recycled. In effect, recycling can be viewed as increasing the effective or usable amount of a substance that we have at our disposal. Moreover, **recycling** often consumes less energy than extracting raw, virgin material.

Recycling has practical limits, however. During the recycling process, the **recovery ratio**—how much of the original material can actually be recovered and recycled—is always less than 100%. For instance, if we begin with a ton of aluminum and manufacture a ton of aluminum cans and other products, use these products for a number of years, and then attempt to gather them all up and melt them down into pure metal, we will end up with something less than a ton of aluminum. Normal wear, rust and corrosion, gears grinding against each other, and so on will result in some loss of metal—perhaps microscopic filings—that are unrecoverable. In the very process of recycling and producing new products, there will be some loss. Heating the metal, cutting it, or crushing it will result in less than 100% recovery (**Figure 10-5**). A recycling recovery ratio of 90% is extremely difficult to achieve for most metals. The only metals for which a higher ratio of recovery is normally possible are precious metals, such as gold, that are relatively nonreactive (so they do not corrode) and are used primarily in jewelry or other products not subject to heavy physical wear. In fact, the current recycling efficiency for many common metals can be as low as 30% or even less.

As a hypothetical example, assume that we have a recycling ratio of about 80%. If we begin with one ton of aluminum, utilize it in products, and subsequently recycle it entirely (that is, as best we can—not all of the metal will even necessarily be re-

Figure 10-5 Recycling metal, such as aluminum shown here, always results in some amount of material loss.

turned for recycling) once a year, then the next year we will have only 0.8 ton of the metal; after two years, we will have only 0.64 ton of the metal; after three years, we will have only 0.512 ton, and so on. If we keep doing this and then sum up all of the metal used from year to year, we will find that, in effect, the one ton of aluminum was stretched into five tons. Turning one ton into five tons through recycling may help, but given increasing demands for metals, it can only slightly delay the inevitable exhaustion of the supply. In a practical sense, recycling alone, without the input of new raw materials, cannot satisfy current demands indefinitely.

Furthermore, even if recycling extends the effective amount of a substance over a number of years, it does not increase the amount of the substance at any one time. Returning to our ton of aluminum, no matter how meticulously we recycle, we can only manufacture a ton's worth of goods from a ton of aluminum at any one time.

Recycling also contributes to pollution, diverts resources such as fresh water, and requires the use of energy. Granted, recycling usually requires much less energy than mining and refining raw ore (producing aluminum from recycled goods uses 90% to 97% less energy than is required to produce the same aluminum from raw ore).

In some cases, recycling used products that were not designed with recycling in mind can be extremely difficult and labor-intensive. For example, the metals used for many manufacturing and industrial purposes must be quite pure—they cannot be contaminated by small amounts of other metals. Thus, one cannot simply take a junk car and "melt it down." The various metals must be carefully separated, and even then large amounts of energy may be required to purify the resulting metal. It is for this reason that various electronic devices, like obsolete computers, can be incredibly difficult to recycle.

It is virtually impossible to recycle some substances in certain uses. Many uses of materials are dissipative—under normal use, they are degraded and dispersed. Examples of such materials include dyes, paints, inks, cleansers, solvents, cosmetics, fertilizers, and pesticides. Metals used in dyes and paints, for instance, cannot be recovered. As the relative need for certain metals (for instance, copper) decreases and demand remains relatively steady, a greater proportion of consumption may well be for dissipative uses. Even when consumption is relatively low, if the uses are dissipative, the metals cannot be recycled—once they are consumed, they are gone for good. Eventually, the supplies of such metals, if subjected to constant consumption without recycling, must run dry.

Substitutability

As the Earth's population continues to grow, recycling may not be enough; we could conceivably reach a point where there is simply not enough copper, for instance, to produce all of the copper piping and wiring demanded. The absolute amount of copper is limited, no matter how carefully we recycle. We can cope with these situations by finding substitutes (**substitutability**). By manufacturing household pipes from plastics, making wires out of metals other than copper, and utilizing fiber optics in place of wires, we cannot increase the supply of copper, but we can lower the demand, at least relative to what it might have been. Notice, however, that even with substitutes, real demand may still increase as the number of consumers increases. Thus, the demand for copper is increasing by approximately 2.7% per year despite many substitutes.

Certain metals are valued for their physical properties and characteristics, which can sometimes be found in other substances. To use an often cited example, the metal tin is relatively scarce from a global perspective. At current consumption rates, the known reserves of tin will be exhausted in 20 to 30 years. Tin is desirable because it is durable, malleable, and light, but each of these properties can also be found in other substances. Thus, tin cans have been largely replaced by aluminum and steel cans and by glass jars and bottles. Replacements can also be found for many other increasingly scarce metals, such as lead, zinc, and mercury.

Substitution only deals temporarily with the scarcity of one or a few metals (or other substances), however. It is not a solution to the generally increasing demand, and therefore scarcity, of all minerals and other exhaustible resources because ultimately the reserves of the substitutes will themselves become depleted. Some people have suggested that plastics could substitute for vast quantities of metals, for example (and indeed great strides have been made in this direction), but most plastics are derived from petroleum products—which are also nonrenewable, rapidly diminishing resources. Once plant-based plastics become common, however, the situation may be different.

In some cases, certain elements or compounds have unique properties, and there is little realistic hope of finding a ready substitute. Platinum and other metals are used as catalysts in industrial processes, mercury is the only metal that is liquid at room temperature, and small but critical amounts of certain elements (such as copper, lead, zinc, mercury, nickel, tin, manganese, chromium, cobalt, and titanium) are absolutely necessary in modern metallurgy. The prospects of developing a totally new system of metallurgical techniques seem remote.

Unfortunately, a few substances that have no ready substitute and cannot be recycled very efficiently also happen to be in very short supply and extremely high demand; they are also critical to life in a direct way. Perhaps the best example of such a substance is phosphorus. Phosphorus is essential to all living organisms: It is used in DNA, in cell membranes, and in the bony tissue of vertebrates, among other things. Without phosphorus, life as we know it could not exist. Not surprisingly, phosphorus is a key component of fertilizer. Phosphorus has no known substitute. It is produced commercially from nonrenewable deposits of phosphate rock, and these deposits are relatively rare and being depleted very quickly. Estimates vary, but assuming current conditions and consumption rates, the practically minable phosphorus supply could be depleted in less than 100 years or, most optimistically, in about 1300 years.

Phosphorus is difficult to recycle because it is used in a diffuse form (for example, spread on the ground as fertilizer), which makes it very difficult to recover. One way of recycling phosphorus would be to crush bones and extract the phosphorus from them, but the recovery rate would probably be fairly low. Another possibility is that wastewater treatment plants are required to remove phosphorus from wastewater, and possibly this phosphorus could be recycled.

Conservation and Durability

Just as the demand for energy can be reduced by conserving energy, so too can we reduce our demand for mineral resources by **conservation**. We can simply use less.

A very simple way to reduce the demand for mineral resources is to produce **durable goods** that are designed to last as long as possible. This can be done by building better quality products to begin with and designing products so that they can be repaired, rebuilt, modified, or refurbished as they grow old—rather than simply discarding and replacing them. In some cases, considerable energy and mineral savings can be accomplished simply by reusing goods, rather than discarding them. Food and drink packaging is a classic example. The refillable glass bottle can typically be reused (washed and refilled) dozens of times. Simple **reuse**, as compared to recycling (which entails breaking down

the product and using the raw materials to make a new product), conserves valuable mineral resources and saves energy and money.

Another strategy is to reduce the size of products where possible, a concept sometimes referred to as **dematerialization**. A smaller, lighter car uses a correspondingly smaller amount of mineral resources and may serve the function equally well. Indeed, a smaller car may be more fuel efficient. Miniaturization, as in electronics, can also save substantial amounts of raw materials and energy— and miniaturized electronic components can be faster, more dependable, and more powerful. Take computers, for example: Computations that in the 1950s required tons of computer hardware housed in its own building can now be accomplished with a system that fits into a briefcase or even a palm-sized machine.

Specific cases of dematerialization must be analyzed carefully; in some cases, first impressions can be deceiving. If a smaller, lighter product lasts only 75% as long as the product that is 25% heavier, the heavier, longer-lasting product will in this case use less material over time. Environmentally, sometimes it may be worthwhile to utilize a more durable product if it lasts significantly longer, rather than a dematerialized product. In other cases, a dematerialized product may last even longer than the heavier product.

Another consideration is that while dematerialized products may require substantially less material to produce, because of the complexity of their designs and incorporated materials (such as various sophisticated alloys), they may be virtually impossible to disassemble and recycle once they have come to the end of their useful lives. This is the case with many modern electronic devices, such as transistors, circuit boards, and batteries. In practice, it is not always cost effective or energy effective to separate and refine the small quantities of various valuable metals that occur in much electronic equipment (although recently advances have been made in recycling metals from electronic circuit boards and other computer hardware). The metals used in manufacturing a portable television, radio, or calculator may be lost to humanity forever. Such products are often intended to be permanently discarded once their useful life is over; indeed, in many cases, they are built in such a manner that they are very difficult or impossible to repair even if they malfunction prematurely.

SUMMARY

- Modern technological society is dependent on mineral resources, but of course minerals have been utilized by humans since prehistory.
- Mineral resources can be classified as metallic minerals and nonmetallic minerals; the latter category includes structural materials (such as building stone), industrial minerals (such as fertilizer components and nonmetallic minerals used in manufacturing), and energy minerals (fossil fuels, not technically minerals, and uranium ore, which is actually a metal).
- An ore is a mineral deposit that can be economically mined.
- Mineral resources are ore deposits that have not yet been exploited.
- Mining and smelting operations can cause severe environmental degradation; in the United States, nonfuel mineral mining produces over a billion tons of waste each year.
- Since the Industrial Revolution over 200 years ago, mineral use has increased by a factor of one hundred or more, and for some specific minerals, the increases have been by factors measured in the thousands.
- Prices of many minerals are low because they do not include externalities—environmental, social, and other costs that are associated with procurement of the mineral; mining may be subsidized by the government or the ore may be virtually free to the miner.
- Mineral scarcity can be dealt with by expanding the resource base (discovering new mineral deposits), recycling, substitution of one mineral or substance for another, and conservation of mineral resources (including the manufacturing of more durable goods that last longer).

KEY TERMS

Conservation
Dead zone
Dematerialization
Durable good
Energy mineral
Externality
Ferrous
Industrial material

Infrastructure
Metallic mineral
Mineral
Mineral deposit
Mineral resource
Nonferrous
Nonmetallic mineral
Ore

Recovery ratio
Recycling
Reserves
Reuse
Structural material
Substitutability
Tailings
Virgin supplies of ore

STUDY QUESTIONS

1. What is a mineral resource?
2. List some of the major types of mineral resources. Which are most important in terms of quantity used? Which are most important in terms of economic value?
3. Why are mineral resources said to be nonrenewable?
4. Distinguish between mineral deposits, ores, and reserves.
5. Summarize some of the ways that mineral exploitation and use can lead to environmental degradation.
6. What are some of the factors that determine the price of a particular metal or other mineral? Have mineral and metal prices generally increased or decreased over the last couple of decades? What accounts for these trends?
7. Describe the broad historical trends in mineral and metal use.
8. Why has the mining industry received special treatment from governments, including the U.S. government, in the past?
9. Discuss the advantages and disadvantages of the following strategies to deal with mineral scarcity: recycling, substitutability, conservation, and durability.
10. What is dematerialization?
11. Do you believe we will ever run out of mineral resources (will demand for minerals ever outstrip supplies)? Justify your answer.

STUDY GUIDE

WHAT'S THE EVIDENCE?

1. The authors assert that the prices of minerals are, in general, artificially cheap (with a few exceptions, such as diamonds and other precious stones whose values may be artificially inflated). What evidence do the authors present to support their assertion that the prices of many mineral resources fall short of their real costs and values? What factors might be included in the prices of minerals to bring the amount the consumer pays more in line with their actual values and the expenses of procurement?

2. The authors maintain that recycling alone cannot satisfy current demands for metals and other minerals indefinitely. What lines of evidence do the authors put forth to support this contention? If recycling alone is not enough, what else might be done to increase the effective supply of mineral resources?

CALCULATIONS

1. If a certain ore body of tin contains 0.7% tin by weight, how much ore will have to be mined and processed (assuming complete recovery of all the tin) to produce one metric ton (1000 kilograms, or 2204.623 pounds, or 1.1023 English tons) of tin? (Express your answer in both metric tons and tons.)

2. If a certain ore body of copper contains 0.4% copper by weight, how much ore would have to be mined and processed (assuming complete recovery of all the copper) to produce one metric ton of copper? (Express your answer in both metric tons and tons.)

ILLUSTRATION AND TABLE REVIEW

1. Referring to Figure 10-3a, calculate by approximately what percentage the consumption of metals increased in the fifty years from 1945 (the end of World War II) to 1995. During this same period, by what percentage did the consumption of industrial materials increase? And by what percentage did the consumption of construction materials increase from 1945 to 1995?

2. Referring to Figure 10-3b, calculate by what percentage world materials use increased between 1970 and 1995. How does this compare to the increase in use by the United States during the same time period?

WWW.ENVIRONMENT.JBPUB.COM

www.environment.jbpub.com

Connect to this book's web site: http://www.environment.jbpub.com/ The site features eLearning, an online review area that provides quizzes, chapter outlines, and other tools to help you study for your class. We've posted the answers to the book's odd-numbered questions. You can also follow useful links for more in-depth information. The site provides sources for the information in this chapter and recommendations for further readings.

Down in the open pit mine.

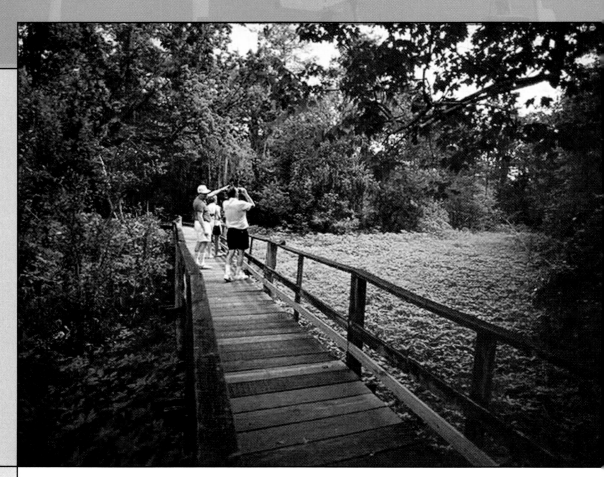

The rich diversity of life on Earth is being eroded away by human activities. This does not usually happen because humans are purposely eliminating other species. Rather, extinction occurs because many human activities, from farming to suburban sprawl, are incompatible with the conditions that many species require. Fortunately, increasing numbers of humans are becoming aware of this "extinction crisis" and there is growing interest in preserving species in many parts of the world. Examples range from local grassroots groups that acquire land for habitat preservation to international agencies that work with governments in developing nations to create national wildlife sanctuaries. Shown here is the Corkscrew Swamp Sanctuary in Florida, owned by the National Audubon Society.

CHAPTER 11

Conserving Biological Resources

Society depends on biological resources in many ways: agriculture and wilderness recreation are just two examples. In this chapter we focus on the current rapid loss of natural biological resources, often called the "extinction crisis." We examine why biological diversity is being lost, why it is worth saving, and how to go about saving it. Few people truly want to see species become extinct, and the effort to slow the loss of perhaps thousands of species each year is gaining national and international support. One indication is the very rapid growth of **conservation biology**, a new subdiscipline of biology that draws from genetics, ecology, and many other fields to find practical ways of saving species.

Biological diversity, or biodiversity, has many definitions, but they all involve the variety of living things in a given area. The U.S. Office of Technology Assessment, for instance, defines **biodiversity** as the variety and variability among living organisms and the ecological complexes in which they occur. One reason for the lack of a precise definition is that life is hierarchical. For example, genes occur in cells, cells occur in organisms, and organisms occur in ecosystems (Chapter 3). At what level do we measure the diversity of life in an area? The number of genes? Organisms? Ecosystems? All can be used.

Measuring What's at Risk

Even if we simplify the definition of biodiversity to describe just the number of species, or **species richness**, no one knows how many species live on Earth. As we discussed in Chapter 4, taxonomists have classified about 1.8 million species. But this sample of 1.8 million species is highly biased toward the northern hemisphere, and may not reflect the true species richness of the world's biosphere. The tropics, the areas with most species presumably, are still not well known, and it's safe to say the oceans contain an unknown number of species we've never seen. Of the 1.8 million species classified, 56% are insects, 14% are plants, and just 3% are vertebrates such as birds, mammals, and fishes.

Estimating the Unknown

It is important to understand species diversity and the abundances of individual species, so that we have some idea of what we're losing and what we're trying to conserve. As discussed in Chapter 4, biologists have devised a number of ways to develop biodiversity estimates from limited information. The methods used include rainforest insect samples, ecological ratios, and species-area curves. These are just three examples of ways in which diversity can be estimated based on small samples. All methods of estimating biodiversity involve a certain amount of extrapolation; so naturally there is disagreement over how many species exist. Estimates range from as low as 4 million land species to as high as 100 million land species, and 1 million to 10 million species in the ocean. Whatever the exact number of species today, the fossil record indicates that we live in a special time of Earth history when global biodiversity is near its all-time peak.

Biodiversity Loss

Extinction is the death of a group. In most cases, people mean species extinction when they refer to extinction, but other kinds occur. Local extinction, or **extirpation**, means that a species has died out in a local area. Grizzly bears, for example, once inhabited Colorado and many other areas of the western United States. They were hunted into local extinction in most areas, but persist in Canada and a few parts of their original U.S. range. **Ecological extinction** means that a species has become so rare that it has essentially no role or impact on its ecosystem. Many of these species are called the "living dead" because their rarity dooms them to eventual total extinction in the wild. The Florida panther, which has a population size of less than 70 individuals, cannot possibly survive without much help because of inbreeding and other problems of small populations discussed below (**Figure 11-1**). Tigers, rhinos, and many other animals suffer similar problems and will likely survive only if drastic action is taken soon. Even then, such species may survive only in zoos and parks outside their original habitat.

Figure 11-1 The Florida panther has a population of fewer than 70 individuals.

Biological impoverishment by local or ecological extinction is much more common than species extinction. But because local and ecological extinctions are rarely reported, this kind of impoverishment goes unnoticed by most of the media and the public. This is one reason why the extinction crisis may be more urgent than is commonly perceived.

Extinction is the ultimate fate of all species. The fossil record indicates that, since life began over 3.5 billion years ago, over 95% of all species that existed are now extinct. Some of these past extinctions occurred during **mass extinctions** (see Figure 4-7 on page 100). These were catastrophic events that killed over 60% of all species on Earth. Most of the five past mass extinctions were caused by climatic disruption of habitat, especially global cooling. This caused sea level changes, and the colder temperatures killed off many tropical species. The last major mass extinction, which included the disappearance of the dinosaurs, occurred about

Objectives

After reading this chapter, you should be able to explain or describe:

Why biodiversity measurement is important

Different types of extinction

The causes of extinction

Traits that make some species prone to extinction

The benefits and problems of the Endangered Species Act

The values of biodiversity

Ways to promote biodiversity preservation

65 million years ago and may have been caused by a large meteorite impact.

Despite widespread interest in past catastrophic mass extinctions, careful analysis of the fossil record shows that over 90% of all extinct species died out during normal background times. In his popular book, *Extinction: Bad Genes or Bad Luck?*, David Raup points out that the average extinction rate throughout Earth's history was probably just a few species per year. Common estimates are that between 2 and 10 species per year died out from natural causes such as volcanic eruptions, floods, and violent weather changes on the local level.

Current Biodiversity Loss: The Sixth Mass Extinction?

The European Age of Expansion in the fifteenth and sixteenth centuries initiated a wave of extinction that has continued to accelerate. Table 11-1 shows the number of recorded species extinctions among vertebrates and plants since the year 1600. In the last four centuries, 1.9% of mammal species and 1.4% of bird species on Earth have gone extinct. While these numbers may not seem especially alarming, they almost certainly underestimate the true nature of the extinction crisis, for a number of reasons:

1. The extinction rate of birds and mammals has been increasing rapidly. Between 1600 and 1700, the extinction rate was only about one species per 10 years. Between 1850 and 1950, the rate was one species every year. Preliminary evidence indicates that the extinction rate has increased yet again since 1950.
2. Some species of birds and mammals became extinct before they were described, so the true

extinction rate is probably higher than that recorded in Table 11-1.

3. Species are not officially recorded as extinct until they have not been seen for 50 years. This lag also means that the true extinction rate of birds and mammals is probably higher than in Table 11-1.
4. Many species of birds and mammals have too few individuals remaining alive to successfully breed. These species are not yet recorded as extinct, but will be in the near future.

As birds and mammals are among the best-described groups, the extinction rates for other groups, including invertebrates (not shown in Table 11-1) are likely to be underestimated even more significantly. To solve such problems, other methods are used to estimate the rate of species extinction. Table 11-2 shows a variety of extinction rate estimates based on the species-area curve and other methods. These indicate that between 1% and 11% of all species on Earth are lost each decade. The average estimate is around 5% per decade, which translates into an extinction rate of at least 50 species per day. This is much higher than the recorded rates of Table 11-1 because many extinctions have gone unrecorded. This is many hundreds of times higher than the normal average extinction rate of about 2 to 10 species extinctions per year, as estimated from the fossil record. Such currently high rates lead many, perhaps most, biologists to predict that as many as half the world's species could be extinct sometime in the twenty-first century.

Perhaps a better way to assess impending extinction is to examine the number of species that are known to be at immediate risk of extinction.

| TABLE 11-1 | Recorded Extinctions of Vertebrates and Plants, 1600 to 2000. |

Many additional species have presumably gone extinct without being recorded by scientists. Many more species are now threatened with extinction in the next few years. Except for mammals and birds, most threatened percentages are probably greatly underestimated.

	Species		Total Species		
	Extinct	Threatened		Percentage Extinct	Threatened
Mammals	87	1130	4600	1.9	24.6
Birds	131	1183	9500	1.4	12.5
Reptiles	22	296	6300	0.3	4.7
Amphibians	5	146	4200	0.1	3.5
Fishes	91	750	19100	0.5	3.9
Plants	90	5611	250000	0.04	2.2

Source: IUCN: International Union for Conservation of Nature and Natural Resources, The World Conservation Union Red List 2000.

TABLE 11-2	Estimated Rates of Extinction		
Estimate		**Global Loss per Decade (%)**	**Method of Estimation**
One million species between 1975 and 2000		4	Extrapolation of past exponentially increasing trend
15%–20% of species between 1980 and 2000		8–11	Estimated species-area curve; forest loss based on U.S. government projections
12% of plant species in Western Hemisphere tropics 15% of bird species in Amazon basin		—	Species-area curve
2000 plant species per year in tropics and subtropics		8	Loss of half the species in area likely to be deforested by 2015
25% of species between 1985 and 2015		9	As above
At least 7% of plant species		7	Half of species lost over next decade in 10 "hot spots" covering 3.5% of forest area
0.2%–0.3% per year		2–3	Half of rainforest species assumed lost in tropical rainforests to be local endemics and becoming extinct with forest loss
5%–15% forest species by 2020		2–5	Species-area curve
2%–8% loss between 1990 and 2015		1–5	Species-area curve; range includes current rate of forest loss and 50% increase

Source: W.V. Reid, "How Many Species Will There Be?" in T.C. Whitmore and J.A. Sayer, eds., *Tropical Deforestation and Species Extinction* [London: Chapman & Hall, 1992]. Reprinted by permission of Chapman & Hall.

Table 11-1 shows the number of vertebrate and plant species that have been designated as threatened with extinction in the next few decades or even years. These are species that have documented declines in population size over the last few decades. They often have population sizes in the range of, or below, several hundred individuals. An alarming 25% of mammals and 13% of birds are now threatened. As with extinction percentages, mammals and birds are the best-studied, so the threat percentages for the other groups are probably gross underestimates. In reality, reptiles, fishes, plants, and most, if not all, other groups, probably have similar percentages of species that are threatened.

Causes of Extinction

Extinction is caused by environmental change. Species can adapt to environmental change if it is not too rapid (Chapter 3), but the changes caused by human population growth and technology have been too massive and too fast to allow many species to adapt.

For convenience, we can categorize all environmental changes that cause extinction into two basic categories: (1) changes in the physical environment and (2) changes in the biological environment (Table 11-3). Biological changes occur in three ways: introducing species, overhunting, and removing species.

It is usually extremely difficult to identify which of the four causes in Table 11-3 was involved in a particular extinction. In most cases, species extinction occurs from more than one cause. For instance, all four causes have led to the decline of the Florida panther. Loss of habitat from development, overhunting, dozens of introduced species, and loss of prey species have combined to reduce the panther population drastically in the last 200 years.

Keeping in mind that extinction-causing environmental changes often occur together, we now discuss each of these four causes of extinction. We will try to focus on examples where one cause has clearly played a major role.

TABLE 11-3	Four Ways that Humans Cause Population Decline and Species Extinction
	Examples
Change Physical Environment:	
1. Habitat destruction	Drain swamp, toxic pollution
Change Biological Environment:	
2. Introduce new species	New predator
3. Overhunting	Big-game hunting
4. Secondary extinctions	Loss of food species

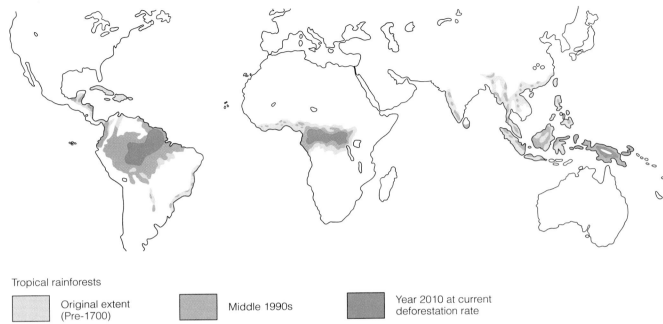

Tropical rainforests

Original extent
(Pre-1700)

Middle 1990s

Year 2010 at current
deforestation rate

Figure 11-2   Past decline of tropical rainforests and projected decline by
year 2010 if current deforestation rates continue.

Habitat Disruption.   Habitat disruption refers to
disturbance of the physical environment in which a
species lives. Such disturbances can range from minor to drastic. Minor disturbances, such as mild

Remaining Natural Vegetation

<25

51–75

25–50

>75

chemical changes from air pollution, tend to affect
only the most susceptible species.

In contrast, extreme physical changes in habitat can eliminate many species from the area.
Deforestation of tropical rainforest by burning and
other agricultural practices is one of the most publicized examples of massive habitat disruption.

Figure 11-2 shows the projected loss of
this critical habitat if current trends continue.
Recall that the tropics contain many more
species than other areas (Chapter 3). By some
estimates, rainforests contain over 50% of the
world's species even though they cover only
7% of the Earth. As of the year 2002, just over one-
half of the rainforest had been deforested. Using the
species-area curve (see Figure 4-6 on page 98), some
experts estimate that this 50% rainforest loss could
mean perhaps a 10% rainforest species loss so far. But
the curve also shows that continued habitat loss will
soon result in a rapid decrease in species numbers.

There has also been an enormous loss of habitat in the United States. A map published by the
Nature Conservancy shows that less than 25% of the
natural vegetation covering most of the central and
northeastern United States remains (**Figure 11-3**).
Only in the more desert-like regions of the western
United States is over 75% of the area covered by
natural vegetation.

Figure 11-3   Remaining natural vegetation in the United States.
(*Source*: B.A. Stein, L.S. Kutner, and J.S. Adams. 2000. *Our Precious
Heritage: The Status of Biodiversity in the U.S.* [New York: Oxford
University Press, p. 229].)

Removal of natural vegetation for farming or development not only destroys the plant species but removes habitat for the animals that require those native plants to survive. Devegetation is usually a piecemeal process, with the habitat being broken into progressively smaller and more isolated fragments. Such habitat fragmentation has occurred in nearly all places where modern agriculture and urban society have moved in. **Habitat fragmentation** is especially disruptive for two reasons. For one thing, remaining areas of habitat become separated and form islands of refuge in a hostile environment. If the islands are too far apart, members of a species may not be able to reach other members in order to reproduce. The species could die out if all the isolated islands of habitat are so small that no island can maintain a self-sustaining population.

Secondly, habitat fragmentation leads to edge effects (**Figure 11-4**). Housecats, dogs, wind and temperature changes, and many other disturbances from the surrounding area can penetrate along the edges of the preserved area, resulting in loss of habitat. Edge effects may be very dramatic. For instance, roads that occupy only 2% of a certain area may actually disrupt about 50% of the habitat! Habitat fragmentation is one of the main causes of decline in global and North American songbird populations.

Habitat disruption is not limited to land. Water pollution, dam building, and other human activities have a huge impact on fishes, clams, and other species in lakes and rivers because such species are confined in limited areas. Even the oceans are suffering widespread disruptions as raw sewage, toxic chemicals, sediments, and other pollutants are poured into them. Extinctions and ecological damage in the oceans are difficult to monitor, but many studies of coral reefs have found signs of extensive damage. Reefs are good biodiversity indicators because they support a great variety of species. Many reefs are in critical condition, meaning that they could be lost in 10 to 20 years. An estimated 10% to 15% of the world's reefs are already dead. At current rates, another 60% could be lost in the next 20 to 40 years. The major causes of death are global warming and sedimentation from logging, farming, mining, construction, and other coastal activities. The suspended sediment blocks sunlight, smothers the reefs, and has many other harmful effects.

Introduced Species. In many parts of the world, populations of **exotic species** have been introduced, sometimes with beneficial effects, as in the case of the horse, but more often with devastating effects. These are species that are introduced into habitats that they have not occupied under natural conditions. More than 1500 species of nonnative insects, including fire ants, gypsy moths, Japanese beetles, and killer bees, have become established in the United States since 1860. Many other kinds of organisms have also been introduced. Well-known U.S. examples include kudzu, walking catfish, and English starlings. Unfortunately, once an exotic species becomes established, it is virtually impossible to remove (**Figure 11-5**).

The economic toll of exotic species is astonishing. The U.S. Office of Technology Assessment es-

Figure 11-4 Even though the fragmented preserves on the right may seem similar in total area to the single preserve on the left, the amount of unaltered habitat is greatly reduced from edge effects.

CASE STUDY 11-1

Invasion of the Zebra Mussels

An oceangoing ship teems with alien life. The organisms come aboard when the ship takes on ballast water from the harbor. A ship will take ballast for several reasons. One is stability: A lightly loaded vessel may ride too high, and is thus more liable to capsize. Another reason is propulsion: The propeller of an underloaded ship may rise half out of the water. To avoid these problems, a ship may distribute ballast water through a network of tanks inside the hull.

In the mid-1980s, a ship leaving a freshwater European port for North America began to take on ballast. As it did, several members of the harbor's population came aboard along with the water. Unfortunately for North America, one of them was a hardy bivalve named *Dreissena polymorpha*, the zebra mussel, which over a 200-year period had spread from the region of the Caspian Sea into much of Europe, plugging water pipes and clamping onto the hulls of boats (**Figure 1**). The unknown ship that carried the zebra mussel across the Atlantic was headed for the St. Lawrence

Seaway. Past Quebec City it sailed, past Montreal, past Toronto, and on through Lake Erie. When the ship finally reached its destination and flushed its ballast somewhere above Detroit, a founding population of zebra mussels tumbled into Lake St. Clair (where it was discovered in 1988)—and a new continent.

Biologists are hastening to quantify the effects of this invasion, as the introduction of any new species is called, but already the costs have been staggering. Propelled by relentless fertility and a talent for spreading themselves abroad, zebra mussels are clogging the intake pipes of power stations and water treatment plants; colonizing navigation buoys in such numbers that they drag them under; fouling fishing nets, marine engines, and hulls of boats; and displacing spawning grounds that are the mainstay of a commercial and sports fishery valued in the billions. In North America, the zebra mussel seems to have found a very hospitable environment. They quickly swept down the length of Lake Erie and penetrated Lake Ontario. By 1990 they could be found in all of the Great Lakes, and within two years zebra mussels were found in the Hudson and Illinois rivers, and thence made their way into the Mississippi River and its drainage area. By 1992, zebra mussels were in the Arkansas, Cumberland, Ohio, and Tennessee rivers as well. The zebra mussels spread so quickly in the United States, that by 1994, they had been reported in or along the borders of Alabama, Arkansas, Illinois, Indiana, Iowa, Kentucky, Louisiana, Michigan, Minnesota, Mississippi, Missouri, New York, Ohio, Oklahoma, Pennsylvania, Tennessee, Vermont, West Virginia, and Wisconsin. As of 2002, they were found in Connecticut and Virginia as well—and they continue to spread.

The zebra mussel is a tiny but troublesome mollusk. It is typically no more than half an inch (1.27 cm) long (although it can sometimes grow up to two inches [50 mm] long) and comes in a

Figure 1 Zebra mussels clog pipes, causing billions of dollars in damage.

timated that over $100 billion in damage was caused by nonnative species during the 1990s. About 17% of the more than 1500 insect species introduced into North America are pests requiring the use of pesticides for control. Unfortunately, many introductions have been purposely done. The English starling, perhaps the most common bird in the United States today, was introduced in a Shakespearean play in the early 1900s. Game ani-

mals have been introduced for hunting or fishing (striped bass in Tennessee), and predators to reduce pests (mongoose in Hawaii). The kudzu plant now overgrows and smothers many acres of land in the southeastern United States at a yearly cost exceeding $1 billion. Yet in the 1930s, the federal government paid farmers to plant it to control soil erosion.

Although purposeful introductions have been sharply curtailed, accidental introductions continue handsome shell marked by alternating bands of light and dark. In addition to the tonguelike "feet" that mussels use to push themselves along the bottom, zebra mussels, alone among freshwater mussels, possess thread that allows the mussel to attach itself to hard surfaces like rocks or steel or other mussels. And unlike native mussel larvae, which disperse themselves by hitching rides on fish, zebra mussel larvae are veligers, meaning that each possesses cilia, tiny hairlike fibers that enable it to suspend itself in water. It is these features that allow zebra mussel larvae in a current to spread with remarkable swiftness.

There seems to be no doubt that the zebra mussel will continue to spread throughout much of North America. They can survive almost anywhere in a range that covers about two-thirds of the United States and the southern part of Canada. In terms of temperature, only Canada's cold northern lakes and the warmer waters of the American South will not sustain them. Their reproductive rates are amazing. A female zebra mussel can produce 40,000 eggs a year, and the male a similar amount of sperm. Even if only a small percentage of these eggs are fertilized and advance to maturity, the rate of proliferation will be impressive. For example, on Hen Island Reef in Lake Erie, the density was 2900 zebra mussels to the square yard (3500 to the square meter) on the first reading; five months later, the count was 19,230 per square yard (23,000 per square meter). Densities as high as 585,000 per square yard (700,000 per square meter) have been reported at a Michigan power plant.

Zebra mussels appear to be driving out the native North American mussels, which could have disastrous effects on local ecosystems. The zebra mussel also threatens native fishes with its insatiable appetite for phytoplankton, the microscopic green plants at the very bottom of the aquatic food chain. Because many fishes feed on the zooplankton that eat phytoplankton, tinkering with phytoplankton

is tinkering with one of nature's building blocks. If the mussels deplete the phytoplankton, the population of zooplankton could crash—with disastrous consequences for certain fishes, for the fishes that prey on those fishes, and for the ducks that eat them and the people who catch them. In this way the zebra mussel, in what amounts to a nanosecond of evolutionary time, could alter the ecosystem of the entire Great Lakes and, ultimately, most of the United States.

How can the spread of the zebra mussel be controlled and its range contracted? One possibility is ozonation, an environmentally benign oxidant that chews away at the soft parts of the organism. Ozonation could be helpful to power and water treatment plants that are trying to prevent zebra mussels from clogging their intakes, but it is extremely expensive. Even then, ozonation can only alleviate specific problems here and there; it cannot do anything to halt the overall proliferation of the zebra mussel. Various chemical, thermal, electrical, and other methods have been suggested as ways to control zebra mussel populations, but any such methods may also have adverse effects on the native organisms and ecosystems. Manual and mechanical removal and destruction of the bivalves is extremely tedious and expensive.

Biological methods have also been suggested, such as using predators, parasites, or disease agents that will attack the zebra mussels. Of course, such methods could end up releasing another organism that could itself cause havoc in the local ecosystems. A fish called the drum, a bottom feeder, does seem to like the mussel, but unfortunately this benefits only the lowly drum itself. Apparently, it is not an extremely tasty fish. At one site in Lake Erie, a diving duck has begun to prey on the zebra mussels. Unfortunately, Europeans have found that only overwintering waterfowl have an effect on mussel control.

to increase, as international trade and travel expand. The very destructive zebra mussel is an example of a species that hitchhiked from one nation to another (see Case Study 11-1). Studies show that the large majority of new species fail to become established, but because so many introductions occur, some of them eventually succeed. Humans promote this both by providing transportation for nonnative species and by disturbing native ecosystems. By

upsetting natural interactions, we make it easier for new species to become established.

Exotic species are especially important as a cause of extinction on islands. Because of their isolation, native island species are often poorly adapted to cope with new species. As a result, new species not only have an easier time becoming established on islands but also have a more devastating impact on native island species than on continental species. Predators and

Figure 11-5   The Old World climbing fern, shown here overtaking cypress trees in Florida, was native to East Asia and Australia before it was introduced to the United States intentionally as a landscape ornamental plant.

competitors have driven many native prey and competitor species to extinction. Others survive only in carefully protected preserves.

An example are ground birds that evolved on many islands, such as New Zealand and Hawaii, in the absence of large predators. The introduction of domestic cats and dogs alone has driven dozens of island ground bird species to extinction or near extinction (**Figure 11-6**). The introduction of the brown tree snake to Guam has effectively eliminated 11 of 18 native bird species on the island. In some areas of Guam, over 5000 snakes per square mile (over 1900 per km^2) have been counted!

Figure 11-6   The New Zealand kakapo once numbered in the hundreds of thousands. But this flightless bird was no match for dogs, cats, and other introduced species. The sixty or so remaining birds survive on two outer islands, which are carefully protected sanctuaries.

Although large continents suffer relatively fewer species extinctions from introduced species, they certainly experience major impacts. The kudzu plant has driven out many local plant populations where it grows. Freshwater lakes and rivers are especially susceptible to destructive invasions, perhaps because they are confined environments, leaving native organisms nowhere to go. Exotic aquatic plants such as hyacinth and hydrilla now cover vast areas of many of Florida's lakes, choking out other plants and blocking sunlight. No less than 50 of the 133 fish species sampled in a recent California study were found to be introduced. Many extinctions have occurred from such introductions so that native fishes, clams, and other freshwater groups are among the groups most strongly affected by exotics such as the zebra mussel. The American Fisheries Society estimates that about three-fourths of the nearly 300 species of freshwater clams in North America are either extinct or in decline.

The most dramatic devastation by an introduced fish is in East Africa in Lake Victoria where a single species, the Nile perch, has already exterminated over 35 species of native fishes in just a few years. If it continues as expected, several hundred more native fish species will become extinct in Lake Victoria from the perch, setting a record for the greatest number of extinctions from a single introduction.

Land species on continents can also become extinct from introductions. The decline of the Florida panther has been accelerated by many invading species. Indeed, South Florida is often said to have more exotics than any other part of the United States. The Florida Department of Natural Resources has estimated that 20% of the 1650 plant species and 23% of the 44 mammal species in South Florida are established nonnatives. In part, this is because South Florida is a tropical environment on the Florida peninsula isolated from the rest of Latin America.

Overhunting.   Overhunting, or overkill, is species decline caused by excessive shooting, trapping, or poisoning of organisms. Usually, this is done for sport or economic reasons. It is very difficult for humans to cause the extinction of pest species, such as roaches or mice, this way because they are so abundant and reproduce so rapidly. However, overhunting has often caused the decline of organisms that were initially rare and/or that reproduce slowly. Recall from Chapter 3 that most species are naturally rare, so many can be eliminated this way. This is especially true of most large animals and plants, which tend to have much lower abundances and slower reproduction rates than small organisms.

Unfortunately, large animals are also the species most often killed for sport or economic motives. Economic motives include killing to protect domesticated animals from predators or to sell parts of the animal itself. Elephant ivory, leopard skins, rhino horns, and tropical bird feathers are but a few examples.

This preference for killing large animals, combined with their low abundance and reproductive rate, means that overhunting has caused drastic reductions in large species throughout the world. Such a process may have begun with prehistoric humans causing extinctions of the woolly mammoth, saber-tooth cats, and other large animals. As killing technology has improved, so has the effectiveness of overhunting. A classic example is the North American bison, which declined from many millions of individuals to under 1100 in just a few decades (**Figure 11-7**). Today, virtually all large species are in decline, including both herbivores such as elephants, rhinos, and pandas, and carnivores, such as large cats and dog relatives. Nor are ocean-dwellers spared, for most whale species are also in decline.

Secondary Extinctions.   Secondary extinctions occur when the extinction of one group causes the extinction of another. Often this involves the loss of a food species. For example, the familiar panda bear of China subsists largely on a diet of bamboo. As the bamboo is being destroyed, the panda may become extinct from that cause alone. Other examples are more subtle, reflecting the complex and unpredictable interactions among organisms that make the effects of human disturbances so difficult to predict. The well-known extinction of the dodo bird, for instance, has caused the Calveria tree to become unable to reproduce. When the dodo ate the seeds of the tree, it digested the outer seed covering and then excreted the seeds without their coverings, allowing them to germinate and grow.

Minimum Viable Populations

Extinction can occur even if the population is not reduced directly to zero. Even if many individuals survive the disturbances, the population may never recover if it becomes too small. The species will fall into an **extinction vortex.**

There are two basic causes for this extinction vortex. One is that small populations may have breeding problems. Too few females may be left, or if the population is too dispersed, individuals may not be able to locate each other to mate. Even if there are enough mates and they can find each other, genetic inbreeding is a major problem in small pop-

Figure 11-7   *Slaughter of Buffalo* (bison) *on the Kansas Pacific Railroad,* from *Plains of the Great West* (1877). (*Source:* Courtesy of the John Hay Library, Brown University.)

ulations. This leads to birth defects and many other malformities. For example, many males in the remaining population of under 70 Florida panthers suffer from testicular and other malformities.

Aside from breeding, the second cause of the extinction vortex is that small populations are much more easily wiped out by random environmental fluctuations, such as an abnormally harsh winter, that would not significantly affect larger groups. The chance of being wiped out increases exponentially with decreasing population size.

The smallest population size needed to stay above the extinction vortex is often called the **minimum viable population** (**MVP**). If a population falls below this size, it is said to be no longer viable, and long-term breeding problems and environmental fluctuations will eventually finish the population off. This is what we meant earlier by the living dead, when habitat fragmentation leaves too few individuals in each isolated fragment.

In the past, calculations of MVP have been oversimplified. Some ecologists initially suggested, for instance, that an MVP of perhaps 500 individuals would permit almost any species to survive for many years. But most ecologists now acknowledge that there is no such magic number for MVP that applies to all species. Organisms vary widely in their ability to rebound from low numbers. Organisms that breed and grow rapidly, for example, will often recover from lower population sizes than slow-growing organisms and therefore often have lower MVPs. In general, most population viability analyses, which seek to estimate how long a certain population will persist, indicate that at least a few thousand individuals must remain alive to ensure long-term survival of most species. In this case, long-term survival means more than a few decades.

Community and Ecosystem Degradation

A growing criticism is that conservationists have focused too much on individual species. While much money has been devoted to saving a few highly publicized species, entire ecosystems are rapidly being degraded or even disappearing. As a result, there is increasing interest in saving endangered ecosystems as well as endangered species.

We saw in Chapter 3 that a community consists of all populations of species that inhabit an area. An ecosystem is a community plus its physical environment. A community or ecosystem becomes extinct when the populations comprising it die out. For example, a forest fire or bulldozer may completely destroy large areas of a certain forest community so that it becomes extinct.

Community Degradation. Communities and ecosystems are more likely to be degraded by human activity than completely destroyed, however. By "degraded" we mean that some species within the community or ecosystem suffer major decreases in abundance. A classic type of disturbance is pollution of a community's water, soil, or air. A basic effect of disturbance is ecosystem simplification, meaning that the number of species in the ecosystem declines. In addition, some of the remaining species become superabundant. These are the species that thrive in the polluted environment, such as certain sewage-eating bacteria or "trash" fishes such as garfishes that have adapted to low-oxygen waters.

Besides simplifying structure, a disturbance can also disrupt the functioning of communities (and ecosystems) by altering the flow of matter and energy through them. The flow of excess nutrients into the ecosystem can cause eutrophication and other disruptions often involving rapid population growth and decay. Conversely, slash and burn agriculture and other disturbances will often lead to "leaky" ecosystems, with excess nutrient flow out of the system (Chapter 3).

Indicators of Ecosystem Health. Some species in a community or ecosystem are more susceptible to disturbances than others. These are often called **indicator species** because they indicate the health of an ecosystem. A decline in the abundance of indicator species is evidence that the entire ecosystem may soon decline. For example, some trees, such as the Fraser firs in the Smoky Mountain National Park, are more sensitive to air pollution than others and will be affected by it first. Similarly, trout and freshwater "jellyfish" are among the organisms most sensitive to water pollution.

Are Diverse Communities More Easily Disturbed?

Until the early 1970s, many ecologists believed that the more species a community had, the more difficult it was to disturb. Although there is still much ongoing research and debate, mathematical models and field data show that the relationship between diversity, or number of species, and stability, or ease of disturbance, in a community is much more complicated than ecologists formerly believed.

Most ecologists now agree that stability is lowest in communities with very low diversity. In monoculture, for example, only one plant, such as corn, is grown for many acres. Such simple communities are easily disturbed, such as by an invading pest that can wipe out much of the community if the corn plants are susceptible. In communities that have more plant species and therefore support more kinds of animals, too, it is very unlikely that all the plant species will be susceptible to a single pest. Thus, increasing diversity leads to redundancy: If a species such as a food species is lost, other species will still exist to support at least part of the community.

However, after a certain point, increasing diversity may make a community easier to disturb. Increasing diversity means that more species are interacting, creating more complex food webs (Chapter 3). Many mathematical models show that such complex webs can become very sensitive because minor changes can cascade through the community. For example, many species can become dependent on certain **keystone species** for food, reproduction, or some other basic need. If the keystone species are removed, many parts of the community can be drastically affected. This seems to be especially true of some highly diverse tropical communities. Because they have been undisturbed for so long, they have built up highly intricate interactions that are easily disturbed.

Stopping Extinctions

Why worry about extinctions? Some people argue that humans have little need of wildlife. They view elephants, exotic tropical insects or plants, and the like as curiosities that have no immediate value and they are not very concerned at the prospect of their loss (**Figure 11-8**). To others, this is an absurd attitude. They take the opposite view and want to preserve all of nature for its own sake. They see humans as the intruders and insist that all extinctions must be stopped as soon as possible. Between these two extremes are many practical realities.

Reasons to Stop Extinctions: The Many Values of Biodiversity

People who worry least about extinction tend to place only direct values on biodiversity. Recall from Chapter 6 that direct values are based on the immediate economic gain made when a resource is destructively harvested. Examples include whaling, logging, and illegal trade in endangered species.

Illegal trade is an especially good example because the entire value of an organism is reduced to the price it brings, either dead or alive, in the marketplace. The World Wildlife Fund estimates that illegal trade in wild animals globally produces $2 to $3 billion per year. Rhino horns, for example, are used as aphrodisiacs, dagger handles, and other adornments. Many other species, such as the imperial Amazon macaw and mountain gorilla, are popular pets or attractions.

But such direct short-term valuation of an organism omits many other potential values, called *indirect values* in Chapter 6. When these are ignored, the true value of a species is underestimated.

Indirect Values. An ethical reason to save species is that humans have no right to destroy other species to the point that they are gone forever. According to this view, animals have intrinsic rights of their own, outside extrinsic human needs. This includes the right to live a life that is relatively free from pain.

The esthetic (aesthetic) and emotional reasons to save species are that biological diversity makes life more enjoyable and enriching. Given the mental and physical rejuvenation many of us experience after a long hike or other outdoor recreation, this argument alone would seem to have much validity. Certainly, such recreation would be less rejuvenating if we had only the same few species to experience all the time.

Indirect economic values include the many nondestructive ways that humans can use species. Ecotourism and sustainable harvesting of exotic foods, medicines, and many other materials are two examples discussed later as practical social incentives for providing jobs while preserving species. When measured over many years, such sustainable economic benefits of biodiversity are generally greater than the short-term economic gain from selling an organism for its skin, horns, or other values.

Environmental services refer to the value of biodiversity in providing us with many of life's essentials. Ecosystems are environmental support systems that provide us with things that we now take for granted: oxygen to breathe (from plants), drinkable water (purified by microbial activity), and many other natural chemical cycles that occur via ecosystem functions (**Figure 11-9**). If we remove too many species from an ecosystem, ecosystem function will be impaired. If we lose too many species in too many ecosystems, environmental services of the entire biosphere could be impaired.

Figure 11-8 What is the value of this endangered plant? This is the Eureka Valley Evening Primrose, known from fewer than five occurrences in Inyo County, California.

Figure 11-9 Wetlands, often called the "kidneys of the water cycle," illustrate many ecosystem functions by filtering groundwater, producing oxygen, and decomposing organic wastes. Seen here is part of a restored wetland in the prairie pothole region of northern Iowa.

The last indirect value is the evolutionary value of species. This is the value of today's species to future generations. Rather than being constant entities, species often change to adapt to new environmental conditions. But the raw material of evolution is variation. As biodiversity disappears, the biosphere will become less able to adapt to change because fewer species will be alive. An impoverished biosphere today means an impoverished biosphere for future generations.

A practical example of the loss of evolutionary value is food production by humans. Today only about 30 plant species provide 95% of the world's nutrition. Just four of these, wheat, corn, rice, and potatoes, provide most of the world's food, and all four have been subjected to centuries of inbreeding. Do we want our diet to be so bland and monotonous forever? Botanists estimate that at least 75,000 edible plant species exist, many superior in flavor and nutrition to those commonly eaten in the United States.

The ability of species to adapt to change also has other important implications for our future food supply. Low diversity makes organisms more susceptible to extinction. Inbreeding of most crop species makes them notoriously susceptible to diseases and insects. By interbreeding wild species with closely related cultivated forms, resistance to disease and general hardiness can often be improved, and genetic engineering now allows us to combine useful traits from species that cannot interbreed naturally (Chapter 13).

It is thus economically advantageous to maximize diversity both by growing a variety of crops and by having as many different crop species at our disposal as possible. Yet if current rates of extinction continue, some 25,000 plant species may die out in the next few years, before we have a chance to study them. Some biologists have proposed that gene banks be set up as a reservoir to save seeds, spores, sperm, and other genetic materials for species that cannot be saved in the wild. Currently, seed banks contain only a small fraction of most plant varieties, and many wild species have yet to be collected and described.

We have focused on plants, but similar arguments can be made for domesticated animal species. Breeding cattle with buffalo ("beefalo") to improve the stamina of cattle breeds and increase their resistance to predation is one of many examples where wild genes have been useful. Similarly, many wild animal species are themselves potentially tasty and healthy food sources, even without being interbred with existing food species.

Which Species to Save?

So many species are at risk that they cannot all be saved from extinction. In 2002, more than 50% of all crayfishes and freshwater clams (mussels) in the United States were at risk (**Figure 11-10**). At least 30% of plant, amphibian, and fish species were at risk, and over 15% of mammal species. Some species may survive with adequate attention, but the cost of saving all of them far exceeds the limited monetary and human resources available for saving species. Thus, painful decisions will have to be made concerning which species we want to save the most.

Conservation biologists often use the concept of **species triage** to refer to the difficult process of selecting which species to try to save from extinction. Triage is a medical term used at accident scenes and in emergency rooms: Some patients are beyond care (fatally wounded),

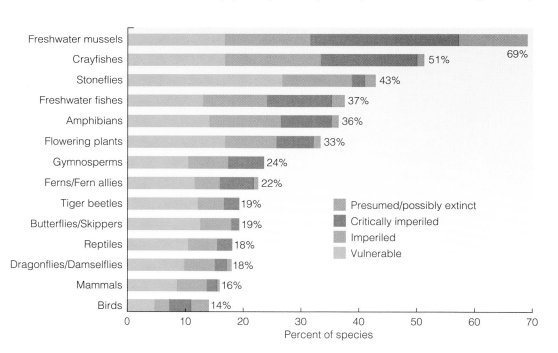

Figure 11-10 Freshwater organisms are among the most imperiled species in North America. These data include only species at risk, not those already extinct. (*Source*: B.A. Stein, L.S. Kutner, and J.S. Adams. 2000. *Our Precious Heritage: The Status of Biodiversity in the U.S.* [New York: Oxford University Press, p.102].)

TABLE 11-4   Characteristics of Extinction Susceptibility

Characteristics	Reason Characteristics Tend to Cause Extinction	Examples
1. Island species	Unable to compete with introduced species.	More than half of the native plant species in Hawaii
2. Species with limited habitats or breeding areas	Some species are found in only a few ecosystems.	Woodland caribou, Everglade crocodile, red-cockaded woodpecker
3. Species that require large territories to survive	Widespread habitat destruction.	California condor, blue whale, Bengal tiger, Florida panther
4. Species with low reproductive rates	Many species evolved low reproductive rates because predation was low.	Blue whale, California condor, polar bear, rhinoceros, Florida manatee
5. Rare species	Few individuals to replenish population.	Tropical insects, rhinoceros
6. Species that are economically valuable or hunted for sport	Hunting pressures by humans.	Snow leopard, blue whale, elephant, rhinoceros, tiger
7. Predators	Often killed to reduce predation of domestic stock.	Grizzly bear, timber wolf, Bengal tiger
8. Species that are susceptible to pollution	Some species are more susceptible than others to industrial pollution.	Bald eagle (susceptible to certain pesticides), pelicans
9. Species with inadaptive behaviors	Behaviors promote death in human environments.	Manatees swimming toward motorboats

and some are walking wounded, so most attention should be focused on those in the middle who are likely to survive *if* they get immediate care. And so it is with species. Weedy species (species that are opportunistic and reproduce rapidly) such as squirrels and deer are in no danger of extinction, while other species are among the "living dead," with so few individuals left that they will likely go extinct. Many experts therefore argue that conservation efforts should be directed at those species that are at risk and can still be saved. Unfortunately, the number of species at risk far outweighs the social resources currently allotted to preserving them.

Which Species Are at Risk?   Species are not equally likely to become extinct. For many reasons, some species are more able than others to survive environmental change. A well-known example of a very resilient group are cockroaches, which have existed for over 300 million years and will probably be alive for millions more. Table 11-4 identifies nine characteristics that make some species more susceptible to extinction than others. Island species are sensitive to the introduction of new species because they have been isolated for a long time. Indeed, most of the bird and mammal extinctions until now have been of island-dwelling species. Species with limited habitats become extinct easily because human activity can quickly eliminate the small space available. For instance, one species of tropical insect will often be adapted to only one part on one kind

of local plant. When the plant is eradicated, so is that insect species, along with others that are adapted to other parts of the plant.

Species with large territories die off quickly because they need lots of area to support them. This has been shown repeatedly when apparently large game reserves are set up and wide-ranging species (such as large predators) still die out. When a single individual needs many square miles to forage, it takes a very large reserve to preserve enough individuals to maintain the species. Many species are naturally rare, even without human disturbance; such species are obviously prone to extinction because they have few individuals to begin with. Low reproductive rates make it difficult for a species to rebound from habitat disturbance, hunting, or other causes of population declines. Economic and sporting value cause species to be sought by hunters. Predators are generally high on the food pyramid, so they are relatively less abundant than many other organisms. Sensitivity to pollution is another trait leading to extinction. Finally, some species may show behaviors that promote extinction. Some birds fly into windshields. Recent data on the Florida manatee indicates that it tends to swim toward motorboats for some reason. Boat injuries are the leading cause of death (**Figure 11-11**).

These nine traits can co-occur: Many species have more than one of the traits and are thus much more likely to become extinct. For example, large animals tend to be rarer, require larger territories,

Figure 11-11   The Florida manatee is a threatened species that suffers from ocean pollution, collisions with motor boats, and many types of human activities. Manatees swim very close to the surface, and it's likely this manatee's scars are from a motorboat's propeller.

and also have lower reproductive rates than small animals. If the large animal is also a predator and is hunted for sport, then it has at least five of the traits that promote extinction. Similarly, the Florida manatee not only swims toward boats but also has low reproductive rates.

Are All Species Equally Important?   With so many species at risk, triage decisions cannot be made on the basis of risk alone. Conservation biologists therefore often ask whether one species is more important than another. Ethically, perhaps one could argue that all species are equal; an insect may have as much right to live as a panther.

But in other ways, in particular, in ecological and evolutionary importance, all species are not equal. Ecological importance reflects the role a species plays in its ecological community. Keystone species play large roles because they affect so many other species. Large predators, for example, often control the population dynamics of many herbivores. When the predators, such as wolves, are removed, the herbivore population may increase rapidly, overgrazing plants and causing massive ecological disruption. Similarly, certain plants are crucial food for many animal species in some ecosystems. Extinction of keystone species will often have cascading effects on many species, even causing secondary extinctions. Many therefore argue that saving keystone species should be a priority.

Evolutionary importance varies among species because all species do not have the same potential for contributing to future biodiversity. Unique species are not closely related to any other living species (**Figure 11-12**). **Unique species** represent unusual gene pools, with many genes and traits not found in other species. The loss of a unique species often represents a much greater loss of genetic and evolutionary potential than the loss of a species with many living close relatives. Pandas, for example, are a unique species of bear. Unlike most bears, which eat many kinds of food, the panda is specialized for eating bamboo. The panda is a result of very unlikely evolutionary events that are unlikely to recur, and the Earth may never again see anything similar to it.

Rather than concentrating on species of ecological and evolutionary importance, to the regret of some biologists, conservation efforts have often focused on high-profile **charismatic species** that attract public support. The Florida panther and the bald eagle are examples of charismatic species. Focusing on such species is not necessarily bad because they are sometimes ecologically and evolutionarily important. Furthermore, because they are usually large animals and often predators, large areas must be set aside for them. These large preserves often contain the habitat of many other species in danger of extinction. For this reason, large charismatic species such as the Florida panther are also called **umbrella species.** Many other species are also protected under the umbrella set aside to preserve the panther.

How to Save Species

Because extinction is caused by change in a species' environment, such as habitat disruption or introduced species, the cheapest and most efficient way to save any species is to preserve its natural environment. This requires establishing protected areas such as wilderness preserves, discussed in Chapter 12. However, protected areas are relatively small in size and number, and many threatened species also occur in areas that are not protected. In the United States, the Nature Conservancy estimates that over half of all federally threatened species occurs on private lands.

Saving species therefore also requires approaches that protect species living in areas that are

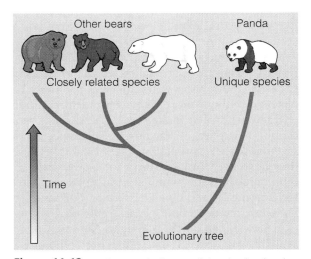

Figure 11-12   Unique species have no living closely related species.

often intensely utilized by humans. This requires legal and economic incentives that reduce human impacts on threatened species. In the rest of this chapter, we focus on the laws that attempt to do this.

Laws Protecting Endangered Species

Early laws to protect such species focused on the trade and sale of their skins and other products. In recent decades, the statute that has been most widely used to protect species and their habitats is the **Endangered Species Act** of 1973. This Act directs government agencies, especially the U.S. Fish and Wildlife Service, to maintain a list of species that are endangered or threatened. Endangered species are in immediate danger of extinction. Threatened species are likely to be endangered soon. The Service is also directed to produce a recovery plan for each listed species and protect its designated "critical habitat" needed for survival.

The Endangered Species Act is very controversial, and Congress often debates whether the Act should be reauthorized in its present form. Many people want to weaken or even eliminate the Act while many others want to strengthen it. Opponents claim the Act has cost billions of dollars and violates the Fifth Amendment's ban on taking property without compensation by preventing property owners from developing their land when it contains habitat of an endangered species. Landowners, logging companies, and other organizations have often brought suit against the government, claiming that the Act violates the right to own private property.

Opponents also point out that the Act has been relatively ineffective. The number of listed species has grown much more rapidly than the number of delisted species. In 2002, there were 1258 species listed as threatened and endangered (over half of these are plant species). In contrast, only 32 species had been delisted (**Figure 11-13**). **Delisted species** include both species that are no longer endangered or threatened, such as the brown pelican, and species that have gone extinct, so even some of the delistings indicate failures, not success. In fact, only 13 of the 32 species were delisted because they recovered. This means that only about 1 species has recovered for every 100 that have been added over the last 30 years! Overall, only about 10% of listed species are classified as improving. All the rest were only stabilized or declining. Furthermore, hundreds, perhaps thousands, of candidate species that are already rare are waiting to be listed, but there is inadequate funding to pay for the listing process.

Supporters of the Endangered Species Act argue that it has not worked well because it is too weak, and that far too little money has been spent on enforcement, less than $100 million annually in the federal budget. State and federal agencies combined spend about $200 million per year to protect endangered species; this is less than 4% of the money spent annually on U.S. lawn care and less than a dollar per person. Furthermore, most of this money is spent on just a few species. In any given year, just a few popular and charismatic listed species receive over half the money. For example, many millions are spent each year on the bald eagle, American alligator, and Florida manatee. Meanwhile, many less popular species receive a few hundred dollars or less. Another sign of imbalance is that in 2002, only 44 insect species were listed compared to 74 mammals, despite the fact that there are thousands more species of insects inhabiting the United States. Supporters of the Act also note that the listing process takes too long because so many species are too close to extinction to be saved when they are finally listed. In most cases, it takes many years for a species to become listed and protected.

A potential solution may be to protect entire ecosystems before species are on the verge of ex-

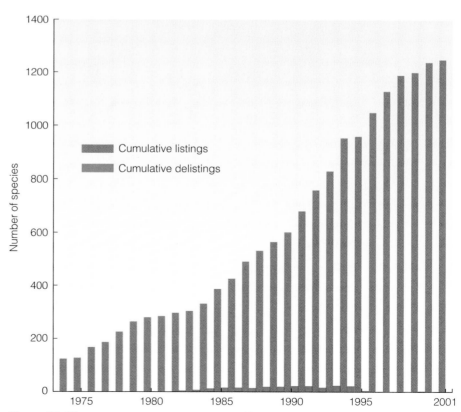

Figure 11-13   The number of species being listed as endangered species is growing much faster than the number being delisted. Many years have too few delistings to show on this graph. (*Source*: Data from U.S. Fish and Wildlife Service, 2002.)

Are Habitat Conservation Plans the Answer? The California Gnatcatcher Example

In 1982, Congress passed an amendment to the Endangered Species Act that was intended to meet some of the main criticisms of the original Act. Opponents had charged that the Act (1) interfered with economic growth, (2) emphasized saving species rather than whole ecosystems, and (3) waited until a species was on the verge of extinction before protecting it. The amendment created a new approach designed to make the Act more flexible, reduce economic costs, and protect many species before they are near extinction.

Figure 1   Encroachment of Gnatcatcher habitat.

The new approach is called a *habitat conservation plan* (HCP). Under an HCP, some of the habitat of an endangered species can be destroyed (called an "incidental take") as long as a plan is drawn up to reduce future losses. An HCP usually evolves as a compromise from discussions among landowners, developers, environmental groups, local governments, and the U.S. Fish and Wildlife Service, which eventually must approve the HCP. Ideally, an HCP will protect all current or potentially endangered species in an area, while simultaneously permitting human use of nearby lands as deemed necessary by social consent.

By 1994, seven HCPs had been approved and another 60 were under discussion from California to Key Largo, Florida. Each HCP is unique and some are more successful than others. California has the most approved HCPs. The first was at San Bruno Mountain south of San Francisco, which emerged as a compromise between housing developers and environmentalists wanting to save the mission blue butterfly. In fact, this conflict at San Bruno led to the legislation creating the HCP concept.

A good example of the complexities an HCP can encounter involved the coastal sage scrub habitat of southern California. This habitat includes some of the nation's most expensive real estate in prime locations around Los Angeles, San Diego, and nearby areas (**Figure 1**). But it is also home to a rare songbird called the California gnatcatcher (**Figure 2**). Only 2600 pairs remain, with 70% to 90% of the habitat

---

tinction, instead of focusing on "one species at a time" and waiting until a species is in imminent danger. Such **habitat conservation plans** have been relatively successful thus far (see Case Study 11-2).

Other laws also protect species. In the United States, each individual state has its own "endangered species act" that protects species that are threatened with extinction within the state boundaries. In most cases, these species are not federally threatened, but are disappearing from the state. State wildlife agencies are usually responsible for recovering these species.

International law states that it is illegal to trade, transport, and sell products made from endangered species. In 1975, 81 countries signed the **Convention on International Trade in Endangered Species** (**CITES**), which outlaws trade in endangered species

products. Though CITES has helped reduce illegal trade in some areas, illegal wildlife trade continues to proliferate (**Figure 11-14**). Unfortunately, thousands of animals die each year, especially tropical birds, in the process of being smuggled into the United States and other nations that are the main markets for illegal wildlife.

Species Recovery: Breeding and Reintroduction

In addition to protecting species in the wild, most wildlife laws also have the goal of **species recovery**, which refers to increasing the population size of a threatened species. One important method for doing this is to breed species in captivity. This allows for the protection of individuals from humans, better medical care, and control over which individualready destroyed. In 1993, after three years of discussion among developers, environmentalists, and government officials, then Interior Secretary Bruce Babbitt announced that the gnatcatcher would be listed as "threatened" instead of "endangered" as proposed by the U.S. Fish and Wildlife Service. The less urgent "threatened" status allowed officials to work out the details of an HCP that permitted development on some of the remaining scrub habitat. But the HCP also called for establishing up to 12 reserves that will benefit as many as 40 other coastal sage scrub species that are also in jeopardy from this disappearing habitat.

Figure 2 The California gnatcatcher has generated national controversy.

Questions

1. Many environmentalists strongly dislike the HCP concept because they believe it is "giving away" species and habitat to development. Can you think of a better way to resolve habitat versus development conflicts? Explain.
2. Since "extinction is forever," why would anyone ever approve of the HCP concept? How could development ever be justified over the irreversible loss of species? Explain.
3. Would you approve of an HCP that permitted one species to go extinct, but allowed five other species to survive? If it allowed ten others to survive? Twenty others? Explain your reasoning.

Figure 11-14 Trade in illegal wildlife continues to grow globally. (a) Seized wildlife parts. (b) Seized illegal shipment of Electus parrots and cockatoos.

(b)

(a)

als breed (to reduce genetic defects in offspring from inbreeding among relatives).

However, breeding in captivity is generally only used as a last resort to save a species. Captive breeding is almost always much more expensive and less effective than preserving habitat and saving species in the wild. Animals tend to be unhappy outside their natural state, especially in older zoos where conditions are poor (**Figure 11-15**). Even under the best of conditions, many animals, such as pandas, do not readily breed in captivity. Furthermore, there are far too few zoos in the world to sustain sufficient populations of all the world's threatened species, even if only large charismatic mammal species are considered. Yet because habitat continues to disappear at alarming rates, captive breeding is a necessity to save many species. As conditions become very desperate, as with the now-extinct dusky seaside sparrow, genetic mate-

rial (sperm and eggs) is being frozen for future use as a gene bank.

The ultimate goal of most captive breeding is for the animals and plants to proliferate and attain sufficient population size for some to be released back into the wild. A **reintroduction** is the release of plants and animals back into habitat that they formerly occupied. The American bison, for example, became extinct throughout most of the West, but is now being reintroduced in many areas. The controversial reintroduction of wolves into some parks and other areas in the United States is another example. Another approach is that of introduction of species into new habitats they did not formerly occupy. Many African savanna species, such as lions, antelope, and rhinos, have been successfully introduced onto game preserves in Florida and Texas, for instance.

Captive breeding can also be used to raise plants and animals for sale. Emu breeding is one example of many exotic species becoming popular in U.S. breeding farms. Other examples include tropical fishes and birds. Some people oppose breeding exotic species for pets and foods for a variety of reasons including animal rights and because it creates markets for captive animals. But supporters point out that such breeding helps reduce the market for the illegal wildlife trade by satisfying the demand for exotic animals and plants without taking them from their native habitat.

Sustainable Uses of Biodiversity

Conservation biologists agree that governmental decrees alone will not solve the extinction crisis. Establishing preserves and prohibiting trade in endangered species will not work as long as people, especially local inhabitants, are poor and must rely on destructive uses of biodiversity to stay alive.

Instead, sustainable uses of biodiversity must be developed and encouraged. Sustainable uses of natural ecosystems, such as tourism and sustainable harvesting of the rainforest, provide economic incentives to save species while also respecting the right of all people to support their families and have a decent quality of life (Chapter 12).

Figure 11-15   Many animals in zoos live in poor conditions that bear no resemblance to their natural habitat. The result is often a lifetime of boredom.

SUMMARY

- Biodiversity is often measured as the number of species.
- Five to 100 million species may now inhabit our planet, but only 1.8 million have been described.
- Five major mass extinctions occurred in the past.
- Extinction rates are best known for mammals and birds.
- About 2% of mammal species are extinct; 25% are threatened.
- Four causes of extinction are habitat disruption, introduced species, overhunting, and secondary extinctions.
- Biological communities can be degraded, as in ecosystem simplification.
- The health of an ecosystem can be inferred from indicator species.
- Biodiversity has many potential values: ethical, esthetic, emotional, economic,

- environmental services, and evolutionary values.
- The best way to save species is to preserve natural habitats.
- Large, rare, specialized, and predatory species are among those most at risk of extinction.
- Unique species, with few living close relatives, are among the most important to save from extinction.
- Species are listed much faster under the Endangered Species Act (ESA) than they are delisted.
- A way to improve the ESA may be to use more flexible habitat conservation plans.
- Species recovery in captivity is an expensive last resort, but it provides protection and ways to increase reproduction.
- The ultimate goal of species recovery is to reintroduce plants and animals back into their natural habitat.

KEY TERMS

Biodiversity
Biological impoverishment
Charismatic species
Conservation biology
Convention on International
   Trade in Endangered Species
   (CITES)
Delisted species
Ecological extinction

Endangered Species Act
Exotic species
Extinction
Extinction vortex
Extirpation
Habitat conservation plan
Habitat fragmentation
Indicator species
Keystone species

Mass extinctions
Minimum viable population
   (MVP)
Reintroduction
Species recovery
Species richness
Species triage
Umbrella species
Unique species

STUDY QUESTIONS

1. How many mass extinctions have occurred before now? What was the average extinction rate before humans? How much higher is the rate now?
2. Name and describe two key reasons why habitat fragmentation is one of the most destructive ways of disrupting habitat.
3. Name the four main causes of extinction. Which are biological? Do most extinctions involve just one cause?
4. What is an extinction vortex? What are two basic causes of an extinction vortex? What is the minimum viable population?
5. Where are exotic species an especially important cause of extinction? Why? Give examples.
6. Are diverse communities more easily disturbed? Explain.

7. Are all species equally important? Give two major examples. Where do unique species fit in?
8. Why is biodiversity important? Discuss some of its many values, and indicate the ones you favor the most.
9. Are marine species being threatened? By what? Discuss possible solutions.
10. Discuss the pros and cons of the U.S. Endangered Species Act. Is it a failure? A success? How should it be improved?
11. What are the traits that make a species vulnerable to extinction?
12. Why are extinction rates almost certainly underestimated?
13. What is the ultimate goal of captive breeding? Why is it best used only as a last resort?

STUDY GUIDE

WHAT'S THE EVIDENCE?

1. The authors suggest that the Earth may currently be undergoing a "mass extinction." What evidence do they present to support this contention? If you agree that the Earth is currently experiencing a mass extinction, do you think it is being caused, either partly or entirely, by humans? Cite evidence to support your opinion.

2. The authors note that the Endangered Species Act is very controversial. What evidence is given to support this statement?

CALCULATIONS

1. If you estimate that 50 species per day are going extinct, how many species will be extinct in a year? In 100 years? What percentage of all species will be extinct in 100 years, if there are 10 million species on Earth?

2. In 2002, there were 1258 species listed as threatened or endangered in the United States and only 13 species that had been delisted due to recovery. What is the ratio of listed to delisted species due to recovery? If this ratio continues, how many delisted species due to recovery will occur if 2000 species are listed in the future?

ILLUSTRATION AND TABLE REVIEW

1. In Table 11-1, which group of organisms has the greatest number of threatened species? Which has the lowest? Which group has the highest proportion? The lowest proportion?

2. In Figure 11-13, which is greater, the number of listed or delisted species? Is the gap between listed and delisted species increasing with time?

WWW.ENVIRONMENT.JBPUB.COM

www.environment.jbpub.com

Connect to this book's web site: http://www.environment.jbpub.com/ The site features eLearning, an online review area that provides quizzes, chapter outlines, and other tools to help you study for your class. We've posted the answers to the book's odd-numbered questions. You can also follow useful links for more in-depth information. The site provides sources for the information in this chapter and recommendations for further readings.

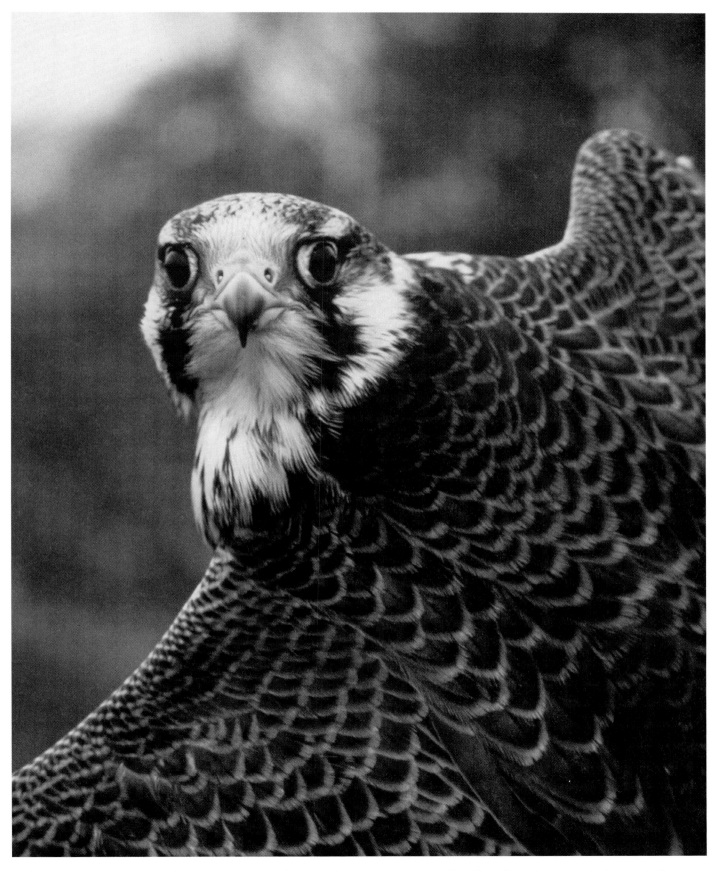

The population of peregrine falcons has recovered. The bird was delisted from the list of endangered and threatened species in 1999.

Wherever human population increases, there is a growing need to manage the way land is used. One reason for this is to reduce human conflicts over incompatible uses of the land. For example, most people who want to live in a quiet housing subdivision don't want a noisy commercial area nearby with lots of traffic. Another reason for land use management is to determine the natural influences on land use. It would be better, for example, to avoid building a new shopping mall on land that is very fertile and more efficiently used as prime farmland. Similarly, it would be a bad idea to put the mall on a major wetland, which would remove the land's ability to recharge and purify groundwater supplies. Unfortunately, in most parts of the United States, such natural influences are largely ignored and land use is mainly determined by immediate economic, social, and political factors.

Dense Urban   Sparse Urban   Forest   Water   Exposed   Cropland

CHAPTER 12

Land Resources and Management

Dramatic urban growth, represented by the red and orange points, can be seen in this series of Landsat images of Atlanta, Georgia.

Land resources in the United States, and most other nations, are typically managed under two general categories: public and private lands. Public lands include national parks, national forests and rangelands, and other lands that are not available for purchase by citizens or businesses. State, county, and city governments also have similar public land holdings including state parks and state forests. In the United States, most public lands are managed for many uses, but especially for resource extraction and recreation. Recently, there has also been growing public pressure for more wilderness preservation on public lands.

Private lands, owned by individuals and businesses, are both a major challenge and a major opportunity for solving current environmental problems. On one hand, private lands are where most intensive land modifications occur, such as farming and building shopping malls. Developers buy land and convert it into commercial or residential real estate that produces large amounts of pollution and provides very little habitat for native species of plants and animals. On the other hand, local land conservation groups can purchase or manage private lands and thereby make them available for wilderness recreation and critical habitat for species that would otherwise be excluded from highly developed areas. Aside from conservation groups, individual landowners anywhere can make enormous contributions by being ecologically concerned stewards of their land. For example, farmers can leave riverbank vegetation intact to reduce run-off and flooding, landscapers and homeowners can leave some native vegetation intact, and so on. Such "bottom-up" or grassroots stewardship is the key to long-term solutions because it allows landowners to have a good quality of life while promoting a healthy sustainable environment.

Managing Public Lands

The largest landowner in the United States is the federal government (**Figure 12-1**). The National Park System includes over 50 major parks and many smaller recreation areas. The National Forest System includes over 150 forests. The National Wildlife Refuge System consists of 535 refuges (as of 2002). The Bureau of Land Management oversees vast areas of the western United States, much of which is used for grazing and mining. These national systems include a huge amount of land, totaling over 42% of the United States. The western states are especially rich in federal land. The federal government owns more than 45% of California, Nevada, Utah, Idaho, Wyoming, Arizona, and Alaska. This federal land does not include the large amount of public land owned by state and local governments.

The management of public lands has been a matter of controversy since the beginning of their establishment more than a century ago. On one side have been preservationists who argue that public natural resources should be preserved as much as possible as pristine wilderness. On the other side have been conservationists, who argue that public resources should be open to mining, logging, grazing, and other human uses (as long as used judiciously). This controversy peaked

Objectives

After reading this chapter, you should be able to explain or describe:

Why land-use planning is important

How public lands are managed

Ways to improve nature preservation on public lands

How private lands are managed

Why urban and suburban sprawl is harmful to the environment

What causes sprawl

Ways to curb sprawl

Legal and economic incentives to curb sprawl

Figure 12-1   Some major systems of public lands that contain some natural habitat. (*Source*: Data from U.S. Geological Survey.)

Federal Lands by Agency
- Bureau of Indian Affairs
- Bureau of Land Management
- Bureau of Reclamation
- Department of Defense
- Fish and Wildlife Service
- Forest Service
- National Park Service
- Tennessee Valley Authority

Albers Equal Area Conic
- Central Meridian: 95° W
- 1st Std. Parallel: 29°30' N
- 2nd Std. Parallel: 45°30' N

in the early 1900s, when it was essentially won by those seeking access to the resources; hence, the conservationist philosophy and the multiple-use principle discussed earlier (Chapter 6) have dominated much public land use. Many historians note that this philosophy was probably inevitable given the rich resources, sparse population, and rapidly growing industrial economy of the United States.

An End to Subsidized Abuse of Public Land?

In recent years, support for more preservationist policies toward public land use has been growing. The multiple-use approach has historically been employed to justify very high levels of resource extraction, which have often resulted in widespread and highly visible environmental damage to public lands. In the western United States, many areas of public land where mineral deposits once existed are now scarred by active and abandoned mines. Each year, most of these mines produce tons of toxic acid and heavy metal drainage that contaminates the soil and water. Similarly, overgrazing of public lands by privately owned cattle damages native vegetation, destroys wildlife habitats, and encourages widespread growth of nonnative weeds. Logging on public lands is a major reason why the remnants of ancient old-growth forests are rapidly dwindling.

Much of this destructive resource extraction has been accelerated by subsidies from federal tax money. Mining, grazing, logging, and other activities have

been conducted at very cheap prices, far below market cost, because taxpayers pay much of the cost. The 1872 Mining Law, which allows miners to lease public land for about $5 or less per acre (0.4 hectare), is a good example. Grazing is much cheaper on public land than on nearby private land. The World Resources Institute estimates that these subsidies cost more than $1 billion per year in public funds spent and taxes not collected from land users. Although public resistance to this subsidized damage is growing, the opponents of subsidies have had difficulty achieving political change. Their efforts have generally been defeated by intense lobbying and the opposition of senators and representatives from the western states. Some of the leading lobbyists have been advocates of the "Wise Use Movement," which consists of a variety of individuals and organizations that seek to maximize private utilization of public lands.

National Parks: Overcrowded and Underfunded

The first major national park in the United States was Yellowstone National Park, established in 1872. As more parks were established, usually through acts of Congress, it became apparent that an organization was needed to manage them. Accordingly, the U.S. National Park Service was founded in 1916. In 2002, the National Park System covered over 80 million acres (32 million hectares) and included 368 sites ranging from major parks such as Yellowstone to urban recreation areas, battlefields, trails, rivers, prehistoric ruins, and homes of presidents. Often, areas in the park system are designated as national rivers, national monuments, national lakeshores, and so on to specify their significance. California has 23 parks, the most of any state. The idea of national parks has been very successful. The spread of such parks worldwide has led some observers to comment that national parks are one of America's great exports. More than 100 nations now contain over 1200 national parks.

But in recent years the National Park System has increasingly suffered from a number of problems. One is overcrowding (**Figure 12-2**). The National Park Service reports that the number of annual visitors to the national parks has risen dramatically from 22 million in 1946 to 133 million in 1966 and over 300 million in 2001. Attendance is increasing faster than the U.S. population and reflects the public's growing interest in the national parks. People who want to camp or rent cabins at the more popular parks, such as Yellowstone, Yosemite, and Great Smoky Mountains, generally must make advance reservations. In 2001, for example, over 9 million people visited the Great Smoky Mountains National

Figure 12-2   Bumper-to-bumper traffic is a common sight in the more popular national parks. This scene shows a traffic jam caused by visitors stopping to view wildlife in Yellowstone National Park, Wyoming.

Park. The parks often turn away hundreds of potential visitors for lack of space. Traffic jams, air pollution from cars, and crime have become problems.

The Funding Problem.   Funding to maintain the national parks has not kept pace with the increasing attendance. The 2001 National Park Service budget of $1.8 billion forced the service to delay maintenance, reduce staff and programs, and allow environmental problems to continue. The service had a $5 billion backlog of high-priority construction projects, $2 billion of unfunded land acquisitions, and $800 million worth of unfunded repairs. Lack of maintenance leads to problems ranging from rusting cannons at Gettysburg to complaints of raw sewage at Mammoth Cave in Kentucky and Sequoia National Park. Roads in many parks have gone unrepaired. The Grand Canyon alone needs an estimated $370 million worth of road, sewer, and other repairs. Park rangers suffer, too. Over 50% of their housing units have problems such as rotting floors, faulty wiring, or insect infestation.

One reason for the underfunding has been efforts to reduce federal government spending. But a number of observers, including some prominent park officials, argue that part of the problem is that Congress has purchased new national parks at the expense of the parks already in existence. Between 1975 and 1999, Congress established 88 new parks. Due to the growing interest in preserving the environment, establishing parks is very popular with many voters, so money for purchasing land has been relatively easy to obtain. In contrast, funding for sewage treatment, road repairs, and other maintenance needs has less popular appeal.

Two possible solutions to the funding problem have attracted considerable support: (1) making the parks more self-sufficient by relying on visitor fees and other private sources, and (2) removing some of the less popular or less unique parks from the park system. Advocates of increased self-sufficiency point out that park visitors spend more than $10 billion annually, but much of this money does not go toward park maintenance. Instead, it goes to concession stand operators in the parks and especially to general government funds used for a variety of purposes. Raising entrance or activity fees is another possibility. The National Park Service spends about $900 million per year on visitor services, but collects only about $150 million in entrance and activity fees.

Removing some less-visited parks from the National Park System is another option. Many critics point out that numerous sites attract few visitors but are costly to maintain. The Yukon-Charley

Rivers National Preserve in Alaska, for instance, has an annual budget of more than $500,000, but receives fewer than 1000 visitors per year. Critics argue that such wilderness can be preserved at less cost either by allowing private conservation groups to purchase it or by managing it outside the National Park System. Some groups are also interested in conducting mining or logging activities in parks in Alaska and the western United States that have few visitors. Congress is considering selling dozens of smaller urban parks, battlefields, and monuments to local governments or private groups. San Francisco's Golden Gate National Recreation Area and Cleveland's Cuyahoga Valley National Park are examples. Removal of parks from the park system is not new; Congress has de-authorized 24 parks since 1916.

Recreation or Preservation? The Dilemma.   Some critics do not agree that the solution to the overcrowding problem is increased funding for more facilities and staff. Instead, they argue that the number of visitors should be reduced altogether. They contend that too many visitors will despoil the natural beauty and contribute to the extinction of endangered species in the parks.

At the heart of the debate is whether the main purpose of national parks is to serve human needs or to preserve natural areas. The multiple-use philosophy that has dominated since the early 1900s has tended to emphasize human uses. But public interest in preserving natural areas is increasing as visitors to many of the major parks encounter huge crowds, traffic jams, commercialism, air pollution, and other unpleasant aspects of urban life. There is also controversy as to what kinds of recreation should be allowed. Should snowmobiles, for example, be allowed in Yellowstone National Park? (See Case Study 12-1.)

As a result, public support is growing for reducing the size of crowds in national parks and perhaps limiting public access. One solution has been to designate some lands in the park system as wilderness areas. Under the **Wilderness Act of 1964**, a designated wilderness area is federal land that is to be managed to retain its primeval character, with no commercial enterprise and no permanent road and no motorized vehicles. The amount of land set aside as designated wilderness has increased greatly since 1970:

1970	10.4 million acres (4.2 million hectares).
1980	79.7 million acres (32.3 million hectares).
1993	95.4 million acres (38.6 million hectares).
2000	105 million acres (42.5 million hectares).

CASE STUDY 12-1

Should Snowmobiles Be Allowed in Yellowstone National Park?

Yellowstone National Park is America's oldest park, established in 1872. It is also one of its most unique parks. Yellowstone's physical beauty contains a wide variety of pristine mountain views and even volcanic activity that produces elegant geysers such as Old Faithful. Elk, grizzly bears, and wolves are among the wildlife that are common in this popularly visited park.

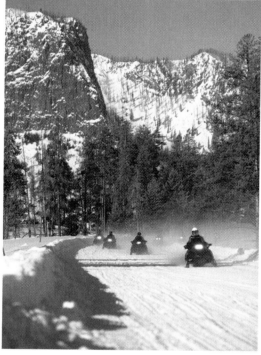

Figure 1 Snowmobiles in Yellowstone National Park have provoked considerable debate.

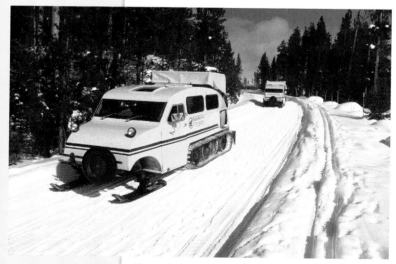

Figure 2 Snow coaches don't provide the same open-air experience of snowmobiles.

However, in recent years, a major controversy has arisen that illustrates the conflicts inherent in the "multiple-use" approach to public lands. Should snowmobiles be allowed in the Park? Each year there are over 80,000 snowmobile visits into Yellowstone (**Figure 1**). Advocates who wish to see the Park focus on wilderness preservation argue that snowmobiles should be banned. They note that wildlife is often disturbed by the loud noises produced by these machines. It is not uncommon, for example, for herds of buffalo to be startled by bands of snowmobiles. In addition, preservationists argue that the wilderness experience is ruined by the noise as well as the fumes and headaches produced by breathing snowmobile exhaust. Some visitors complain that they can't hear the splashing of Old Faithful because of the buzzing noises. In addition to many wilderness groups, many and perhaps most of the Park rangers seem to agree that some type of snowmobile restrictions are needed. In fact, surveys show that most of the general public is against snowmobiles in the Park.

On the other side are advocates of snowmobile use in the Park. Aside from the users themselves are many businesses in nearby towns that rent snowmobiles and sell related items. The mayor of West Yellowstone, for example, is an outspoken critic of a snowmobile ban arguing that it would "devastate" the town's economy. The town of West Yellowstone has many snowmobile rental shops and thousands of hotel beds that are used by snowmobilers in the winter.

After five years of study and several rounds of public hearings, the Clinton administration proposed a ban on snowmobiles in the park. The snowmobile visits every winter were to be phased out over a few years. But on the day he was sworn into office, President Bush imposed a moratorium on the ban, so snowmobiles continue to be widely used.

In the meantime, several compromises have been proposed. One is the use of quieter and less-polluting four-stroke-engine snowmobiles. However, because the machines are more expensive and less powerful than the traditional two-stroke models, renting them has not been easy. Also, snow coaches, which can carry several people, have been suggested as an alternative, but they have limited appeal to those who want the freedom and exhilaration that come from a solo ride on a snowmobile (**Figure 2**).

These figures include land in the National Forest and Wildlife Refuge Systems, as well as in the National Park System. In the National Forest System, the largest system by far in the lower 48 states, about 15% of the land is set aside as wilderness.

Interest in preservation has also led to two other trends: (1) increasing the size of parks by acquiring adjacent land through purchases or donations and (2) restoring parks to a more natural condition. Donations often come from conservation groups. In 1995, for example, a local conservation group, The Foothills Land Conservancy, purchased a few hundred acres of important black bear habitat and donated it to the Great Smoky Mountain National Park (**Figure 12-3**).

Park restoration efforts include allowing some forested areas to return to the old-growth condition and reintroducing native species that were exterminated by humans, such as the recent reintroduction of wolves into Yellowstone National Park. This project also illustrates the conflict that can arise from the multiple use of public land. Many environmentalists welcome the return of the wolf, while some farmers argue that the wolves pose a threat to humans and livestock. At some parks, there are plans to place a greater limit on visitation. In Yosemite, for example, some buildings have been torn down and visitor access to some areas may be reduced.

Figure 12-3   The recent donation of critical black bear habitat to the Great Smoky Mountain National Park is a good example of how private citizens can enhance the preservation of public lands.

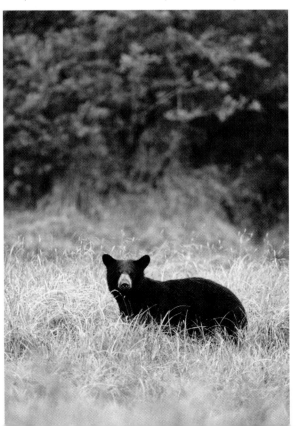

The trend toward increasing preservation is not without controversy. President Clinton was strongly criticized for designating public lands as National Monuments. Many environmentalists applauded the efforts to protect what they regarded as unique but threatened natural treasures, but many other people, particularly in the western United States, resented what they saw as a federal tactic to interfere with the use of their lands.

National Forests

The United States contains over 737 million acres (298 million hectares) of forests. About two-thirds of this land is classified as commercial timberland, meaning it is available for timber production. Only about 15% of this commercial timberland is owned by companies directly involved in producing forest or paper products. Instead, the timberland occurs on federal, state, and other public land. Most of the federal forests are in the National Forest System and are managed by the U.S. Forest Service (see Figure 12-1). Indeed, nearly half of the large commercially valuable timber trees in the United States are found on Forest Service land. This includes the last virgin stands of old-growth forests in the continental United States, which are found in the national forests of the Pacific Northwest.

Utilization or Preservation? The Dilemma (Again). The dilemmas that arise from multiple use of public land are especially common in national forest management because the National Forest System was originally designed to allow considerable utilization of natural resources. While national parks were mainly designed for recreation and to protect wildlife, water, and all natural resources, national forests were designed not only to make available these resources but also to provide U.S. citizens with access to the land for livestock grazing and the extraction of wood and mineral resources. This difference is evident in the fact that the National Park Service and the Fish and Wildlife Agencies are in the Department of the Interior while the Forest Service is in the Department of Agriculture. As a result, national forest management has historically been less inclined toward preservation. Because the National Forest System is so large, this philosophy helps explain the relative lack of wilderness preservation in the United States.

With the growing public interest in environmental preservation, the traditional methods of forest management, such as use of pesticides and nonnative plants (**Table 12-1**), are often criticized as unnecessarily destructive to wildlife and their habitat. Of special note is the controversial practice of

TABLE 12-1	Common Forestry Practices and Their Effects on Biodiversity	
Forestry Practice	**Purpose**	**Effect on Biodiversity**
Planting of exotics (nonnative species) or genetically "improved" tree species	Improved yield of commercial tree species	Replacement of native species
Pesticide spraying	Protection of forest tree species of commercial value	Reduction in vast numbers of nontarget insect species; secondary effects on nontarget organisms; major disruption of ecosystem
Clear-cutting/reforestation	Maximum utilization of existing tree biomass/maximum speed of new forest growth	Artificial cycle of disturbance; loss of species richness; loss of structural and functional diversity
Clear-cutting/even-aged management	"Efficient" regulation of the forest; maximum profit from growing trees	Shortened successional cycle; loss of forest structural diversity
Slash-burning	Site preparation for new forest; esthetics ("neatening up" the forest)	Major loss of structure, biomass, and nutrients from forest ecosystem
Tree thinning	Increased growth of commercial tree species	Reduced structural diversity in forest
Brush removal/herbicide spraying	Removal of species believed to be delaying reestablishment of commercial tree species	Truncated succession with loss of important successional processes; loss of species richness
The Bottom Line		
Maximum production of commercial sawtimber and pulpwood	Maximum forest output; maximum profit	Reduced structural and species diversity of forest; deterioration of forest ecosystem

Source: Used with permission from *Saving Nature's Legacy* by Reed F. Noss and Allen Y. Cooperrider. Copyright © 1994 by Defenders of Wildlife. Published by Island Press, Washington, D.C. and Covelo, California.

clear-cutting, also called "even-aged" management (**Figure 12-4**). In this tree-harvesting system, an entire stand of trees is completely removed, and the site is prepared for planting a new "crop" of trees. Clear-cuts are unattractive, as the land is stripped bare of trees, and can contribute greatly to increased sedimentation and erosion, which not only deplete the soil but often pollute local streams. Very large clear-cuts are no longer practiced in the United States. However, even patchworks of clear-cuts can be harmful and unattractive, especially since they often require large road networks to be built into the forest.

An alternative to clear-cutting is **selective cutting**, also called "uneven-aged" management. In this method, only certain trees in the stand are cut down, so the land is not stripped bare. Selective cutting is generally more expensive than clear-cutting, however, and is not without its own environmental problems. Roads must often still be built, and selective cutting usually alters normal forest development because shade-intolerant trees such as most pines and the Douglas fir cannot grow. Nevertheless, the "New Forestry" is a movement to incorporate biodiversity preservation into forest management, and selective cutting is preferred over clear-cutting. Replanting and careful restoration of harvested forest are promoted. Indeed, the forest products industry is the largest planter of new trees in the country; one-third more wood is grown each year in the United States than is harvested or lost to fire, insects, or disease.

Figure 12-4 Clear-cutting strips the land bare of trees, promoting erosion and water pollution and devastating habitat. This photograph was taken in Willamette National Forest, Oregon.

Subsidized Abuse? The Tongass National Forest Example. We noted earlier that taxpayer dollars are often used to subsidize rapid, and often harmful, depletion of natural resources. In the case of national forests, each year the Forest Service loses money on timber sales in most of its forests. If the cost of roads built by the government for the logging companies is included, the Forest Service shows a net loss of millions of dollars, according to data from the Wilderness Society and other environmental groups. Such groups therefore argue that the emphasis on tree production is outmoded at a time when forests often contain endangered species and shrinking, rare ecosystems. These shrinking forest ecosystems are not limited to the northern spotted owl and the old-growth forests of the Pacific Northwest, but include many other areas containing dozens of endangered aquatic species (such as freshwater clams) and birds (such as the red-cockaded woodpecker of the southeastern United States).

The public lands of Alaska are especially important because they contain the largest natural areas remaining in the United States and also the largest amount of untapped resources. As a result, the stakes are high for both developers and environmentalists, and so is the controversy. For example, the Tongass National Forest in southeastern Alaska contains the last intact stretch of America's only temperate rainforest, which once extended down the Pacific Coast to California (**Figure 12-5**). Currently, timber companies are pushing hard for the right to log the heart of the Tongass Forest and have won considerable support in Congress. Alaskans living near the forest point out that although past logging restrictions have cost jobs, they have also improved tourism and commercial fishing, which were harmed by the clear-cutting operations. A similar controversy is brewing over the Arctic National Wildlife Refuge, the largest wildlife preserve in the United States, where oil companies want to drill. Critics point out that the damage to wildlife, especially caribou, from the roads and development could be enormous. And according to geological estimates, there may be little oil in that area anyway. Whatever happens in the Tongass and Arctic Refuge, it seems likely that the debate over preservation versus extraction of Alaska's wealth will persist for many years to come, demonstrating the controversies inherent in the multiple-use concept.

NEPA and Environmental Impact Statements

In 1969, a fundamental federal law was passed called the **National Environmental Policy Act** (**NEPA**), which required that all major federal actions, including all activities on federal public lands,

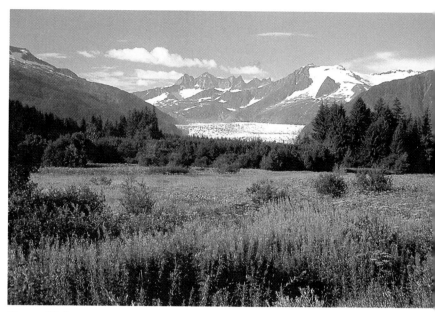

Figure 12-5   The Tongass National Forest of southern Alaska contains the last large area of America's temperate rainforest.

be reviewed for their environmental impacts before those activities could go forward. A key part of NEPA was the **Environmental Impact Statement** (**EIS**), which has become a basic tool to protect the environment on federal lands. Most states also require such a review on state public lands. The EIS requires that anyone proposing action on public lands, such as logging or construction of roads or buildings, first perform a study that documents the need for the project, reasonable alternatives to it, an overview of the natural environment affected, and the environmental consequences of the project.

Like many federal laws, the EIS was well received and is widely regarded as a good idea. However, in practice, there have been many problems. One issue is the vast amount of paperwork that is often generated. As a result, in 1979, the process of **scoping** was introduced; this means that local government agencies and the general public are invited to participate in producing the EIS for a project. Members of the community in a proposed Forest Service logging project, for example, are invited to town meetings with Service officials to discuss the pros and cons of its impact on the community. As a result, the EISs of many projects produce considerable controversy because nearly all projects have both harmful and beneficial impacts. For example, local environmentalists often complain that EISs prepared by private companies for work on federal lands are biased because the private companies hire the technical firms that produce the EIS and the data are often selectively used.

Managing Private Lands for Biodiversity Preservation

The future resolutions of many environmental problems rely heavily on how private lands will be managed. Environmentalists hope that many of them will be managed for the preservation of native species biodiversity. Because extinction is caused by change in a species' environment, such as habitat disruption or introduced species, the best way to save any species is to preserve its natural environment. Until recently, the establishment of preserves has occurred mainly on public lands such as those discussed above. However, such public preserves, in the United States and most of the world, are currently far too inadequate to prevent the extinction of many species.

One inadequacy is simply that the amount of land (and ocean) protected is insufficient. One common estimate is that about 10% of any given area, such as the United States, would need to be preserved as unmodified habitat in order to save the large majority of threatened species. Overall, only 5.9% of the world's total land area (excluding Antarctica) is protected, far below the 10% goal established by the World Conservation Union. Another inadequacy is that even where larger amounts of land are preserved, the habitat is of poor quality: It is often highly fragmented and/or located on mountains, in deserts, or in other habitats that were set aside simply because the land had little commercial value for humans. Finally, many of the preserves are ineffectively managed. Especially in poor nations, there is little money available to prevent poaching and habitat loss.

Fully aware of these inadequacies of preserved public lands, some environmental groups have tried to safeguard additional private lands. The past few decades have seen the rapid growth of private organizations in many nations, usually called nature conservancies or land trusts, which purchase, lease, or find some other way to legally protect private lands from development.

The Nature Conservancy is an example of a group that has made a huge impact by doing this. Founded in 1951, the **Nature Conservancy** has a membership of approximately one million. By using a large fund of donations (almost $400 million in 2001 alone), the Conservancy has created the largest system of private natural areas and wildlife sanctuaries in the world. It includes more than 12.6 million acres (5.1 million hectares) in the United States as of 2002 and over six times that much outside the United States. By focusing on buying species-rich threatened habitats, buffer zones, and corridors, the Conservancy has greatly enhanced the survival chances of many species in the United States. Another example is the **Land Trust Alliance**, which is an association of hundreds of local land trusts in the United States.

Managing land for preservation is actually quite complex and involves two key steps: (1) selecting and designing preserves on the basis of ecological principles and (2) using legal and economic principles to establish and maintain the preserves. The first step is largely scientific, the second largely social.

Selecting Preserves

Creating a preserve involves more than just setting aside a section of land or water. Preservation of habitats containing many nonnative species or abundant native species often does little to maximize regional or global biodiversity. Instead, preserves should be selected that save native species that are rare and in danger of extinction.

Historically, natural habitat preservation in North America and Australia has usually arisen from the desire to preserve unique or scenic landscapes. Most public parks, for example, were established for this reason. Yellowstone National Park, Great Smoky Mountains, and the Everglades National Park are but a few examples. On a smaller scale, this is also true of most state parks, which are usually characterized by their presence near mountains, canyons, and so on. The first national park, for example, was Yellowstone National Park created in 1872 for its scenic beauty. It has also been noted that most of these early parks were created on lands that were available simply because they had low commercial value: Mountains and canyons, for example, have poor soils for farming and are isolated from population and transportation centers. In Europe, many of the first game preserves were created to preserve large game animals, whose numbers were decreasing. One of the earliest was established in Poland in 1564 to attempt to preserve wild cattle. In developing tropical nations, many preserves have also been established to protect disappearing species.

As many native species become rare at increasing rates, it is clear that simply preserving either a few unique habitats or the habitats of a few charismatic species will not suffice. If the goal is to save many threatened species, a more comprehensive approach is needed. A common approach has been to use popular charismatic species as umbrella species to preserve large areas, but this approach has its limitations. Many rare species are narrowly distributed in areas that are not currently occupied by umbrella species. Also, the past emphasis on lands with low commercial value has disproportionately favored preserving some habitats and neglected others. For example, mountainous and canyon habitats

Figure 12-6 Global hot spots for plant biodiversity. (*Source*: G.K. Meffe and C.R. Carroll, *Principles of Conservation Biology* [Sunderland, MA: Sinauer, 1994], p. 123. Reprinted by permission of Sinauer Associates, Inc.)

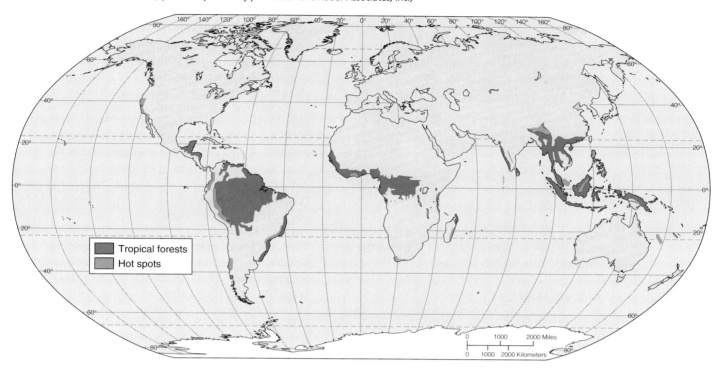

are relatively well represented by national parks in the United States, whereas river valley and wetland habitats are not. River valleys and wetlands have been favored for human settlement because they have a high commercial value, with fertile soils and river access for industry and travel.

A way for future biodiversity planning to correct these past omissions is to identify species hot spots that should be preserved. **Hot spots** are areas of exceptionally high species richness, especially concentrations of localized rare species that occur nowhere else. Hot spots are generally areas of great geographic diversity that have promoted lots of evolutionary change. **Figure 12-6** shows a number of global hot spots for plant species. Ecologist Norman Myers has estimated that these cover just 0.5% of the Earth's land surface, but contain the only habitat of about 20% of the Earth's plant species. Similar analyses of birds indicate that about one-fourth of the world's bird species are confined to just 2% of the Earth's land surface. Hot spots for birds do not necessarily occur in the same areas as plant hot spots. In the United States, a number of hot spots for various groups have been identified (**Table 12-2**). Unfortunately, many hot spots in the United States and the world are already nearly

TABLE 12-2	Hot Spots of Species Richness in the Continental United States
Areas of Greatest Species Richness	
Vascular plants	California, followed by Texas, Arizona, Oregon, and Florida (in that order)
Trees	Southeastern Coastal Plain and Piedmont, northern Florida
Mollusks	Tennessee River system (Tennessee, Alabama) and Coosa River system (Alabama)
Butterflies	Western Great Plains and Central Rocky Mountains (Colorado)
Fishes	Cumberland Plateau in the Tennessee and Cumberland River drainages
Amphibians	Southern Appalachians and Piedmont
Reptiles	Gulf Coastal Plain (eastern Texas)
Birds (breeding)	Sierra Nevada, southeastern Arizona–southwestern New Mexico
Mammals	Sierra Nevada and, secondarily, southern Cascades and desert Southwest

Source: Used with permission from *Saving Nature's Legacy* by Reed F. Noss and Allen Y. Cooperrider. Copyright © 1994 by Defenders of Wildlife. Published by Island Press, Washington, D.C. and Covelo, California.

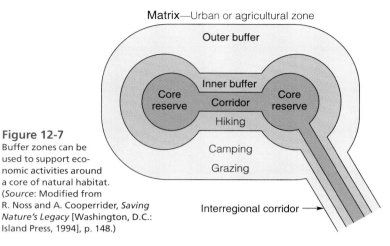

Matrix—Urban or agricultural zone

Figure 12-7
Buffer zones can be used to support economic activities around a core of natural habitat. (*Source*: Modified from R. Noss and A. Cooperrider, *Saving Nature's Legacy* [Washington, D.C.: Island Press, 1994], p. 148.)

destroyed. By many estimates, most of the plant hot spots in Figure 12-6 and Table 12-2 will suffer major habitat losses within 30 years.

Designing Preserves

After a location is selected, the design of the preserve is critical. Species preservation is maximized by three key characteristics: size, shape, and connectivity. Larger size increases the number of species contained in the preserve. Rounder shape minimizes edge effects because the perimeter (edge) is smaller relative to the area inside than with other shapes. Connectivity is the opposite of fragmentation. Increasing the connections (corridors) between po-

tential fragments allows members of the same species to immigrate and interbreed.

Buffer zones are another important preserve characteristic (**Figure 12-7**). A **buffer zone** is moderately utilized land that provides a transition into the unmodified natural habitat in the core preserve where no human disturbance is allowed. For example, campgrounds and limited cattle grazing may be permitted in the outermost buffer zone, with hiking in the innermost buffer zone. Buffer zones are a major departure from traditional preserves that were viewed as islands of natural habitat in a hostile matrix of agricultural or urban landscape.

Buffer zones are very important for both psychological and practical reasons. Sharp boundaries, as were used around traditional preserves, tend to promote the idea that nature should be fenced in and humans fenced out; such boundaries suggest that humans are separate from nature rather than being part of nature. On a practical level, conservation biologists have found that inhabitants of areas surrounding preserves must derive some benefits from the preserve. Many preserves have been established in developing nations only to have the endangered species poached into extinction because local inhabitants needed food or money. Buffer zones integrate some of the area surrounding the preserve into the local economy without harming the core preserve. By permitting moderate recreational, forestry, farming, and other activities, buffer zones provide jobs and income with no ill effects on species in the core preserve. Indeed, endangered species in the core preserve occasionally wander into the buffer zones, providing thrills for tourists, income for inhabitants, and larger forage area for the endangered species.

Preserve Networks

The small, fragmented nature of many preserved areas has contributed to a steady decline in biodiversity in the United States. Large species that need sizeable areas have especially suffered. For example, Yellowstone National Park contains only about 100 grizzly bears (**Figure 12-8**). This is far too few to form a sustainable minimum viable population, and the bears are showing signs of inbreeding and population decline.

Many conservation biologists suggest that the solution is to: (1) expand the size of existing preserves, (2) create buffer zones, and (3) create habitat corridors that connect the preserves. Such preserve networks would allow habitat preservation in the United States to have the desirable characteristics of preserve design discussed above. Many of the proposals for extensive networks have been

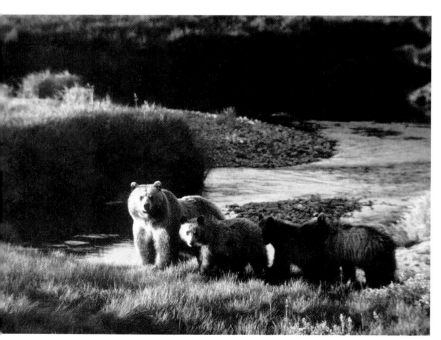

Figure 12-8   Yellowstone National Park, like many parks, is far too small to sustain viable populations of large mammals such as grizzly bears for many years.

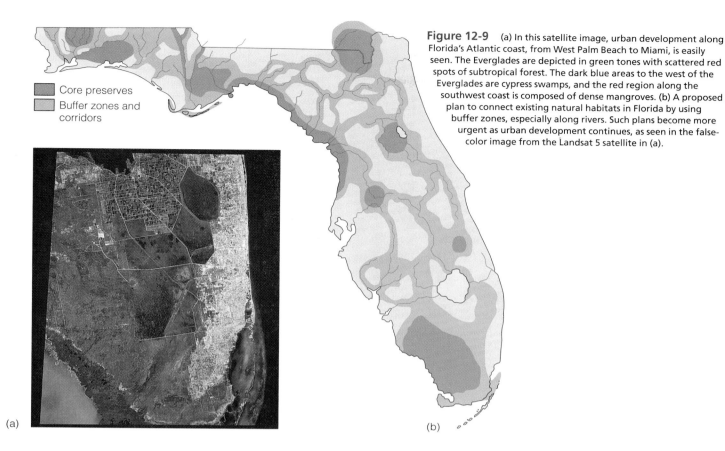

(a)

(b)

Figure 12-9 (a) In this satellite image, urban development along Florida's Atlantic coast, from West Palm Beach to Miami, is easily seen. The Everglades are depicted in green tones with scattered red spots of subtropical forest. The dark blue areas to the west of the Everglades are cypress swamps, and the red region along the southwest coast is composed of dense mangroves. (b) A proposed plan to connect existing natural habitats in Florida by using buffer zones, especially along rivers. Such plans become more urgent as urban development continues, as seen in the false-color image from the Landsat 5 satellite in (a).

criticized as unrealistic; one plan, for example, includes much of Florida (**Figure 12-9**). An even more extensive plan, called the **Wildlands Project**, outlines a similar goal with networks throughout the entire United States. However impractical such plans may seem, most wildlife experts agree that corridors, buffers, and other network concepts will eventually be necessary in order for wide-ranging species to survive and for many regional ecosystems to stay healthy.

Sustainable Management of Preserves

Conservation biologists agree that governmental decrees alone will not solve the extinction crisis. Establishing preserves and prohibiting trade in endangered species will not work as long as people, especially local inhabitants, are poor and must rely on destructive uses of biodiversity to stay alive.

Instead, sustainable uses of biodiversity must be developed and encouraged. Sustainable uses provide incentives to save species while also respecting the right of all people to support their families and have a decent quality of life. This is sometimes called the **bottom-up approach** to biodiversity conservation because it permits local citizens to play a role in planning and establishing preserves. In contrast, with the top-down approach, national and regional governments simply establish preserves regardless of the needs of local citizens.

Ecotourism. One very important sustainable use of preserves is **ecotourism**, which the Ecotourism Society defines as "responsible travel to natural areas that conserves the environment and sustains the well-being of the local people." Ecotourism uses the travel facilities created by the native population. It avoids international hotel chains in favor of properties owned and managed by locals. Adherents dine not on foods flown in from abroad, but on regional dishes, supporting the community's economy. Ecotourism respects the local environment and the native culture of an area. To be done correctly, ecotourism must not harm the local ecosystems in any significant way.

Examples of this fast-growing industry abound throughout the world. This growth is driven by widespread environmental concern combined with the fact that ecotours can be an enjoyable escape from the routine of industrialized society. Travel agencies arrange ecotours that range from a few days in an air-conditioned boat cruising the Amazon to weeks of hiking in the Himalaya Mountains in Nepal. Many of these tours do accomplish at least some of the goals of ecotourism.

Figure 12-10   Ecotourism can benefit animals such as this orangutan if it leads to money to buy quality natural habitat.

A good example is the Orangutan Rehabilitation Centre in Bukit Lawang, Indonesia. This project rehabilitates orangutans illegally held in captivity or displaced by deforestation and returns them to natural habitats (**Figure 12-10**). Because of the great international interest in this disappearing primate species, the Indonesian government is promoting the Centre as a destination where tourists can view orangs in their habitat. Consequently, it has become one of the most popular ecotours in Southeast Asia and is bringing in much revenue to help support the Centre.

While ecotourism can help preserve natural habitat and promote a sustainable economy, it has limitations. One is that many natural areas are not suitable for ecotourism. Many ecosystems are simply not appealing enough to the general public to provide a large pool of visitors. Aesthetic sites, such as coral reefs, rainforests, and pristine mountain hikes are popular, as are areas with charismatic species, such as elephants or grizzly bears. But many ecosystems lack the spectacular scenery or unique creatures necessary to draw sufficient visitors. In all likelihood, large portions of deserts, swamps, tundra, and many other areas cannot be preserved via ecotourism. Indeed, some experts suggest that only a few isolated, relatively unique ecosystems can be completely supported by ecotours.

Even where ecotourism is feasible, practical problems arise. Money generated by ecotours does not always contribute to the local economy, especially in poor developing nations. Some fees from

the Orangutan Rehabilitation Centre, for example, have gone to government officials in Jakarta. Also, in many cases, native species and ecosystems have been harmed by crowds of people and all the activity. Motorized tundra buggies that take crowds to see Hudson Bay polar bears have been criticized as seriously disturbing the bears. Similarly, the Manuel Antonio National Park in the heart of the Costa Rican rainforest now gets 1000 visitors per day and 300 once-wild monkeys have become garbage feeders. Even native cultural diversity can be affected. Some boat tours along the Amazon now make stops so ecotourists can trade with native Indians who previously had little contact with civilization.

Despite these problems, ecotourism has enormous potential because the travel industry is the second largest industry in the world. (Agriculture is the largest.) In 2001, tourism generated over $245 billion and employed 207 million people worldwide. If properly done, ecotourism can reduce local poverty, conserve nature, and educate the public about both biological and cultural diversity.

Sustainable Harvesting.   Another sustainable use of preserves is sustainable harvesting. Sometimes called *extractive forestry*, **sustainable harvesting** includes the harvesting of nuts, fruits, and many other products that can be extracted from an ecosystem without harming it. Specific examples include collecting Brazil nuts and tapping rubber trees in the rainforest to sell these products to industrialized nations. Animals such as parrots can also be sustainably harvested, in theory, although most pet bird groups recommend buying captive-bred birds. Sustainably harvested products can also be used for health care at home. Most of the primary health care in many developing countries is provided by traditional healers using native plant remedies. The World Health Organization estimates that nearly 3 billion people use wild-harvested medicines, with an annual value of up to $15 billion. With so many undescribed plants, many potentially useful medicines will be lost if the plants they are derived from are allowed to become extinct before their potential value is discovered. This has led to a new field, called **bioprospecting**, where biologists and chemists are rapidly compiling a huge database of the commercial potential of various species. One of the most effective types of bioprospecting uses the knowledge of native healers, or shamans, about the medicinal properties of local plants. But chemical prospecting is not limited to rainforest species or even land species. Many marine organisms have yielded chemical compounds now widely used as medicines, foods, and other products.

Although chemical prospecting has been very successful in finding potential uses for biodiversity, many local inhabitants still do not benefit from it. Typically, drug and agricultural companies and other businesses that profit from the genetic diversity of developing nations are owned and operated by citizens in industrialized nations in the Northern Hemisphere. This inability of the developing nations to profit from their own biodiversity was a major issue at the 1992 Earth Summit in Rio de Janeiro. After heated debate, 158 nations signed the Convention on Biological Diversity (CBD), which acknowledged that developing nations deserve a greater share of the profits generated from genetic resources by agriculture and genetic technology. In essence, this means that genetic resources are similar to intellectual property rights that can be patented, with profits returning to local populations where the genetic information originated. Such **genetic patent rights** would allow native people who conserve their biodiversity to receive some of the enormous profits often derived from pharmaceuticals, agriculture, and other uses of their genetic resources.

Studies often show that the long-term value of a rainforest is greater when sustainably harvested for medicines, rubber, foods, and other products than when it is logged, burned so that grass can be planted for grazing cattle, or otherwise destructively exploited for short-term profits. The long-term value from sustainable harvesting is even greater if the preserve can also be used for ecotourism. The total revenue from Amboseli National Park in Africa, for example, has been estimated at about 40 times the revenue that would be produced from farming the same land. Each lion was estimated as worth $27,000 in tourist revenue, over twice its value either dead or alive in the illegal wildlife market (**Figure 12-11**).

Sustainable harvesting of natural healthy temperate forests is important too. Only a portion of the commercial forest area in the United States is in national forests. Much of the remainder is owned by private companies that grow trees for commercial logging. Many rare species occur in these private forests, and many more could exist there if current forestry practices were discontinued (Table 12-1). An important result of the emphasis on wood production is the loss of old-growth, or virgin forests. More than 95% of such forests have been cut down in the lower 48 states. Old-growth forest takes many decades to grow because it is a late-successional, climax stage in forest growth (Chapter 3). Old-growth timber is generally denser, higher-quality wood. Old-growth forests are also rich in species not found in early-growth forests. Because timber takes so long

Figure 12-11   A lion is generally much more valuable as a source of income to local inhabitants if it is alive.

to grow, timber companies tend to raise only early-successional trees, such as certain pine species, that grow quickly. These can be cut down in a few years and replanted to start the cycle anew. This practice may maximize wood production, but it reduces overall biodiversity in U.S. forests.

Marine Preserves

Although the extinction of land species has attracted more attention than the extinction of species living in water, aquatic species, in both freshwater and marine environments, are also being decimated by widespread human disturbances. Protection of aquatic species has also lagged behind that of land species, for many reasons. One is that people cannot see the extent to which biodiversity loss is occurring there. In addition, in some ways, aquatic species are more difficult to protect. Large bodies of water, especially oceans, are classic commons, and are not clearly owned by any one nation. Consequently, regulating protected areas can be difficult. Pollution discharge in one area, for example, may travel hundreds of miles in an ocean current.

Nevertheless, there is growing interest in protecting these crucial marine habitats, especially nearshore bays and reefs that are being overvisited by tourists and contaminated by water pollution. Governments are establishing **marine protected areas**, which are areas of ocean that are set aside as preserves for marine life. In theory, fishing, construction, tourism, pollution, and other human dis-

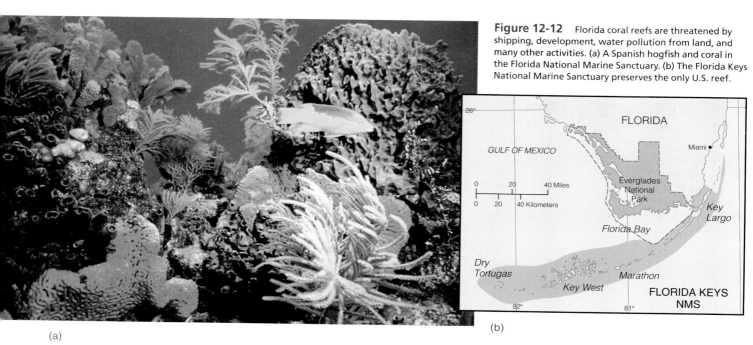

(a)

Figure 12-12 Florida coral reefs are threatened by shipping, development, water pollution from land, and many other activities. (a) A Spanish hogfish and coral in the Florida National Marine Sanctuary. (b) The Florida Keys National Marine Sanctuary preserves the only U.S. reef.

(b)

turbances are closely regulated and restricted in these areas, but the restrictions are often difficult to enforce, especially in poor developing nations with relatively few resources.

In the United States, marine protected areas are included in the **National Marine Sanctuary** program. This program was created in 1972, a full century after Yellowstone was dedicated as the first national park, illustrating how marine protection has lagged behind. Since 1972, twelve sanctuaries have been established under the jurisdiction of the National Oceanic and Atmospheric Association. One of the largest is the 220-mile-long (354 km long) Florida Keys sanctuary where fishing and coral collecting are restricted because of widespread ecosystem disruption by humans (**Figure 12-12**).

Many marine biologists argue that many more such sanctuaries desperately need to be established, particularly throughout the Caribbean and western Pacific, especially around Southeast Asia, where poor nations have allowed massive destruction of coral reefs and other crucial marine habitats. Buffer zones, ecotourism, sustainable harvesting, and many other economic incentives, if properly applied, can be used to promote marine preserves just as effectively as with forest and other land preserves.

Managing Lands to Reduce Urban Impacts

The last two hundred years have seen a strong global trend toward urbanization. In about 1800, only 3% of humanity could be classified as urban dwellers. By 1950, this number had risen to approximately

30%, and in 2000, 46% of all people lived in cities. If present trends continue, it is projected that in 2025, the urban population will amount to 60% of the world's total population.

As a city's population grows, it spreads out along the edges into the surrounding countryside, giving rise to **suburban sprawl**. Slowly, residential suburbs push farther and farther away from the city center, and shopping centers and malls, theaters, and medical facilities, not to mention schools, churches, and post offices, are built to service the suburban residents. Soon, people living in the suburbs may rarely venture into the heart of the city at all; they can find everything they need without leaving suburbia. City centers tend to decline as the suburbs develop and expand. An important general rule is that lateral expansion increases at a much faster rate than population growth of a city. In fact, the population of many cities in the United States has declined while the area occupied by sprawling development has increased! As the population moves away from the center, the core of the city may deteriorate until eventually low land prices attract businesses and developers once again. A large city may go through stages of growth, stagnation, and rejuvenation.

As the suburbs of large cities expand outward, adjacent cities and their suburbs may meet and begin to merge, forming a vast network of adjacent urban and suburban communities (sometimes referred to as a *conurbation*), until they become a single vast urban area, the **megalopolis**. The term *megalopolis* was first coined to describe the almost continuous urban and suburban sprawl in the northeastern

United States from Boston, through New York City and Philadelphia, down to Washington, D.C. Examples of megalopolises can now be found throughout the world; for instance, in the Netherlands, Amsterdam, Leiden, The Hague, Rotterdam, and Dordrecht all run together, and in Japan, the Tokyo–Yokohama complex forms a megalopolis.

In the United States, suburban sprawl is a very widespread problem. In 2001, 50% of the U.S. population lived in the suburbs, while 30% lived in cities. A mere 20% of the population, the lowest percentage in U.S. history, lived in rural areas. More ominously, the amount of land being converted to "built-up" areas is increasing exponentially, so that nearly 3 million acres per year are now being developed in the United States. The rate of development is now much faster than the rate at which land is preserved as public or private land (**Figure 12-13**).

Causes of Sprawl

Surveys consistently show that most people tend to have a negative opinion of sprawl. Then why does it happen? Suburban sprawl has many causes. One of the most important is the rise of automobile use and the proliferation of roads that make commuting easy. Economics influences both of these factors in the United States, where gasoline is much cheaper than in nearly all other nations. In addition, road construction is encouraged by federal and state gasoline taxes that are strictly dedicated to the state and federal departments of transportation and provide enormous funds for the purpose of building roads.

Of course ease of transportation is only part of the cause. Suburbanites must have some incentive to move to the outer fringes of cities. When asked, many suburban-dwellers say that the main attraction to living in suburbia is the desire to own a parcel of land that can be gardened. (Gardening is the most popular outdoor activity in the United States.) Another reason often cited is the desire to escape the greater perceived crime rates of downtown areas.

Cheap land at the suburban fringe is yet another incentive because it allows developers to build residential subdivisions at relatively low cost. This land is often cheap because it is agricultural land and the farm industry in the United States is in decline in many areas(**Figure 12-14**). This is especially true of the small family-owned farms surrounding many urban areas. The expansion of large-scale corporate farming has made it very difficult for small farms to compete. As a result, it is usually much more profitable for individuals to sell farmland near cities for development into residential and commercial subdivisions than to farm the land. Thus,

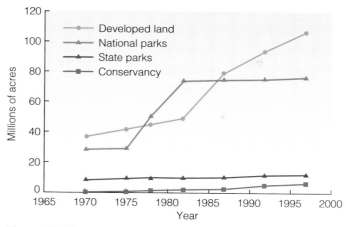

Figure 12-13   In the United States, acreage being developed is increasing much faster than acreage being preserved as public (parks) or private land (conservancy). (*Source*: Data from U.S. Census Bureau , 2001.)

many homeowners move to the suburbs because the housing is most affordable there.

Harmful Consequences of Sprawl

Sprawl has many costs. These include economic, environmental, and human health costs. Starting with economic costs, sprawl is the most expensive way for humans to live because it requires a widespread infrastructure of roads, sewers, telephone and television cables, police and fire units, and many other services. It is much more costly per person to provide these services to individual residences than to a few communities (such as large apartment buildings) that house many people.

Figure 12-14   There are no easy answers to the question of which is more desirable: affordable homes for a growing population or farmland to feed the growing population?

Sprawl is also the most harmful way to live, in terms of harmful impact per person on the environment. By widely dispersing human habitation on the landscape, the effects of each person in producing air and water pollution and consuming resources (such as gasoline and wildlife habitat) are greatly enhanced. For example, a single gallon of gasoline produces over 20 pounds of carbon dioxide plus considerable amounts of smog and other harmful pollutants. Yet 80% of the automobiles on the road have only one person in them, most of them commuting to work, to a store or to some other local chore that can only be accomplished in a car because walking is not feasible. Similarly, water pollution is greatly increased by the extensive pavement and concrete that much sprawl necessitates: At least 50% of the surface area of the average suburb or commercial center is covered with buildings and pavement. In fact, 5% of the surface area of the lower 48 states is covered with pavement and concrete. Such "sealed surfaces" do not permit the infiltration of rainfall into the ground; thus, surface water runoff carries increasing amounts of litter, oil, and many chemicals into local streams. This also slows the recharge of groundwater supplies and increases flooding. Extensive areas of roads and concrete are the most devastating of all human activities to both plants and wildlife, producing what is essentially a "biological desert" in which nothing grows. If ugly landscapes can be considered visual pollution, then we should note the ugliness created by traffic, strip malls, and monotonous housing developments caused by sprawl.

Sprawl increases the consumption of resources per person in many ways. There is one car per person in the United States. The extensive driving of these cars consumes many billions of gallons of gasoline. The average American adult drives over 200 miles per week. Nationally, Americans drive over 1.5 trillion miles per year, which is equal to nearly 15,000 trips to the Sun at 93 million miles away.

Sprawl also takes a very heavy toll on human health. Many thousands of people in the United States die prematurely from illnesses caused by smog that originates mainly from automobiles (Chapter 16). Water pollution arising from sprawl contains sewage and many other harmful substances that take a human toll in waters used for drinking and recreation (Chapter 15). Perhaps a less obvious health cost is the lack of exercise produced by spending so many hours in a car. The average American spends nearly an hour per day in a car, which adds up to nearly three years for a person who has a lifespan of 70 or so. Instead of walking to the store or the neighbor's house, suburbanites drive everywhere and as a result get out of shape. In 2002, the American Medical Association issued a report stating that 60% of Americans are overweight and many Americans suffer from heart disease and other health problems; the main cause is apparently a lack of exercise produced by their suburban lifestyle. Finally, there are mental health costs as well. Many studies done by social scientists have concluded that the typical suburban lifestyle produces considerable social isolation that leads to a "loss of community." Unlike true villages of past cultures, modern neighbors have little need for cooperative interactions and they very often do not even know the names of people living 100 feet away. These studies also show that the resulting social isolation generally has harmful effects on both mental and physical health.

Curbing Sprawl

The best solution to suburban sprawl is compact development. Instead of dispersing human habitation across the landscape, **compact development** minimizes the amount of space occupied by setting boundaries on city growth (**Figure 12-15**). You might think that compact development would lead to a lower quality of life for urban dwellers. However, many studies by urban planners show that there is a vast amount of inefficiently used land within most cities. By de-

Dispersed development

Compact development

Commuter rail transit   Light rail transit   HOV or bus lane   Major highways and freeways

Urban center   1990 Urban area   2020 Urban area

Figure 12-15 Compact development is a more efficient way to accommodate population growth because it encourages development that builds "up" instead of "out," as seen in dispersed development. (HOV = High Occupancy Vehicle)

signing dwellings, transportation, and other human needs to more efficiently use this land, compact development can accommodate much population growth at a high living standard.

There are at least four ways that cities can increase the efficiency of land use to meet the needs of a growing population. Perhaps the most obvious is the use of multifamily housing. Instead of single-family homes on individual land parcels, many more people can be housed per acre in condominiums, apartment buildings, and other "highrises." A second way is the use of **clustered housing,** where developers cluster houses fairly close together, but leave relatively large amounts of open or undeveloped land that can be used for a social commons, park, and wildlife habitat. Clustered housing contrasts with the standard approach of placing houses on uniformly separated land parcels, with very little land left over for recreation, wildlife, or other nonresidential uses. A third way is called **multiple-use development,** where the same piece of land is simultaneously used for different purposes. Some cities, for example, now use riverside land for recreation, flood buffers, water filtration by plants, and wildlife habitat. Some recreational city parks have been designed with native plants and natural features to provide much better wildlife habitat than the typical urban park designed only for humans.

A fourth way of more efficient land use is **sequential-use development,** where land that is no longer needed in some capacity is utilized in another way. For example, old landfills have been converted into parks, residential neighborhoods, and shopping areas. In Knoxville, Tennessee, a local nature center is transforming an old mining quarry into a natural landscape to be used as wildlife habitat and also for recreation (see Case Study 12-2). A very important kind of sequential-use development is **brownfield development.** A "brownfield" refers to old, run-down parts of a town, such as an old factory or downtown area with empty stores. Brownfield development thus refers to improving such abandoned parts of a town by renovating them into new industries, stores, or multifamily housing. For instance, abandoned stores in many downtown areas of the United States are being modernized into comfortable condominiums that are bought by professionals who want to avoid commuting from the suburbs.

Even in regions where a city must expand into adjacent natural lands, there are ways to minimize the negative impacts of land use on the environment. One very important approach is to use **natural siting criteria** when determining where buildings, roads, and other structures will be constructed. This means that the location (siting) of a structure is selected to be where the geological and biological factors are most favorable. If you have a choice between building a new mall, for example, on the side of a mountain that contains many rare species versus building the mall in a flat area with only widespread common species, you make the latter choice as much less environmentally harmful. This is sometimes called the "working with nature" approach to designing and engineering human settlements. Unfortunately, this approach is very rare, especially in the United States. Many studies done by urban planners show that human cultural needs, such as real estate values and commuting time, are the overwhelming criteria used to determine where a structure will be located. In the vast majority of the cases analyzed, geological and biological criteria played no role at all in determining where structures were built. As a result, erosion, landslides, and flooding have increased, unique and natural vistas are lost, and many species are removed from the area.

Another way to minimize the impact of sprawl is sometimes called **wildscaping** (a contraction of "wilderness" and "landscaping"), referring to the retention of native vegetation, soil, and other natural features of the land when building on it. Most construction projects begin with the complete removal, using bulldozers, of the soil and native vegetation. A main reason for this is economic: The presence of vegetation, especially trees, impedes the extensive movement of people and machines during construction. When construction is finished, the area is then replanted, but usually with ornamental plants that are not native to the area and thereby are lower quality wildlife habitat. Wildscaping is therefore usually more expensive because equipment use is restricted and special care is taken during construction of the houses or buildings. However, a growing number of cities, homeowners, and even golf courses are using wildscaping techniques to preserve natural habitat and reduce environmental impacts where construction is deemed necessary (see Case Study 12-3).

A final way of minimizing sprawl impact is **mitigation,** which is the replacement of a destroyed ecosystem with a new one that is created by humans. If a developer is permitted, for example, to pave over five acres of wetland habitat, then the developer is usually required by state and federal law to replace those five acres by creating or restoring a healthy wetland somewhere nearby. In fact, most laws require a 2 or 3 to 1 ratio so that 2 or 3 acres must be created to replace each one that is destroyed. While this sounds like a good compromise between social and ecological needs, many environmental-

CASE STUDY 12-2

The Restoration of Mead Quarry

Visit www.environment.jbpub.com for more information.

The restoration of Mead Quarry by Ijam's Nature Center of Knoxville, Tennessee, is a good example of sequential land use. Mead Quarry was first established as a marble quarry in the early 1880s. It was then sold to Williams Limestone Company in 1945 and operated as a limestone quarry until the company filed for bankruptcy and abandoned the property in the mid-1970s. The site has played a significant role in Knoxville's history since it was established, and throughout the years, marble cut and polished at the quarry has been used in the construction of several notable buildings in downtown Knoxville, as well as other buildings and monuments in Maine, New York, Ohio, and Indiana. Since the mid-1970s, the site has been used as a dumping ground for everything from stolen cars to household garbage to toxic chemicals and worse.

In 2001, Knox County acquired the 50-acre Mead's Quarry property that is located adjacent to Ijam's. The property, which has been subsequently leased to Ijam's on a long-term basis, has provided an unprecedented opportunity for the preservation of spectacular natural beauty in South Knoxville. Alongside Ijam's existing 100-acre park, the site's natural high bluffs, a 25-acre lake, an abundance of wildlife, and noteworthy geological features combine to provide unique recreational and educational opportunities for the local community.

Ijam's Nature Center has a longstanding history of environmental stewardship within the local community through considerable wetland protection and conservation initiatives. Through strong partnerships with other local organizations, they are able to make a difference in the quality of life of community residents. The restoration of the Mead's Quarry property will continue this trend by preserving stunning green space within only three miles of downtown Knoxville.

Restoration Plans

In order to restore the property to a state of natural beauty for the enjoyment of the local community, the site will undergo significant cleanup activities to remove all toxic waste and other dangerous materials. Ijam's will coordinate a series of volunteer-led workdays to remove lightweight debris by hand. Larger piles of industrial waste that cannot be cleared by the volunteers, including potentially dangerous materials such as asbestos or glass, will be removed by industrial frontloading equipment.

In addition to the removal of discarded waste, the major components of the restoration plan include the following:

Vegetation removal and native plantings. Ijam's education staff will undertake conservation activities to restore the environmentally damaged area and strengthen the wildlife habitat already in existence. The focus will be to increase biodiversity through exotic plant removal and native planting and to improve water quality through protection of riparian zones. The planting of forage and grasses will stabilize cleanup sites and excavation. Native trees and shrubs will be planted from Ijam's stock to restore the site. Habitat improvements will also include ground leveling and refilling of disturbed areas. Landscape plantings in the lime-rich soils will greatly beautify and soften the site. Selective reforestation will be undertaken in appropriate areas and nest boxes will be installed for bird populations.

Trail Construction. The first proposed hiking trail will be 0.5 mile in length. Commencing at the main entranceway, the trail will follow an existing roadbed along the southern perimeter of the property and ascend up a gradual incline at the west end of the lake. The trail will incorporate a natural overlook affording spectacular views of the property. Expansion of the trail to complete a circular walkway around the whole site will be undertaken at a later date.

Figure 1 The property has already shown itself to be a source of neighborhood and community pride. Since the acquisition of the property, there has been a groundswell of support from within the community to assist with a volunteer-led cleanup operation.

ists and ecologists are critical of mitigation because it is very difficult to "create" an ecosystem that will function as well and persist as long as a natural ecosystem.

Legal and Economic Incentives to Curb Sprawl. The traditional method of regulating urban development has been zoning regulations. These regulations specify the type of development, such as commercial or residential, that is allowed in an area. Zoning regulations are typically made by county, and sometimes city, governments based on the needs of the community. In the past, most zoning has not been aimed at restricting suburban growth. Rather, it has focused on keeping areas of commercial, residential, and agricultural uses from becoming intermixed. It would be possible, however, to zone certain areas for "non-development," where the land would be managed as wildlife habitat and light recreation. A common example of this is **greenbelt zones** that prohibit any removal of native vegetation around a city (**Figure 12-16**). These zones are especially important in cities surrounded by hills or mountains, such as Boulder, Colorado, because the vegetated mountains are such an important part of the scenic view visible from throughout the city. Currently, the use of zoning to regulate sprawl varies greatly among communities, so there is growing interest in producing state laws that provide a coherent set of standards. Some states, such as Tennessee, have already passed such laws, requiring cities to designate boundaries that limit their spread. As with all laws, however, it remains to be seen how effectively these state urban growth boundary laws are enforced.

As more cities struggle to control sprawl, they have devised a variety of economic incentives to slow its growth. Probably the most generally effective method is a gasoline tax. Increasing gasoline taxes makes long commutes less attractive, so there is more demand for less dispersed housing and commercial space. These taxes also cause drivers to take fewer unnecessary trips, buy cars that are more fuel-efficient (and less polluting), and encourage commuters to car pool and use mass transit. Perhaps more important is that the funds collected by increased gas taxes can be used by the city to improve mass transit systems, for brownfield development, and to buy land (or the "development rights" to that land) that can be used as parks and other undeveloped spaces. Even without an increase in gasoline taxes, a current cause of sprawl is that much of the money now collected as gas taxes is automatically transferred to state and federal departments of transportation. As a result, these departments have an enormous amount of funding available for build-

ing roads and very little incentive to promote alternative transportation. In most states, education and most other state agencies must compete for dwindling funds every year, while the transportation agencies have almost sole access to the vast and increasing funds created by gas taxes. A key solution therefore is to see that more gas tax money is distributed for purposes besides road building.

Another option is to increase taxes on developed land to collect funds for brownfield development and buying undeveloped land. At the same time, the government could reduce taxes on land that is farmed or otherwise not used for development. A final option is the **developer pays** approach, where cities require the developer of a residential subdivision to pay for the costly infrastructure of roads, sewer lines, utility lines, and other services that are often paid by taxpayers. Such taxpayer subsidies only encourage the growth of sprawl by making it relatively cheap for the developer to build housing and acquire very large profits in the process. If the developer is required to pay for these, then there will be less incentive to build large subdivisions, and more incentive to spend development money on compact development, brownfield development, and other ways of more efficiently using land that is already available. Under the developer pays approach, housing in new suburban subdivisions will be more expensive, because it will not be subsidized

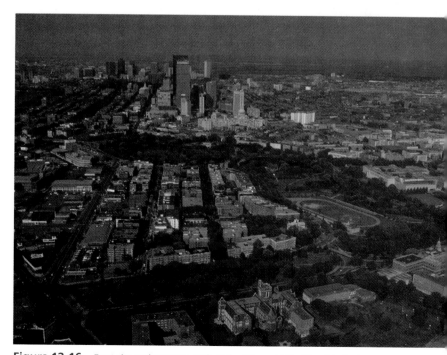

Figure 12-16 Even city park systems, such as the "Emerald Necklace" in Boston, Massachusetts, can serve as a greenbelt within a city, prohibiting development and greatly increasing the visible beauty of the city.

CASE STUDY 12-3

Should We Rethink Lawns and Golf Courses?

Lawns and golf courses are excellent examples of how small-scale activities that we take for granted can accumulate to cause widespread environmental harm. Both lawns and golf courses have become traditions in the United States and other developed countries, ironically, because of our desire to enjoy nature (**Figure 1**). Yet, as so often happens, traditional activities can become environmentally damaging if they are carried out on a huge scale by many individuals.

If all the lawns in the United States were assembled together, they would cover an area of 25 million acres (10 million hectares), which is equivalent to the combined area of Vermont, New Hampshire, Massachusetts, Connecticut, Rhode Island, and Delaware. This makes lawn grass the largest single crop in the United States, in terms of area. We care so much for our lawns that we spend $25 billion on the lawn care industry each year.

Many psychologists believe that lawns, and our care for them, are a way to feel close to nature. The smell and images of greenery are pleasant to many of us. Unfortunately, this particular way of staying close to nature is actually very "unenvironmental." Suburban lawns account for much of household water use, which is a serious problem in water-poor areas such as the southwestern United States and especially southern California. Furthermore, lawns average more than *three times* more pesticides per acre than an equivalent area of cropland. U.S. lawns absorb more synthetic fertilizers than the entire nation of India applies to all its food crops. Finally, mowing the lawn for an hour with a gas-powered mower can produce as much air pollution as driving a car for 340 miles (544 km).

What can we do? Growing a variety of grass types in good topsoil and keeping the grass high discourages weeds and reduces the need for pesticides and herbicides. Many air quality management districts have sponsored buy-back events in which people exchanged their gas-powered mowers for a rebate coupon, typically worth $100.00, good toward the purchase of an electric mower from a specific manufacturer. The EPA is phasing in stricter emission standards for new models. F. Herbert Bormann edited a book called *Redesigning the American Lawn: A Search for Environmental*

Figure 1 Unless carefully managed, golf courses can use more pesticides and fertilizers, and cause more water pollution, per acre than most farms.

by taxpayers; however, this is another example of how consumers may be obliged to pay the "true environmental costs" of items for the free market system to promote sustainability (Chapters 6 and 19).

A final important option to control sprawl is by the actions of private preservation groups, espe-

cially the land trusts and conservancies noted above. These groups do a great public service by preserving land before it is developed. However, it is often too costly to own preserved land outright because of maintenance costs. Therefore, a more common preservation tool is the use of **conservation ease-**

Harmony (1983). This book's answer is that if we wish to enjoy nature, we should stop trying so hard to cultivate the grasses and other plants that are not native to our backyards. The huge investment in water, pesticides, fertilizer, and mowing in the typical suburban lawn is only needed because we attempt to grow non-native plants. Instead, it is suggested that we step back and let some of the native plants, such as crabgrass, chickweed, and many others, become established. We call such plants "weeds," but they are really just native plants that recur because they are better suited to the environment. Bormann and his Yale colleagues suggested that a totally environmental "Freedom Lawn" would consist entirely of native plants, require less care, and still provide the same feeling, if not more, of being close to nature.

On a per acre basis, golf courses are even more harmful than lawns. Golf courses can use nearly seven times as much pesticide per acre as cropland, twice as much as the average for lawns. A single golf course can cover 200 acres (81 hectares) and usually requires even more fertilizer per acre than a lawn. The United States already has nearly 18,000 golf courses occupying an area larger than the state of Delaware, and furthermore, the rate of new golf course construction is increasing. From 1995 to 2000, an average of 340 courses were built or older courses expanded each year. This is more than twice the rate of the mid-1980s. Florida, California, and Michigan have the most courses, with over 1000 each.

All these pesticides have had visible effects. Bird kills related to golf course pesticides have been reported: In some cases, hundreds of birds died. Human health effects could be very serious. A 1993 study by the University of Iowa Medical School examined data on the cause of death of 618 golf course superintendents between 1970 and 1992. It showed that the superintendents had exceptionally high frequencies of cancers of the lung, brain, intestine, prostate, and some lymphomas. These are often related to pesticides, and the study recommended more restricted use of pesticides and better protective clothing for golf course workers.

What can be done? As with lawns, there are more "environmentally friendly" ways to design golf courses. For new and existing courses, environmental planning, wildlife and habitat management, chemical use reduction and safety, water conservation, and water quality management by course superintendents can help courses serve as ecologically valuable green spaces.

Questions

1. Do you think lawns and golf courses should be redesigned? Perhaps you are like many Sierra Club members: One in every six members is a regular golf player. Some are not concerned, and others have had to rethink their hobby.
2. What about lawns? If you owned a home, would you have a traditional yard and mow the lawn on weekends?
3. What are some alternatives to green lawns? How would you redesign golf courses to be less environmentally harmful?

ments, which are legal agreements with a landowner where the development rights to that land are surrendered. In other cases, these rights are leased by the preservation group.

Such land trust agreements are often complex, involving expiration dates when development rights may be repurchased and problems when the owner sells the land. However, land trusts will be a fast-growing and very important method of preserving land for many decades to come.

STUDY GUIDE

SUMMARY

- Management of public lands must balance the needs of people versus the need to maintain healthy ecosystems for future generations.
- Public land management in the United States varies with the goal of the government agencies involved.
- There is a continuing debate over how many tax dollars should be spent on subsidizing resource extraction of public lands.
- A major tool for protecting public lands is the Environmental Impact Statement (EIS), which requires that planned activities be assessed for their impacts.
- Management of private lands presents both a challenge and an opportunity for conservation.
- Unlike the past, when preserved land was selected for its scenic beauty or low commercial value, future parks should include biodiversity preservation as a selection criterion.
- Preserves should be designed to have buffer zones, connecting corridors, maximum size, and other variables that promote biodiversity preservation.
- Ecotourism and sustainable harvesting are good examples of how preserves can be maintained for sustainability while providing income for local human inhabitants.
- Suburban sprawl is a rapidly growing problem in many parts of the world, especially the United States.
- Causes of suburban sprawl include: cheap land prices at the urban fringe, a strong reliance on the automobile, and taxes and economic incentives that promote suburban development.
- Harmful impacts of sprawl include air and water pollution, habitat loss for wildlife, and human health costs from lack of exercise and social isolation.
- Compact development reduces sprawl and can be promoted by: clustered housing, multiple-use development, sequential-use development, and brownfield development.
- Ways to minimize sprawl impacts include: using natural siting, mitigation, and wildscaping.
- Incentives to control sprawl include: greenbelt and other zoning, lower taxes on farmland, buying development rights, the "developer pays" approach, and conservation easements.

KEY TERMS

Bioprospecting
Bottom-up approach
Brownfield development
Buffer zones
Clustered housing
Compact development
Conservation easements
Developer pays
Ecotourism
Environmental Impact Statement (EIS)

Genetic patent rights
Greenbelt zones
Hot spots
Land Trust Alliance
Marine protected areas
Megalopolis
Mitigation
Multiple-use development
National Environmental Policy Act (NEPA)
National Marine Sanctuary

Natural siting criteria
Nature Conservancy
Scoping
Selective cutting
Sequential use development
Suburban sprawl
Sustainable harvesting
Wilderness Act of 1964
Wildlands Project
Wildscaping

STUDY QUESTIONS

1. What are some controversies over the use of public land?
2. What is the bottom-up approach? How are buffer zones, ecotourism, and sustainable harvesting related to this?
3. What is an EIS? What does it do and what are some problems with it?
4. How are genetic patent rights important in promoting biodiversity conservation? Give specific examples. What is bio-prospecting?
5. What is a hot spot? How can hot spots be used in selecting the location of preserves?
6. What are two widely suggested solutions to the national park funding problem?
7. Compare and contrast selective cutting and clear-cutting.
8. Discuss the problems of the U.S. Forest Service and those of the National Park Service. What differences and similarities do you see in the problems of the two agencies?
9. Name four ways that land can be more efficiently used to reduce sprawl.
10. What is ecotourism? Discuss its advantages and disadvantages in saving species.
11. What is wildscaping? What is mitigation? How are they useful in coping with sprawl?
12. What is a conservation easement? Why is it often better than buying land to preserve?
13. Give some examples of sustainable harvesting.
14. Name some ways that traditional forestry can harm the forest ecosystem.
15. What is the Wildlands Project? Is it realistic? Why or why not?

WHAT'S THE EVIDENCE?

1. The authors make several assertions that sub-urban sprawl is bad for human health and the environment. What is the evidence to back this up?
2. The authors suggest that abuse of federal public land is sometimes subsidized by tax-payers. What is the evidence for this? Is this evidence objective?

CALCULATIONS

1. By what percentage did the amount of designated wilderness land increase between 1980 and 2000? (Refer to the in-text table on page 285.)
2. If an average American drives 200 miles per week, how many miles will that person drive in a year? In 10 years? In a driving lifetime of 50 years?

WWW.ENVIRONMENT.JBPUB.COM

www.environment.jbpub.com

Connect to this book's web site: http://www.environment.jbpub.com/ The site features eLearning, an online review area that provides quizzes, chapter outlines, and other tools to help you study for your class. We've posted the answers to the book's odd-numbered questions. You can also follow useful links for more in-depth information. The site provides sources for the information in this chapter and recommendations for further readings.

Harvesting rice in India. With approximately a billion chronically undernourished people in the world today, world hunger is an enormous problem. Most of the hungry live in the developing countries, and without assistance, they have little hope of improving their situation. But assistance does not always need to mean massive inputs of relief supplies or monetary funds; assistance can take the form of small-scale one-to-one exchange of information and sustainable strategies.

CHAPTER 13

Feeding the World

We are part of the global biosphere, and as such we are ultimately dependent on other living organisms for virtually all aspects of our lives. Many of the ways we depend on other organisms are obvious. Trees shade us; green plants and other photosynthetic organisms produce the oxygen (O_2) that we breathe and absorb carbon dioxide; worms and microorganisms prepare the soil for us and recycle necessary nutrients; insects pollinate our crops. We do not fully understand all of the complex interrelationships among bacterial and other microorganismal, plant, fungus, and animal species that make up the biosphere, and we have no way of knowing what roles certain seemingly insignificant species perform now or will play in the future.

Modern industry and medical technology are dependent on animals and plants for a plethora of important drugs and other substances. All of our fossil fuels were formed by organisms over millions of years, as were many commercially important rock and ore deposits (such as limestone and phosphate deposits). Biological resources provide a large percentage of the raw materials that we utilize (such as wood, natural rubber, and leather).

Perhaps most importantly, we are totally dependent on other organisms as a source of nutrition (see the diagram at the beginning of Section 3, p. 141). We raise plants and animals to eat; we harvest wild plants and animals as food (for example, fishes). In the present chapter, we focus on this most fundamental use of biological resources by humans. Since the majority of the world's food needs are met today by the cultivation of crops, we include a discussion of the soil resources that form the basis of **agriculture**.

Food as a Biological Resource

Virtually all of the food that humankind depends on is derived from other organisms. Although we eat many different types of plants and animals, actually only a very small number of species provide the vast majority of our foodstuffs. Only 20 different species of plants supply 80% of the world's food supply, and just three kinds of plants constitute 65% of the food supply—namely rice, wheat, and maize (corn). Global **grain production** currently (data for 2001) stands at about 1.843 billion metric tons per year. Likewise, animal protein (meat production) is dominated by only a few terrestrial species: poultry, sheep and goats, pigs, and cattle. Globally, about 237 million metric tons of such animal meats are harvested each year. Also extremely important, at approximately 117 million metric tons per year globally, is the fish and aquatic organism catch (including 36 million metric tons of aquaculture production—farmed fishes).

Hunger

Terms such as **hunger** and *malnutrition* can be hard to define, although their meaning can be painfully obvious when one sees a starving human. To maintain good health, a person must have adequate nutrition—proper amounts of protein and various vitamins and minerals—and an adequate supply of kilocalories as an energy source (in popular usage, kilocalories are often referred to simply as "calories," but a kilocalorie or Calorie is actually equal to 1000 calories). The amount of kilocalories and nutrients individuals require also depends on their age, sex, and other characteristics. Statistics of global hunger are often based on the number of kilocalories available or ingested per person per day. According to the United Nations, the recommended daily intake per person is 2350 kilocalories. The U.S. National Research Council recommends a kilocaloric intake of 2700 and 2000 for the average adult male and the average adult female, respectively.

From a simple kilocaloric point of view, many people do not consume enough to live and work actively. It is estimated that as many as 1.1 billion people are undernourished and underweight. In some countries, the average daily kilocaloric intake is less than required for people to carry out productive work; many of these nations are in sub-Saharan Africa and Asia. To give a couple of extreme examples, based on 1997 statistics (the latest currently available), in North Korea the average daily per capita kilocalorie supply was only 1837, in Afghanistan it was 1745, and in Somalia it was a mere 1566. The last number was well less than half of the 3699 kilocalories per person recorded in the United States for the same time period. Studies have demonstrated that in many countries, malnutrition is correlated with high death rates, particularly among children. Children suffering from malnutrition are characterized by stunted growth, reduced mental functions and learning capacities, and lowered general activity levels. Damage caused by severe malnutrition may be irreversible. One study found that increasing the per capita consumption of food from slightly under 2000 kilocalories to 2700–3200 kilocalories was associated with an approximately 50% decrease in the death rate (**Figure 13-1**).

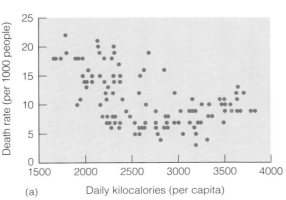

Figure 13-1   (a) The relationship between the daily energy supply per capita in 1989 and the crude death rate per 1000 people in 1990 for 119 countries. Note that both too little and too much food, as measured using kilocalories, results in an increased death rate.
(b) Severely undernourished children are more likely to die from childhood illnesses. Here a relief worker treats a child suffering from starvation. (*Source*: (a) From H. R. Pullian and N. M. Haddad, "Human Population Growth and the Carrying Capacity Concept." In *Bulletin of the Ecological Society of America* [September 1994], p. 150.)

(b)

Objectives

After reading this chapter, you should be able to explain or describe:

What constitutes the major categories of human food sources on a global scale

The concept of hunger

Agricultural food production, both at present and projected into the future

The differences between "traditional" and "modern" agriculture (including the "Green Revolution")

Genetically modified food

Aquaculture

The nature and importance of soil

Besides an adequate intake of kilocalories, a healthy person requires proper amounts of vitamins and minerals. Vitamin and mineral deficiencies can cause many symptoms, including general poor health; blindness (vitamin A deficiencies are found in 100 to 140 million children globally, and cause blindness in up to 500,000 preschool children annually); mental retardation; learning disabilities; decreased work capacity; decreased resistance to illnesses, diseases, and infections; and premature death. Over a billion people, primarily in developing countries, currently suffer from vitamin and mineral deficiencies, while another billion are at risk.

Feeding the World Today

By current estimates, 800 million people do not eat enough every day for a normal, healthy life; at least 500 million people are chronically hungry, including about 160 million children five years old or younger. Approximately 8 to 11 million people die each year due to hunger or hunger-related causes (about 24,000 to 31,000 a day), and many more people are weakened by hunger and therefore succumb to other factors (such as disease) more easily. Even

Figure 13-2 Canadian soldiers with United Nations aid trucks carrying food are stopped by Bosnian Serb forces near Srebrenica, Bosnia-Herzegovina, April 1993.

in the United States it is estimated that some 20 to 30 million people (almost half of them children) may be suffering from chronic malnutrition. Why are so many people in need of food, especially when some countries have grain surpluses? In part, the large inequities in food distribution are related to political and social problems; all too often, food is withheld as a weapon (**Figure 13-2**), especially in internal struggles in developing nations. But political considerations are only part of the problem.

In fact, perfect management of the world's current food production might just barely feed the global population, but this would be only a temporary measure, for soon the global population will outrun current food supplies. Per capita grain production actually peaked in 1984 at 342 kilograms per person, and since then, the population has grown faster than grain production such that in 1997 it was only 322 kilograms per person; lower global grain production and more people to feed brought this number down to 299 kilograms in 2001. Based on the peak total grain harvest of 1997 (not the peak per capita grain harvest), researchers have calculated the percentage of the world's population that could be fed using various diet models. In the United States, the typical citizen consumes about 25% to 30% of his or her kilocalories from animal products, such as meat and cheese—a notoriously inefficient way of deriving energy from foodstuffs. If the entire world followed the American dietary example, less than half of the present global population could be fed adequately. Typical Latin Americans consume about 10% of their kilocalories from animal sources. Using the Latin American diet as a model, about 4 billion people could be fed based on the record harvest of 1997. Only with everyone maintaining a strictly vegetarian diet, and assuming perfect food distribution systems (which is unrealistic given current political and transportation problems) could the current population of over 6 billion be adequately fed. If we could considerably decrease the waste factor (for example, up to 40% of all food typically spoils or is eaten by insects, rats, or other pests, and much food is thrown away as "leftovers," made into food for nonessential house pets, and so forth), perhaps 8 billion people worldwide could be fed a subsistence diet.

But such measures would be of little consequence to the billions of extra mouths that will need to be fed by the end of the twenty-first century—especially if the world's population reaches 8 to 10 billion. There seem to be few options for dealing with this situation. One possibility is to increase global food production dramatically and ensure that the food is equitably distributed. But whether major gains in global food production will be possible is a point of heated debate. Some analysts argue that innovative agricultural techniques, combined with advances in biotechnology (see Case Study 13-1, page 318), will allow us to significantly increase food production. Other researchers believe that global food production has already peaked (see the following discussion) and that even maintaining current levels will be difficult. With few realistic prospects for significant increases in world agricultural output, some analysts argue that the only way out of the predicament is to reduce world population growth significantly (see Chapter 2).

Food for the Future

If the world population continues to grow, we will need more food in the future. World food production could be increased through two basic, but certainly not mutually exclusive, strategies: Increase the amount of land under cultivation, or increase the yield per unit of land under cultivation. Today slightly over 1.5 billion hectares of land worldwide are under some form of cultivation—an average of 0.25 hectare per person. According to various theoretical estimates, there are between 2 and 4 billion hectares of **cultivable land** on Earth (much of the variation in the estimates is due to the use of different criteria for defining "cultivable land"—how fertile the soil must be, how much water must be available, and so on).

Differing scenarios involving various combinations of land cultivation and yields on a global scale have been proposed (see **Figure 13-3**). If the population increases at an exponential rate of the 1990s and present average yields continue (that is, they do not increase), then before the year 2050, all of the theoretical 4 billion hectares of cultivable land will have to be put into production, and by the end of the twenty-first century, the population will quickly outstrip its food supply. This scenario assumes, unrealistically, that the present high crop yields of prime land can be maintained, even on marginally arable land. However, it also assumes a higher population growth rate than is currently expected for the first half of this century. It also assumes that no cultivable land is lost during the next hundred years, which is also unrealistic because, as a later section explains, the absolute amount of cultivable land on Earth is declining due to various factors.

If the population increases according to various World Bank projections, which are probably more realistic at this point, we can sustain the Earth's population through the twenty-first century within the range of estimates of land available for cultivation. However, unless crop yields can be significantly increased or food losses and wastage significantly de-

creased, an additional billion or more hectares may have to be placed under cultivation by the end of the century in order to feed all the globe's people. Such a dramatic increase in cropland will severely strain natural ecosystems and may be quite difficult pragmatically and politically.

Some optimistic studies have concluded that if we really wanted to, we could grow enough food to support a global population of 50 billion people. Such conclusions, however, are based on totally unrealistic assumptions. For one thing, they assume that all potentially arable land would be cultivated, including land occupied by forests and land that is of marginal fertility or is so arid that massive irrigation would be necessary. The human population would have to live in areas, such as the polar regions, where agriculture is totally impossible, while the potentially arable land beneath our current cities and towns would be put under the plow. Furthermore, these studies assume the yields on all this land would ei-

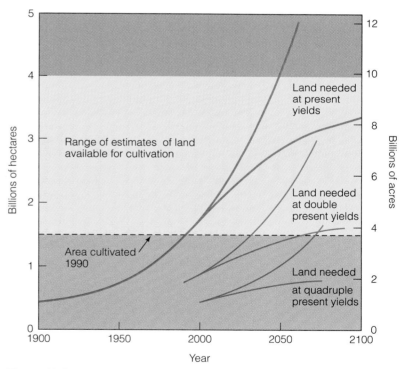

Figure 13-3   Possible land futures. The pairs of lines show the amount of land needed to maintain present per capita food production if the world population continues to grow exponentially at current rates (upper line of each pair) or if the world population grows according to early 1990s World Bank forecasts (lower line of each pair). The heavy solid pair of lines shows land requirements at present crop yields; the two lighter pairs of lines show land requirements at double and quadruple present crop yields. The yellow shaded area is the range of estimates of land that could possibly be cultivated for food; note that much of this land is currently covered with forests. (*Source:* D. H. Meadows, D. L. Meadows, and J. Randers, *Beyond the Limits,* [Post Mills, VT: Chelsea Green, 1992], p. 51. Reprinted from *Beyond the Limits,* copyright © by Meadows, Meadows, and Randers. With permission from Chelsea Green Publishing Co., White River Junction, Vermont.)

ther match or (with technoagricultural advances) surpass those that have been attained under ideal conditions on the most fertile land in the past. Such super high yields are only a pipe dream at present. Finally, these projections ignore the detrimental and nonsustainable aspects of modern agriculture (discussed in a later section) and the consequences (climatic and otherwise) of destroying the world's remaining forests. The question of where the massive quantities of energy, **fertilizers, pesticides, herbicides,** and freshwater necessary for modern, intensive, high-yield agriculture will come from is not addressed, nor are the attendant problems of pollution caused by chemical use, topsoil loss and exhaustion, soil salinization, and waterlogging addressed. To suggest that we could feed close to 50 billion people is irresponsible—indeed, it is a lie.

A more conservative estimate is that we could potentially feed about 8 billion people, but even this estimate assumes better yields than occur today in many agricultural areas. Although some increase in yields may be possible (the best farmers in Iowa, for instance, can produce four times the world's average corn yield per acre), in recent years global yields have shown signs of leveling off, and we can-

not necessarily count on further large increases in the future.

Taking all these factors into account, some experts insist that we are just about at the limit of the number of people we can realistically expect to feed. And they note that even now, with 15% or more of the world's population underfed, we do not do a very good job of feeding a mere 6.3 billion—and furthermore, from 1997 to 2001, per capita grain production has declined. Still, it must be pointed out that this is a very controversial area of research. New crop varieties continue to be developed, and new methods of farming and more efficient equipment have yet to be used around the world. Even if crop yields have leveled off in the last few years, that does not necessarily mean they will remain at these levels for all time. They may begin to increase once again as successful new cultivation techniques spread around the world.

Agricultural Food Production and Supplies

Grain (rice, wheat, maize, sorghum, barley, oats, rye, millet, and so on) forms the backbone of the world's food supply, and thus global annual grain production takes on an extreme importance. Total world grain production nearly tripled between 1950 and 1997 (**Figure 13-4**), in large part because of the "Green Revolution," which is discussed later.

Not all grain grown is used for human consumption. Since 1960, about a third to over two-fifths of the world's grain supply each year has been used to feed livestock and poultry. To raise livestock and poultry on grains, farmers must routinely include a protein supplement in the animals' diet. The most important such supplement is the protein-rich soybean, which is also a valuable source of oil and protein for humans.

It is far more efficient to consume grains and other plants directly rather than to feed them to animals and then consume the animals. As the world population continues to increase and grain production remains stable or declines, it may become necessary to allocate less grain to animals if we want to ensure that all humans are fed adequately.

Not all grain produced in a given year is necessarily consumed in that year. An important concept is that of **carryover grain stocks** from one year to the next. By the best estimates, each year from 1996 through 1998, some 29 to 43 million metric tons of grain were produced globally over what was actually consumed. However, in 1993 through 1995, consumption *exceeded* production by 13 to 49 million metric tons a year. More recently, consumption exceeded production by 2 million metric tons in 1999, by 34 million metric tons in 2000, and by

Figure 13-4   (a) World grain production, 1950–2001; and (b) world grain production per person, 1950–2001. (*Source*: United States Department of Agriculture.) The background images show wheat and barley harvesting in Palouse, Washington.

an astounding 54 million metric tons in 2001. Several such years have the effect of depleting the world's carryover grain stocks. The size of the world's grain carryover stock is often used as an indicator of global food security. Often, if the stocks drop too low, as they did previously in the early 1970s, grain prices may fluctuate widely with step increases. In such cases, it becomes even more difficult for the poor to obtain adequate food until a good harvest is restored.

Some observers point out that, historically, there has always been grain to buy—as expressed in stockpiles of grain around the world. They contend that the primary reason hunger exists in the world today is because poor nations cannot afford to buy food. The best way to ensure that all people are adequately fed, some researchers maintain, is to expand food production in temperate regions (such as the United States) where advanced agricultural technologies and transportation systems are already in place. Furthermore, there must be a global free trade policy for food. A worldwide free market system for food would discourage inefficient, often government-subsidized, food production in marginal areas. For instance, in India, overpumping of groundwater supplies using free government-provided electricity has resulted in lowered water tables, salinization, and waterlogged soils. Indonesia, in the name of "self-sufficiency," has cleared 1.5 million acres (607,000 hectares) of tropical rainforest to grow soybeans for use as chicken feed. The problem is that the cost of Indonesian soybeans is higher than the price for soybeans on the world market. Likewise, India produces milk at a cost above world market prices.

Even those who espouse a "free market solution" to world hunger must acknowledge that the problems of debt (particularly on the part of developing countries), trade imbalances, and restrictions on free market policies are complex and not easily solved. Politically, it would be very difficult to have a genuine global free market system for food products. Furthermore, even if free trade/free market policies could accomplish the equitable distribution of food around the world, there must be enough food to go around. At present, the food supply does appear to be adequate, but as we have already discussed, if the global population continues to grow, it is far from certain that there will be enough food to feed everyone in the future.

Land, Fertilizers, and Water Devoted to Agricultural Production. An enormous amount of land is devoted to agricultural use (and not all agriculture is devoted to food production—for instance, cotton and other crops are grown to manufacture textiles, and animals are raised for leather products). Globally, 3.3 billion hectares of land are used for grazing animals and slightly over 1.5 billion hectares is devoted to cropland, of which an estimated 275 million hectares are artificially irrigated to grow crops.

Taking a closer look at one of the more important agricultural resource statistics, worldwide grain area harvested (a subset of world cropland), we find that it has increased from 587 million hectares in 1950 to approximately 680 million hectares at the end of the century. The high was reached in 1981 with 732 million hectares. Every year, new land is put under cultivation as forests are cut and dry areas are irrigated, but other land is removed from cultivation because of such factors as soil exhaustion, degradation, and the building of residences and shopping malls. At present, there is a rough balance between land newly put under cultivation each year and land removed from cultivation. But some observers fear that within a few decades the cropland area may begin to diminish substantially.

More importantly, though, due to global population growth, the amount of grain area harvested per person has steadily declined from 0.23 hectare in 1950 to the present 0.11 hectare. Despite this steady decrease in per capita area devoted to grain, per capita world grain production is higher now than in 1950. This was accomplished through intensive, often mechanized, farming techniques using specially bred varieties of crops and massive doses of artificial fertilizers, pesticides, herbicides, and in some cases artificial irrigation (the "Green Revolution" is discussed further later).

In many regions, **irrigation** is essential to grow crops that would not otherwise survive. Interestingly, since about 1961, the global irrigated land area has grown at about the same rate as the world's population. In 1961, 139 million hectares were under irrigation (.0451 hectare per person), and in 1999 there were 274 million hectares of irrigated land (.0457 per person). Unfortunately, as is discussed further below, much irrigated land is watered unsustainably. Some analysts predict that the growth in irrigated land cannot continue for very long into the future.

The Effects of Agriculture. Most agriculture alters and manipulates natural ecosystems, transforming them into artificial ecosystems that are inherently unstable and can only survive with constant human attention. Maximum food production is the only goal.

The Pioneer Stage and Its Effects. In nature, during the process of **ecological succession**, a clear patch of land will be colonized by successive groups of plants and animals. The first settlers will generally be smaller, fast-growing, pioneer plants. Then larger, slower-growing, and longer-lasting plants will progressively replace the original colonists. The final stage of succession is the **climax community**, which in many terrestrial areas consists of mature forest composed of large trees interspersed with younger trees and other plants and animals.

In clearing land for agricultural use, farmers essentially begin the cycle of succession anew. But the farmer does not allow succession to follow its natural course and reach a climax stage. Instead, the land is artificially maintained at the **pioneer stage**; furthermore, the pioneer plants that are allowed to grow on the land are carefully picked, maintained, and managed. Corn (maize), wheat, or rice may be planted as the pioneer plant; other weeds are eliminated (many major food crops are essentially cultivated weeds). When the crop has matured, it is harvested; the next season the land is cleared, and the system begins again. Most agricultural systems emphasize the pioneer stage of succession because this is when an ecosystem is most productive (although not most efficient in energy use). In this stage, virtually all the energy and nutrients utilized by the plants go into growth. In contrast, in the climax stage, much of the energy and most of the nutrients go into maintaining the system; the only new growth that occurs replaces plants that die.

The long-term arrestment of ecosystems at the pioneer stage leads to major problems, however. Pioneer ecosystems are inherently unstable, and this instability is exacerbated by the human habit of planting only one variety of plant per field at a time (**monoculture**). In a mature, climax ecosystem, the complex relationships and interactions among many species of plants and animals promote long-term stability. In the climax community, there are natural checks and balances on predator-prey relations (including insect attacks on vulnerable plants), disease, population explosions of particular species, and so on. Climax communities are also less susceptible to the ravages of climatic fluctuations such as droughts or floods. In the artificial environment of a crop field, pests, disease, and the vagaries of climate must be mitigated to a greater extent by human intervention. Pests might be controlled by applying poisonous chemicals to a field. Watering or irrigation may compensate for a lack of rain.

Pioneer stages also extract a heavy toll of nutrients from the soil without replenishing them. Replenishment occurs naturally during later stages of ecological succession, but when humans harvest and remove their pioneer crops, the nutrients are lost from the land and must be restored by artificial fertilization. In contrast, the climax community is characterized by complementary, even symbiotic relationships between organisms. The nutrients extracted by one organism are eventually passed on to and restored by another organism. The cycle of growth, death, decay, and regrowth—all on the same parcel of land—ensures continued recycling of raw materials.

The characteristics of pioneer communities that make modern monoculture farming so productive and successful on a short-term basis cause continued environmental degradation in the long term. Rapid nutrient uptake (absorption) without recycling destroys the soil's fertility. Lack of a balanced vegetation—or no vegetation at all between harvesting and the next planting season—to hold the soil in place and absorb moisture can lead to massive erosion of valuable topsoil, flash floods, dust storms, and droughts. Monocultures are notoriously susceptible to attack by disease and pests uncontrolled by natural predators or other mitigating agents.

The Effects of Irrigation.   In areas where irrigation is necessary, a whole new set of problems is encountered, in particular, **salinization** and **waterlogging**. All soils contain various mineral salts. Under natural conditions, in areas that are characterized by relatively high rainfall and good drainage, these salts are washed out of the soil and travel, through water flow, to the sea. This is why the sea is salty—it is the final resting place where salts from the land surface accumulate. In contrast, arid regions tend to have higher natural concentrations of salts in the soil and in any groundwater or standing bodies of water simply because there is not a constant flow of water to remove the salt. Irrigating arid land dissolves the salts in the soil, and as the water evaporates, the salts are drawn toward the surface. Many artificially irrigated lands are poorly drained, and as a result, the salts simply remain in the upper levels of the soil rather than being flushed out and carried to the sea. Furthermore, the poorly drained land and soils themselves can become waterlogged, and the water table can rise over time, as the groundwater and soils become progressively saltier. Accumulated mineral salts are toxic to most plant life, and as land becomes increasingly salinized, it may reach a point where it can no longer support most crops or other plants (**Figure 13-5**).

Traditional Methods of Coping with Agriculture's Effects.   A traditional and time-honored way of circumventing the problems inherent in agriculture is to occupy the land for a year or two, often utilizing slash and burn or swidden techniques (cutting and burning the natural vegetation in order to clear the land and release the nutrients into the soil), and then move on. In this way the natural ecological cycle of succession can occur once again; the land is allowed to regenerate and replenish itself. This method is feasible as long as the human population in any one area is relatively small, and they are willing to pick up and move on a regular basis.

A variation on this theme is to maintain a permanent place of residence, but utilize alternating fields in different years. In late ancient and medieval Europe, many farmers used a "two-field system" in which only half of their land was planted with crops in any particular year; the other half lay fallow. Native wild plants would colonize the fallow land, and farm animals were allowed to graze there; their manure helped to restore fertility to the soil. Fallow fields could also serve as a home for wildlife, such as birds, that could help keep insects and other pests in check. Each year, the crops would be planted on the previous year's fallow land. A related method is to rotate crops and the pasturing of livestock from field to field and season to season. With proper rotation, nutrients that were utilized by one crop can be restored by the next crop. For instance, periodically planting a field with legumes (members of the pea and bean family) will restore nitrogen to the soil since these plants' roots attract soil bacteria that have the ability to remove nitrogen from the air and produce nitrogen compounds upon which other forms of life are dependent. Crop rotation also tends to decrease the threat of pests and disease. If the same crop is planted in the same field year after year, a colony of a harmful pest or disease agent (be it a rodent, insect, fungus, or other life-form) can take up permanent residence in or near the field. Such a colony has less chance of establishing itself if crops are rotated from year to year.

Another traditional way to avoid the problems inherent in some agricultural practices is to promote diversity. This can take many forms and is not unrelated to the concept of crop rotation. In many traditional, aboriginal agricultures, numerous varieties of many different crops are planted each season; the aboriginals of Amazonia, for instance, used at least 70 varieties of manioc (a group of tropical plants with edible roots, also known as cassava or tapioca). In some cases, many different types of plants are cultivated within a small area, even planted together in the same space—mimicking some of the characteristics of a climax community. In Central America, the farmers have traditionally interplanted maize (corn), beans, and squash. The three crops benefit one another, and the system leads to greater long-term productivity and sustainability than planting a single crop at a time. The more varied diet such interplanting promotes is also nutritionally preferable for humans. In addition, using a variety of crops is a form of insurance—one does not put all of one's eggs in a single basket. Different crops and varieties have different tolerances for adverse pest, soil, disease, and climatic conditions (**Figure 13-6**). Even if unexpected rains or droughts occur or an abnormal fungus or insect plague strikes, it is less likely to destroy the entire harvest if a variety of plants have been cultivated.

Modern Agriculture's "Solutions": Fertilizers, Pesticides, and Herbicides.   "Modern" agricultural techniques of the late nineteenth through twenty-first centuries have avoided many of the traditional methods used to cope with the inherent weaknesses in artificial agriculture. This has resulted in the "quick fix" of bumper crops, but has been at the expense of the land, nonrenewable mineral and energy resources, and long-term sustainability.

Figure 13-5   Salt-affected agricultural land near Katanning, Western Australia.

Figure 13-6   Planting two or more crops in alternating strips is referred to as strip cropping. Here, alternating strips of alfalfa with corn protect this field in Iowa from soil erosion.

Much modern agriculture is synonymous with the circumvention of biological agents in the restoration of depleted soil fertility; crop diversity, crop rotation, and even manure use are abandoned. Instead, minerals and chemicals are mined, processed, and applied directly to croplands in the form of fertilizers; simultaneously, irrigation efforts are intensified. Sometimes this has been referred to as "force-feeding" the land.

The main nutrients applied to the soil are phosphorus, potassium, and nitrogen. Phosphorus and potassium are mined from mineral deposits; phosphorus in particular is potentially in short supply as high-grade naturally formed phosphate deposits are exploited faster than new deposits are discovered (like the fossil fuels, there is only a finite supply of

Figure 13-7   Spraying pesticide on leaf lettuce in Yuma, Arizona.

geologically formed high-grade phosphate deposits, many of which are the result of accumulations of ancient animal bones millions of years ago). Nitrogen was initially supplied from manure or from bird droppings known as "guano." Some isolated islands contain huge mountains of bird droppings, and these were mined for their nitrogen content. Now, however, artificial nitrogen-bearing fertilizers can be manufactured synthetically using the abundant nitrogen of the atmosphere.

The heavy use of fertilizers was accompanied by an increasing emphasis on monoculture (planting huge fields with a single variety of a single crop). Monoculture allows the farmer to tailor the fertilizers to the specific needs of the particular crop and increases efficiency in mechanical harvesting and processing of the crop. But monoculture brought with it increasing problems from pests and diseases that found a happy point of attack in the huge, ecologically unstable fields. This meant that such pests and diseases needed to be controlled. The preferred way to control them was through the use of more chemicals—synthetic pesticides and herbicides that could be designed to kill everything except the crop being cultivated. In the United States, pesticide use in agriculture rose from about 340 million pounds (154 million kg) per year in 1965 to about 890 million pounds (404 million kg) in the early 1980s. U.S. pesticide consumption then dropped slightly; it is currently 750 to 800 million pounds (340 to 363 million kg) per year. Globally, there has a been a steady increase in pesticide use, with the average amount applied to agricultural land more than tripling, from about 0.5 kilogram per hectare in 1961 to between 1.5 and 2.0 kilograms per hectare at the end of the century. But such techniques led to obvious problems. Despite the massive addition of fertilizers, the soils slowly became exhausted. While the major nutrients extracted by the plants were being temporarily restored, many trace elements necessary for the ultimate sustainability of agriculture were not; examples of such trace elements include zinc, iron, boron, copper, molybdenum, and manganese. Furthermore, good healthy soil is more than just a handful of dry minerals and fertilizers. It is full of organic debris, humus, and living organisms including worms, beneficial insects, fungi, and bacteria. These organisms help mix and aerate the nutrients. The texture, structure, and quality of the soil are necessary for the roots of plants to take hold and the soil to retain water—which helps minimize both droughts and waterlogging of the earth below. Dumping massive amounts of toxic substances (in the form of pesticides and herbicides) literally kills the soil (**Figure 13-7**). Dead soil loses its structure and no longer functions properly. An unstable, dead soil may quickly erode away, perhaps further spreading the noxious chemicals that killed it.

The Green Revolution.   Shortly after the end of World War II, modern, chemically based agriculture began to be used on a large scale in the industrialized countries. In the 1960s, the Food and Agriculture Organization of the United Nations began a massive program to increase world food production, especially in the developing countries. This effort was based on the use of modern agricultural techniques applied to high-yielding, western-designed crops. The resulting food production gains of the 1950s through 1980s (**Figure 13-8**) are often termed the **Green Revolution.**

The immediate, short-term gains of the Green Revolution were truly impressive. For instance, between 1950 and 1986, world grain production rose from approximately 631 million metric tons per year to 1665 million metric tons per year. This growth rate in grain production was greater than the rate of human population growth across the planet, with the net result that per capita grain production rose from 247 kilograms per person to 337 kilograms per person during the same period (see Figure 13-4). Given the increase in the Earth's population from just slightly over 3 billion in 1960 to the present 6.3 billion, the Green Revolution may have staved off immediate starvation for billions, but this has come at a price. The massive application of "modern" agricultural techniques has resulted in numerous problems, and it is unclear whether current food production levels will be maintainable for much longer. The peak per capita production of grain in 1984 (342 kilograms) has thus far (through 2001) not been duplicated, and 2001 per capita grain production (299 kilograms) was the lowest since 1972 (when it came in at 295 kilograms). In many areas, **soil fertility** is declining rapidly as nutrients are extracted from the soil, but not returned in kind. Massive irrigation, even in countries where forms of irrigation have been successfully carried out for millennia, is causing waterlogging and salinization at unprecedented rates. In arid and semiarid regions, this often leads to **desertification (Figure 13-9),** the spread of desertlike conditions caused by human exploitation and misuse of the land. In China, perhaps over a million hectares of agricultural land have had to be abandoned since 1980 because of problems with salinization and waterlogging. Similarly, 2.9 million hectares were removed from use in the Soviet Union between 1971 and 1985, and in India, the newly irrigated land that is put into production each year is counterbalanced by damaged land that must be removed from production. In Egypt, irrigation has been a necessity since ancient times, yet traditionally the fields were not used continuously; the fields were allowed to lie fallow, and thus the accumulating salts could be naturally washed out and the land rejuvenated. Since about 1960, intensive modern irrigation has caused salinization of about a third of Egypt's cultivated land, and an estimated 90% is suffering from the effects of waterlogging.

Just as serious as nutrient depletion, salinization, and waterlogging are the effects of toxics (in the form of pesticides and herbicides) and soil erosion. Pollution from herbicides and pesticides is a global

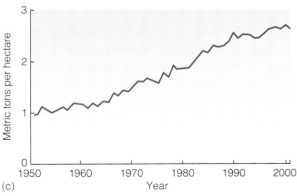

Figure 13-8   (a) Composite world grain yields, 1950–2001; (b) World corn and rice yields, 1950–2001; (c) World wheat yields, 1950–2001. All yields are in terms of metric tons per hectare. (*Source*: United States Department of Agriculture.)

Figure 13-9   Deserts and areas at risk of desertification. (*Source*: Based on United Nations data.)

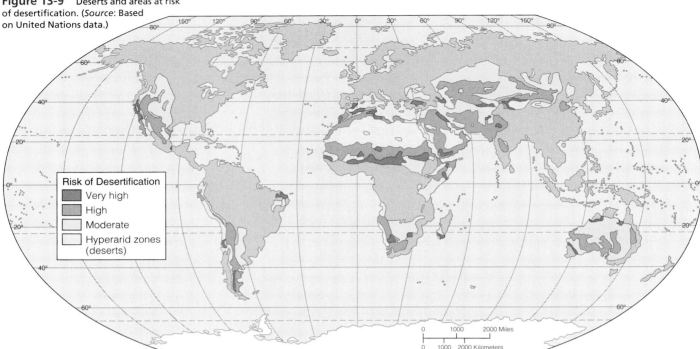

problem: It destroys not only the living organisms in the soil, but other wildlife and vegetation as well, and is directly harmful to the human population. Denuded soils quickly erode, and topsoil loss is a serious global problem (discussed further below). A study in Tanzania found that land with natural vegetation cover suffered virtually no topsoil loss and absorbed almost all the rainfall, compared to land that was either artificially cultivated or left bare. It has been estimated that currently about 26 billion metric tons of topsoil are lost worldwide from erosion every year. It can take up to 1000 years to form a layer of soil 0.4 inch (1 centimeter) thick, yet it can be lost in just a few years from poor agricultural management. At current rates of erosion, some researchers estimate that the once fertile land of the U.S. corn belt could be nearly depleted of topsoil before the middle of the twenty-first century.

The Green Revolution has also been held responsible for the contamination and depletion of groundwater supplies in many parts of the world. Agrochemicals—artificial fertilizers, pesticides, herbicides, nitrates (derived from fertilizers), and other chemicals (many of which are highly toxic and carcinogenic)—applied in abundance to fields have penetrated down and polluted groundwater supplies. An estimated 50 million people in the United States are potentially exposed to pesticide-contaminated groundwater used for drinking. Contaminated groundwater is often virtually impossible to clean up.

Underground aquifers are cool, dark, and well protected; they have poor water circulation, contain little in the way of life-forms, and thus form an ideal place for contaminants to be stored and remain stable (and toxic) for centuries or millennia. Deep in the aquifer there are no natural mechanisms to break down or neutralize contaminating toxics, so they remain indefinitely. Many authorities feel that in most cases, a badly contaminated groundwater supply must simply be dismissed as a future source of fresh water.

The Green Revolution has also greatly stressed the supplies of freshwater. Many of the miracle crops of the Green Revolution were hybrid varieties that, although they may have been more productive, required much more water than traditional varieties of wheat and rice. In addition, in some areas crops for export were introduced that required even greater amounts of water; a case in point is sugarcane, which can require 10 times as much water as wheat. Furthermore, arid land was put under artificial irrigation using modern wells that tapped deep underground aquifers. In the past few decades, many areas have routinely withdrawn water from aquifers much faster than the aquifers are recharged by rain. The net result of these practices is that water tables are declining around the world.

Beyond the Green Revolution—Integrated Pest Management and Biological Control.   The acme of the Green Revolution, with its heavy dependence on synthetic chemical compounds—fertilizers, pesticides, and herbicides—and its use of water-consuming, genetically identical monocultures grown in factory-style intensity with the help of heavy equipment powered by fossil fuels, has now passed. The gains of the Green Revolution were impressive, but they were achieved unsustainably. The down side of the Green Revolution has been environmental damage to an extent previously unknown in recorded history.

Fortunately, "new" (in fact, based on traditional practices), ecologically sound, and sustainable methods of growing food are being used by an increasing number of farmers around the world. New methods of "no-till" planting, where only a narrow slit is cut through the sod and crop residues in order to plant the new crop, can help protect the soil from erosion and degradation (**Figure 13-10**). Drip irrigation techniques can drastically reduce water usage and losses and help avoid such problems as soil salinization and waterlogging. Also popular are the concepts of **Integrated Pest Management (IPM)**, the use of **biological controls**, and organic farming.

The basic philosophy behind IPM and biological control is that the farmer does not try to totally eliminate pests, as was often the idea behind using massive amounts of poisons as part of the Green Revolution, but simply attempts to control pests so that they do not cause serious damage. IPM advocates "natural" controls, such as the use of the pests' biological predators. IPM systems also use cultural practices such as **crop rotation**, allowing fields to lie fallow periodically, and interplanting to help control various pests and weeds.

To an increasing extent, farmers are returning to the use of natural fertilizers such as crop wastes that are plowed back into the soil or left to rot on top of the soil, natural compost, animal manures, and even human wastes. In some areas, farmers have taken up true **organic farming**, which avoids the use of any synthetic chemicals—fertilizers, pesticides, or herbicides.

IPM, biological control of pests, and organic farming are proving to be productive and economically feasible. In some cases, the yields have been slightly lower (although sometimes they are higher), but since the farmers did not have to purchase extra synthetic chemicals, their costs were lower and their profits the same as or higher than they would have been using more conventional methods. In fact, one study of nine crops in 15 U.S. states found that the farmers using IPM systems had a collective profit of $579 million more than their projected earnings using other methods. Thus, these new tech-

Figure 13-10   No-till planting in the residue of the previous crop reduces erosion and returns nutrients to the soil. This field is in northwest Iowa.

niques are economically viable; they will not drive farmers out of business, nor cause a dramatic drop in food production. Most importantly, however, they do not deteriorate the land and general environment to the extent that the techniques of the Green Revolution did. In fact, at their best, organic farming and IPM, combined with very limited use of synthetic chemicals, appear to be sustainable—a claim the Green Revolution could never approach.

Biotechnology and Genetically Modified (GM) Crops

The Green Revolution was based on many new "miracle" strains of crops that grew faster and produced higher yields. Many hope that we can continue to increase food production through **biotechnology** and **bioengineering**—the artificial use and manipulation of organisms toward human ends, including genetic manipulations that can in effect produce new types of organisms (see Case Study 13-1). Genetically modified (GM), transformed, or **transgenic crops** (transgenic varieties) are already a reality. In 1994, the Flavr Savr tomato, developed by Calgene, Inc. (since taken over by Monsanto Corporation), of Davis, California, became the first genetically engineered whole food product to hit the markets. Essentially, the Flavr Savr was designed so that an altered gene blocked production of a certain enzyme that controls ripening and softening. Normally, tomatoes are harvested before they are ripe, shipped, and then artificially ripened (such as

Genetically Modified Foods and Crops

In less than a decade, genetically modified (GM) crops have literally exploded on the world stage. In 1996 (when the first edition of this book was published), a mere 1.7 million hectares were planted worldwide, but by the turn of the century some 50 million hectares were being planted with GM crops, an area the size of Spain. Most of the use of GM crops is concentrated in the United States (the leader in such operations), Argentina, Canada, and China. It is estimated that in the United States, between 60% and 70% of all food contains some GM component.

But GM crops did not explode just in terms of dramatic rises in plantings; they also exploded in terms of worldwide controversy over the advisability of relying on such technologies. In many European countries, much of the populace is extremely wary of genetically modified foodstuffs, and as of early 2002, Britain, for instance, did not have any GM crops being grown commercially. In Europe, generally there is much more concern with labeling of GM foods and regulation of their importation. In the United States, there is no requirement to label foods containing genetically modified components.

Genetically modified organisms (GMOs), also known as genetically engineered, bioengineered, or transgenic organisms, are plants (or in some cases animals or microorganisms) that contain genes extracted from other types of organisms, be they viruses, bacteria, plants, animals, and so forth, inserted artificially into the subject organism. Humans have been selectively breeding, and thus artificially modifying, the genetic makeup of domesticated organisms for thousands of years, but so-called genetically modified organisms ("transgenic organisms" is actually the more accurately descriptive term for organisms that contain genetic material transferred from another species) are different. Traditional breeding involves, by necessity, the crossing of organisms that are closely enough related that they can interbreed. The engineering of transgenic organisms involves the mixing of genes from organisms that are widely separated evolutionarily, such as the inserting of bacterial genes in a plant or animal, or even animal genes in a plant or vice versa. For this reason, sometimes transgenic food crops are referred to as "Frankenfoods" (after the fictional Dr. Frankenstein's monster which was manufactured from a combination of body parts originally belonging to different individuals).

Why engineer transgenic organisms? Direct desired benefits include increasing crop yields, developing more advantageous characteristics of crops (be it better taste, higher nutritional value, or longer shelf life), resistance to pests, and tolerance of herbicides and pesticides (so that the crop can be treated with a herbicide that will kill everything other than the desired crop). Through transgenic engineering, plants can also be developed that will yield precursors of plastics, vaccines, and other products not necessarily associated with the plants in nature. And it is not only plants that are genetically engineered. The so-called "Super Salmon" has been genetically engineered with a growth hormone gene that causes the fish to grow extremely quickly (four to six times as fast as the original, unaltered fish) and reach very large sizes.

One early example of a generically engineered plant is the GM soybean developed by the Monsanto Company. This variety of soybean was developed to be immune to the Monsanto herbicide known as "Roundup." Farmers could plant the genetically modified soybeans and then control weeds simply by spraying Roundup on the crop. All plants other than the soybeans with the immunity to Roundup would be eliminated. This, it was argued, would reduce the need to use other, more toxic and dangerous herbicides. Various herbicide-resistant varieties of soybeans, corn, cotton, and canola (rapeseed) are among the most common types of transgenic crops currently planted (in 1999, the latest year for which there are accurate statistics as of this writing, accounting for 71% of the area worldwide planted with transgenic crops). Another major development was the insertion into corn, cotton, and other crops of bacterial genes (from the common bacterium *Bacillus thuringiensis*) that produce a toxin poisonous to various insect species. Such genetically modified Bt crops (named after the initials of the bacterium) have a "built-in" resistance to insect damage, and as of 1999, Bt-corn and Bt-cotton accounted for 22% of the transgenic crop area planted globally. Plants can also be modified to both produce Bt toxins and be herbicide-resistant simultaneously, known as "trait-stacked" varieties that contain more than one genetic modification; in 1999, 7% of the transgenic crop area consisted of corn and cotton varieties that produced Bt and were herbicide-resistant.

GM crops such as those described above have incredible potential, so their proponents argue, to increase production and decrease costs (less money need be spent on pest control), and thus transgenic crops have been rapidly adopted in a few countries, led by the United States. However, there is a darker side to GM foods, the critics have argued. One example often pointed to is the use of so-called "terminator technologies." In the late 1990s, Monsanto and other companies were worried, in part, that their investments and what they regarded as their valid intellectual property rights would be lost if genetically modified plants that they had worked hard to develop could simply be purchased once and then re-grown year after year by saving some seeds from the previous year's harvest, as is done by many traditional farmers. To eliminate such a possibility, they worked to engineer sterility into the seeds—that way, each year the farmers would be forced to buy more seeds from the parent company. Such technology, however, came under intense public scrutiny and, under pressure from the critics, Monsanto decided in 1999 not to commercialize terminator technology.

Still, this incident did not make for good global public relations when it came to GM products. It also brings up larger philosophical and ethical issues, such as should private individuals and companies be allowed to patent life forms or parts of life forms? On the one side, some people believe it is wrong to "play god" and patent organisms, or simply unethical to develop a beneficial strain of a crop that could help feed the poor in developing nations, yet make it available only to those willing to pay the asking price, thus potentially eliminating those who could most benefit from such a development. On the other hand, it can be argued, why shouldn't a company reap the benefits from the risk and investment involved in attempting to develop new GM products? Indeed, in 1980, the U.S. Supreme Court cleared the way for the patenting of newly developed types of organisms and this helped make it more attractive to private corporations to invest in genetic engineering.

Many questions have also been raised, especially by environmental and consumer advocate groups in Europe, concerning the potential health and environmental risks of GM foods and crops. In contrast, according to advocates of the benefits of GM foods and crops, no such "hypothetical" risks to humans have yet to be definitively demonstrated for GM organisms, and what minimal risks might be involved in the usage of GMOs are far outweighed by the benefits. Let us briefly examine both sides of the issue.

It has been suggested that transgenic foods may produce toxic or allergic reactions in humans (although no studies to date have definitively demonstrated this to be the case), and at the least critics argue that consumers should be allowed to choose if they want to purchase and ingest GM foods. Why should a strict vegetarian, for instance, be unwittingly subjected to ingesting a GM product that might contain an animal gene? Of course, this would not only mean labeling all GM foods, but keeping separate and distinct GM crops and non-GM crops and processing so that the end products could be labeled correctly. (In the fields, non-GM crops planted too close to GM crops may become inadvertently pollinated with GM pollen, thus turning a non-GM crop into a GM crop, to the detriment of the farmer attempting to grow a GM-free product. This has already become a concern in some areas where GM crops are grown.) Given that there are no demonstrated human health risks involved with commercially available GM foods, such separation and labeling is unnecessary and only an added expense that will scare consumers, so GM proponents argue. In opposition, GM opponents note that there have been very few studies of GM foods and their safety, and the few studies that have been carried out were aimed primarily at determining whether the GM food in question is "substantially equivalent" to its natural counterpart, and if it is "substantially equivalent" in composition to the natural form, then it is considered safe. However, "substantial equivalence" is a poorly defined concept and potentially a very slight difference in composition could have major health or environmental ramifications.

Indeed, depending on what level of evidence one accepts, there is at least some evidence that some GM foods may pose health risks. In a widely publicized and controversial study, Dr. Arpad Pusztai of the Rowett Research Institute in Aberdeen, Scotland, reported stunted growth, damaged internal organs, and damaged immune systems in rats after feeding them experimental varieties (not commercially grown) of genetically modified potatoes for several years. Part of the problem, at least according to critics of GM food, is found in the very techniques that are used to engineer the organisms. Viruses are typically utilized to insert foreign genes into an organism, and this process can result in other, unintended insertions of genetic material with unanticipated consequences either in the short term or the long term.

Genetic engineering can also be used to increase the nutritional value of crops. Thus, for example, in early 2000, a form of GM rice, commonly known as "Golden Rice" (it is yellow in color), was developed to address vitamin A deficiencies in parts of the developing world where rice is consumed as a staple. The golden rice is engineered to contain beta carotene, the precursor of vitamin A. Over a hundred million rice-eating children in developing countries suffer from vitamin A deficiencies, leading to suppression of the immune system, blindness, and death. Clearly, addressing this problem is a worthy goal. However, critics suggest that Golden Rice is not the panacea that some would have us believe. Beta carotene is best transformed into vitamin A in a healthy person, not someone already weakened from a lack of proper nutrition, and furthermore a dozen times as much Golden Rice as is normal for rice consumption daily would have to be eaten to receive an adequate supply of beta carotene. Still, proponents point out, Golden Rice is certainly an improvement nutritionally over regular rice.

As a result of the disparate GM climates on various continents, different parts of the world have treated GM food very differently. In the United States, there is no labeling requirement. Europeans in general, it can be suggested, have a different attitude toward food than Americans. Europeans have traditionally cared more about the food they eat and have been more "purist" concerning what they ingest. Furthermore, there were the "mad cow disease" scares of the late 1990s that eroded European public confidence in their health care officials, and GM food was also viewed as an imposition from America. Regardless of whether any harm could be demonstrated to be caused by GM foods, as a precautionary measure and to allay public concerns, beginning in the late 1990s, the European Union implemented the labeling of GM foods and restricted the import of GM crops. This had a severe negative effect on corn imports from the United States to Europe, given that much of American corn is composed of GM varieties. By 1999, the concerns of Europe had spread to many other large importers of American crops, including Japan and South Korea. Various food companies in Europe and Japan implemented policies of removing any GM ingredients from their products. These developments have cost U.S. agriculture hundred of millions of dollars as exports dropped and transgenic crops were devalued because of a lack of markets.

There are also many environmental concerns about GM organisms, and the current evidence suggests that these are perhaps better founded than the human health concerns. There have been laboratory studies indicating that pollen produced by Bt-corn can harm or kill Monarch butterfly larvae, for instance. Other studies have suggested that the toxins produced by Bt-corn can accumulate in soils and may have negative ecological effects. There is also the concern that using crops genetically engineered to produce their own "insecticide toxins" could, analogous to the overuse of antibiotics to treat diseases, induce the evolution of insects and other crop pests that are resistant to the toxins. Such super pests might then not only attack the crop, but begin attacking other plants as well. If such toxin-resistant pests have not evolved yet, that does not mean there is no potential for their evolution; after all, GM crops have only seen widespread commercial use for less than a decade. Then there is the concern that genetically engineered herbicide resistance could be spread to wild plants, such as wild relatives of the crops (many crops are essentially cultivated weeds), or the plants may even escape into the wild and ultimately produce "super weeds" that cannot be controlled by standard herbicides. Much damage has already been caused by the introduction of exotic species from one region of the world to another; the escape of GM organisms and their genes would effectively constitute additional cases of the introduction of exotic species, in this case from the laboratory to nature.

Another concern, not distinct from the above considerations, is the basic unpredictability of the effects of genetic engineering. A recent case in point involves potatoes. When potatoes were genetically modified to repel aphids, it was found that they actually attracted other pests, including the potato leafhopper that feeds on the plant's leaves. It has also been found that the stems of herbicide-resistant Monsanto soybean plants are more prone to cracking open in hot climates than non-GM soybean plants.

The use of GM crops might not be "all or nothing." Grafting is common in certain types of horticulture, and genetically modified root stocks that are pest and disease resistant can have non-GM stalks and fruit- or nut-bearing portions grafted on to them. This has been done, for instance, with walnuts. Such hybrid plants may combine the best of both—the advantages of GM with a final non-GM product that will cause no concern to consumers.

GM organisms not only serve as foodstuffs, but have many other purposes as well. For instance, microorganisms can be engineered to digest and get rid of toxic substances. Another potentially important application that has been touted in some circles is to genetically engineer plants, such as corn (maize) to produce substances that can be extracted and processed to produce "green" (that is, environmentally friendly) plant-based and fully biodegradable plastics as an alternative to using scarce petroleum resources to make plastics. Such plastics would be produced from a renewable resource (plants) and could be disposed of in the same manner as organic compost and other wastes. Yet, such a technology is far from a viable reality at the moment. To produce plant-based plastics with current methods requires large amounts of energy for extraction and processing—energy that typically comes from fossil fuels. Based on one analysis, more fossil fuel will typically be used to produce plant-based plastic than to produce an equivalent amount of petroleum-based plastic, in which case it makes more sense environmentally to go with the petroleum-based plastic. Furthermore, given the problems of global climate change, biodegradability is not necessarily an advantage. If the plastic is allowed to break down, it will produce greenhouse gases such as carbon dioxide and methane. Also, devoting crops to the production of plastics may detract from the world's food supply, and there are already many people that do not have enough to eat. An exception might be if plastic precursors can be simultaneously grown in an otherwise inedible part of an edible plant, such as in the leaves of corn (maize), in which case the same field can produce both food and plastics simultaneously.

Passions run deep on both sides of the GM organism debate. Should farmers in Britain, France, Sweden, and India be forced to destroy their crops of rapeseed or cotton, as has been the case in some instances, because they were sowed using "illegal" or "unapproved" GM seeds? Irrational fear of GM crops can lead to nonsensical actions, contend GM proponents. Yet opponents of GM crops argue that it is better to be "safe than sorry" when we do not know the potential consequences of the widespread use and consumption of genetically modified organisms. Some proponents of GM foods and crops suggest that the opponents of GM are hurting primarily the poor farmers in developing countries who may ultimately be the ones to most benefit from the "Gene Revolution" (following a half century after the "Green Revolution"). Wealthy European consumers, it has been suggested, can easily afford to indulge in non-GM products, but the poor and starving need all the help they can get. GM plants and technologies should be designed, these proponents advocate, specifically to increase yields and decrease costs in tropical and developing countries, and then distributed at fair prices. In fact, over the last few years, this is exactly what has been happening in China. China currently has the second largest GM program (the United States is number one), but in China the emphasis has been to engineer insect and disease resistance, rather than focus on herbicide resistance. As of 2002, it was reported that the Chinese had introduced over 120 different genes into approximately 50 different species. Already either in use or in the late trial stage are Chinese GM varieties of rice, cotton, wheat, tomatoes, sweet peppers, potatoes, rapeseed, peanuts, cabbage, melons, maize, chilies, papaya, and tobacco. Literally millions of Chinese farmers benefit from GM crops, especially Bt-cotton. Around two million Chinese plant Bt-cotton on 7000 square kilometers worth of fields, and since the introduction in 1997 of the Chinese version of Bt-cotton, the use of toxic pesticides on these fields has dropped by 80%, costs have decreased by 28%, and the farmers have accordingly increased their earnings. Nay Sayers, however, suggest that it is only a matter of time before resistant insect strains evolve and all of the GM gains will be lost. So far, however, the Chinese are benefiting from GM technology.

Questions

1. List some of the reasons that genetically modified (GM) organisms have been, and are being, developed. How are they enhancing food production? Or are they?

2. If you live in the United States, you probably eat some foods that are composed, at least in part, of GM organisms. Does this bother you? Are you aware of when you are eating foods with a GM component?

3. It has been suggested that GM crops and organisms might "escape" into nature, or crossbreed with close natural relatives, and cause damage to ecosystems. What is the evidence, thus far, for such a potential threat? What precautions do you think should be taken with GM crops to avoid such problems?

by using ethylene gas) once they reach their destination. Flavor is lost with such procedures, though, and an estimated 30% of the tomato crop is still destroyed by rotting or damaged during shipping. The idea was that the Flavr Savr could be left on the vine longer, so it would ripen naturally and develop a better flavor, resist spoiling during the shipping process, and have a longer shelf life once it arrived at a supermarket or home. However, the Flavr Savr was not a success. Within a few years, it was off the market due to reported problems with taste, damage during shipping, poor growing in soils and climates outside of California where it was developed, and the general public wariness concerning GM foods. Still, many other fruits and vegetables are plagued by the same ripening and spoilage problems as tomatoes, and might benefit from similar genetic engineering.

Genetic engineering can also change the taste or other properties of plants by modifying their sugar and starch content. Peas, corn, tomatoes, and other crops can be made sweeter, or the starch content of potatoes can be raised, making them more suitable for potato chips.

In the long run, genetic engineering's most important contribution may be to increase the resistance of crops to insect and disease vectors. The common bacterium *Bacillus thuringiensis* (commonly abbreviated as Bt) naturally produces a substance that is toxic to certain types of pest caterpillars. For several decades, Bt and its derivatives have been used as a natural pesticide on crops with good results; it is relatively nontoxic to birds, mammals, and various nonpest insects. Through biotechnology, the Bt bacterial genes can be implanted into the crops themselves so that they produce the toxin. Such a transgenic organism is in effect mostly plant but also part bacterium. Spiders and other creatures also produce toxins that kill insect pests. Plants that resist various viral, bacterial, and fungal diseases can be designed by implanting

genes from various viruses, bacteria, plants, and animals into crop plants. Many people may not realize it, but sizable percentages, from a third to a half or more, of such crops as soybeans, corn, and cotton are composed of transgenic varieties in America.

Clearly, transgenic crops require reduced loads of standard pesticides and have an advantage in resisting diseases. However, researchers point out that genetically engineered crops must be utilized carefully. Many insects and diseases can evolve very rapidly. If too much reliance is placed on one or a few types of transgenic crops, natural pest populations may rapidly evolve immunities to the toxins given off by the crops. Already there are reports of insects that can tolerate fairly high levels of Bt toxins. This situation is analogous to cases where insect populations evolve the ability to withstand the assaults of standard insecticides. To prevent immunities from developing in the pest insects or disease vectors, IPM techniques can be used in conjunction with transgenic crops. For example, different types of transgenic and standard crops might be combined in the same field. The unaltered, nonresistant stands of plants would act as a feeding and breeding ground for insects that are not immune to the toxins engineered into the transgenic crops. Thus, the more damaging individuals—those carrying natural immunities—would never be allowed to dominate the population. By not overusing the resistant strains of crops, their effectiveness will be maintained.

Efforts are underway to genetically alter plant crops so that they will have increased tolerances to stresses such as drought, cold, heat, or high soil salinities. So far, however, less progress has been made in this area than in developing insect- and disease-resistant strains. Stress-tolerant crops could be a real boon in developing countries that have only marginally arable lands, suffer from soil salinization, or lack adequate irrigation systems. Some observers worry, however, that in the long run, stress-tolerant crops could cause more harm than good by encouraging the continued cultivation of marginal, fragile, or already damaged lands until they are completely destroyed.

From a human dietary perspective, an extremely important potential of genetic engineering is to improve the nutritional content of familiar foods. People whose staple is rice often suffer from vitamin A deficiency; thus, researchers engineered a rice variety containing substantial quantities of beta carotene, a precursor of vitamin A. (This so-called "Golden Rice" has come under some criticism, however. It is suggested that the average person would have to consume twelve times the normal intake of rice to get the necessary amount of beta carotene, and furthermore, beta carotene is best converted into vitamin A in a healthy person, not the undernourished that Golden Rice is aimed at. It might be more useful to help people with vitamin A deficiencies to grow various green vegetables that not only are rich in beta carotene, but other nutrients as well—which are lacking in rice, Golden or otherwise.) Likewise, levels of various proteins might be increased both in crops that are eaten directly by humans and in those used as feed for farm animals.

Genetic engineering is also being used to meet the specialized needs of consumers in industrial countries: coffee with a lower caffeine content; rapeseed (oilseed, canola) varieties that produce specialized oils for use as lubricants, in cosmetics, for soaps, and in cooking. A mustard family plant has been designed to produce the biodegradable plastic known as polyhydroxybutyrate (PHB), which is similar to polypropylene (derived from petroleum). One concern over this kind of research is that currently many specialty oils, waxes, and rubbers are derived from tropical forests and are among the major exports of the developing countries. Successful development of "oil crops" could restrict the market for these goods. At the same time, however, by growing "plastic crops," the developing countries could produce their own plastics without relying on oil or petrochemical facilities. Additionally, biodegradable plastics could alleviate many of the disposal problems associated with traditional nonbiodegradable plastics (see Chapter 18).

Nonagricultural Food Sources

It is sometimes suggested that we could feed extra billions of people by harvesting the natural, renewable biological resources that grow wild on land and in the seas. Humans have been hunting wildlife and collecting naturally growing edible plants for millennia, but the reserves of such sources are virtually depleted. Certainly, traditional hunting and gathering on land is not a viable option for feeding

anything but an infinitesimally small proportion of the current global population.

The ocean, however, is often viewed in a different light. Using modern techniques, tens of millions of tons of fish and other seafood are harvested from the sea each year. The oceans are so vast, couldn't we make a dent in our food shortages (especially protein shortages, for fish is high in protein) by drawing more from this resource?

Many people have a mistaken impression of the magnitude and abundance of life in the seas. They may be familiar with the productive shallow-water coastal and reefal areas, which are unrepresentative of the vast majority of the oceans. Most of the open ocean is a "biological desert" that is very sparsely populated by life-forms. Only certain areas where nutrients upwell and collect near the surface are highly productive. Unfortunately, we are already closely approaching, and have perhaps now surpassed, the maximum sustainable yield of fishing from the oceans: estimated at approximately 80 to 100 million metric tons of fish a year. In the last decade, the global fish catch from the oceans has been in the range of 80 to 95 million metric tons per year. Fishing on the open oceans has often been done on a massive scale utilizing driftnets. These huge nylon nets—some are up to 30 miles (50 km) long and 40 feet (12 m) deep—indiscriminately catch everything in their path, including squid, fishes, dolphins, seals, sea turtles, and even water birds. The oceans are literally mined; fishes and other organisms are removed much more quickly than they can replenish themselves. Many animals in addition to the desired fishes are caught inadvertently; these carcasses are simply thrown overboard as waste. Fortunately, there has been mounting international pressure to limit the use of driftnets. Since 1992, the United Nations has imposed an international moratorium on the use of driftnets over 1.55 miles (2.5 km) long. But even if driftnets are totally banned, some fishing fleets will continue to utilize them illegally on the high seas. Furthermore, many broken pieces of nets or damaged and abandoned nets ("ghost nets") are floating unattended through the oceans. These ghost nets continue to catch and kill sea organisms indiscriminately.

Already, certain species of ocean organisms have been overexploited by humans, perhaps to the point of **commercial extinction**, that is, it is no longer economically viable to harvest them. Populations that have dropped to such low levels may never fully recover; indeed, they may become extinct. Classic examples of such overexploitation include the Peruvian anchovy fishery, the Alaskan king crab fishery, and the exploitation of whales. The Peruvian anchovy industry more than tripled its catch from 1960 to 1970, peaking at about 13 million metric tons in 1970. By 1973, it had collapsed to less than 2 million metric tons, possibly due to both overexploitation and adverse climatic conditions. It has never recovered to the levels of peak production. The Alaskan king crab story presents a similar scenario: Peak production in 1980 was 84,000 metric tons, but this dropped to 7000 metric tons in 1985 and has not fully recovered since. Whales have been hunted to commercial extinction over several centuries, beginning on a large scale in the 1700s and early 1800s and continuing into the present century.

In addition to mining our oceans, we are polluting them at a tremendous rate. Some seafood species are being killed off altogether, and others contain such high levels of toxic chemicals that they are unfit to eat. Many coastal cities continue to dump their raw or inadequately treated sewage and waste directly into the oceans. This pollution is destroying the wildlife. In 1988, 10,000 seals died in the North Sea, apparently from a viral infection that they could not fight off because the pollutants in the water had weakened their immune systems. Beluga whales inhabiting the St. Lawrence River have been reported to contain such high levels of heavy metals, PCBs, and other pollutants in their flesh that their corpses are classified as toxic waste when they die.

The surfaces of the oceans are manifesting the symptoms of the "pollution disease." Around the world, surface algal blooms, sometimes called the "red tide," are appearing with increasing frequency. Currently, dozens of red tides occur each year in Hong Kong's harbor, where they were unknown before the mid-1970s. In a red tide, tiny algae that live as phytoplankton in the upper layers of the oceans proliferate out of control. Apparently sewage, fertilizer runoff, and other pollutants that are released into the water are fit nutrients for the algae, which grow uncontrollably on the surface of the water. As the algae grow, they deplete the oxygen in the water that is necessary for the survival of other organisms. Shellfishes, crabs, shrimp, a variety of fishes, and numerous other organisms can literally be suffocated under the red tide (**Figure 13-11**). An algal bloom off the coast of Norway reportedly killed more than 1.34 million pounds (609,000 kg) of salmon and trout. If poisoned shellfish are eaten by other organisms, including humans, they can cause food poisoning in the consumers.

Perhaps even more pernicious than overfishing and pollution are the effects that global warming and the weakening of the ozone layer may have on marine life (Chapter 17). Abnormally warm ocean temperatures appear to be killing coral reefs; it is estimated that 11% of the world's reefs had been lost as of 2001, and up to 40% could be lost by 2010. Reefs are particularly important because an estimated 500 million people live within a 100 kilometers of a reef, and depend on the reefs and their biota for food and employment, either directly or indirectly. An estimated 25% of the fish catch in developing countries comes from coral reef areas, helping to feed a billion people. In 1988, when ozone levels reportedly declined 15% because of the ozone hole over Antarctica, phytoplankton levels also decreased by 15% to 20%. Phytoplankton are the small, photosynthetic organisms that inhabit the top layer of the oceans; they form the basis of the oceanic food chain and also help the oceans to absorb carbon dioxide (the prime greenhouse gas). If increasing global heating and destruction of the ozone layer adversely affect the phytoplankton, this will obviously have a detrimental effect on the entire ocean ecosystem. Some researchers have even suggested that life in the oceans may collapse. As phytoplankton die, the oceans will take up less carbon dioxide, which will lead to increased global warming, which in turn will accelerate the destruction of the ocean ecosystem. Ironically, both too much phytoplankton (producing red tide) and too few phytoplankton (destroying the base of the marine food chain) are extremely detrimental to oceanic ecosystems.

What about the possibility of increased "fish farming" through **aquaculture** (used to refer to

Figure 13-11   Red tide seen along a coastline.

aquatic organism farming in general, or freshwater "seafood" farming in particular) or **mariculture** (saltwater seafood farming)? Organisms such as salmon, shrimp, and edible seaweed are already being raised under controlled conditions in many countries. Aquaculture systems can be extremely productive and efficient at producing animal protein—generally more efficient than terrestrial farms (see Case Study 13-2). The drawbacks of aquaculture are that it is very labor-intensive, it can involve very intricate management of delicate ecosystems, and it is not suited to all locations. One must have adequate water of the right purity, salinity, and so forth, and temperatures need to be maintained within close tolerances. The start-up and maintenance costs of aquaculture can be relatively high. Some of the best locations for aquaculture are coastal areas that are being destroyed by pollution and development. For these sorts of reasons, many experts have little hope that aquaculture will ever significantly relieve the world's hunger.

Figure 13-12   Lack of growth (the light areas), for this field of new wheat in Palouse, Washington, indicate topsoil loss.

The Soil of the Earth

Soil is one of our most precious commodities: It is vital for the health and well-being not only of human civilization, but of most terrestrial ecosystems. Without soil we could not grow food, our single most important activity.

Yet we are quickly squandering our natural inheritance of soils. Various estimates suggest that we are losing soil to **erosion** at a rate of 25–75 billion metric tons a year globally. The annual loss of topsoil from agricultural lands averages about 7.6 tons per acre (17 metric tons per hectare) in the United States and Europe and as much as 30 to 40 metric tons per hectare in parts of Asia, Africa, and South America. In contrast, erosion rates in undisturbed natural forests are on the order of 0.004 to 0.05 metric ton per hectare per year; in nature, soil loss is generally more than offset by soil formation. It has been estimated that soil erosion costs the United States some $44 billion a year in direct damage to agricultural lands and indirect damage to infrastructures, waterways, and health (**Figure 13-12**). Globally, the direct and indirect costs of soil erosion may be close to $400 billion a year.

As soils are depleted on prime agricultural lands, crop yields decrease. For every inch (2.54 cm) of topsoil lost, average corn and wheat yields drop by about 6%. Generally, at least 6 inches (15 cm) of topsoil are needed to grow crops. Once the layer of topsoil has become too thin, the land is no longer useful for agricultural production.

We can sustain some soil loss, for soil is continually produced on the surface of the Earth. The generation and maintenance of soils are functions that are performed for "free" by healthy natural ecosystems. Under the best of conditions, however, soil formation is a very slow process. Some scientists have estimated that soil is forming in the United States at an average rate of only about an inch (about 2.5 cm) per century, which is equivalent to about 3.7 metric tons per hectare per year; other researchers suggest that the average rate may be closer to 1 metric ton per hectare per year. Around the world, some studies suggest that average rates of topsoil formation may be as little as an inch (2.5 cm) in 500 or a 1000 years. The inescapable conclusion is that we are losing our soils more quickly than they are forming.

At present, modern civilization is living off—and eroding into—the capital of the past (the soils accumulated over many thousands of years), rather than using the soils in a sustainable manner. We need to learn to live off income; that is, to deplete the soils no faster than they are forming under natural conditions.

What Is Soil?

Soil is a combination of weathered, disintegrated, decomposed rocks and minerals (technically known as regolith) plus the decayed remains of plants and animals (organic matter or humus), small living animals, plants, fungi, bacteria, other microscopic or-

CASE STUDY 13-2

Aquaculture

As catches from natural fisheries have stabilized or even decreased, production from fish farms (including freshwater and marine fishes, mollusks, crustaceans, and other aquatic edible animals) has skyrocketed over the last two decades (**Figure 1**). At present (estimate for 2000, the latest year for which data is available at this writing), about 36 million metric tons of seafood protein are produced by aquaculture annually. For comparison, aquaculture produced just 6.9 million metric tons in 1984, and 24.5 million metric tons in 1995. Today, aquaculture continues to be the fastest growing form of food production in the world; 31% of the world's food fish is produced by aquaculture currently, and this is bound to increase in years to come. China is the leader in fish farming, producing an estimated 68% of the world's output. In terms of volume, but not value of final product, the next largest producer is India, followed by Japan, Indonesia, and Bangladesh. Overall, nearly 90% of all aquaculture is carried out in Asia. Although still small by global standards, Chile has been quickly increasing its aquaculture production, from 32,477 metric tons of fishes in 1990 to 274,216 metric tons in 1999. Eighty-five percent of the Chilean production is farmed salmon and trout.

Fish farms produce not only fishes, such as flounder, salmon, trout, carp, tilapia, and catfishes, but also oysters, clams, shrimp, prawns, and many other aquatic organisms (**Figure 2**). Over 200 species are farmed using aquaculture, although the majority of production is only a dozen or so species. More than half of world production consists of relatively low-value freshwater fishes, such as carp and tilapia, which are primarily raised for local consumption. High-value species, such as salmon, shrimp, and certain mollusks, are grown primarily for export. About two-thirds of fish farming activities take place along inland rivers and in lakes, ponds, and artificial tanks, while the remainder are located along the coasts, in bays, and sometimes even in the open ocean.

An important reason for the steady expansion of aquaculture is that fishes and other aquatic organisms are typically very efficient at turning feed from plants into animal meat. Fishes, crustaceans, and mollusks are cold-blooded, so they do not burn excess calories to keep warm. The water they live in helps support their body weight, so they expend less energy than comparable terrestrial animals. As a result, only two pounds (or less) of feed is typically required to produce a pound of fish, far less than is needed to produce an equivalent amount of beef or pork.

There are numerous drawbacks to fish farming, however. Aquatic farming naturally requires tremendous quantities of clean water, a substance in increasingly short supply. Of course, the water is not consumed in the same way that water is when crop plants are irrigated, but fouling of the water environment can be a real problem. Excess organic wastes may pollute the water to the extent that all aquatic organisms suffer; for instance, excess wastes may induce algae blooms, resulting in oxygen depletion and suffocation of fishes, mollusks, and crustaceans. Like any animal farmers, fish

Figure 1 The amount of seafood protein produced by aquaculture annually.

Figure 2 An aquaculture facility in Louisiana raises catfish. The color differences between ponds are due to the number and type of algae in each pond.

farmers must generally purchase grain products to feed their stock. This, on a global scale, means there is less grain for other purposes (such as feeding terrestrial livestock or even people). Furthermore, some aquatic species, such as shrimp or salmon, are carnivorous or omnivorous. In aquaculture, such species are typically fed pellets with a high-protein content. But where does the protein come from? In some cases, it comes from plant-based proteins, but in many cases the feed pellets used in aquaculture are made, at least in part, from fishmeal derived from relatively low-value fish caught in the wild, such as anchovies and herrings. In such cases, aquaculture is not supplementing and adding to the wild fish catch, but actually consuming part of the wild catch. In contrast to carnivorous species that must be artificially fed a rich diet, marine mollusks (like oysters and clams) raised in pens along coasts and in bays can feed on nutrients that naturally occur in the water and thus require little in the way of artificial inputs.

Specialized and expensive equipment may be necessary in aquaculture operations, especially if the stock is being raised in artificial tanks, ponds, or holding areas. Hormones, antibodies, vaccines, and other medical supplies may be required. The dense populations of fish that are typical of modern fish farms are extremely vulnerable to infectious diseases. Disease outbreaks among farmed aquatic species, especially in monoculture situations (where a single species is being raised in a small, confined area), are a constant threat. In 1999, for instance, farmed shrimp in Ecuador suffered from an outbreak of white spot virus that resulted in a loss of nearly half a billion dollars of product. Inbreeding, resulting in genetically weakened strains, can also be a problem, particularly if cultivated organisms escape and interbreed with a wild population.

Another controversy that has developed in recent years is the use, or potential use, of transgenic (bioengineered or genetically modified) species in aquaculture. For instance, transgenic Atlantic salmon (sometimes colloquially referred to as "Super Salmon") that have been genetically engineered with growth hormones can reportedly grow up to four to six times as quickly as the wild variety, reach larger sizes, and are more efficient at turning feed into fish flesh. As of this writing, the transgenic salmon have yet to be approved by governing authorities in all of the countries where

their use has been proposed, including Canada, Chile, New Zealand, and the United States. Concerns raised relative to the transgenic salmon are comparable to the concerns expressed more generally relative to genetically modified organisms: Are they totally safe for human consumption? Is there a possibility that the modified fish could have a negative impact on natural fish populations and ecosystems? If, or when, transgenic fish escape into the wild, will they outcompete their nonmodified relatives? Will they destroy natural prey populations? (Many critics suggest that it is inevitable that some will escape—farmed salmon have already been known to escape to the wild, sometimes in large numbers. In December, 2000, up to 100,000 farmed salmon held in pens off the coast of Maine escaped when a storm damaged their cages.) Will transgenic fishes interbreed with nonmodified populations and essentially genetically destroy the wild populations? These are all issues that will increasingly come to the fore in future years. At present, there are major efforts underway to derive the benefits of transgenic salmon and other fishes while at the same time protecting wild varieties. Examples include raising and keeping transgenic species in secure, self-contained, land-based facilities so that there is no possibility of escape into the wild, or developing strains where the final adults raised for food purposes are sterile so that even if they inadvertently escape into the wild, they will not reproduce or mate with wild varieties.

Another major problem with fish farming is the space it requires. The best settings for fish farms are along coasts, rivers, and lakes, but these same areas are considered prime waterfront property, and land values are often very high. Furthermore, in some areas, coastal mangrove forests and other wetland areas have been cleared in order to build fish farms. But these coastal wetlands are the breeding grounds for wild fishes, so clearing such areas often causes natural (wild) fish populations to decline.

Still, fish farming continues to expand, and with good reason. It is one of the most efficient means of turning plant products into animal meat. And with proper management, fish farming can have a very low impact on the environment. As the human population increases in future years, we can expect that more and more of the animal protein in our diet will come from the cultivated aquatic realm.

Questions

1. How can aquaculture serve to protect wild populations of aquatic organisms?
2. What are some potential ways that fish farming can harm wild populations?
3. Do you believe that "Super Salmon" and other genetically modified fish varieties should be farmed? If so, under what conditions? If not, why are you opposed to the raising of such varieties?
4. List the advantages and disadvantages of aquaculture. Do you think that the advantages outweigh the disadvantages? Explain.
5. Do you believe that aquaculture operations should be promoted in the United States?

ganisms, water, and air. Typical soil is about 50% mineral and organic matter by volume and about 50% water and air. There are literally thousands of different types of soils around the world, but they all serve the same vital functions in the ecosystems in which they are found. Soils hold nutrients and water in place such that surface fauna and flora can grow and thrive. Without healthy, porous soils, most rainwater quickly runs off the surface of the land instead of soaking in. Soils supply the vital nutrients, such as usable nitrogen, phosphorus, sulfur, carbon, hydrogen, oxygen, and various trace elements and important compounds, to the plants that grow in the soil. As organisms die and are de-composed in the top layer of the soil, the nutrients are recycled back to the above-ground organisms.

Healthy soil is alive, a complex ecosystem unto itself. Without its living components, soil lacks its characteristic properties: texture, fertility, and the ability to dispose of wastes and recycle nutrients. An amazing number of organisms can live in a handful of soil. Larger animals that live in soils include earthworms, mites (relatives of spiders and ticks), millipedes, and insects. Under a square yard (0.84 m²) of pasture in Denmark, researchers found 40,000 small earthworms and related organisms, almost 10 million roundworms, and more than 40,000 mites and insects. This is not even taking truly microscopic organisms into account. An ounce (28.35 g) of good forest soil can contain well over 28 million bacteria, about 3 million yeast cells, and 1.4 million individual fungi organisms. An ounce of good agricultural soil can contain billions of bacteria, 11 million fungi, 1.4 million algae organisms, and 850,000 protozoa.

A typical well-developed soil is not a homogeneous mass. It consists of layers or **soil horizons** that are approximately parallel to the surface of the Earth (**Figure 13-13**). The horizons have different biological, physical, and chemical attributes, such as the amount of living and dead organisms and organic matter they contain, water and air content, texture, structure, color, and mineral content. In any particular part of the world, the soil horizons develop to a characteristic degree based on the underlying bedrock, under the influence of the climate, flora, and fauna over time. In some places, many horizons develop, whereas elsewhere only one or two horizons are distinguishable. From top to bottom, a typical soil profile (a vertical section through the soil at a particular locality) may exhibit the following basic soil horizons: the uppermost organic matter and humus (heavily decomposed organic matter), the **topsoil** (a mixture of mineral matter and humus, alive with microscopic and macroscopic organisms), the subsoil (composed mainly of minerals), a layer of partially disintegrated rock, and the underlying bedrock.

Figure 13-13 A typical soil is composed of five horizons.

Global Assessment of Soil Degradation

For many years, the amount and degree of **soil degradation**, leading to land degradation, has been a topic of extreme controversy. For several decades, certain environmentalists have enumerated cases of **deforestation**, overgrazing, desertification, and clear destruction of once fertile lands. At the same time, other experts pointed out that crop yields and live-stock production have increased significantly since World War II and concluded that land degradation is not a major global problem. This dispute was es-sentially unresolvable without a global database on soil degradation.

A major study entitled the Global Assessment of Soil Degradation (GLASOD) was sponsored by the United Nations Environment Programme and co-ordinated by the International Soil Reference and Information Centre in the Netherlands. GLASOD, unlike many earlier studies, only looked at soil/land degradation that has occurred due to human inter-vention since World War II, specifically, from 1945 to 1990. Hundreds of soil scientists measured the de-gree, area, and causes of land degradation since 1945, and the results were compiled by continental regions and globally. The findings were alarming. Globally, more than 4.85 billion acres (1.96 billion hectares), or 17% of the vegetated land surface of Earth, have been degraded by humans to some ex-tent in less than half a century.

Figure 13-14   Areas of concern for soil degradation. (*Source*: World Resources Institute, *World Resources 1992–93* [New York: Oxford University Press, 1992], p. 117. From *World Resources 1992–93*. Copyright © 1992 by The World Resources Institute. Reprinted by permis-sion of Oxford University Press, Inc.)

On a worldwide basis, livestock overgrazing, deforestation, and agricultural activities account for over 90% of the soil degradation since 1945. Overgrazing is responsible for 35% of land degra-dation, deforestation for 30%, and agricultural activities for 28%. Of course, these percentages vary greatly by continent. Unfortunately, damage from soil and land degradation is still occurring. As **Figure 13-14** shows, every inhabited continent has areas that are at serious risk.

The U.S. Midwest and Great Plains, the bread-basket area, have experienced various degrees of soil degradation. The Soil Conservation Service has determined that approximately a quarter of all U.S cropland is eroding away faster than is sustainable. Central America is suffering from extreme soil degradation and is an area of serious concern for the future. Most of this damage is a result of defor-estation and overgrazing, but improper agricultural practices are also an important factor. In South America, areas of intensive deforestation are par-ticularly in danger of continued soil degradation, as is the mountainous region on the west coast.

Europe, particularly the middle and eastern portions, is suffering from extreme soil degrada-tion that is predicted to continue into the future. Much of this is due to pollutants, including indus-trial and urban wastes and pesticides that have dam-aged the soil. Asia is also suffering from severe and continuing soil degradation, especially in India, China, and Southeast Asia. Asian soil degradation is caused primarily by deforestation, agriculture, and overgrazing. Africa suffers from continued soil

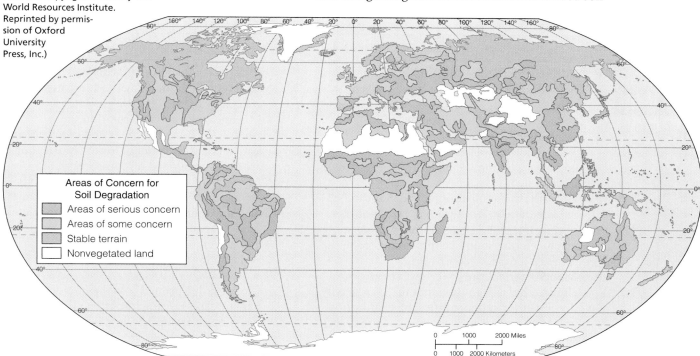

degradation and desertification, especially along the north coast, in the sub-Saharan Sahel region, and in South Africa. Overgrazing, wind erosion, and poor agricultural practices are to blame for much of Africa's problems. Compared to the rest of the world, Australia is an area of only moderate soil degradation. Most of Australia's problems with soil deterioration result from overgrazing.

Stopping Soil Degradation

As the world population continues to increase, stabilization and restoration of soil resources will become increasingly important. Since World War II, agricultural yields have increased greatly, but crop yields would have been even higher with healthier soils. Unfortunately, for decades, modern agricultural technologies have masked the soil deterioration by increasing yields even as the soil has become degraded. If unsustainable agricultural practices continue, the soil will become so degraded that despite our fertilizers, pesticides, and high-yield crop varieties, we will still not be able to produce a good harvest. Once the soil is dead or eroded away, the land becomes barren.

Techniques, such as no-till sowing of crops, drip irrigation, crop rotation, and leaving land fallow, can mitigate or prevent soil degradation, but farmers often do not practice them for simple economic reasons. Farmers often find that it does not make short-term economic sense to invest very heavily (if at all) in soil conservation, preservation, or restoration efforts. Of course, in emphasizing the maximization of short-term profits, the farmer is destroying the capital upon which the business depends, but in a fertile area, the soil may take half a century to become severely degraded—longer than the individual farmer may stay in business. Classically, the next generation would move on to new land, but the world is now running out of new land to place under the plow.

Aside from economic considerations, even if an isolated farmer desires to practice soil conservation techniques, often this is impossible because the action required to stop soil degradation may go beyond the scale of a single farm. As we have discussed, many factors are causing soils to degrade around the world. Soil conservation measures, such as watershed management and river and catchment basin maintenance, may be required on a regional or national level. Such projects can only be implemented by governmental authorities on a local, national, or even international level. An isolated farmer may be virtually helpless if the government does not support and implement sound conservation policies.

Ultimately, it seems that the problem of global soil degradation can only be solved through a variety of actions addressing a multitude of causes on every level from the individual to the international community. The long-term needs of society, which essentially means sustainability in practical terms, must take precedence over all other concerns, be they personal short-term economic gains, debt payment on the part of a poor government, or political jockeying in the international arena. Without healthy soil, civilization as we know it cannot survive.

STUDY GUIDE

SUMMARY

- Virtually all food for humankind comes from other organisms, with just three kinds of plants (rice, wheat, maize [corn]) constituting 65% of the global food supply.
- Depending on estimates, globally at least 500 million people are chronically hungry and over a billion are undernourished and underweight.
- Globally, the number of people that can be fed adequately depends on the components of the diet (grain versus animal products, for instance); how many people can be fed in the future is a topic of controversy.
- Modern agriculture of the last fifty years, including the "Green Revolution," was heavily dependent on monocultures, mechanization, irrigation (where needed), and the use of fertilizers, pesticides, and herbicides.

- Newer, ecologically sound and environmentally friendly farming methods are now increasingly being used, such as Integrated Pest Management and crop rotation.
- Genetically modified foods and crops potentially hold great promise for the future, but are also surrounded by controversy.
- Aquatic organisms (both wild and farmed) are an important component of the human diet, and aquaculture (aquatic organism farming) is increasing at a fast pace.
- Soil is vital for terrestrial food production, and also for the well-being of most terrestrial ecosystems.
- Globally, due primarily to the intervention of humans, soil is being lost at an alarming rate—much faster than it is being produced by natural processes.

KEY TERMS

Agriculture
Aquaculture
Bioengineering
Biological control
Biotechnology
Carryover grain stocks
Climax community
Commercial extinction
Crop rotation
Cultivable land
Deforestation
Desertification

Ecological succession
Erosion
Fertilizer
Grain production
Green Revolution
Herbicide
Hunger
Integrated Pest Management
 (IPM)
Irrigation
Mariculture
Monoculture

Organic farming
Pesticide
Pioneer stage of succession
Salinization
Soil
Soil degradation
Soil fertility
Soil horizon
Topsoil
Transgenic crops
Waterlogging

STUDY QUESTIONS

1. Describe the present global food situation. Is everyone fed adequately?
2. What is hunger? Approximately how many hungry people are there in the world?
3. What three plant species supply most of the world's food?
4. What was the "Green Revolution"?
5. Describe the effects of modern intensive agriculture.
6. Given that the world population continues to grow, more food will be needed in the future. What are two basic strategies that can be pursued to increase world food production?
7. Do you believe that the world could adequately feed a population of 10 billion? Justify your answer.
8. What are the high and low estimates of the number of people that could be fed in the future? What types of assumptions are these estimates based on?
9. What is the major nonagricultural food source? Is it being used sustainably?
10. How can biotechnology, bioengineering, and the increasing use of genetically modified organisms (GMOs) help us deal with increasing food scarcity?
11. What are some of the criticisms of GMOs?
12. How does Integrated Pest Management attempt to control crop pests?
13. What is soil?
14. Discuss the types and extent of soil degradation that are occurring globally.
15. Why are soil and potential soil degradation such important issues?
16. What is being done to help stop global soil degradation?

WHAT'S THE EVIDENCE?

1. The authors state that perfect management of the world's current food supply could adequately feed the world's population. Do you agree? What is the evidence for this statement? Is "perfect management" realistic or possible?
2. The authors contend that global soil loss and depleted soil fertility, in part the result of modern agricultural techniques, are vital issues that must be addressed in this century. Are you convinced they are as important as the authors suggest? Cite evidence to support your answer.

CALCULATIONS

1. Currently, about 3.7 billion acres (1.5 billion hectares) of land worldwide are under cultivation. Assuming that currently just enough food is produced to feed everyone on Earth, if the world population increases from 6.3 billion to 8 billion, and yields per unit area of land remain constant, then approximately how much more land will need to be cultivated to feed everyone?
2. Using the same assumptions as in Question 1, how much more land will need to be cultivated if the world population increases from 6.3 billion to 10 billion?

STUDY GUIDE

1. Study carefully the projections shown in Figure 13-3 (Possible land futures). If present crop yields per area of land remain the same, and the world population grows according to World Bank estimates, approximately how much land will need to be cultivated in 2050 to feed the world's population? What percentage increase is this over the amount of land currently under cultivation?

2. If present crop yields per area of land double, and the world population grows according to World Bank estimates, approximately how much land will need to be cultivated in 2050 to feed the world's population? How does this compare to the amount of land currently under cultivation?

3. Referring to Figure 13-8a (World grain yields), beginning in 1950, approximately how many years did it take to double average world grain yields?

4. Referring to Figure 13-14 (Areas of concern for soil degradation), a continuous band of desert and areas at moderate to very high risk of desertification stretches from the coast of western Africa east into the Asian heartland. About how many thousands of miles long is this continuous band?

www.environment.jbpub.com

Connect to this book's web site: http://www.environment.jbpub.com/ The site features eLearning, an online review area that provides quizzes, chapter outlines, and other tools to help you study for your class. We've posted the answers to the book's odd-numbered questions. You can also follow useful links for more in-depth information. The site provides sources for the information in this chapter and recommendations for further readings.

Growing crops in the desert: Each of these circular features are crop fields southwest of Riyadh, Saudi Arabia.

Pollution and Waste are symptoms, not causes, of the environmental crisis.

Paul Hawken, businessman and environmentalist

Population

Consumption

1. The Environment and
Humans (Chapters 1-2)

Lithosphere

Biosphere

Hydrosphere

Atmosphere

2. The Environment of
Life on Planet Earth
(Chapters 3-5)

Lithosphere

Biosphere

Hydrosphere

Atmosphere

3. Resource Use and
Management
(Chapters 6-13)

Lithosphere

Biosphere

Hydrosphere

Atmosphere

4. Dealing with
Environmental
Degradation
(Chapters 14-18)

Lithosphere

Biosphere

Hydrosphere

Atmosphere

5. Environmental Issues: Social Aspects and Solutions
(Chapters 19-20)

A brown smog covers Denver. Pollution is an inevitable byproduct of modern life. Nearly all human activities produce wastes that will find their way into the air, water, and soil. In contrast to the dominant "throwaway" mindset of consumption-oriented cultures, a sustainable society will minimize the waste produced, and recycle as much waste as possible.

CHAPTER 14

Principles of Pollution Control, Toxicology, and Risk

Up to this point, we have focused on environmental problems caused by resource depletion, or excess inputs to society. We now turn to the second major category of environmental problems, those caused by too many outputs by society (see the diagram at the beginning of Section 4, page 335). It is also important to know which chemicals in our daily lives impose the greatest risks to our health. Despite many problems (both of an ethical sort and questions involving the applicability of animal data to humans), data gathered from laboratory animals provide the most widely used way of making social decisions about whether chemical products, including pesticides, should be used.

What Is Pollution?

Pollution generally refers to excess outputs by society into the environment. In this case, "excess" means something produced in amounts high enough to be harmful to us, other life, or valued objects such as cars and buildings. Almost anything can be harmful if it is concentrated enough in a particular context, so all matter and energy can cause pollution if locally produced in sufficient amounts. Because pollution is such a widespread environmental problem, many aspects of it deserve special consideration:

- *Pollution as matter cycling and energy flow.* All of the environment, including land, sea, air, and life, ultimately consists of matter cycles and energy flows. Pollution represents local concentrations in the matter cycle or energy flow. The rapid burning of fossil fuels, for instance, releases into the atmosphere tons of carbon that was stored underground as coal and petroleum. Similarly, concentrations of energy can be a form of pollution. Heat pollution is a serious form of air and water pollution. Astronomers speak of "light pollution" from nearby cities that disrupts their view of the stars.

- *Pollution as an accelerated natural process.* We associate pollution with belching factories, but many natural processes have been causing pollution for billions of years. For example, volcanoes release gases that are harmful to life, affect global climate, and cause acid rain. Many organisms produce highly toxic chemicals.

    Humans, however, cause pollution at a much greater rate than nature for two reasons, related to both the quantity and the quality of our waste:

    1. The quantity of waste produced is vast. For example, the over 6.5 billion metric tons of carbon we release annually from combustion, especially the burning of fossil fuels, far exceeds what would be released from forest fires and other natural combustion sources.

    2. The quality of our waste accelerates pollution because it includes so many new substances. Although often they are not more harmful than natural materials, these substances are not easily decomposed by natural processes and are therefore more persistent. This quality aspect of waste began in 1856 when the first human-made industrial chemical was created: an artificial dye. In 1895, the first bladder cancer associated with artificial dyes was reported. Between 1957 and 2003, the American Chemical Society recorded over 15 million new chemicals. Now the Society registers new chemicals at the rate of over 70 per hour. Only about 500 of the chemicals invented each year ever reach a wide market. Still, more than 70,000 chemicals are in everyday use worldwide. Yet less than 1% of the chemicals on the market have ever been completely evaluated as potential health or ecological hazards. Most have been subjected to very little, if any, testing.

- *Pollution as a stepwise process.* A primary reason that humans produce so much waste is that large concentrations of matter and energy are produced as byproducts ("waste") during each step of many human activities (**Figure 14-1**). Because "everything must go somewhere" as either energy flowing or matter cycling, this waste often ends up polluting water, air, or land. Increasingly, engineers are using **Life Cycle Analysis (LCA)** to pinpoint steps that can be either eliminated or made more efficient. Both measures reduce waste generation. LCA analyzes the entire "life cycle" of a product (such as an appliance), from mining of raw materials to its eventual disposal by the consumer, and measures how much energy and matter are used at each step.

Objectives

After reading this chapter, you should be able to explain or describe:

What is meant by "pollution"

How pollution is produced

How pollution is controlled

Toxic chemical testing

How society copes with risk

The costs and benefits of pesticides

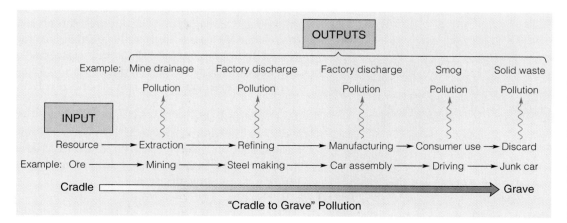

Figure 14-1 "Cradle to grave" pollution refers to the many ways that a single product, such as a car, can pollute during its lifetime. Each step, from mining through final disposal, often results in the release of dozens of toxic pollutants into the air, water, and land (Life Cycle Analyses elucidate such processes). Nearly all products, including food and other agricultural products, create such stepwise pollution.

Sometimes called "cradle to grave" analysis, LCA is most widely practiced in Germany, where laws require that cars and many appliances be recycled. For example, all of the various parts in a car or appliance have their own manufacturing steps. To recycle the entire car or appliance, the life cycle of each part must be studied. In Germany, the parts are labeled (with computer bar codes) when they are assembled to make it easier to dismantle and recycle them when the car or appliance is discarded.

- *Pollution = population × consumption.* Recall from Chapter 1 that environmental impact (I) = population (P) × consumption (C). In this case, the amount of pollution produced by society depends on two basic factors, the number of people (P) and the amount of waste produced per person, mainly determined by consumption.

Total pollution therefore increases as both populations and waste per capita (per person) grow. In the past, waste per capita has been strongly correlated with the rapid growth of resource-intensive industry and technology. "Green" technologies emphasize source reduction as a main goal. The need for reduction is especially great in the United States, which has among the world's highest per capita air, water, and land pollution. Each U.S. citizen accounts for more than 2000 pounds (907 kg) of waste per day when transportation, agricultural, industrial, and household wastes are included. As a result, with 5% of the world's population, the United States produces perhaps 50% of the world's solid waste and 20% to 35% of most other pollutants.

Figure 14-2   A shell mound in Georgia where prehistoric Native Americans discarded shellfish remains and other wastes.

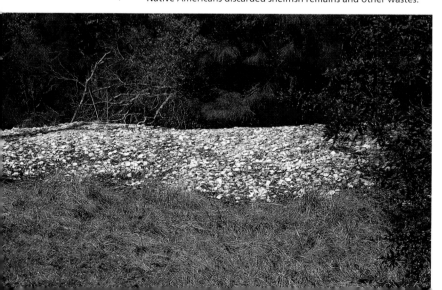

History of Pollution

Natural pollution is as old as the Earth. Ancient volcanoes, for instance, spewed vast amounts of gases into the atmosphere. Many of these gases affected global climate, causing at least some of the mass extinctions that devastated the biosphere in the past. Nor are humans the only animals that pollute. Huge herds of grazing animals have been known to pollute rivers with urine and feces; droppings from large flocks of birds can contaminate the local environment near nesting areas.

Early humans also polluted the environment. The remains of the first "landfills" are among the most common finds in caves and other sites of fossil humans. Often these contain the bones of hundreds of fossil animals (**Figure 14-2**). The first written records indicate that Egyptians, Greeks, and other early civilizations sometimes had to cope with polluted drinking water. As cities grew and population density increased, so did pollution. Roman laws prohibited dumping in certain areas. One of the earliest references to air pollution dates from 1307, when air pollution in London was so bad that King Edward I banned coal burning. Modern wastewater treatment and the flush toilet did not become common until the 1920s in Europe and the United States. Until that time, pollution of lakes, rivers, and groundwaters from human waste posed enormous threats to drinking water quality. Not until 1912 was the first federal law enacted to set standards for drinking water in the United States.

While pollution has always occurred, the scale and rate of pollution in the last few decades are vastly greater than in the past. In today's world, large human populations and fossil fuel-driven technologies are daily producing huge volumes of many different kinds of pollutants on regional and even global scales. These pollute all parts of the environment, from land to water to air to living things.

Controlling Pollution

We often associate environmental deterioration with smokestacks and pipes pouring out toxic substances. Unfortunately, controlling pollution involves far more than just passing laws that require factories to stop toxic emissions. Realistic efforts to control pollution must consider the many complexities involved.

Myths of Pollution Control

We begin by examining some of the misconceptions about pollution and how it can be controlled. These myths are widely held and often lead to costly political and economic decisions.

Myth of Purity in Nature.   We often hear phrases such as "pristine waters" that imply a certain purity in natural places untouched by humans. In reality, virtually nothing is "pure" in nature. A single drop of water contains about 500 trillion molecules. Even before humans existed, no drop of rain or river water would have contained 500 trillion molecules of only H_2O. Millions or, more likely, billions would have been molecules of various dissolved gases, organic solids, mineral solids, and many other natural "pollutants." Similarly, the "pure" country air we breathe contains a huge number of gases and particles. Furthermore, even highly polluted water or air usually contains only tiny concentrations, far less than 1% of toxins or other pollutants. This is why pollution is measured in parts per million or even billion. Even raw sewage entering treatment facilities is usually at least 99.9% water.

Myth of Zero Pollution. Although eliminating all pollution sounds ideal, zero pollution is an unrealistic goal for four reasons. First, modern society cannot exist without producing pollutants. Everything must go somewhere, so human activities will inevitably produce waste matter and energy. Second, as we will see, the costs of removing all pollutants from any given activity increase exponentially after a given point, making total purity economically impossible in nearly all cases. Third, the benefits of pollutant removal decrease exponentially after a point, making total purity unnecessary. Fourth, not even nature is totally pure. Even if zero pollution were economically feasible, should we necessarily seek an environmental standard higher than nature?

Myth of Zero Risk. Most of us have many misconceptions about risk, including that of a risk-free existence. In reality, every activity, from eating peanut butter to taking a walk, involves some risk. Therefore, in all activities, including control of pollution, we can at best seek to minimize, not eliminate, the many risks we face. Unfortunately, our efforts to minimize risk are greatly hindered by our inaccurate perceptions of risk.

Deciding How Much Control: Being Realistic

The pollution myths have had a strong influence on policy, leading to some unrealistic and costly decisions. Pollution control in the 1970s and early 1980s emphasized "rights-based" and "technology-based" approaches. Rights-based control assumes that all individuals, even the most pollutant-sensitive, have a right to be exposed to the least pollution that society can provide. For example, the Clean Air Act of 1970 set pollutant concentrations at levels low enough to protect even the most sensitive citizens. Technology-based control sets pollutant levels ac-

cording to technological ability; technology-based laws often specify that companies reduce pollution by using the "best available technology."

In the past, rights- and technology-based approaches have been very successful in improving many aspects of air and water quality. But many people have argued that they tend to be too stringent and costly because they strive for zero pollution and zero risk, which are unattainable ideals. Determining pollution control based only on the most sensitive individuals or the best technology is a very narrow approach that can lead to overprotection for most of society at an unnecessarily high cost, according to critics.

Rise of the Benefit-Cost Approach. Concerns about a declining U.S. economy have led to increasing interest in lowering the costs of pollution control in recent years. As **Figure 14-3** shows, the portion of U.S. businesses' total expenditures that goes for pollution control has been climbing since the early 1980s; the 2001 percentage of 19% represented more than 2% of the gross national product (GNP). Some critics argue that much, or all, of these costs were a needless drain on the U.S. economy.

Others argue that the money spent on pollution control has yielded many benefits to human health and the environment. Nonetheless, the trend toward increasing costs, combined with general economic problems, has encouraged a search for less costly approaches. The environmentalism of the 1960s and 1970s has gradually been replaced by a sustainability movement, beginning in the early 1980s. One of the traits of the sustainability movement is its focus on addressing fundamental economic changes, as opposed to the legal mechanisms often emphasized by environmentalism (Chapter 1). In the case of pollution control, this has meant replacing the rights- and technology-based approaches with a benefit-cost approach.

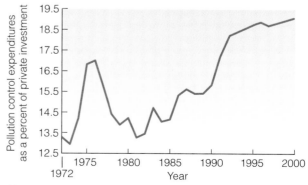

Figure 14-3 The cost of pollution control has risen from about 13% of business investment in 1972 to about 19%. (*Source*: U.S. Commerce Department.)

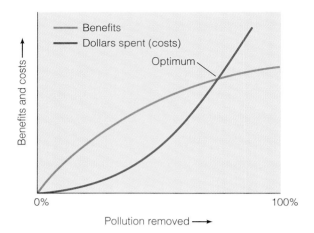

Figure 14-4   As pollution is removed, the relative benefits of further pollution control decrease, and the economic costs of control increase.

Benefit-cost analyses (BCA) address the benefit-cost ratio. With pollution control, this means finding the greatest relative health and environmental benefits for the least amount of economic cost. The cost of pollution control increases slowly at first so that the initial reduction in pollution is relatively cheap (**Figure 14-4**). But after a certain point, cost increases rapidly as increasingly refined equipment and more energy must be applied to extract the more problematic pollutants, such as finer particles or less soluble gases.

Fortunately, society usually does not have to pay the very high costs of further removal because the detrimental effects of pollution tend to follow a curve that is the inverse of the cost curve. As pollution is initially removed, rapid benefits accrue, but the "law of diminishing returns" sets in as pollution removal continues, and continued health benefits are much less. The point where the cost and benefit curves intersect can be considered the point where benefit:cost (ratio = 1) is optimized, denot-

ing maximum relative benefit for the cost incurred. Ratios greater than 1 indicate that additional controls could result in a net overall gain to society. Ratios less than 1 indicate "overcontrol," where controls are costing society more than the benefits it receives.

While benefit-cost studies are extremely useful in social decisions about pollution control, they are not without problems. Benefit-cost results may be very misleading, or even worthless, if some benefits or costs are omitted or inaccurately estimated. This is very likely when dealing with environmental problems because it is very difficult to estimate the value of many environmental resources. Recall that society has historically tended to underestimate the true environmental costs of most activities. For example, we can measure the health costs of particulate matter relatively easily using medical data. But we should also include the aesthetic damage done to wilderness areas if plants are killed. What is the value of your favorite fishing lake? How much would you pay to keep it from being polluted? Such environmental benefits of pollution control are not only difficult to place a dollar value on, but people value them differently. To some people, a wilderness area has less value than a shopping mall (**Figure 14-5**). Others place no value on the mall.

Figure 14-5   Which do you value more—the shopping mall or the wilderness? Or do you value them the same? Determining how much people value something is extremely difficult, yet it is essential if natural resources are to be conserved and managed.

SOURCE

Input reduction
(conservation)

1. Efficiency improvements

High input

2. Reuse and recycle

3. Substitution

Society
Population and
Consumption

Lower throughput

SINK

Output
management

4. Remove
(pollution control)

5. Remediate

6. Restore

Figure 14-6   Input reduction, or conservation, can be accomplished in three ways. Green technologies are used to achieve efficiency improvements, reuse/recycling, and substitution.

This "valuation problem" is one of the most fundamental obstacles to solving pollution and other environmental problems. A partial solution that economists often use is called the **Contingent Valuation Method (CVM)**; it attempts to "objectively" measure the dollar value of changes in environmental quality, often by using questionnaires and other surveys that ask people what they would pay for various environmental improvements. One survey, for example, found that U.S. citizens would be willing to pay $1.30 to $2.50 per household for air pollution controls on a power plant that was reducing air quality in the scenic Grand Canyon. CVM has been used for more than 20 years and has considerably improved our understanding about environmental values, such as how much people are willing to pay to remove pollutants. Even so, many hurdles remain because personal and social values are inherently subjective, change often, and are difficult to measure.

Input versus Output Controls.   The flow of matter and energy through society can be reduced in two basic ways, input reduction or output management (**Figure 14-6**). Input reduction is usually the most desirable way to reduce pollution because it also slows resource depletion and usually saves money, such as by increasing efficiency and reusing/recycling materials.

In contrast, output management deals with the flow of matter and energy only after society is finished with it—when it becomes waste. This rarely promotes resource conservation. It also is generally very expensive compared to input reduction. Wastewater treatment facilities and scrubbers that trap particles and gases in smokestacks are examples of pollution control through output management. Whereas input reduction saves money by increased efficiency and reuse/recycling, output management increases costs even further because the pollution must be cleaned up after it is produced. In

the case of pollution control, these added costs include (1) cost of the pollution control equipment, or **abatement devices**, which are relatively expensive, and (2) cost of waste disposal. An average coal-burning electricity generating plant, for example, produces many tons of toxic ash, sulfur, and other waste that is extremely expensive to remove and dispose of in landfills or incinerators. This demonstrates why input reduction has much higher benefit-cost ratios than output management. If we do not produce the waste to begin with, such as by using low-sulfur coal, there is no need to pay for equipment to eliminate sulfur or pay for its disposal.

Besides pollution control, Figure 14-6 also illustrates two other kinds of output management, remediation and restoration of the polluted environment. **Remediation** seeks to counteract some of the effects of pollution after it has been released into the environment. For example, lakes polluted with acid rain are sometimes treated with lime to neutralize the acid. Consider how expensive this is compared to stopping acid rain by burning low-sulfur coal or even trapping the sulfur before it is released. Remediation of toxic waste dumps has proven so expensive that the amount of Superfund money originally estimated by Congress is much too small. A rapidly growing type of remediation that promises to lower costs in some cases is bioremediation, which uses microbes to digest pollutants in groundwater, oil in oil spills, and many other toxic substances.

Restoration is more ambitious than remediation. Rather than just counteracting some of the effects of pollution in the environment, restoration seeks to reinstate the environment to its former condition. Restoration of the Kissimmee River and Everglades in Florida and restoration of the tallgrass prairie in the Midwest are examples. Restoration is, not surprisingly, the most expensive of the three output management methods. The eventual cost of partly restoring the Everglades has been estimated at $2 billion. Furthermore, complete restoration is almost never accomplished in any project. Once the environment is harmed, it is nearly impossible to fully return it to its former state.

Despite the high costs, pollution reduction attempts have traditionally emphasized output management. Most of the state and federal laws passed during the 1970s and early 1980s, such as the Clean Air and Water Acts and the Superfund Act, mandated pollution controls and pollution cleanup. But as noted above, the early 1980s saw increasing interest in reducing costs, such as by the use of benefit-cost analyses. Such analyses repeatedly indicate that pollution can be reduced much more cheaply by greater application of input reduction methods. Reducing the need for output control with increased efficiency, reuse/recycling, or other input reduction methods can result in more pollution removed for less money. Growing numbers of businesses are recognizing this.

Implementing Pollution Controls

Implementing pollution controls is often the most difficult step because it usually entails an initial monetary cost and/or change in behavior. Even though society as a whole may decide that the benefits of controlling a certain pollutant outweigh the costs, the average citizen is often reluctant to pay the added costs and sacrifices necessary to bring this about. Even the cheapest forms of pollution control, such as increased energy efficiency, require an ini-

tial cost, perhaps buying a new, smaller car, that only yields benefits after payment.

Social scientists tell us that there are three basic ways to promote change in human activities such as those that pollute:

1. *Persuasion.* Ask people to change polluting behaviors.
2. *Regulation.* Pass laws requiring less pollution.
3. *Incentive.* Reward behavior that reduces pollution.

Persuasion can be an effective way of changing people's behavior, especially if it is accompanied by education. People are much more likely to change if they have information about the harmful consequences of their actions. An example would be an advertisement urging people not to pour household chemicals into the sewer system. Persuasion is also relatively inexpensive compared to regulation and incentive methods, but because it has no means of enforcement, persuasion alone is rarely sufficient. By itself, it is generally the least effective of the three methods, but it can be used to supplement the others.

Legal Aspects of Pollution Control.   Table 14-1 lists the major federal laws regulating pollution in the United States. Most of these laws are discussed in

TABLE 14-1	**Some Federal Laws Regulating Chemicals in the United States**			
Year First Enacted	**Law**	**Regulatory Agency**	**Regulated Products**	**Regulatory Model[a]**
1938	Food, Drug and Cosmetics Act	FDA	Food, drugs, and cosmetics, medical devices, veterinary drugs	Risk (food, cosmetics) Balancing (drugs, medical devices)
1947	Federal Insecticide, Fungicide, and Rodenticide Act	EPA	Pesticides	Balancing
1960	Federal Hazardous Substances Act	CPSC[b]	Household products	Risk
1970	Occupational Safety and Health Act	OSHA	Workplace chemicals	Technical feasibility
1970	Clean Air Act	EPA	Air pollutants	Risk (stationary sources). Technical feasibility (moving vehicles)
1972	Clean Water Act	EPA	Water pollutants	Technical feasibility
1974	Safe Drinking Water Act	EPA	Drinking water contaminants	Technical feasibility
1976	Toxic Substances Control Act	EPA	Industrial chemicals not covered elsewhere	Balancing
1980	Superfund Amendments and Reauthorization	EPA	Contaminants at waste sites	Risk/technical feasibility

[a] "Risk" means the agency considers only risk information when reaching decisions. "Balancing" means that both risks and benefits are considered. "Technical feasibility" means that the law requires the agency to consider not only risks, but also the availability of technology to control risk. Some laws invoke more than one model.

[b] CPSC: Consumer Product Safety Commission.

Source: From Table 8, p. 203, of Joseph V. Rodricks, *Calculated Risks* (New York: Cambridge University Press, 1992). Reprinted with the permission of Cambridge University Press.

the appropriate chapters that follow, but some key generalizations should be noted here. First, these laws can be subdivided into two basic categories: those using the input approach and those using the output approach. The input approach is used in laws that control the production of toxic chemicals, such as FIFRA (Federal Insecticide, Fungicide, and Rodenticide Act) and TSCA (Toxic Substances Control Act). These laws try to restrict the creation of materials that can become pollutants. Other laws use the output approach, trying to reduce pollution after it has been produced. Examples include the Clean Air and Clean Water Acts, which limit pollution discharges from factories and other places, and the Superfund Amendments, which mandate pollution cleanup. Also note that the laws are administered by different agencies (although the EPA administers most of them) and use different regulatory models.

Regulation by laws is most useful where polluters are few in number and the pollution can be easily monitored. For example, hazardous waste disposal and large factories are relatively easy to oversee. It is also important where the pollution is very dangerous and must be tightly controlled; again this includes toxic and hazardous waste.

Economic Aspects of Pollution Control. The United States has seen a rapid increase in regulatory laws at the local, state, and federal levels. Many businesses and citizens complain that there are too many laws and that they are too complicated, with environmental lawyers and administrators being the main benefactors. Furthermore, laws are not very effective at controlling activities that involve many people acting independently. It is difficult and costly to regulate pollution where there are many polluters, such as people littering or disposing of used oil. In such circumstances, incentive methods are often better.

Rather than using force and other legal threats, incentive methods provide economic rewards for nonpolluting activities. This controls pollution at much lower costs because society does not have to pay for many law enforcement officials to police the many individual polluters. Instead, individuals carry out nonpolluting activities on their own because they are monetarily rewarded for doing so. Two common examples are (1) deposits paid when potential waste is purchased and (2) "pay as you throw" schemes that make the polluter pay for waste discarded.

Requiring deposits for such items as aluminum cans will greatly increase the rate of return for can recycling. Generally, the larger the deposit, the

higher the rate of return. The use of deposits has been so successful with cans that it is being increasingly applied to other waste items. For example, discarded car engine oil is a hazard to water supplies in many areas. Initial studies show that requiring a deposit when new oil is purchased, and reimbursing the buyer when the waste oil is returned, can drastically reduce oil waste. "Pay as you throw" schemes are also becoming common, especially with solid waste disposal. Both industry and households are being charged for the amount of waste that is picked up by waste disposal companies.

In summary, regulatory means of pollution control are most effective where polluters are relatively few in number and can be readily monitored. Incentive means are better for pollution from many sources and generally are less costly than regulation. Persuasion is a generally weak but cheap form of social implementation that can supplement the other two.

International and National Aspects

Pollution in both ex-communist countries and developing nations illustrates the economic principles just discussed. Ex-communist countries are generally the most polluted in the world largely because the old communist governments encouraged pollution by subsidizing fossil fuels. This kept prices very low so people had no incentive to conserve. Studies show that raising energy prices to Western European levels will decrease air pollution by more than 50%. This again demonstrates the effectiveness of input reduction.

Developing countries illustrate input management of pollution in another way, that of investment in "clean" (green) technologies. Such technologies emphasize hydropower and other renewable resources for electricity generation. Clean technologies not only decrease pollution as those countries industrialize, but according to studies by the World Bank, the overall cost of these technologies is lower than the cost of current technologies due to improved efficiency and less fuel use.

National Pollution. As a leading industrialized country, the United States is the world's leading polluter in many respects. For instance, the United States produces over 25% of the world's greenhouse gas emissions. Nevertheless, some types of water and air pollution in the United States have been reduced considerably since the early 1970s due to more stringent efforts at all levels of society. Unfortunately, the amount of solid waste in the United States continues to increase exponentially, driven by a steady increase in consumption.

Toxic releases: Top Five U.S. industries

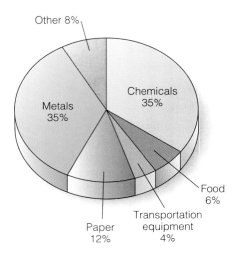

Total release=2,389 million pounds (1,084 million kilograms)

Figure 14-7   About 70% of toxic releases reported in the 2000 TRI came from the chemical and metal industries. (*Source:* Environmental Protection Agency.)

An annual overview of toxic pollution in the United States is provided by the **Toxics Release Inventory (TRI)**. The Emergency Planning and Community Right-to-Know Act of 1986 requires industries to report to the EPA the amounts of their toxic releases. The EPA uses these reports to produce the TRI. Several hundred toxic substances must be reported. Nevertheless, environmental groups have criticized the TRI for not including many kinds of toxics. Another criticism has been that some polluters, such as coal-burning power plants and sewage plants, are not required to file reports.

Despite these shortcomings, the TRI data have proved very useful in tracking the flow of toxics in the United States. For example, the 2000 TRI data show that most toxics were released by the chemical and metal industries (**Figure 14-7**). Direct re-

TABLE 14-2	**Direct Releases of Toxic Substances between 1990 and 2000**	
	Percentage change 1990–2000	
Total releases	−39%	
Air releases	−38%	
Water releases	−14%	
Underground injection	−48%	
Land releases	−38%	
Source: Environmental Protection Agency, 2002.		

leases into all parts of the environment have declined significantly due at least in part to community efforts and the publicity generated by the release of the TRI data (**Table 14-2**).

Toxicology: The Science of Poisons

Toxicology studies the effects of chemicals that are harmful or fatal when consumed by organisms in relatively small amounts. A conventional pollutant may be deadly in parts per million, whereas a toxic pollutant can be deadly in parts per trillion or even less. Toxic pollutants include not only pesticides, which are purposely made to be toxic to certain target organisms, but thousands of other chemicals that have toxicity as a side-effect. Remember that more than 70 new chemicals are registered every hour and more than 70,000 chemicals are in everyday use worldwide. An estimated 2% of the thousands of new chemicals available each year are sufficiently harmful to be considered toxic.

Effects of Toxic Substances

Toxicologists trace the path of a toxic substance through the body in five steps identified by the acronym ADMSE: absorption, distribution, metabolism, storage, and excretion. Absorption of toxic substances can occur in three ways: ingestion with food or drink, inhalation, or skin contact. The body rarely absorbs 100% of any substance that enters it. For example, about 50% of the lead in food is absorbed—and much less if the lead is in a chemically nonreactive form. After toxic substances enter, they can be distributed via the blood to cells where they are metabolized. Metabolism refers to the body's biochemical response to the toxic substance. Sometimes metabolism can detoxify the poison, rendering it less harmful in the body. In other cases, the opposite occurs, and harmless substances are made harmful.

A **carcinogen** is a cancer-producing substance. Cancer is the uncontrolled multiplication of cells, often forming tumors and spreading throughout the body. Cancer causes about 22% of U.S. deaths annually. There are numerous kinds of cancers, some of which can be inherited. In many cases, however, environmental factors, such as diet and chemicals in the workplace, can contribute to or directly cause cancer. Tobacco and diet are the two most important environmental factors, accounting for perhaps two-thirds of avoidable (not inherited) cancer deaths (**Table 14-3**). Notice that these, like most causes of cancer deaths, are largely related to our personal lifestyle choices. Pollution and food additives, which are cancer causes often associated with environmental problems, are a very small percentage of cancer deaths. Such estimates are very approximate because it is obviously very difficult to do controlled experiments with humans. Nevertheless, similar results from many rigorous statistical studies lead most cancer experts to consider these figures valid as rough estimates of noninherited cancer causes.

Two types of chronic toxicity can cause birth defects. A **mutagen** is a substance that causes genetic mutations in sperm or egg cells. A **teratogen** is a substance that affects fetal development, such as when the mother drinks alcohol or has other harmful chemicals in her body. Teratogens are common because development is a very complex sequence of cellular and biochemical interactions and is therefore easily disturbed. A general rule of thumb is that sensitivity to chemicals decreases with age, at least until old age is reached. Very young embryos are most sensitive to chemicals because cell multiplication, organ formation, and many other delicate growth processes are occurring. This is why pregnant women are discouraged from smoking, drinking alcohol, and taking drugs.

Toxic Risk Assessment

One of the most important changes in environmental policy in the 1980s was the incorporation of risk measurements into environmental decision making. This was based on the realization that everything involves costs (including risk) and that the benefits of substances must be weighed against the risks to humans and other organisms. During the 1990s, interest in risk grew even more.

For this policy to work, we must have some way to measure the risks. One way is through the use of epidemiological statistics, which examine correlations between environmental factors and the health of humans and other organisms. The strong relationship between smoking cigarettes and developing lung cancer, for instance, is statistically highly significant. This approach has two disadvantages. First, such statistical correlations examine harm "after the fact," so that much damage has already occurred by the time we determine that a chemical was harmful. Second, the best evidence in science is usually based on controlled experiments, where certain variables are held constant. Information on toxicity therefore often requires the direct testing of a chemical's effect on organisms under controlled conditions. Of course, humans are not usually tested directly, so animals ranging from bacteria to primates are used (**Figure 14-8**). However, animal testing raises the issue of animal rights (see Case Study 14-1).

Toxic risk assessment in the United States is a four-step process:

1. *Toxic identification.* Determines if a chemical is toxic, usually by testing microbes.
2. *Dose-response assessment.* Determines the strength of toxicity, often by testing rodents.

TABLE 14-3	Estimated Causes of Cancer Deaths
Factor	**Percentage of Total Cancer Deaths**
Tobacco	30
Alcohol	3
Diet	35
Reproductive and sexual behavior	7
Occupation	4
Food additives	<1
Pollution	2
Industrial products	<1
Sunlight, ultraviolet light, other radiation	3
Medicines, medical procedures	1
Infections or inherited factors	13
TOTAL	100

Source: Data from R. Doll and R. Peto, "Avoidable Risks of Cancer in the U.S.," *Journal of the National Cancer Institute* 66 (1981): 1191.

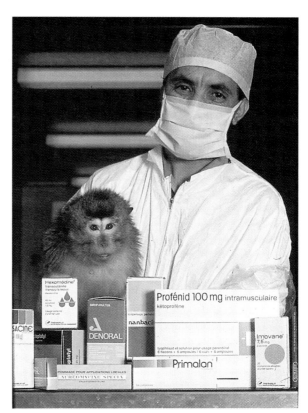

Figure 14-8 Animals are used to test the effects of substances on humans. Seen at left with the researcher is a rhesus monkey.

CASE STUDY 14-1

Should We Use Animals to Test Products? Which Animals?

Visit www.environment.jbpub.com for more information.

Many people question whether animals should be used to test products. Animal activist groups, such as People for the Ethical Treatment of Animals (PETA) and the Animal Liberation Front (ALF), oppose animal experimentation. While some argue that no animals should be used for experimentation, others oppose the use of only certain kinds of animals (such as monkeys and apes). Others agree that animal experimentation is needed, but call for more humane treatment: The amount of pain inflicted, which is often ignored, should be considered.

While most animal rights advocates take a nonviolent approach, others have acted violently. The National Association for Biomedical Research documented 107 attacks on institutions conducting animal research between 1981 and 1994. The monetary damage from these attacks was estimated at more than $7.7 million, which does not include the loss of scientific data that could have saved human lives or even animal lives in the case of data used for veterinary medicine.

Perhaps the most desirable solution to the animal-testing controversy is to find ways of testing products that provide the same toxicological information but do not use "higher" animals (such as mammals) or at least do not inflict pain on or kill them. Fortunately, several new methods are becoming available, although none is fully satisfactory for all purposes. These methods include computer models, "in vitro" tests of cell cultures, and use of microbes and invertebrates. Computer models based on previously gathered data can be very powerful predictors if properly utilized. Knowing the biochemical pathways of the way one substance is metabolized, for instance, can allow us to predict how a similar substance will affect an organism. Unfortunately, computer models are always limited by the quality of the data and the accuracy of the assumptions made. Just one inaccurate assumption out of hundreds can nullify the model and produce invalid results. Consequently, many toxicologists are skeptical that models will ever replace high-quality data gathered under controlled laboratory conditions. In vitro tests of cells or tissues can avoid animal pain or death where the cells can be grown in the lab, but such tests are very incomplete compared to whole-animal data because they provide little information on how organ systems in the animal will interact if exposed to the chemical.

Fewer objections are raised to the use of microorganisms and invertebrates such as shellfishes or insects, but chemical testing on them is often criticized for two reasons. One is scientific: The less closely related a species is to humans, the less similar to humans are its biological responses to chemicals. Hence, vertebrates, especially monkeys and apes, provide the closest analogs to humans. In contrast, the genes, biochemistry, and physiology of microbes and invertebrates are often so different from our own, that the data produced are useless for human toxicology. The other problem is ethical: Can we say that a mammal, for example, deserves more humane treatment than, say, a shellfish? If so, is this just because we are mammals? Or is it because mammals are "smarter" or feel more pain? And where do we draw the line? Is it valid to subject some mammals, such as mice, to pain, but not others, such as monkeys?

These questions cut to the very heart of the animal-testing issue, and there are, as we often find in environmental issues, no simple answers. Perhaps a "techno-solution" can eventually be found, such as the use of very complex computer models or cloning of cell tissues that will provide precise toxicological information. But until that time, animal testing seems certain to remain controversial as the debate over the needs of humans versus the rights of animals continues.

Questions

1. Do you think that all animal testing should be banned? Why?
2. Should apes be used for testing? Why or why not?
3. How can computer models reduce the use of animal testing? Give specific examples.
4. What are two criticisms of chemical testing of invertebrates?

3. *Exposure assessment.* Determines how often humans are exposed to the chemical.

4. *Risk characterization.* Considers the scientific data from steps 1–3 along with societal values to reach a final decision.

The first two steps use animals; dose-response assessment may test higher animals, including dogs or monkeys.

Toxic Identification.   In the initial step of toxic identification, researchers seek evidence if a certain substance is toxic. Given the number of new chemicals developed each year, it is impossible to thoroughly test each chemical for toxicity. Instead, government agencies and manufacturers use hierarchical sequences of tests, called **bioassays**, to test each chemical. A relatively short, inexpensive initial test is administered to search for any possible indications of harmful effects. If harmful effects are indicated, then progressively more thorough tests are carried out until statistical analysis shows that the chemical is harmful (or not) with a high degree of certainty.

An example is the test sequence for cancer, which begins with a short-term test to see if the chemical may cause mutations. The most widely used mutagenicity test, called the *Ames test*, subjects certain strains of bacteria to the chemical. These bacteria are specially bred to be incapable of normal reproduction by cell division. If the bacteria begin to multiply and form visible colonies after the test chemical has been applied, it is likely to be mutagenic because only bacteria that have mutated back to a form capable of normal cell division can reproduce. As some mutagens can cause cancer, a series of more thorough tests for carcinogenesis is begun. These involve longer-term tests (several months), using mice and rats, to see if tumors develop in specific organs. If these tests also show signs of carcinogenic properties, the chemical is subjected to an even lengthier and more costly set of tests, the chronic carcinogenesis bioassay. This involves hundreds or thousands of animals over a period of several years.

These procedures for identifying toxic substances have two basic problems: (1) they do not test humans directly, and (2) they cannot detect very small risks. With animal tests, we can never be totally certain that a chemical is safe for humans. The complex biochemistry of even closely related species often differs in subtle ways, called **intrinsic factors,** so that a chemical that has no effect on one species may have acute or long-term effects on another. The inability to detect very small risk arises because even the long-term testing of thousands of animals cannot generate statistically certain evidence on chemicals that have slight effects.

Dose-response Assessment.   If a chemical shows evidence of toxicity in the bioassays, the next step is to measure the strength of toxicity. **Supertoxins** can cause death in very tiny amounts; less than a drop can kill an adult human. Conversely, the chemical may be slightly toxic, with only very large doses being harmful. Toxic strength is essential to know because the benefits may outweigh the harm if the substance has nontoxic uses and is toxic only in amounts much greater than normally encountered, as is the case with aspirin. Any substance can cause harm in sufficient amounts, even vitamins and other nutrients that we require in moderation.

Dose-response assessment is based on the application of various doses of the test toxic substance to a number of organisms, usually animals, to see what the response is: no harm, impairment, or death. The statistics generated are then used to make probabilistic statements on the chemical's effect on humans. Lethal dose curves, which examine the dose needed to cause death in the test organism, are among the most common types of statistics. The usual pattern in any population of organisms is a bell-shaped curve. Some individuals are much more sensitive to the toxic substance than others, some are more tolerant, and most are intermediate (**Figure 14-9**). The mid-point represents the dose that kills 50% of the population, called **LD-50** meaning Lethal Dose 50%. LD-50 has become the standard reference for summarizing the toxicity of substances.

LD-50 testing suffers from the same two problems as toxic identification: These tests use nonhuman data and cannot detect very small effects that accumulate over a long time. The problem of small effects is most pronounced with cancer-causing substances. For noncarcinogenic substances, it is usually

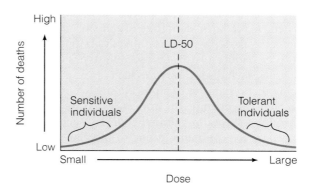

Figure 14-9   Sensitive individuals die at small doses, but a few tolerant individuals can sustain large doses. Most of the population dies at doses around LD-50, which is the dose where 50% of the population has died.

assumed that there is a **toxic threshold** below which no harm is done. For carcinogens, it is often conservatively assumed that exposure to any amount will create some increased chance of cancer in a population, so there is no threshold. For example, a study of mice may indicate that only 100 cancers may be produced in a population of one million mice over the next few years. Even though the calculated risk is very small (0.01%), it may be considered serious enough to make a chemical product unmarketable.

Another obstacle to extending animal results to humans is differences in body sizes, often called **scaling factors**. For example, dosage is generally measured as the amount of a substance administered to the organism divided by the body weight of the organism. When LD-50 data on mice are applied to humans, the data must be adjusted for body weight differences between mice and humans.

Exposure Assessment.   While much public attention is given to identifying and measuring the toxicity of chemicals, the amount of exposure is equally important. Even very toxic substances can do no harm if we are not exposed to them in sufficient amounts. The first step in exposure assessment is to determine possible pathways by which toxic chemicals may reach humans. The three most important pathways are inhalation (air transport), eating (food transport), and drinking (water transport). Pathways can involve many steps, and many substances have a number of different, often complex pathways.

In addition to mobility, the persistence of a substance is also important. Persistence refers to whether the substance remains intact long enough to be transported long distances. In the case of exposure by food, a crucial property is the tendency toward bioconcentration. **Bioconcentration** (sometimes called *bioaccumulation* or *biomagnification*) is the tendency of a substance to accumulate in living tissue; it involves two steps. First, an organism takes in a substance, but does not excrete or metabolize it. Then, a predator eats that organism and in the process ingests the substance already accumulated by its prey (**Figure 14-10**). This second step is common in fishes, shellfishes, carnivores such as eagles and cats, and other organisms that prey on smaller organisms so that the chemical is concentrated by passage through the food pyramid. Mollusks, such as scallops, are exceptionally prone to bioconcentration because they feed by filtering millions of tiny plankton from the water. **Heavy metals,** such as mercury, attack the nervous system and are among the many substances that bioconcentrate in living tissue.

Recently, a relatively unstudied impact of toxic pollutants on wildlife and humans has been arousing concern. Often called **environmental hormones,**

these pollutants send false signals to the complex hormonal systems that regulate reproduction, immunity, behavior, and growth. The result can be infertility, behavioral and growth abnormalities, and low resistance to disease. Some ecologists suspect that hundreds of chemical products may have caused population declines in many species of birds, fishes, and other wildlife, but the effects are usually so subtle and complex that they are difficult to prove (see Case Study 14-2).

A final key aspect of exposure assessment is that we can be exposed to many chemicals simultaneously. Consequently, we need to understand how substances can affect us when they interact with one another. There are three basic kinds of interactions: antagonistic, additive, and synergistic. Antagonism occurs when substances work against each other and cancel each other's effects. For example, deadly snake venoms, such as nerve toxin, can be neutralized by chemical antidotes that destroy the toxic molecules. Additivity occurs when the toxic effects are added together. If chemical A harms the liver and chemical B harms the kidneys, then exposure to both will harm the liver and kidneys. **Synergism** is perhaps the most important because it occurs when the toxic effects are multiplied through interaction, making this the most dangerous type of interaction of all. For example, smoking cigarettes increases your chance of getting lung cancer, as does inhaling radon gas. But a smoker who is exposed to radon gas has a much higher chance of developing lung cancer than one would predict by adding the chances of getting cancer from smoking and radon gas. Tobacco ash particles in the lungs serve as attachment areas for radioactive radon particles, greatly enhancing the chance of getting lung cancer.

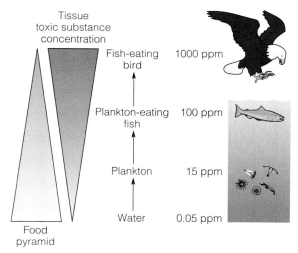

Figure 14-10   Organisms occurring higher on the food pyramid tend to have increasingly greater concentrations of toxic substances in their tissue. (Note: ppm = parts per million.)

Risk Characterization. The fourth step of toxic risk assessment is risk characterization, which combines scientific data from the first three steps with societal values to reach a decision about what to do with a new chemical. Is it too dangerous to be marketed? Do the risks outweigh the benefits? Because human emotions, economics, and many other social variables are involved, this is often the most complex step of all.

Risk refers to any potential danger. The first three steps of toxic assessment provide data on the risk of newly invented chemicals. Unfortunately, even when data are available, our perception of risk is often distorted and does not accurately reflect the actual risk. Consequently, our efforts to try to guard ourselves against perceived risks can lead to under- or overprotection. The average American is much more likely to die from eating peanut butter than from being struck by lightning (**Table 14-4**), yet most of us worry much more about lightning.

There are two basic reasons for such misperceptions of risk. One is that our emotions play a large role in how we interpret facts. The most "dreaded" (highly perceived) risks are hazards that are neither observable nor controllable, such as radioactive waste (**Figure 14-11**). Observable and controllable hazards, such as motorcycles, are perceived as less

TABLE 14-4	Annual Risk Calculated as Number of Deaths per 100,000 People in the United States
Activity/Exposure	**Annual Risk (Deaths per 100,000 Persons at Risk)**
Motorcycling	2000
All causes, all ages	1000
Smoking (all causes)	300
Smoking (cancer)	120
Fire fighting	80
Hang gliding	80
Coal mining	63
Farming	36
Motor vehicles	24
Rodeo performer	3
Fires	2.8
Chlorinated drinking water (chemical by-products)	0.8
4 tbs peanut butter/day (aflatoxin)	0.8
3 oz charcoal broiled steak/day	0.5
Floods	0.06
Lightning	0.05
Hit by meteorite	0.000006

Source: P. Slovic, *Risk Analysis 6* (New York: Plenum, 1986): 405–15. Reprinted with permission of Plenum Publishing Corporation.

Figure 14-11 The most dreaded threats (maximum perceived risk) are often those that are least observable and least controllable. The least dreaded may be higher actual risks, but we dread them less because our control and observability are greater. (*Source*: Adapted from M. Granger Morgan, "Risk Analysis and Management." *Scientific American* [July 1993]: 41. Copyright © July 1993 by Scientific American, Inc. All rights reserved.)

CASE STUDY 14-2

Environmental Hormones: A Growing Threat?

We are exposed to thousands of synthetic chemicals every day. We eat them, drink them, breathe them, and use them at work, at home, and in the garden. They are present in soil, water, air, and food. Direct environmental exposure occurs through the foods we eat and industrial and household products including detergents, drugs, lubricants, cosmetics, pesticides, and plastics. Drinking water may also be contaminated by chemicals and their breakdown products found in industrial discharge and sewage effluent. A study released in 2002, of streams and rivers in the United States, found surprisingly high levels of caffeine in most samples. When caffeine is urinated, it is very stable and remains in natural waters for quite some time, serving as a chemical tracer for human waste products. Indirect exposure to chemicals occurs when chemicals are released into the air and water. Examples include airborne particles from industry, which can land on grass or hay, be eaten by livestock and passed along to humans.

Some environmentalists warn that this constant exposure to synthetic chemicals may be posing significant health risks for humans as well as wildlife. Usually the first health risk that comes to many people's minds is cancer, but in fact the most serious risks, some contend, could be the loss of fertility and problems of fetal development. The problems occur because many synthetic organic chemicals disrupt natural hormones, such as estrogen, and when organisms are exposed to even extremely small amounts at critical times in their development these chemicals can produce severe damage, especially in unborn fetuses. Sometimes called *endocrine-disrupting chemicals*, they alter hormonal functions in several ways:

1. They can mimic or partly mimic the sex steroid hormones estrogens and androgens by binding to hormone receptors.
2. They can alter production and breakdown of natural hormones.
3. They can affect the manufacture and function of hormone receptors.

Perhaps the best-documented example involves DES (diethylstilbestrol), an artificially synthesized mimic of estrogen. DES was given to many pregnant women in the 1940s through the 1960s to prevent miscarriages. Then it was discovered that DES can interfere with the development of the fetus, particularly causing female sterility and other fertility problems,

as well as a rare form of vaginal cancer in young adulthood, in the children of mothers who took DES.

While some natural plant-produced chemicals (called phytoestrogens) can alter hormonal function, the main focuses of concern are the many endocrine-disrupting synthetic chemicals that are commercially manufactured. Examples include: pesticides (insecticides such as DDT and chlordane, and various herbicides and fungicides); products associated with plastics such as bisphenol A; pharmaceuticals (drug estrogens—birth control pills); ordinary household products (such as breakdown products of detergents); industrial chemicals (PCBs, dioxin, and benzopyrene); and heavy metals (lead, mercury, and cadmium).

In recent years, such endocrine-disrupting chemicals have been blamed for reduced levels of intelligence, increases in violence, hyperactivity, and birth defects, an epidemic of prostate and breast cancers, and even increasing neglect of children on the part of their parents. Such assertions are extremely controversial, however, and scientists really aren't sure how many of these environmental hormones we are exposed to and how that exposure affects us. Specifically, there is no consensus among researchers as to the strength of the evidence, the validity of many studies, how much (or how little) of an endocrine-disrupting chemical may produce significant results, or whether studies performed on laboratory animals or wildlife are also applicable to humans.

A classic example is a study of the alligator population in Lake Apopka, Florida. Significant quantities of a pesticide were spilled into the lake and several years later it was found that 60% of the male alligators had undersized penises and only about 25% of the normal level of testosterone. The conclusion was that the pesticide was affecting their hormonal levels, essentially "feminizing" the males. Dozens of other studies, from Lake Michigan minks to Baltic Sea fishes, have purported to show similar results—including reduced penis sizes in males, "feminized" male behaviors, female sterility, and other problems. While such studies on wildlife document a direct impact of environmental hormones on ecosystems, it is unclear how well these data can be extrapolated to humans.

Another large group of published studies focuses on the impacts of environmental hormones on sperm counts around the world. Since the early 1990s, these studies have been reporting that human sperm counts have decreased by as much as 50% in cities and other areas of high exposure to synthetic chemicals over the last half century. Early studies were, justifiably, criticized for many sampling errors, such as combining data from different areas at different time periods and failing to take potential geographic variations in sperm counts into account. However, more recent studies continue to imply an impact on fertility. For example, a study released in 2002 by Lynn Fraser of Kings College in London showed that even small concentrations of three kinds of environmental estrogens had a much stronger (about 100 times greater) effect on sperm behavior in mice than natural estrogens found in female mice. Fraser hypothesized that this could reduce fertility by reducing the sperm's ability to fertilize the egg.

Again, critics of such studies note that differences between mouse and human reproduction may invalidate their application to humans. To help resolve the debate, the United Nations, the World Health Organization, and many other international organizations sponsored a global comprehensive review by many scientists of the publicly available scientific literature on environmental hormones. This review was released in 2002 as a report entitled *Global Assessment of the State-of-the-Science of Endocrine Disruptors*. The report found there is sufficient evidence that adverse effects have occurred as a result of exposure to environmental hormones in some wildlife species. The report also concludes that while the evidence supporting the contention that human health has been adversely affected by exposure to environmental hormones is generally weak, further research and information is needed. Therefore, because of continuing concerns and scientific uncertainties, studies on the potential effects posed by these chemicals should remain a global priority requiring coordinated and strengthened international research strategies. There is, in particular, an urgent need for studies of vulnerable populations, and especially of infants and children, since exposure during critical developmental periods may have irreversible effects.

Until such studies are completed, public concern over the hormonal impacts of synthetic chemicals will continue to have many social ramifications. For example, in 2002, the EPA released a pesticide assessment report saying that over 6000 pesticides had been successfully assessed for their impacts on humans and animals. The EPA was sued by the National Resources Defense Council (NRDC), which stated that the EPA's assessment was incomplete because the testing focused on pesticide acute toxicity and failed to test whether the pesticides harm the body's hormone controlled endocrine system.

Questions

1. How great is your concern over synthetic chemicals in the environment? How about in the food you eat? What would make you more concerned than you currently are? Do you think such issues need to be addressed more forcefully than they currently are by government and/or industry?

2. One large group of synthetic organic chemicals are the "organochlorines"— including chlorinated pesticides (such as DDT), PCBs, dioxins, chloroform, CFCs, vinyl chloride, carbon tetrachloride, and so forth (see also Case Study 14-3). While organochlorines have often been considered the mainstay of the chemical industry and chlorine is even used to kill pathogens in drinking water, many chlorinated organic compounds are known to be carcinogenic or otherwise toxic, while many others are suspected of having harmful effects. In 1991, Greenpeace launched a campaign to phase out the industrial use of chlorine. Do you think such a chlorine ban is called for? Should organochlorines perhaps be evaluated one by one for harmful effects and then their use restricted if necessary? How practical can such an approach be given that there are over 10,000 different organochlorines being manufactured? Phasing out the use of chlorine in most industrialized processes could cost at least tens of billions of dollars. How well documented should the potential harmful effects of organochlorines be to justify such an expense?

CASE STUDY 14-3

To Worry or Not to Worry? Chlorine, Dioxin, and PCBs

Visit www.environment.jbpub.com for more information.

An excellent example of the complexity of the benefit-cost decisions facing society today is the intense debate over chlorine and its derived chemical compounds. Chlorine is a very reactive element. It can combine with many other elements to produce thousands of substances including plastics, pesticides, cosmetics, antifreeze, drugs, paints, and solvents. Furthermore, these compounds are often very stable because the chemical bonding with chlorine is so strong. Chlorine-related industries employ 1.4 million people in the United States and Canada.

But the same properties that make chlorine so useful also make it dangerous. Chlorine compounds are often toxic, even in small amounts. Furthermore, their high stability allows them to persist for a long time in the environment and often bioaccumulate in living tissue. CFCs are chlorine-based chemicals that pose a threat to the ozone layer because they persist for many decades in the atmosphere. An especially controversial group of chemicals are the "organochlorines," which are formed from chlorine and carbon. Many are considered to be potentially toxic and they are extremely common. Half of the 362 synthetic chemicals found in a recent survey of the water, sediment, and aquatic organisms of the Great Lakes were organochlorines.

Two of the best known organochlorines are dioxin and PCBs. Dioxin is a family of over 75 related compounds, some of which are extremely toxic to laboratory animals in very small amounts. Dioxin has long been suspected to be a potent human carcinogen, but it is difficult to prove with laboratory animals. An opportunity to directly test its effects arose in 1976, when an industrial accident spewed large amounts of dioxin into the air in Seveso, Italy. Studies of local citizens still living in the area 15 years later showed exceptionally high rates of liver and other cancers. People who moved away just after the accident showed no such effects, indicating the importance of long-term exposure.

Dioxin also appears to have many other harmful effects in addition to cancer. The U.S. Environmental Protection Agency released a major study showing that even small amounts of dioxin significantly affect the immune, reproductive, and nervous systems. For example, lowered sperm counts, birth defects, and lowered disease resistance may occur. This study provoked so much debate that a $4 million reassessment study was done, with contributions from more than 100 scientists from universities, governments, and private labs around the world. The reassessment confirmed many previous results and provided insight into how dioxin harms humans: It disrupts cell and tissue growth and function by binding to cellular proteins. The reassessment was unsuccessful in specifying just how much dioxin exposure is needed to harm humans, but most data indicate that humans are in the middle range of sensitivity; some organisms are much more sensitive than humans to dioxin's effects, others less so. Continued research was called for and will no doubt occur for years to come.

risky than nuclear power, even though statistically the risk is much higher. The second reason for risk misperception is the way we receive information about environmental hazards. Nearly all the news media tend to focus on dramatic spectacles such as oil spills rather than long-term problems because the media have found that people tend to be more interested in dramatic events. While oil spills do represent short-term local risks, the most dangerous risks to society as a whole are often more long term in nature, such as global warming and species extinctions.

Public misperception of risk results in wasteful and inefficient spending of money on environmental problems. Much more money is spent on problems of lower overall risk such as sewage and hazardous wastes than on high-risk problems such as global warming and indoor radon. Some have proposed that such spending inefficiencies could be avoided by using benefit-cost analyses that incorporate accurate risk information. In practice, this has proven very difficult, however, largely because so many of the benefits and costs, including risk, are very difficult to measure. The statistical risks of death in Table 14-4, for example, omit many other crucial considerations needed for a complete benefit-cost analysis. What about the risks of nonlethal injuries, which are omitted from simple "death count" tabulations? Or, the risks to wildlife and ecosystems? Or the social groups that will suffer the highest risk, and obtain the least benefit, from a proposed activity? This last consideration is crucial because risks, costs, and benefits are almost never equally distributed. As a result of such com-

Although complete understanding of dioxin's impacts remains elusive, we have a fairly good understanding of how humans are exposed to dioxin and, thus, how to reduce exposure. While dioxin does occur in nature, such as from volcanoes and forest fires, most of it is a product of modern industry. Even though just 30 pounds (13.6 kg) of dioxin are released each year in the United States, it is so pervasive that everyone has been exposed. Most exposure comes from eating meats and dairy products from animals that have eaten plants contaminated by dioxin emissions. What kind of industries emit dioxins? Waste combustion, especially medical and city waste, accounts for about 95% of all dioxin emissions. This is one reason why many communities oppose waste incinerators. Other sources of dioxin are the chemical manufacture of chlorine compounds, such as pesticides, and pulp and paper mills. Bleaching paper with chlorine is an especially important source. In response to growing pressure, many paper companies are switching away from chlorine to other substances such as chlorine dioxide, which produces less dioxin. Oxygen compounds such as ozone that produce no dioxins can also be used, but they are more costly.

Another well-known organochlorine group are PCBs (polychlorinated biphenyls). These were once common in electrical insulation, paints, and many other industrial products. Their properties and effects on humans are similar in many ways to those of dioxin. In fact, combustion of PCBs will produce dioxin. Manufacture of PCBs in the United States ceased in 1978. They have been slowly disappearing since then, but are still common because of their persistence. Industrial waste sites and other old dump sites often have high concentrations of PCBs; they are one of the main challenges of hazardous waste cleanups.

In response to the problem of chlorine compounds, Greenpeace and other groups have campaigned to ban all use of chlorine by industry. The evidence of potential harm to humans and other life can be used to support such a ban. But society also has to consider the overall benefits and costs. The costs of banning chlorine in industry are very high; according to some estimates, a switch to other compounds would cost consumers more than $91 billion. Chlorine-based materials comprise over half the pharmaceuticals on the market, for instance. Similarly, cleanup of dioxin and PCBs is exacting a huge cost right now. As much as $100 billion may ultimately be spent in the United States to remove PCBs from the environment. Is it worth it? Many people argue that while no one can deny that some harm is done by such compounds, the huge monetary costs to society will greatly outweigh the health and ecological benefits of removing these compounds. Once again, we see the problem of measuring true environmental costs. It is difficult to make monetary decisions when we cannot measure such things as just how many human lives are lost and how many ecosystems are seriously harmed by these chemicals.

plexities, society has spent enormous resources in legal, political, and public debates over the costs and benefits of many substances (see Case Study 14-3).

Pesticides: Pollutants Made to Kill

A **pesticide** is a chemical manufactured to kill organisms that humans consider to be undesirable. Specific kinds of pesticides refer to the targeted pests: insecticides, herbicides, rodenticides, and fungicides. Herbicides are the most commonly used, followed by insecticides. More than 100,000 chemicals are recorded as potential pesticides.

More than 1.1 billion pounds (500 million kg) of pesticides are applied each year in the United States at a cost exceeding $4 billion. It is estimated that this saves $64 billion in crops. Household pest eradication (mainly mice, termites, and other insects) is a $5 billion a year industry. Worldwide, more than $20 billion per year is spent on pesticides. Chemical pesticides have many disadvantages, however, ranging from harming people and ecosystems to evolved immunity of pests. About 45,000 accidental pesticide poisonings occur in the United States each year, resulting in about 50 deaths per year. However, many more people may die of cancer caused by pesticide exposure. A National Academy of Science study estimated that 20,000 people per year get cancer because of pesticides; even if only 10% of these die, the death rate would be 2000 per year.

Furthermore, because pests can develop immunities to pesticides and new pests are introduced by human transport, increasing use of new pesticides

a

has not reduced the number of pests. Farmers still lose about one-third of their crops to pests, just as they did 50 years ago before advanced pesticides were invented. Therefore, the current trend in the United States and some other parts of the world is toward nonchemical controls, such as using natural enemies. In undisturbed ecosystems, natural enemies control pest populations 5 to 10 times more effectively than chemical pesticides, with little or no harm to other organisms.

Three Stages of Pesticide Evolution
The history of pesticide use by humans has occurred in three basic stages.

Inorganic Pesticides. The first stage of pesticide use began more than 4500 years ago when the Sumerians used sulfur to kill insects. Other inorganic chemical pesticides include copper, arsenic, and lead. These substances were used in many parts of the world until the early 1900s, but are now generally banned as pesticides because of their high persistence and, especially, their nonspecific toxicity. They can readily harm or kill nearly all forms of life, including humans.

Synthetic Organic Pesticides. The second stage, synthetic organic pesticides, revolutionized pest control when they were first widely used around the time of World War II. Initially, they were heralded as a "cure-all" that would result in the total elimination of pest species, but this has decidedly not been the case. Organic pesticides have caused a growing number of environmental and health problems and have steadily declined in effectiveness as pest species have become immune to them.

There are three basic kinds of synthetic organic pesticides (Table 14-5). **Chlorinated hydrocarbons** include chlordane, DDT, and other familiar pesticides. These are nerve toxins that cause paralysis, convulsions, and death. They are broad-spectrum toxins, attacking any organisms with central nervous systems. While these are the cheapest groups of pesticides, they also cause the most long-term harm to biological communities because they are easily bioconcentrated in the food pyramid. Chlorinated hydrocarbons are "organochlorines," which often have high persistence and tend to be concentrated in tissue. DDT in particular has one of the highest fat solubilities ever measured, and the body tissue of organisms often becomes concentrated to over 1000 times the surrounding levels. For this reason, DDT and most of the other chlorinated hydrocarbons have been banned in the United States. As a result, the concentration of such persistent pesticides in human tissue has steadily decreased in recent years.

The other two groups of organic pesticides, **organophosphates** and especially **carbamates**, are less persistent, lasting only a matter of days, weeks, or months in the environment compared to decades for chlorinated hydrocarbons. However, these two groups are also more expensive and much more toxic to humans. An adult human would have to ingest more than a pound (0.45 kg) of DDT in one sitting to cause death, but much smaller amounts of organophosphates or carbamates, which are rapidly absorbed by human skin, lungs, and the digestive tract. Like chlorinated hydrocarbons, organophosphates and carbamates attack the nervous system of the pest, but in a different manner. They attack an enzyme that controls nerve impulses, setting off a stream of uncontrolled impulses that cause the organism to twitch violently before death from organ failure. Common examples of organophosphates are malathion and parathion; sevin is a common carbamate (Table 14-5).

Biochemicals: Bacterial Toxins and Synthetic Hormones. The third and most recent stage of pesticides is the commercial production of chemi-

TABLE 14-5	Classification and Properties of Synthetic Organic Pesticides		
Classification	Use (Insecticide Unless Otherwise Noted)	Relative Cost of Class	Relative Persistence of Class
Organophosphates			
Diazinon	Mites		
Malathion			
Parathion	Mites	Expensive	Short
TEPP			
Carbamates			
Sevin			
Baygon		Expensive	Short
Chlorinated hydrocarbons			
Aldrin			
Benzenehexachloride			
Chlordane			
Dieldrin			
DDD			
DDT			
Endrin		Cheap	Long
Lindane			
Mirex			
Toxaphene			
2,4-D	Herbicides		
2,4,5-T	Herbicides		

Source: U.S. Department of Agriculture.

cals that are ordinarily found in nature and can be used to control pests. Such naturally occurring biochemicals have the advantage of being very nonpersistent. Unlike many artificial chemicals, biochemicals are readily degraded into nontoxic forms by microbial digestion and other natural processes.

Bacterial toxins are pesticides that are made by such bacteria as *Bacillus thuringiensis*. These are toxic to a number of insect pests, such as Japanese beetles. Synthetic hormones are biochemicals that mimic hormones of pests, especially insects. Two basic types are especially important. One type interferes with growth-regulating hormones. For example, insects must periodically shed, or molt, their hard outer skeleton in order to grow (**Figure 14-12**). Synthetic growth hormones inhibit the hormones that control molting. Being unable to shed their outer skeleton, the juvenile insects die. **Pheromones** are the second type of synthetic hormone; these are the biochemical scents used by males or females to attract the opposite sex. Artificial pheromones are used to attract individuals to traps, where they are killed. Pheromones are so powerful that only a drop can attract hundreds of individuals from many miles away.

Synthetic hormones have two major advantages over the nonpersistent synthetic organic pesticides. First, they are generally more specific and more likely to affect only the target species. For example, mammals and birds are generally unaffected by insect hormones. Second, only small doses are needed because they have such a powerful effect on the target species.

Problems with Chemical Pesticides

In 1962, Rachel Carson published *Silent Spring*, the first book that warned against the ecological and health effects of widespread pesticide use. Since then, many studies have confirmed much of what she said. Specifically, three main problems have emerged that severely limit the long-term utility of pesticides: nontarget toxicity, secondary pest outbreaks, and increasing immunity. These problems are especially severe with synthetic organic pesticides, which continue to be widely used because biochemical pesticides are not yet available for many pests. Also, biochemical pesticides are often less effective. Trapping pests by attractants is not only much more labor intensive but it generally reduces the pest population less than a potent poison would. A heavy poison dose will kill more than 90% of the pest individuals, whereas dozens of traps will kill only a few percent. Nor are the biochemical pesticides completely free of any of the following problems.

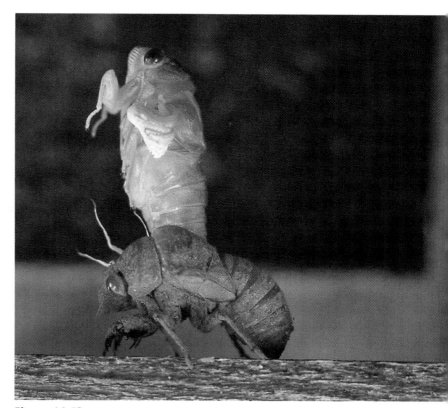

Figure 14-12 Biochemicals that prevent insects from molting or shedding their outer skeleton are effective at reducing reproduction in a certain insect species without harming other species. Seen here is a cicada emerging from its nymphal case.

Nontarget Toxicity. The widely used organic pesticides can have devastating effects on nontarget species. Just one of countless illustrations occurred in 1958, when massive doses of the chlorinated hydrocarbon dieldrin were applied on Illinois farmland to eradicate the Japanese beetle. Cattle and sheep were poisoned, 90% of local cats died, 12 species of wild mammals and 19 species of birds were decimated, with robins, starlings, pheasants, and other birds virtually eliminated. Even synthetic hormones can affect nontarget species. For instance, synthetic growth hormones can affect nonpest insects because many insects have hormones that are biochemically similar. Nontarget toxicity is enhanced in the more persistent and fat-soluble organic pesticides such as chlorinated hydrocarbons. This often results in increased toxicity in organisms high on the food pyramid from bioconcentration. Pesticides come in many different forms, such as dusts and sprays, which can be carried great distances by air and water. Over 7200 miles (11,585 km) of U.S. rivers and 2.5 million acres (1 million hectares) of lakes have restricted or banned fishing, largely because of pesticide contamination.

(a)

(b)

Figure 14-13   Pesticides and herbicides pose considerable health risks to the farm workers who must use them. (a) In less-developed countries, workers often cannot read the warning labels or they lack access to (b) the protective clothing that should be worn when handling the chemicals.

Nontarget toxicity is also directly hazardous to humans. Pesticides cause an estimated 10,000 deaths and 1 million injuries each year worldwide (**Figure 14-13**). These occur from a variety of toxic effects, ranging from immediate to long-term tissue accumulation. Immediate effects are most likely from exposure to the highly toxic organophosphates and carbamates. Temporary symptoms include nausea, blurred vision, and convulsions. Permanent symptoms from exposure to higher concentrations include damage to major organs (heart, lungs, and especially liver and kidneys) and brain and other nervous system damage. Chlorinated hydrocarbons tend to cause chronic toxicity leading to cancer, birth defects, and mutations. An example is Agent Orange, a chlorinated hydrocarbon herbicide used in Vietnam to defoliate forests. Now believed to cause cancer and birth defects, it has been banned in the United States since 1985.

How much pesticide do we ingest in our food? Unfortunately, a majority of the tested fruits and vegetables eaten in the United States contain pesticide residues. One special cause of concern is that many of these residues include pesticides banned in the United States, such as DDT and many other chlorinated hydrocarbons. Despite the ban on use in the United States, they continue to be manufactured here and shipped to developing countries, where they are widely used because of their low cost. U.S. citizens continue to be exposed to these pesticides because much of the produce grown in these developing countries is imported by the United States in what is sometimes called the **circle of poi-**

son. Most of the residues on supermarket produce occur in such low concentrations that many experts are not concerned. Nevertheless, efforts to regulate pesticide use can fail, and increasing numbers of consumers are willing to pay more for foods grown organically without pesticides.

Secondary Pest Outbreaks.   Natural populations are constrained by biological interactions such as predation and competition. When a pesticide strongly reduces key predators or competitors, the species they had kept in check often experiences a "population explosion." Cotton pests in Central America are a prime example. In 1950, synthetic organic pesticides were applied to reduce the boll weevil, resulting in much greater crop yields. By 1955, however, a number of secondary pests that were less affected by the pesticide, such as the cotton aphid and cotton bollworm, began to reduce crop yields again. These pests had been limited by competition from the boll weevil. The typical response to such secondary pest outbreaks is to apply a different kind of pesticide. By the 1960s, farmers were combating eight pest species instead of the original two with an average of 28 applications of various pesticides per year.

Evolved Immunity.   Pests become immune to pesticides with striking speed. In the last few decades, insects, plant diseases (such as fungi), and weeds have shown an exponential increase in the number of pesticide-resistant species (**Figure 14-14**). At least 10 species of rodents are immune to some brands of rat poison. Perhaps the most impressive pest in this regard is the common housefly, which has become immune to almost every pesticide invented so far, even a synthetic growth hormone designed for it. The rate of adaptation is so high largely because pests are generally fast-growing organisms that re-

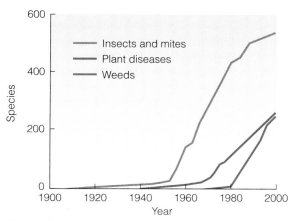

Figure 14-14 Pesticide-resistant species have increased exponentially since 1908. (*Source*: Updated from *Vital Signs 1994* by Lester R. Brown et al. with permission of Worldwatch Institute, Washington, D.C., Copyright © 1994.)

produce often and are adapted to a wide range of conditions. In fact, these "weedy" traits are why they are pests to begin with. Their fast growth ensures that tolerant traits are passed on to offspring quickly, so that evolution occurs quickly. Their broad adaptations ensure that at least a few tolerant individuals will likely occur somewhere in the population. The result is what is often called the

pesticide treadmill: To keep up with the rapid pace of evolving pesticide immunity, new pesticides must constantly be developed.

Reducing Chemical Pesticide Use

The problems with chemical pesticides have caused a recent shift in agriculture toward **Integrated Pest Management (IPM)**, which seeks to reduce the use of chemical pesticides by relying on other methods of pest control, such as natural predators and breeding pest-resistant crops (**Figure 14-15**). IPM is becoming increasingly common in the United States and many parts of the world. In fact, since 1980, U.S. pesticide use has leveled off and begun to decrease (although some of this decrease can be attributed to more lethal pesticides, so that less is needed).

IPM is yet another example of the efficiency of source reduction. Reducing the amount of pollutant produced is much easier than trying to control pollutant dispersal or cleaning up the pollution after it is released into the environment. Less than 1% of sprayed pesticides actually reaches pests. The rest is carried off by water and other transport methods to pollute groundwater, soil, and other parts of the environment. Source reduction—in this case, spraying less pesticide—is the best way to reduce such pollution.

Besides reducing pesticide pollution, IPM is often cheaper than chemical pesticides. Furthermore, the use of nonchemical methods actually improves the efficiency of chemical pesticides when they must be used. Immunity to chemicals develops much more

(a)

(b)

Figure 14-15 Integrated pest management can include the careful use of an invasive pest's natural predators or parasites to help reduce the pest's damage. (a) A female *Catolaccus grandis* wasp, a native of Mexico, homes in on a boll weevil larva. The wasp will deposit an egg in the larva rearing cell, and soon the larva will become a meal for the newly hatched wasp. (b) An *Aleiodes indiscretus* wasp is parasitizing a gypsy moth caterpillar.

slowly when the pests are exposed to the chemicals less often. Most experts predict that these benefits will cause IPM to be increasingly used in the United States and worldwide. Indeed, the main obstacles now are a shortage of teachers trained in IPM and a lack of funds to finance the education of farmers in IPM.

Pesticide reduction through IPM is often associated with farms and agriculture, but cities and households are significant contributors to pesticide pollution. U.S. suburban lawns use more chemical pesticide per acre than U.S. cropland. Similarly, golf courses pour tons of pesticides, fertilizer, and other pollutants into local waterways. Should homeowners and golf courses reduce pesticide pollution by substituting more natural plants for lawns?

Legal Aspects of Toxic Substance and Pesticide Control

Three major federal statutes are of special importance to toxic substance input reduction because they limit the actual production of toxic substances. In contrast, most other federal laws limit the release and environmental exposure of toxic substances after they are produced.

The **Toxic Substances Control Act** (TSCA) was designed to regulate toxic substances not covered under other laws. Its basic goals are to require industry to produce data on environmental effects of chemicals and to prevent harm to humans and the environment, while not creating unnecessary barriers to technology. These objectives exemplify the ambiguity inherent in many regulations because the protection of humans and the environment often conflicts with the promotion of technology. What is an "unnecessary" barrier, for example?

The centerpiece of the TSCA is the "premanufacturing notice," which requires the chemical industry to notify the EPA 90 days before a new chemical is to be manufactured for sale. This notice must contain information on the chemical identity, molecular structure, and test data obtained from the toxic identification, dose-response assessment, and risk characterization procedures discussed earlier. The EPA uses this (and other) information to decide whether the chemical can be produced, should be banned, or requires further testing. The TSCA has caused hundreds of chemicals to be withdrawn that would otherwise have been sold.

Unfortunately, more than 60,000 chemicals were already in use in 1976 when the TSCA was enacted. This was far too many to test, so nearly all of these chemicals were simply "grandfathered in":

They continue to be sold without any testing. But these pre-1976 chemicals make up more than 99.9% of the 6 trillion pounds (2.7 trillion kg) of chemical products made in the United States each year.

The **Federal Insecticide, Fungicide, and Rodenticide Act (FIFRA)** requires anyone wishing to manufacture a pesticide to register that product with the EPA. By 2000, more than 25,000 pesticides had been registered. As with other toxic substances under the TSCA, the EPA may ban the pesticide, permit manufacture (sometimes under limited use), or require further testing if it is deemed to present an "unreasonable risk" to humans and the environment. But, again, the agency's mandate is ambiguous: What is an "unreasonable" risk? The importance of ensuring the safety of pesticides before they are used is clear when (1) less than 1% of all fruits and vegetables are inspected by the Food and Drug Administration, (2) some pesticides are not tested for, and (3) tests take an average of 28 days, so that the produce is often sold before the tests are completed. Unfortunately, because about 25% of U.S. produce is imported from countries that use EPA-banned pesticides, we are still exposed.

Even if manufacture and use in the United States are permitted, each pesticide must be reregistered every five years and may be banned if information has emerged that it is exceptionally harmful. Recall that DDT and many other chlorinated hydrocarbons were taken off the market in the United States. More recently, Diazinon, one of the most widely used lawn and garden insecticides, has been banned in the United States. Shipment to retailers stopped in August 2003, although retailers can sell existing stock until December 2004 when the pesticide's registration expires. A major problem with FIFRA is that pesticide manufacturers are not required to list "inert" chemicals when they label the pesticide for consumers. Such "inert" chemicals often constitute more than 80% of the pesticide and are sometimes very harmful in their own right. But the manufacturers insist that they must withhold this information to prevent competitors from discovering their formulas. A national telephone network, 800-858-PEST, has been established to obtain pesticide information.

The oldest statute regulating production of toxic substances is the **Food, Drug and Cosmetics Act**, which authorizes the Food and Drug Administration (FDA) to test products to see if they are safe to market for human use. The FDA is often criticized for being overcautious and taking too long to test medical products that could save lives. On the other hand, the agency has a very difficult job because a highly toxic product could take many lives.

Toxic Torts

The complexities of regulating toxic chemicals are enormous. Errors will inevitably be made in a large economy. When errors happen, the next line of protection—one that is increasingly being used—is the so-called **toxic tort**, a lawsuit by individuals against manufacturers for harm that has occurred. This area of law is very chaotic and varies from state to state because torts are governed by state law. Basically, the harmed individual(s) must prove that the company was negligent in not providing adequate warning or in marketing a product that was unnecessarily dangerous.

For example, one can sue a chemical manufacturer for negligence if it produces toxic chemicals that cause harm to unknowing consumers. As we have noted, in practice, it is very difficult to prove that a certain chemical causes cancer or other chronic illness. Nevertheless, many individuals have won millions of dollars in compensatory damages (for medical bills and other expenses to recover their original quality of life) and, less often, punitive damages, which are awarded to punish the company. An important environmental benefit of toxic torts is that chemical companies are now taking special care to police the toxic properties of the chemicals they market. Even if a toxic chemical gets by the EPA review process, the companies know that they are liable to lawsuits if negligence or exceptional toxicity can be shown.

Industrial Accidents. Accidental releases of industrial chemicals into the environment form a special class of toxic torts because of their huge potential for catastrophic ecological and human harm. This was made evident in the "Bhopal disaster" in India in December, 1984, when approximately 15 metric tons of deadly pesticide gas were accidentally released (**Figure 14-16**). The gas covered more than 30 square miles (77 km^2) and exposed up to 600,000 people. Local inhabitants estimate that at least 7000 people died immediately, but much suffering continues. Medical reports indicate that at least 50,000 people suffer a variety of serious ailments and perhaps one person dies every two days from the effects. Union Carbide, which owned the pesticide plant, was held legally responsible for the accident and paid various types of compensations to some individuals. Such cases are often the subject of "class-action" lawsuits, where a single group, like the citizens of Bhopal, sues the party responsible for the accident. But many thousands of sick

Figure 14-16   Victims of the Union Carbide Bhopal disaster, December 1984.

Bhopal citizens have complained that they have as yet received nothing.

Despite the warning of Bhopal, the potential for similar catastrophes is growing as chemical industries expand in developing nations, and industrial accidents are statistically increasing. In the eight years following Bhopal, there were 106 major accidents, compared to just 74 in the eight years before. One of the most tragic examples occurred in Hungary in 2000 when cyanide used in a gold mine overflowed into a major river, the Tisza, and killed all aquatic life for dozens of miles downstream. Even the famous Danube River, which receives water from the Tisza, was impacted by this spill. Nor is the United States exempt. There are over 10 billion pounds (4.5 billion kg) of extremely hazardous chemicals stored in the United States. Between 1990 and 2000, several incidents of industrial accidental gas releases that exceeded Bhopal in quantity and toxicity occurred in the United States. For example, on December 13, 1994, an explosion occurred in the ammonium nitrate plant operated by Terra Industries, Inc. in Port Neal, Iowa. Four persons were killed as a direct result of the explosion, and 18 were injured and required hospitalization. The explosion released approximately 5700 tons of anhydrous ammonia into the air, approximately 25,000 gallons of nitric acid to the ground and lined chemical ditches and sumps, and liquid ammonium nitrate solution into secondary containment basins. The chemicals released as a result of the explosion contaminated the groundwater under the facility.

- Pollution refers to excess concentrations of substances.
- Life Cycle Analysis (LCA) can eliminate waste and pollution at many steps in product use.
- Pollution = population × consumption.
- Three myths of pollution control are: purity of nature, zero pollution, and zero risk.
- The benefit-cost approach seeks to reduce pollution to acceptable levels while minimizing economic cost.
- Input reduction is more efficient than output control in reducing pollution.
- Persuasion, regulation, and economic incentives are three basic ways to implement pollution controls.
- Economic incentives are more effective than laws where many polluters are involved.
- The United States produces more pollution per person than any other nation.
- The Toxics Release Inventory (TRI) provides an annual overview of toxic pollution.
- Toxicology studies how chemicals harm living organisms.
- Bioassays determine if a chemical is toxic.
- Dose-response assessment determines the strength of toxicity; a standard measure is LD-50.
- Inaccuracies in using dose-response data for humans include the use of test animals and scaling factors.
- Exposure assessment examines the pathways by which toxic substances can reach humans, including bioconcentration.
- Risk characterization helps scientists decide if the benefits of a chemical outweigh the costs (risks).
- Humans widely misperceive risks.
- Benefit-cost analyses are rarely simple to carry out in practice despite the widespread desire to apply them.
- Pesticides have evolved through three stages: inorganic, synthetic organic, and biochemical.
- Problems with pesticides include: nontarget toxicity, secondary pest outbreaks, and evolved immunity.
- Integrated Pest Management minimizes pesticide use by relying on natural enemies and pest-resistant crops.
- The major U.S. laws regulating toxic chemicals include: The Toxic Substances Control Act, the Federal Insecticide, Fungicide, and Rodenticide Act, and the Food, Drug and Cosmetics Act.

KEY TERMS

Abatement devices
Benefit-cost analysis (BCA)
Bioassays
Bioconcentration
Carbamates
Carcinogen
Chlorinated hydrocarbons
Circle of poison
Contingent Valuation Method (CVM)
Environmental hormones
Epidemiological statistics
Federal Insecticide, Fungicide, and Rodenticide Act (FIFRA)

Food, Drug and Cosmetics Act
Heavy metals
Integrated Pest Management (IPM)
Intrinsic factors
LD-50
Life Cycle Analysis (LCA)
Mutagen
Organophosphates
Pesticide
Pesticide treadmill
Pheromones
Pollution
Remediation

Restoration
Risk
Scaling factors
Supertoxins
Synergism
Teratogen
Toxicology
Toxic Substances Control Act (TSCA)
Toxic threshold
Toxic tort
Toxics Release Inventory (TRI)

STUDY QUESTIONS

1. What is *Life Cycle Analysis* (LCA)? How does it reduce waste generation?
2. Why is nothing in nature truly pure?
3. What are some problems associated with benefit-cost studies? What does a benefit-cost ratio over 1 indicate?
4. How does output management increase pollution control costs compared to input management?
5. What method uses questionnaires to try to objectively estimate the dollar value of changes in environmental quality? Explain how this works.
6. Name the three basic ways to promote change in human behavior, such as reducing pollution.
7. Name the three common myths of pollution control.
8. What is *toxicology*?
9. Name three kinds of synthetic organic pesticides. Which kind has the longest persistence and is easily bioconcentrated? Give two examples.
10. What is *bioconcentration*? What kinds of substances tend to bioconcentrate? Give examples.
11. What is an environmental hormone? How can these hormones harm organisms?
12. Define *risk*. How does perceived risk vary from actual risk? Why do people sometimes perceive risk inaccurately?
13. Name and describe the three basic kinds of chemical interactions among substances in the environment.
14. What causes secondary pest outbreaks?
15. What causes evolved immunity?
16. What does the Toxic Substances Control Act require industry to do? What is the centerpiece of the act? What happened to the chemicals already being produced before the act was enacted in 1976?
17. What is a *toxic tort*? Is this area of law a simple one? Explain.

WHAT'S THE EVIDENCE?

1. The authors state that benefit-cost analyses are difficult to implement properly. List the evidence presented to reach this conclusion. Do these reasons seem valid to you? Why or why not?
2. The authors indicate that pesticide use imposes many problems on society. What are these problems and what is the evidence for them?

CALCULATIONS

1. It is estimated that pesticides cause up to 1 million injuries to humans each year. What percentage of the human population is this, if there are 6.3 billion people?
2. Less than 1% of sprayed chemical pesticides usually reaches pests. If someone sprayed 1 million gallons (3.785 million liters) on a field, how many gallons (and liters) would reach the pests, assuming that 1% of the spray did? (Recall 1 gallon = 3.785 L.)

ILLUSTRATION AND TABLE REVIEW

1. Table 14-2 shows releases of toxic substances between 1990 and 2000. What kind of toxic release had the greatest decline? What kind had the smallest decline?
2. Figure 14-3 shows that, after 1975, the percentage cost of pollution control in the United States reached its lowest point in what year? What year did it reach its highest point?

WWW.ENVIRONMENT.JBPUB.COM

www.environment.jbpub.com

Connect to this book's web site: http://www.environment.jbpub.com/ The site features eLearning, an online review area that provides quizzes, chapter outlines, and other tools to help you study for your class. We've posted the answers to the book's odd-numbered questions. You can also follow useful links for more in-depth information. The site provides sources for the information in this chapter and recommendations for further readings.

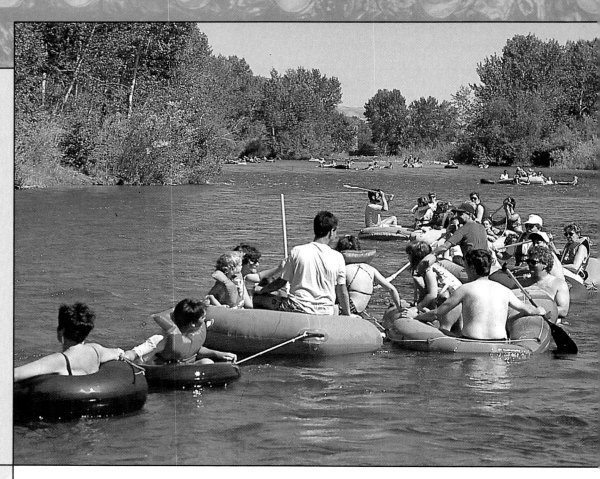

In the early 1900s, Boise, Idaho, started using the Boise River for waste disposal. By 1962, untreated wastes , including rotting carcasses from slaughterhouses, grease, and raw sewage clogged the river. Such environmental degradations helped instigate landmark environmental laws during the 1970s. One of the most successful has been the Clean Water Act of 1972, which provided federal funding for wastewater treatment and spurred many other improvements in "end-of-pipe" waste removal. The payoff for the improved surface water quality has been increased economic activity and a cleaner environment. In Boise, for example, thousands of people now swim and canoe in the river.

CHAPTER 15

Water Pollution

Water is an excellent solvent, meaning that it readily dissolves substances. As a result, humans have long dumped their wastes into natural waters because the waste was quickly diluted and dispersed. Such natural processes of purification are effective as long as human population density is relatively low compared to the amount of water available. But the rapid increase in human populations has so overwhelmed the natural purification processes that pollution of both fresh and oceanic waters has become a global crisis.

Polluted water is water rendered unusable for its intended purpose. Even if water is unfit for human drinking, it is not polluted if that water is normally used for watering a golf course and is still safe for that purpose. Unfortunately, many of the pollutants that find their way into the waters of the environment are often extremely toxic and render the water unusable for any purpose: Humans cannot drink it, farmers cannot water crops with it, and it is too corrosive or otherwise harmful to machinery for industrial use. Contaminated water is rendered unusable for drinking. Disease organisms, heavy metals, and many other toxic substances causing death or illness can contaminate water. Potable water is water that is safe to drink.

Damages and Suffering

Water pollution is highly costly both in the effects it has on the health of humans, other organisms, and ecosystems and in the economic damage it inflicts on industries including agriculture. The effects on human health are greatest in developing nations that do not yet have widespread modernized wastewater treatment. In the year 2000, at the beginning of the new century, the World Health Organization estimated that 82% of the world's population, a total of 4.9 billion people at the time, had access to an improved water supply (such as a household connection, a public standpipe, a protected well or spring, or rainwater collection). This was up from 79% a decade earlier, so improvements have been made. Still, at the dawn of the twenty-first century, 1.1 billion people are without an improved water supply. Even more alarming, however, is the fact that in 2000, fully 40% of the world's population, approximately 2.4 billion, lacked access to improved sanitation excreta disposal facilities (such as a public sewer, a septic system, or even an adequate latrine). Inadequate sanitation (**Figure 15-1**) and poor water supplies lead to deadly outbreaks of cholera, typhoid, hepatitis, and many other diseases. There are approximately 4 billion cases of diarrhea each year, causing an estimated 2.2 million deaths (mostly among children under the age of 5 years). Approximately 500 million people are at risk of trachoma; this disease causes blindness in about 6 million annually. About 200 million people are infected with schistosomiasis (also known as bilharziasis—infections with parasitic trematode worms), with 10% of those suffering severely. All told, the global consequences of poor water supplies and inadequate sanitation are staggering.

Modernized societies have a much lower incidence of waterborne diseases. In the United States, improvements in sanitation beginning around 1850 have led to a steady decline in waterborne disease. By 1920, outbreaks of cholera and typhoid were rare. But problems can still occur. In 1993, for example, a breakdown of the Milwaukee water supply system caused about 403,000 people to become ill and contributed to the deaths of at least 9 people (and possibly as many as 50 people or more).

Aside from its effects on human health, water pollution in the United States causes widespread damage to aquatic ecosystems such as rivers, lakes, wetlands, and coastal marine areas. A survey by the Environmental Protection Agency (EPA) found that more than 17,000 (or about 10%) of the nation's streams, rivers, and bays are significantly polluted. Furthermore, about one-quarter of the nation's usable groundwater is contaminated. Groundwater is the only source of drinking water for about half of the U.S. population. The EPA estimates that total economic damages from water pollution in the United States exceed $20 billion per year.

Water Purification in Nature

The water cycle, driven largely by the energy of the Sun, purifies water in many ways. Rain, streams, lakes, and groundwaters therefore tend to be relatively clean of dissolved matter and are usually drinkable in the natural state. Water is purified in nature through a number of processes (**Figure 15-2**). For convenience, these can be grouped into two basic categories: physical and chemical processes.

- *Physical Processes.* The four physical processes involve forces such as gravity. These processes are important in removing the larger particles and debris and thus often account for the removal of the bulk of the material in water.
  1. **Dilution** is the reduction in concentration of a pollutant when it is discharged into water. Dilution increases as

Figure 15-1 Inadequate sanitation is a major cause of disease for at least one-third of the world's people. Seen here is an open sewer and slum housing, Belize City, Belize.

Objectives

After reading this chapter, you should be able to explain or describe:

The concept and risks of polluted water, including the common major categories of water pollution

Natural water purification processes

Sources of water pollutants

Where most water pollutants ultimately end up

Technical ways to decrease and treat water pollution

Legal methods used to address water pollution

Figure 15-2 Five natural chemical and physical purification processes.

one moves away from the point of pollutant discharge (Figure 15-2). Not surprisingly, the faster the water moves, the more effective the dilution is because more water flows past the point of discharge in a given period of time. Indeed, fast-moving streams can remove the waste of up to about five times more people than slow-moving streams.

2.   **Sedimentation** is the settling out of suspended particles (Figure 15-2). Sedimentation varies with particle size and water velocity, occurring most readily with larger particles and in slow-moving waters. Very fine particles (such as clay) can stay suspended for long periods and travel great distances.

3.   **Filtration** is the percolation of water through sand and other settled sediment to remove suspended particles (Figure 15-2). Even very fine particles can be filtered out if the water percolates through fine sediment.

4.   **Aeration** is the release of gaseous impurities into the atmosphere (Figure 15-2). Water, like all liquids, contains many substances in the form of dissolved gases. Aeration is greatly accelerated when more water is exposed to the air such as when the water trickles over rocks in shallow streams. It is also more effective in warm and fast-moving waters.

•   *Chemical Processes*. Natural waters contain many dissolved minerals and gases that chemically interact in complex ways. Most of these reactions are biochemical, involving enzymes and many other chemicals produced by organisms, especially microbes. Humans have put these microbes to work as major degraders of waste in sewage treatment plants. A number of inorganic (nonbiological) chemical reactions are also important in purifying water. Many elements, such as iron and phosphorus, may react with one another to form molecules that are insoluble in water. Upon reaction, the molecules will therefore precipitate out and settle on the bottom.

Pollution: Overwhelming Natural Purification

As human populations and per capita use of water have both increased, the quantity of wastes released into natural waters has increased exponentially, overwhelming natural processes for removing wastes from the waters in many areas. In addition to this massive increase in quantity, the quality of the waste has become increasingly toxic in recent decades. The rapid growth of chemical and other industries has created a huge outpouring of thousands of new chemicals every year in the United States alone. Moreover, these chemicals are often highly resistant to degradation and persist in the water for some time. The result has been increasingly severe water pollution affecting increasingly larger bodies of water. Entire ocean basins such as the Mediterranean Sea are now becoming significantly polluted.

Water pollution can be classified by (1) its composition (what it is), (2) its source (where it originates), and (3) its fate (where it goes). The next sections examine these three categories in that order.

Composition and Properties of Water Pollutants

The U.S. Public Health Service classifies water pollutants into eight general categories by composition, as shown in **Table 15-1**. In reading about these pollutants, recall from earlier chapters that such pollutants often interact synergistically to become exceptionally toxic and may also undergo bioaccumulation in living tissue. Influx of two weak toxic substances into a lake, for example, may lead to a massive fish kill when their joint effects are combined.

Oxygen-Demanding Wastes.   Oxygen-demanding wastes include materials produced by plants and animals, such as body wastes in sewage and unused body parts from food preparation. That something as "natural" as body waste is a highly destructive pollutant is a good illustration of how anything is harmful in too great a quantity (Chapter 14). In this case, low levels of plant and animal matter provide food for the microbes in natural waters that live by decomposing such matter. Too much plant and animal waste, however, leads to a rapid increase in the rate of decomposition. Because decomposition requires oxygen, the **biological oxygen demand** (BOD)—defined as the amount of oxygen needed to support biological activity, including oxygen absorbed or used ("demanded") by organisms and chemical processes (such as natural oxidation)—in a particular stream, lake, or other body of water increases. The absorption, dissolution, and movement of oxygen through water takes much longer than through air (the oxygen content of water can be increased by bubbling air through water, or by allowing water to "fall" through air, as when water runs quickly down a rocky stream or over a waterfall). Thus, when there is a very large and sudden demand for oxygen in an aquatic system, the local oxygen supply may be depleted faster than it is replenished. Where BOD levels from decomposition are higher than the local dissolved oxygen content in the water, there is not enough oxygen left for other organisms, such as fishes, causing them to die (**Figure 15-3**). Dead fishes and other organisms may then further add to the BOD as their bodies decompose, causing in turn the deaths of even more organisms deprived of oxygen. Unfortunately for fishermen, game fish such as trout tend to be less tolerant of low oxygen content than other fishes.

Infectious Agents. Four major kinds of disease-causing agents are involved in water pollution: bacteria, viruses, protozoa, and parasitic worms. These are usually found in association with oxygen-demanding wastes in sewage and other waters carrying plant, animal, and especially human wastes. Before there were modern sewage treatment plants, outbreaks of cholera, typhoid fever, and diphtheria were major causes of human death, and as noted above, they still are in areas with poor sanitation. Modern sewage treatment plants can easily remove most of

TABLE 15-1	Classes of Water Pollutants, with Some Examples
1. Oxygen-demanding wastes	Plant and animal material
2. Infectious agents	Bacteria and viruses
3. Plant nutrients	Fertilizers, such as nitrates and phosphates
4. Organic chemicals	Pesticides, such as DDT, detergent molecules
5. Inorganic chemicals	Acids from coal mine drainage, inorganic chemicals such as iron from steel plants
6. Sediment from land erosion	Clay silt on stream bed, which may reduce or even destroy life-forms living at the solid-liquid interface
7. Radioactive substances	Waste products from mining and processing of radioactive material, radioactive isotopes after use
8. Heat from industry	Cooling water used in steam generation of electricity

Source: U.S. Public Health Service.

these pathogens simply by adding chlorine, which is highly toxic to many life-forms. However, some pathogens, including the protozoan *Giardia* and a number of kinds of hepatitis viruses, are resistant to chlorine. These are often associated with human feces and can cause diarrhea, cramps, and other illnesses if wastewater treatment plants are ineffective. The illness of over 400,000 people in Milwaukee in 1993 was caused by a combination of factors, including a contaminated water supply, poor water treatment, and an especially resistant parasite called *Cryptosporidium* (see Case Study 15-1).

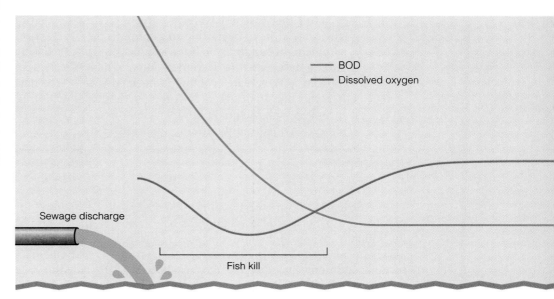

Figure 15-3 Biological oxygen demand (BOD) decreases away from sewage discharge point. Fish kills can occur for many miles downstream from lack of oxygen.

CASE STUDY 15-1

The Milwaukee Incident

In 1993, a series of events culminated in a major outbreak of a waterborne illness, commonly known as "traveler's diarrhea," among the residents of Milwaukee, Wisconsin. The culprit was the small parasitic organism named *Cryptosporidium*, which assembles along the inner membrane of the intestines. This organism infected half of the 800,000 individuals served by the city's water system. The symptoms varied in severity and generally lasted two weeks. Individuals with compromised immune systems, including those suffering from AIDS or cancer, and the elderly suffered the most. More than 50 deaths (or more than 100, according to some estimates) have been attributed to the outbreak, including deaths directly and indirectly linked to the parasite.

The source of the outbreak, namely the municipal water system, was identified within several days, and Milwaukee's mayor, John O. Norquist, recommended that all users of city water boil their water prior to drinking or washing food. Although the contamination was remedied eight days later, questions remained. How did the drinking water become contaminated with this infectious agent? And why did the municipal water treatment facilities fail to remove this organism?

Upon investigation, it was found that the *Cryptosporidium* most likely came from either (1) infected cattle on farms upstream, whose runoff flowed into the Milwaukee River and thence into Lake Michigan, the source of the city's water, or (2) Milwaukee's own sewage system discharge, which also flowed into Lake Michigan just a few miles from the city's source water intake.

In either case, the water treatment facilities should have been able to eliminate the problem. In fact, one facility was functioning adequately, but the second facility, which processed the greater volume of infected water, was not. In an effort to correct an earlier problem with the quality of its water, this facility was using polyaluminum chloride to lower the corrosiveness of the water, but this measure inadvertently decreased the effectiveness of the sand filters used to eliminate *Cryptosporidium* as well as other solids from the treated water.

Although incidents of drinking water contamination affecting large numbers of people are not common in the United States, the Milwaukee incident (by far the largest on record) was not the first case nor the last case. Between 1984 and 2001, a dozen significant outbreaks in the United States occurred. In 1987, for instance, 13,000 residents of Carrollton, Georgia, were affected by a *Cryptosporidium* infestation of the local municipal water system even though all state and federal standards were reportedly met at the time.

What should be learned from these events? They reveal the imperative need to address and deal with nonlocalized sources of contamination of our water supplies. They also illustrate the delicate balance of water treatment processes. The complex intricacies of these processes should not be altered randomly or haphazardly. Instead, thorough evaluations must be performed to determine any possible ramifications from system changes.

Questions

1. Does the Milwaukee incident scare you? Do you ever worry about the safety of your local water?
2. Do you know where your drinking water originates? Is your town or city dependent on one or more water treatment facilities?
3. It has been suggested that a prime way for a terrorist organization or enemy nation to attack the population of the United States is through the drinking supplies of major cities (contaminating the supplies and/or damaging water treatment facilities). Do you think this is a realistic assessment? How important and vulnerable are our water supplies?

Plant Nutrients. Plant nutrients are another example of how too much of a "good thing" can be a major pollutant. Plants in aquatic ecosystems (such as algae) are adapted to a fairly narrow set of environmental conditions, including the chemical nutrients in the water. Ecologists have found that there is generally one nutrient, called the "limiting nutrient," that limits plant growth. If large amounts of this limiting nutrient suddenly become available, plant populations will increase rapidly. The result is sometimes called an "algal bloom," which appears as a green or red scum on the water (**Figure 15-4**). As the plants die, microbial decomposition will increase the BOD, and the animal life will suffocate, as discussed above. This process of abundant plant growth is called **eutrophication** (Greek *eu* = good, *trophikos* = food, so "good food" is a rough translation).

In freshwater lakes, rivers, and streams, the limiting nutrient is usually phosphorus. In many marine waters, it tends to be nitrogen. Fertilizers and human sewage are rich in both phosphorus and nitrogen, so pollution of fresh or salt waters by either of these causes eutrophication. Eutrophication (and BOD increase in general) is usually greatest in standing bodies of water, where water circulation is slow.

Organic Chemicals. Some organic (carbon-containing) chemicals are natural, such as petroleum or coal; these can be toxic to many species. But many other organic chemicals are human-made. Most of the many thousands of chemicals invented each year are organic because of carbon's unique bonding abilities. As a result, thousands of synthetic organic chemicals (SOCs) are now used in industry, agriculture, and other activities. Synthetic organic chemicals are among the most abundant and most toxic pollutants. They also tend to bioconcentrate. Among the most common organic pollutants are organochlorines, including some pesticides.

Inorganic Chemicals. The inorganic chemicals, which are not carbon based, form a broad group. Two of the most destructive water pollutants in this category are metals and acids. Metals are naturally occurring elements such as arsenic, zinc, lead, and mercury. In high concentrations, metals often have a very toxic effect on living organisms because metal atoms will chemically bind to protein molecules such as enzymes, interfering with their functioning. The well-known example of mercury poisoning affecting over 3500 people near Minimata, Japan, in the 1950s was the result of the residents eating fish from the waters into which a local industry was discharging mercury; chronic illnesses, even deaths, and birth defects resulted from the contamination. A more common metal that can cause health damage is lead. It is often carried in tap water in homes with old plumbing. Like mercury, lead attacks the nervous system.

A second major type of inorganic pollution is the release of acids into natural waters. This can occur through "acid rain" (see Chapter 16) or from discharge of acidic waters including drainage of coal and metal mines. Aquatic organisms, such as different kinds of fishes, vary greatly in their tolerance to acidity.

Figure 15-4 The discharge of nutrients into natural waters can lead to the rapid growth of algae and other plants.

Sediment Pollution. Sediment from erosion is the sixth major type of pollution (Table 15-1). We do not often think of sediment as pollution, but human activities can cause thousands of times more sediment to be introduced into natural waters than would otherwise occur. This sediment has many detrimental effects to human use, causing the water to be murky and aesthetically unappealing. In economic terms, it fills in channels and reservoirs and damages power-generating equipment. It is also very detrimental to aquatic organisms. For example, sediment can impair gills and other organs of fishes and shellfishes and reduce the amount of sunlight available to underwater plants, thereby slowing down production at the base of the food pyramid. Many coral reefs in the Florida Keys, the Caribbean, and other parts of the world, for instance, are being destroyed by dredging operations that stir up large amounts of sediment and kill the tiny coral animals.

Radioactivity. Radioactive substances may be discharged into natural waters by nuclear power plants. Radioactivity has many detrimental effects on living organisms, ranging from nearly immediate death to cancer, mutations, and sterility. The radioactivity of the substances routinely released by power plants is generally considered to be so limited as to be harmless, but this practice is still controversial.

Thermal Pollution. Heat as a water pollutant often comes from hot industrial waters, such as those discharged by power-generating plants. These

Figure 15-5 Pesticides are a major source of pollution of surface waters and groundwater.

Sources of Water Pollutants

Virtually all human activities produce some kind of environmental disturbance that can contaminate surrounding waters with the eight pollutants just discussed. Eating (body wastes), gardening (pesticide and sediment runoff), and many other activities create byproducts that can find their way into the water cycle. For convenience, we can assign the large majority of the sources for the eight pollutant groups to three broad categories of waste:

1. industrial,
2. agricultural, and
3. domestic wastes.

Industrial and domestic wastes are often point sources, which discharge pollutants at easily identified single locations, such as discharge pipes. Agricultural sources of water pollution tend to be nonpoint sources, which discharge pollutants from many locations; an example is pesticide runoff from crops (**Figure 15-5**).

Industrial Wastes. Wastes from industry serve as major sources for all eight types of water pollutants. Many major industries contribute significantly to water pollution, but some of the most important are the (1) manufacturing, (2) power-generating, (3) mining and construction, and (4) food processing industries.

Manufacturing industries contribute many of the most highly toxic pollutants, including a variety of organic chemicals and heavy metals. In many cases, both the product, such as paint or pesticides, and the byproducts from the manufacturing process are highly toxic to many organisms, including humans. A key problem with such toxic wastes is not just the many kinds produced, but the sheer volume of each kind. Many billions of pounds of waste are produced each year, especially by the chemical and metal industries, which are the largest producers of toxic and hazardous wastes by far.

Power-generating industries are the major contributors of heat and radioactivity. Nearly all power plants, whatever the fuel, are major sources of thermal (heat) pollution. Radioactivity from nuclear power plants can pollute waters in a variety of ways, including discharge of mildly radioactive wastewater and groundwater pollution by buried radioactive waste.

The mining and construction industries are major contributors of sediment and acid drainage. Sediment pollution occurs because both industries can denude the land of vegetation. Construction in particular results in a drastic rise in the rate of land erosion and transportation of sediment into streams.

heated waters can seriously disturb aquatic ecosystems. Fishes have distinct temperature tolerances for spawning, egg development, and growth. By changing the water temperature, certain organisms may be eliminated and others may prosper. In many cases, a temperature change of only a few degrees can completely exclude a species from breeding in the area. As with low-oxygen waters and acidity, game fishes tend to be more sensitive to temperature changes than most other fishes. Plants, plankton, and shellfishes are also highly sensitive to temperature changes. For example, green algae grow best between 86° and 95° F (30°–35° C), while blue-green algae grow best between 95° and 104° F (35°–40° C). As blue-green algae are a poorer food source than green algae, an increase in the water temperature and the resulting shift from green to blue-green algae will have a big impact on animals higher on the food web. Heated water can also affect large organisms. The endangered manatees of Florida often crowd into the warm discharge waters of power plants during cold winter periods.

In addition to exceeding the temperature tolerances of organisms, heated water affects the oxygen supply. Warmer water is able to hold less oxygen than cooler water. The ability to dissolve oxygen ("solubility") decreases exponentially with increasing temperature. Even if some organisms are not directly affected by increases in temperature, they may be affected indirectly because they cannot tolerate the lower oxygen content of the water.

Acid drainage is mainly a product of mining coal and metallic ore minerals. Because acid drainage occurs only in regions where mining is important, this environmental problem is often overlooked. Yet more than 12,000 miles (19,300 km) of streams in the United States have been seriously affected by acid drainage from mining operations.

Food processing industries include slaughterhouses, canning factories, and many other plants that produce large amounts of animal and plant parts that become oxygen-demanding wastes in nearby waters. These are also sources of waterborne diseases.

Agricultural Wastes.   Agricultural wastes are generated by the cultivation of crops and animals. Globally, agriculture is the leading source of sediment pollution, from plowing and other activities that remove plant cover and disturb the soil. Agriculture is also a major contributor of organic chemicals, especially pesticides.

The other three major agricultural pollutants have biological aspects. Oxygen-demanding wastes are largely body wastes produced by livestock. Livestock produce five times as much waste per pound as humans and are therefore the major cause of this type of pollution (**Figure 15-6**). Infectious agents are nearly always found in body wastes, so livestock are also major producers of this type of pollutant. Agriculture is the major source of plant nutrient pollution through runoff carrying fertilizers applied to crops.

Domestic Wastes.   Domestic wastes are those produced by households. Most domestic waste is from sewage or septic tank leakage that ends up in natural waters. In the past, some cities dumped untreated or barely treated sewage directly into rivers, lakes, or coastal waters. The bulk of domestic waste pollution consists of body wastes and other oxygen-demanding wastes. In addition, domestic sources may be a major contributor of infectious agents and plant nutrients. Infectious agents are a common hazard of all human waste. Plant nutrients occur in the form of nitrogen and phosphorus. These come not only from human waste, but also from fertilizers used extensively in household lawns and gardens.

Fate of Pollutants

Ultimately, most pollutants find their way into natural waters. This is inevitable given the dissolving power of water and its tendency to flow toward rivers and basins. The natural waters that ultimately absorb the pollutants can be divided between freshwater and marine water. The freshwaters, in turn,

Figure 15-6   Runoff from livestock yards such as this one are a source of water pollution.

can be either surface water (rivers and streams, or lakes) or groundwater.

Freshwater: Rivers and Streams.   Rivers and streams drain water that falls on upland areas. Moving water dilutes and decomposes pollutants more rapidly than standing water. But then why are more than 10% of U.S. rivers and streams significantly polluted? A primary reason is that all three major sources of pollution—industry, agriculture, and domestic (cities)—are concentrated along rivers. Industries and cities have historically been located along rivers because the rivers provide transportation and have traditionally been a convenient place to discharge wastes. The very fact that fast-moving waters carry off wastes and purify themselves more readily has encouraged people to pollute them in ever-greater amounts. Agricultural activities have likewise tended to be concentrated near rivers. River floodplains are exceptionally fertile due to the many nutrients that are deposited in the soil when the river overflows.

One of the unique aspects of river pollution, as compared to other types of water pollution, is that "everyone is downstream from someone else." If a city or factory discharges waste, it is sure to find its way into someone's drinking water downstream (**Figure 15-7**). This is not so bad if the next water user is far enough downstream that the natural purification processes have time to act. But many types of pollution are discharged into rivers, and the purification

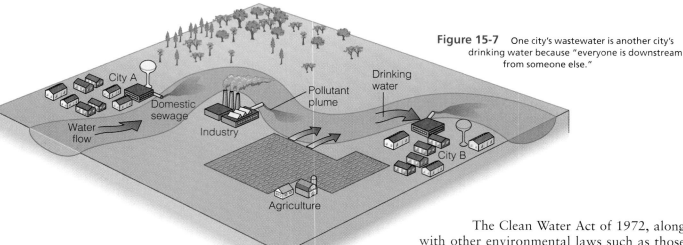

Figure 15-7 One city's wastewater is another city's drinking water because "everyone is downstream from someone else."

processes remove them at varying speeds. Some heavy metals, for example, are removed relatively quickly because suspended clay and organic particles have a slight electric charge and adsorb ("attach") the metal atoms. When the clay or organic particles settle out of the water, they take the metal atoms with them. But pollutants that are very persistent in the water can actually accumulate downstream, causing great hazards for those living at the "end of the line." For instance, New Orleans draws much of its water from the Mississippi River, just before it empties into the Gulf of Mexico (**Figure 15-8**). In the past, the city has had many problems with water quality as some of the hundreds of industrial chemicals in the water were detected in amounts high enough to be harmful. Consequently, many residents and visitors drink only bottled water.

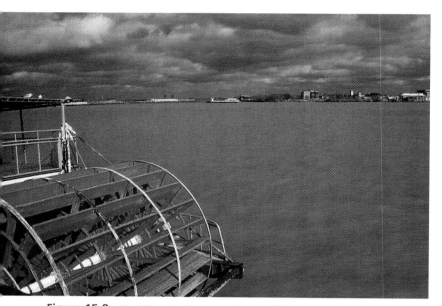

Figure 15-8 The Mississippi River contains many kinds of pollutants by the time it reaches the "end of the line" at New Orleans.

The Clean Water Act of 1972, along with other environmental laws such as those regulating pesticide use, has led to a substantial improvement in water quality in many U.S. rivers, especially the larger ones. Heavy metals and many synthetic organic compounds including pesticides have, on average, substantially decreased since the early 1970s.

Despite these improvements, many U.S. rivers remain in serious trouble, and some are declining rapidly due to various causes. One is nonpoint pollution. Some rivers, such as the Everglades river system, suffer from agricultural and other sources of polluting water runoff. The second is the problem of physical alteration of rivers. Alterations can take four forms, often called the "four horsemen" of river destruction: dams, diversion of water (canals), channel alteration, and land development. These alter flow patterns, increase sediment pollution, and vary water temperature, among many other changes that reduce water quality and destroy aquatic ecosystems. While many of the problems of point source pollution have been addressed, rivers (and streams) in many areas will continue to decline until nonpoint pollution and physical alteration of rivers are addressed.

Freshwater: Lakes. Lakes are subjected to many of the same pollutants as rivers. But lakes are more easily polluted than rivers for at least three reasons. First, lake waters circulate much more slowly. Deep waters in many lakes are circulated only during seasonal temperature changes. Second, lakes are often "dead ends," being basins into which water flows. Pollutants therefore accumulate in lakes, having nowhere else to go. Third, lakes often contain less water than rivers, especially if you consider the amount of water that flows through a river over time. Smaller lakes, of course, are polluted even faster than larger ones.

Lake Erie is a classic example of lake pollution. Because of its very large size, Lake Erie (like the other Great Lakes) was thought to be extremely resistant to pollution. However, the close proximity of many large industries and cities eventually began to overwhelm the lake's natural purification systems. By the late 1960s, sewer systems serving over nine million people were emptying into the lake on the U.S. side alone. Tons of industrial waste from steel mills, paper mills, auto plants, and many others were added daily. By the early 1970s, the lake was declared "dying, if not dead" by many experts, based on the fact that game fishes and many other organisms were absent from large parts of the lake and increasingly rare in the remainder. Much of the problem was the result of widespread eutrophication, with rampant algal growth and decay causing oxygen to be depleted in the lake. A major contributing factor was that even though Lake Erie is nearly 300 miles (483 km) long, it is very shallow, with an average depth of less than 65 feet (20 m). The lake is now recovering, thanks to concerted action taken since the early 1970s. Pollution from industry and sewers has been greatly reduced by water treatment before discharge (**Figure 15-9**). Game fishes have begun to reappear in areas where they had disappeared, oxygen levels are up, and there are many other hopeful signs that water pollution problems are being corrected. Unfortunately, however, many of these gains are being offset by increasingly severe biological damage from dozens of introduced species.

Freshwater: Groundwater. Most freshwater is groundwater. As human populations increase and industrialize, demand for this resource is rising. Unfortunately, there has been an accompanying rapid increase in pollution of groundwater. About half of the U.S. population relies on groundwater as their only source of drinking water, yet it has been estimated that about one-fourth of the usable U.S. groundwater is in danger of contamination or is already contaminated.

A primary reason why groundwater is so easy to pollute is that it moves so slowly. The water must migrate through pores in the aquifer rock. Groundwater flow rates vary, but the average is only a few inches (several centimeters) per day. Compare this to a river where water usually moves many hundreds or thousands of feet (up to thousands of meters) per day. This slow movement causes water shortages because of the long times required to recharge aquifers. The slow movement also leads to pollution for the same reasons that lakes are more easily polluted than rivers: Natural purification processes such as aeration and dilution are slowed. Once pollution occurs, groundwater tends to stay polluted for a long time, often for decades. Some

TABLE 15-2	Major Sources of Groundwater Pollution	
Unintentional		**Intentional**
Surficial percolation:		1. Deep-well injection
1. Landfills		
2. Waste disposal ponds		
3. Spills		
4. Agriculture and land use		
Underground leakage:		
5. Septic tanks		
6. Buried wastes		
7. Underground storage tanks		
Overpumping:		
8. Saltwater intrusion		

aquifers will take many centuries to become purified if only natural processes are relied upon.

Table 15-2 lists nine major sources of groundwater pollution. Only one of these is intentional: deep-well injection, which involves pumping waste into a deep aquifer. Ideally, this aquifer is separated from higher "drinking water" aquifers by an impermeable barrier such as a shale layer. However, the barrier may be fractured, permitting leakage upward. The eight major sources of unintentional groundwater pollution in the table can be subdivided into three basic categories. One of these is saltwater pollution from overpumping the aquifer (see Figure 9-15, page 235).

The other two categories are easily distinguished: surficial percolation and underground leakage. Some

Figure 15-9 Lake Erie may cover a large area, but it is relatively shallow and was readily polluted by surrounding cities and industries. Seen here is Sandusky Harbor, Ohio.

Figure 15-10 (a) Leaking septic tanks can soon pollute local groundwater. (b) Nearly all landfills eventually form leachate that diffuses into the groundwater.

(a)

(b)

Figure 15-11 Major contaminants of U.S. groundwater: Nitrates (from fertilizers) and pesticides are mainly from agriculture. Volatile organic compounds (VOCs) and petroleum products come from oil and chemical storage tanks underground. The rest come largely from hazardous waste sites and leaking sewage and septic tanks. (*Source*: Environmental Protection Agency.)

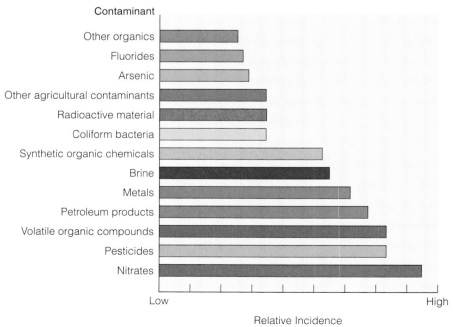

examples are shown in **Figure 15-10**. The four surficial sources involve downward percolation of rainwater that has percolated through landfills, waste disposal ponds (such as where industries put toxic chemicals), spills, and various agriculture and land-use products, including fertilizers, animal waste, and salt sprinkled on roads to melt ice. The three underground sources include leakage from septic tanks, buried wastes, and underground storage (such as gasoline and fuel oil tanks). Septic tank pollutants are primarily oxygen-demanding wastes, while buried wastes tend to be highly toxic chemical byproducts of industry.

Petroleum product leakage from underground fuel tanks is one of the most common types of groundwater pollution. The EPA has estimated that more than 50% of such tanks leak before they are removed. Volatile organic compounds and petroleum products, both mainly from leaking petroleum storage tanks, rank third and fourth among groundwater pollutants in the United States (**Figure 15-11**). Nitrates (fertilizers) and pesticides rank first and second, respectively, illustrating that nonpoint sources are also a problem for groundwater; both enter the aquifer mainly as runoff from farms and other cultivated areas.

Much progress has recently been made in reducing many sources of groundwater pollution. Disposal of waste by deep-well injection has been greatly curtailed in most areas by strict regulations. Reduction of solid waste through recycling and other means has slowed the use of landfills. Even where landfills are being used, they are often designed with linings and other barriers that reduce percolation into the groundwater. Federal legislation passed in the early 1990s requires many gas station owners to replace their older underground tanks with new tanks designed to minimize leakage. Some of the most dangerous sources are the tens of thousands of hazardous and toxic waste disposal sites in the United States such as disposal ponds, buried wastes, and some landfills. Tens of billion of gallons of water percolate through these sites each day. Like polluted rivers and lakes, nonpoint sources will remain a major challenge for years to come.

Marine Waters: Chemicals and Sediments. In 1990, a group of scientists appointed by the United Nations issued a major report that concluded that the oceans have contamination and litter that can be observed from the poles to the tropics, from the beaches to the deep sea. But this pollution is not evenly distributed; most coastal areas are polluted, but the open oceans are relatively unpolluted. One reason for this distribution is that most ocean pollution comes from land-based activities. Runoff (44%) and atmospheric pollution (33%—mainly from factories and other land-based sources) comprise 77% of marine pollution.

The disproportionate pollution of coastal waters is unfortunate for both ecological and economic reasons. Coastal waters contain some of our most productive ecosystems. Because of rich nutrient runoff from land, nearshore estuaries are the "nurseries" of vast numbers of young organisms. Furthermore, young organisms, such as larvae, are the most sensitive to pollution. The widespread destruction caused when estuaries and coastlines are developed aggravates this problem. The economic consequences are also significant. More than 99% of the global catch of marine fish comes from within 200 miles (322 km) of shore; much of the decline in the worldwide fishing industry is attributable to loss of coastal habitat.

Agricultural runoff is the major source for sediment, plant nutrients, and pathogens. Industry is the main source for synthetic organic chemicals, such as toxic chemical waste. Domestic (city) waste is the main source of oxygen-demanding waste, especially through discharge of sewage. Unfortunately, in some parts of the world, city sewage is often not treated before being discharged into the ocean. To save money, many coastal municipalities simply discharge raw waste directly into the sea. For example, much of the sewage discharge from cities in the Mediterranean is untreated. Long pipes are often used to transport the waste a few miles out to sea, so local impacts of the waste were thought to be negligible. However, the rapid growth of coastal populations has begun to overwhelm the purifying abilities of many local marine waters, and currents often bring polluted waters back to shore.

Marine Waters: Oil and Litter. Oil spills probably receive more attention in the media than all the other ocean pollutants combined. Most oil pollution, however, does not come from spills. According to one estimate, globally nearly half comes from land-based sources such as atmospheric pollution, city and industry waste, and runoff. Marine transportation

provides another 45%, but less than one-third of this is from oil spills. Most comes from routine ship operations. For example, tankers often discharge oily bilge and ballast water at sea. According to a 2002 National Academy of Sciences report, 29 million gallons (110 million liters) of petroleum are released into U.S. coastal waters annually, and nearly 85% of this oil originates from land-based runoff, polluted rivers, airplanes, ships, jet skis, and small boats. Less than 8% is from tanker or pipeline spills, and about 3% results from offshore oil exploration and drilling. Interestingly, the same study estimated that 47 million gallons (178 million liters) of oil enter U.S. ocean waters each year naturally, such as through cracks in the seafloor. Still, a major oil spill can be significant and one such incident has the potential to greatly affect the above annual averages. For instance, the grounding of a tanker off the Shetland Islands in 1993 spilled an estimated 26 million gallons (98 million liters), and the grounding of the *Exxon Valdez* in Prince William Sound, Alaska, in 1989, spilled over 10 million gallons (38 million liters). Estimates of the amount of oil spilled during the Persian Gulf War in 1991 range from 25 million to 130 million gallons (95 million to 492 million liters).

Despite their relatively modest contribution to overall oil pollution, oil spills can cause massive devastation to the local environments where they occur. This is why they receive so much media coverage. Long-term events that continue for many years include tar balls washing up on local beaches and scarcity of marine life. In some cases, species that are highly sensitive to oil have never returned. Long-term economic effects include a decline in tourism and fishing. Fishing losses occur not only from death and a decline in breeding but also because the fishes and shellfishes that survive may contain high concentrations of oil in their tissue, often for many years.

Marine litter is a major form of water pollution. Such litter includes materials of many kinds, usually discarded from boats. In the past, it was common practice to dump trash at sea; municipalities used boats to routinely dump tons of garbage. They assumed that the sea is so vast that the garbage would effectively "disappear." But the rapid growth of human populations and material goods worldwide has begun to overwhelm the ocean's ability to absorb and decompose the solid waste. Millions of tons of shipboard litter alone are tossed into the sea each year. This great volume is made worse by the widespread use of plastics, which are extremely durable, often taking many decades to decompose. The impact of plastic is evident in the litter cleared from the beaches each year (**Figure 15-12**). As an example, a brief three-hour cleanup of 157 miles (253 km) of Texas coastline was typical of many U.S. coastlines: It collected 279 metric tons of litter. The litter collected in a typical beach cleanup is about two-thirds (64%) plastic; much of this is plastic cigarette filters, followed by other pieces of plastic. Metal, glass, and paper make up most of the remainder. Plastic is not just a problem for popular beaches. Plastic's durability, combined with its ability to float, allows plastic litter to travel very long distances. Recent surveys of remote locations, thousands of miles from large landmasses, have found increasing amounts of plastic debris.

Besides the obvious ugliness of marine litter, it is harmful to marine life. It is estimated that up to 2 million seabirds and 100,000 marine mammals die each year from eating or being entangled in plastics. The extent of this problem is evident from a study on Midway Island in the Pacific, which found that 90% of the albatross chicks had plastic in their digestive tracts. A random global sampling found that 25% of the world's seabirds have such undigested plastic particles. Sea turtles and marine mammals (such as porpoises and seals) often die from eating plastic, especially bags that they apparently mistake for jellyfish. In addition, birds, turtles, and mammals can be killed when discarded plastic bands encircle their bodies and strangle them as they grow. Discarded fishing nets have also become a major cause of death, entangling and strangling the animals.

Slowing Pollution: Reduction, Treatment, and Remediation

In Chapter 14 we discussed various ways of slowing pollution. Most of these methods are currently used to slow water pollution, including (1) source reduction of waste, (2) treating wastewaters before they are discharged into natural waters, and (3) remediating (cleaning up) natural waters after they have been polluted. These are listed in order of generally increasing cost. It is usually much more costly to clean up a polluted lake than to produce less pollutant at the source. At present, most of society's efforts are directed at the second method—treating polluted wastewaters—so most of our discussion will focus on that.

Source Reduction: Efficiency, Recycling, and Substitution

Source (input) reduction decreases pollution by producing less waste. Reducing the use of water-polluting materials is the least costly way of slowing pollution because it saves the cost of treating pol-

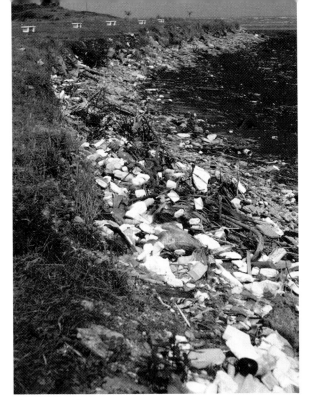

Figure 15-12   Tons of litter accumulate each year on an average beach.

luted discharge waters or cleaning up polluted natural waters. Furthermore, it is the only effective way of slowing pollution from the many nonpoint sources, such as agricultural runoff.

Efficiency.   One way of reducing the introduction of polluting materials into natural waters is improved efficiency, or using less of a polluting product. An example is oil. Between 1974 and 1986, the total number of oil spills worldwide decreased from 1450 to 118 per year. While much of this decrease was from better transportation methods (such as improved oil containment), some of the decrease was from oil conservation. Following the OPEC embargo in the early 1970s, world exports decreased about 25% between 1977 and 1986 as higher oil prices led to increased energy efficiency. In other cases, concerns over the water-polluting effects of materials have led to improved efficiency. This is one reason for reduced growth of pesticide use in the United States. Soil conservation efforts are yet another example. These not only promote sustainable agriculture, but also reduce sediment pollution into nearby waters; pollution from fertilizer is often reduced as well because natural soil fertility is retained.

Recycling.   Recycling plastics, chemicals, and many other materials helps to reduce litter and nearly all types of water pollution because less waste is produced. Recycling of city and industrial wastewater, as in "closed loop reclamation," is a rapidly growing method of reducing water pollution while also extending water supplies.

Substitution.   Another method of source reduction is the substitution of other materials for water-polluting materials in manufactured products. A classic example is the reduction of the amount of phosphate in detergents. As the limiting plant nutrient in many aquatic ecosystems, phosphate pollution is a primary cause of eutrophication. Until the early 1980s, a typical detergent contained about 9% phosphorus by weight, and more than 500 million pounds (227 million kg) of phosphorus were being dumped into U.S. wastewaters yearly. The use of phosphate substitutes by manufacturers has significantly reduced phosphate pollution from domestic and industrial wastewaters. Other examples are biodegradable packaging (instead of plastic) and pesticides that decompose before they reach natural waters.

Treating Wastewater

Until recently, most efforts at pollution control have neglected the generally more effective and less costly methods of source (input) reduction just discussed. Instead, pollution control has emphasized output controls. For water pollution, this has meant an emphasis on wastewater treatment by industry, cities, and households.

Wastewater Treatment.   The overwhelming importance of proper wastewater treatment is illustrated by those countries that lack it. Inadequate water sanitation accounts for much of the world's sickness and death. Even early civilizations realized that the human waste that accumulated under crowded conditions led to rapid disease transmission. The first known sewer system was built in Mesopotamia more than 5000 years ago. It was made of clay pipes that were used to carry wastes away from the cities. Some major Roman sewers are still used today. The flush toilet was invented in the late 1800s, removing the need to carry the waste to the sewer canals in buckets. Carts that picked up and carried the buckets were called "honeywagons," a term still used to refer to garbage trucks in many regions.

Although the early sewer systems greatly relieved the health problems of local urban populations, they increasingly damaged the health of the local aquatic ecosystems and contaminated drinking water downstream because they dumped the raw sewage directly into rivers or lakes. Therefore,

scientists began to develop ways of treating the wastewater; by the 1920s, modern treatment had come into widespread use in North America and Europe.

Treatment today is often carried out by municipal treatment (sewage) plants, which serve as the endpoint for waters carried by city sewer systems. These sewer waters often carry both domestic and industrial wastes. The composition of these highly polluted sewer waters varies from city to city, depending largely on what industries are allowed to dump waste into the sewer system and how much they are required to treat it before they discharge it into the sewers. But even the most highly polluted wastewaters are usually more than 99.9% water. Once the polluted sewer water reaches the municipal plant, its treatment is generally divided into three basic stages: primary, secondary, and advanced (or tertiary). Each stage is progressively more expensive, and only areas with special needs or problems use advanced treatment.

Primary and Secondary Treatment.   **Primary treatment** uses physical processes, especially screening and settling, to remove materials (**Figure 15-13**). A typical primary treatment sequence has three basic steps: (1) A bar screen removes branches, garbage, and other large objects. (2) The grit chamber holds the wastewater for a few minutes while the sand and other coarse sediments settle out. (3) The primary settling tank holds the wastewater for about two to three hours, allowing finer sediments and organic solids to settle out (or float, where they are skimmed off). These settled sediments and solids, called *sludge*, contain large numbers of bacteria, fungi, protozoa, and algae. These are piped into a sludge digester that uses bacteria to decompose the sludge. After removal of water in the sludge drying bed, the digested sludge is usually taken to a landfill. As landfills become full (Chapter 18), alternatives for sludge disposal are being explored, such as incineration and use as a soil conditioner. Unfortunately, sludge usually has high concentrations of toxic synthetic organic chemicals, heavy metals, and pathogens that require costly treatment before the nutrient-rich sludge can be used for crops. The toxic chemicals and metals also cause problems when the sludge is incinerated.

Primary treatment typically removes about 60% of the suspended solids and 35% of the oxygen-demanding waste. As recently as the early 1970s, this is all the treatment received by the sewage of about one-fourth of the U.S. population. Primary treatment alone is inadequate to prevent many long-term health and ecological problems, however, so secondary treatment has been required by U.S. law since 1977. When it follows primary treatment, **secondary treatment** uses biological (microbial) processes to remove up to 90% of both the suspended solids and oxygen-demanding waste.

In secondary treatment, the water discharged from primary treatment is subjected to three basic steps:

1. An aeration tank mixes oxygen, wastewater, and bacteria together. The bacteria digest the sewage.
2. A secondary settling tank allows the many fine organic particles in the digested wastewater to form more sludge, which is then piped back to the sludge digester.
3. A chlorination tank, which adds chlorine to the water, is the last treatment. Chlorine is a highly reactive chemical and is very effective for killing disease-causing organisms, although it produces potentially toxic chemicals that have become very controversial.

Figure 15-13   Conventional wastewater treatment plants utilize a two-step process of primary and secondary treatment. Note the disposal of sludge.

Advanced (Tertiary) Treatment.   Despite its effectiveness in removing oxygen-demanding waste and suspended solids, secondary treatment is relatively ineffective in removing (1) plant nutrients, (2) toxic chemicals (especially synthetic organics and metals), and (3) some pathogens. Secondary treatment may be adequate if treated water is discharged into natural waters where natural purification eventually eliminates these pollutants before the water is used for drinking. However, some situations require advanced treatment to remove some of these pollutants beforehand. As much as 50% of the nitrogen and 70% of the phosphorus often remain in the wastewater after secondary treatment. Therefore, treatment plants that discharge the treated waters into such sensitive ecosystems as Lake Tahoe use advanced treatment to remove the nutrients.

Dozens of kinds of advanced treatment are available. They use a wide variety of methods, from physical processes such as microfiltration, heating, electricity, and evaporation, to chemical processes such as oxidation and precipitation. An example of a physical process is carbon adsorption, which is commonly used to remove synthetic organic molecules, such as PCBs or pesticides. The wastewater goes through filters of fine carbon particles. The carbon is so finely ground that a single handful has the surface area of an acre, allowing the organic molecules to adsorb ("attach") to the carbon. An example of a common chemical process is chemical precipitation, which is often used to remove metals. Lowering the acidity of the wastewater, for instance, can cause metals to precipitate out of the water into a solid mass that can be easily removed. Advanced treatment is often used as a part of closed loop wastewater reclamation.

Septic Tanks.   In areas without a sewer system (mostly rural areas), individual homes use septic tanks. A **septic tank**, which is usually made of watertight concrete and often has a capacity of about 1000 gallons (3785 L), is buried underground to receive household sewage. Waste from the house (such as from the flush toilets) undergoes settling in the tank for several days, and sludge forms at the bottom. The more liquid part of the flow passes across the top of the tank into the absorption field where the polluted water diffuses outward into the

Figure 15-14   Septic tanks permit sewage to settle and then disperse into an absorption field of soil where bacteria digest it.

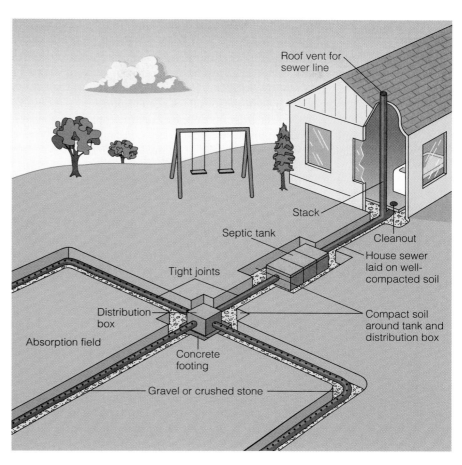

soil (**Figure 15-14**). The waste is decomposed by bacteria in the tank and by soil bacteria in the absorption field.

Soils with finer sediment (such as fine sand or silt) are most suitable for septic tanks because they slow down the rate of diffusion, giving bacteria more time to work. They also provide more surface area for bacteria. In fact, septic tanks should not be built on land with coarse sediment (such as gravel) or underlain by fractured rocks (such as limestone) because the rapidly moving wastewater will pollute the groundwater. On the other hand, if the soil is too fine or the water table is high, the septic tank will back up because outward diffusion in the absorption field is impeded. After about three years, the sludge must be pumped out of even properly working septic tanks. The absorption field usually becomes saturated in a decade or so, causing persistent problems of backing up. This is why septic tanks are generally unacceptable for houses on lots of less than half an acre (one-fifth hectare) of land; the absorption field becomes saturated in just a few years.

Remediation: Cleaning Up Polluted Waters

Although cleaning up polluted waters is much more expensive than reducing pollutant input or treating waters before discharge, it is sometimes the only option. Groundwater can be treated with microbes to decompose organic pollutants. This is accomplished either by (1) injecting "seed" microbes into the polluted groundwater or (2) injecting nutrients and oxygen into the groundwater to stimulate growth in microbes already there. This is often used to treat groundwater contaminated with gasoline. Groundwater can also be pumped to the surface, treated, and pumped back underground.

Polluted surface waters (lakes, rivers, oceanic) can be dredged to remove contaminated sediment. Lake acidity caused by acid rain can be treated with chemicals. Chemical bases such as lime are disseminated in the lake to neutralize acidity.

Legal and Social Solutions

Major federal legislation aimed at water pollution control began in 1899 with the Refuse Act, which prohibited dumping of waste into major waterways.

Through the 1960s, Congress passed various laws that gradually expanded federal responsibilities for maintaining water quality and conducting research. Many of these laws were ineffective and were so widely ignored that water quality in the United States continued to deteriorate. The increased environmental awareness that led to many landmark laws in the early 1970s produced two federal laws, the Clean Water Act and the Safe Drinking Water Act, which were far more effective than earlier laws for reasons discussed shortly. These acts are reauthorized every few years.

Many other laws have also played key roles in improving water quality. These include the Coastal Zone Management Act, Federal Insecticide, Fungicide, and Rodenticide Act, Toxic Substances Control Act, and the Ocean Dumping Ban Act.

Cleaner Rivers and Lakes

The **Clean Water Act (CWA)** of 1972, along with its 1977, 1987, and 1996 amendments, has been very successful in its goal of improving the water quality of lakes and rivers. The act (1) provided for enforcement by the EPA with stiff penalties, (2) created a system for identifying new point sources, (3) established water quality standards for discharged wastewaters, (4) set pretreatment standards for industrial wastes prior to discharge, and (5) provided federal funding to build wastewater treatment plants.

While all five provisions have been important to the CWA's success, the funding for treatment plants and enforcement by the EPA have been especially crucial. Before the Act was passed, many local communities could not afford wastewater treatment and often dumped highly toxic material into local waters despite local and federal laws against it. Through federal subsidies, the CWA provided up to 75% of the cost of building treatment plants. As a result, many thousands of sewage treatment plants were constructed in the 1970s; more than 10,000 plants have been built since 1981 (**Figure 15-15**). In addition, the EPA was authorized to (1) establish specific water quality criteria (pollutant concentrations) that had to be met, (2) monitor surface waters to ensure compliance, and (3) punish violators. The EPA has established limits for over a hundred priority pollutants.

Cleaner Drinking Water

The Clean Water Act focused on improving water quality of rivers and lakes by reducing pollution in wastewater discharge. The **Safe Drinking Water Act (SDWA)** of 1974 was explicitly aimed at improving drinking water. It set more rigorous standards that applied to over 60,000 public water supply systems with 25 or more customers.

Public Water Supplies.   About 10% of the water used in the United States is distributed by the public water systems as the "tap water" found in all major U.S. cities and towns. This water is usually withdrawn from local groundwaters, rivers, or lakes and is often stored in reservoirs that allow sediment to settle out and other pollutants to decompose through natural processes. Before distribution to public taps, the water is sometimes treated, especially if it is from a river or lake. A variety of treatments are used including (1) coagulation to cause suspended particles (including most bacteria) to settle out, (2) filtration by clay or sand filters to remove remaining suspended particles and chemicals (see Case Study 15-2 for an example where a city attempts to avoid the need for costly filtration), and (3) disinfection, usually by chlorine.

These methods are very effective in removing most pollutants, and tap water is generally safe. Nevertheless, the increasing age of treatment facilities, the growing number of chemical pollutants, the potential dangers of chlorine compounds, and other problems have caused many citizens to opt for other sources of drinking water besides tap water, as we discuss later.

Safe Drinking Water Act.   Like the Clean Water Act, the Safe Drinking Water Act authorized the EPA to monitor drinking water supplied by the public systems and other drinking water sources. Since 1974, standards have been added for more contaminants as they become progressively more common. For instance, a 1988 study listed over 2100 contaminants found in U.S. drinking water since the SDWA was passed in 1974. In 1986, a series of strengthening amendments was added to the SDWA. These amendments (1) required the EPA to speed up the creation of standards for 85 new contaminants, (2) increased civil and criminal penalties for violations of all standards, and (3) required the use of lead-free solder in plumbing pipes. Before 1986, water pipe solder often contained about 50% lead, which resulted in widespread lead contamination of tap water in many homes.

The drinking water standards required by the SDWA fall into two categories. **Primary standards** specify contaminant levels based on health-related criteria. **Secondary standards** are based on non-health criteria. These improve the quality of water in aesthetic and other ways. For example, taste, color, odor, and corrosivity have secondary standards. Because secondary standards do not directly relate to life-threatening characteristics, they are generally not legally enforced. They represent guidelines that water systems try to follow when possible. Another familiar secondary standard is **hardness**, which is the amount of calcium, magnesium, and other ions in the water. "Hard water,"

Figure 15-15   Until recently, construction of conventional wastewater treatment plants has been heavily subsidized with government funds. Seen here is the Hill Canyon Wastewater Treatment Plant in Thousand Oaks, California.

which has high amounts of these ions, causes two annoying problems for water users. One is that these ions react with soap to produce gummy deposits, such as a "bathtub ring." These deposits make all cleaning, from dishes to baths to laundry, more difficult. In addition, when hard water is heated, it forms rocklike "scales" (mineral deposits) that can clog pipes. Adding water softeners to hard water removes the ions and can reduce these problems.

Primary standards are actively enforced because human lives are directly affected. Contaminants that are covered by primary standards are classified into four basic categories: inorganic chemicals, organic chemicals, radioactive matter, and microbes (pathogens). Note that these are also four of the eight main classes of water pollutants discussed above. The remaining four—heat, sediment, oxygen-demanding wastes, and plant nutrients—are harmful to ecosystems, but are usually not toxic to humans in small amounts and are therefore omitted from drinking water standards.

Table 15-3 lists **maximum contaminant levels** (**MCLs**) for some inorganic chemicals. MCLs are the highest concentration allowed by the EPA; water with concentrations exceeding this level is considered harmful to at least some people if they consume it over a lifetime. The EPA has similar listings of MCLs for organic chemicals, radioactive matter, and microbes. Examples of common organic chemicals with MCLs are pesticides, petrochemicals, and chlorine byproducts. The most common radioactive contaminant is dissolved radon gas, which enters groundwater from surrounding rocks. Radon is colorless, tasteless, and odorless, and it is potentially very dangerous where it occurs (see Chapter 16).

Microbial contaminants, the fourth category, include disease-causing microorganisms, mainly certain bacteria and viruses. Because running separate standard tests for each of the many species of disease-causing microbes would be too expensive, a simple test for one class of microbes, coliform bacteria, is carried out. **Coliform bacteria,** especially *Escherichia coli,* live in the human intestine in huge numbers, where they aid digestion and perform other functions essential for our health (**Figure 15-16**). Because they have a short life span, large numbers of dead coliform bacteria occur in human feces, about 1.4 billion individuals per ounce (50 million per gram) of feces.

The abundance of coliform bacteria in water is thus an excellent indication of how much human feces has recently entered the water. As human feces are the main source of disease-causing microbes in many waters, the abundance of coliform bacteria also provides an estimate of how abundant disease-causing microbes are. Raw sewage contains about

TABLE 15-3	Maximum Contaminant Levels (MCLs) for Certain Inorganic Chemicals	
Contaminant	**Principal Health Effects**	**Maximum Contaminant Levels (mg/L)**
Arsenic	Dermal and nervous system toxicity effects	0.10
Barium	Circulatory system effects	2.0
Cadmium	Kidney effects	0.005
Chromium	Allergic dermatitis	0.1
Fluoride	Skeletal damage	4.0
Lead	Central and peripheral nervous system damage; kidney effects; highly toxic to infants and pregnant women	0.015
Mercury	Central nervous system disorders; kidney effects	0.002
Nitrate and nitrite	Methemoglobinemia (blue-baby syndrome)	10.0/1.0
Selenium	Gastrointestinal effects	0.05
Thallium	Liver/kidney effects	0.002

Source: Environmental Protection Agency, 2002.

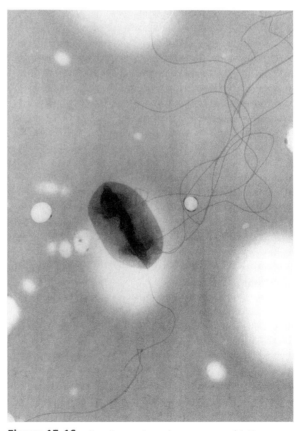

Figure 15-16   *E. coli* are extremely common symbiotic organisms in the human intestine.

The New York City Watershed Protection Plan

Visit www.environment.jbpub.com for more information.

During the nineteenth and most of the twentieth century, New York City has been supplied with clean and tasty drinking water from upstate watershed areas covering nearly 2000 square miles in the Catskill Mountains and Hudson River Valley (**Figure 1**). Forests, soils, vegetation, and wetlands provided natural filtration and purification of the water that was collected and brought by aqueduct to the millions of residents of the city. No treatment of the water was necessary other than standard chlorination to ensure that there would be no contamination by living waterborne diseases.

In the last decades of the twentieth century, the quality of New York City's water began to deteriorate as the watershed area became increasingly developed. More people living in the watershed area, with more septic and sewage systems and wastewater treatment plants, badly strained the natural filtration and purification processes of the land and its ecosystems. Increasing amounts of substances such as petroleum products, road salts, fertilizers, herbicides, and pesticides contained in the runoff from lawns, farms, roads, golf courses, parking lots, and other human-made constructions only exacerbated the situation. By 1990, the U.S. Environmental Protection Agency (EPA) made it clear that either the watershed would have to be protected or New York City would have to build and operate a water filtration system. At the time, it was estimated that to build such a water treatment plant would cost anywhere from $3 billion to $8 billion, but to implement a watershed protection program would cost only around $1.5 billion.

On one level, it seemed obvious to some people at the time that theoretically the less expensive option, to protect the watershed areas, was the best. Not only would it save money, they argued, but it would also help protect natural ecosystems and open space, improving the quality of life in upstate New York.

Generally speaking, however, protecting watershed areas is in many ways more difficult, at least politically, than simply building a water filtration/treatment plant through which the water can be processed. Protecting watersheds involves working with numerous individuals, industries, organizations, municipalities, local and regional communities, and state and federal governments. Developing and implementing a plan that will be accepted by all of the parties involved is a major undertaking, and at the outset success is by no means certain. In fact, some parties involved may resent being asked to limit development or change their lifestyles or ways of doing business for the benefit of an enor-

Figure 1 The Catskill Mountains watershed and the Hudson River Valley supply New York City with drinking water. The Catskills are southwest of Albany in this satellite image taken during the fall.

1 virus for every 92,000 coliform bacteria. Similar ratios have been calculated for disease-causing bacteria, worms, and other pathogens. Fortunately, most of these disease-causing organisms have a much lower survival rate than coliform bacteria outside the human body.

Although the EPA has oversight authority, each individual state is largely responsible for monitoring and enforcing the standards on a day-to-day basis. As a result, the effectiveness with which water quality is maintained varies considerably from state to state. Water quality in Maine, Massachusetts, and New Jersey often ranks highest in yearly lists produced by environmental groups. Maine devotes a large part of its drinking water budget to training water system operators and reportedly allows no community to deviate significantly from the SDWA.

mous, wealthy city of which they are not a part. In the middle 1990s, the city of New York owned less than 7% of the critical areas of the watersheds, and the state of New York only owned about 20%; thus, some 73% of the critical watershed areas were out of direct control by the city and state.

Still, after many years of intense and sometimes contentious negotiations, in 1997, a watershed management and protection plan was agreed on that involved the city of New York, state and federal officials, and a variety of villages, towns, environmental groups, and organizations in the watershed areas. Based on this plan, the U.S. EPA waived filtration requirements for New York City water through at least 2002. Aspects of the plan included New York City spending a quarter billion dollars directly to acquire and preserve land in the watershed areas, and development of new regulations to protect the watersheds and the quality of the water. New York City also agreed to provide funding for upgrades to wastewater treatment plants in the watershed areas, funding for farmers to implement voluntary measures to reduce runoff and contamination of waters, and payments to communities in the watershed areas to subsidize water protection procedures and environmentally friendly development that will help preserve the quality and integrity of the watershed areas.

Whether the program will actually accomplish its goals of avoiding the construction of a filtration system and protecting New York City's water quality is still an open question. Critics point out that it is dependent in part on the cooperation of numerous individuals, villages, and towns, as well as voluntary measures to be taken by farmers and others. Also, there is no guarantee that the city and state will be able to acquire enough acreage in the critical watershed areas to adequately ensure acceptable water quality. Critics also suggest that the plan does not adequately limit development in watershed lands, and may place too much emphasis on pollution control rather than more effective (at least in the long run) pollution prevention. Ultimately, some critics suggest, filtration will still be needed. However, the measures already being put into place to help protect New York City's watershed areas are having a positive benefit in terms of protecting open space, water bodies, and natural ecosystems—even if ultimately the measures are not enough to avoid filtering the water. And as an added benefit, dialogue over environmental concerns has been created among many officials at all levels (local, state, and federal), communities, industries, and organizations. Even if the plan to protect New York City's watershed areas is not fully successful, it is hoped that the lessons learned from the experience can be used to develop watershed protection programs for other communities.

Questions

1. Do you agree with the initial decision that was made, to try to protect the watershed areas rather than build and operate a water filtration system?

2. What are some of the obstacles that must be overcome, politically, socially, and technologically, in order to adequately protect the watershed areas upon which New York City is dependent?

3. If, despite spending hundreds of millions of dollars attempting to protect the watersheds, it is found that a water filtration system still needs to be built, will the money have been wasted? Justify your opinion.

Point-of-Use Treatment, Bottled Water, and Alternative Sources. For many reasons, increasing numbers of U.S. citizens are concerned about their drinking water. Annual polls by the Gallup Organization regularly show that many Americans have a "great deal" of concern about their drinking water. This has led to a rapid increase in the use of (1) point-of-use treatment, (2) bottled water, and (3) alternative sources such as drilling home wells.

Point-of-use treatment refers to a variety of home treatment devices that alter water quality as it enters the home. Three of the most common are listed in **Table 15-4**: carbon filters, distillation, and reverse osmosis. The advantages and costs vary, depending on the area and the quality of the device. Countertop carbon filters are the most common. Not listed in the table are water softeners, which are home treatment devices that replace with sodium the

TABLE 15-4 | Treatment Methods and Alternative Water Sources: Advantages, Disadvantages, and Costs

Treatment Method/ Alternative Source	Contaminants Removed/Advantages	Disadvantages	Estimated Cost
Carbon filter	Removes halogentated and some other organics, radon, and residual chlorine.	Bacteria held in carbon may wash into water.	Countertop under $50 and up. Undersink $300 and up plus maintenance costs. Whole house under $1200.
Distillation	Removes heavy organics, inorganics, metals, and microbiological contaminants.	Not effective on volatile organic compounds, trihalomethanes, and radon. May corrode pipes.	$200–$2000 plus maintenance costs.
Reverse osmosis	Best method of removing inorganics, also reduces the level of many organics and metals.	Wastes water; 75% to 90% of water is lost in straining process.	$90–$700 and maintenance costs.
Bottled water	Natural spring water from an isolated source could be water of good quality.		$10–$15 per week depending on the size of the household.
Drilling a new well	Possible location of an uncontaminated nearby source.		Averages $10–$15 per foot plus casing and pump costs nationwide, but can range anywhere from $5–$50 per foot depending on the area.

calcium and magnesium atoms that make water "hard." Hard water is particularly common in the midwestern states. Limestone and other alkaline rocks are major contributors to hardness in surface and ground waters.

The use of bottled water (Table 15-4) in the United States has surged in recent decades. In 2001, Americans spent $6.5 billion to purchase over 5.4 billion gallons (20.5 billion liters) of bottled water, approximately an 11% increase over the figures for 2000. Is it worth it? Some people buy bottled water for improved taste, smell, or color compared to their tap water. But others buy it in the belief that it is healthier than tap water. This is not always true; a recent study carried out by the Natural Resources Defense Council found that bottled water is not necessarily cleaner or safer than most tap water. While most tested bottled water was found to be of good quality, at least some bottled water was found to be contaminated with such substances as bacteria, synthetic organic materials, or arsenic. A primary reason for such discrepancies in bottled water quality is that at a national level, the Food and Drug Administration regulates bottled water, while the Environmental Protection Agency regulates tap water quality. The FDA has considerably less rigorous standards, both in terms of testing and purity, for bottled water than the EPA has for tap water.

Furthermore, the many bottlers selling in only one state are regulated by state authorities, not the FDA, and one out of five states fails to adequately regulate bottled waters.

Another possibility is to find an alternative source to the local municipal water by drilling a private well. Costs of drilling vary considerably, depending on how deep clean groundwater is to be found, the type of rocks to be drilled through, and other variables. In some regions, groundwater is extremely high quality and requires no treatment at all for safe, tasteful drinking. But because of possible unseen contamination, such water should routinely be tested by professional laboratories. About 15% of Americans rely on private wells or springs.

Future of U.S. Water Pollution Control
The Clean Water Act of 1972 was reauthorized in 1995, as it is every five years or so. The main purpose of the reauthorizations is to rectify deficiencies in the laws that have become apparent over time. The Clean Water Act, along with the Safe Drinking Water Act of 1974, has already played an enormous role in improving water quality, and these laws and their modifications will largely determine how water pollution is controlled in coming years.

One of the most widely discussed deficiencies of water pollution control in the United States, and nearly all industrialized nations, is the emphasis on "end-of-pipe" methods. Hundreds of billions of dollars have been spent on water pollution control since the Clean Water Act was established in 1972. This includes money spent on cleanup methods, the building and upgrading of urban sewage treatment plants, and the treatment of industrial and other wastes not directly associated with city sewer systems.

While the end-of-pipe approach has been successful in reducing localized sources of water pollution (often called "point" sources because the pollution is discharged at a single point), many problems such as pollution of native ecosystems and groundwaters arise from nonlocalized or nonpoint sources, such as runoff from farm fields and urban areas that contains pesticides, oils, and other pollutants. The EPA has estimated that runoff from agricultural lands accounts for over two-thirds of the pollution in "impaired" rivers. A major focus of future water pollution control has therefore been on reducing runoff and other nonlocalized sources (**Figure 15-17**). This is much more difficult, but it can be accomplished through input reduction such as using less pesticide, fertilizer, and many other substances that end up in our waters. Thus, instead of concentrating on output reduction by end-of-pipe methods, it is often much more effective to stop pollution at the source. Input controls are cheaper than end-of-pipe methods, so this approach will also address the problem of rising costs of water pollution control.

Other key problems the United States must address include (1) the obsolescence and age of many urban wastewater treatment plants (many were built in the 1970s); (2) increasing concern over toxic chemicals, such as chlorine-related compounds (resulting from the use of chlorine to disinfect water) and many newly invented chemicals; (3) the lack of an integrated network of water testing facilities to provide comprehensive data on water quality; and (4) pollution of waters that have been largely neglected until recently. While cleanup efforts have focused on lakes and rivers, groundwater, marine waters, and wetlands have continued to suffer increasing pollution and need immediate protection.

Clean Water outside the United States

Poor water quality is a major cause of death in many developing nations. Providing modern water treatment is a primary goal of many development programs of the World Bank, the United Nations, and many other organizations. India, for example, is

Figure 15-17   This state-of-the-art waste management system helps control runoff from a 900 head hog farm.

engaged in a $200 million project to clean up the 1500-mile-long (2400 km) Ganges River, which provides water for one-third of India's massive population. However, this will scarcely dent the tons of sewage, industrial waste, pesticides, and other pollutants that flow into the Ganges daily.

Water quality is also very poor in many countries in Eastern Europe and in the territory of the former Soviet Union. The Baltic Sea is deteriorating rapidly. Dioxin concentrations are so high that once-popular Baltic cod liver is no longer edible; many fish have large tumors. The main reason is that surrounding cities have no or only primitive wastewater treatment. The sediment in St. Petersburg's harbor has a thousand times the normal concentration of many toxic metals such as lead and cadmium. The Finnish Baltic Marine Environmental Commission estimates the cost of cleaning up the Baltic at $1 billion per year for 20 years. Each year Russia's 2300-mile-long (3700 km) Volga River receives billions of tons of sewage and industrial waste, including toxic waste from more than 3000 factories. Currently, 70% of the fish in the Volga contain mercury. Much of this pollution ends up in the Caspian Sea, which is also rapidly becoming polluted. A growing environmental movement, called Save the Volga, has appeared, but, as in other ex-communist countries, lack of money seriously impedes cleanup efforts.

STUDY GUIDE

SUMMARY

- Polluted water can be defined as water that is rendered unusable for its intended purpose.
- In the year 2000, over a billion people globally lacked an improved water supply and over two billion lacked improved sanitation facilities, leading to outbreaks of cholera, typhoid, hepatitis, and many other diseases.
- Water is purified in nature by physical processes (dilution, sedimentation, filtration, aeration) and chemical processes (biochemical reactions, many produced by microbes, and precipitation).
- Basic categories of water pollutants are: oxygen-demanding wastes, infectious agents, plant nutrients, organic chemicals, inorganic chemicals, sediment pollution, radioactivity, and thermal pollution.
- Sources of water pollutants are: industrial, agricultural, and domestic wastes.
- Ultimately, water pollutants can end up in natural waters, such as rivers and streams, lakes, groundwater, and marine waters.
- Water pollution can be reduced and mitigated by source reduction, wastewater treatment, and remediation (cleaning up polluted waters).
- In the United States, the Clean Water Act and Safe Drinking Water Act have been instrumental in addressing water pollution issues.

KEY TERMS

Aeration
Biological oxygen demand (BOD)
Clean Water Act (CWA)
Coliform bacteria
Contaminated water
Dilution
Eutrophication

Filtration
Hardness
Maximum contaminant levels
    (MCLs)
Potable water
Polluted water
Primary standards

Primary treatment
Safe Drinking Water Act
    (SDWA)
Secondary standards
Secondary treatment
Sedimentation
Septic tank

STUDY QUESTIONS

1. What is polluted water?
2. Approximately how many people in the world are without an improved water supply? Without improved sanitation? What types of problems can this cause?
3. Describe the basic water purification processes as they occur in nature.
4. What are the eight basic categories of water pollutants as classified by the U. S. Public Health Service?
5. What is the effect of oxygen-demanding wastes as water pollutants?
6. Give an example of an effect caused by high concentrations of a metal in living organisms.
7. Give examples of how sediment from erosion can be a pollutant.
8. What are two harmful effects of increased temperatures on aquatic organisms?
9. What are the major sources of water pollution? Specify the water pollutants produced by each activity, such as agriculture.
10. What are two reasons that cities historically have been built along rivers?
11. Give three reasons that lakes are more easily polluted than rivers.
12. Why is groundwater so easily polluted? What are some of the major sources of groundwater pollution?
13. Discuss why ocean pollution is a growing concern. Describe some of the major types of ocean pollution and their sources, and suggest potential solutions.
14. What are the three basic ways of addressing water pollution concerns?
15. Describe the concept of source reduction and how it applies to water pollution issues.
16. Describe the process of wastewater treatment. What do primary treatment and secondary treatment refer to in this context?
17. The Safe Drinking Water Act (SDWA) of 1974 specifies maximum contaminant levels for what four classes of health-threatening water pollutants?
18. Under the SDWA, what distinguishes a primary standard from a secondary standard? What is hard water and what problems does it cause?
19. Distinguish between localized ("point") sources of water pollution and nonlocalized sources of water pollution. Which has been a primary emphasis of water pollution control efforts in the past? Why is it important to address both classes of water pollution?

WHAT'S THE EVIDENCE?

1. Many people buy bottled water thinking that it is "better" quality water. What does "better" mean in this context? Healthier? Better tasting? What is the evidence that bottled water is superior to local tap water? How are the answers to such questions locally specific?

2. The authors state that remediation—cleaning up polluted waters—can be more expensive than reducing pollutant input or treating waters before discharge. Are you convinced that this is the case? Can you cite general examples to support the authors' contention? (Suggestion: You might also consider the lost economic and other benefits of polluted waters, the increased "value" that clean waters can have, and the costs of treating disease caused by polluted waters.)

CALCULATIONS

1. If a beach cleanup gathered 307 tons (278.6 metric tons) of litter along 157 miles (253 km) of Texas coastline, how much litter per mile was this? How much litter per km? How many tons of plastic were collected if 64% of the total was plastic?

2. If a beach cleanup gathered 600 tons (544.5 metric tons) of litter along 200 miles (322 km) of Florida coastline, how much litter per mile was this? How much litter per km? How many tons of plastic were collected if 58% of the total was plastic?

ILLUSTRATION AND TABLE REVIEW

1. Referring to Table 15-3, which contaminant would you conclude is the most dangerous at very low levels? Which does the EPA allow at the highest concentrations?

2. Referring to Table 15-3, the maximum allowable concentrations of barium is how many times that of mercury?

WWW.ENVIRONMENT.JBPUB.COM

www.environment.jbpub.com

Connect to this book's web site: http://www.environment.jbpub.com/ The site features eLearning, an online review area that provides quizzes, chapter outlines, and other tools to help you study for your class. We've posted the answers to the book's odd-numbered questions. You can also follow useful links for more in-depth information. The site provides sources for the information in this chapter and recommendations for further readings.

Strict legislation has significantly reduced air pollution in many cities, including New York City (seen here on a clear evening).

CHAPTER 16

Air Pollution: Local and Regional

Air pollution is one of the most widespread environmental problems because it occurs at many scales: indoor, local urban and regional, and global (**Figure 16-1**). Regional and global air pollution are especially difficult to control because polluters are often very distant from those damaged by their emissions.

Based on Environmental Protection Agency (EPA) statistics (2000), over 160 million tons of pollution annually is emitted into the air from the United States. Each year, outdoor air pollution causes tens of billions of dollars worth of direct damage in the United States, including crop and livestock damage, cleaning costs to clothes, and weathering of statues and buildings. Ground-level ozone alone causes $500 million dollars worth of reduced yields to agriculture and commercial forests annually. Many billions of dollars more are lost due to increased health care expenses. At the end of the twentieth century, the EPA reported that many cities were still polluted by too much smog and ozone despite major reductions since 1970. For instance, there were 39 days during 1999 that the nation's capital, Washington, D.C., failed to meet acceptable air-quality standards as set by the EPA. Houston suffered from 50 such days in 1999, and Los Angeles 27. Although it is still unacceptable to have 27 such bad air days in Los Angeles, this is a tremendous improvement compared to 1992, when L.A. suffered 175 such days. In the United States, 121 million people live in areas where the air was unhealthy for at least a portion of the year. Globally, approximately a million premature deaths are attributable to particulate air pollution yearly.

Yet all is not bleak. Air quality has considerably improved in the United States over the last three decades (see Case Study 16-1). Since 1970, EPA tracking of principal pollutants has shown that, in the aggregate, they have dropped by 29% (2000 statistics). The air we collectively breathe is cleaner today than it was in the 1970s. As noted by the EPA, these positive results have been achieved even as the U.S. population (up by 38% over this period) and economy continued to grow (the U.S. Gross Domestic Product increased by 158% over the same period), more people drove cars (vehicle miles traveled increased by 143%), and more energy was consumed (showing a 45% increase).

Indoor air pollution is potentially an even greater health hazard than outdoor air pollution. Radon and especially cigarette smoke cost many billions of dollars more. In terms of overall impact, global air pollution is perhaps the greatest single environmental challenge humans will face in the next few hundred years. Climate changes caused by pollution could lead to massive alterations of agriculture, large-scale migration of humans and other organisms, flooding of the world's coastal cities, and many other changes.

In this chapter, we discuss local and regional air pollution and indoor air pollution. Global air pollution, including ozone depletion and climate change, are discussed in the next chapter. We will see that local and regional air pollution are largely a problem of cities and industrialization: Large concentrations of people and machines produce large inputs into the atmosphere, especially by combustion of the fossil fuels that we now use to run civilization. Alternative fuels, such as solar energy, would therefore alleviate much air pollution. In contrast, global air pollution is less restricted to cities; the burning of tropical forests is an example.

Local and Regional Air Pollution

Until the late 1980s and early 1990s, local and regional pollution generally received more attention than global or indoor pollution. The United States and most Western industrialized nations have significantly reduced most types of local and regional air pollution since 1970, when strict amendments to the Clean Air Act were passed. This has been achieved by reducing the small group of so-called criteria pollutants (also known as "principal air pollutants"), which are the source of most local and regional air pollution.

The six basic **criteria pollutants** as currently recognized by the EPA are particulate matter, sulfur dioxide (SO_2), nitrogen dioxide (NO_2), ozone (O_3), lead (Pb), and carbon monoxide (CO). Carbon monoxide, lead, nitrogen dioxide, and sulfur dioxide are emitted directly into the air from various sources. Ozone and particulate matter are somewhat more complicated. Particulate matter (a number of different types of which can be distinguished) may be emitted directly, or it can form when var-

Figure 16-1 Global air pollution occurs when so much pollution, such as carbon dioxide, is released into the atmosphere, that global changes occur. These are covered in Chapter 17. Urban and regional air pollution are more local; pollutants travel less distance and stay in the air for shorter time periods. Acid rain, smog, and other examples are discussed in this chapter. (*Source*: Modified from F. Mackenzie and J. Mackenzie, 1995, *Our Changing Planet* [Englewood Cliffs, N.J.: Prentice-Hall, 1995], p. 237.)

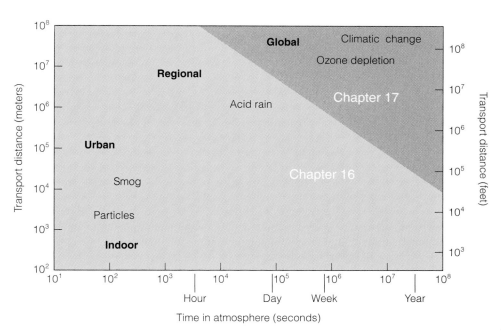

CASE STUDY 16-1

Clean Air in Los Angeles

Smog, and air pollution more generally, in cities and from major industrial operations has been a significant issue of concern for well over a century. London, for instance, became known for its "fog" (often, in modern times, actually smog). In 1892, in London 1000 people died as a result of a major smog incident, and in 1952, some 4000 people succumbed to another London "killer fog (smog)." In the latter case, the smog and smoke were so dense that traffic was stopped and buses could only operate by being led by a person on foot carrying a lantern. Other smog incidents in London, in 1955 and 1962, were held responsible for 1000 and 750 deaths respectively. An air inversion trapping industrial emissions in the Meuse Valley of Belgium in 1930 killed 60 people and made thousands sick. In 1948, in Donora, Pennsylvania, emissions from local industrial facilities killed 20 people, numerous animals, and made 6000 (half the local population) sick. In 1950, smog in Poza Rica, Mexico, killed 20 and hospitalized hundreds. These are just a few examples of how serious the health effects of air pollution could be in the early and middle twentieth century before there were any consistent and strict laws or regulations in support of clean air.

One city in the United States that became well known for its high levels of air pollution, and smog in particular, is Los Angeles. As early as the summer of 1943, there was a major smog incident in the Los Angeles area. Residents suffered from irritation of the eyes and nose, nausea, vomiting, and breathing disorders. Visibility was limited to three blocks during the incident. At the time, the problems were blamed on a local butadiene (a chemical used in the manufacture of synthetic rubber) plant, but although it may have contributed to the problem it was not the sole cause. Even after the plant shut down, the problems continued due to the high local population density with much congested traffic and industry. In 1945, the city of Los Angeles first began a pollution control program, and also in the late 1940s, the state of California addressed air pollution issues through legislation. Still, the quality of the air in the Los Angeles area remained unhealthful. In the early 1990s, Southern California's air quality was the worst in the United States. Air pollution in the region reached unhealthful levels on half the days each year, and it violated four of the six federal standards for healthful air—those for ozone, fine particulates, carbon monoxide, and nitrogen dioxide. In 1991, the South Coast Air Basin exceeded one or more federal health standards on 184 days.

Nevertheless, air quality had improved dramatically since the 1970s. From 1955 to 1992, the peak level of ozone—one of the best indicators of air pollution—declined from 680 parts per billion to 300 parts per billion. The California Air Resources Board documented that population exposure to unhealthful ozone levels was halved between 1984 and 1994, when the lowest level on record was registered.

All these improvements were achieved at a time when human activity in the Los Angeles area was increasing at a rapid rate. Since the 1950s, the population of Los Angeles proper has almost doubled, from 1.97 million in 1950 to 3.69 million in 2000, and the greater Los Angeles area is home to some 15 million people (depending on how the area is defined) and

ious pollutants such as sulfur dioxide, nitrogen oxides (NO_x), ammonia, or other gases react in the atmosphere. Ozone forms as a result of the reaction of volatile organic compounds (VOCs) and nitrogen oxides in the presence of sunlight. For this reason, it is often very important to monitor the levels of VOCs in a given area. These criteria pollutants have two main sources: (1) burning of fossil fuels by motor vehicles and stationary sources such as power plants and (2) industrial processes other than fossil fuel combustion. We will discuss each of these six criteria pollutants in turn, but first we should note some key patterns in the United States:

1. Most pollutants, with the exception of nitrogen oxides and ground-level ozone in some areas (formed as a result of increased emissions of nitrogen oxides), have decreased significantly since 1970 due to increased pollution controls.

2. Transportation is the largest source of carbon monoxide, volatile organic compounds, and nitrogen oxides. This category includes motorboats, mowers, and small motors, which until recently have been largely unregulated by air pollution controls.

contains over 10 million automobiles and other vehicles; the city has become one of the most prosperous in the world.

The hazy days of the Los Angeles area are caused in part by a natural weather phenomenon known as an inversion layer. Such inversions often form off the coast of Los Angeles as the Pacific Ocean cools the atmosphere just above it. After ocean breezes blow the air mass inland, the layer of cool air may slip underneath a layer of warmer air. The inversion layer traps air pollutants in the cool air near the ground where people live and breathe. The mountains that surround the region compound the problem by preventing the pollutants from dispersing.

Natural materials, such as dust, pollen, fibers, and salt, are important components of haze. But industries and motor vehicles contribute to the problem by adding carbon particles, metallic dust, oil droplets, and water vapor. Automobiles, factories, and other sources also release such raw pollutants as hydrocarbons, water vapor, carbon monoxide, and heavy metals. When these chemicals are exposed to intense sunlight, they react to yield a vast number of secondary pollutants.

Over the years, regulations were gradually developed that concentrated on reducing the major sources of air pollution: particles from trash incineration, emissions from industry, and pollutants from motor vehicles. The EPA prepared standards for airplanes, trains, and ships that travel through the region. Local governments tried to reduce traffic by improving the transportation infrastructure and expanding mass transit. Automotive emissions in particular decreased because of more stringent tailpipe standards as well as programs encouraging carpools and use of public transportation.

Since the middle 1990s, the dramatic improvements of Los Angeles air quality seen in previous decades have not been maintained at the same pace. For instance, up through 2000 (the last date for which data are currently available), peak ozone levels have been creeping upward. However, this trend is somewhat misleading because even as peak recorded ozone levels have increased slightly, the number of days during the year when ozone (and other) standards as established by the EPA have been exceeded has dropped. In 1992, the Los Angeles area failed to meet acceptable air-quality standards on 175 days of the year, but this number has been reduced every year since then to only 27 days in 1999 (latest year for which there is data as of this writing). Thus it appears that, overall, the air quality of Los Angeles continues to improve.

Questions

1. Is there significant local air pollution where you live, work, or attend college? Do you keep track of the local air quality? Are there days that you can "feel" that the air is unhealthy?

2. Some people are resigned to the "fact" that there is bound to be more air pollution (as well as other drawbacks) in large metropolitan areas, yet they would rather put up with such issues in exchange for the opportunities and conveniences of the city. How do you feel about this matter? Must cities necessarily have unhealthful air? How does the example of Los Angeles influence your opinion on this issue?

3. Stationary fuel combustion (mostly coal-burning power plants) is the largest source of sulfur oxides, which cause acid rain.

4. Industrial processes besides fuel burning are the largest source of particulates.

Take special note that fossil fuel burning is the main cause of much of the air pollution that we face. This is one reason why many environmentalists advocate switching to alternative fuels, such as solar or wind. These energy sources are not only renewable, but would greatly reduce many air pollution problems.

Particulates

Particulates are any particles of dispersed matter, solid or liquid, that are larger than individual molecules. This category is very complex because particulate matter varies widely in composition and size. Soot, which arises from incomplete fuel combustion in cars and especially coal-burning factories, is generally the most common particulate in urban areas. It comprises at least 50% of the particulate air pollution in most cities of the world. Soot is especially common in developing countries, where air pollution is poorly controlled. If you have visited large cities in developing countries, you know that

(a)

(b)

Figure 16-2   (a) Smog in Beijing, China. Particulates and other pollutants greatly reduce visibility. (b) A satellite image of smog covering Beijing.

white clothes can quickly become dirty from the accumulation of dark particles. Particulates also scatter light, reducing visibility (**Figure 16-2**).

Particulate matter (PM) is often classified by size. $PM_{2.5}$ refers to "fine" particulate matter that is equal to or less than 2.5 microns (micrometers) in diameter (one micrometer is one-millionth of a meter). PM_{10} refers to "coarse" particulates that are large than $PM_{2.5}$ but less than or equal to 10 microns in diameter. As a general rule of thumb, coarse particulates (PM_{10}) are primarily the result of direct emissions, whereas much of the fine particulate matter ($PM_{2.5}$) is composed of secondarily produced par-

ticles (as when various gaseous pollutants interact in the atmosphere).

Aside from their effects on clothes, machinery, buildings, and visibility, particulates are the single most damaging air pollutant to lungs. Larger particles are usually trapped in the hairs and lining of the nose and throat and then coughed or sneezed out. Small particulates however, can penetrate deep into the smaller canals and pockets of the respiratory system, where they cannot be easily coughed up or otherwise removed (**Figure 16-3**). Accumulation of sufficient amounts of these fine particles will cause impaired breathing from blockage and irritation.

Figure 16-3   Anatomy of the lung. Fine particulates can penetrate deep into the bronchioles and alveoli where oxygen and carbon dioxide are exchanged.

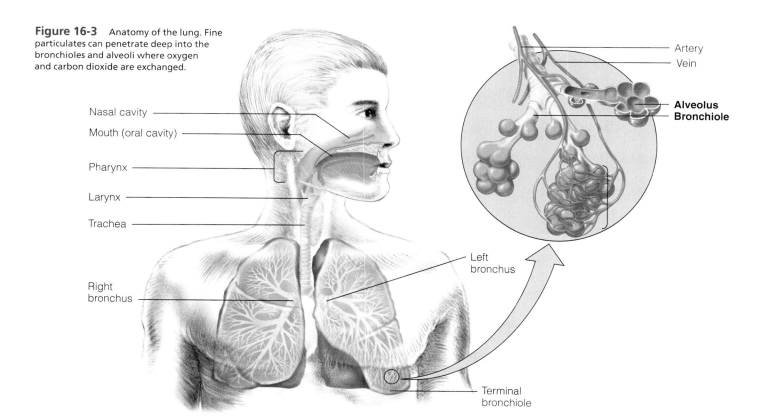

Examples include "black lung" of coal miners, asbestos fibrosis, and urban emphysema. Many statistical studies show that increases in particulate concentration in the atmosphere are correlated with increased visits to hospitals for respiratory problems, cardiac disorders, bronchitis, and many other illnesses associated with breathing.

An additional problem is that toxic chemicals may become attached ("adsorbed") to the surface of soot particles. These chemicals, such as the toxic metals cadmium or nickel, can be absorbed by the blood and cause heavy metal poisoning, usually affecting the nervous system. Organic chemicals, such as those used in pesticides or manufacturing, are also often adsorbed to soot; these can cause cancer. Because particulate inhalation is cumulative, as are the toxic effects of adsorbed chemicals, the effects are usually most pronounced in the elderly. Particulate matter can also have detrimental affects on plants, animals, and natural ecosystems. In addition, PM can cause damage to painted surfaces and building materials. Despite these effects, particulates as a whole are generally not considered as detrimental to the environment as most other forms of air pollution. For one thing, particulate pollution caused by humans is actually small compared to pollution from natural sources. More than 90% of the particulates in the air globally come from salt spray, dust storms, and especially volcanoes; humans introduce less than 10%. Second, unlike gaseous air pollutants, particulates settle out of the air in a few days by gravity and rain. They thus tend to be local problems. Third, particulates are among the easiest and cheapest air pollutants to remove at the source.

Reducing Particulate Pollution. As with all pollutants, input reduction is always the cheapest and most effective method. In the case of particulates, this means burning coal, for instance, that has a low ash content. However, even low-ash coal produces tons of emissions. These are usually removed from the air flow after combustion. Particulates are generally removed from the air flow in a stepwise fashion. Large particles are removed first by a so-called cyclone collector (**Figure 16-4a**). This spins the emissions in a vortex, causing the heavier particles to collide with the sides and slide down into a collecting bin at the bottom. This step typically removes more than 90% of the largest particles and is relatively cheap and maintenance-free.

Next, the smaller particles are removed by using either electrostatic precipitators or baghouses. **Electrostatic precipitators** take advantage of the small electric charge that most particles carry. A high

(a) Cyclone

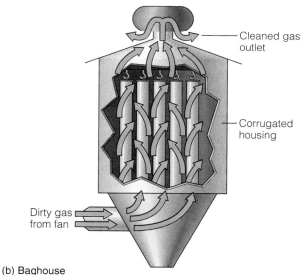

(b) Baghouse

Figure 16-4   (a) A typical cyclone collector for particulates. (b) A typical baghouse filter for particulate pollution. (*Source:* U.S. Department of Health, Education, and Welfare.)

voltage is created between metal walls, causing the particles to collect on the walls, where they are washed or shaken off into a bin. These are very efficient, removing up to 99.9% of small particles. **Baghouses** are a series of fabric bags that act as filters (**Figure 16-4b**). In efficiency and cost, they are similar to precipitators. Despite certain disadvantages, such as fire and explosion potential, baghouse filters are used almost as frequently as precipitators.

Trends in Particulate Pollution.   It can be difficult to get a precise handle on particulate emissions since there are several different types and sizes of particulate matter involved; some is emitted directly and some particulate matter forms secondarily (as described above). In addition, in the late 1990s, the EPA changed its procedures for estimating and monitoring different types of particulate emissions. Still, it is clear that particulate emissions in the United States have decreased dramatically in the last few decades; it is estimated that direct emissions of PM_{10} declined by 47% during the period 1981 to 2000 (with a drop of 6% during the 1991–2000 period), and directly emitted $PM_{2.5}$ declined by 5% during the 1991–2000 period. Much of this reduction is the result of the Clean Air Act Amendments of 1970, which, among other things, required the use of control devices to limit coal particle emissions. Across the United States, however, there can be major regional discrepancies in levels of particulate matter. The eastern United States, and in particular the southeast and mid-Atlantic states, tend to have higher levels of $PM_{2.5}$ than the western states due to the sulfate concentrations resulting from coal-burning power plants.

Particulate air pollution has also decreased over the last few decades in nearly all of the other wealthier nations. In contrast, the formerly communist countries of Europe and the developing nations continue to have high levels of particulate pollution largely because they lack the money to invest in controls. Historically, as per capita income increases, particulate emissions decrease because nations have more money to invest in pollution controls.

Sulfur Oxides

Sulfur oxides are produced when fossil fuels (in particular, coal and oil) that contain sulfur are burned or during metal smelting and other industrial operations. This oxidizes the sulfur to form SO_x compounds, such as SO_3 or, most commonly, SO_2. Coal generally contains much more sulfur than the other fossil fuels. Consequently, the majority of SO_2 air pollution in the United States is generated by coal-burning electric power plants.

In terms of overall damage to humans and the environment, SO_x may be the most serious local and regional air pollutant. Sulfur dioxide, or SO_2, is a gas that is toxic to living things. Plants are especially sensitive, exhibiting stunted growth, discoloration, and reduced crop yields. Plant damage

has been recorded over 50 miles (80 km) downwind of large sources. Effects on humans range from irritation of mucous linings in the eyes and lungs to death from respiratory and heart failure. Death can occur in just 30 seconds at concentrations as low as 3 parts per million.

SO_2 is also the main cause of **acid rain** (also known as *acid precipitation*), which is rainfall that is more acidic than normal. Acidity is measured on the **pH scale**, where 7 is neutral and 1 is very acidic (**Figure 16-5**). This is a logarithmic scale, so a decrease of 1 in pH indicates a 10-fold increase in acidity. Even unpolluted rainfall is slightly acidic, with an average pH of about 5.6, because water combines with naturally occurring carbon dioxide in the air to form the mild carbonic acid. When sulfur oxides are added to the air, water combines to form the much stronger sulfuric acid, causing acid rain. Technically, it becomes "acid rain" when its pH is below 5.0. In extremely industrialized urban areas, local "acid fogs" can be produced; likewise, in colder regions, "acid snow" may form. In the past, there have been fogs in the Los Angeles basin with a pH of 1.7, more acidic than lemon juice.

Such extremely high acidity usually occurs in confined areas near the sulfur oxide sources. More often, the sulfur oxides disperse downwind, often traveling three or four days and many hundreds of miles before falling as acid rain. Dry acid deposition

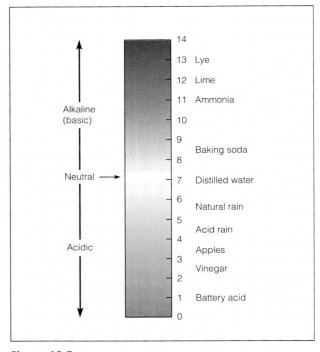

Figure 16-5   On the pH scale, 7 is neutral. Values above 7 are increasingly alkaline, and values under 7 are increasingly acidic.

can occur when relatively dry sulfate-containing particles, or sulfuric acid salts, settle out of the air. When combined with moisture, these particles may form a very strong acid. This problem is aggravated by the tall smokestacks built by many coal-burning power plants to prevent local particulate pollution by ejecting emissions high into the atmosphere. As a result, acid rain tends to be concentrated in areas downwind from the large industrial and urban centers that produce the pollution. In North America, acid rain is concentrated in the Great Lakes region, New England, and southern Canada, where the pH has dropped below 4.5 (**Figure 16-6**). These areas are downwind of industrialized areas around the southern Great Lakes such as Cleveland and Pittsburgh. In Europe, acid rain tends to be concentrated in northern Germany and Scandinavia, which are downwind of the industrial, urban areas of England and southern Germany.

Acid rain and acid deposition damage both the physical and biological environment. Physical damage includes paint discoloration; corrosion of metals; and dissolution of marble, mortar, and other building materials that contain lime, which reacts with acid. This has caused many buildings and statues to become pitted and eroded. In the United States alone, acid rain annually causes tens of billions of dollars in equipment repairs and replacement costs. Archaeologists calculate that acid rain has caused more damage to the marble buildings of ancient Greece in the last 50 years than in the entire preceding 2400 years. Whenever possible, the most valuable historic monuments, such as statues and parts of temples on the Acropolis in Greece, are being moved into museums, and plastic replicas are left in their place. Biological damage from acid rain is most visible in forests and lakes. Coniferous trees, such as pines, are especially sensitive to changes in soil acidity. Symptoms include yellowing and loss of needles (**Figure 16-7**). Many of the coniferous forests in New England, Canada, and northern Europe have shown these symptoms—some drastically. For example, it is estimated that more than half of western Germany's forests have been affected by acid rain.

Lakes suffer biological damage as their waters become more acidic. Generally speaking, lake ecologists have found that larger organisms are more sensitive to acidity:

Game fishes   most sensitive
Invertebrates
Rough fishes
Algae
Microbes   least sensitive

There is much variation within this pattern, however. Invertebrates such as insects and snails vary widely, but many suffer damage below pH 6. Fishes have a lower average pH tolerance because

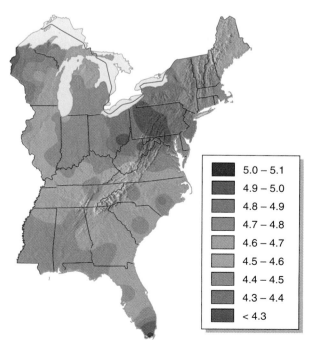

■	5.0 – 5.1
■	4.9 – 5.0
■	4.8 – 4.9
■	4.7 – 4.8
■	4.6 – 4.7
■	4.5 – 4.6
■	4.4 – 4.5
■	4.3 – 4.4
■	< 4.3

Figure 16-6   Average pH values of rain and snow in the eastern United States as of 1999, the latest data available.

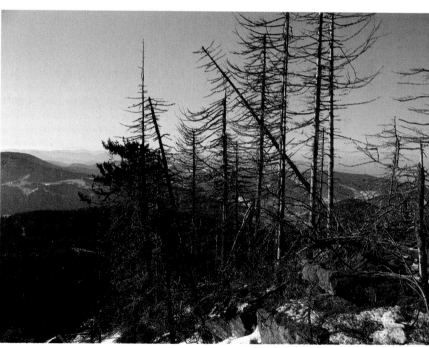

Figure 16-7   Acid rain damage.

so-called rough fishes, such as garfishes, can tolerate pH values even below 4. However, many game fishes, especially trout, are affected at pH values as high as 6. Larvae and eggs are usually the most sensitive in all organisms.

Most of the water that fills a lake is runoff that has percolated through the surrounding rocks and soil. In many cases, these rocks and soils are more **alkaline** than the rain and therefore neutralize some of the rain's acidity. For example, where lakes are surrounded by limestone and other sedimentary rocks that are extremely alkaline, even highly acid rain is often readily neutralized. Unfortunately, New England, eastern Canada, and northern Europe have surface rocks and soils formed from igneous and other crystalline rocks that are much lower in alkaline chemicals. This is a major reason why acid rain affects lakes so strongly in these areas.

Reducing Sulfur Oxides. As with all pollutants, the most effective and cheapest way of reducing sulfur oxides is through input reduction. In this case, this means reducing the amount of sulfur in coal, which produces the vast majority of sulfur oxide problems. The sulfur content of power plant coal typically ranges from 0.2% to 5.5% (by weight). By switching to low-sulfur coal, sulfur emissions can be reduced from 30% to 90%, depending on the original and the new sulfur contents. Unfortunately, about 85% of U.S. low-sulfur coal reserves are in the western United States, while two-thirds of coal consumption is in the east. Coal's large bulk makes it expensive to transport, so the cost of shipping low-sulfur coal is a major deterrent to its use, given the many tons burned per day by each plant. A growing alternative is to clean high-sulfur coal using a variety of methods. For example, sulfur often occurs in heavy minerals such as pyrite (FeS_2) that readily settle out when the coal is washed in water. Coal cleaning is relatively cheap because it reduces other costs by increasing coal-burning efficiency and reducing waste.

Another option is to use scrubbers. **Scrubbers** are devices that cleanse emissions, usually with water, before they are released into the air. Finely pulverized limestone, mixed with water, is sprayed into the emissions. The gaseous sulfur mixes with the calcium carbonate to form a "sludge" of calcium sulfate (gypsum). Scrubbers can remove about 90% of SO_2 from the emissions, but using them is costly. One problem is the high cost of installation: A scrubber adds 10% to 20% to the total cost of a new power plant. Operation costs are also high, adding about 1–2 cents per kilowatt-hour of electricity produced. The problem here is that scrubbers often corrode

and clog, reducing plant efficiency. Finally, and perhaps most important, is the high cost of removing the highly toxic and acidic gypsum sludge that is produced in vast amounts. For example, a large 1000-megawatt plant burning 3% sulfur coal produces enough sludge each year to cover a square mile (2.59 km²) of land to a depth of over one foot (0.31 m). The cost of disposing of this material is rising rapidly as landfills and other disposal sites fill up. Attempts have been made to use the gypsum in construction, but it is not generally economical compared to gypsum from other sources. Yet another problem with scrubbers in arid areas is that they use up to 1000 gallons (3785 L) of water per minute in large plants.

Trends in Sulfur Oxide Pollution. Thanks to the Clean Air Act Amendments of 1970 and the EPA's Acid Rain Program beginning in 1995, U.S. SO_2 emissions decreased by 31% over the 1981–2000 period (with a 24% decrease during 1991–2000). Even more dramatically, nationwide ambient concentrations of SO_2 dropped by 50% between 1981 and 2000 (with a 37% drop during 1991–2000). Much of this reduction was achieved from the use of the control methods described above in industry and stationary combustion sources such as power plants. World Bank data show that sulfur oxide pollution is most common in countries of intermediate wealth. Poor countries do not yet have the industrial and electrical capacity to burn large amounts of coal, while the wealthiest countries can afford pollution reduction.

Nitrogen Oxides, Ground-level Ozone, and Volatile Organic Compounds

Nitrogen oxides, ground-level ozone, and volatile organic compounds (VOCs) are discussed together because they are major causes of smog and other photochemical pollutants in urban areas. Note that ground-level ozone is a major pollutant, forming the primary constituent of smog, that remains a major concern in the United States and around the world. Ozone also occurs naturally in the stratosphere, where it serves important functions, such as protecting the surface of Earth from high levels of ultraviolet radiation. Both increased concentrations of ground-level ozone and depleted levels of stratospheric ozone pose major environmental and health concerns (see further discussion of stratospheric ozone in Chapter 17).

Nitrogen oxides are usually referred to as NO_x because they are composed of many different oxides, especially NO and NO_2. VOCs include a wide variety of hydrocarbon molecules. A **photochemical pollutant** is produced when sunlight initiates chemical reactions among NO_x, VOCs, and other components of air. These reactions are very complex and can produce many different kinds of photochemical pollutants. Nevertheless, we can summarize the basic photochemical reaction as:

Volatile organics + NO_x
+ sunlight = photochemicals

These photochemicals create a variety of environmental problems. Direct effects on humans and animals include irritation of eyes, lungs, and other mucous membranes. Plants may suffer stunted growth and death. Photochemicals are examples of **secondary pollutants,** meaning that they are produced by reactions among other air pollutants. **Primary pollutants** are those that are directly emitted, such as SO_x by coal-burning plants.

Smog originally meant air pollution that was a mixture of smoke and fog, but it now refers to the brown haze of photochemical pollutants in urban areas that irritates membranes and reduces visibility. **Ozone,** or O_3, is the most abundant photochemical pollutant in smog. Oxygen is very reactive, so the O_3 molecule is particularly destructive, leading to much of the lung irritation in humans and plant damage. Indeed, ozone alone is thought to be responsible for a large percentage of all air pollution damage to crops, with a total loss of hundreds of millions of dollars to U.S. agricultural productivity annually. Ground-level ozone damages the foliage of plants and is associated with increased susceptibility of plants to diseases, pests, and environmental stresses (such as lack of water or unusually harsh weather). The effects of ozone exposure may not be fully manifested for many years or decades, and thus there is the potential for long-term forest and ecosystem decline that may not be readily apparent initially. The reactivity of ozone also leads to damage of the physical environment. Long-term exposure to ozone causes rubber products to crack, for example. Ozone occurs naturally in air in small concentrations and is even crucial for protecting the Earth from solar radiation at very high altitudes (Chapter 17). In highly urbanized areas, however, where concentrations of ozone can be over 10 times higher than the natural near-surface level, its reactive properties are harmful.

In addition to their role in smog, nitrogen oxides contribute significantly to acid rain. While sulfuric acid from SO_x contributes about two-thirds, nitric acid, formed from NO_x and water, contributes about one-third of the acid in acid rain.

Sources of Photochemical Pollutants. NO_x and VOCs are largely produced by fuel combustion.

NO_x comes mainly from fossil fuel burning in transportation and power generation; VOCs come from transportation and industry. NO_x is formed when nitrogen in the air and in fuel combines with oxygen during combustion. VOCs are largely gasoline and other fossil fuel (hydrocarbon) molecules that are not completely burned during combustion.

Because so much NO_x and VOC pollution is associated with transportation, urban smog varies considerably with traffic patterns. **Figure 16-8** shows how morning "rush hour" traffic in Los Angeles causes a peak of nitrogen oxides in the early morning. Photochemical reactions on these oxides and VOCs take about two hours to produce the ozone peak in late morning.

Reducing Photochemical Pollution. NO_x is formed from nitrogen in air and in the fuel during combustion; therefore, a major method of reducing NO_x has been to alter the combustion process. Because coal-fired power plants emit about 25% of the NO_x pollution in the United States, emphasis has been placed on designing new technologies that control or reduce NO_x production during coal combustion. The so-called low excess air process restricts the amount of air allowed into the combustion chamber. This oxygen-starved environment reduces the oxidation of nitrogen during combustion so that nitrogen in the fuel is released as the harmless N_2 gas. NO_x emissions from coal combustion can be reduced up to 50% by this method.

Reducing NO_x and VOCs from vehicle emissions by changing the combustion process has proved much more difficult than reducing power plant emissions. The problem is that NO_x emissions are maximized under the same combustion conditions that minimize VOCs and most other pollutants.

This has been a continuing problem for vehicle manufacturers, who are required by law to meet certain emission standards. The primary solution has been the development and installation of emission control devices that remove pollutant gases after fuel combustion occurs. These are discussed in the next section. However, a final source of VOC pollution should be mentioned: Up to 20% of an auto's emissions can come from evaporation of gasoline from the gas tank and carburetor. This source has been reduced through increasing use of vapor recovery systems in the tank and carburetor that capture and recycle the fumes through the engine. Also, gas stations are increasingly using vapor recovery nozzles at the pumps.

Trends in VOC and NO_x Pollution. Growing use of vapor recovery systems, emission controls, and

Figure 16-8 (a) Rush hour traffic is a familiar sight to nearly all city-dwellers. Such traffic peaks are the main source of many health and other air pollution woes. Seen here is traffic in Calabasas, California, just northwest of Los Angeles. (b) Typical concentrations of NO$_x$ and ozone in Los Angeles over the course of a day. Morning traffic causes a late-morning peak of ozone. (*Source*: (b) U.S. Department of Health, Education, and Welfare.)

(a)

(b)

ability of the red blood cells to carry oxygen to the organs: The CO molecule attaches to the blood cells more readily than oxygen molecules. Because the brain requires a very large amount of oxygen, dizziness, headaches, and mental impairment are among the first symptoms.

Concentrations as low as 50 ppm will cause headaches and other effects over the course of a few hours. All effects are felt sooner if you are active; the body's cells are depleting oxygen faster and are in greater need of it. This is one reason why jogging during rush hour traffic is often discouraged. Fortunately, the effects of CO are usually reversible, except for extreme exposure. As usual with air pollution, the very young, elderly, and people with certain lung and heart ailments are most at risk.

other methods has led to a general decline in VOC pollution since about 1970; between 1981 and 2000, VOC emissions decreased by 32%. Most of this decline has resulted from reduced transportation (vehicle) emissions.

In contrast, NO$_x$ is the only major air pollutant that has not shown a significant decline because of the engine design difficulties described above. Even so, emission controls and other methods have made some strides in reducing NO$_x$ emissions per car. However, nearly all of this gain has been offset by the increasing number of cars being used, both in the United States and worldwide. Thus, total NO$_x$ emissions in the United States actually increased by about 4% between 1981 and 2000.

Carbon Monoxide

Carbon monoxide (CO) is a deadly gas in high concentrations, killing more than 200 people every year in the United States. Its deadly effects are all the more dangerous because it is so difficult to detect: It is colorless, odorless, and tasteless. Carbon monoxide has little effect on plants or materials, but in humans and animals it interferes with the

Sources of Carbon Monoxide Pollution. CO is produced by incomplete combustion, when fossil fuels, wood, tobacco, and other organic materials burn under less than ideal conditions. As a result, the carbon is not fully oxidized to CO_2. For example, an engine with a poor air supply or low burning temperature will produce more CO than an engine that has a better air supply or burns at a higher temperature. Thus, the same high temperatures that cause nitrogen to form NO$_x$ will minimize CO.

About 60% of carbon monoxide emissions in the United States are from transportation, mainly vehicle exhausts; in American cities, 95% of the CO can come from automobile exhausts. Similarly, cars and trucks are the main source of CO in nearly all cities of the world. The rest comes primarily from fuel combustion at power plants and industry and solid waste combustion. It is ironic that so much attention is given to these sources while many cigarette smokers deliberately produce much higher levels: Directly inhaled cigarette smoke may contain more than 400 ppm CO, compared to only 5–100 ppm in urban air from very congested, stalled traffic. Even "second-hand" cigarette smoke may raise indoor CO to 20–30 ppm.

Reducing Carbon Monoxide Pollution. Control of CO is closely related to control of NO_x and VOCs. Vehicle emissions are a major contributor of all three. There are three basic ways to reduce vehicle emissions: precombustion, postcombustion, or changing the combustion process. These methods have been successful in reducing CO emissions (and also VOCs) since 1970 despite a large increase in the number of vehicles driven. CO emissions decreased 18% in the United States over the period 1981–2000, and national average ambient air levels for CO decreased by 61% over the same period.

Most automobile manufacturers have traditionally favored postcombustion control because it is easiest and most economical. Postcombustion control devices remove the pollutants from the exhaust gas before it is released into the air. Examples include thermal reactors, which are "afterburners," and exhaust recirculators, which divert the exhaust back into the engine. Both of these reduce emissions by further oxidizing the CO and VOCs, and are thus less useful for reducing NO_x for the reasons noted above. Furthermore, they significantly reduce fuel economy and performance.

For these and other reasons, the auto industry's preferred postcombustion control is the catalytic converter. A "catalyst" is a substance that facilitates and speeds up a chemical reaction. A **catalytic converter** is a device that carries out a number of chemical reactions that convert pollutants to less harmful substances. Because CO, NO_x, and VOCs each have different chemistries, these reactions are quite complex. CO and VOCs must be oxidized to remove them, whereas further burning (oxidation) will only make NO_x pollution worse. The solution has been to make "two-stage" catalytic converters. In the first stage, NO_x is converted to harmless nitrogen gas (N_2) by removing oxygen. The second stage acts on CO and VOCs in the opposite way—by oxidizing CO to CO_2 and VOCs to CO_2 and H_2O. Both stages use finely ground particles of substances such as platinum as the catalyst.

Changing the combustion process is another way to control emissions. There are many alternatives to the standard gasoline internal combustion

engine, and some of these produce fewer emissions. Diesel engines are currently the most popular alternative, although electric motors may be the engine of the future. Diesel engines inject fuel directly into the cylinder. By omitting the need for spark plug ignition, diesels are not only more fuel-efficient, but have very low emissions of CO and VOCs. On the negative side, diesel combustion in the cylinder produces high temperatures that create high levels of NO_x pollution. Furthermore, high levels of unburned carbon, or "soot," are emitted. Such particulates can be mutagenic and carcinogenic. The NO_x and soot problems can be reduced by exhaust recirculation and filters, but have created maintenance problems. Overall, diesels are declining in popularity.

Other alternative engines that have been marketed include the stratified charge engine, gas turbines, and the rotary engine. For economic and other reasons, they have not become popular with consumers, so car companies have tended to meet air quality laws with gasoline internal combustion engines linked to catalytic converters. Instead, the car of the future may be the electric car, which runs on a battery or fuel cell. By eliminating combustion in the car, electric cars can nearly eliminate CO, VOC, and NO_x emissions in dense urban traffic. Note that the net environmental impact of electric cars is largely determined by how the electricity to run them is generated. If a coal-fired plant is used, the net impact could be more of many pollutants, since coal is not only a very "dirty" fuel, but much usable energy is lost in converting coal to electricity. Urban smog and CO pollution would be reduced, but concentrated air pollution would be shifted to the local area around the power plant. On the other hand, electricity generated by direct solar, wind, or other cleaner fuels would greatly reduce net air pollution everywhere.

Control of CO, NO_x, and VOC emissions by precombustion methods is also growing. Examples include "reformulating" gasoline before it is burned and using alternative fuels such as methanol, ethanol, compressed natural gas, propane, or hydrogen. All have various advantages and disadvantages. A promising method of reformulating gasoline is **oxygenation**, which blends oxygen-rich liquids into it. By adding oxygen, CO is converted to CO_2 and VOCs are oxidized better.

Perhaps the most commonly discussed alternative fuel of the future is **methanol**, which can be produced from natural gas, coal, or biomass. Most commercial methods now use natural gas. Methanol has a much higher octane rating than gasoline, allowing it to burn more completely and reducing CO and VOCs; it also burns at lower temperatures, reducing NO_x. Ozone pollution is reduced by half. But methanol also has disadvantages including higher emissions of formaldehyde; it is also an eye irritant and a possible carcinogen. A practical difficulty is its low volatility, which makes it difficult to start engines on a cold morning. Most important is its low energy content, which is about half that of gasoline; this means gas tanks would have to be twice as large to give cars the same mileage range. A solution is to use a mixture of 85% methanol and 15% gasoline, which eliminates the cold-start problem and improves mileage. Carmakers are introducing "flexible fuel" cars that can run on either gasoline or a mix with methanol. In addition to reducing local and regional air pollution, methanol reduces global emissions of CO_2, but only if it is made from natural gas. If methanol is made from coal, overall CO_2 emissions more than double.

Lead

Lead (Pb), like other metals, can accumulate in the body and is especially destructive to the nervous system. Lead poisoning can cause brain damage leading to learning disabilities, seizures, and death at high concentrations. Children and pregnant women are most at risk. In 1977, the EPA estimated that 600,000 children in the United States had blood levels of lead that were near, or exceeded, those high enough to cause behavioral changes. Airborne lead occurs mainly as particles that do not travel far from the source. About one-sixth of inhaled lead contributes to poisoning: One-third of all inhaled lead particles become trapped in the lungs, and about half of those are eventually absorbed by the bloodstream.

Fortunately, human exposure to airborne lead has decreased more than 95% since 1970. The main reason was the gradual elimination of leaded gasoline during the 1980s and early 1990s. Instead of lead additives to reduce "knocking" in the engine, other anti-knock additives are now used. However, unacceptably high atmospheric lead concentrations can still be found around some smelters and battery manufacturing operations. Also, independent of atmospheric-borne lead, ingestion of lead from water in lead-contaminated pipes and from lead-based paint continues to be a major health threat to children, especially in urban areas. Pipes and paint are now required to be relatively lead-free so the threat occurs mainly in older homes.

International Trends

In contrast to the United States, where most types of local and regional air pollution are declining, the worldwide total of the six criteria pollutants has generally been increasing. A main reason for these international trends is the increasing industrialization of developing nations and their use of fossil fuel technology. As coal-burning power plants, motor vehicles, and industries become more common in places like China, inevitably ever-greater amounts of air pollution are released (see Figure 16-2). The U.S. Department of Energy has estimated that the global fleet of automobiles, trucks, and buses could grow from the current 700 million to 1.1 billion in 2020.

There are many ways to reduce these trends. Most involve promoting industrialization that reduces or even eliminates fossil fuel use, which is the main source of most air pollution. Mass transportation can have a major impact by reducing the use of cars. The average American annually produces more than twice as much carbon dioxide, for example, compared to the average Japanese, in part due to the very low use of mass transit in the United States. Reformulated gasoline and more fuel-efficient technology will also help. Ultimately, however, the most effective way to reduce fossil fuel pollution is the use of alternative fuels: electric cars, hydrogen fuels, wind power, and so forth. The widespread use of such green technologies in developing nations and their adoption in developed nations can simultaneously improve the environment and the economy.

Weather and Air Pollution

Local weather conditions can strongly influence the impact of air pollution. Temperature is especially important. For instance, cars release more air pollution when they are first started on cold winter mornings than in the summer. Because incomplete combustion causes much air pollution, a cold engine does not burn fuel as efficiently and it releases greater amounts of CO and other pollutants. You may have noticed that car exhaust smells different on cold winter mornings. The most important local weather phenomenon that affects air pollution is **thermal inversion**. This occurs when a layer of warm air overlies cooler air, "inverting" the usual condition in which air becomes cooler as altitude increases (**Figure 16-9**). The layer of warm air, which can be several yards to several miles thick, acts as a trap for rising air pollution. Such inversions can last up to several days, but are most common at night. Low-lying valleys, such as the Los Angeles area, are most prone to inversions because it is easier for warm air to become trapped. Before air pollution controls were common, inversions could be quite disastrous in urban industrial areas. The most severe urban inversion episode in history occurred from

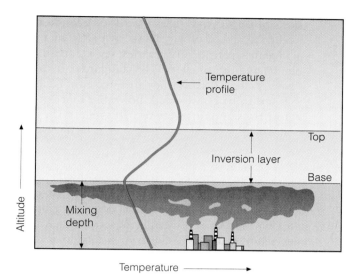

Figure 16-9 The warm inversion layer prevents pollutants from rising into the air above it.

December 5 to December 10, 1952, in London and caused 4000 deaths. Air pollution controls have reduced the frequency of such disasters, but urban air quality is still generally at its worst during periods of thermal inversions.

In addition to weather affecting air pollution, the reverse can occur: Air pollution can affect weather. For example, industrialized urban areas tend to experience more rain and snow than surrounding areas because of the pollution they produce. Cities release great amounts of thermal pollution. The concentrated burning of fossil fuels and other energy sources, along with the concrete and metal that retain the Sun's energy, turns cities into "heat islands." Rising hot, moist air causes rain when it cools, and this effect is enhanced by the high amounts of particulate air pollution from cities. Particulates such as soot form tiny nuclei on which the cooling moisture can condense into droplets. Besides affecting local weather conditions, air pollution can affect global climate. This obviously has a much greater potential effect, as we see in the next chapter.

Social Solutions

We have discussed many technological ways of reducing air pollution. Since most air pollution originates as fossil fuel combustion, these solutions have focused on those fuels. We have seen three ways to control combustion: (1) precombustion, such as higher fuel efficiency or cleaning coal before use; (2) postcombustion, such as scrubbers; and (3) changing the combustion process, such as in electric cars. We have also discussed how input reduction by us-

ing less fossil fuel, especially by using alternative energy technology, is often the most effective solution. Technological reductions will not become widespread, however, unless people have some incentive to use them. In the United States, legal and economic incentives have been used most often.

Legal Solutions

Legal incentives have been the main reason for the decline in many of the local and regional air pollutants since about 1970. Federal legislation actually began with the Air Pollution Control Act of 1955. This act mainly provided funds for research, but its enactment was a key event because it initiated federal participation in air pollution control. Until 1955, air pollution was considered a local problem and was therefore treated by a hodgepodge of local and state laws that varied widely. The **Clean Air Act** of 1963 was the first of a series of acts and amendments that exerted increasing federal pressure on air polluters. Not until the Clean Air Act Amendments of 1970, however, were widespread legal standards established that were actually enforced. These standards included the National Ambient Air Quality Standards (NAAQS) and New Source Performance Standards (NSPS). The NAAQS established standards of allowable concentrations in surrounding air, whereas the NSPS established standards of allowable emissions, or rates at which polluters could emit. The newly established Environmental Protection Agency (EPA) was charged with the task of enforcing these standards.

Air Quality Standards. In applying NAAQS, the EPA established air quality standards for various major pollutants, such as CO, NO_2, O_3, SO_2, PM_{10} (particulate matter), and lead. These are easily measured, occur in significant amounts, and are health threats. They indicate the general "health" of the air. If one or more of these are present in large amounts, then we can infer that other, associated air pollutants are also common. Allowable concentrations in California are often lower, illustrating how federal law may allow states to set stricter standards for air quality and many other environmental regulations.

For many years the **Pollutant Standards Index** (**PSI**) was used, particularly in cities, to monitor air quality. This index ranged from 0 (best air quality) to 500 (worst air quality). The PSI was based on the major NAAQS pollutants listed in the last paragraph. In 1999, the EPA revised and updated its air

TABLE 16-1	The Air Quality Index	
Air Quality Index (AQI) Values	**Levels of Health Concern**	**Colors**
When the AQI is in this range:	*. . . air quality conditions are:*	*. . . as symbolized by this color:*
0 to 50	Good	Green
51 to 100	Moderate	Yellow
101 to 150	Unhealthy for sensitive groups	Orange
151 to 200	Unhealthy	Red
201 to 300	Very unhealthy	Purple
301 to 500	Hazardous	Maroon

quality index, renaming it the Air Quality Index (AQI) based on levels of ground-level ozone, particulate matter, carbon monoxide, sulfur dioxide, and nitrogen dioxide. Like the earlier PSI scale, the AQI ranges from 0 (best quality) to 500 (worst). A value of 100 for a particular pollutant is equivalent to the level set for the pollutant as the national air quality standard, and thus levels below 100 are considered acceptable or safe (0 to 50 is generally considered "good"), while levels above 100 indicate various degrees of health concern. An overall AQI value for a particular day at a given location corresponds to the highest AQI value for any of the monitored pollutants. The current AQI includes a color coding scheme (**Table 16-1**).

Emission Standards. The EPA sets New Source Performance Standards (NSPS) to limit air emis-

sions by certain industries. The affected industries include fossil fuel power plants, incinerators, nitric acid plants, petroleum refineries, sewage treatment plants, and ore smelters. The standards are called "New Source" because they are directed at facilities that are to be constructed, not those that existed before the standards were set. This reflects the economic reality that it is usually much cheaper to design a new plant to produce lower emissions than to modify an existing plant that uses older technology. Of course, this begs the question of what to do with older facilities. By 1977, for instance, only about one-third of the air quality control regions in the United States were meeting the set standards. This deficiency in the 1970 Clean Air Amendments led to the 1977 Clean Air Amendments, which attempted to deal with these so-called nonattainment regions.

Perhaps the most successful policy to result from these amendments was **emission offsets**: To receive a construction permit, a major new pollution source in a nonattainment area must first find ways to reduce emissions from existing sources. Emission offsets have been very efficient in economic terms because they use the free market to find optimal solutions. For example, someone seeking a construction license may install emission controls on a heavy polluter, even if someone else owns it. Alternatively, the license seeker may buy out the old plant and shut it down.

Clean Air Act Amendments of 1990. In 1990, the U.S. Congress passed a new set of Clean Air Act Amendments (CAAA). These toughened air pollution controls for the first time since 1977. Some of the amendments' major provisions are listed in **Table 16-2**. These provisions focus on three types of local and regional air pollution that were considered inadequately addressed in earlier amendments:

1. Smog is a focus of provisions 1–6 as listed in Table 16-2. They were needed because nearly a hundred major cities had not complied with existing regulations. The new amendments required lower tailpipe emissions of NO_x and VOCs by 1994 and even applied to small industrial sources of smog-producing chemicals, such as bakeries. Carmakers were required to redesign pollution controls and encouraged to build electric and other so-called zero- or low-emission cars. Fuel manufacturers were required to produce methanol and the other "reformulated" gasolines. The EPA estimated the total cost of complying with

TABLE 16-2	Three Main Air Pollutants, Their Sources, and Major Provisions of the 1990 Clean Air Act Amendments that Seek to Control These Sources*	
Source	**Provision**	
Smog		
1. Gasoline	"Reformulated" fuels required in major cities	
2. Tailpipe emissions	Reduce NO_x and VOC emissions by about half	
3. Tailpipe emissions	Emission controls must last 100,000 miles	
4. Tailpipe emissions	Low-emission vehicles to be developed	
5. Gas stations	Vapors must be captured during refueling	
6. Cities	Reduce smog until federal standards are met	
Acid Rain		
7. Power plants	Reduce nitrogen emissions by one-third	
8. Power plants	Reduce sulfur emissions by one-half	
9. Power plants	Can exchange "pollution credits"	
Airborne toxics		
10. Businesses	Reduce toxic air pollutants by 90%	

*These provisions are required by federal law, unless they are repealed by later congressional actions.

these six provisions at up to $12 billion yearly.

2. Acid rain is a focus of provisions 7–9. Their goal is to reduce SO_2 and NO_X emissions from coal-burning power plants.

3. Airborne toxics (air pollutants that are dangerous in small amounts, such as airborne mercury, arsenic, and many carcinogens) are a focus of provision 10. In 1988, more than 2.7 billion pounds (1.2 billion kg) of toxic air pollutants were released in the United States. Only seven such pollutants were regulated until 1990; the new amendments regulated 189 kinds of toxics and more have been added since then.

These amendments have been very controversial. They are costly to many businesses, local governments, and citizens. Car prices, fuel prices, local taxes, and business operating costs increased as the provisions were phased in. Some employers who had too many employees who drive to work alone instead of car pooling or taking mass transit were fined. As a result, the amendments have encountered considerable resistance from the public and local and state governments.

Economic Solutions

The atmosphere is a classic example of an environmental "commons." It is a shared resource that people will exploit or pollute as long as they are free to do so: Polluters gain many benefits, while paying few of the costs of damage to the commons. Legal restrictions are one method of reducing pollution of a commons. Another method is to make polluters pay more of the true costs of damage to the commons. This can be done in many ways, such as taxation, fees, and other government charges on emissions or amount of fuel burned.

Many economists favor economic solutions because they impose costs on pollution and then allow the market to find a cost-effective way of paying for them. We have already discussed how emission offsets do this. The 1990 Clean Air Act Amendments also created a free-market trading system for emission offsets. Congress identified 110 utilities and assigned each a number of SO_2 allowances based on its history of fuel use. Each allowance represents 0.91 metric ton of sulfur; plants can buy, sell, or trade these allowances. A plant might thus pay for more efficient emission controls and sell its allowances to a plant that would rather buy the "right to pollute." The idea of a "right to pollute" may seem unattractive, but this method will keep the net regional air pollu-

tion below any predetermined limit in an efficient way. The first use of the 1990 CAAA allowances occurred in May 1992, when the Tennessee Valley Authority bought the pollution rights to 10,000 tons (9100 metric tons) of SO_2 emissions from Wisconsin Power and Light. In the mid-1990s, the EPA's Acid Rain Program was implemented, setting a cap of 8.95 million tons (8.1 million metric tons) on the total amount of SO_2 that may be emitted by power plants across the country (this is about half the amount emitted in 1980) combined with an emissions trading program. In contrast to emission offsets, which make polluters pay for their outputs, are gasoline, coal, BTU, and other "green" taxes that make polluters pay for the amount of polluting fuel burned. These "input reduction" solutions are the most efficient means of all because they not only reduce pollution but also conserve resources. A polluter who has to pay higher prices for fuel will use it more efficiently and consume less.

Indoor Air Pollution

Indoor air pollution as an environmental problem has been receiving increasing attention in recent years. Indeed, indoor air pollution is usually a greater direct threat to human health than outdoor air pollution. The cost of health problems from indoor air pollution in the United States is estimated in the tens of billions of dollars. Although there are many kinds of indoor air pollutants, just two, radon and smoking, cause the vast majority of harm.

Indoor air pollution is a major threat to human health for two main reasons. First, the indoor environment tends to concentrate pollutants. Some toxic and cancer-causing pollutants can reach air concentrations that are 100 times greater than outside air. Second, on average, people in industrialized societies spend more than 80% of their lives indoors, including time spent in offices on the job. **Sick building syndrome** is a popular medical term that refers to chronic ailments such as headaches, nausea, allergic reactions, and other symptoms that are caused by indoor air pollutants where we work or live.

Sources and Types of Indoor Pollutants

There are many sources of indoor air pollution (**Figure 16-10**). These sources can be divided into three categories: underground diffusion, combustion, and chemical emissions. In nearly all cases, the harmful effects of indoor pollution can be greatly reduced by increasing ventilation in a building. If properly done, this will not significantly reduce the heating (energy) efficiency of many buildings.

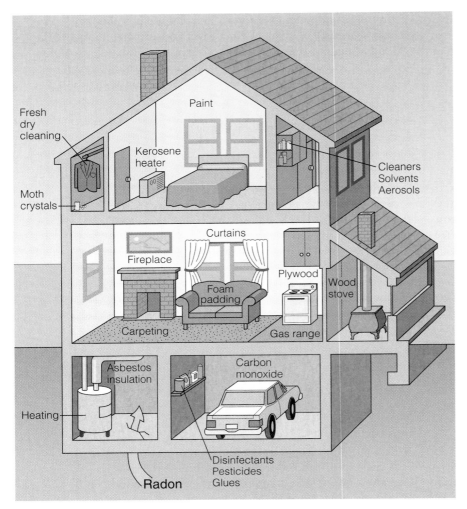

Fresh
dry
cleaning

Paint

Kerosene
heater

Moth
crystals

Cleaners
Solvents
Aerosols

Curtains

Fireplace

Plywood

Wood
stove

Foam
padding

Carpeting

Gas range

Asbestos
insulation

Carbon
monoxide

Heating

Disinfectants
Pesticides
Glues

Radon

Figure 16-10   Sources of indoor air pollution in the home. (*Source:* U.S. Environmental Protection Agency.)

Where underground diffusion of radon is found to be a danger, there are a number of relatively simple and inexpensive ways to greatly reduce indoor concentrations. The precise methods depend on the type of building construction, but the large majority involve increasing ventilation air flow through the building. Increasing air flow through basements, crawl spaces, and other lower parts of buildings is especially effective. For example, large window fans can be used to blow air through basement windows. Another common method is to insert pipes into the basement floor. The pipes draw the radon from the soil and transport it outside where the other end of the pipe exits.

Combustion: Especially Smoking.   Stoves, fireplaces, and heaters are common appliances that emit air pollution from combustion. Carbon monoxide is a poison caused by incomplete combustion; it can cause headaches and even be lethal in closed areas where stoves or heaters are not functioning properly. Certain cancer-causing chemicals, such as hydrocarbons, are also released by combustion. These problems are relatively minor where appliances are functioning efficiently and the building is well ventilated. In developing countries, however, inefficient equipment and poorly designed housing often result in harmful levels of indoor air pollution from heating and cooking. Particulate air pollution from ash particles of wood or coal burned in the home is a major problem. A study by the World Health Organization estimated that indoor particulate levels were more than 20 times higher than considered safe in many developing countries where wood is burned in the home.

Radon: Death by Diffusion.   Radon gas, which is among the most harmful indoor pollutants when found in high concentrations, generally enters the home through underground diffusion. Radon is a radioactive decay product of uranium; it is most concentrated in houses built on soil and/or rocks that are rich in naturally occurring uranium minerals and their products.

According to a National Research Council study, radon and related radioactive gases are linked to an estimated 15,000 to 22,000 deaths per year in the United States. These deaths occur mainly from lung cancer that is caused by inhaled radioactive particles that become lodged in the lungs. However, there is a synergistic effect between smoking and radon exposure; the same study concluded that only about 1200 to 2900 radon deaths are among nonsmokers. Radon testing of homes is now common, but efforts to establish the true danger of radon have led to considerable controversy. Most U.S. homes are built on soils that emit relatively harmless amounts of radon, and proving the cause of lung disease is often very difficult.

The most harmful combustion source of indoor air pollution is tobacco smoke (**Figure 16-11**). Many medical studies have found that smoking causes 400,000 to half a million deaths per year in the United States, especially from heart disease and cancer. Furthermore, there is strong evidence that nonsmokers are also affected: "Second-hand smoke" causes an estimated 3000 to over 50,000 deaths per year in the United States (numbers vary widely depending on the study cited—second-hand smoke is not only implicated in lung cancer, but also heart disease, asthma, bronchitis, and other diseases, especially in children). In addition to carcinogens, tobacco smoke contains carbon monoxide, particulates, and other harmful substances. Finally, smoking can enhance the deadly effects of radon. Smoke particles trapped in the lungs act as attachment sites for radioactive radon particles, increasing the chances for lung cancer. This is a classic, if deadly, example of a synergism where two effects enhance one another.

Chemical Emissions: Source of Many Pollutants. Chemical emissions can come from dozens of sources indoors. Many of these chemical emissions are carcinogenic, but only in concentrations that are generally much higher than occur in the large majority of homes. They do, however, often cause headaches, dizziness, and nausea, especially in people with a high sensitivity to certain chemicals.

Mothballs, bleach, shoe polish, cleaning solvents, air fresheners, and many other chemical products emit gases that can be toxic if sufficiently concentrated by poor air circulation. Many materials release carcinogenic fumes, such as benzene and formaldehyde. Benzene is emitted by plastics, synthetic fibers, and cleaners; formaldehyde is emitted from foam insulation and wood products such

as particle board, plywood, and some glues. Asbestos has received much notoriety as a cause of lung cancer and the lung ailment asbestosis. Once used for fireproofing, insulation, tile, and cement, this fibrous material is now being removed from old buildings at great expense. The need for much of this great cost is hotly debated, in part because only some types of asbestos are harmful (see Case Study 16-2).

Noise Pollution

Noise is often defined simply as "unwanted sound." Usually, it is unwanted because the sound is either too loud for comfort or is an annoying mixture of sounds that distract us, such as a distant conversation. However, noise is partly subjective and depends on one's state of mind and hearing sensitivity: Many studies show that sounds that harm or annoy some people do not bother others. This subjective aspect is important because it adds to the difficulty of formulating policies to control noise. Some people are not as concerned about certain noises as others and are unwilling to pay for controls. For example, spending money for airport noise reduction is often controversial.

Noise is a kind of air pollution because it is usually transmitted to our ears through the air that surrounds us. Indeed, when we hear any kind of sound, noise or otherwise, our eardrums are receiving vibrations being transmitted to us by colliding air molecules. Air molecules thus transmit sound waves in much the same way that water molecules transmit ocean waves. Whereas a large storm at sea provides the energy that sets the water waves in motion, a large explosion, machinery, or some other source provides the energy that sends the sound waves toward you.

Measuring Noise

Loudness increases with intensity, meaning the amount of energy carried by sound waves. Loudness is measured using the **decibel (dB) scale.** (*Deci* refers to the units of 10 used, and *bel* refers to Alexander Graham Bell, who was a teacher of the deaf as well as the inventor of the telephone.) The scale is exponential in that each unit of 10 increase in sound level (dB) is a 10-fold increase in sound intensity, or energy. The scale ranges from 0 to over 180, where 10 is the sound made by a leaf rustling and 180 is a rocket engine. Most daily noises in a busy building or on a city street average around 50–60 dB; in a quiet room, they are about 30–40 dB. Hearing damage begins around 70 dB for long exposure to a sound such as a loud vacuum cleaner. Outside noises at this level result in widespread complaints

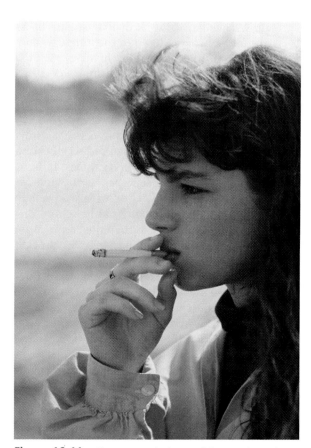

Figure 16-11 Cigarette smoke is one of the most hazardous indoor air pollutants.

CASE STUDY 16-2

Is Asbestos a Classic Example of "Environmental Hypochondria"?

Visit www.environment.jbpub.com for more information.

Despite the enormous scope and variety of environmental issues, a few have generated much more intense controversy than many others. One of those that have inflamed heated passions is the exposure of humans to asbestos (**Figure 1**). Some forms of asbestos can be quite dangerous to people; for instance, exposure to asbestos can cause mesothelioma, a cancer of the membranes around the lungs and internal organs that can lead to death. So, if asbestos is so dangerous, why not simply ban it? Indeed, a number of countries have done just that, including Argentina, Chile, Croatia, Iceland, Latvia, and Saudi Arabia. It is expected that by 2005, asbestos will be banned throughout the European Union.

Asbestos has many uses, however, and some people argue that the use of asbestos actually protects and saves more lives than it harms. Another major reason for the debate over asbestos is that the incredibly high economic costs of eliminating all asbestos currently in use—over $100 billion in the United States—do not clearly yield benefits that are worth the costs in some people's opinions. Furthermore, if asbestos is banned, substitutes will have to be found for many current asbestos uses, and such substitutes may be more expensive than asbestos. Indeed, critics argue that most types of asbestos pose a much smaller hazard than many daily activities such as walking across a street. They argue that asbestos is a classic example of "environmental hypochondria": an illusory environmental hazard produced by anxiety.

"Asbestos" was a term originally used mainly by geologists to refer to fibrous minerals found in certain metamorphic rocks. Because many of these fibers are heat-resistant, asbestos became widely used, especially during the 1950s and early 1960s, for "fireproofing" in building materials. Mines, shipyards, schools, houses, offices, and many structures used asbestos to reduce the chance of fire. Asbestos is also an excellent material for such high-temperature uses as brake pads on vehicles and engine gaskets. In the early twentieth century, it was realized that people working directly with asbestos on a regular basis suffered from health problems, and by the early 1970s, there was significant scientific evidence that simply breathing asbestos was harmful. The primary problem is that some kinds of asbestos are very crumbly and readily disperse into the air. When inhaled, the needlelike asbestos fibers cannot be broken down in the lungs, so they remain as a constant irritant. In these cases, the body's reaction to this irritation over 10 to 40 years can produce stiffening of lung tissue called "asbestosis." This makes breathing difficult and can lead to lung cancer.

These findings resulted in many lawsuits against makers of asbestos products, costing many billions of dollars in legal fees. Billions more dollars have been spent removing asbestos from buildings. The costs are especially burdensome for the many public school districts with limited funds that must remove asbestos from school buildings. In 1989, the U.S. Environmental Protection Agency attempted to ban all uses of asbestos, but was unsuccessful in large part due to pressure from the asbestos industry and asbestos users. New uses of asbestos were banned, but older uses were allowed to continue with regulation. Today (2002) asbestos remains in use in America in various roofing materials, gaskets, brake pads, and miscellaneous other uses. Another problem is that various levels of asbestos have been

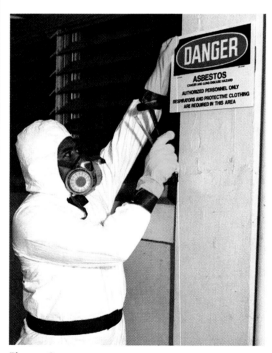

Figure 1 A worker wearing a protective suit and mask prepares for asbestos abatement.

found as a contaminant in other, supposedly asbestos-free, products such as shingles and insulation, cardboard, crushed rock used in landscaping, fertilizers, and many mineral products where asbestos may naturally occur in very low concentrations.

Is the energy, time, and money spent so far addressing the asbestos issue worthwhile? For decades now this has been hotly debated. As the 1990s progressed, there was a growing number of critics of the "anti-asbestos" campaign. In 1995, *U.S. News & World Report* stated that there is "now a broad consensus among scientists and physicians that asbestos in public buildings is not much of a threat to health" (February 20, p. 61). A 1995 study by the U.S. Office of Technology Assessment said much the same thing. A 1990 EPA report (the "Green Book") stated that asbestos removal is often not a building owner's best course of action—it is better in some cases to stabilize and cover the asbestos so that it does not enter the air or cause other damage, rather than attempt to remove it.

A primary reason for this rethinking of the asbestos threat was that more detailed research indicated that most of the asbestos in buildings is not the very dangerous variety. There are many kinds of asbestos, including two very different mineral groups. Most of the proof for toxicity centers on the "amphibole" type of asbestos. For example, one study showed that of 33 factory workers who breathed large doses of this type of asbestos in 1953, 19 workers had died of asbestos-related illness by 1990.

However, about 95% of the asbestos used commercially is not of this variety. Instead, it belongs to a second mineral group, the "chrysotile" type of asbestos, which evidently is nowhere near as toxic. It is very difficult to gather toxicity data because no one can measure exactly how much exposure each worker received, but the only clear proof that chrysotile asbestos is harmful is when it is breathed in industrial settings with exceptionally high levels for long periods of time. The result is that many voluntary risks that people want to engage in are apparently much more harmful than this commonly used asbestos. The Harvard University Energy and Environmental Policy Center estimated that

cigarette smoking or walking across the street is more than 200 times more dangerous than asbestos in school buildings. A person is three times more likely to be struck by lightning than to be harmed by asbestos in school buildings.

In recent years, concern over asbestos exposure and the fight to ban all asbestos in the United States have been renewed. In June 2002, Senator Patty Murray (Democrat, Washington State), with the endorsement of the Sierra Club, introduced the Ban Asbestos in America Act of 2002 into the United States Senate. If it is passed in its proposed form, Senator Murray's bill will not only ban asbestos, but will require EPA studies of asbestos-containing products (including asbestos-contaminated products), additional research into diseases caused by asbestos exposure, a public education campaign focusing on the dangers of asbestos exposure, and a review of the health effects of not only asbestos, but also "unregulated minerals," along with a review of relevant current laws and regulations established for the protection of workers and consumers.

Questions

1. How serious do you think the asbestos risk is? Is it something you worry about? How have perceptions of the "asbestos risk" changed with time? Do you think asbestos should be banned in the United States? Justify your answers.

2. People often perceive more risk in an activity when they have little control over it. Does this explain why asbestos may be perceived as being riskier than it may really be? Or do you agree with the most severe assessments of the risks of asbestos? Explain.

3. People also tend to perceive more risk when they cannot observe the threat. Is the asbestos threat observable? Explain.

4. List and discuss some other environmental problems that might be criticized as examples of "environmental hypochondria." Or, in your opinion, is there no such thing as "environmental hypochondria" since all environmental and health issues are important and we should not "put a value on human life"?

by citizens. At about 130 dB, irreversible hearing loss can occur almost instantaneously.

Health Damage from Noise

The most obvious health damage from loud noises is hearing loss. This occurs from physical damage to the fragile mechanisms in the ear. The damage can occur very rapidly from very loud noises or slowly from long-term exposure to moderately loud noises. Noise can also lead to a variety of ailments that arise from stress and anxiety induced by loud or distracting noises. Such stress-related ailments can be psychological, such as emotional trauma, or physical, such as headaches, nausea, and high blood pressure. Many studies have shown that such stress-related effects reduce productivity in the workplace.

Control of Noise

Noise can be reduced in three ways: (1) at the source, (2) as it travels to the person, (3) by protecting the person. Reduction at the source is often the easiest and most cost-effective means, reflecting the old engineer's adage that "noise means inefficiency." This is because noise is often caused by vibrations, friction, and other signs that equipment is not functioning efficiently. The solution is to design machinery that minimizes vibrations and friction, such as reducing the number of moving parts. Similarly, the machinery needs to be well maintained, lubricated, and so on. If equipment still produces loud noises after these steps are taken, as often occurs with industrial machines, the other two methods must be used.

Reduction of noise as it travels through the air to the person is achieved by sound-absorbing or **acoustical materials**. Examples include tiles and baffles made of wool or many kinds of synthetic materials that do not transmit sound vibrations efficiently. Car mufflers and jet engine noise deflectors are examples of devices made to reduce noise after it has left the engine source.

The third method of noise reduction, protecting the person, usually employs earplugs or earmuffs. These are made of materials that absorb sound waves before they reach the eardrum—not cotton, which provides little protection when inserted in the ear. Earplugs, and especially earmuffs, can greatly reduce noise up to 50 dB. Their efficiency depends heavily on an "acoustical seal," meaning that they must be inserted tightly, with as little exchange of air into the ear as possible.

Electromagnetic Fields

The flow of electricity through wires produces an electromagnetic field (EMF) that can extend through the air for many feet (**Figure 16-12**). Concern over the health effects of such fields, in particular, the possibility that they may cause cancer, has been growing since the late 1960s. Numerous studies have often yielded conflicting results, however, so the threat is controversial. Extensive studies carried out in the late 1980s by the Electric Power Research Institute and the Congressional Office of Technology Assessment concluded that health effects of electromagnetic fields could be neither proved nor ruled out. More recent studies have come to similar, non-conclusive, results.

U.S. utility companies have spent over $1 billion on such studies, with no conclusive evidence of harm yet shown. These, and similar high costs for other controversial threats such as radon and asbestos are a main reason that the 1995 Congress showed strong interest in benefit-cost analyses. Such analyses try to determine true risks objectively and to estimate appropriate benefits for money spent on them. But many environmentalists note that benefit-cost analyses are often flawed because of the difficulty of measuring benefits and risks accurately. They argue that unforeseen harm could result if spending on EMF and other potential threats is reduced too soon.

Statistical studies, which test for a correlation between cancer and people exposed to high EMF, usually show a weak correlation or none at all. This indicates that EMF "bioeffects" depend on many variables. Biological studies, which directly examine the effects of EMF on living cells and tissues in the laboratory, produce similar results. In some cases, biochemical regulation of cell multiplication

Figure 16-12   Power transmission lines produce enormous electromagnetic fields as electricity flows through them.

does seem to be affected, possibly from changes in the flow of electrically charged calcium ions through cell membranes. But the dose-response relationship is clearly very complex and is affected by many variables: (1) frequency and wavelength of the EMF, (2) duration of exposure, (3) kind of cells exposed and type of organisms, (4) and even orientation of the EMF with the Earth's electromagnetic field. Considering that individuals of nearly all species vary in their susceptibility to environmental stresses, it will likely be very difficult to precisely determine EMF effects, if any.

One of the most interesting and useful findings so far is that fields generated by toasters, electric blankets, and other indoor electrical equipment may be a greater threat than the high-voltage transmission lines that receive so much publicity. We spend much more time in close proximity to these indoor appliances, and they can sometimes generate a surprisingly strong electromagnetic field.

SUMMARY

- Air pollution is both a health issue and an economic issue—air pollution costs tens of billions of dollars a year in the United States; in the United States, over 100 million people live in areas where the air is considered unhealthy for some portion of the year.
- Currently, the U.S. EPA recognizes six basic criteria pollutants: particulate matter, sulfur dioxide, nitrogen dioxide, ozone, lead, and carbon monoxide.
- Volatile Organic Compounds (VOCs) are another important class of pollutants.
- Most air pollutants in the United States have declined over the last thirty years (exceptions are ground-level ozone and nitrogen oxides in some areas).
- Transportation is the largest source of carbon monoxide, nitrogen oxides, and VOCs.
- Coal-burning power plants in particular are a major source of sulfur oxides.
- Sulfur dioxide is the major cause of acid rain, which can cause damage to both biological (such as natural ecosystems) and physical materials (for instance, degradation of building materials).
- Photochemical pollutants, commonly known as smog, are produced by chemical reactions among nitrogen oxides, VOCs, and other components of the air.
- Ozone, the most abundant photochemical pollutant in smog, is very reactive and destructive, causing damage to human health, crops, natural ecosystems, and physical entities (such as causing rubber products to crack).
- Carbon monoxide, a gas dangerous to human health, can be produced by incomplete combustion, such as by fossil fuel powered vehicles.
- Vehicle emissions can be addressed through such means as catalytic converters, alternative engine designs, and alternative fuels that burn cleaner.
- Lead poisoning can cause brain damage, especially in children, and other disabilities; however, with the elimination of leaded gasoline over a decade ago, airborne lead levels have decreased by over 95% in the United States.
- In the United States, the Clean Air Act and its amendments have gone a long way toward addressing outdoor air pollution.
- Indoor air pollution, caused by sick building syndrome, radon, or tobacco smoke can be a threat to human health in specific instances and is usually addressed on an individual basis.
- Noise pollution, "unwanted sound," can also be an issue in specific cases.
- A controversy remains about whether electromagnetic fields produced by electricity flowing through wires poses a health hazard.

KEY TERMS

Acid rain	Electrostatic precipitators	pH scale
Acoustical materials	Emission offsets	Pollutant Standards Index (PSI)
Alkaline	Methanol	Primary pollutants
Baghouses	Noise	Scrubbers
Catalytic converter	Oxygenation	Secondary pollutants
Clean Air Act	Ozone	Sick building syndrome
Criteria pollutants	Particulates	Smog
Decibel (dB) scale	Photochemical pollutant	Thermal inversion

STUDY QUESTIONS

1. Distinguish among global air pollution, local and regional air pollution, and indoor air pollution, giving examples of each.
2. Name the six basic criteria pollutants as currently recognized by the U.S. EPA. What are the two main sources of these pollutants?
3. What are some key patterns and historical trends related to criteria pollutants in the United States since about 1970?
4. What are particulates? List some of the terms used to describe them. Why are they the most damaging pollutant to lungs?
5. Where does most acid rain occur in North America? Why does this pattern exist?
6. Describe the methods by which sulfur oxides can be reduced.
7. What does 1 on the pH scale mean? What is the average pH of rainfall? Which fishes are most sensitive to acid waters?
8. What is a photochemical reaction? How does smog form?
9. Distinguish secondary pollutants from primary pollutants.
10. What are VOCs, and why are they important?
11. Distinguish ground-level ozone from stratospheric ozone. Which is "good" and which is "bad"?
12. Why can carbon monoxide be a major health hazard to humans?
13. What are some basic ways to address carbon monoxide pollution?
14. What is a thermal inversion? Why, when combined with local air pollution, can it have deadly results?
15. What does a value of 100 on the Air Quality Index (AQI) mean?
16. What did the National Ambient Air Quality Standards (NAAQS) and the New Source Performance Standards (NSPS) establish?
17. What was the focus of the 1990 Clean Air Act Amendments? Why are the 1990 Clean Air Act Amendments so controversial?
18. Why is indoor air pollution a generally greater direct threat to human life than smog and many other forms of air pollution?
19. What are the three main categories of sources of indoor air pollution? How can the harmful effects of these sources be reduced?
20. What is noise? Why can it be considered an air pollutant? Discuss ways that noise can be reduced.
21. Describe the controversy over electromagnetic fields (EMFs).

WHAT'S THE EVIDENCE?

1. The authors contend that the most harmful combustion source of indoor air pollution is tobacco smoke. What is the evidence for this statement? Do you agree with this assessment? Why or why not? Are you a smoker? Do your personal habits relative to smoking or not smoking influence your position on the issue?
2. Suppose you have a friend who lives in a house directly adjacent to high-voltage transmission lines. One day you are in a conversation with your friend and she expresses concern that the EMF (electromagnetic field) from the power lines might adversely affect her health. What would you say to her? Is there definitive evidence that EMFs from power lines are harmless? Is there strong evidence that such an EMF can be detrimental to human health? Should she worry? Would you worry if you were in a similar situation?

CALCULATIONS

1. On the decibel scale, how much greater is the sound intensity (energy) in a noise of 50 dB compared to a noise of 10 dB?
2. On the decibel scale, how much greater is the sound intensity (energy) in a noise of 60 dB compared to a noise of 40 dB?
3. On the pH scale, a substance such as lye (pH 13) is approximately how much more alkaline (basic) than ammonia (pH 11)?
4. If the global fleet of automobiles, trucks, and buses increases from a current 700 million to 1.1 billion in the year 2020, what percent increase will this represent over this period?

ILLUSTRATION AND TABLE REVIEW

1. Referring to Table 16-1, what level of health concern is indicated by an AQI (Air Quality Index) of 85? Of 175?
2. Based on Figure 16-8 (Los Angeles traffic), what time of day do nitrogen dioxide levels in Los Angeles peak? How about ozone levels? When is not a good time to go jogging in Los Angeles?

WWW.ENVIRONMENT.JBPUB.COM

www.environment.jbpub.com

Connect to this book's web site: http://www.environment.jbpub.com/ The site features eLearning, an online review area that provides quizzes, chapter outlines, and other tools to help you study for your class. We've posted the answers to the book's odd-numbered questions. You can also follow useful links for more in-depth information. The site provides sources for the information in this chapter and recommendations for further readings.

The TOMS (Total Ozone Mapping Spectrometer) satellite image of atmospheric ozone concentrations in the southern hemisphere, September 17, 2001. Dark blue corresponds to the thinnest ozone. Light blue, green, and yellow indicate progressively thicker ozone. The ozone hole is about 26 million square kilometers—roughly the size of North America. In the near future, assuming no large volcanic eruptions, the hole will likely grow to about this size each summer.

Sep 17 2001

CHAPTER 17

Global Air Pollution: Destruction of the Ozone Layer and Global Climate Change

Some environmental problems, by their very nature, are global in scope. Perhaps the two most immediate environmental threats that we face as a global community are climatic change due to the artificial introduction of large amounts of greenhouse gases into the atmosphere and abnormally high incidences of ultraviolet radiation on the surface of Earth due to the destruction of the ozone layer. These problems are interrelated: Both are caused by the introduction of large quantities of human-produced gases into the atmosphere, and in many cases, the same gases are implicated in both problems. Chlorofluorocarbons (CFCs), for instance, are not only a major culprit in the destruction of the ozone layer but also an extremely effective greenhouse gas.

The story of the human assault on the ozone layer now appears to be drawing to a close (although the story is not quite over yet—ozone-depleting compounds will continue to be detectable in the atmosphere for years to come). The problem was discovered about 1974, and by 1990, all the major governmental authorities of the world had agreed on a solution. For the first time in history, political leaders, scientists, corporations, and consumers worked together to find an international solution to a global problem that potentially threatened all life on the planet. The "ozone story" sets a precedent as the world tackles even more complex environmental issues. We now know that nations can work together if they must. Perhaps the next major global problem that will be systematically and successfully addressed is global warming caused by the accumulation of greenhouse gases.

In this chapter, we first tell the ozone story. Then we address the issue of global warming.

Ozone Depletion

Ozone in Nature

Ozone (O_3) is an important natural component of the **stratosphere**. (Ozone is also an important pollutant at ground level, see Chapter 16.) Ozone occurs in scant amounts between about 6 and 31 miles (10 and 50 km) above sea level and is most strongly concentrated at an altitude of 12 to 16 miles (20–25 km)—the **ozone layer**. Ozone is formed in the stratosphere (**Figure 17-1**) when high-energy **ultraviolet (UV) radiation** splits normal oxygen molecules (O_2) into atomic oxygen (O). The atomic oxygen may then combine with a standard diatomic oxygen molecule (O_2) to form triatomic ozone (O_3). Under natural conditions, ozone is not only formed in the atmosphere, but also removed by various reactions. An ozone molecule can absorb UV radiation and split into O_2 and O. The atomic oxygen can then either recombine with an O_2 molecule to form ozone once again, combine with another atomic oxygen to form diatomic oxygen, or combine with some other substance in the stratosphere. In nature, excluding human interference, a dynamic equilibrium exists between ozone production and ozone destruction, such that the stratosphere always contains a small amount of ozone. The amount of stratospheric ozone is so small that if it were all brought down to sea level, it would form a blanket over the surface of Earth only 0.118 inch (3 mm) thick.

Nevertheless, this stratospheric ozone is essential to the preservation of current forms of life on the Earth's surface. The ozone layer acts as a shield that absorbs biologically dangerous UV-B radiation (**Figure 17-2**). (Ultraviolet radiation is commonly divided into two bands: UV-A has wavelengths of 320 to 400 nm [nm = nanometers; 1 nm = 10^{-9} m, or one billionth of a meter or approximately 3.937×10^{-8} inch], and the higher-energy UV-B has wavelengths less than 320 nm [sometimes the very shortest wavelength UV, approximately in the range of 200–280 nm, is labeled "UV-C" as in Figure 17-2].) When an ozone molecule is hit by UV-B wavelengths, it absorbs the radiant energy and photodissociates into O_2 and O while giving off heat. This keeps the UV-B radiation from reaching Earth's surface and also causes a temperature inversion in the stratosphere that helps to maintain relatively stable climatic conditions on and near the ground.

Human Assaults on the Ozone Layer

The stratospheric ozone layer has remained in a dynamic equilibrium for much of geological time, but by the early 1970s, scientists had discovered evidence suggesting a decline in the ozone concentration due to human interference. In particular, humans had been injecting enormous quantities of ozone-destroying substances into the atmosphere, especially a class of chemicals known as **chlorofluorocarbons (CFCs)** that found important uses as solvents, aerosols, fire retardants, and refrigerants.

Other human-produced substances that can attack stratospheric ozone include the hydrochlorofluorocarbons (HCFCs—substances with only 2%–5% the ozone-destroying potential of CFCs, and therefore often used as substitutes for CFCs) and the halons, a class of fluorocarbons that contain bromine atoms. The halons release their bromine atoms, which then destroy stratospheric ozone. Halons have been commonly used in fire extinguishers. Methyl chloroform, methyl bromide (widely used as a pesticide and also produced naturally by oceanic plankton; due to its ozone destroying properties, bans on the production and use of methyl bromide have been proposed), and carbon tetrachloride are some other important and widely used chemicals that have ozone layer-destroying properties.

By the mid-1970s, the possible destruction of the ozone layer by artificial chemicals had become a topic of heated controversy. Large chemical companies and many manufacturers down-

| High-energy UV radiation | Oxygen molecule | 2 Oxygen atoms | Oxygen atom | Oxygen molecule | UV radiation | Ozone molecule |

Figure 17-1 Schematic representation of the formation of stratospheric ozone. Ozone forms when a free oxygen atom, released by the splitting of an oxygen molecule by cosmic radiation, combines chemically with another oxygen molecule.

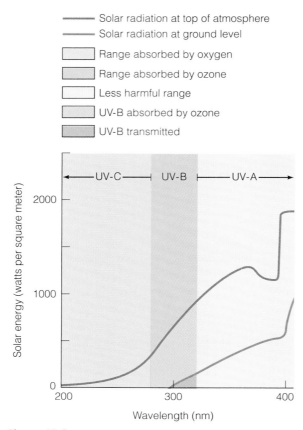

Legend:
- Solar radiation at top of atmosphere
- Solar radiation at ground level
- Range absorbed by oxygen
- Range absorbed by ozone
- Less harmful range
- UV-B absorbed by ozone
- UV-B transmitted

Figure 17-2   Absorption of ultraviolet light by the atmosphere. (Note that wavelength is in nanometers; one nanometer equals one-billionth of a meter.) (*Source*: D. H. Meadows, D. L. Meadows, and J. Randers, *Beyond the Limits* [Post Mills, VT: Chelsea Green, 1992], p. 146. Reprinted from *Beyond the Limits*, copyright © 1992 by Meadows, Meadows, and Randers, with permission from Chelsea Green Publishing Co., White River Junction, Vermont.)

played or denied the possibility, while many environmentally oriented citizens called for immediate, even if economically expensive, action to save the ozone layer. The controversy abated somewhat in the late 1970s after some action was taken to protect the ozone layer (for instance, in 1978, the United States banned the use of CFCs as **aerosol spray** propellants), but it erupted again in the 1980s, when an ozone hole was detected over Antarctica.

Historical Background.   The ozone layer's modern troubles began innocuously enough in 1928, when chemists discovered a new class of chemicals that could replace sulfur dioxide and ammonia as the basic fluids in refrigerators. This new class of chemicals, which was produced by DuPont under the trade name of "Freon," became known more generally as the chlorofluorocarbons (CFCs). CFCs were considered a triumph of modern chemistry; they were safe, nonflammable, stable, unreactive chemicals that proved ideal for many purposes. They were put to use in car air conditioners as well as in re-

frigerators. They were used as a nontoxic propellant for aerosol cans and as a blowing agent in producing Styrofoam and other plastics. They had insulating properties and could also be used as solvents and cleaning agents in the electronics industry. Ironically, the chemical stability that makes CFCs so ideal for use on the Earth's surface is the very characteristic that allows them to reach the stratospheric ozone layer. Since CFCs do not appreciably react on the Earth's surface, they are free to float up into the stratosphere. But under stratospheric conditions, the CFCs readily break down and become a powerful catalyst of ozone destruction.

Applications of CFCs multiplied after their initial discovery, and after World War II, the market for these substances expanded rapidly. The global production rates of Freon 11 (= CFC-11, $CFCl_3$) and Freon 12 (= CFC-12, CF_2Cl_2), trade names for two of the most common CFCs, rose from less than 50,000 metric tons a year in 1950 to 725,000 metric tons in 1976. An estimated 90% of these chemicals was released directly into the atmosphere during, or immediately after, their use; the remaining 10% was generally released once the equipment in which the chemicals were used (such as a refrigerator) was discarded. Since their invention, approximately 20 million metric tons of CFCs have been released into the atmosphere. By the early 1970s, CFCs constituted an $8 billion per year industry in the United States alone.

In 1973–1974, however, two chemists at the University of California at Irvine, Mario Molina and Sherwood Rowland, discovered the effect that CFCs accumulating in the stratosphere have on the ozone layer (for their research they shared the 1995 Nobel prize in chemistry with Paul Crutzen, another researcher on stratospheric ozone destruction). Molina and Rowland found that CFCs are almost inert in the **troposphere** (the atmospheric layer underlying the stratosphere and covering the immediate surface of Earth). CFCs are not even washed out by rain since they are not soluble in water, and so they drift on air currents up into the stratosphere. In the stratosphere, ultraviolet radiation breaks down the CFC molecules, releasing atomic **chlorine**. A free chlorine atom reacts with an ozone molecule, removing one oxygen atom and thus converting the O_3 into O_2. The chlorine monoxide molecule (ClO) is not stable, however; it readily reacts with a free oxygen atom to form a molecule of diatomic oxygen (O_2) and a free chlorine atom once again. Thus, once released from the CFC, the single chlorine atom acts as a catalyst that can destroy thousands (perhaps 10,000 to 100,000) of ozone molecules. Eventually, the chlorine atom will be washed out of the atmosphere by precipitation.

According to Molina and Rowland's calculations, ozone depletion due to CFCs threatened the very existence of life on our planet. They and other environmentally oriented people quickly lobbied for a ban on CFCs. But the large chemical companies and a few scientists argued that the theory of CFC-induced ozone depletion was speculative and did not justify controlling CFC releases. Part of the problem was that initially there was little empirical evidence to support the Molina-Rowland calculations, largely because of the difficulty of measuring ozone concentrations in the stratosphere. Furthermore, this was not the first time that concerns about ozone depletion had been raised; earlier some researchers had worried about the effects of nuclear bomb detonations on the ozone layer. Still, as news of possible ozone depletion hit the streets, the public responded. By the first half of 1975, the use of CFC-based aerosol spray cans had dropped by 25%, and by 1978, the United States had effectively banned CFCs as an aerosol propellant. Yet CFCs continued to play a major role in other industries, both in North America and around the world.

Despite confirmation by other scientists, including a committee of the U.S. National Academy of Sciences, that the threat to the ozone layer was real, little more action was taken for several years. Then in 1983–1985, a discovery was made that stunned many previously unconcerned observers: Based on 30 years of measurements, the British Antarctic Survey found that each spring a hole developed in the ozone layer over the South Pole; the hole closed again later in the year. Could this be a result of CFCs released into the atmosphere? NASA's *Nimbus-7* satellite confirmed the depletion in the ozone layer; curiously, satellites had not detected this ozone hole previously because the computers had been programmed to disregard as errors any data that recorded such low ozone values. Scientific expeditions were quickly mounted to confirm or refute the reality of the Antarctic ozone hole and to try to determine its cause if it was real. The upshot was that the spring hole was real (the loss of ozone was about 50% in general, but rose as high as 100% in some spots) and that it was caused by chlorine carried up to the stratosphere in CFC molecules. Molina and Rowland had been correct all along.

International action was soon taken. In 1985, representatives of nearly two dozen nations met to consider what should be done to protect the ozone layer. The result was a series of conferences and meetings that culminated in September 1987 with the **Montreal Protocol**. At a meeting hosted by the Canadian government in Montreal, representatives from over two dozen countries agreed to freeze CFC production (at 1986 levels) and then gradually decrease CFC and halon production to 50% of 1986

levels by the year 1999. Very quickly, however, it became clear that the protocol did not go far enough. Evidence of a thinning of the ozone layer over the Northern Hemisphere and continued enlargement of the Antarctic ozone hole indicated that more drastic measures were needed. Furthermore, the Montreal Protocol did not cover many ozone-destroying chemicals, such as methyl chloroform and carbon tetrachloride. And not until March of 1988 did DuPont officially acknowledge the damage CFCs were causing to the ozone layer and agree to cease CFC production once substitutes were found.

In 1989, the U.S. Congress imposed a stiff excise tax on all ozone-depleting chemicals sold, produced, or imported into the United States. The idea was to discourage the use of CFCs and other ozone-damaging chemicals and to encourage the search for safer substitutes. Yet such a tax could have little effect globally.

By the late 1980s and early 1990s, stratospheric ozone loss was well documented. The largest losses occurred over Antarctica. In the Arctic, ozone losses have amounted to about a 10% reduction. Significant losses have also been recorded over the mid-latitudes both north and south of the equator. Ozone losses of 3% to 10% or more were recorded over parts of Australia, New Zealand, South Africa, and South America. According to recent data collected by the *Nimbus-7* satellite, some of the largest decreases in ozone occurred over the Northern Hemisphere, including North America, Europe, and large portions of Asia, where ozone values had fallen below "normal" (pre-early 1970s) values by as much as 13% to 14%. Lesser ozone losses were documented over the tropics. Averaged over the planet as a whole, it appears that from 1979 to 1991, Earth lost 3% of its stratospheric ozone.

In June 1990, 93 countries meeting in London agreed to end the production of CFCs, most halons, and carbon tetrachloride by the year 2000 (with the exception of certain developing countries, which had until 2010 to stop CFC production). Further, methyl chloroform production was to end by the year 2005. An international fund of more than $200 million was also established to assist the developing countries in switching to CFC substitutes, which are often more expensive than the CFCs they replace. One of the main stumbling blocks to controlling worldwide CFC production has been that most of the CFCs have been dumped into the atmosphere by the rich, industrialized countries. Thus, the rich of the world created the problem, but then expect everyone (rich and poor alike) to help in solving it. Meanwhile the developing countries, including China and India with their immense populations, desire the same refrigerators, air condi-

CASE STUDY 17-1

Methyl Bromide and the Ozone Layer

Visit www.environment.jbpub.com for more information.

Besides CFCs (chlorofluorocarbons), there are many other substances that can harm the stratospheric ozone layer to various extents, including hydrochlorofluorocarbons (HCFCs), methyl chloroform, carbon tetrachloride, methyl chloride, and halons (to name just a few). Halons are a class of fluorocarbons that contain bromine atoms; when released in the stratosphere, bromine (like chlorine) has the ability to destroy ozone molecules. Another substance that has been implicated in the destruction of stratospheric ozone is methyl bromide. Indeed, now that production and use of CFCs and certain other chemicals have been greatly curtailed, according to some researchers, at present the biggest threat posed to the ozone layer may be methyl bromide.

Methyl bromide, a widely used pesticide and fumigant, is applied to everything from alfalfa and tomatoes, strawberries, grapes, walnuts and wheat—all told, it is used for over a hundred different crops, as well as in forests, nurseries, wood protection, to protect harvests, and in quarantine procedures. It has the advantage of killing insects, weeds, and other pests in the soil even before the crop is planted. It is also commonly used to exterminate termites. In the late 1990s, the United States was using over 40 million pounds (18 million kilograms) of methyl bromide each year, three-quarters of which went to fumigate soil prior to planting crops. Worldwide, an estimated 144 million pounds (65.3 million kilograms) a year were being used. The problem was that methyl bromide may also be a potent destroyer of the ozone layer. Some studies indicate that a molecule of methyl bromide can break down ozone about 40 to 50 times as quickly as a CFC molecule. It has been suggested that methyl bromide may currently account for 10% of all stratospheric ozone damage.

Amendments to the Montreal Protocol recognized methyl bromide as an ozone-depleting substance and have mandated a phaseout of its use gradually between 1999 and 2005 in developed countries. In developing countries, the phaseout has been extended to 2015. Certain exemptions in both developed and developing countries have been allowed, including the use of methyl bromide for certain quarantine procedures, pre-shipment fumigation of some commodities, and other so-called critical uses.

There is a great deal of controversy about whether methyl bromide should in fact be banned. Unlike the CFCs, methyl bromide is found in nature—it is not simply an artificial, human-made substance. Methyl bromide, as well as methyl chloride (another ozone-destroying compound), are both produced by many natural processes. Indeed, the plankton of the oceans produce prodigious quantities of both methyl chloride and methyl bromide; perhaps 30% to 60% of all methyl bromide released into the atmosphere according to various United Nations estimates. One study estimated that salt marshes around the world may produce 10% of all methyl chloride and methyl bromide released into the atmosphere; these same substances are also produced by tropical and subtropical coastal forests and vegetated islands, and by both biological and nonbiological degradation of organic matter (such as dead vegetation) in soils. It has been estimated that 10% of all methyl bromide injected into the atmosphere is from the burning of vegetation, and in recent years (before the Montreal Protocol restrictions on methyl bromide use took effect), 20% resulted from fumigation uses.

So why didn't naturally occurring methyl bromide and methyl chloride completely destroy the stratospheric ozone layer long ago?

tioners, electronics, and convenience items that have become so common in the West. Only if the developed world contributes to the developing world can global cooperation be anticipated.

Given the urgency of the ozone crisis, the Montreal Protocol continued to be amended during the 1990s during a series of meetings held in Copenhagen (1992), Montreal (1997), and Beijing (1999). As a result, CFC production was banned in

industrial countries as of 1996, with the exception of a small amount of production solely for export to developing countries or for use in specialized and essential situations, such as in asthma inhalers. Furthermore, even this production is being phased out. Reportedly all CFC production ceased in Russia as of December 2000, and one of the single largest CFC manufacturing plants in the world, located in the Netherlands, is scheduled to close at the end of

Apparently there has been an equilibrium in nature, with inputs of these gases being offset by natural processes that destroy them. For instance, in the lower atmosphere there is an abundance of hydroxyl free radicals (-OH) that can react with and remove these gases from the atmosphere. The fear on the part of many experts is that the anthropogenic release of massive amounts of methyl bromide will upset the delicate balance of nature and lead to significant stratospheric ozone depletion.

But is it really necessary to ban the manufacture and use of methyl bromide? At least some studies indicate that the human contribution of methyl bromide to the atmosphere is far outweighed by natural contributions. And as for the methyl bromide used in agriculture, some researchers (although admittedly funded by the methyl bromide industry) contend that most of the methyl bromide used as a pesticide never enters the stratosphere where it could harm the ozone layer. Rather, it is absorbed by the soil or by the oceans where it may be broken down into harmless substances. Banning methyl bromide, it is contended, is only harming the agricultural industry and entailing enormous costs (the cost of finding substitutes, and losses in food production and damage to stored harvests) without significantly benefiting the ozone layer. Unfortunately, no one substitute has been found that can serve all the purposes to which methyl bromide has been applied, and thus a range of new options and techniques must be developed as methyl bromide is phased out. Alternatives to applications of methyl bromide include employing integrated pest management, utilizing beneficial microbes that attack pests, rotating crops, and treating soils with heated water to kill pests prior to planting of crops. To fumigate foodstuffs and other organic matter (such as wood products) prior to shipping, hot steam or dry air applications, hot water dips, cold treatments, and subjection to a pure nitrogen atmosphere (therefore suffocating aerobic organisms) have been suggested. Still, critics contend, this variety of processes may be more cumbersome and less effective than treatment with methyl bromide.

Questions

1. Can you draw any parallels between the current controversy over methyl bromide and the controversy in the 1970s over CFCs? (Remember that in the 1970s, many CFC industry advocates denied that CFCs could be damaging the ozone layer. At that time, it was often argued that ceasing CFC production would hurt the economy and lead to food shortages as refrigeration units, dependent on CFCs, lapsed and allowed food to spoil.)

2. Given that CFCs and methyl bromide are two different classes of chemicals, is it possible that CFCs really do deserve to be banned, whereas the dangers of methyl bromide to the ozone layer have been exaggerated by some researchers? How important is it to point out that methyl bromide is produced in large quantities by natural organisms? Can you trust research that is funded by the methyl bromide industry? How about research funded by the EPA or United Nations agencies that may be biased against any chemical suspected of contributing to ozone layer deterioration?

3. If you had to decide, would you allow methyl bromide to be used unchecked, would you limit but not eliminate the use of methyl bromide, or would you totally ban the production and use of methyl bromide? Justify your position.

2005. As the twentieth century came to a close, among the leading producers of CFCs were China and India (both developing nations). The Montreal Protocol and its amendments currently require CFC production in developing nations to be completely phased out by 2010.

Over the past two decades, it has become apparent that it is not only CFCs that can damage the ozone layer; many other substances also have the potential to damage stratospheric ozone, including chemicals that were originally seen as replacements for CFCs. Ozone-damaging substances include hydrochlorofluorocarbons (HCFCs), various halons (such as Halon-1202), methyl bromide (see Case Study 17-1), n-propyl bromide, hexachlorobutadiene, 6-bromo-2-methoxy-naphthalene, and many other substances. In recent years there have been, in conjunction with the phaseout and banning of

CFCs, restrictions and phaseouts of these and other ozone-harming substances. As more chemicals are found to damage the ozone layer, we can expect further restrictions.

The Current Situation. It may appear that the ozone story has come to a happy closure; however, despite the Montreal Protocol and its subsequent amendments, the story is not over yet. Even with the complete phaseout of CFCs and other chlorine-bearing chemicals, stratospheric chlorine concentrations are expected to remain high well into this century as already released CFC molecules rise from the troposphere into the stratosphere (once released, a CFC molecule can take 15 years to make its way up from the Earth's surface to the stratospheric ozone layer).

Even with the accelerated phaseout mandated by amendments to the Montreal Protocol, the problem will persist. Some of the more common CFCs have life expectancies of between 75 and 110 years, and the chlorine catalyst in the stratosphere can also be quite long-lived. The estimated background level of chlorine in the atmosphere, derived from natural sources such as volcanoes, is about 0.6 ppb (parts per billion). Currently, due to human releases of chlorine-bearing chemicals like CFCs, the concentration of chlorine is about 3.5 to 3.9 ppb; fortunately, this may be a plateau and it is expected that levels will decline in the future. Another, perhaps more direct, way to gauge the amount of stratospheric ozone depletion is to monitor the annual spring ozone hole over Antarctica. In 2001, NASA reported that the Antarctica ozone hole, about 26 million square kilometers in size (approximately equivalent in size to North America) was approximately the same size as the previous three years. The degree of ozone depletion within the hole can also be measured, using Dobson units (essentially a measure of the thickness of the ozone layer above a certain location). A minimum of 100 Dobson units was recorded in September 2001 (springtime in Antarctica) as compared to 98 in 2000, or the record low of 88 in 1993; "normal" levels of ozone for the Antarctic are about 275 Dobson units. These data are consistent with the hypothesis that stratospheric ozone depletion has peaked and is now leveling off— and hopefully ozone will then have a chance to build up in the stratosphere once again. Scientists estimate that the ozone layer will remain depleted to some degree until at least 2040 or 2050, with ozone losses as high as 10% to 30% at times over the northern latitudes where most of the world's population resides. Although over the next fifty years stratospheric ozone should build up again,

measurable amounts of CFCs will continue to reside in the atmosphere well into the twenty-fourth century. Thus, even if the Montreal Protocol and its amendments are upheld by all nations (perhaps an overly optimistic assumption), the Earth will be subjected to an abnormally thin ozone layer for decades to come.

Natural Causes. A number of naturally occurring substances also have the ability to destroy stratospheric ozone. Most important of these are the hydrogen oxides (HO_x) derived from water vapor, methane (CH_4), hydrogen gas (H_2), and nitrogen oxides (NO_x). The nitrogen oxides, which are very effective destroyers of stratospheric ozone, not only occur naturally but are being supplemented by anthropogenic sources. In particular, nitrogen-containing fertilizers release N_2O into the air; the N_2O can diffuse into the stratosphere and form nitric acid (NO), which has the ability to react with and destroy ozone molecules. However, there is evidence that in the presence of methane and chlorine (perhaps released from a CFC), NO_2 will form from NO; additionally, solar radiation can break the NO_2 into NO and O. The atomic oxygen (O) can combine with O_2 to form ozone (O_3). Nitrogen compounds will also bind with chlorine atoms, preventing the chlorine from destroying the ozone. Thus, under certain circumstances, nitrogen compounds may actually enhance the ozone layer rather than deteriorate it. Similarly, it has been suggested that stratospheric methane may benefit the ozone layer by helping to remove chlorine atoms. Another substance that breaks down stratospheric ozone is methyl bromide. Indeed, some researchers feel that this is a major threat to the ozone layer. Methyl bromide produced by humans is widely used as a pesticide, but it is also a substance found in nature (for instance, plankton of the oceans produce enormous quantities of methyl bromide—see Case Study 17-1).

There are some suggestions that naturally occurring chlorine oxides, such as chlorine monoxide (ClO), may promote the deterioration of stratospheric ozone. During major volcanic eruptions, significant amounts of chlorine may be injected into the stratosphere, but this has never been conclusively demonstrated to be a significant agent of global stratospheric ozone destruction. More troubling, however, is the recent discovery that tiny particulate matter in the stratosphere, known as stratospheric aerosols, may help promote ozone destruction by providing convenient surface areas where the ozone-destroying reactions can readily take place. In 1982, the eruption of the El Chichùn volcano in Mexico launched 6 or

Figure 17-3 The eruption of Mt. Pinatubo, in June of 1991, sent solid material 30 to 40 km into the atmosphere. Within 10 days, the aerosol cloud from the eruption formed a nearly continuous band that stretched 11,000 km from Indonesia to Central Africa.

7 million metric tons of sulfate aerosols into the stratosphere, which may have caused ozone levels in the Northern Hemisphere to temporarily drop by as much as 10%. In June 1991, **Mt. Pinatubo** in the Philippines erupted, injecting an estimated 20 million metric tons of sulfate aerosols into the stratosphere (**Figure 17-3**). Some atmospheric scientists believe that this further exacerbated the ozone depletion problem for several years. Additionally, sulfate aerosols released during the burning of fossil fuels are accumulating in the stratosphere.

Effects of Increased Ultraviolet Radiation Reaching the Earth's Surface

Why have people been so concerned about the depletion of the ozone layer? Is a little bit more exposure to ultraviolet radiation really all that bad? A small amount of UV radiation is necessary for the well-being of humans and other organisms. UV radiation promotes the synthesis of Vitamin D in humans and, in small amounts, acts as a germicide to control populations of microorganisms. Yet an increased incidence of UV-B radiation at the Earth's surface can have numerous adverse effects on the health and well-being of humans and other organisms and can also cause the degradation of nonliving materials. For every 1% decrease in the ozone layer, the intensity of UV-B radiation on the Earth's surface increases by 2%.

The most widely publicized problems associated with increasing amounts of UV-B are the projected increases in skin cancer and cataracts among humans. DNA (the genetic material found in all living organisms) is extremely sensitive to, and can be damaged by, UV-B radiation. A solid body of evidence indicates that skin cancer in humans can be caused by UV-induced damage to DNA. The United Nations Environment Programme has estimated that the extra UV-B exposure due to a 10% loss of global ozone could result in a 26% increase in the incidence of nonmelanoma skin cancers. A rise in malignant **melanoma** skin cancers, which are much more dangerous than nonmelanoma cancers, would also be expected. In one of the world's southernmost cities, Punta Arena, Chile (near Antarctica), incidents of skin cancer rose by 66% between 1994 and 2001. The U.S. Environmental Protection Agency (EPA) has estimated that continued depletion of the ozone layer could cause an additional 800,000 cancer deaths in the United States over the next century.

Some researchers have downplayed the increased risks of skin cancer by pointing out that the incidence of skin cancer among the general population increases as one moves from higher latitudes toward the equator; presumably, this phenomenon is independent of any loss of ozone in the stratosphere (perhaps in warmer areas, more people spend more time in the Sun). The National Academy of Sciences, in a 1975 study, found that in the United States the doubling distance for skin cancer was about 600 miles (965 km) south. From this, one can extrapolate that the risk of skin cancer increases about 1% for every six miles (9.65 km) closer to the equator one lives. Thus, a 26% increase in skin cancer resulting from a 10% loss in ozone would be the same as if everyone simply moved about 156 miles (250 km) closer to the equator (of course, an impossibility for those already living along the equator). Such reasoning seeks to trivialize the effects of ozone depletion. It does not consider the effects of ozone depletion and increased UV-B radiation exposure on delicately balanced ecosystems, and it is also of little comfort to people who contract cancer from increased UV-B exposure.

Increased exposures to UV-B have also been linked to increased incidences of **cataracts** (where the lens of the eye becomes opaque), damage to corneas, and retinal disease in humans. It has been estimated that a 10% decrease in the ozone layer could cause more than 1.5 million new cases of cataracts each year. Evidence also suggests that excess dosages of ultraviolet light can suppress the human immune system, allowing the spread of infectious diseases. Furthermore, it has been hypothesized that ultraviolet light may help activate the AIDS virus.

Ultimately, the widespread ecosystem damage caused by increased dosages of UV-B may be more important than its immediate effects on humans. Abnormally high levels of UV radiation inhibit photosynthesis, metabolism, and growth in a number of plants, including food crops such as soybeans, potatoes, sugar beets, beans, tomatoes, lettuce, wheat, sorghum, and peas. The UV-B radiation can destroy cells and also cause mutations. Many tree species are particularly sensitive to UV levels, and increasing amounts of UV may result in a major decline in forest productivity. Insect activity is disrupted by elevated levels of UV radiation, and, of course, the other organisms in a typical terrestrial ecosystem depend on the insects.

Many of the plants and animals in freshwater and marine ecosystems are extremely sensitive to UV levels, especially since UV-B can penetrate several yards (several meters) of water. Phytoplankton (plants and algae that form the bases of many food chains) and fish larvae may be especially susceptible. Reportedly in 1988, when the ozone levels over Antarctica declined by 15% overall, surface phytoplankton levels also decreased by 15% to 20%. Experimental evidence demonstrates that elevated levels of UV-B radiation will damage fish, shrimp, and crab larvae (and certainly many other species that have not yet been tested).

UV-B exposure can also damage nonliving material objects. For instance, UV exposure will cause or accelerate the breakdown and degradation of various types of paints and plastics, such as polyvinyl chloride (PVC), an important construction material that is used for vinyl siding, garden hoses, and other products that are regularly exposed to sunlight. Such material damage could cost billions of dollars per year.

Although extremely poorly understood at present, major fluctuations in the ozone layer could have significant climatological effects on the Earth's surface. The stratospheric ozone naturally absorbs UV radiation and reemits it as infrared energy (essentially heat) into the troposphere; this contributes to the Earth's energy budget. At present, the absorption of UV radiation warms the stratosphere, creating a temperature inversion at the boundary between the troposphere and the stratosphere. Some scientists suggest that a decline in the ozone layer could cause a cooling of the stratosphere, which in turn might cause a cooling of the troposphere and ultimately a slight cooling of the Earth's surface (even though more UV radiation would directly reach the ground). Such a cooling effect could be very small, however. According to one calculation, a 20% reduction in ozone concentrations in the stratosphere would cause only a 0.25° C (0.45° F) decline in surface temperatures. But before jumping to the conclusion that ozone loss will have one good effect—countering global warming due to the greenhouse effect—one must consider that many of the chemicals that are contributing to stratospheric ozone declines are also potent greenhouse gases. The increased greenhouse effects of CFCs and other ozone-destroying gases may more than offset any surface cooling that would otherwise have been brought about by destruction of the ozone layer.

Changes in the concentrations of ozone at different levels in the atmosphere could also affect the temperature gradient of the atmosphere, which in turn could affect atmospheric circulation patterns and climatic patterns on the Earth's surface in unpredicted, but potentially adverse, ways. The relative temperature differences among the various layers of the atmosphere are extremely important in determining convective processes that produce clouds and precipitation patterns. Interfering with this system by depleting the ozone layer could possibly have disastrous effects.

Global Climate Change

The Greenhouse Effect

The basic idea behind the **greenhouse effect** is as follows. Short-wavelength, high-energy, solar **radiation** shines from the Sun onto the Earth. Some of this incoming solar radiation is reflected by the atmosphere back into space, some passes though the atmosphere and is absorbed as it heats the air, and about half reaches the Earth's surface (**Figure 17-4a**). The surface heats up and in the process gives off longer-wavelength, lower-energy (infrared, or **heat**), radiation. This **infrared radiation** passes up into the atmosphere, but instead of being radiated 100% back into space, much of it is absorbed by the atmosphere and reradiated to the surface (**Figure 17-4b**). This phenomenon occurs because many trace gases (the **greenhouse gases**) in the atmosphere are relatively transparent to the higher-energy sunlight, but trap or reflect the lower-energy infrared radiation. Thus, the greenhouse gases act as a one-way filter, letting energy in the form of sunlight in, but not

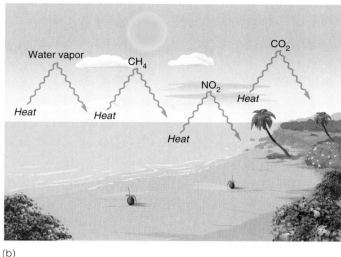

Figure 17-4 (a) Greenhouse gases in the atmosphere allow energy from the Sun in, (b) but trap or reflect much of the lower-energy infrared radiation (heat) from the Earth's surface.

allowing the infrared heat to escape at the same rate. This process is crudely analogous to the way glass in a greenhouse allows sunlight to shine in, but stops much of the longer-wavelength heat from escaping. Even on a cold winter day, the inside of a greenhouse can become quite warm if the Sun is shining.

Likewise, the Earth's surface would be a frozen mass if it were not for the natural greenhouse effect of the atmosphere. Without this phenomenon, average global temperatures might be on the order of $-17°$ C ($1.4°$ F). Note, however, that with current levels of greenhouse gases, some infrared heat does continue to escape. In geologically recent Earth history, a relative steady-state balance has been achieved that maintains the average global surface temperature at about $14°$ C ($57.2°$ F). If no heat escaped, the surface would continue to heat up to unbearable temperatures. The perceived problem is that, due to inadvertent human intervention, greenhouse gases are accumulating very quickly in the atmosphere, and some predict this will lead to catastrophic global warming.

Greenhouse Gases.   The best-known greenhouse gas (**Table 17-1**) is **carbon dioxide** (CO_2). Carbon dioxide is breathed out or otherwise given off by living organisms (including plants) as they undergo aerobic respiration, and is taken in by green plants during photosynthesis. A small amount of carbon dioxide in the atmosphere is an absolute necessity for life on Earth; without it, the planet would be too cold to support living organisms and plants would lack an essential raw ingredient. But an excess of car-

bon dioxide will cause the Earth's surface to heat up abnormally.

Clearly, we humans are very quickly adding to the amount of carbon dioxide in the atmosphere (**Figure 17-5**). In 1850, the average annual concentration of carbon dioxide was about 250 ppm (parts per million; also referred to as ppmv, parts per million volume); this increased to 316 ppm by 1959, and by 2001, it had grown to over 370 ppm. The CO_2 content of the atmosphere continues to increase.

TABLE 17-1	Greenhouse Gases: Sources and Estimated Relative Contributions to Global Warming	
Note that the estimates of relative greenhouse gas contributions involve significant uncertainties.		
Gas	**Major Sources**	**Percentage of Contribution**
Carbon dioxide	Fossil fuels, deforestation	50
Chlorofluorocarbons, other halocarbons	Refrigeration, solvents, insulation, foams, aerosol propellants, other industrial and commercial uses	20
Methane	Rice paddies, swamps, bogs, cattle and other livestock, termites, fossil fuels, wood burning, landfills	16
Tropospheric ozone	Fossil fuels	8
Nitrogen oxides	Fossil fuels, fertilizers, soils, burning of wood and crop residues	6
Source: Based primarily on estimates made by World Resources Institute.		

Figure 17-5   Average annual atmospheric concentrations of carbon dioxide, 1850–2001. (*Source:* Compiled from various sources, including Worldwatch Institute and United Nations Environment Programme, *Vital Signs 2002* [New York: W. W. Norton, 2002, p. 53.].)

The primary way we are increasing the CO_2 content of the atmosphere is through the burning of fossil fuels (**Figure 17-6**): coal, oil, and gas. For millions of years, the carbon in these fuels has been out of atmospheric circulation, buried deep under the surface of the Earth. But suddenly, over a period of just two and a half centuries (and especially during the last few decades), we have released massive amounts of this fossilized carbon back into the atmosphere. In 2001, for instance, over 6.5 billion metric tons of carbon were released worldwide by the combustion of fossil fuels—slightly over 1 metric ton of carbon per person on Earth. Each metric ton of pure carbon

released forms approximately 3.66 metric tons of CO_2, so 23.8 billion metric tons of CO_2 were released into the atmosphere simply by the burning of fossil fuels in 2001. **Carbon emissions** from the burning of fossil fuels are produced disproportionately by the people of the industrialized countries. Indeed, the wealthiest 25% of the world's population burn nearly 70% of all fossil fuels. The United States is the biggest single contributor, producing nearly 24% of the world's CO_2 emissions (primarily from burning fossil fuels), while containing less than 5% of the world's population.

An additional 1 to 2 billion metric tons of carbon are emitted into the atmosphere each year from deforestation. Like the fossil fuels, which were once living organisms, the extant forests hold vast stores of carbon. When the trees and other plants die and are either burned or allowed to decay, this carbon is converted to CO_2. Furthermore, trees serve the vital function of removing CO_2 from the atmosphere as they grow. Unless deforested areas are quickly replanted, the CO_2 that is emitted directly into the atmosphere remains there, and a vital mechanism for removing excess CO_2 from the atmosphere is destroyed.

Although carbon dioxide is blamed for 50% to 70% of the current abnormal global warming (depending on the authority consulted and how "global warming" is calculated), it is not the only major greenhouse gas. The other major culprits are chlorofluorocarbons (CFCs), **methane** (natural gas, CH_4), tropospheric ozone, and **nitrogen oxides** (NO_x).

The chlorofluorocarbons that promote global warming are the same CFCs that are destroying stratospheric ozone. Indeed, they are up to thousands of times more efficient at absorbing heat and promoting global warming than CO_2. At present, it is estimated that CFCs may account for 15% to 25% of the human contribution to global warming. This number would have been even higher if steps had not been taken to reduce the CFCs released into the atmosphere. Thus, there are two good reasons to reduce our reliance on CFCs: to save the ozone layer and to reduce global warming.

Methane accounts for an estimated 15% to 20% of current global warming. Currently, methane does not occur in anywhere near the concentrations of CO_2 in the atmosphere, but it is up to 30 times more effective than CO_2 in trapping heat. Like CO_2, methane is a naturally occurring gas that has been present on the Earth's surface since its formation. Methane is produced by bacteria in swamps and

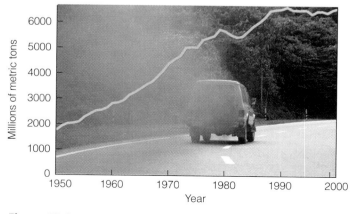

Figure 17-6   World carbon emissions from fossil fuel burning, 1950–2001. (*Source:* Worldwatch Institute and United Nations Environment Programme, *Vital Signs 2002* [New York: W. W. Norton, 2002].)

other areas as organic matter is decomposed. Termites give off methane as they break down wood, and many animals produce large amounts of methane in their digestive tracts. Ruminants such as cows in particular release enormous quantities of methane into the atmosphere. These natural sources of methane have been intensified by human habits. We raise billions of cows and other livestock, which serve as natural methane factories. Unfortunately, virtually all of this methane is lost to the atmosphere, although some have suggested that we should try to capture it and burn it as a fuel. Rice farming is another major source of methane—in typical rice paddies, large quantities of organic matter rot under a shallow layer of water to produce methane.

The increased use of fossil fuels has also added significant quantities of "fossil" methane to the atmosphere. During coal mining, underground stored reserves of methane are often inadvertently released into the atmosphere. Likewise, methane occurs above oil in many oil wells. Although this methane is now generally seen as a valuable resource, in the past (and in some cases to this day) it was considered not worth bothering with and simply released into the air. The current increased use of methane as the fossil fuel of choice (it burns cleaner than coal or gasoline) also opens up the possibility that more stored reserves of methane will make their way into the atmosphere, escaping from leaking tanks or pipelines. As much as 15% of all methane released annually into the atmosphere may be leaking from poorly designed and maintained natural gas lines in Eastern Europe and the former Soviet Union.

Another aspect of natural methane could have extremely serious repercussions. An estimated 14% of all the organic carbon of the world is buried (mostly as partially degraded vegetation) and frozen in the permafrost of the tundra. With increased global warming, which many scientists predict will affect the poles to a greater extent than the Earth as a whole, vast tracts of tundra could begin to melt. Bacteria would then decompose the organic matter, releasing prodigious amounts of methane into the atmosphere. The added methane would accelerate global warming, which in turn would accelerate the production of more methane—setting off what some fear could be a runaway cycle of increased warming and methane production. Some people even fear that if global temperatures rise high enough, much organic matter found in the soil of temperate regions (where the ground is not frozen) may also be converted to methane or CO_2.

A similar effect may occur in the oceans. The oceans, in particular the cold regions, hold huge stores of organic matter, containing more carbon than all the coal reserves on land. About 36,000 billion metric tons of carbon are stored in the oceans, compared to about 4000 billion metric tons in fossil fuels. As the oceans slowly warm up, this organic carbon could also decompose, forming methane that will bubble to the surface.

Compared to CO_2, CFCs, and methane, nitrogen oxides (NO_x) and tropospheric ozone are relatively minor greenhouse gases, together accounting for perhaps 10% to 15% of global warming. Nitrogen oxides and tropospheric ozone are important air pollutants in other contexts, contributing to petrochemical fog and acid rain (see Chapter 16); thus, it is important from several standpoints that emissions of these substances be decreased. Nitrogen oxides are formed when chemical fertilizers break down, when coal is burned in power plants, and in general when any fuel is burned at high temperatures (nitrogen and oxygen, the primary components of our atmosphere, combine at high temperatures to form nitrogen oxides). Large quantities of nitrogen oxides can also be released during the production of certain synthetic substances, such as nylon stockings and pantyhose.

Other gases also contribute to global warming. Water vapor (H_2O), which can act as a greenhouse gas, is a potentially important factor, but virtually nothing can be done to control the amount of water vapor in the atmosphere. The Sun naturally heats the surfaces of the oceans and lakes, and plants naturally transpire, all of which inject large quantities of water vapor into the atmosphere. Of course, this water vapor collects and precipitates out again as rain or snow.

Is Earth Really Warming Up?

Many researchers would respond to this question with a definitive "yes." Fairly accurate records of global temperatures have been kept since 1880, and they show a fairly steady increase in temperature from about 1880 to 1940. From 1940 to the mid-1960s, temperatures underwent a slight cooling and stabilization, but since the 1960s, there has been a dramatic (though somewhat erratic from year to year) increase (Figure 17-7), with record highs in the last decade. Based on data through 2001, the warmest year on record was 1998 (global average temperature of 14.54° C [58.172° F]) followed by 2001 (14.43° C [57.974° F]). Global temperatures did experience a drop in the early 1990s, but this drop has been attributed to the 1991 eruption of Mt. Pinatubo in the Philippines—the smoke and ash spewed into the atmosphere had a temporary cooling effect.

The eruption of Mt. Pinatubo and its effect on global temperatures raise an interesting point. The

Figure 17-7   Global average temperatures, 1880–2001. (*Source*: Data from Goddard Institute for Space Studies, 2002.)

cooling effect of atmospheric pollutants, such as soot and acid particles, may be masking much of the global greenhouse warming. The eruption of Mt. Pinatubo shot a fine mist of sulfuric acid and sulfate aerosols into the atmosphere; this material reflects sunlight back into space and so cools the Earth. Soot, acid particles, and other aerosols are also injected into the atmosphere through human activities—burning fossil fuels, clearing forests, smelting metals, and so forth. For a long time, many climatologists were a bit puzzled when their theoretical models of the effects of increasing concentrations of greenhouse gases predicted much greater temperature increases than have been actually observed. Only recently has it been widely recognized just how powerful the cooling effect of many anthropogenic airborne pollutants can be. These atmospheric particles not only reflect sunlight back into space, but also form the nuclei of water droplets, promoting the formation of clouds that reflect still more sunlight. According to some recent calculations, atmospheric pollution and particulate matter may be counteracting, or masking, much of the effect of atmospheric greenhouse gases. As progress is made toward reducing global air pollution, greenhouse warming may greatly accelerate.

Other evidence for global warming exists besides direct measurements of atmospheric temperatures. Satellite data of ocean-surface temperatures indicate that the oceans have been warming at a rate of about 0.1° C (0.18° F) per year since the early 1980s. Another way to measure temperature changes over time is by using geophysical borehole data. Heat diffuses from the Earth's surface into the soil and rocks, so by plotting a profile of temperatures at

increasing depths, it is possible to reconstruct broad temperature trends at the surface. Borehole temperature profiles taken at localities ranging from Alaska to Africa confirm that there has been a distinct warming trend in recent decades.

The retreat of nonpolar glaciers provides another line of evidence for global warming over the last few decades. According to data compiled by the United Nations Environment Programme's World Glacier Inventory, most nonpolar glaciers and ice masses (particularly in Asia, Africa, and South America) have been retreating since the turn of the century; in some cases, they have been experiencing particularly rapid melting and decreases in snow cover over the last 40 years. Many glaciologists believe that tropical glaciers (found on high mountain tops) may be especially sensitive to slight increases in global temperature.

The recent rise in global temperature over the last few decades (Figure 17-7) may not seem very dramatic, but from a broader perspective, it is quite rapid indeed. Temperature and climate may be changing more quickly than the ability of many natural species to evolve to the new conditions, with the result that ecosystems are damaged. In 1950, the global average temperature was 13.87° C (56.966° F), in 1955, it was 13.88° C (56.984° F), in 2000, it was 14.30° C (57.74° F), and in 2001, it was 14.43° C (57.974° F; preliminary calculation as this book goes to press). Based on this data, the current global warming rate is about 0.4° to 0.5° C (0.72° to 0.9° F) every 50 years.

How Much Will Earth Warm Up?   If we keep dumping greenhouse gases into the atmosphere, unquestionably the surface of Earth will warm up. About this there is no doubt—all one has to do is analyze the surface temperatures of Venus and Mars, which are determined in large part by the carbon dioxide in these planets' atmospheres. The real question is how much the surface temperatures on Earth at different locations will be affected by the equivalent of, say, a doubling of CO_2 over preindustrial levels. This topic has been the subject of much debate and speculation, but slowly a consensus seems to be emerging. Many different researchers have come to the conclusion that a doubling of CO_2 levels in the atmosphere will lead to an average global warming of 1.3° to 4.5° C (2.3° to 8.1° F) above the 14° C (57.2° F) average global surface temperature over the last 10,000 years. It is important to remember that these numbers refer to average global temperature changes; in many specific areas, the increases could be much greater, especially during the winter. In 1990, a group of about two hundred scientists working under the auspices of the United Nations Environment Programme and World Meteorological Organization, the **Intergovernmental Panel on Climate Change (IPCC)**, concluded that given the present situation of increasing greenhouse gas emissions into the atmosphere, there is at least a 50% chance that an average global increase in temperature of 1.6° to 5.5° C (3° to 10° F) will take place by about 2050. Their most likely estimate was a 2.8° C (5° F) rise by 2050. Since 1990, the IPCC has issued revised assessment reports on global climate change. According to the latest report of the IPCC (2001), it is expected that over the next century carbon emissions from the burning of fossil fuels will dominate changes in atmospheric CO_2 levels, and by 2100, carbon dioxide concentrations could be in the range of 540 to 970 ppm. This could lead to global average temperature increases of 1.4° to 5.8° C (2.52° to 10.44° F) between 1990 and 2100, and a projected rise in sea level of anywhere from 9 to 88 centimeters (3.54 to 35 inches). Note that there is a broad range in these figures because of many uncertainties in the raw data, the interpretations of the data, and the models used to make future projections.

Figure 17-8 Variations in average global temperature over the past 20,000 years. Also shown is the IPCC's prediction for the next century. (*Source*: A. J. McMichael, *Planetary Overload* [Cambridge: Cambridge University Press, 1993], p. 84. Cambridge University Press. Reprinted with the permission of Cambridge University Press.)

Questioning Global Warming.   Although they are in the minority, numerous scientists question the concept of global warming both in theory and in practice.

Ice Ages.   Over the last few million years, the Earth's surface temperature has undergone natural fluctuations, expressed as the **Ice Ages** (**Figure 17-8**). Even in historical times, there have been warmer and colder periods, such as the warmer "Medieval Little Optimum" around A.D. 900–1100. At this time, Earth's surface was warm enough that ice melted back in the North Atlantic Ocean and the Norsemen colonized Greenland. A couple of hundred years later, the **Little Ice Age** set in and the surface became noticeably colder; the colonies in Greenland failed and mountain glaciers advanced. Only in the last few hundred years has Earth begun to warm up again. No one really understands the causes of preindustrial temperature fluctuations (perhaps they are correlated with sunspot activity or other natural phenomena such as slight variations in the tilt and direction of Earth's axis and the shape of its orbit), but they certainly had nothing to do with industrially produced greenhouse gases.

Over the course of billions of years, there has been a general decline in atmospheric CO_2 concentrations. During the Mesozoic Era (the "time of the dinosaurs," about 245 to 65 million years ago), average global temperatures may have been 5° C (9° F) or more warmer than at present, and CO_2 concen-

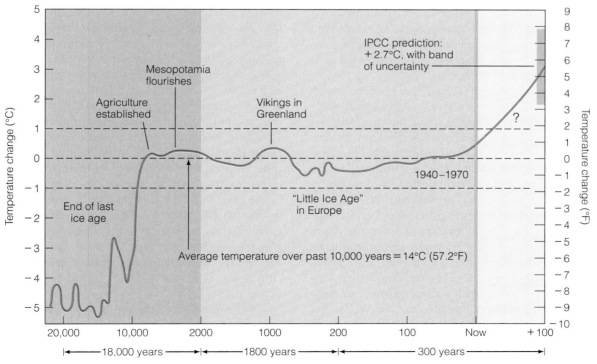

trations may have been appreciably higher than to-day, perhaps several times higher. In contrast, CO_2 concentrations may have been only 200 ppm about 40,000 years ago during the last great glaciation. Since then, they rose naturally to the preindustrial levels of about 250 to 280 ppm. Despite the recent rise in CO_2 concentrations, a few researchers suggest that the natural long-term decline of atmospheric CO_2 concentrations over tens of millions of years, if not reversed, could eventually spell the end of life on our planet in the distant future.

The Pleistocene Ice Ages of the last few million years occurred in cycles of approximately 100,000 to 120,000 years, with each ice age lasting about 100,000 years followed by an approximately 10,000- to 20,000-year interglacial period. The last ice age ended just over 10,000 years ago, so if the pattern continues, we could soon enter another ice age. Indeed, some scientists (though a minority) strongly believe that we are on the verge of entering the next ice age.

These scientists do not deny that atmospheric CO_2 levels are on the rise; in fact, they believe this is evidence in support of their hypothesis. According to the theories of John D. Hamaker and Larry Ephron, during a glacial period, the soils of the Earth are fertilized by ground-up mineral matter. At the beginning of an interglacial period, plant life thrives, and as a result, CO_2 levels in the atmosphere are relatively low. But toward the end of an interglacial epoch, trees and other plants die off as the mineral nutrients in the soil are exhausted. As forests in particular recede, less CO_2 is removed from the air and atmospheric concentrations increase. The immediate result of the increased CO_2 levels is greenhouse warming.

But what about the ice age? Ice age proponents believe that most of the warming will occur in tropical and temperate regions, while the high latitudes will remain cold (or perhaps become even colder: Note that many scientists disagree with this analysis and believe that the higher latitudes will warm up relative to the lower latitudes). According to certain ice age proponents, greenhouse warming will reinforce the temperature differences between the tropics and the poles. With increased warming, large quantities of water will evaporate from the low latitudes and move toward the higher latitudes, where it will precipitate out as rain and snow. Increased cloud cover, along with increased snow and ice cover, in polar and temperate latitudes will reflect sunlight and eventually cause a cooling trend to set in. Once the cooling begins in earnest, glaciers will grow on land and slowly spread into progressively lower latitudes. Another ice age will be upon us.

What do the proponents of this ice age theory suggest that we do about the situation? Since greenhouse warming will initially trigger the next ice age, many advocate the same measures as do those concerned about global warming—decrease the consumption of fossil fuels, limit the amounts of greenhouse gases that are spewed into the atmosphere, and so on. But some ice age proponents go much further. They advocate spreading minerals (in the form of ground rock) over forest floors all over the world so as to revitalize tree growth. They also advocate planting billions and billions of new trees to promote CO_2 removal from the atmosphere.

Imprecise Data. Some researchers question the whole notion of global warming, not because they believe that another ice age is imminent, but because they see no clear trend in global temperature changes or they question the whole theory of global warming per se. They point to such data as the slight cooling and stabilization of mean global temperatures during the 1940s to 1960s as indicative of a lack of the predicted global warming. They point out that many of the global warming data are imprecise and that we have a mere century's worth of accurate data. With such data it is very difficult to separate the signal from the noise, especially since daily temperatures can fluctuate by tens of degrees and seasonal temperatures can fluctuate much more widely, while the global warming trend that is being sought is only on the order of 0.5° C (0.9° F) over several decades. Local temperature readings during the past century may also be skewed, depending on where they were taken; large urban areas have progressively become "heat islands," so that temperature readings taken in cities may not reflect real Earth surface conditions. One study found what appears to be a relative rise in the night temperatures throughout the United States since about 1930 without a concurrent rise in daytime temperatures: Such data have been used to argue both for and against the seriousness of possible global warming. Rises in nighttime temperatures without an initial rise in daytime temperatures could be expected to occur as greenhouse gases build up, for they would let sunlight in during the day, but not let heat (infrared radiation) escape back into space at night. Those who downplay the significance of global warming suggest that even if this phenomenon of nighttime temperature rises is real, in and of itself it is probably relatively benign.

Some researchers also argue that even if global warming is occurring, its initial effects might be absorbed by the oceans as they heat up, so atmospheric global warming might be delayed for years, decades, or possibly even centuries. Even if this hypothesis is correct (and data to accurately test it do not exist at present), global ocean warming could potentially have effects just as devastating as global atmospheric warming (especially given the role that oceans play in determining climatic conditions on the Earth's surface).

Another aspect of the oceans is that they store a tremendous amount of carbon. The oceans appear to have the capacity to continue to absorb and store much more carbon, but they can do so only at a certain maximum rate. At present, we have far exceeded the capacity of the oceans to absorb our excess CO_2 as fast as we emit it. The oceans now absorb approximately half of the excess CO_2 emitted by humans; the remainder collects in the atmosphere. Nevertheless, without the oceans, CO_2 would collect even faster in the atmosphere than it does currently.

General Circulation Models.   Among the primary research tools of climatologists are **general circulation models (GCMs)** of the atmosphere. Using GCMs, some research groups have predicted that a doubling of the atmosphere's CO_2 levels (or the effective greenhouse equivalent of CO_2, such as smaller amounts of CFCs and methane) from about 300 ppm to 600 ppm will raise the average surface air temperature by about 4° C (7.2° F). Proponents of global warming have placed much emphasis on these predictions, but skeptics have pointed out the limitations of the current generation of GCMs.

GCMs are complex mathematical models that, with the help of supercomputers, simulate the Earth's climatic patterns. In typical GCMs, the Earth's surface is conceptually covered with a number of huge boxes (each perhaps a few hundred miles [several hundred kilometers] long on a side, and a few miles [several kilometers] high) laid next to each other and stacked one atop another to a height of several tens of miles (tens of kilometers). Each model uses hundreds of thousands of such boxes. For each box, a set of initial conditions is specified, such as air temperature and pressure; H_2O, CO_2, and other gas concentrations; and wind speed and direction. The model also includes expected inflows and outflows from one box to another and from/to the Earth's surface (e.g., CO_2, dust, and other substances spewed into the atmosphere, water evaporated from the oceans, rain and snow precipitated out) and outer space (e.g., incoming solar radiation); equations are used to calculate how matter and energy flow in, through, and out of each box. Once the mathematical model is devised and the initial data are inserted, the model is allowed to run reiteratively, generating predictions as to future climatic conditions on different parts of the Earth's surface.

The actual workings of the atmosphere are so complex, and so many variables and uncertainties are involved, that even the best GCMs are extremely crude in comparison. As a model is run for longer periods of time (to make predictions farther into the future), any errors inherent in the system will accumulate and multiply. Current GCMs are incapable of perfectly predicting present or, retrospectively (also known as retrodiction), past climates with any great accuracy. Critics conclude that predictions of future climatic conditions made by these GCMs are not to be believed. Some adherents of GCM methodology contend that the purpose of these models is not so much to precisely prognosticate the future, but to simply improve our understanding of the atmospheric phenomena being simulated. Even if we must take the actual predictions of the GCMs with a large grain of salt, they can help us make rational choices when it comes to human endeavors that affect the atmosphere and climate.

The use of GCMs is not the only way to model predicted global warming from the greenhouse effect. Based on the greenhouse effects experienced on Venus and Mars, taking into account the atmospheric CO_2 pressures on these planets and their distances from the Sun, it has been predicted that increasing the effective CO_2 content of Earth's atmosphere to about 600 ppm will cause an average surface warming of only about 0.4° C (0.72° F), or about one-tenth of that predicted by many GCMs. Of course, this prediction is, in turn, disputed by many GCM theorists.

The Consensus View.   Despite the questions raised concerning the reality of global warming, it should be firmly emphasized here that the vast majority of scientists who have studied the problem are convinced that global warming is real and will change the distribution of resources across the Earth. There is no doubt that the concentrations of greenhouse gases in our atmosphere are increasing dramatically because of human activities. It is known that these greenhouse gases trap heat that will increase the temperature of the Earth's surface. It is also thought that as the Earth heats up, the warming will be unequally distributed, with higher latitudes generally warming up more relative to the lower latitudes; the least change will occur in the equatorial regions. If the rise in temperature is great enough, sea level will surely rise due to thermal expansion of the oceans and the melting of ice.

What is unclear to many scientists is exactly what effects global warming will have. Exactly how much will the surface of Earth warm? How high will sea levels rise as glaciers melt and the oceans expand? What effect will global warming have on cli-

The Benefits of CO$_2$

Some researchers argue that increasing the concentration of carbon dioxide in the atmosphere will have benefits that will far outweigh any deleterious "greenhouse effects." As is well known, CO$_2$ is an essential raw ingredient on which all photosynthetic green plants rely. There is evidence that throughout much of Earth's history the atmospheric carbon dioxide content was higher than it has been during humankind's existence; thus, many plants may have evolved to grow optimally in somewhat higher concentrations of CO$_2$. Experiments have shown that doubling the CO$_2$ content of the air (for instance, from 330 ppm [the level in the middle 1970s] to 660 ppm CO$_2$) will raise the productivity of many plants by about a third on average. That is, the plants produce more organic matter more efficiently. Yet some authorities contend that even if the rate of photosynthesis for many plants increases in the presence of elevated levels of CO$_2$, not all of the cultivated crops that humans currently depend upon may respond in this way (**Figure 1**).

Rising CO$_2$ concentrations also lead to a decrease in the transpiration (water loss by evaporation from the leaves) rate in many plants. Experiments have demonstrated that on average, the amount of water lost by plants in this manner is decreased by one-third if CO$_2$ content is doubled from 330 to 660 ppm. Overall, this means many plants become much more efficient producers of organic matter as CO$_2$ concentrations increase; they produce more with less loss of water. Because of these benefits, many commercial nurseries routinely use a CO$_2$-enriched environment to grow plants.

But even if increased concentrations of atmospheric CO$_2$ accelerate plant growth (al-

Figure 1 To determine how wheat might respond to elevated levels of atmospheric carbon dioxide, scientists near Phoenix, Arizona, are raising the levels artificially around the plants in a study called Free Air Carbon Dioxide Enrichment.

matic patterns, including wind directions, rainfall patterns, and ocean currents? Has global warming actually begun, or is it still too early to detect it relative to the "background noise" of daily and seasonal temperature fluctuations? These questions may be partially to completely answered in the next half century, as we begin to directly observe more and more of the effects of global warming.

Consequences of Global Warming

Sometimes it is suggested that a little bit of global greenhouse warming might not be such a bad thing, especially for people who live in regions visited by harsh winters (see also Case Study 17-2). But, in fact, global warming will not cause overall weather patterns to become consistently warmer and more equitable—indeed, just the opposite may occur in certain regions. As Earth heats up, climatic patterns will shift, and in some places local weather conditions will become much more violent. Cold air currents may be displaced such that, ironically, regions that are currently relatively warm may experience cold snaps and abnormal winter storms. In recent decades, abnormal frosts that have destroyed Florida tomato crops and record low temperatures experienced in Chicago have been attributed to global warming. Due to shifting air currents caused by increased heating of the Earth's surface, cold air masses were displaced from the Arctic.

Climate and Weather. Atmospheric circulation, which ultimately causes what we see as weather (in the short term) and climate (in the longer term), is caused by the differential heating of air masses on though there is currently no unambiguous evidence that the increase in CO_2 in the atmosphere so far has produced this result), the ultimate outcome may not be beneficial from a human perspective. Weeds may proliferate more rapidly at the expense of other types of plants. Insects and other pests that feed on plants may also increase in numbers. Some studies demonstrate that certain insects feed more rapidly on plants in a high CO_2 environment. Apparently, the fast-growing plant leaves contain lower concentrations of essential nutrients (particularly nitrogen), so the insects must eat more in the same period of time. Other insects are incapable of using this strategy (perhaps they already feed as quickly as they can), and so are not as healthy under elevated CO_2 conditions. It might just be that the latter groups of insects are necessary components of the local ecosystem, perhaps the pollinators that ensure plant reproduction.

Human crops may also grow faster when subjected to higher concentrations of CO_2, but the growth may be concentrated in parts of the plant that are not consumed or used otherwise. Thus, the rapid growth may yield little net gain for humans.

More critical than any potential increase in plant production are the disruptions that will certainly occur from increasing surface temperatures on Earth. Insect-plant relations will be assaulted. In higher latitude temperate regions, the freezing conditions of the annual winter season are extremely important for containing the populations of insects that feed on human crops and other plants. If global warming occurs to the extent that it prevents or mitigates freezes and frosts in certain regions, insect pest populations will proliferate. Of course, the increases in violent storms and abnormal weather patterns that are expected from global warming will not benefit plant production either, nor will encroaching oceans from rising sea levels along currently productive coastal areas. These latter disruptions, which might actually decrease the overall net growth of many types of plants, are simply a function of global warming per se, whether this global warming is caused by CO_2 or other greenhouse gases (which do nothing to enhance the growth of plants).

Questions

1. Can the results of greenhouse experiments necessarily be extrapolated up to the scale of the entire Earth?

2. Even if increased concentrations of CO_2 may accelerate the growth of some types of plants, is this always beneficial? What types of unforeseen consequences might result from increased atmospheric concentrations of CO_2?

3. In your opinion, do the predicted "benefits" of increased atmospheric CO_2 concentrations justify—as some people argue—not taking action to counter the buildup of CO_2 in the atmosphere?

the Earth's surface. Although establishing the exact effects global warming will have on these patterns is extremely difficult, some educated predictions can be ventured. As more heat is retained in the system, more air will move across the Earth's surface, producing winds, clashing warm and cold fronts, and generally causing more violent weather conditions. Hurricanes, tornadoes, and other dangerous storms will increase in intensity. Some researchers attribute the record number of extremely damaging storms in the last decade to global warming.

Global warming will also dramatically change overall climatic patterns. With increased warming, evaporation from the oceans and other large water masses may increase, which will lead to higher levels of precipitation. But the increased precipitation will not necessarily occur where it falls now—with the changing air currents, the areas of rainfall will be displaced. The American Midwest, often referred to as the "breadbasket" of America (and the world), may experience such intense droughts that it will become a desert. The rain that would have fallen in this area may well be pushed north into Canada. Rainfall may also shift from one season to another; some agricultural regions may receive more rain on average than at present, but the bulk of it will come during the winter months when it is of little use for crops. Changing rainfall patterns, coupled with the generally more violent weather, will cause increasing incidences of flash floods. Ironically, some areas will experience droughts and floods simultaneously; during the height of a drought, a violent cloudburst will cause rivers to swell and flood, but it will not replenish reservoirs, which require gentle, protracted rains in order to recharge.

Due to changing atmospheric circulation patterns, increased heat, and altered rainfall patterns, large portions of temperate regions could experience major declines in soil moisture. The agricultural productivity of these areas could decline, resulting in severe food shortages. An increase of only 2° C (3.6° F) might cause a decline in grain yields of up to 17% in North America and Europe. Unfortunately, the global food situation is already precarious (see Chapter 13); any disruptions caused by greenhouse warming will not be easily rectified. Some crops will have a hard time growing under even slightly warmer conditions. Many fruit trees and berries need to be subjected to chilling before they will produce fruit.

It is sometimes naively suggested that global warming will simply push many major agricultural belts in the Northern Hemisphere further north; prime cropland will simply be redistributed. To a certain extent, this may be true. Alaska, Finland, and Denmark, for instance, may experience major increases in crop production. In contrast, the classic breadbaskets of central North America, Europe, the former Soviet Union, and portions of China will become much less productive. But one must remember that established farmers in the central United States are not likely to pack their bags and move north into Canada as global warming sets in. Furthermore, many nonclimatic factors must be considered. Even if warmer weather conditions might theoretically favor the cultivation of the northern Canadian prairies and boreal forests, the soils in these areas are not suitable for growing most food crops.

Another consideration is that global warming will not take place all at once. New climatic zones will not simply replace the old. Decades to centuries of very violent, wandering weather patterns may occur before the climate readjusts to a more or less stable equilibrium. It has been suggested that this is already taking place, based on data from insurance companies. For instance, insurance claims from violent weather in 1998 alone were greater than for all of the 1980s combined. Severe storms and unexpected cold snaps or heat spells are very damaging to any agricultural enterprise. The loss of food production could be one of the most serious immediate consequences of global warming.

Not only will atmospheric circulation patterns change with global warming, but the paths of ocean currents will also be modified. Currently, for instance, cold water sinks in the Arctic Ocean and then travels southward along the bottom of the seas toward the equator. In the tropics, the water is warmed and then moves north once again as a surface current, forming what can be thought of as a giant convection cell. With global warming, the northern polar regions may heat up enough to disrupt the convection cell; adequate quantities of cold water will no longer sink to the bottom to drive the surface currents back to the pole. The Gulf Stream in particular may slow down, stop flowing, or even change its direction. This could dramatically change local weather and climate patterns. England's weather, for instance, is heavily influenced by the warm waters of the Gulf Stream passing by its shores. If global warming disrupts the Gulf Stream, British winters may become much colder and harsher. Also, disrupting deep-water currents may affect the circulation of nutrients in the oceans, exacting a heavy toll on extremely delicate ocean ecosystems.

Sea Levels. One of the most discussed effects of global warming is the predicted rise in sea levels. One study by the EPA predicted that sea levels could rise by as much as 7 feet (2.2 m) by the year 2100, although some more recent studies, based on IPCC data, predict only a 19-inch (48-cm) rise by that time. But any significant rise in sea level, even a mere 19 inches, could have devastating effects on the human population and global ecosystems.

Approximately one-third of all people live within 37 miles (60 km) of the sea, often at elevations close to sea level, so a rise in sea level would affect them directly, driving them to higher ground. On the east coast of America, such cities as Boston, New York, and Washington, D.C. (to name just a few) would be prone to flooding. Of course, productive agricultural areas could be destroyed as well, freshwater supplies could be contaminated with seawater, coastal wetlands could be wiped out, and other, unpredicted changes could occur. The IPCC estimated that a 3-foot (1-m) rise in global sea levels (a distinct possibility within the next few centuries given current trends in global warming) would flood almost 250,000 miles (402,250 km) of coastlines around the world.

Global warming can promote rises in sea level through several mechanisms. It is often suggested that increases in mean global temperature will cause mountain glaciers and polar ice caps to melt (**Figure 17-9**). As the ice melts, the water will eventually empty into the oceans, raising their level. The gravest concern is that the Antarctic ice cap, which is 2 miles (3.2 km) thick in places and contains an estimated 80% of all the ice in the world, may undergo significant melting. If the entire Antarctic ice cap were to melt, the mean average sea level would be raised by almost 300 feet (90 m). (In fact, the average level of the oceans is about 200 feet (60 m) higher today than at the end of the last ice age about 10,000 years ago—much of this rise is attributed to the melting of great continental glaciers.) Realistically, though, not many scientists believe that the Antarctic ice cap as a whole will melt significantly or very quickly. This ice cap has lasted for about 10 million years, withstanding many warm and cold spells. If it did begin to melt on a large scale, the melting would probably take place very slowly; the core of the ice cap is extremely cold (much colder than ice in many Northern Hemisphere glaciers, for instance) and ice forms a very good insulator. Even with extreme global warming, melting all of Antarctica's ice would take many thousands of years.

Of course, only a very small percentage of Antarctica's ice needs to melt to raise sea level by a few feet, causing global devastation. In recent decades, ice shelves have been calving and breaking up, apparently due to warmer temperatures. The breakup and melting of floating ice shelves and icebergs has only a minor effect on global sea levels per se (since the ice is already in the water). But when Antarctic coastal shelf ice melts and breaks away, land-based ice creeps or slides down into the sea, replacing the ice that broke away; consequently, more water is added to the oceans, and the sea levels rise.

Global warming could also cause a rise in sea levels through thermal expansion of water. Liquid water expands and contracts slightly with changes in temperature. Given the tremendous volume of water in the world's oceans, even a slight expansion of the water from global warming will cause a measurable rise in sea level.

When we speak of changes in sea level, we usually refer to global averages—but to the people or other organisms living near the shore, the height that the sea reaches during exceptional high tides or major storms may be even more important. On a local or regional scale, global warming will contribute to this aspect of "sea level" also. Increasingly violent storms such as typhoons and hurricanes, accompanied by high winds and changes in barometric pressure, will drive water high onto land. Changes in atmospheric pressure can actually pull the water up, locally raising sea level. Such storm surges will only increase as global warming continues.

Climate Zones and Biodiversity.   In another major effect of greenhouse warming, the climatic zones on the Earth's surface will tend to shift from the equator toward the poles. The effects will be devastating for many ecosystems, but forests will be especially hard hit. As climatic zones migrate away

from the equator, trees will find themselves living in environments for which they are not adapted. Very small changes in average temperatures could spell death for large tracts of forest. The decline of red spruce populations in the eastern United States since 1800 has been attributed to global warming. One analysis suggests that if the effective concentration of CO_2 in the atmosphere doubles, the ranges of such trees as the eastern hemlock, yellow birch, beech, and sugar maple would have to migrate 300 to 600 miles (480 to 965 km) north. If the warming occurs rapidly, as is expected, many of the trees will simply die.

Biodiversity, already under assault from humankind, could be further reduced by rapid global warming. Many species of plants and animals require very narrow ranges of temperature and moisture. As greenhouse warming modifies their habitats, they will not be able to adapt or migrate quickly enough, and so will succumb; other species, however, may be able to expand their ranges, sometimes with devastating results from a human perspective (see Case Study 17-3). Species that live on mountaintops or in the high-latitude Arctic and Antarctic regions will have nowhere to relocate. Other animals that could possibly migrate to higher latitudes will be impeded by human barriers (such as roads, farms, and urban areas). Mangrove swamps and other coastal wetlands will be flooded by rising sea levels. Some wildlife will find themselves trapped in artificial wildlife preserves that were originally designed to protect them. Already, global warming may be exacting its toll on some relatively immobile species.

Figure 17-9   The Antarctic home of these emperor penguins may be threatened by global warming.

Global Warming and Disease

A major theme of this book is that the complex connectedness of environmental systems leads to unpredictability of human impacts. Our individual actions can accumulate, to have unseen consequences that are not evident for many years.

A good example is a 1995 landmark study released by the World Health Organization on the effects of global climate change on human health. Until the mid-1990s, the vast majority of research on global climate change had focused on physical impacts, such as a rise in sea level. But there were growing signs that the potential effects on human health are no less serious. This possibility came to public attention in 1993, when a controversial paper published in the medical journal *The Lancet* concluded that the 1991 cholera outbreak in South America was related to localized warming of Pacific Ocean waters from global climate changes. The paper argued that the warming had caused the rapid growth of plankton that harbor the cholera bacterium leading to thousands of deaths.

Over the last decade, numerous studies have linked global climate change with disease threats and adverse health effects, not just for humans but for other organisms as well.

Some of the projected adverse effects on health and well-being are physical in nature: strong, hot, and prolonged heat waves in some areas, and warmer nights in particular. Prolonged heat can enhance the production of smog in populated areas, and one predicted impact of continued global warming is that many cities will experience "killer heat waves" that will increase deaths from bronchitis, asthma, and many other ailments. Indeed, in some areas, it has been projected that the number of deaths directly caused by heat waves could double between 2000 and 2020. Global climate change is also increasing the number and severity of weather fluctuations and extremes, such as both severe droughts and massive rainstorms. These erratic weather phenomena can cause harm to humans outright, such as by drowning, dehydration, or starving (and of course also severely impact ecosystems upon which humans are dependent), but they can also promote the spread of infectious diseases.

The promotion and spread of diseases is perhaps the most obvious, and pernicious, biological effect caused by global warming. Since an outbreak of cholera was first linked to global warming a decade ago (as described above), increased incidences and new outbreaks of numerous diseases have been linked to global warming, including West Nile virus, schistosomiasis, malaria, dengue fever, yellow fever, various forms of encephalitis, Rift Valley fever, filariasis, onchocerciasis (river blindness), trypanosomiasis (sleeping sickness), tuberculosis, and hantavirus.

Many of the infectious pathogens or their carriers are opportunists that thrive on the sequential extremes in environmental conditions brought about by global climate change. Some diseases, such as tuberculosis, increase in situations of overcrowding, as when people are displaced by flooding and cramped into a smaller area with perhaps inadequate sanitary facilities. Microorganisms or viruses cause many diseases, but animal carriers, or "vectors" transmit the pathogens. For example, various species of mosquito (vectors for malaria, dengue fever, yellow fever, various forms of encephalitis, filariasis, Rift Valley fever, and West Nile virus), black flies (river blindness), tsetse flies (sleeping sickness), ticks (lyme disease and some forms of encephalitis), water snails (schistosomiasis), and rodents (hantavirus) can all spread disease among humans. Mosquitoes and other vectors are very sensitive to temperatures; cold, such as during the winter season or at higher (and cooler) latitudes and altitudes, will kill adults, larvae, and eggs. As global warming progresses, temperatures

Figure 1 The *Anopheles gambia* mosquito is one of the leading malaria vectors in the world.

are rising and remaining at much warmer levels longer and over larger geographic areas, allowing mosquito and other disease-carrying populations to expand. For example, *Anopheles* mosquitoes that spread malaria only do so effectively in a sustained manner in regions where temperatures are maintained above about 15.56° C (60° F). Currently, malaria kills several thousand people daily, and over 45% of the world's population lives in areas where there is a risk of malaria; based on some models of global warming, by the end of the twenty-first century, 60% of the world's population may live in areas with a risk of malaria (**Figure 1**). Similarly, *Aedes aegypti* mosquitoes, those that carry yellow fever and dengue fever, must be in regions where the temperatures seldom fall below 10° C (50° F) in order to spread the diseases—as global warming occurs, their range will expand.

In the tropics, the altitude on mountains where it remains below freezing year-round has climbed nearly 500 feet since about 1970; these temperature changes correlate with the melting of mountain glaciers around the world and also allow insect populations to spread to higher upland elevations that were once "safe" from many diseases. In the past many people, and in particular nineteenth century European colonists, lived in the uplands and mountains of the tropics to escape both the heat and the insects, along with the diseases they brought. Now insect-borne diseases are spreading to elevations of above 1 mile (1600 meters), unheard of 50 to 100 years ago, in the mountains of east and central Africa, Asia, and South and Central America. In higher latitudes, such as temperate regions, many diseases and their vectors were controlled by the cold weather of the winter months that killed off large numbers of insects that could otherwise rapidly multiply and spread disease. Now, with milder winters in New England and elsewhere over the last decade, this natural mechanism of insect (and other disease-carrying organisms) population control has been failing.

To make matters worse, mosquitoes and other vectors may become more active in warmer environments (as long as it is not too warm to kill them). Mosquitoes, for instance, tend to breed more quickly and bite more victims more often in warmer temperatures, increasing the chance of picking up a pathogen and transmitting it to healthy victim. Furthermore, the disease-causing microorganisms within the mosquitoes also reproduce more quickly in warmer temperatures. The malaria-causing parasitic microorganism, *Plasmodium falciparum*, takes 26 days to develop inside the mosquito at 20° C (68° F), but only half that time at 25° C (77° F). This further increases the effectiveness of the carrier mosquito to spread the disease. Increasing winter temperatures and nighttime temperatures will also effectively increase the mosquito season, meaning that diseases can be spread for a longer period.

Global climatic disruption, resulting in sustained droughts and intense rainstorms with flooding can cause rises in mosquito and other insect populations, and the diseases they carry. Abnormal weather may negatively affect species such as birds, fishes, amphibians, and insects that are predators on certain insects and thus help to control their population numbers. Ladybugs, for instance, are predators on mosquitoes. Both floods and, somewhat ironically, droughts can increase the spread of insects such as mosquitoes that breed in water. Flooding may allow dormant, but viable, eggs to hatch. Drought often results in stagnant pools of what water remains (including artificial pools, perhaps even a barrel of water created by humans as a storage mechanism) that can be perfect incubating grounds for insect larvae. Droughts can also lead to the congregation of people and animals near the few remaining sources of water, and the overused water itself may become contaminated and spread disease. The crowding of people or other organisms allows disease to spread more readily.

Flooding can also promote the spread of water-borne (non-insect-carried) diseases, and so can drought conditions. Flooding can flush sewage and pathogens into drinking water supplies as well as fertilizers from farms and even yards. Algal blooms can produce toxins directly harmful to humans, as well as contaminating fishes and shellfishes. Algal blooms can also encourage the growth of pathogens, such as *Vibrio cholerae*, the cause of cholera. Droughts can promote water-borne diseases as supplies of safe drinking water are depleted and contaminated supplies are used. Furthermore, lack of water can lead to a lack of proper hygiene, further promoting the spread of diseases.

Nobody really knows how much increase in human disease can be expected from global warming. Of course, the answer will depend in part on how much global warming actually takes place, which in turn depends on what measures we take in the coming years and decades to stop greenhouse gases from accumulating in the atmosphere. Already, billions of people are at risk, and the numbers will surely increase substantially. One study projected that an increase of global temperatures by just a few degrees could result in 50 million or more new malaria cases a year. It is not only human populations that are being affected by increases in diseases due to global climate change—nonhuman populations and ecosystems are also feeling the effects. With slight increases in temperature, slime molds have been found growing more quickly on eelgrasses (a group of important aquatic plants that grow in, and help support, productive coastal areas). Eastern oyster disease has been found in the waters of Maine where previously it was too cold for the disease to survive. It seems certain that more diseases linked to warming temperatures will be found in nonhuman populations if global climate change continues.

Questions

1. How many people do you think are aware, or ponder, that a trip to the supermarket in their gasoline-powered automobile (which spews out greenhouse gases) can contribute to an increase in worldwide malaria and other diseases? Does this knowledge have any impact on your personal choices for transportation?

2. If major wide-scale problems, such as global warming, result from the accumulation of millions of personal choices each day, how can such problems be solved?

3. Does a nation such as the United States, which contributes more than its "fair share" of greenhouse gases to the atmosphere, bear any responsibility to help treat increased incidences of malaria and other diseases in developing countries?

In Canada, long-term studies have documented decreased amounts of ice and longer ice-free seasons on some lakes, and this is affecting the relative populations of various aquatic species. In the Caribbean and around the world, corals have been dying; some biologists believe this is the result of rising sea temperatures. Overly warm water appears to stress the corals (corals are a type of sedentary animals) to the point where they expel the algae which live symbiotically within them, become "bleached" (they lose their color without the algae) and ultimately

die (**Figure 17-10**). In addition, such factors as pollution, mining, overfishing, and overharvesting of reefal organisms are also damaging coral reefs. It has been estimated that over a quarter of the world's coral reefs may already be damaged beyond repair. Coral reefs contain a wealth of biodiversity and are the breeding grounds for many ocean organisms, so their loss is particularly significant.

General Strategies for Dealing with Global Warming

There are three basic attitudes toward, and ways of dealing with, the potential of massive global warming: (1) The Waiting Strategy holds that since all the data are not yet in, we should wait before taking substantive measures to combat global warming (perhaps the predictions are simply wrong). (2) The Worst Case Scenario Strategy says we should assume that the most pessimistic predictions are valid and act accordingly. (3) The Compromise Strategy says we should follow a path that is environmentally safe and promotes the general well-being of the planet, regardless of whether global warming turns out to be a valid phenomenon.

Each of these strategies has proponents and opponents. The Reagan and first Bush administrations strongly espoused the Waiting Strategy. Their position was that the reality and magnitude of global warming, and its causes, had not yet been definitively

Figure 17-10 The bleaching of this brain coral is one possible symptom of global warming.

"proven" scientifically, and without such proof, taking serious steps to reduce CO_2 emissions would be premature and inappropriate. National policies, it was argued, must be based only on hard facts. Any actions that would significantly reduce emissions of greenhouse gases would impose a heavy financial burden on the country; enormous amounts of money would be spent, economic output could decrease, and jobs might be lost. Such pain might be for naught if the predicted global warming turned out to be either false or grossly overestimated.

The problem with the Waiting Strategy is that if global warming is real, dealing with the eventual problems (both economic and otherwise) will be more painful and costly than taking measures at an earlier stage. This line of thinking leads some people to argue that even though all the data are not yet in, we should not take the chance that the more pessimistic scenarios will turn out to be true. Those advocating the Worst Case Scenario Strategy believe that as a form of insurance, we should do everything we can to prevent potential global warming. Thus, we should significantly decrease the burning of fossil fuels, halt deforestation, reduce cattle production, reforest areas that have been clear-cut, and so on. If all the appropriate actions are taken and extreme global warming never occurs, we may never know whether human actions prevented a disaster or whether the predictions were simply incorrect. However, taking no action at all would be taking an unacceptable risk.

The Compromise Strategy, sometimes referred to as the "no regrets strategy" or the "tie-in strategy," takes a middle-of-the-road approach. Its advocates argue that we should develop policies and take actions that will help prevent the worst of global warming while benefiting the environment and society otherwise. For instance, increasing conservation and raising energy efficiencies, curbing the use of fossil fuels, and placing more reliance on renewable energy sources will benefit humankind in the long run whether global warming is real or not. Independent of the greenhouse effect, acid rain and general air pollution will be reduced, decreased coal mining and oil drilling will help preserve the landscape, and reduced dependence on oil in particular will lessen dependence on foreign energy supplies. The main problem with the Compromise Strategy is that if the most dire predictions turn out to be true, the moderate response may be inadequate—too little, too late.

International Efforts to Combat Global Climate Change. Through the 1990s, many governments pursued a weak compromise strategy and took rel-atively little action to combat global warming. As a result of the 1992 Earth Summit sponsored by the United Nations in Rio de Janeiro, Brazil, over 160 countries signed the United Nations **Framework Convention on Climate Change (UN FCCC)**. Although the convention established no legal obligations or specific target dates, it required signing countries to use their best efforts to control emissions of greenhouse gases. The convention suggested that emissions of greenhouse gases be stabilized at 1990 levels by the year 2000. By the beginning of 1994, enough nations had ratified the convention to make it a legal document, and by 2002, fully 181 nations and the European Union were parties to the convention. According to its provisions, certain nations were required to produce a national action plan (NAP) explaining the policies they were following to reduce greenhouse emissions and projecting the levels of emissions they intend to allow in the future. The NAPs would then be submitted to a Conference of Parties (established by the convention) for review. The hope was that the participants would then be able to establish specific, legally binding emissions levels.

According to many scientists at the time, the Climate Convention did not go nearly far enough; for instance, the IPCC estimated that greenhouse gas emissions should be reduced to 60% of 1990 levels just to stabilize the climate. In 1995, delegates of the countries that ratified the 1992 Rio agreement met in Berlin and agreed to negotiate a new set of targets for reduced emissions of greenhouse gases within the next few years. These talks resulted in the Kyoto Protocol of 1997 (named after the Japanese city where meetings were held—note that the basic protocol was developed in 1997, but not ratified), which among other actions, collectively called for industrial and former Eastern communist bloc countries to reduce their greenhouse gas emissions by 5.2% below 1990 levels by and during the period 2008–2012 (with different targets of emissions reductions for different countries). The 1997 Kyoto Protocol includes provisions for the trading of emissions permits, the use of carbon sinks such as forests, and credits based on carbon-reducing and saving initiatives, as ways of garnering credits for reducing greenhouse gas emissions. The Kyoto Protocol also commits developing countries to the monitoring and ultimate reduction of their greenhouse gas emissions.

Between 1998 and 2000, governments met for negotiations to attempt to finalize the rules and provisions of the Kyoto Protocol so that it could be finally ratified and implemented. William (Bill) Jefferson Clinton, a Democrat, was president of the Untied States from 1993 to 2001 (with the environmentally oriented Al Gore as vice president) during the initial development of the Kyoto Protocol. In January 2001, the Republican George Walker Bush (son of the former president George Herbert Walker Bush, 1989–1993) was inaugurated president, and one of his first major international acts was to announce in March 2001 that he would not sign the Kyoto Protocol and the United States effectively withdrew from the negotiating process. Some commentators believe that Bush's announcement may have spurred other countries to reach some form of international agreement on the issues all the more quickly. In July, 2001, a meeting held in Bonn, Germany, resulted in 178 nations reaching agreements on various elements of the Kyoto Protocol, and further meetings (with various agreements and compromises reached) were held in Morocco in October–November 2001. By early June, 2002, Japan and all 15 of the European Union nations had ratified the Kyoto Protocol; still, not enough countries had ratified the treaty for it to come into force.

Technological and Other Measures to Combat Global Warming. The principal ways to control the greenhouse effect are to cut down on emissions of CO_2 and other greenhouse gases into the atmosphere and to restore or provide additional sinks for CO_2 (including artificial means of sequestering carbon dioxide). Interestingly, virtually all measures that curb greenhouse gas emissions have additional benefits as well. Thus, reducing emissions of CFCs and related gases will help protect the ozone layer as well as helping to mitigate the greenhouse effect. CO_2 emissions can be readily controlled by increasing energy efficiency and decreasing global dependence on fossil fuels (for instance, switching to more environmentally friendly fuels, such as clean-burning hydrogen gas). Of course, decreasing our use of fossil fuels offers other benefits as well, including reducing air pollution and acid rain, protecting the landscape, and extending the time that our finite supplies of these fuels will last. Of particular importance to the United States and other developed countries is that a decrease in dependence on fossil fuels (and oil in particular) will mean a decrease in dependence on foreign supplies from volatile regions of the world, such as the Middle East. Switching to other fuel sources and increasing energy efficiency will not only benefit the environment but also save billions and billions of dollars in military expenses and actually prevent open conflicts and save lives in a direct sense. Alternative, nonpolluting, cleaner energy sources, such as solar and wind technologies, need to be further developed and implemented—although certainly great strides have been made along these lines in recent years. Some researchers suggest that at least on a global scale, nuclear power may also play a much larger role in the future since this form of energy emits very little in the way of greenhouse gases (although typically greenhouse gases are emitted when fossil fuels are used to power the mining operations that obtain the uranium fuel). However, there are other drawbacks to increasing our dependence on nuclear power.

One way to discourage the burning of fossil fuels is to impose a "**carbon tax**" on their consumption. A carbon tax is levied in proportion to the amount of carbon emitted during combustion of a particular fossil fuel; accordingly, coal would generally be taxed at a higher rate than natural gas. Such a tax would encourage more efficient energy use and the development of non-carbon-emitting energy technologies; it would also bring the price of burning fossil fuels more in line with the environmental costs they entail. The amount of money that could be raised with carbon taxes is truly staggering: The Congressional Budget Office has estimated that a tax of $28 per ton (0.907 metric ton) of carbon content in fossil fuels would raise over $30 billion a year. Although not strictly a carbon tax, every penny added to the federal taxes on 1 gallon (3.785 L) of gasoline contributes about $1 billion annually in revenue to the U.S. government.

An argument often heard against carbon taxes is that they are regressive. That is, relatively poorer people pay a much higher proportion of their income for energy than do richer people; therefore, any carbon tax will extract a relatively higher proportion of the income of the poor than of the rich (even if the rich pay a higher amount of carbon tax in absolute terms). A solution might be to couple a stiff carbon tax with lower Social Security and income taxes for the poor.

Another suggestion is to institute a standardized system of emissions permits, on both a national and a global level, that would allow a company or organization to emit only a certain amount of CO_2 and other pollutants into the atmosphere. These permits could be sold or traded, so a firm that did not emit its allotted amount of pollutants could sell the "right" to emit the pollutants to another firm. Such a system has been used to control air pollution in California for a number of years and was established at a national level by the 1990 Clean Air Act Amendments.

Besides the burning of fossil fuels, another important source of greenhouse gases is the CO_2 released during deforestation (**Figure 17-11**). A "simple" solution to this problem is to stop cutting and burning forests, but, of course, real life is not

Figure 17-11 Deforestation can be clearly seen in this Landsat satellite image of part of the Amazon Basin, Brazil. The original natural forest is dark green; the pale greens, yellows, and browns mark leveled forest.

so simple. Deforestation occurs because people or governments find it to be economically profitable. Furthermore, sometimes forests catch on fire and burn despite people's best attempts to put out the fire. This, of course, releases enormous amounts of carbon dioxide into the atmosphere. And it is far from clear how much carbon, in the long-run (given that forest fires and various decay processes that release carbon back into the atmosphere occur naturally), can be permanently or even semi-permanently sequestered and stored in forests. Various environmental groups have recently de-emphasized the role of forests as a way of mitigating global warming in the long run, suggesting that planting and preserving forests should not be seen as a substitute for increasing energy efficiency measures and developing alternative fuels. Nevertheless, saving the Earth's forests is a worthy pursuit in its own right, and not only may help abate the greenhouse effect, but will contribute to the preservation of biodiversity, help protect soils, ameliorate climatic extremes, and serve many other useful and necessary functions.

In addition to reducing emissions of greenhouse gases, we can develop sinks that will absorb CO_2. This has given rise to a new field, sometimes re-

ferred to as "carbon sequestration" which aims to "geo-engineer" the Earth and its atmosphere by using various natural and artificial means to remove atmospheric carbon dioxide. In recent years, the United States Department of Energy and the EPA have increased their interest in, and promotion of, various carbon sequestration techniques.

Perhaps one of the easiest ways that has been widely promoted is to plant trees where none currently exist. Trees, like all photosynthetic plants, convert gaseous CO_2 into a solid form such as cellulose (essentially wood fiber), where it is stored until the tree is burned or allowed to decompose. Planting billions and billions of trees, it is suggested, not only will help reduce CO_2 levels in the atmosphere but will restore many of our devastated forests, add scenic beauty, help stabilize eroding soils, provide habitats for wildlife, and help ameliorate climatic extremes. A couple of well-placed shade trees can help keep a house cool in the summer, thus significantly cutting down on the amount of energy needed to air-condition or otherwise cool the house. But, as noted previously, it is doubtful that simply planting trees (or other plants, for that matter—even plants that have been bio-engineered to absorb abnormally large quantities of CO_2) will solve the problem of global climate change. In order to keep the carbon dioxide out of the atmosphere once a tree (or even a part of a tree, such as a leaf or limb) or other plant dies, it must be further sequestered and preserved. If it burns or decomposes, the CO_2 will once again be released. Thus, some researchers have advocated the permanent landfilling of suburban yard waste rather than allowing it to be composted and otherwise decomposed to retrieve the valuable nutrients stored in the organic material.

Another method that has been proposed for removing carbon dioxide from the atmosphere is to fertilize the oceans with iron, nitrogen, phosphorus, and other limiting nutrients so that the growth of photosynthetic plankton will be promoted, thus absorbing carbon dioxide. The theory is that once the plankton die, they will fall to the ocean floor and thus store the carbon. In the 1990s, an experiment was carried out in the waters of the equatorial Pacific where 500 kilograms of iron (in the form of iron sulfate) was dumped into the ocean. This caused a temporary and localized phytoplankton bloom that consumed hundreds of tons of carbon dioxide. However, most or possibly all of the CO_2 used by the phytoplankton may have already been dissolved in the ocean water, and it is not clear that any significant amount of carbon dioxide was transferred from the atmosphere to the water. It is also unclear what the environmental and ecological repercussions of the massive fertilization of phytoplankton would be, and some researchers suggest that it could reduce oxygen levels (with possible detrimental effects) in the deep oceans.

Rather than fertilizing phytoplankton, another suggested approach is to inject carbon dioxide directly into deep ocean areas a thousand or more meters below the surface. The water can absorb the CO_2; however, it will make the water more acidic and this could have detrimental effects on deepwater organisms. Another unknown is how CO_2 enriched water at that depth will interact with non-enriched seawater. Furthermore, the technology to pump carbon dioxide deep below the ocean surface, perhaps using long pipelines from tankers, may end up being quite expensive.

Carbon dioxide can also be pumped into geologic formations, either below the ocean floor or below the surface of the dry land. For instance, in 1996, four million metric tons of carbon dioxide removed from natural gas deposits in the North Sea was pumped into a thick sandstone layer nearly a kilometer below the ocean floor rather than releasing the CO_2 into the atmosphere. Similarly, it has been suggested that carbon dioxide could be pumped into other porous rock formations around the world, or even into depleted underground coal, oil, and gas deposits. Cost factors could be prohibitive, however, and more importantly, there are concerns over how much such carbon sinks may "leak" in both the short term and the long term. There also may be a much more limited capacity for sequestering carbon artificially than is released, particularly in certain regions, and the costs of transporting carbon dioxide to other regions with more storage capacity, at least in the short term, could be extremely expensive. Furthermore, sequestration of carbon may at best be a finite solution—ultimately, the global capacity to store excess carbon will be reached. In the very long term, our best bet may be to pursue energy efficiency measures and cleaner fuel sources rather than continuing to rely on fossil fuels and then attempting to find ways to safely dispose of the carbon dioxide.

Global Light Pollution: A Developing Problem?

Besides stratospheric ozone depletion and the potential of global climate change, there is another issue that is gaining increasing attention and appears to have the potential to affect the atmosphere around the world. Many people in urban, suburban, and even relatively rural areas have never seen a truly dark night sky due to light pollution. **Light pollution** is generally considered to be the excess "waste" light given off by outside light sources at night, such as street lights, illumination around buildings, automobile headlights, and so forth, plus such sources as light emanating outdoors from the interiors of illuminated buildings (**Figure 17-12**). In many communities, the sky at night is never fully dark, but rather glows (perhaps in shades of pink, orange, white, yellow, or gray) due to all of the excess light in the local atmosphere. The light further reflects off of clouds, dust, and atmospheric pollution. This phenomenon of light pollution is often referred to as "**sky glow**," and many people now grow up thinking that this is how the sky naturally appears!

For many years astronomers in particular, both amateur and professional, have bemoaned light pollution. Light pollution can make it very difficult to see celestial objects at night, even with sophisticated, high-powered telescopes. It has been estimated that under natural conditions (without light pollution), on a clear, cloudless night, one can observe with the unaided eye the Milky Way and over 10,000 stars. However, today in many urban areas, the residents can no longer identify the Milky Way and, even on the clearest night, a big city dweller might see only around 100 stars. It is reported that during a power outage in Los Angeles during the 1990s many residents were actually afraid of the strange illuminated "cloud" in the night sky that they had never seen before. In fact, they were observing the Milky Way for the first time.

But does it really matter if we can see the stars at night? Is this just a minor aesthetic nuisance for some people? Isn't it better to have well-illuminated streets and buildings so that people can see, traffic accidents can be avoided, and burglaries and violent crime reduced? For millennia, humans have "feared the dark" and attempted to stave off the night; it is not without reason that the invention of the incandescent light bulb was heralded with such fanfare. But, some people contend, more light is not necessarily better. Unnecessary lights waste energy (estimated at a value of two to eight billion dollars a year in the United States, depending on how it is calculated). Much of the electricity comes from fossil fuels ultimately, and thus adds to the greenhouse effect. Excessively bright lights in automobile drivers' eyes and unwanted glare may actually cause some accidents, and artificial over-illumination at night may otherwise impair visual acuity and performance. Another concern is the right to privacy: Bright lights from elsewhere shining on one's property may be seen as violating privacy at night.

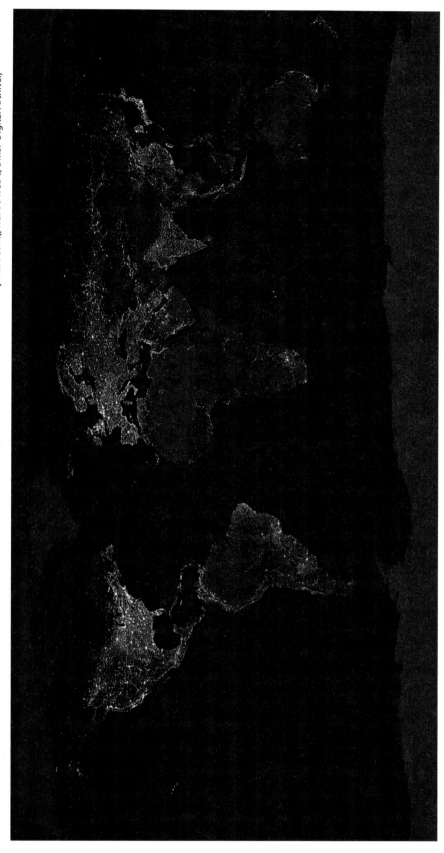

Figure 17-12 Earth at night, dramatically illustrating the extent of artificial lighting in populated areas around the globe. The above image is a composite of hundreds of photographs taken by orbiting satellites. (*Source:* C. Mayhew & R. Simmon [NASA/GSFC], NOAA/ NGDC, DMSP Digital Archive.)

There may be many more subtle negative ramifications of light pollution, effects that we are only beginning to become aware of. Many people feel that they cannot get a good night's sleep in some cities. Artificial illumination shining into the bedroom may be so intense that normal blinds and shades do not block it completely. There is research indicating that artificial light (even at very low levels over extended periods of time) can disrupt natural hormone production and circadian rhythms. Circadian rhythms are physiological cycles with an approximately 24-hour periodicity in humans. This may lead to a poor night's sleep, and ultimately to severe sleep deprivation over time, which can reduce waking-hour mental and physical performance and quality of life. The disruptions can also induce illness and disease (including, possibly, the risk of certain types of cancer).

Effects on Animals and Plants

The effects of light pollution on flora, fauna, and entire ecosystems may be profound, but we are just beginning to understand such issues. Artificially produced sky glow can affect areas that are far away from obviously visible artificial lights, and the sky glow may in turn disrupt wild animals and plants. The migrations of birds may be impacted, and birds may be attracted to artificially lit areas (much to their detriment, in some cases being killed by electrocution on electrified towers). The cycles of many nocturnal species, such as various mammals, frogs and other amphibians, and even aquatic worms, may be disrupted. In Florida, sea turtles lay their eggs on the beaches, and when the baby turtles hatch, they instinctually move toward the open ocean at night, apparently drawn by the light of the moon and stars reflecting off the ocean. Now, with the development of hotels and other well-lit structures along the coast, the hatchlings become disoriented and may migrate inland only to die. In another example of the disruptive effects of light pollution, it was found that tiny aquatic creatures known as phantom midges migrate to the surfaces of lakes at night to feed on algae. Now there is evidence that sky glow can prevent the phantom midges from coming as close to the surface at night as they would under natural conditions, and there is fear that this could disrupt a delicate ecological balance in some lakes. If the midges do not eat enough of the algae at night, an excess of algae may grow, disrupting other organisms in the lake and ultimately the entire local ecosystem. Light pollution may also affect the health and life cycles of various plants, which of course are the producers of our terrestrial ecosystems upon which all other life is dependent. At this point in time we have too little information to fully comprehend what the long-term effects of light pollution may be.

Curbing Light Pollution

Light pollution can be addressed in many ways. The obvious "solution" is to simply turn off unnecessary lights, but in many urban settings night illumination is considered necessary for safety, security, and economic reasons (people work, shop, or engage in leisure activities after dark—especially in the winter). In many situations, however, current light illumination is poorly directed and as a result a good percentage shines into the sky causing light pollution without benefiting anyone. Shielded lights are available that shine down in a directed manner so as to illuminate only the areas meant to be lit. Appropriate lighting, not too bright or harsh for the purpose at hand, can be utilized. Ordinances can be passed against one property owner shining unwanted light on another property (violations of which are sometimes known as "Light Trespass").

SUMMARY

- Two major environmental issues, by their nature, are global in scope: stratospheric ozone depletion and global climate change (global warming).
- Ozone naturally occurs in the stratosphere and shields Earth's surface from dangerous levels of ultraviolet (UV) radiation.
- Declining stratospheric ozone concentrations were first documented in the 1970s and are linked with increased levels of human-produced gases such as chloro-

fluorocarbons (CFCs), hydrochlorofluorocarbons (HCFCs), and halons.
- Initial international action to protect the ozone layer culminated in the 1987 Montreal Protocol, which was amended several times during the 1990s.
- As a result of international agreements, CFCs and other ozone-depleting chemical production and use have been greatly curtailed and, it is hoped, will be almost completely eliminated within the next decade or two.

SUMMARY–cont'd

- It is estimated that ozone depletion has peaked and the ozone layer will return to "natural" conditions by the middle of the century.
- Increased UV radiation on Earth's surface can adversely affect human health (causing skin cancer and cataracts), cause widespread damage among ecosystems, and also affect climates by modifying temperatures at different levels in the atmosphere.
- Greenhouse gases, such as carbon dioxide, methane, and nitrogen oxides trap infrared radiation (heat) and thus increase the temperature near the surface of Earth.
- Over the last two centuries or so, humans have measurably increased the atmospheric concentrations of carbon dioxide and other greenhouse gases, such as by burning fossil fuels.
- Although the reality of human-induced global warming continues to be questioned by some researchers (and Earth has experienced natural fluctuations in climate over geologic time), the consensus view among climate scientists is that global warming is real.
- Apparent and projected consequences of global warming include changing weather patterns (including more violent weather and altered rainfall patterns), shifting of climatic zones (which will affect ecosystems and the biodiversity they contain), melting of glaciers covering both continents (e.g., Antarctica) and in mountains, and rises in sea level.
- The best ways to address human-induced global climate change are to reduce the release of carbon dioxide and other greenhouse gases into the atmosphere (for example, by reducing the use of fossil fuels) and remove carbon dioxide from the atmosphere (for instance, encourage reforestation of deforested areas).
- On an international level, negotiations over reducing emissions of greenhouse gases have been slow—the Kyoto Protocol (1997) aimed at reducing greenhouse gas emissions has not yet been ratified by enough nations to come into force.
- Light pollution caused by artificial illumination may be a growing problem that has the potential to adversely affect not only human health and well-being but also may disrupt natural ecosystems.

KEY TERMS

Aerosol spray
Carbon dioxide
Carbon emissions
Carbon tax
Cataracts
Chlorine
Chlorofluorocarbon (CFC)
Framework Convention on Climate Change (UN FCCC)
General Circulation Model (GCM)

Greenhouse effect
Greenhouse gas
Heat
Ice age
Infrared radiation
Intergovernmental Panel on Climate Change (IPCC)
Light pollution
Little Ice Age
Melanoma
Methane

Montreal Protocol
Mt. Pinatubo
Nitrogen oxides
Ozone
Ozone layer
Radiation
Sea level rise
Sky Glow
Stratosphere
Troposphere
Ultraviolet radiation (UV)

STUDY QUESTIONS

1. What is the history of ozone layer deterioration? When was the problem first discovered? How long did it take for the world community to be convinced that action should be taken?
2. What are CFCs?
3. Why were CFCs invented? What applications did they have?
4. How do CFCs damage the ozone layer?
5. Describe how humans will be able to manage without CFCs and other ozone-destroying chemicals.

STUDY QUESTIONS–cont'd

6. Name some other chemicals that are destructive to the ozone layer. Are they naturally occurring or artificial?

7. What is the importance of the ozone layer? What effects does significant thinning of the ozone layer have?

8. Discuss the Montreal Protocol and its subsequent amendments.

9. How do the problems of ozone destruction and global warming differ from one another? In what ways are they similar?

10. What is the nature of the controversy surrounding potential global warming? Is the evidence for substantial global warming convincing to you? Why or why not? What is the evidence supporting global climate change?

11. What are some of the major greenhouse gases?

12. Describe how the greenhouse effect promotes global warming.

13. Which countries are currently the biggest contributors to potential global warming?

14. How fast have greenhouse gases been accumulating in the atmosphere during this century? Based on the work and projections of various climatologists, how much might the Earth heat up? Would the heating be evenly distributed over the Earth's surface?

15. List some of the potential consequences of significant global warming.

16. What can, is being, and should (in your opinion) be done to combat global warming? How serious do you think the problem really is? (Be sure to fully justify your answer.)

17. If global warming is real, do you think we should "learn to live with it," try to reduce global warming by reducing greenhouse gas emissions, or address the issue by developing more sinks for carbon dioxide and other greenhouse gases? Or should we use some combination of strategies to deal with global climate change?

18. What is light pollution? Do you think this is a serious issue in the area where you live? Why or why not? What, if anything, should be done to address light pollution? Justify your answer.

WHAT'S THE EVIDENCE?

1. The authors state that the consensus view of scientists who have studied the issue is that Earth's atmosphere is warming and at least a component of this warming is the result of human activities. Is consensus the way issues are ultimately decided in science, or should evidence and "facts" be the ultimate foundation upon which scientific opinions are based? In this case, what is the evidence for the assertion that global warming is real and at least in part caused by humans? Do you find the evidence convincing?

2. The authors suggest that one consequence of global climate change will be the spread of disease among human populations. What evidence and rationale is given in support of this contention? Do you find it persuasive?

STUDY
GUIDE

CALCULATIONS

1. Given that the concentration of CO_2 in the atmosphere increased from 317 ppm in 1960 to 371 ppm in 2001, what percentage increase occurred over this period?

2. Given that the concentration of CO_2 in the atmosphere increased from 250 ppm in 1850 to 371 ppm in 2001, what percentage increase occurred over this period?

ILLUSTRATION AND TABLE REVIEW

1. Referring to Figure 17-7 (Global average temperatures, 1880–2001), what is the lowest average global temperature recorded during this period, and in what year did it occur? What is the highest average global temperature recorded during this period, and in what year did it occur? What is the difference between these two extremes?

2. Referring to Figure 17-8 (Variations in average global temperature over the past 20,000 years), approximately how much have average global temperatures increased since the end of the last ice age? When did the last ice age end?

3. Referring to Figure 17-8, how much have average global temperatures fluctuated between approximately 3000 years ago and 150 years ago? Is this variation, from the warmest temperatures to the coldest temperatures, more, less, or comparable to the variation that has occurred over the last 150 years?

4. Referring to Figure 17-6 (World carbon emissions from fossil fuel burning, 1950–2001), global annual carbon emissions in 2000 were approximately how many times greater than those in 1950?

WWW.ENVIRONMENT.JBPUB.COM

www.environment.jbpub.com

Connect to this book's web site: http://www.environment.jbpub.com/ The site features eLearning, an online review area that provides quizzes, chapter outlines, and other tools to help you study for your class. We've posted the answers to the book's odd-numbered questions. You can also follow useful links for more in-depth information. The site provides sources for the information in this chapter and recommendations for further readings.

Locally, regionally, nationally, and globally, solid and hazardous wastes continue to be of concern, and as the global population increases, even more wastes are produced each year. This is not a problem with an easy solution. Probably the best way to address waste output is to decrease resource input; for instance, by increasing the efficiency of manufacturing processes and decreasing unnecessary packaging of consumer items. Shown in this photograph is a waste disposal facility for a Superfund site (circa 1992) in New Bedford, Bristol County, Massachussets. The sign specifically warns people to keep out of the facility due to the high levels of hazardous wastes, including PCBs, found on the site.

CHAPTER 18

Municipal Solid Waste and Hazardous Waste

The amount of waste generated each year is staggering. No one really knows how much waste humans generate, but much of it originates from the developed countries (poverty tends to generate less waste). Just as the United States is a leader in energy consumption and pollution, this country also produces the most waste per capita. Estimates of the amount of solid waste generated by the United States range from around 5.4 to 9 billion metric tons a year or more (depending on how "solid waste" is defined and how the estimates are calculated). Using the former figure, this amounts to about 19.2 metric tons of waste a year for every American or slightly more than 115 pounds (52 kg) of waste a day. Now the average person does not literally throw away 115 pounds of trash each day. Most waste in the United States comes from mining, agricultural, and industrial operations; but ultimately these operations exist to feed and provide for the necessities and desires of the consuming public. The trash that individuals put out for the garbage collector, the **municipal solid waste** (old newspapers, packaging materials, empty bottles, and so forth), makes up somewhere in the range of 2% to 4% of America's waste (**Figure 18-1**), but in 1999 (the most recent data that has been compiled), this still amounted to 208.7 million metric tons. The average American directly disposed of a little over four and a half pounds of trash and garbage each day, significantly more than just a few decades ago (**Table 18-1**) and also more than is generated per person in various other countries (**Figure 18-2**). However, of this 4.62 pounds discarded daily, 1.28 pounds were recovered for recycling or composting, such that only about three and a

Objectives

After reading this chapter, you should be able to explain or describe:

What constitutes solid waste

Basic approaches to solid waste management

Problems with incineration

Issues surrounding landfills

Ways to mitigate solid waste problems

Types of reuse and recycling, including their benefits

Hazardous waste

The Superfund

Toxics Release Inventory

TABLE 18-1	Generation, Materials Recovery, Composting, and Discards of Municipal Solid Waste in the United States, 1960–1999								
Pounds per Person per Day									
	1960	1970	1980	1990	1994	1995	1997	1998	1999
Generation	2.68	3.25	3.66	4.50	4.51	4.40	4.49	4.52	4.62
Recovery for recycling	0.17	0.22	0.35	0.64	0.89	0.94	0.97	0.98	1.02
Recovery for composting*	Negligible	Negligible	Negligible	0.09	0.18	0.20	0.25	0.27	0.26
Total Materials Recovery	0.17	0.22	0.35	0.73	1.06	1.14	1.22	1.25	1.28
Discards after Recovery	2.51	3.04	3.31	3.77	3.44	3.26	3.27	3.27	3.33
Population (thousands)	179,979	203,984	227,255	249,907	260,682	263,168	267,645	270,561	272,691

*Composting of yard trimmings and food wastes. Does not include mixed MSW composting or backyard composting. Details may not add to totals due to rounding.

Source: Environmental Protection Agency.

third pounds per day was actually discarded (refer to Table 18-1). From this point of view, final discards per person per day in the United States were about the same in 1999 as they had been in 1980 even as waste generation rose by a pound per person per day over the same period.

Defining Solid Waste

Solid waste, broadly defined, includes a number of items that are not generally thought of as "solid." Certainly, household **garbage, trash, refuse,** and **rubbish** are all solid waste, but so too are solids, various semisolids, liquids, and even gases that result from mining, agricultural, commercial, and industrial activities. Often substances such as liquids and gases are confined in solid containers and disposed of with more conventional solid wastes. Sewage, effluent, and wastewater from commercial enterprises, organizations, and private homes are not solid waste, but once wastewater is treated and various residues are removed from the water to form sludge, the sludge is usually treated as a form of solid waste. Solid waste may be divided into two broad categories depending on its origination: municipal solid waste (produced by various institutions, businesses, and private homes) and **industrial solid waste.** Another useful distinction is between **hazardous waste** (see page 462) and **nonhazardous waste.**

It is interesting to look at the composition of typical American municipal solid waste (**Figure 18-3**). The largest single component is paper products; other significant components include yard waste (which can easily be composted), food waste, plastics, and metals. If we analyze the waste in terms of products discarded (Figure 18-3b), the single largest component at 33.1% by weight is containers and packaging (including paper, plastic, and metal packaging), followed by nondurable goods (including such items as newspapers, telephone directories, clothing and textiles, and various consumer electronic products that have a short lifespan), durable goods, yard waste, and food waste.

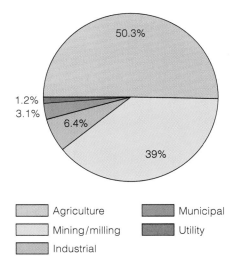

Agriculture — Municipal
Mining/milling — Utility
Industrial

Figure 18-1 A breakdown, by percentage, of the approximately 5.4 to 9 billion metric tons of solid waste generated annually in the United States. (*Source:* Environmental Protection Agency.)

Figure 18-2 Annual municipal waste generated per person in selected developed nations. (Note that the data in this figure are for comparative purposes only; more recent data across all countries was not available. Due to differences in methods of estimating and accounting, the annual waste generated per person in the United States as shown here does not precisely agree with the data given in Table 18-1.) (*Source*: Data compiled from World Resources Institute, *World Resources 1992–93* [New York: Oxford University Press, 1992], p. 319.)

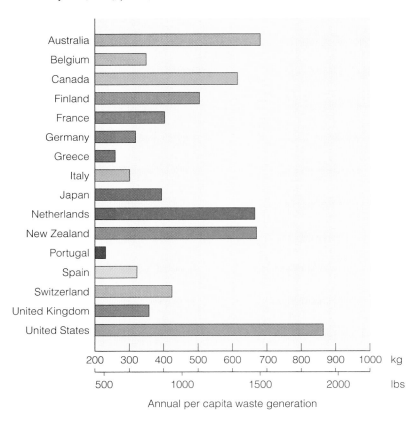

Alternative Paradigms for Waste Management

For most of human history, wastes were usually disposed of by a "dilute and disperse" strategy. Early gathering and hunting cultures simply left their trash where it fell, and moved on. In the Pre-industrial and early Industrial Age, settlements, communities, and factories were often located near waterways. Streams and rivers not only supplied fresh water from upstream, but also provided a convenient way to get rid of wastes. Waste materials were dumped into the river and washed away. As the water flowed downstream, the wastes and pollutants became less concentrated and were also naturally filtered out or absorbed by organisms and sediments. Wastes were often simply dumped into the seemingly infinite reservoirs of the atmosphere and the oceans. Most rivers eventually run to the sea, and coastal cities could dump raw sewage and waste directly into the ocean. Tall smokestacks could inject waste, in the form of gases and fine particulate matter (smoke), into the skies.

At the beginning of the Industrial Age, two hundred twenty years ago, many philosophers and social thinkers could not conceive of a time when, on a global scale, we would begin to run out of clean air and fresh water—but this is in fact just what is happening (as we examined in Chapters 15–17). Today, the pollution of industrial humanity is found throughout the globe, even in the most remote and isolated areas. It has become clear that the dilute and disperse approach to waste man-

Figure 18-3 The typical composition of municipal solid waste in the United States according to the weight of various types of materials. (*Source*: Environmental Protection Agency.)

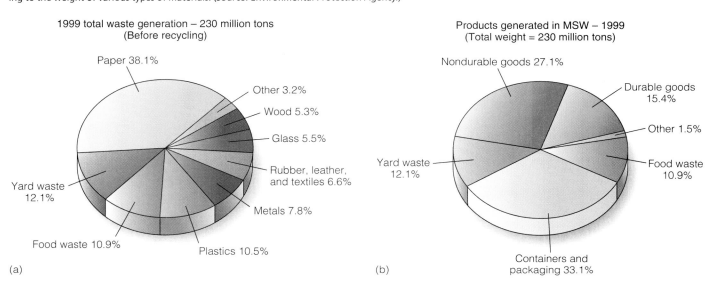

agement is no longer valid as a general paradigm. The capacities of Earth's natural ecosystems and reservoirs (sinks) for absorbing human-produced wastes are being reached and surpassed. In many cases, the natural systems have been so overloaded that not only are they incapable of dealing with all the human-produced waste hurled into them, but also they are actually damaged by the excess waste, with the result that they can no longer function effectively and efficiently.

Humans have produced innumerable new types of waste that nature was never equipped to deal with, such as new elements (for instance, plutonium and other artificially produced elements and isotopes), synthetic compounds (CFCs, assorted types of plastics, synthetic pesticides, fertilizers, and so forth), and extremely concentrated quantities of substances that are found in nature only in very dilute form, such as heavy metals given off by factory operations. Even under the best of conditions, with certain types of waste, the dilute and disperse strategy will do only that—dilute and disperse—since natural systems lack the capability to absorb and nullify these wastes. Long-lived radioactive isotopes or synthetic compounds simply accumulate throughout the globe.

By the early twentieth century, it was clear that the dilute and disperse strategy had reached its practical limits, and a new concept was introduced—"concentrate and contain." Concentration of waste has been utilized to a certain extent for many millennia. North American natives concentrated oyster shells in huge middens, and the ancient cities of the Old World developed large urban trash heaps (which are now much sought after by archaeologists). In many cases, concentration was only the first step toward ultimate dispersal. Trash might be collected and brought to a dump to be burned, and the ashes then thrown into a river or harbor. The relatively modern concept of concentration and containment is to collect waste and then permanently isolate it from the rest of the environment. Sanitary landfills and hazardous and radioactive waste disposal sites are tangible expressions of the concentration and containment strategy of waste management. Often the dilute and disperse and concentrate and contain strategies are used in tandem, as in some incinerator operations. Rubbish may be burned in an incinerator, dispersing into the atmosphere a good percentage of its original volume (60% or more), but the remaining bottom ash and collected fly ash constitute a highly concentrated, and often toxic, waste product that must be isolated from the environment, usually by being permanently

contained in a specially designed **landfill**. (Bottom ash collects in the bottom of the incinerator, and fly ash is removed from the smoke in the incinerator's chimney.)

There are numerous problems with the concentrate and contain strategy, especially as it is applied to modern industrial wastes. Many modern wastes are synthetic or artificial in nature and are extremely hazardous, especially in concentrated form. Therefore, they must be contained securely for very long periods of time, essentially forever. Yet we have learned from experience that perfect containment is very difficult to achieve; even the best-designed landfills may leak, containers can break, accidents and natural disasters happen, and there is always the risk that toxic substances in containment will be purposefully released, for instance, by a terrorist group. Besides these practical problems, containment is also subject to deeper criticisms. Is it fair and right for one generation to use resources for its own advantage and then remove them forever from use at the expense of all future generations? It can be strongly argued that ethically we should not be depriving future people of valuable material resources by converting those materials into hazardous or toxic waste that must be taken out of circulation permanently.

As a result of such considerations, a third strategy toward waste management has been promoted, variously referred to as "resource recovery," "industrial ecosystems," "sustainable waste management," and most optimistically "resource management." The basic philosophy underlying this strategy is that there should be no such thing as material waste, although the energy powering the system may produce waste heat. Under sustainable waste management, unnecessary waste (that is, unnecessary use of materials and energy) is first reduced at the source. Once materials (goods) are used for the designated purpose, all remains are reused or recycled. At the moment, however, we are a long way from achieving this goal. Even if the best reuse and recycling programs currently available were put into place globally, human-produced waste would still exist. A major change in the technology of manufacturing will be necessary if we are to eliminate most or all waste. But in the future, there may be no such thing as waste—only potential resources. Even our twentieth- and twenty-first century dumps and landfills may be viewed as raw resources (antique "urban ore") by the planners of the twenty-fifth century. Until that day comes, however, we will have to continue to address the question of waste management.

Traditional Means of Waste Management

For the past century, humans have dealt with solid wastes in three basic ways: (1) by burning the waste, thus essentially injecting much of it, converted to gases and smoke, into the atmosphere; (2) by storing wastes, including the leftover ash from burning, in dumps, impoundments, and most recently sanitary landfills; and (3) by injecting or burying wastes in rock cavities deep underground (a method proposed for the disposal of industrial and conventional toxic or hazardous waste, as described below, and also for radioactive waste—see Chapter 7). Each of these waste disposal methodologies has its proponents and its critics. In the next few sections, we will consider each briefly.

Incineration.   In the industrial technique of **incineration** (**Figure 18-4**), trash and garbage are burned in a large furnace at high temperatures to get rid of as much of the refuse as possible. Of course, burning trash is a time-honored procedure, but use of large incinerators dates back only to the late nineteenth century.

During their first 50 years of existence, incinerators were in and out of fashion. Many early incinerators were relatively inefficient, caused massive pollution, and left large quantities of ash and other nonburnables. But by World War II, some seven hundred new and improved incinerators were operating throughout the United States, on both a large and a small scale. Some apartment buildings even had small incinerators to burn the residents' trash. Nevertheless, incinerators continued to cause problems. Aesthetically, incinerators were an offensive intrusion on the skyline, and people who lived near them complained of the odors and said

Figure-18-4   A trash incineration plant.

the smoke and gases caused respiratory problems. As early as the 1950s, many incinerators were shut down. The increasing environmental awareness of the late 1960s and early 1970s continued to erode people's confidence in, and tolerance of, incinerators. The Air Quality Act of 1967 and the Clean Air Act Amendments of 1970 established new emission standards that many existing incinerators did not meet; most operators simply closed their incinerators rather than adding costly emission control devices.

Shutting down the incinerators meant that the trash and garbage had to be disposed of some other way. The preferred alternative was sanitary landfills. By the late 1970s and early 1980s, many cities and municipalities were finding that their landfills were running out of space, and there were fewer and fewer sites available, at least politically. An apparent solution to this predicament was presented by a new breed of incinerators known as *resource recovery plants*.

The basic idea behind **resource recovery facilities** (or "plants") was that trash and garbage would be burned and the heat generated could be recovered and applied to some useful end, such as generating electricity. Resource recovery facilities, also commonly referred to as *waste-to-energy facilities*, currently take two basic forms: refuse-derived fuel facilities (RDFs) and mass-burn incinerators. In an RDF, the solid waste is fed into the plant on a conveyer belt. Then, through a series of mechanical operations such as pulverization, shredding, sieving, density separation, and magnetic separation, the waste is sorted into various components—paper, wood, and other combustibles; iron, steel, and other magnetic metals; aluminum; glass; and so on. In this manner, recyclable materials can be recovered from the waste, while the shredded combustibles can either be burned on-site to generate electricity or formed into a fuel (often in the form of pellets) that can be sold to conventional coal-fired plants to be burned in place of coal. RDFs are often expensive to operate because of the complicated machinery involved, and also because of the fact that there are often large fluctuations in the markets for the recyclable materials and fuel produced by such a plant. Furthermore, the refuse-derived fuel cannot always be sold easily because it may not be of consistent quality.

Mass-burn incinerators take a more direct approach to the waste (**Figure 18-5**). The unsorted trash and garbage are simply fed into a furnace that burns the refuse at very high temperatures (around 1800° to 2000° F, or 980° to 1100° C). The heat from the burning refuse is used to produce steam that drives a turbine to generate electricity. Whatever is not burned in the incinerator is removed and simply disposed of (for instance, in a landfill).

Problems with Incineration.   Incinerators not only burn trash, reducing it to ash and gases, they give off—indeed manufacture—huge amounts of toxic pollution. As an example, a typical 1990s state-of-the-art incinerator (one that may well be in use today) that meets all regulatory requirements burning 2042 metric tons of refuse a day could emit 4.5 metric tons of lead and 15.4 metric tons of mercury per year, as well as 263 kg of cadmium, 263 kg of nickel, 2040 metric tons of nitrogen oxides, 774 metric tons of sulfur dioxide, 705 metric tons of hydrogen chloride, 79 metric tons of sulfuric acid, 16 metric tons of fluorides, and 89 metric tons of small dust particles. Perhaps most dangerous of all, incinerators produce and expel dioxin.

Dioxin (actually not a single substance as the term is commonly used, but a group of over seventy-five related compounds) is considered by some experts to be one of the most toxic chemicals ever manufactured by humans. Exposure to dioxin can cause skin eruptions (which resemble acne), headaches, dizziness, digestive disorders, birth defects, and various forms of cancer. Dioxin was an unintended contaminant of the defoliant Agent Orange that was used during the Vietnam War. It is formed as a byproduct in the production of some herbicides, certain bleached papers, and other products. Dioxin can form inadvertently in many situations. In particular, dioxin can be formed when organic wastes containing chlorine are burned. Plastic containers, paper, and other trash typically found in municipal solid wastes can often give rise to dioxin when burned in incinerators.

Not until the late 1970s was it even realized that dioxin production from incinerators might be a problem. At first, many experts believed that any dioxin in trash would be destroyed by the high temperatures of incineration. In fact, this was essentially true: Dioxin is destroyed in the furnace of the incinerator, but dioxin is then synthesized from the chemicals present in the cooler parts of the incinerator beyond the furnace and rises up the smokestack.

Modern incinerators are equipped with all sorts of mechanisms to capture the dioxins, heavy metals, and other toxic substances that would otherwise spew out of their stacks. The gases are subjected to various scrubbers, electrostatic precipitators, and fabric filters that catch the fly ash and toxic components of the gases. Even in the best cases, however, a commercial incinerator that meets all regulations can release substantial quantities of dangerous substances through its stacks. Furthermore,

after incineration, a substantial bulk of both bottom ash and fly ash remains to be disposed of—usually in a landfill or the equivalent. Some bottom ash may contain toxic levels of contaminants, but in other cases, it may be safely deposited in an ordinary landfill or even used to make building materials such as cinder blocks. The fly ash, which contains the dioxin, heavy metals, and other toxic substances that were removed before the gases left the stacks, is typically much more hazardous than the bottom ash.

This hazardous material must be carefully isolated from the environment. In some cases, it is chemically stabilized and hardened to a rocklike consistency before being deposited in a specially designed landfill. Unfortunately, abuses have occurred, to everyone's detriment. For instance, incinerator ash containing toxic substances has been improperly dumped in older landfills from which the toxics escaped into the general environment.

Due to the regulations involved, incineration of waste is generally the most expensive form of waste disposal, costing much more per ton of waste than landfilling (landfilling generally costs between $60 and $270 per metric ton, whereas land-based incineration costs range from about $400 to $550 per metric ton). At present, the construction of a typical mass-burn incinerator that can dispose of 1815 metric tons of garbage a day will cost about $250 million. Assuming a rate of garbage production of 4 pounds (1.8 kg) per person per day, such an incinerator would service about a million people. To avoid many of the regulations that are applied to

Figure 18-5   The interior of the RESCO (Refuse Energy System Company) refuse-to-energy plant in Bridgeport, Connecticut. The crane is transporting waste from the refuse pit to feed the high-temperature furnace.

land-based incinerators, large amounts of material (especially waste liquids such as oils and solvents) are regularly burned on ship-based incinerators in the open oceans (ocean-based incineration can cost half the price of land-based incineration). This may cut costs by skirting some regulations that do not apply, or cannot be enforced, on the high seas, but, of course, the hazardous materials spewed out by ship-based incinerators still enter the biosphere.

Dumps and Landfills.   One of the most ancient, and most common, ways of disposing of solid waste is to simply pile it up in a convenient but out-of-the-way place. Accordingly, most cities, towns, and other populated areas have generally had open **dumps** on their outskirts. There, trash and garbage would simply be left to sit or to be picked through by garbage pickers and sifters. In other cases, liquid or semiliquid wastes were contained in artificial or natural lakes or impoundments. But open dumps and impoundments tend to be associated with many problems, especially as they grow larger. Aesthetically, they can be smelly and unsightly. They can present serious health hazards, serving as breeding grounds for disease vectors and attracting pests. Dumps can significantly contribute to local air pollution, especially if the trash is periodically ignited and allowed to burn—as is done at some open dumps to reduce the volume of waste. Chemicals can also leak from an impoundment or leach from a dump, perhaps as it is rained upon, and the resulting runoff can seriously pollute the local surface and groundwater.

Despite the problems associated with open dumps, they remain an extremely common form of solid waste disposal, especially in developing countries. But even in the United States, open dumps are not hard to come by. A few of these dumps are officially sanctioned, but most open dumps in this country are relatively small and illegal—essentially, accumulations of rubbish in relatively deserted areas. The percentage of solid waste that actually ends up in American open dumps is quite small (Case Study 18-1 describes a type of solid waste that often ends up in large open piles).

Replacing the ancient dumps have been modern techniques of solid waste disposal, such as incineration and the lineal descendant of the open dump, the modern **sanitary landfill**. In the simplest sense, a modern sanitary landfill is essentially a closed dump (**Figure 18-6**). The solid and semisolid wastes are confined to a specific area, usually a giant hole in the ground; after being compacted, they are covered over daily with a layer of soil. Thus, the hole in the ground is filled ("landfill"), and the waste is isolated from the general environment by the periodic dirt covering, making the whole complex relatively "sanitary" as compared to open dumps.

The earliest recorded predecessors of sanitary landfills date to the early 1900s in the United States and to the 1920s in Britain. By the 1930s, sanitary landfills were catching on throughout the United States. In 1945, around 100 American cities had sanitary landfills; by 1960, approximately 1400 cities had adopted this method of solid waste disposal. Today, over half of U.S. municipal solid waste is disposed of in landfills.

Through their hundred years of existence, the philosophy and design of sanitary landfills have undergone major evolutionary shifts. From the beginning, the central idea was that once a landfill was completely filled, it could be covered over with a final layer of clay and dirt, and the "reclaimed" land devoted to some other purpose. This is the concept of sequential land use: What was "wasted" land in the past, perhaps a natural swamp or gully, becomes a place to deposit waste, and once the land has been completely filled, the reclaimed land can serve as the site for homes, a park, a factory, or any of numerous other uses. But over the decades, there has been a dramatic shift in thinking as to what land is suitable for landfills. In the early decades of landfill development, the most suitable and desirable locations were thought to be in swamps, marshes, and other low-lying wetlands. Wetlands were viewed as classic "waste land" that served no useful purpose: They were too wet to build on or grow crops, and were believed to breed only disease, insects, and

Figure 18-6   A sanitary landfill in the process of being capped.

other pests. By filling the wetlands with garbage interbedded with layers of soil, the land could be raised in elevation and made dry and usable. Of course, we now understand the importance of wetlands in the overall global scheme; far from filling them in, efforts are underway to preserve remaining wetlands. To make matters worse, building sanitary landfills upon wetlands certainly does not isolate the waste from the rest of the environment; in fact, the opposite is true. The natural wetlands provide an easy conduit for liquids, including extremely hazardous and toxic substances, to drain from the landfill and enter surface and groundwaters. Thus, a swamp or other wetland is just about the worst possible place to site a sanitary landfill.

Today, landfills are sited in areas where contamination of surface and groundwaters will not occur. Some of the best sites for sanitary landfills are in arid regions with little rainfall to produce leachates from the landfill. **Leachate** is formed as water percolates through the refuse, either from the top, such as from rain falling on the landfill, or laterally due to groundwater flow intercepting the landfill. Leachate is essentially an aqueous solution containing any chemicals and particles that can be dissolved, leached, or removed from the trash, including in some instances live disease-producing microorganisms. Leachate can be an extremely hazardous and concentrated brew of substances.

Modern sanitary landfills are usually begun by digging a huge hole at a suitable site. In some cases, a preexisting hole such as an abandoned coal or copper mine can be used, but usually not because the surrounding rock will be too permeable. Once the hole (which may be 50 feet [15 m] deep) is excavated, the next step is to line the sides of the cavity with thick layers of dense clay, a thick, sealed plastic liner, and sand and gravel. Many older sanitary landfills are not lined or have inadequate linings, but newer landfills are invariably lined. The liner serves as a barrier to the uncontrolled migration of the leachate out of the sanitary landfill. In the newest landfills, the leachate is not allowed to collect in pools at the bottom of the landfill; if it did, the pool would eventually overflow the sides of the landfill. Instead, pipes and drains in the bottom of the landfill collect the leachate so that it can be removed.

Most landfills now try to stay as dry as possible, so the leachate is removed, collected, and treated. The leachate may be treated like sewage, either being treated at a special plant on site at the landfill or sent to the local municipal sewage facility. Typically, the water is separated from the leachate, purified, and then released into a local river or the ocean. The remaining solid sludge may be dumped into the landfill once again, or it may be burned and then the ashes placed in the landfill. In some cases, the sludge can be used as a fertilizer, or it may be safely dumped into the ocean. In other cases, the leachate and sludge contain enough toxic substances that they are considered to be a hazardous waste and therefore must be disposed of in a suitable hazardous waste facility.

Besides leachate, sanitary landfills produce **methane gas** as certain organic components in the landfill are decomposed by bacteria. This methane must be removed from the landfill on a regular basis, or it could accumulate and form a hazard if it ignited. In some cases, pipes perforated with holes are inserted into wells drilled down into the landfill; the gas can then be collected from these pipes. Another method is that a lattice of perforated pipes may be included within the landfill as the refuse is accumulating. At some landfills, the methane is simply burned or released directly into the atmosphere (of course, contributing to the greenhouse effect—see Chapter 17), but at other facilities, the methane is collected, purified, and then sold as fuel.

Once a sanitary landfill is full, it is capped with a thick layer of compacted clay and soil. It can then be seeded, landscaped, and developed to serve any number of purposes. Parks, golf courses, and similar facilities are ideally suited for areas underlain by old landfills, but even houses or industrial buildings can be situated on these sites. The materials in the landfill may continue to compact for a number of years after closing, however, causing surface and subsurface subsidence—thus, it is unwise to build major structures on a fresh landfill. Moreover, methane gas will continue to form in the landfill for many years, as may leachate. The methane must continue to be removed long after the landfill has been closed. Likewise, the former landfill and the area adjacent to it should be closely monitored to ensure that no pollutants begin to escape.

About 57% of U.S. municipal solid waste is currently disposed of in landfills, 28% is recycled, and the remainder is incinerated or otherwise burned. Despite this continued high reliance on landfills, over the past two decades, many landfills have closed down, and it is increasingly difficult to open new landfills (due both to their expense and to the NIMBY [Not-In-My-Back-Yard] syndrome). In 1978, there were about 20,000 operating landfills in the United States; by 1988, the number had dropped to about 8000, in the early 1990s, it was less than 6000, and by 1999, the count had dropped to just over 2200 landfills. During the same period, however, the average size of landfills has increased.

CASE STUDY 18-1

Tires

Tires—car tires, bus tires, truck tires, and tractor tires—although only about 1.8% of the American municipal solid waste stream by weight, pose a major disposal problem (**Figure 1**). An estimated 300 to 500 million (numbers vary, depending on the source) used tires are stockpiled in America today, many of them in large, unsightly outdoor heaps. Every year, Americans dispose of an estimated 270 million car, truck, bus, and tractor tires. Many communities no longer allow tires to be discarded with other rubbish; tires are classified as "special waste," and disposing of them may cost $2 or more a piece. One of the problems is that tires contain pockets of air and are not easily flattened completely, so in older landfills where the tires were not properly disposed of (perhaps by weighting them) they can "float" up to the top of the landfill. In floating to the top, they can disrupt and destabilize the other components of the landfill, causing serious damage. They can even float up and break the clay caps that have been placed over landfills. At some older landfills that once accepted tires, the tires that have surfaced must periodically be skimmed off. Today, over 30 states ban whole tires from landfills.

Piles of tires are not just eyesores, but contribute to other environmental and health problems. The tires can catch on fire, producing billows of noxious black smoke and fumes.

The tire pile is mostly air, which promotes the fire, and the rubber in the tires burns at very high temperatures. As they burn, tires give off dangerous oily liquids. Putting water on the fire in an attempt to extinguish it can cause these liquids to spread, causing more environmental damage. Extinguishing the fire can be extremely difficult. Some tire fires have burned out of control for weeks or even months. A major tire fire in Hagersville, Ontario (Canada), in 1990 consumed 14 million tires and temporarily drove 4000 people from their homes. In 1999, a tire fire near Westley, California, burned for 30 days and spewed an estimated 8 million pounds of pollutants into the atmosphere.

Piles of tires also provide a breeding ground for rats, mosquitoes, and other disease-spreading pests. Water can collect in tires, providing the perfect incubation site for insects. Rats and other rodents make their homes in tire piles, especially if the piles are located near a garbage dump (as is often the case) that provides a ready source of food.

Fortunately, not all scrapped tires are simply sent to landfills or stored in unsightly heaps. Of the estimated 273 million scrap tires (84% from cars, 15% from heavy and light trucks, and 1% from aircraft, heavy equipment, and other sources), weighing over 3.25 million metric tons (3.6 million tons), generated in the United States in 2000, an estimated 196 million went to some kind of market for reuse, although that still left over 75 million tires that were simply "trashed." In 2000, the tires that went to market fell into the following major categories: used as fuel, 125 million; used in civil engineering projects, such as building erosion barriers or retaining walls, 30 million; tires processed into ground rubber, 18 million; tires exported out of the country, 15 million; tires cut up and made into new products, 8 million. Unfortunately, only a limited number of practical uses have been devised for used tires.

Perhaps the most obvious way to deal with the tire problem is to cut down on the number of tires entering the waste stream by making each tire last longer. Proper inflation, balancing, and periodic rotation can all extend the life of tires. Some brands and types of tires last significantly longer than other types, and retreading can extend the life of a used tire. Due to their initial expense, many bus and truck tires are commonly retreaded, but most individuals

Figure 1 A typical tire dump.

prefer to buy new tires for their passenger cars since new tires tend to cost little more than re-treads. In 1998, it was estimated that less than 28 million retread tires were sold in the United States and Canada combined. But even doubling, tripling, or quadrupling the life span of the average tire will not solve the tire disposal problem. With so many cars on the road, tens of millions of tires will continue to be added to America's waste stream each year.

Reclaiming and recycling the rubber from tires is difficult and expensive. The various synthetic rubber, glass, and steel layers in radial tires (there are 1.13 kilograms [2.5 pounds] of steel embedded in a typical steel belted radial passenger tire) make reclamation all the more difficult and generally impractical. Whether whole or cut up, tires have only limited uses. Some playground equipment is composed largely of used tires (**Figure 2**); tires are used on boat docks and piers as bumpers; sandals, floor mats, and other products can be made from cut-up tires; and rubber chips produced from chopped-up tires can sometimes replace traditional mulch in gardens. Whole, shredded, and chipped tires have been successfully used in the construction and

maintenance of landfills. Tires, weighted with concrete, have been used to start artificial reefs on coastlines (although not always with satisfactory results); likewise, tires have been used as a primary building material for breakwaters and erosion barriers on beaches. There has been some use of tires packed with cement as a structural material inside the walls of buildings. A promising large-scale secondary use for tires is in "asphalt rubber." Simply put, tires are ground up and mixed with asphalt to produce a material, asphalt rubber, which can be used to surface, repair, and restore roads. Million of tires have been used in asphalt rubber in both the United States (especially on federal highways) and Europe. Most secondary uses of tires, however, are rather limited (for instance, only so many playgrounds or breakwaters are needed) and do not solve the problem of what to do with the hundreds of millions of stockpiled tires and the tens of millions more entering the waste stream annually.

Tires can be burned as fuel in specially designed power plants; indeed, as you will have noted from the statistics cited above, this is the single largest market for used tires. Currently, there are approximately 70 facilities in the United States that use tire-derived fuel. Tires can also be burned with other materials in more conventional power-generating incinerators or as a source of heat and energy for industrial operations such as paper production. Critics of tire burning are concerned about the toxic air pollution and ash that it may produce. In many areas, tire burning has not found favor with the public.

Figure 2 This playground and park utilized 3000 recycled tires.

Questions

1. Make a list of possible uses for used tires—include both current uses that you are aware of as well as possible new uses.
2. Referring to your list of possible uses for old tires, which uses are the most environmentally friendly? Which are the least? Are some uses of tires worse than simply stockpiling them?
3. Do you advocate the burning of tires as a source of fuel? Or is it better to keep the carbon (and various other substances found in tires) sequestered rather than release it into the atmosphere?
4. How can we reduce the number of tires that are disposed of annually?

On a national level, according to the EPA, landfill capacity (ability to receive more wastes) has remained relatively constant and does not appear to be a problem, although locally a particular community may fill a landfill to capacity and then need to seek alternative waste disposal means (perhaps opening a new landfill, or shipping wastes to another landfill combined with more intensive recycling).

New EPA regulations are increasing the pressure on the nation's landfill capacity. The requirements include stricter guidelines for landfill liners, leachate collection systems, and groundwater and methane monitoring. This will increase the cost of opening and operating a landfill to an estimated average of $125 million and may force closure of many older landfills that cannot economically be upgraded to meet the new requirements. As of the mid-1990s, the estimated average remaining life expectancy of an American landfill was about 14 years.

Some researchers contend that the whole question of landfills running out of capacity is a political issue rather than a scientific problem for the United States. They argue that the only reason it is harder to site new landfills is the NIMBY syndrome—people do not like the thought of "disgusting" landfills in their communities. One economist has estimated that all of America's projected solid waste for the next 500 years could be easily fit within a single landfill 20 miles (32 km) on each side and 100 yards (91.5 m) deep. This would certainly be a huge landfill, but it would represent only a small dot on a map of the continental United States. Another researcher has suggested that the ideal location for such a national landfill (or landfills) would be in the thick shales that cover most of western South Dakota.

Deep-Well Injection. Various types of industrial or hazardous wastes are sometimes disposed of by injecting them into deep wells drilled into the crust of the Earth. The basic idea is to pump wastes into rocks that are situated well below all freshwater aquifers, so as to avoid contaminating any groundwater supplies. A typical well might be hundreds to several thousands of yards (or meters) deep. For decades, the oil industry has used deep-well injection to dispose of liquid wastes, such as salty brines, that are often pumped out of the ground in association with oil production—the brines are simply injected back into the rock. But using **deep-well injection** to get rid of other types of wastes is associated with many potential problems and as a result is opposed by many environmentalists.

Natural earthquakes may divert groundwater flow and allow previously isolated injected waste to contaminate other bodies of groundwater. This could ultimately lead to the spread of the waste and contamination of drinking water. Furthermore, the increased fluid pressure that results from injecting liquid waste deep underground may actually initiate earthquakes. Apparently, the increased fluid pressure allows rocks to move or slide along pre-existing joints and fractures. Earthquakes caused by the injection of wastes into deep wells have been reported in Colorado, Texas, Utah, and California. In another case, the chemical wastes that had been pumped down a well blew back up and spewed into Lake Erie. Despite such accidents, proponents of deep-well injection point out that taken as a group, most operations have no such problems.

Another concern over deep-well injection is that what constitutes a "deep well" at present may not seem so deep in the future. As we deplete the readily available freshwater aquifers, we need to drill deeper and deeper for usable water. In the future, we may need to drill as deep or deeper than the injected waste, but in the process the injected waste may contaminate our efforts. Also, not all wastes should be injected into all rocks. Certain wastes may react adversely with some types of rocks or with the natural pore fluids that already occur in the particular rock.

In areas where deep-well injection of waste is used, once the waste is injected into the rocks and the well plugged, the job is not over. The injection field must be permanently monitored to ensure that the waste does not migrate out of the confining rocks into which it was injected, perhaps through the natural pores of the rock or along natural or artificial cracks, holes, or fractures. After many years, even a properly sealed well may begin to leak.

Garbage as a Source of Revenue for Poorer Communities

Some communities have found that an easy way to attract money and jobs is to take the garbage from other areas. A classic recent example is Kimball County, Nebraska, which allowed a hazardous-waste incinerator to be built in its jurisdiction because this represented a $60 million investment in the financially depressed region. Likewise, New York City shipped treated sludge to a 128,000-acre (51,800-hectare) ranch in Hudspeth County, Texas, which created 35 badly needed jobs for local residents.

On a global scale, waste is shipped from richer to poorer nations (**Figure 18-7**). In many cases, hazardous wastes in particular are shipped from industrialized nations, where tight regulations make the materials expensive to dispose of, to developing nations with looser regulations or lax enforcement of existing regulations. Such overseas shipments are hard to monitor, especially since the material may not be properly identified as hazardous waste. Millions of tons of hazardous waste are being shipped from the United States and Western Europe to Asia, Africa, Latin America, and Eastern Europe every year. Sometimes host countries are happy to take the trash; before German unification, East Germany willingly accepted wastes from West Germany and Denmark for a fee. In other instances, the disposal is totally surreptitious. In 1988, for instance, the Nigerian government discovered that hazardous wastes from Italy were being dumped in a Nigerian port. To stem the illegal movement of waste, the United Nations Environment Programme oversaw the development of a treaty, adopted by many nations in 1989, to regulate the international movement of hazardous wastes.

Although such waste exchange arrangements can be mutually advantageous, as already pointed out, critics argue that the short-term economic gains are more than offset by the long-term costs and hazards of accepting other people's waste. Ultimately, the waste may pollute and otherwise degrade the environment, causing a self-perpetuating cycle of decreasing property values and the movement of wealth from the area. In the case of a typical landfill or incinerator/landfill operation, eventually the space will be filled, and where will the accepting community be then? When the waste no longer flows in, neither will money and jobs; the community may be left with a pile of garbage that still needs to be monitored.

Dealing with Rubbish and Other Waste in the Near Future

Many nations, experts, and organizations (including the U.S. Environmental Protection Agency) agree on a common approach, at least in principle, to the problem of waste management. Often known as the "waste management hierarchy," this approach involves a list of options ordered from the most desirable to the least desirable along the following lines:

1. **Source reduction**. Reduce the generation of waste in the first place.
2. **Reuse** of products. For instance, washing and reusing beverage containers directly.
3. Recovery and **recycling**. Using "waste" as the raw materials of industrial processes, such as collecting old aluminum cans, melting them down, and using the recovered aluminum to manufacture new products.
4. Waste treatment and incineration. This may include recovering energy as the trash is burned.

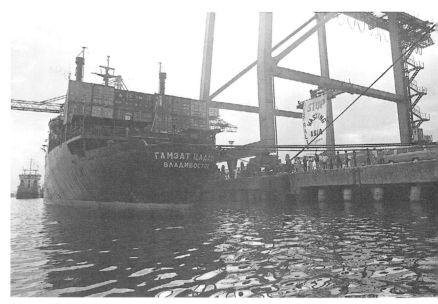

Figure 18-7 Containers of Australian computer waste being impounded by Filipino authorities in Manila.

5. Storage and disposal. The residual ash and solids of incineration or other waste treatment must ultimately go somewhere. In some cases, the material can be used to manufacture items, such as cinder blocks from certain types of incineration ash; as a last resort, it may be permanently disposed of, the ultimate resting place usually being a landfill.

Certainly, the most immediate, and ultimately most beneficial, solution to the solid waste problem is source reduction—utilizing less raw material in the first place. Source reduction can be accomplished in many different ways, including the use of alternative raw materials (such as substituting lighter and stronger aluminum for other materials in some goods), altering the product being produced, and changing manufacturing procedures. Ultimately, source reduction can be accomplished by simply consuming less. Many waste management experts suggest that the best way for the average citizen to help solve the waste management problem is to simply not purchase or use unnecessary goods. But source reduction alone can never solve all our waste problems. Humans require a minimum number of material products to survive, so appropriate source reduction must be combined with other objectives farther down on the waste management hierarchy.

Reuse is next on most people's lists of ways to deal with the garbage crisis. In some ways, reuse is really just a form of source reduction. Instead of making 50 soft-drink bottles, use one bottle

50 times. Yet, reuse has not captured the popular imagination to the same degree as recycling.

Lately, recycling has become increasingly popular, and most people view this as a positive development (recycling is discussed in more detail later in the chapter). But some environmentalists argue that as a solution to the refuse problem, recycling is deceptive and misleading and perhaps even contributes to the problem by diverting energy and resources that could be used for more beneficial endeavors—namely, actual waste reduction and reuse of goods. Instead, it is argued, we should change our basic consumption habits. We can use and reuse while consuming at a lower rate. For instance, virtually all food-carrying containers could be manufactured so as to be reusable—from the bottles that your beverages come in, to the containers for your cereal (although they may have to be manufactured from something other than cardboard). Reusing, if properly organized, should be no more inconvenient than a recycling or ordinary garbage collection program. Massive reuse would save energy and resources.

Waste treatment, such as incineration, and permanent disposal in landfills are on the bottom of many people's waste management hierarchy. Due to the air pollution and other dangers inherent in some forms of incineration, some environmentalists contend that landfilling and other forms of permanent storage should be placed before incineration in the hierarchy.

In contrast, many proponents of the "waste-to-energy" industry argue that incineration should be much higher on the hierarchy than it already is. Waste-to-energy incineration plants are touted as a form of "recycling"—heat from burning trash is converted to usable energy. Other people suggest that the whole notion of a "waste management hierarchy" is counterproductive. They argue that individual communities should be allowed to pursue an "integrated approach" to waste management, choosing the technologies that best suit local needs. A strong argument against this line of thinking is that waste management is no longer simply a local problem. The world must soon begin addressing this issue in concert if we are not to be overwhelmed by trash and pollution on the land, in our waters, and in the world's atmosphere.

Waste Disposal and Recycling in the United States

According to recent EPA statistics (for the year 1999, the latest currently available), in the United States as a whole approximately 57.4% of all municipal solid waste (by weight) is disposed of in landfills, about 14.8% is incinerated or otherwise combusted, and the remaining 27.8% is recycled. These numbers represent a significant change in municipal solid waste disposal approaches just over the last few decades; today a much higher percentage is being recycled (**Figure 18-8**). Besides reducing the waste stream, using recycled secondary materials decreases our energy use, water use, mining wastes, and air and water pollution (**Table 18-2**).

In the United States, four prominent categories of materials are commonly recycled: glass, metals (especially aluminum, steel, and iron), paper, and plastics. But not all recycling is equal. In just these four categories, we have both **closed-loop** (or nearly closed-loop) recycling, represented by the glasses and metals (these substances can be recycled completely and indefinitely), and various types of **open-**

TABLE 18-2	Environmental Benefits Derived from Substituting Secondary Materials for Virgin Resources			
Environmental Benefits	**Aluminum**	**Steel**	**Paper**	**Glass**
Reduction of:				
Energy use	90%–97%	47%–74%	23%–74%	4%–32%
Air pollution	95	85	74	20
Water pollution	97	76	35	—
Mining wastes	—	97	—	80
Water use	—	40	58	50

Source: Based primarily on data found in R. C. Letcher and M. T. Scheil, "Source Separation and Citizen Recycling," in W. D. Robinson, ed., *The Solid Waste Handbook* (New York: John Wiley and Sons, 1986).

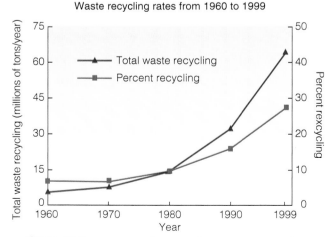

Figure 18-8   Waste Recycling Rates from 1960 to 1999 in the United States. (*Source*: Environmental Protection Agency.)

loop recycling with paper and to an even greater extent with plastics (these substances can only be recycled to a limited extent). As cases in point, we will take a brief look at the current status of recycling efforts for these four categories of materials.

Glass. In many ways, pure glass is the ideal substance to recycle for it is virtually 100% recyclable. Glass bottles and jars can be melted down to make new glass bottles and jars, and this process can be repeated over and over again in a virtually endless cycle without damaging the raw constituents of the glass—primarily silica-based sand (quartz) and a few other ingredients. Thus, under ideal conditions, glass recycling can form a closed-loop system: Only energy needs to be added to make new glass products from old. Recycling glass, as compared to producing new glass from virgin materials, entails enormous energy savings, decreases the amount of mining that needs to be carried out, and significantly reduces the amount of air and water pollution generated.

In the United States, approximately 27% of glass containers in municipal solid waste is recycled. Billions of glass jars and bottles are recycled every year, but more could be done. Theoretically, virtually all glass containers could easily be recycled. Not all glass products can be easily and directly recycled, however. Objects that contain glass and other substances (for instance, metals), such as lightbulbs, television tubes, and mirrors, are difficult to recycle.

Even when glass cannot be recycled directly back into similar glass products, it can be used for construction materials, such as fiberglass (although this usually requires stringent limits on the chemical composition of the glass). Crushed glass has also been used as a substitute for crushed stone in asphalt used in paving roads and other surfaces (sometimes known as "glasphalt").

Metals. Many metals offer the same recycling advantages as glass. Pure metals can be melted and reused over and over without damage to their substance. For instance, precious metals such as gold and silver, as used in jewelry or tableware, have been recycled in this manner for thousands of years.

In the United States, steel is the most commonly recycled material—more steel is recycled every year than all other materials combined. Over 100 billion pounds (45 billion kg) of steel are recycled in the United States annually (much of this, of course, is from sources other than municipal waste), and over half of all steel cans from the municipal waste stream are recycled; overall, an estimated two-thirds of all steel products are recycled. Steel is relatively easy to separate out for recycling, because it is magnetic, and producing new steel from scrap steel instead of virgin ore entails tremendous savings in energy and mining expenses, as well as considerable reductions in the pollution produced.

Aluminum is also easily and widely recycled. Beverage cans are the best-known example, but aluminum foil, pie plates and other disposable containers, and even parts of cars and appliances can be recycled. Producing aluminum from ore (bauxite) is a very energy-intensive process, so recycling can result in energy savings of up to 90% or more (and also reduces pollution given off during manufacturing by comparable amounts; see Table 18-2). Aluminum cans, trays, and other products are easily identifiable and 100% recyclable, forming the ideal closed-loop system. Currently, about 55% of all aluminum cans in the United States are recycled. Certainly, this is admirable, but there is little reason why we cannot recycle better than 90%.

Many other metals are also recycled in the United States, although mostly by specialized scrap metal dealers. Some of the more common metals include copper (from copper pipes, wiring, and other products), lead (especially from automobile batteries), zinc, nickel, titanium, and chromium. Over 90% of lead-acid automobile batteries are recycled. An emerging industry is the recycling and processing of old computers, including the circuit boards and chips, for valuable metals such as copper, nickel, cobalt, silver, gold, palladium, and platinum.

Paper. About 38% of America's municipal solid waste is composed of paper and paperboard products. Americans are profligate users of paper, using over 600 pounds (272 kg) of paper a year per person. For comparison, annual per capita paper use in parts of the former Soviet Union is 25 pounds (11.3 kg), and in China, it is only 2 pounds (0.9 kg)—of course, not everyone can get a daily newspaper in China. Fortunately, a very high percentage of all paper (perhaps over 40%) is recycled in some form.

Paper, especially cardboard and other corrugated material, is one of the more commonly recycled products in the United States. Indeed, about half of all corrugated paper materials are recycled. Many businesses that depend heavily on cardboard boxes for shipping and packing goods have been recycling the used cardboard for years, so networks for recycling cardboard are well established.

Printing and writing paper, computer paper, glossy paper, and newsprint are all recycled to various degrees in different parts of the country. Currently, some 59% of all U.S. newspapers are re-

CASE STUDY 18-2

Paper versus Plastic

Which is less damaging to the environment, paper or plastic? Which fills up our landfills? Which should you use over the other? Many grocery stores, for instance, give the shopper the option of paper or plastic grocery bags. Many people instinctively answer that paper is less degrading to the environment—after all, paper is made from natural plant fibers, is biodegradable, and is recyclable. In contrast, most plastic is manufactured using scarce petrochemicals (although there are a few plant-based plastics being developed), plastic is not always easily or economically recyclable, and, once manufactured, plastic may last virtually indefinitely. Virtually no common plastic products are truly biodegradable.

Yet in the United States, paper and cardboard products are the single largest component of municipal waste. Even though paper and wood products may theoretically be biodegradable, in most landfills they do not biodegrade; conditions within a landfill are generally not amenable to microorganismal growth. Furthermore, with continuing concerns over global warming, biodegradability is not necessarily a good thing. When an organic substance biodegrades, it is broken down, ultimately, into primarily water and carbon dioxide, and in many cases, significant methane is also given off. Carbon dioxide and methane are powerful greenhouse gases, and some researchers feel it is better to leave the carbon sequestered in solid form (in the form of nonbiodegraded paper, for instance) rather than release it into the atmosphere. Thus, for the environmentally aware consumer, the choice between disposable plastic and paper products is not always clear-cut. Here we will review some of the pluses and minuses of using paper versus plastic.

Tremendous numbers of trees are chopped down every year to be pulped and turned into paper products. Yet, as everyone knows, paper is "recyclable," so why do we have to cut down very many new trees at all? Why can't we just recycle a larger proportion of the massive amounts of paper that end up either deposited in landfills or burned in incinerators? A major hurdle to paper recycling is that the paper fibers are modified as they are used and reused. To be recycled, discarded paper must be deinked and converted to pulp before it can be manufactured into new paper. In the process of pulping, the paper fibers are broken. Therefore, paper made from recycled fibers will be composed of shorter fibers than the original paper, resulting in a lower quality paper of inferior strength. As paper is recycled over and over again, the fibers become progressively shorter, and the paper's quality and strength diminish. Thus, the value of paper made from progressively recycled fibers and its potential uses drop off sharply. At some point, the fibers become too short to produce serviceable paper for most purposes. For instance, newsprint or other paper that is too weak cannot be run through typical high-speed presses without tearing or other problems. Thus, even under ideal conditions, paper fibers cannot be recycled indefinitely, and a percentage of virgin wood pulp must be injected from time to time. For this reason, newspapers cannot always be printed on 100% recycled paper; rather, paper containing about 80% recycled material and 20% virgin fibers might be used. Yet newsprint made of 80% recycled fiber is certainly preferable to newsprint made of 100% virgin fiber. Also, there are many important uses for very low-quality, short-fiber paper—such as the ever-useful toilet tissue.

Of course, not only can paper be recycled, but many paper products can actually be reused. Paper bags, cardboard boxes, even paper from photocopy machines can be reused. Grocery bags, for instance, typically need not be discarded after one use. Manila envelopes and boxes can often be used to package and send things through the mail several times. Many times, photocopies are on one side only; why not photocopy on both sides of the paper (this not only saves paper, but saves space as well) or use the backs of old one-sided photocopies as notepaper? Rather than being insulted that someone writes to you on the back

cycled. Yet newspapers are still a major component of landfills. Fifty-three percent of high-grade office paper and 23% of magazines were recycled in 1999.

Recycling paper is not exactly comparable to recycling materials such as glass or metals, for paper recycling is ultimately not a closed-loop system. Paper cannot be used over and over endlessly, for the fibers eventually degrade and become unusable for most purposes (see Case Study 18-2). Realistically, many paper fibers can be recycled no more than six of "scrap paper," feel proud that the writer is environmentally conscientious enough to reuse the paper.

Compared to paper, plastics generally have fewer uses—and accordingly make up a lesser volume of our solid waste. But many of the common uses of plastic are identical to some of the uses of paper. Both plastics and paper are used heavily in packaging, indeed often in tandem (for instance, the plastic bottle with a paper label, the cardboard box or drink container that is plastic lined, the cardboard backing to an otherwise plastic container). Plastic bags can be used over and over again, often more times than a comparable paper bag. But once the plastic bag is torn or broken, it may be discarded—plastic cannot yet be recycled as easily as paper. Over the years, plastic bags, containers, and other items have become lighter and less bulky while retaining their strength, so once they end up in the landfill, they occupy even less room. On the other hand, when plastics (and paper in some cases) are burned in an incinerator, they may give off many toxic substances, including the deadly dioxins. Recently, much research has been devoted to developing reuse and recycling options for standard plastics, and truly **biodegradable plastics** may play a significant role in the future.

Many modern environmentalists suggest that when it comes to the question of paper versus plastics, there is no clear-cut answer. In fact, this may be the wrong question to ask. What is really much more important is to cut down on unnecessary usage and then reuse whenever possible. When you go to the grocery store bring with you your own stash of used grocery bags and use them again. Or, better yet, invest in a couple of durable cloth bags that you can carry with you and use over and over and over for many different purposes, whether it be buying groceries or carrying books. Ultimately, for many people there may be bigger and more important issues that they should consider instead of fretting over whether to ask for paper or plastic bags at the grocery store. How did you get to the store in the first place? Did you walk or ride a bicycle? If so, excellent. Did you take public transportation? Very good. Did you drive a super-low emissions and super-high mileage hybrid car? Well, good. Or did you drive your gas-guzzling SUV (sport utility vehicle) or out-of-tune clunker? If the latter is the case, there are more important things for you to worry about, from an environmental point of view, than your use of paper versus plastic. Put your priorities into order and find a more fuel-efficient, and more environmentally friendly, mode of transportation when you take that trip to the store. You need to look at the entire picture and keep various concerns in perspective.

Questions

1. So, the obvious question is, you are in a grocery store and offered a choice of either paper or plastic bags—which do you choose? Justify your answer. How much does your individual answer to this question depend on specific factors in your personal life at the moment (for instance, whether you can more easily reuse a paper or a plastic bag)?

2. Overall, from a pro-environmental point of view, do you think more emphasis should be placed in the future on paper, on plastics, or should the current balance by maintained?

3. Perhaps you would like to drive a state-of-the art, fuel-efficient, and super-low emissions hybrid vehicle, but you just cannot afford it. You are grateful to have your gas-guzzling, smoke-spewing clunker. What else can you do to help preserve the environment? Is there reliable and safe public transportation offered in your area? If so, do you use it? If not, perhaps you're stuck driving the clunker. Frankly and honestly assess your own situation and lifestyle. Is there more that you could personally be doing to help the local and global environment? Is there more that you are willing to do?

to eight times. To offset this effect, virgin material is often mixed with recycled fibers such that even "recycled" paper may contain a significant percentage of virgin fiber. When purchasing "recycled" paper, it is extremely important to distinguish between "preconsumer" and "postconsumer" recycled content. Preconsumer waste is primarily waste produced during the manufacturing of the paper, such as scraps from cutting sheets of paper; paper companies have traditionally reused this preconsumer waste.

Postconsumer content is made from other paper that was actually used for some purpose by the public and then returned to be recycled.

The paper recycling business has historically been a roller coaster. Sometimes old newspapers can be sold for a profit, but at other times, a hauler must be paid to take bundled newspapers away. In the late 1980s and early 1990s, some warehouses were full of paper waiting to be recycled. Well-meaning legislators had passed laws requiring the collection of paper for recycling without requiring the use of recycled paper, which would have developed a market for the product. Now, however, the federal government as well as many state governments require all paper used in government offices to have a recycled component. In the past, due to the public's distrust of recycled paper, paper companies were often unwilling to produce it. Today, however, the public is generally more accepting of paper containing recycled content. Indeed, domestically and globally there is a continuing strong demand for paper and paper products, and recycled paper is helping to fill the need. During the five years from 1990 to 1994, over 85 paper mills that could recycle paper products were built in the United States. In the early 1990s, old newspaper was of negligible value, but in the middle and end of the decade, recyclable newspaper averaged around $55 to $64 per metric ton and in some areas of the country, prices reached $160 a metric ton. Old newspapers and other scrap paper have become such a valuable commodity across the country that in some areas, theft of papers from curbside recycling bins has become a significant problem.

Using scrap paper to manufacture other paper is not the only way that paper can be recycled. For instance, recycled paper is already used to manufacture construction materials, such as wallboard, roofing paper, padding, and insulation. Some farmers successfully use shredded newspaper in place of straw for bedding material for animals. Although not exactly recycling in the traditional sense, paper can be either burned directly as fuel or converted into fuel pellets.

Plastics.   Traditionally, plastics have been the bane of environmentalists. Plastics make up only about 10.5% of U.S. municipal solid waste by weight, but pose many problems. Most plastics are synthetic compounds composed of polymers containing hydrogen, carbon, and oxygen (usually manufactured from petroleum and its derivatives). Typical plastics in use today are not biodegradable or otherwise readily broken down in nature; thus, they not only clog our landfills, but produce unsightly litter across

the landscape. When burned or incinerated, otherwise inert, and thus nontoxic, plastics can give off many toxic substances, including carcinogens such as dioxin.

The large-scale recycling of plastics faces many practical obstacles. For one, the many different types of plastic, which look very similar, must be sorted according to specific plastic resins. Some common products, such as squeeze bottles for food, are composed of several layers of different types of plastics, making it very difficult to separate them for recycling. Furthermore, even under the best conditions, most plastics cannot be recycled back to their original use. Common plastics can only be recycled in an open-loop system. Unlike a glass bottle, a plastic soda bottle cannot simply be melted down and made into another plastic soda bottle. Not only does the quality of the plastic diminish, but the temperatures at which plastics are melted and remolded are not always high enough to sanitize the plastic for use as a food or beverage container. In order to recycle plastics, secondary uses must generally be found for them.

To make it easier to identify different types of plastics for potential recycling, many plastic manufacturers have adopted a voluntary coding system that appears on containers, bags, and other products. This code takes the form of a number within a recycling triangle and generally an acronym below the triangle (Table 18-3). Of the categories of plastic listed in Table 18-3, PETE and HDPE are currently the most commonly recycled plastics, but the amounts of all plastic types that are recycled continues to increase, and demand for good recyclable plastic is strong. As of 1999, about 40% of plastic soft drink containers were recycled in the United States, but when it came to other plastics the rate was less impressive—a 9.7% rate of plastic recovery and recycling for plastic containers and packaging overall, and a 3.8% rate for plastics in discarded durable goods.

Despite the significant progress that has been made in recycling plastics, it should be emphasized that recycling plastics once is quantitatively and qualitatively different from the reuse of paper fibers half a dozen times (after which the fibers can theoretically be composted, as described below) or the virtually endless recycling without degradation of some types of glass and metal. Eventually, second-generation plastic products, such as lawn furniture made from recycled milk jugs, must be disposed of. Furthermore, today's typical plastics are manufactured from scarce fossil fuel resources in an energy-intensive fashion, often producing significant amounts of pollution as a byproduct. For these reasons many environmentally concerned citizens continue to look askance at plastics.

Composting. Nearly half of all yard wastes (leaves, grass clippings, and so forth) are composted. Still, many experts agree that **composting** is generally underutilized in the United States (**Figure 18-9**). About 23% of our solid municipal waste, by weight, consists of yard wastes and food wastes (a mere 2.2% of food wastes were composted in 1999), which are easily composted to make fertile humus or topsoil. It is ironic that such organics are often either buried in landfills, where they are permanently removed from the ecosystem and generally do not biodegrade to any considerable extent due to insufficient oxygen and/or moisture, or incinerated, when they could be used to produce natural fertilizer or topsoil—substances that are sorely needed for our agricultural lands (see Chapter 13). Furthermore, if not contaminated with heavy metals or other toxic substances, sewage sludge (consisting, in large part, of dead bacteria—sewage sludge is not the same as raw sewage) can be composted to make a natural fertilizer. In addition, the paper, wood, leather, textiles, and other biodegradable substances that make up about 50% of our municipal waste stream could also be composted. Composting these latter substances may not produce high-grade humus or topsoil, but it will provide a low-grade compost that could be used for nonagricultural purposes such as landfill cover, strip-mine reclamation, reforestation projects, or clean fill in areas along roads or construction sites. A fully implemented program of reuse, recycling, and composting, combined with separation and removal of hazardous waste for special disposal, would leave very little material to be placed in a landfill. Under such a system, landfills would be receptacles primarily for synthetic, nonrecyclable, and non-biodegradable substances such as certain types of plastics and synthetic textiles.

Composting has occurred in nature for hundreds of millions of years. It is a free service provided by the microorganisms of the soil. When leaves and other organic material fall to the ground in a natural ecosystem, they are "eaten" by microscopic decomposers that attack the organic matter and break it down into smaller molecules, which can then be recycled in the ecosystem. This is the natural process by which humus and soil form.

Figure 18-9   A community compost center.

TABLE 18-3	**Coding System for Plastics**
	Polyethylene terephthalate (also known as PET)—this is a transparent plastic (although it may be colored) commonly used to make two-liter soda bottles and other containers such as peanut butter jars. PETE is used in approximately 25% of all plastic bottles. It can be recycled into such items as strapping (for packaging), fiberfill for winter clothing, carpets, surfboards, sailboat hulls, and the like.
	High-density polyethylene—this plastic is commonly used to manufacture plastic milk jugs, bleach and detergent bottles, motor-oil bottles, plastic bags, and other containers. HDPE is used in more than 50% of all plastic bottles. It can be recycled into trash cans, detergent bottles, drainage pipes, base cups for soda bottles, and other similar items.
	Vinyl or polyvinyl chloride (also known as PVC)—this is used in the manufacture of vinyl siding, plastic pipes and hoses, shower curtains, and so forth; it is also used to make some cooking-oil and shampoo bottles, as well as bottles for some household chemicals. It can be recycled into fencing, house siding, handrails, pipes, and similar items.
	Low-density polyethylene—a plastic commonly used to make cellophane wrap, it is also used to manufacture bread bags, trash bags, and other types of containers. It can be recycled into grocery and garbage bags.
	Polypropylene—a lightweight plastic commonly used in packaging some foods (for instance, some margarine and yogurt containers) and for certain types of lids and caps. It can be recycled into car-battery cases, bird feeders, and water pails.
	Polystyrene—this is the substance that is commonly known as "styrofoam," used to make coffee cups, plastic peanuts for packing, egg cartons, meat trays, plastic utensils, video-cassettes, and so forth. It has been recycled into tape dispensers and reusable cafeteria trays.
	Other: Plastic resins other than the six basic categories listed above, as well as objects produced from several different types of plastics mixed together. Mixtures of plastics can sometimes be recycled into "plastic lumber" and used to manufacture benches, lawn furniture, picnic tables, marine pilings, and other types of outdoor equipment.

Virtually anyone can facilitate composting; it can be carried out on a small or large scale, in a low-tech or high-tech fashion. Many individuals and families have compost piles or bins in their backyards. Grass clippings and other yard refuse, perhaps along with kitchen scraps, animal manure, and some dirt, are piled together. As long as the pile is properly aerated (it helps to stir the compost or turn it over occasionally) and has sufficient moisture (in temperate climates, the rain provides all the moisture needed), the microorganisms will do their job. The organic matter will decompose to form good compost. During the composting process, the aerobic microorganisms generate heat (the interiors of backyard compost piles can reach temperatures of up to 160° F [70° C]) that naturally kills many pathogenic bacteria and weeds, thus "sterilizing" the resultant compost. In large-scale commercial or community composting operations, care must be taken that the compost does not overheat, killing the very microorganisms that promote the composting process.

Depending on the quality, compost has numerous uses. High-quality compost can be used to help replace lost topsoil on farms. The amounts of compost that could be used for this purpose are enormous; it takes an estimated 59 metric tons of compost to add just an inch (2.54 cm) of topsoil to an acre (0.4 hectare) of land. But then the potential for producing good compost is also large. Composting the organic wastes of a community of one million people could produce an estimated 544.5 metric tons of compost daily. Compost is also widely used in the nursery and landscaping industries; it is necessary for many types of land reclamation projects; and, of course, sanitary landfills need compost or soil between the trash layers and as a cap on top once the landfill is closed.

Recycling—Closing the Loop. Millions of Americans carefully sort their trash, then dutifully deliver the various categories of recyclables to a recycling center, or place the separated trash outside for curbside recyclable collection. Certainly, recycling will help address the solid waste problem in the United States, but sorting trash and taking it to a recycling center is only the first step. The materials must actually be recycled. Unbeknownst to many citizen recyclers, not all the material that is collected for recycling is actually recycled. The problem is not that the material could not potentially be recycled, but that markets for recyclables do not always exist.

While communities and the general public have been extremely eager to sort and collect recyclables, the capacity for processing recyclable materials has not always expanded accordingly. For instance, in the early 1990s, the market for used paper was so poor, millions of tons of it were accumulating in warehouses. Fortunately, as discussed above, this situation is now vastly improved. Similarly, mountains of plastic bottles and sorted glass sometimes await recycling. In September 1992, the company Waste Management of Seattle, Inc., had a reported 5445 metric tons of glass for which it could not find a ready market. In other cases, segregated paper and other trash intended for recycling have been incinerated or landfilled because there was no market for the material and storing it was uneconomical. However, it looks like the market for recyclables is now, in the early twenty-first century, gaining strength and momentum. If this trend continues, these problems will be a thing of the past.

Still, such situations have presented a dilemma for some recycling advocates who have had to decide whether to tell the public that their "recycling" efforts were for naught. Some leaders of the recycling movement have been known to practice deception, believing that it is important to encourage the recycling habit, even when the material is not recycled. They hoped that if the beginning of the recycling loop—sorting and collection—remained intact, the remainder of the loop—processing plants and markets—would eventually develop. Based on the current strength of the recyclable market, early indications are that this strategy has worked.

Strengthening the Market for Recyclables. Various ways of strengthening the market for recyclables have been suggested, not all of which have proved equally effective. Some local governments have mandated the collection of sorted recyclable materials, but collecting cannot lead to recycling if there are no markets for recyclables. In some instances, such short-sighted laws have actually driven trash collectors into bankruptcy when they were forced to spend extra time and money picking up recyclables but could not sell them.

Indeed, there is a popular misconception that the collection, sorting, and selling of recyclable trash should be a profit-earning venture for a community or at worst a break-even situation. In fact, recycling may be costly, at least until markets develop. But even if markets must be "artificially" created through tax incentives and other means, recycling can still pay off if the ultimate cost is less than or equal to the cost of waste disposal otherwise. Indeed, disposal of waste by any means is expensive. According to one estimate, the United States spends approximately $30 billion on municipal waste disposal annually.

If we are to increase the rate of recycling, both increased research and development and economic incentives will be needed. Although reuse and recycling have been carried out on a limited scale for ages, recycling of twenty-first century goods on a twenty-first century scale requires new technologies. New ways to sort, clean, and reuse materials will certainly be welcome. Perhaps even more importantly, technical innovations may allow products manufactured from recycled materials to be of equal or better quality than products produced from virgin materials. To close the recycling loop, products manufactured from recycled materials must be purchased by consumers. This means that the recycled goods must be competitive on the open market, from both a quality and a cost perspective. At present, some products manufactured from virgin materials cost less than they would were it not for various tax breaks and subsidies that apply to the virgin materials; instead, tax incentives and subsidies could favor products manufactured from recycled materials. Local, state, and national governments can give an economic boost to recycling efforts by mandating the use of certain minimum levels of recycled materials in various new products, thus creating a ready market for recycled materials.

Industrial Ecosystems

Natural ecosystems do not produce substantial quantities of material waste; the "waste" of one organism is the lifeblood of other organisms. Likewise, an **industrial ecosystem** does not produce waste; the effluents of one industrial process form the raw materials for another industrial process. Already, some companies are finding that they can sell their effluents for a profit, whereas previously they had to pay to dispose of the same effluents as "waste." For example, a steel-processing company may "recycle" the sulfuric acid utilized in the steel mills to manufacture iron sulfate compounds that can be sold to magnetic tape companies. Otherwise, the "used" sulfuric acid might simply be disposed of. A key component to the success of industrial ecosystems is communication. The company with a particular effluent must be able to locate the potential users of the material. Accordingly, information networks are being established in many countries (for example, France, Germany, and Belgium) and in various regions of the United States to put potential consumers of effluents in touch with the appropriate producers.

If the industrial ecosystem is to properly mimic natural ecosystems, virtually all materials must be reused/recycled. This means that not just manufacturing effluents but goods discarded by society at large must reenter the system as raw materials.

Due to the human penchant for treating symptoms rather than root problems, however, in some instances we are backsliding from the ideals of the industrial ecosystem. Take the typical automobile as an example: It is probably much more difficult to recycle all of the components found in a car today than it was in the 1920s. A modern automobile is manufactured from numerous different alloys of metal, including iron, steel, aluminum, nickel, lead, and copper, various polymers of plastic, rubber, cardboard, and so forth. The typical car has a very short life span and produces enormous amounts of waste during its operation; it not only spews pollution into the atmosphere, but runs through tires that typically end up in landfills (refer back to Case Study 18-1, page 450).

Eventually, the car is junked, and although many of its materials could be recycled, the sad truth is that often they are not. It requires a lot of time and effort to strip a car and separate its components for reuse or recycling. Furthermore, because of impurities, goods manufactured from the recycled materials may be inferior to goods manufactured from virgin materials.

As increasing numbers of vehicles have been put into service, they have produced escalating levels of pollution. A typical response to this problem has been to design ever more sophisticated cars that produce less pollution; that is, the symptom—pollution from car exhausts—is attacked. A much simpler, though surely less popular, alternative would be to attack the root problem—namely, too many cars.

In the 1970s, catalytic converters began to be widely used on the exhaust systems of automobiles to reduce various pollutants in the exhaust before they enter the atmosphere. But catalytic converters also have drawbacks; while helping to solve one environmental problem, they exacerbated another. Standard catalytic converters rely on the use of scarce platinum group metals. Before the extensive use of catalytic converters, platinum group metals were typically recycled at efficiencies of 85% or higher due to their rarity and value. Early catalytic converters were not designed with recycling in mind, so their introduction substantially reduced the rate of recycling among platinum group metals, deviating from the ideal of the industrial ecosystem. Fortunately, systems are now being developed to remove catalytic converters from old cars and recover the platinum group metals. Likewise, the metals from discarded computers and other electronic equipment are increasingly being recycled (but see Case Study 18-3, page 464).

Hazardous Waste

The Comprehensive Environmental Response, Compensation, and Liability Act (CERCLA), the federal statute passed in 1980 that established Superfund (see page 466), defines a hazardous substance as "any substance that, when released into the environment, may present substantial danger to public health, welfare, or the environment." CERCLA goes on to define extremely hazardous substances as substances that "could cause serious, irreversible health effects from a single exposure." Hazardous waste is essentially waste composed of such hazardous substances. Hazardous waste can originate from the home (such as household cleaners or pesticides), local or national government operations, agricultural use, industry, or other sources (Table 18-4). Hazardous waste often includes substances that are chemically reactive, corrosive, flammable, explosive, or toxic to living organisms (toxic materials are harmful or fatal when consumed by organisms in relatively small amounts).

There are three main categories of hazardous waste: organic compounds, inorganic compounds and elements, and radioactive waste. Organic compounds are carbon-based substances that also contain a substantial percentage of hydrogen and oxygen. Some organic compounds are naturally

formed molecules, some are derived from once-living material such as petroleum and other fossil fuels, and some are totally artificial substances synthesized in the laboratory. Many fertilizers, pesticides, organic dyes, plastics, and other substances are organic compounds. Theoretically, at least, organic compounds should be degradable into simple, nontoxic substances such as carbon dioxide and water (plus some residue). In reality, however, many are not easily decomposed in nature and persist for extremely long periods of time. Sometimes incineration can be applied successfully to break down and detoxify hazardous organic compounds; another approach is to use microbes that can chemically attack and degrade some such substances. Bacteria have been used with success to help clean up oil spills.

Inorganic hazardous compounds have little or no carbon, but they commonly contain heavy metals such as lead, mercury, cadmium, copper, arsenic, iron, aluminum, manganese, chromium, beryllium, nickel, selenium, zinc, silver, and others. Some heavy metals are necessary to life in small, or trace, amounts, but all are toxic to organisms in larger doses. All of these elements occur naturally in the environment, usually in low concentrations, and are cycled through the Earth's chemical and bio-

TABLE 18-4 — Health Effects of Selected Hazardous Substances

Chemical	Source	Health Effects
Pesticides		
DDT	Insecticides	Cancer; damages liver, embryo, bird eggs
BHC	Insecticides	Cancer, embryo damage
Petrochemicals		
Benzene	Solvents, pharmaceuticals, and detergent production	Headaches, nausea, loss of muscle coordination, leukemia, linked to damage of bone marrow
Vinyl chloride	Plastics production	Lung and liver cancer, depresses central nervous system, suspected embryotoxin
Other organic chemicals		
Dioxin	Herbicides, waste incineration	Cancer, birth defects, skin disease
PCBs	Electronics, hydraulic fluid, fluorescent lights	Skin damage, possible gastrointestinal damage, possibly cancer causing
Heavy metals		
Lead	Paint, gasoline	Neurotoxic; causes headaches, irritability, mental impairment in children; damages brain, liver, and kidneys
Cadmium	Zinc processing, batteries, processing fertilizer	Cancer in animals, damage to liver and kidneys

Source: Based primarily on data compiled by World Resources Institute and published in *World Resources 1987* (New York: Basic Books, 1987).

logical systems. The problem is that modern industry has mined and released much higher concentrations of heavy metals than the natural geochemical and biological systems are capable of readily handling. Therefore, once released into the environment, these substances tend to slowly accumulate, often to toxic levels. Once dispersed in the general environment, like all hazardous substances, they are very difficult to remove. In order to dispose of toxic inorganic compounds safely, they are sometimes combined with various chemicals that transform the material into a stable, cementlike block that can be buried or otherwise disposed of. The idea is that the solid block will resist leaching and thus not allow the inorganic compounds to escape into the environment. However, the best way to deal with such inorganic compounds is to not allow them to mingle with the general environment at all. Rather than being discarded, waste substances containing heavy metals should be recycled whenever possible (see Case Study 18-3).

Radioactive waste is the third generally recognized category of hazardous waste. We have already discussed certain types of radioactive waste in Chapter 7. Besides nuclear power plants, radioactive waste is generated by nuclear weapons manufacturing facilities, medical facilities (radioactive equipment and treatments), some research labs, and many common commercial, industrial, and home functions (for instance, the radioactive elements of home smoke detectors). High-level radioactive waste, from nuclear power plants and weapons facilities, is disposed of according to special regulations and handling procedures. But much low-level radioactive waste is considered to be below regulatory concern, and no special precautions are taken in disposing of it. In some cases, radioactive material has been incinerated along with other rubbish or been disposed of in municipal dumps and landfills.

Hazardous waste is a major problem because of both its inherent dangers and the amount that continues to be produced each year. No one really knows how much hazardous waste is produced in the world every year, but one estimate is that the United States alone produces more than 260 million metric tons annually. World production must be at least twice this amount. Much of this waste comes from the chemical, paper, and petroleum industries.

The best way to control the influx of hazardous substances into the waste stream is simply to reduce the amounts used in the first place and then reuse and recycle such substances whenever possible rather than disposing of them. Many manufacturing processes use or generate hazardous substances. In some cases, the amounts of these substances can be reduced by making the processes more efficient. Better information networks would enable hazardous material generated by one manufacturer to be passed on to another company that could use it (see the earlier discussion of industrial ecosystems). Sometimes biodegradable and environmentally benign substances can be substituted for artificial, synthetic, or less environmentally friendly chemical substances; an example would be using natural, biologically based fertilizers, pesticides, and herbicides in place of synthetic chemicals. Even around the house, the average consumer can buy less toxic, more environmentally friendly paints, sprays, cleansers, and other chemicals.

Hazardous waste disposal is now recognized as a special problem that requires particular techniques and solutions. All too often in the past, however, hazardous wastes were disposed of improperly with little concern for long-term consequences, even though their disposal was technically legal at the time. What are now recognized as hazardous substances were often thrown away along with ordinary wastes. Now we are haunted by hazardous waste dump sites across the nation, and indeed around the world. One of the first to come to the attention of the nation was **Love Canal** (**Figure 18-10**).

From 1947 to 1953, Hooker Chemical and Plastics Corporation used Love Canal, located in the town of Niagara Falls, New York, as a disposal site. An estimated 19,873 metric tons of chemical wastes were deposited in the abandoned canal, mostly contained in 55-gallon (208-liter) steel drums.

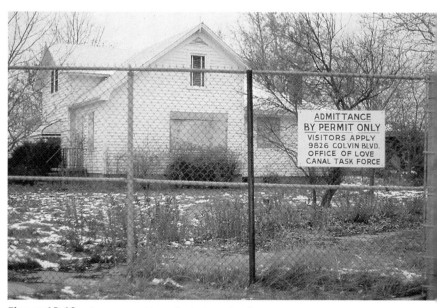

Figure 18-10   An evacuated home in the Love Canal community, Niagara Falls, New York.

CASE STUDY 18-3

E-Waste and Technotrash

The electronic age of computers, the World Wide Web and the Internet, e-commerce, mobile phones, and instant global communication have brought about a revolution in the way we (at least those of us connected to the system) interact, work, and relax (perhaps surfing the net or playing the latest online game). In 2001, an estimated 520 million people were using the Internet; of that number, 166 million were in the United States, 47 million in Japan, and 34 million in China. As of 2001, 292 million of the people using the Internet had native languages other than English, despite the fact that the primary language of the Internet remained English. Generally, Internet users had either European or Asian languages as their mother tongues. It has been suggested that by 2007, the most widely used language by Internet surfers could be Chinese. And the Internet continues to grow at an amazing pace; between 1999 and 2002, it is estimated to have doubled in size in terms of number of host computers and users, and during the same period, material posted on the web expanded from an estimated billion pages to some 10 billion pages.

Mobile (cellular) telephone subscriptions and use have seen a similar meteoric rise in the last decade and a half. In the late 1980s, there were only a few million cellular phone subscribers worldwide, and now there are about a billion (2002), and the number continues to climb. For many people, especially in developing countries, mobile phones are their only phones. At present (2002), there are approximately as many cellular phone connections as there are fixed-lined telephone connections.

Electronic communications hold the promise of reducing waste and energy consumption per amount of information that can be stored and transferred. For instance, the equivalent of tens of thousands of pages of text can easily be fit onto a single CD-ROM and millions of pages can be stored on a small hard drive that can be held in the palm of a hand. The savings in paper and other raw materials can be enormous—or so some contend. But while modern electronics may reduce certain types of wastes, entirely new forms of hazardous wastes, variously referred to as e-waste (electronic waste) or technotrash (technological waste) are being produced at a prodigious rate. To manufacture the semiconductors and other components used in computers and various electronic devices, massive amounts of water and chemicals (many highly toxic) are required, resulting in equally massive amounts of wastewater and many hazardous wastes. It is difficult to get accurate statistics, but manufacturing the components of a typical personal computer generates on the order of thousands of liters of wastewater, several kilograms of hazardous waste byproducts, and it requires hundreds of kilowatt-hours of electricity. The wastes of the electronics industry do not just disappear; they must go somewhere. Unfortunately, they now contaminate many sites. The place credited with the birth of the U.S. semiconductor industry, Santa Clara County, California, now contains more U.S. EPA Superfund sites than any other county in the country. Some workers in the semiconductor industry have been allegedly exposed to hazardous and toxic materials under less than safe conditions. IBM (International Business Machines) and National Semiconductor, for instance, have faced lawsuits involving allegations of cancer and birth defects caused by the exposure of workers to toxic chemicals, and in at least one case, IBM settled out of court for an undisclosed sum over an instance of a child being born with birth defects to parents who had been employed in an IBM plant in New York in the 1980s.

Computers, cell phones, and other electronic equipment typically have a very short half-life—often on the order of only about 18 months or less. Such devices are not discarded only, or even primarily, because they are broken and cannot be repaired, but perhaps more typically because they have become obsolete as more and more speed and power is packaged into smaller and less expensive computer components and other electronic devices. Indeed, some electronic gadgets, such as cell phones, are for all practical purposes "disposable." Yet these electronic devices are difficult and dangerous to recycle. It can be extremely difficult to extract minute quantities of valuable rare metals from integrated circuit boards and chips; modern electronic devices are not designed to be disassembled. Furthermore, they contain many harmful substances. A typical cathode ray tube computer monitor contains 1.8 to 3.6 kilograms of lead (a dangerous heavy metal linked to nervous system disorders), cadmium (linked to increased risks of cancer and damage to reproductive systems) is found in many batteries used in electronic devices, and flat panel screens contain mercury (a highly toxic substance that damages the nervous system, especially in children). It is estimated that some 40% of all lead in U.S. landfills is from computer monitors; accordingly, some states, such as Massachusetts and California, have banned cathode ray tube monitors from landfills and incinerators.

So where do obsolete computers, discarded cell phones, and other trashed electronic devices end up? Millions of tons of electronic waste are dumped in U.S. landfills every year. But this is not the only place that electronic wastes end up. According to one study, in 2002, approximately 10 million personal computers were discarded by Americans (if anything, this may be a conservative estimate, and given the pace of new technology, hundreds of millions of personal computers could become obsolete in the next few years), and of those 10 million (or more) discarded computers, two-thirds were shipped to Asia (for instance, China, India, and Pakistan) to be dismantled by hand by rural villagers for recycling and disposal. Estimates suggest that from 50% to 80% of electronic waste collected in the United States to be "recycled" is actually exported to other, poor, countries. Such recycling may, superficially, seem like a good thing, but the poorly paid workers are routinely exposed to hazardous and toxic chemicals while their villages, rivers, and groundwater supplies are poisoned. As documented in a 2002 report by the Basel Action Network and the Silicon Valley Toxics Coalition, in Guiyu, China, which has become a major center for computer and electronics "recycling," poorly or totally unprotected laborers smash cathode ray tube monitors to extract the copper and other metals; they strip wires and sort wires and plastics from computers by hand; they burn unusable plastics and circuit boards in open pits (releasing carcinogens into the environment); they de-solder circuits using crude burners and open fires; and they use strong acids in uncovered vats to extract gold, silver, and other precious metals from electronics components. This activity has polluted the local air, soil, river (local river water was measured to contain 190 times the levels of some pollutants allowed under World Health Organization guidelines), and groundwater. In fact, drinking water must be shipped in from 18 miles away. Guiyu is not an isolated instance; similar areas occur elsewhere in Asia, including in India and Pakistan.

In 1989, an international treaty commonly known as the Basel Convention (Convention on the Control of Transboundary Movements of Hazardous Wastes and Their Disposal) sought to restrict the transfer of hazardous wastes, including electronic wastes, across national borders. The treaty seeks to minimize hazardous waste generation and encourage disposal of wastes close to the points of origin. In 1994–1995, the Basel Convention was amended to prohibit the export of wastes from developed industrial (rich) nations to non-developed (poor) nations, even for supposed "recycling" purposes. Although not yet fully legally binding (all of the amendments have yet to be ratified by a sufficient number of countries), many countries abide by the Basel Convention and its amendments. China has banned the importation of electronic waste, yet it still flows in a steady stream to Guiyu and other areas of the country. This, critics suggest, is in large part because the United States—the world's single largest producer of hazardous waste—does not fully honor the Basel Convention and its amendments (in fact, although the United States signed the convention in 1989, it has not yet ratified it) and the U.S. government has effectively made export part of its overall strategy for dealing with electronic wastes.

However, as the hazards of e-waste are being exposed and brought to the attention of consumers who never realized the dangers posed by their discarded electronic equipment (which many assumed was being safely recycled), there is increasing pressure from environmental and human rights organizations (such as the Basel Action Network, Greenpeace China, Silicon Valley Toxics Coalition, Society for Conservation and Protection of the Environment, and Toxics Link India), as well as the public in general, in America and overseas, to implement genuine recycling programs in the country of origin rather than simply shipping such waste to developing countries to be "recycled." The recycling methods are very crude and dangerous, and much of the material ends up being disposed of in an environmentally unsound manner. Japan has instituted an "Appliance Recycling Law" that requires manufacturers (after a fee is paid by the consumer) to take back and recycle various appliances such as air conditioners and refrigerators. As of this writing, it is anticipated that computers and other electronic devices will be added to the list covered by the Japanese law. Likewise, various European nations have instituted "take-back programs" for electronic equipment; in some cases, the cost must be borne by the original manufacturer, whereas in other cases, it is paid by the consumer or the local municipality (and thus the taxpayers).

Ultimately, however, the solution to e-waste and technotrash may not be "after-the-fact" recycling, whether in the nation of origin or abroad, but the design of electronic components and devices that can be manufactured in an environmentally friendly manner. The products should be easily disassembled, so that the components can be reused and recycled, and be free of as much toxic and hazardous substances as possible.

Questions

1. Do you think manufacturers should be held responsible for the entire life-cycle of their products, from cradle to grave (sometimes known as the principle of "Extended Producer Responsibility" or EPR)? In other words, should their responsibility for the product continue after the purchase of the product by the consumer? Should a manufacturer be required to "take back" a product at the end of its life cycle? Should this be the case just with electronic products, only certain types of products, or for all consumer items? Justify your answer.

2. Should a manufacturer be held legally responsible and liable if the products it sells are found to be causing human and environmental damage, whether during routine use or after being discarded?

3. Should manufacturers be required to design electronic equipment with ease of disassembly and recycling of components in mind, and produce electronic components utilizing less toxic and poisonous substances? What if such procedures significantly increase the cost, perhaps by a factor of two or three times or more, of the end product? Or do you think such procedures could be encouraged on a voluntary basis? Would you purchase a more expensive, but more environmentally friendly, computer or cell phone?

4. Should manufacturers be encouraged (or required) to design longer lasting, more easily repairable and upgradeable, products? Based on what you know about computers and other electronic devices, do you think this is technically feasible, economical, or practical? Why or why not?

5. The principle of "environmental justice" is that no group of people, on the basis of economic status, social status, or race, should be forced to bear the environmental risks of others. Is the shipping of wastes, such as e-waste and technotrash, from rich, developed nations like the United States to poor developing nations a form of "environmental injustice" (the concept that a group of people, community, or country is the recipient of the wastes, pollution, and environmental problems created by another group that is generally better off materially)? Put simply, do you believe it is morally wrong for rich nations to ship their wastes to poor nations? Why or why not?

Numerous hazardous wastes, including powerful carcinogens, were buried in the canal—substances such as dioxin, benzene, chloroform, and dichlorethylene. Later, an elementary school and numerous homes were built in the area, only to be abandoned in the late 1970s when toxic chemicals began to seep from the ground. Ultimately, it cost hundreds of millions of dollars to clean up the mess.

Superfund

After the chemical contamination of Love Canal was exposed in the 1970s, the United States realized that something had to be done about hazardous waste sites. In 1980, Congress passed the Comprehensive Environmental Response, Compensation, and Liability Act (CERCLA), authorizing the EPA to recommend which hazardous waste sites should be given the highest priority in cleanup efforts. Accordingly, in that year the EPA set out to identify all hazardous waste sites in the United States, determine what dangers to health and well-being each posed, and list what measures would be required to clean up each site. The agency compiled a National Priorities List (NPL) of high-priority sites that merited action. The cleanup and related operations were to be funded by taxes on the chemical and petroleum industries, as well as assessments on various corporate polluters. Initially, $1.6 billion was allocated to a Hazardous Substance Response Fund, commonly referred to as "**Superfund**," to be used for cleanup activities.

After over twenty years, the Superfund program has grown and evolved. Tens of thousands of hazardous waste sites have been identified, and more continue to be located. In the first twenty years of the program, over 6400 actions were taken to reduce health and environmental threats from such sites, and cleanup was completed at over 750 sites (the work being carried out by the responsible parties at 70% of the sites). Through the Superfund program, private parties paid or funded over $18 billion in settlements. No one really knows how much it will cost to clean up all of the hazardous waste sites identified by the EPA. As of mid-2002, there were over 1200 sites listed on the National Priorities List (sites are added to and removed from the list periodically). According to one estimate, the average cleanup cost for a single site ranges from $21 million to $30 million; on that basis, the NPL sites could cost $27 to $40 billion to clean up. Some analysts have estimated that the cost of cleaning up all hazardous waste nationwide could range from $750 billion to $1 trillion and require more than 50 years of sustained effort. To solve some of the continuing financial problems involved with cleaning up hazardous waste, "no-fault" programs have been proposed. Under no-fault policies, all companies would be required to help finance cleanup efforts, regardless of whether they were directly involved in creating the problem in the first place. Of course, many "clean" companies and organizations regard such policies as inherently unfair: Why should they help pay for damage caused by other companies? Or should the federal government simply pick up the entire tab, which means that the public will ultimately pay for the cleanups? There are no easy answers as to how to fund the enormous cleanup costs that we face in the future.

The necessity and effectiveness of Superfund cleanup operations have also been questioned. Some researchers suggest that the health risks of many hazardous waste sites have been overstated, and in some cases, cleanup efforts may disturb toxic materials and create more of a hazard than if the site were simply left alone. Various EPA studies have suggested that problems such as indoor pollution and outdoor air pollution pose a greater health risk than hazardous waste—the logical conclusion being that limited resources might better be expended on cleaning up these other environmental problems rather than focusing on hazardous waste sites.

The United States is not the only country that faces huge price tags to clean up its hazardous waste problems; similar situations exist around the globe. The small country of the Netherlands will have to spend at least $1.5 billion to clean up its hazardous waste sites. The former West Germany needed an estimated $10 billion to clean up its own hazardous waste sites; now with the unification of East and West Germany, the price will sharply increase. Some of the worst toxic dump sites and environmental hazards are in Eastern Europe and the countries of the former Soviet Union. Developing nations are also plagued by toxic dumps and rampant pollution; in many cases, the dumping of hazardous waste is inadequately regulated. In Mexico City, for instance, wastewater contaminated with heavy metals and toxic organic chemicals is discharged into the municipal sewer system. From there, with very little treatment, it is transported to agricultural areas for use in irrigation, with the result that toxic substances have been found in vegetables and other crops. Likewise, in China, some 400 million metric tons of industrial wastes and mining tailings are annually dumped on the outskirts of cities or into lakes, streams, and rivers with devastating results—148,260 acres (60,000 hectares) of land in China may be covered with hazardous material.

The Bhopal Disaster

One of the most infamous industrial chemical disasters occurred at the Union Carbide pesticide-producing plant located in **Bhopal**, India, on December 3, 1984. Apparently, a faulty valve allowed water to leak into a tank containing approximately 15 metric tons of liquid methyl isocyanate, an extremely toxic substance. The water and methyl isocyanate reacted chemically, producing high temperatures and pressures that vaporized the liquid; ultimately, the material escaped as a deadly cloud above the city of 800,000 people. Unfortunately, there were no emergency plans or procedures in place, and no one really knew how to deal with the situation. The results were devastating—several thousand people died immediately, and over the next few years, thousands more (estimates run as high as 10,000) died as a result of methyl isocyanate poisoning. In addition to the deaths, an estimated 50,000 to perhaps 500,000 were injured to some degree as a result of the disaster.

The Bhopal disaster should not have occurred. A number of safety devices at the Union Carbide plant were not operational at the time of the disaster; indeed, some had been turned off for months.

The Bhopal disaster suddenly made people aware of the inherent dangers of many industrial processes, especially if plans are not in place to deal with extraordinary disaster. In the United States, the Bhopal incident caused Americans to look with increasing suspicion at the chemical industry and helped inspire Congress to pass the Emergency Planning and Community Right-to-Know Act, which established the important Toxics Release Inventory.

Toxics Release Inventory

In October 1986, Congress passed the Emergency Planning and Community Right-to-Know Act as Title III of the Superfund Amendments and Reauthorization Act. Under this law, larger companies and organizations must report their production, use, and release of hundreds of potentially hazardous chemicals listed by the EPA. Based on this information, the EPA produces an annual report, known as the **Toxics Release Inventory (TRI)**. With the release of the 1987 TRI, for the first time Americans could get an overall picture of how much toxic material the country's industries were emitting. Based on the material gathered for the TRI, in 2000 (the latest year for which figures are currently available), 7.1 billion pounds (3.2 billion kg) of over 650 EPA-listed toxic chemicals were released or disposed of by U.S. industry. This was down from 7.8 billion pounds in 1999, and overall chemical releases have decreased by about 48% since 1988. In 2000, 27% of these chemical releases went into the air, 4% went to water, and 69% to land. Manufacturing accounted for 32% of the releases, and electric utilities another 16%.

Over the past decade and a half, the TRIs have had some very positive benefits. They have enabled citizens and communities, armed with data, to put pressure on companies to cut emissions of hazardous substances. Likewise, people in high-level management positions were themselves appalled by the amounts of toxic chemicals their own factories were releasing. Many industries are now taking voluntary actions to reduce the amount of toxic material released into the environment. Despite the good that the TRI has done, it has not gone without criticism. Many different types of businesses, especially smaller businesses such as dry cleaners and photographic processors, are not required to submit information to the EPA. Furthermore, the list of toxic chemicals compiled by the EPA and added to periodically is not complete. Perhaps most critically, the TRI is merely a compilation of certain toxic chemicals released directly by large facilities. Originally, companies were only required to report releases if they manufactured or processed 25,000 pounds [11,340 kg]

or more, or otherwise used 10,000 pounds [4536 kg] or more, of listed chemicals. To correct this deficiency, at least in part, in 1999, reporting thresholds for certain substances were significantly lowered. However, the TRI still does not include all of the relatively small releases by individuals, small businesses, municipalities, and other organizations; yet these can add up. Also, many hazardous and toxic chemicals are incorporated into end products (paints, stains, coolants, and so forth) that are used by a consumer, but eventually released into the environment—again, these are not accounted for in the TRI. While certainly an important step toward accounting for our chemical waste, the TRI may represent merely the proverbial "tip of the iceberg."

Technologies for Dealing with Hazardous Waste

Engineers and scientists have long grappled with the problem of what to do with hazardous wastes. Common solutions have involved isolating such wastes from the environment in permanent waste disposal sites, such as a landfill or deep in the ground through deep-well injection. Another approach is to treat hazardous and toxic wastes so as to eliminate or reduce their harmful properties. For decades this was attempted by simply burning the waste, but in many cases burning at temperatures that are not sufficiently high actually compounds the problem by creating and releasing molecules that are more harmful than the initial substance (for example, dioxin and heavy metals may be released during standard incineration, as discussed earlier).

In recent years, several laboratories have concentrated on developing more efficient ways of destroying and neutralizing toxic chemicals. Westinghouse Electric Company is working on a "plasma torch" that can reportedly generate temperatures up to 10,000° F (5500° C). The torch would be trained on toxic organic chemicals and break them down into relatively simple and harmless gases. The Westinghouse plasma torch is small enough so that it can be mounted on the back of a truck and brought to hazardous waste dump sites. In this way, extra transportation of hazardous waste can be avoided, lessening the chance that an accident might happen on the way to a disposal facility. A potential drawback of plasma torch technology is that it may consume large amounts of energy to destroy a significant amount of hazardous waste.

The Solar Energy Research Institute in Colorado and Sandia National Laboratory in New Mexico, both U.S. government organizations, are researching ways to destroy toxic chemicals using solar energy. One method resembles in principle some solar energy collectors described in Chapter 8. Water contaminated with toxic organic chemicals is mixed with catalysts and then pumped through tubes while the Sun's light is focused on the mixture. The ultraviolet light of the Sun has the ability to break down and destroy up to 90% or more of the contaminants. In another method, the Sun's light is focused on a special vessel containing toxic organic chemicals. The solar energy can raise the temperature inside the vessel to very high levels, converting the more complex chemicals into simple substances such as carbon monoxide and hydrogen gas. It has even been suggested that these products could then be used as a fuel; either the CO and H_2 can be used to form methanol, or the hydrogen gas can be burned directly. In a related process, magnified sunlight focused on a container full of dioxin broke down 99.999% of the dioxin into simpler, relatively harmless substances.

Bioremediation. Bioremediation, or biotreatment, is the use of bacteria and other small organisms (such as single-celled and multicellular microbes and fungi) to clean up or reduce unwanted concentrations of certain substances. Essentially, the microorganisms eat or digest unwanted chemicals, transforming them into simpler, less harmful, or useful forms. Certain bacteria, for instance, will attack oil and related organic substances, converting the material into primarily water and carbon dioxide.

The biological action of microorganisms has been a necessary part of nature ever since life orig-

inated nearly four billion years ago. Microorganisms convert kitchen waste into compost, and bacteria have been used commercially for years to treat sewage and even to concentrate metals such as copper and nickel from low-grade ores. Recently, microorganisms have been finding much wider applications, especially in oil spills, at toxic and hazardous waste sites, and as biological "scrubbers" to remove dangerous pollutants from factory emissions before they leave the facility. Certain strains of bacteria and fungi can digest even extremely dangerous chemicals, such as DDT (the insecticide), TNT (the explosive), PCBs, dioxins, tolulene, naphthalene, xylene, carbon tetrachloride, toxic nitrates, asphalt products, creosote (a wood preservative), and many other substances. If a bacterium cannot be found in nature to perform a certain job, researchers may be able to genetically engineer one in the laboratory. Of course, the substances produced as organisms digest dangerous compounds may also be extremely dangerous. For example, DDT may be degraded to an even more toxic substance than the original pesticide. One must be careful in applying bioremediation techniques in specific cases.

After the Exxon Valdez oil spill of March, 1989 (**Figure 18-11**), simple bioremediation techniques were used on a very small scale with quite encouraging results. Liquid farm fertilizers were sprayed on a portion of an oily beach, and after approximately two weeks, the oil on the beach was significantly reduced, compared to a test area that had been left untreated. No microorganisms per se had been added to the beach—essential nutrients had

Figure 18-11   A cleanup crew works on a beach shortly after the Exxon Valdez oil spill of March, 1989.

simply been spread to encourage the growth of already present bacteria that would digest the oil. The success of this simple experiment has had wide ramifications for the prospects of bioremediation. Prior to this time, the EPA generally did not sanction the use of microorganisms for cleaning up oil spills or hazardous waste sites. Since the initial success of the Exxon Valdez experiment, bioremediation has been put to work cleaning up hundreds of sites overseen by the EPA.

Today, dozens of companies are developing and marketing biotreatment products—everything from bacteria and nutrients to help clean out a home septic system, to mixtures that are specifically designed to deal with gasoline spills and other hazardous materials. Bioremediation tends to be cheaper than traditional treatment methods such as incineration of contaminated soil, and it has the distinct advantage that it can often be applied on site. For example, imagine that some toxic substance, such as a petroleum product, has leaked into the ground. Instead of digging up hundreds of tons of contaminated soil and transporting it to an incinerator, large augers can drill into the soil and mix in bacteria and nutrients.

The bacteria then attack the substance, breaking it down into harmless chemicals. Such techniques have already been used to treat soil contaminated with creosote at a lumber-treatment plant and at a 40-acre (16-hectare) site near Los Angeles where the soil had been contaminated with marine fuel.

Biotreatment is also finding many useful applications to avoid releasing pollutants into the environment in the first place. Microbes are of particular use in treating volatile organic compounds (VOCs) that are often released as gaseous fumes from factories and small businesses, such as dry cleaners, print shops, auto body shops, and even bakeries. In one technique, the fumes are passed through layers of well-humidified and aerated peat moss and compost that contain bacteria. The gases are caught on wet films covering the compost and peat moss and can then be broken down into harmless gases (mostly water and carbon dioxide) by the bacteria. This is a relatively low-tech, inexpensive, but very effective method of substantially reducing the release of harmful VOCs. Bacteria simply do the work of chemical degradation for free. Such systems are finding wide use in the Netherlands and Germany.

STUDY GUIDE

SUMMARY

- Solid waste consists of municipal solid wastes (garbage and trash, refuse, and rubbish from homes and various businesses) and industrial solid waste.
- Alternate traditional paradigms for solid waste disposal are the "dilute and disperse" (e.g., burning and dispersing the smoke) versus "concentrate and contain" (e.g., landfilling) strategies.
- A third paradigm is "resource recovery," including reuse and recycling.
- Incineration is the burning of trash and garbage—the heat may be used to generate electricity or otherwise be put to use in a resource recovery facility.
- Problems with incineration include the dioxins and other toxic substances that may be given off.
- Landfills isolate and contain solid waste, but the isolation may not be complete—for instance, hazardous leachate may drain from the landfill or methane gas may be produced.
- There is increasing emphasis among waste management experts to address current solid waste issues through source reduction, reuse

of products (rather than dispose of them), and recycling of materials.
- Some materials, such as glass and metals, can be recycled virtually indefinitely in closed-loop systems.
- Other materials, notably paper and plastics, can only be recycled a limited number of times (open-loop systems) before the material is too degraded to use further.
- Hazardous wastes are substances that, when released, may present a substantial danger to public health, welfare, or the environment and include chemically reactive, explosive, and toxic substances; such materials originate from industries, businesses, governments, and households.
- Hazardous waste accumulations can be a major issue locally; in 1980, the U.S. Congress passed legislation that set up a "Superfund" to be used for cleanup activities at hazardous waste sites.
- In the United States, the Toxics Release Inventory (TRI) provides data on the billions of kilograms of toxic chemicals released or disposed of by industry each year.

KEY TERMS

Bhopal
Biodegradable plastic
Bioremediation
Closed-loop recycling
Compost
Deep-well injection
Dioxin
Dump
Garbage
Hazardous waste
Incineration

Industrial ecosystem
Industrial solid waste
Landfill
Leachate
Love Canal
Methane gas
Municipal solid waste
Nonhazardous waste
Open-loop recycling
Recycling

Refuse
Resource recovery facility
Reuse
Rubbish
Sanitary landfill
Solid waste
Source reduction
Superfund
Toxics Release Inventory
Trash

STUDY QUESTIONS

1. Define *solid waste* and its subcategories.
2. Discuss the alternative paradigms that have historically been used for waste management. What are the pros and cons of each? Which are still in common use?
3. Discuss both the benefits and the drawbacks of the following techniques of waste disposal: incineration, resource recovery facilities, dumps and landfills, and deep-well injection.
4. Define *hazardous waste*. What different types of hazardous waste are there?
5. What are some of the new technologies that are being developed to deal with hazardous waste?
6. How can garbage be a source of revenue for some communities and nations? What are the positive and negative aspects of such an enterprise?
7. Describe the "waste management hierarchy." Is there universal agreement on this hierarchy?
8. Describe the advantages and disadvantages of source reduction, reuse, and recycling. What can and should be done to encourage recycling?

9. Why do some environmentalists object to too heavy an emphasis on recycling?
10. How do glass, metal, paper, and plastic recycling differ from one another?
11. How are most municipal wastes disposed of in the United States? How much is recycled?
12. Describe the extent of the solid and hazardous waste problem in the United States. How serious are these problems compared to other environmental issues discussed in this book?
13. Briefly describe the process and benefits of composting. The process of composting can produce greenhouse gases, so is composting always a good thing? Justify your answer.
14. Comment on the controversies surrounding the paper versus plastics debate.
15. Briefly describe the concept of an industrial ecosystem.
16. Briefly describe the Love Canal and Bhopal incidents. What practical effects, especially legislatively, did they have?
17. What is Superfund? What has it accomplished? Why has it been controversial?
18. What is the Toxics Release Inventory? What purpose does it serve?

WHAT'S THE EVIDENCE?

1. Today, over half of all municipal solid waste in the United States is disposed of in landfills, and many people are concerned that we are quickly running out of landfill capacity. Does the evidence support this assertion? Name some of the factors that pertain to, and affect, landfill capacity.
2. Some people argue that it is wrong to ship wastes, especially wastes that contain haz-

ardous and/or toxic components, from developed nations to developing nations. Part of the rationale behind this point of view is that even if the developing country gains short-term economic benefits from accepting the waste, in the long term it will suffer. Is there hard evidence to support this assertion? What is your position on this issue? Justify your answer.

STUDY GUIDE

CALCULATIONS

1. Based on the data presented in Table 18-1, the per capita daily garbage output of the average American increased by what percentage from 1960 to 1980?
2. Based on the data presented in Table 18-1, the per capita daily garbage output of the average American increased by what percentage from 1980 to 1999?
3. Based on the data presented in Table 18-1, in 1960 in America, what percentage of municipal solid waste was recovered for recycling?
4. Based on the data presented in Table 18-1, in 1999 in America, what percentage of municipal solid waste was recovered for recycling?

ILLUSTRATION AND TABLE REVIEW

1. Referring to Figure 18-3 (The typical composition of municipal solid waste in the United States according to the weight of various types of materials), what are the categories of products that make up the first and second largest components of municipal solid waste?
2. Referring to Figure 18-3, what are the categories of materials that make up the first and second largest components of municipal solid waste?
3. Referring to Figure 18-8 (Waste Recycling Rates from 1960 to 1999 in the United States), how many more million tons per year of municipal solid waste were recycled in America in 1999 as compared to 1960?

WWW.ENVIRONMENT.JBPUB.COM

www.environment.jbpub.com

Connect to this book's web site: http://www.environment.jbpub.com/ The site features eLearning, an online review area that provides quizzes, chapter outlines, and other tools to help you study for your class. We've posted the answers to the book's odd-numbered questions. You can also follow useful links for more in-depth information. The site provides sources for the information in this chapter and recommendations for further readings.

Sampling water from a cyanide-poisoned lagoon.

Environmental Issues: Social Aspects and Solutions

The economy is like a huge digestive tract, with a flow of inputs and outputs. Sustainable economies will mimic more efficient, not bigger digestive tracts.

Herman Daly, environmental economist

A culture of permanence will not come quickly. We can expect no instant revolutions in social values. All we can realistically hope for is painfully slow progress . . . punctuated by rapid advances. When most people see a large automobile and think first of the air pollution it causes rather than the social status it conveys, environmental ethics will have arrived.

Alan Durning, How Much Is Enough? (1992)

1. The Environment and Humans (Chapters 1-2)

Population

Consumption

2. The Environment of Life on Planet Earth (Chapters 3-5)

Lithosphere

Biosphere

Hydrosphere

Atmosphere

3. Resource Use and Management (Chapters 6-13)

Lithosphere

Biosphere

Hydrosphere

Atmosphere

4. Dealing with Environmental Degradation (Chapters 14-18)

Lithosphere

Biosphere

Hydrosphere

Atmosphere

5. Environmental Issues: Social Aspects and Solutions (Chapters 19-20)

Lithosphere

Biosphere

Hydrosphere

Atmosphere

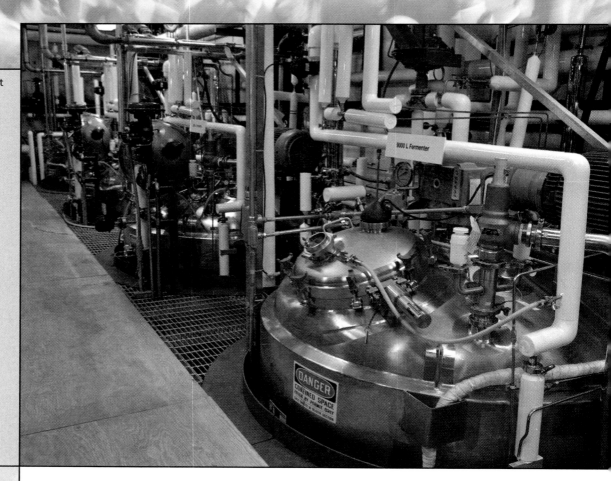

Economics plays a key role in most environmental problems. The laws of supply and demand not only determine the availability of natural resources, but also play a role in many activities that affect the environment, such as selling pollution permits, developing cleaner technologies, or starting an organic farm. In fact, as nearly all people want a clean environment, it is often the economic costs, not social willpower, that prevents society from solving environmental problems. Until recently, the study of economics has focused rather narrowly on short-term human needs alone. This is changing, as awareness of global environmental issues has deepened in recent years, and economics is playing an increasing role in efforts to produce a truly sustainable society. Shown here is a pilot biomass-to-ethanol plant at the National Renewable Energy Laboratory built to test economical ethanol production technologies.

CHAPTER 19

Environmental Economics

Economics is often defined as the study of scarcity and how we cope with it. Economics is thus concerned with environmental issues in two basic ways. The most obvious is resource scarcity. In the resource section of the book, we saw how humans are depleting a variety of resources, or inputs to society. Economics is also concerned with outputs, or pollution. Thus, it addresses fresh air, clean water, and other basic needs when they become scarce from pollution and other forms of environmental degradation. Economic realities are an inevitable part of solving all environmental problems, so we have discussed them throughout this book. In this chapter, we provide an overview that covers many basic principles of economics and the environment.

Economics versus Environmental Science?

Economics and environmental science are frequently depicted as natural enemies. Indeed, economists and environmental scientists are often critical of each other. A main reason for the conflict is that the two disciplines have fundamentally different ways of looking at the world, especially in three areas (**Table 19-1**):

- *Time span of outlook.* Environmental science has historically been concerned with species interactions, natural cycles, and other aspects of nature that reflect the long-term impact of resource depletion and pollution on the biosphere. Such a long-term view is useful when considering such problems as global warming or radioactive waste, whose impacts last hundreds to millions of years. In contrast, economics is a social science that deals with human time scales and focuses on short-term problems, such as unemployment and taxes, that are measured in yearly or even monthly terms.

- *Priorities.* Traditional economists have mainly been concerned with society, especially how society produces, distributes, and consumes goods and services. They tend to view nature as a resource to be used for these purposes. Environmental scientists have tended to place higher priority on preserving the natural environment than producing more goods and services.
- *Social solutions.* Economists are social scientists and know that incentives are often the most effective way to regulate human behavior. Environmental scientists have tended to emphasize persuasion and laws such as the Endangered Species Act as solutions to environmental problems.

The popular news media tend to reinforce the idea that economics and environmental science are fundamentally opposed by presenting such stories as "development versus wildlife preserves." One of the best-known examples is the spotted owl controversy, which was widely discussed in terms of jobs (logging the forest) versus environment (preserving the forest habitat of the spotted owl).

In a limited sense, these apparent conflicts are real. A logger who loses his job to an endangered species would certainly see a conflict. In a broader sense, however, the conflict between economics and the environment is really a false dichotomy. **A false dichotomy** is when a complex question becomes polarized into an "either–or" issue. The dichotomy of jobs or environment is false because usually the ultimate choice is much more complex than that. In the long run, environmental preservation typically produces *more* jobs. For instance, in the spotted owl debate, if the old growth forest were harvested, many loggers would still be out of a job in a few years as the remaining forest disappeared. On a larger scale, in many countries where the tropical rainforests have been destroyed, the local inhabitants have lost their way of life and live in abject poverty. The same could be said for fishermen in the overharvested fishing grounds of offshore New England, farmers who have exhausted the soils in Africa or Haiti, and countless other examples.

Sustainability Produces More Jobs

These examples show that controlled use of resources often provides more employment in the long term. Fishermen, loggers, farmers, and many others can only remain employed if their resource base is not depleted. In other cases, the loss of environmentally damaging occupations will usually be more than compensated by new occupations that promote environmental preservation. We have discussed many examples throughout this book where activities that help preserve the environment usually create more jobs than they eliminate. Many of these new jobs are in pollution cleanup. Since the early 1970s, the U.S. environmental cleanup business, including pollution control and toxic waste remediation, has grown dramatically. This growth has created millions of new jobs in over 75,000 new firms with annual sales well over $100 billion each year.

Even more jobs are created by indirect environmental activities such as increased efficiency, recycling, and renewable resource use. For example, the Worldwatch Institute reports that to produce 1000 gigawatt-hours of electricity, the following numbers of workers are needed:

Unsustainable fuels:
- Nuclear plant–100 workers
- Coal-fired plant–116 workers

Sustainable fuels:
- Solar thermal plant–248 workers
- Wind farm–542 workers

In another example, data from the New York City Department of Sanitation show that recycling produces at

TABLE 19-1	Three Areas Where Economics and Environmental Science Differ		
	Time Span	**Priorities**	**Social Solutions**
Economics	Short term	Society	Incentives
Environmental Science	Long term	Environment	Laws, persuasion

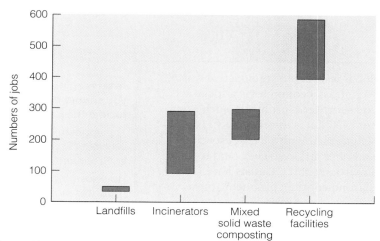

Figure 19-1   The number of jobs created for each 1 million metric tons of waste processed in New York City. A range is shown for the different processes because some facilities (such as different incinerators or different recycling facilities) are more efficient than others. Recycling creates the most jobs, even when the variation is included. (*Source*: Reprinted from Worldwatch Paper #104, "Jobs in a Sustainable Economy," by Michael Renner with permission of Worldwatch Institute, Washington, D.C., copyright © 1991.)

least 10 times more jobs per 1 million metric tons of waste than landfills do (**Figure 19-1**). Environmental laws, increased business interest in efficiency, and increasing public demand for thousands of "green" consumer items such as organic foods and recycled clothing, will likely cause further growth in environmental jobs.

Environmentally sustainable activities produce so many jobs because they are often labor-intensive. They rely on people instead of large amounts of

Figure 19-2   Older, obsolete industries sometimes blame environmental regulations for closing them down or causing layoffs. But such regulations actually cause only a tiny fraction of total job losses. Seen here is an abandoned limestone mill in Bainbridge, Pennsylvania.

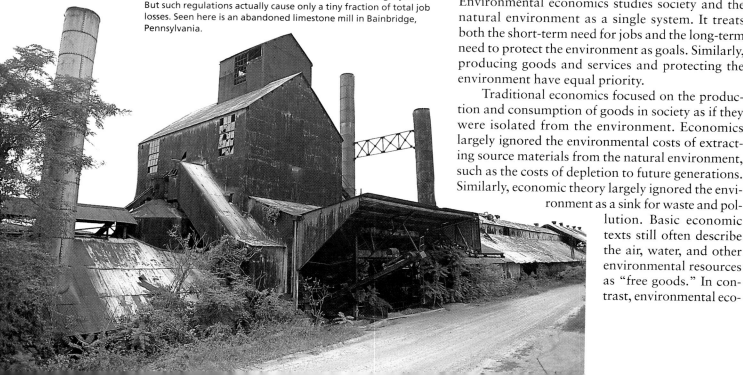

energy and machinery. In contrast, many highly polluting, resource-depleting industries are just the opposite, using much energy and few people. In the United States, for example, just five manufacturing industries, oil refining and coal products, chemicals, primary metals, paper, and food products, account for over 80% of both energy use and toxics release. Yet they account for less than 25% of total employment. The discrepancy is particularly great for the chemical industry, which produces about 35% of the toxic releases and uses 20% of the energy, while accounting for less than 6% of the total manufacturing jobs.

Despite the greater job-generating capacity of many environmentally related activities, some older industries promote the belief that environmental preservation causes job losses. Companies threaten to close or relocate (often to another country) if they are forced to comply with environmental laws. This is sometimes called **greenmail**, or job blackmail. Such industries are often inefficient, rely on depleted resources, or have declining employment from increased automation or international competition (**Figure 19-2**). In fact, annual reports consistently show that only about 0.1% of all U.S. job layoffs are related to environmental regulations.

Environmental Economics: Blending the Dichotomy

The increasing integration of jobs and environmental preservation has led to the development of a new discipline, called **environmental economics**. (This discipline and its subdisciplines are also sometimes called *ecological* or *green economics*.) Environmental economics studies society and the natural environment as a single system. It treats both the short-term need for jobs and the long-term need to protect the environment as goals. Similarly, producing goods and services and protecting the environment have equal priority.

Traditional economics focused on the production and consumption of goods in society as if they were isolated from the environment. Economics largely ignored the environmental costs of extracting source materials from the natural environment, such as the costs of depletion to future generations. Similarly, economic theory largely ignored the environment as a sink for waste and pollution. Basic economic texts still often describe the air, water, and other environmental resources as "free goods." In contrast, environmental economics incorporates environmental considerations into its theory and practice. The environmental costs and benefits of mining, pollution, and other production activities are included in economic calculations. A reason for this different approach is that when traditional economics arose, human populations and technological impacts were much smaller than they are now (**Figure 19-3**). Society has greatly increased in size, while the environment has remained finite. The human impacts are now so great that they can no longer be ignored.

What Is Sustainable Growth?

Traditional economics is based on the principle that economic growth is extremely desirable. This view arose because economic growth, by increasing the quantity and quality of goods and services, has led to a general improvement in the human condition. As long as human populations and the technological capacity to consume and produce were relatively low, environmental costs of growth were not of much concern. Environmental economics agrees that economic growth can be desirable, such as where people are poor, but argues that the goal should be growth that can be sustained for a long time, not growth at any cost. **Sustainable growth** both preserves the environment and permits a nation's economy to increase in size, creating jobs and improving the human condition.

Sustainable growth differs from unsustainable growth in four basic ways (**Table 19-2**):

1. *Resource use.* As we have seen, traditional economics was based in part on the idea of limitless resources. Accordingly, unsustainable growth maximizes resource use and thus the rate of resource depletion. Economist Kenneth Boulding has called this approach **cowboy economics** because it views the environment as a source of endless space and riches just as the pioneers viewed the Old West. In contrast, sustainable growth recognizes that resources are finite and that the rate of resource depletion must be slowed if we are to have growth in the long term. Accordingly, sustainable growth emphasizes minimizing resource use by reducing inputs to society through increased efficiency, reuse/recycling, and substitution of renewable resources (Chapter 6).

2. *Pollution control.* Reducing inputs not only reduces the rate of resource depletion but also reduces the pollution associated with mining, processing, manufacturing, using, and disposing of goods. In contrast, unsustainable growth, with its emphasis on maximizing input, concentrates on output management,

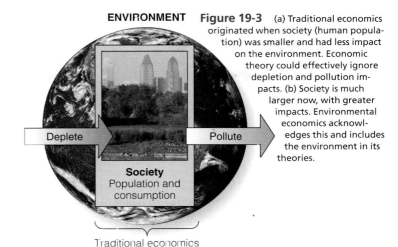

Figure 19-3   (a) Traditional economics originated when society (human population) was smaller and had less impact on the environment. Economic theory could effectively ignore depletion and pollution impacts. (b) Society is much larger now, with greater impacts. Environmental economics acknowledges this and includes the environment in its theories.

(a) Small population and little technology. Society has low impact on environment

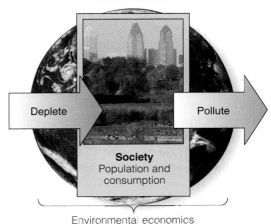

(b) Larger populations and increased technology. Society has great impact on environment

that is, controlling pollution after the resource is used. Consequently, unsustainable growth relies on abatement devices, such as wastewater treatment and smokestack scrubbers, that are much more expensive than reducing the waste stream on the input side.

TABLE 19-2	Unsustainable Growth versus Sustainable Growth	
	Unsustainable	**Sustainable**
Resource use	Inefficient	Efficient
Pollution control	Output reduction	Input reduction
Resource fate	Matter discarded	Matter recycled
Resource type	Nonrenewable	Renewable

3. *Resource fate.* Sustainable growth emphasizes reuse and recycling of matter instead of discarding matter in the one-way flow that is characteristic of the unsustainable **throwaway society.** Reuse and recycling not only reduce pollution from solid waste but reduce depletion of virgin resources.

4. *Resource type.* Sustainable growth substitutes renewable resources where possible. For example, paper containers are preferable to plastic made from petroleum. Future generations will be able to manufacture more paper from trees, but the petroleum from which plastic is made will likely be gone well within the next century. Also, renewable resources tend to be biological in origin and therefore are often more biodegradable. Renewable energy is preferred over nonrenewable forms.

Efficiency improvements, reuse/recycling, and substitution, the hallmarks of sustainable growth, are among the major kinds of **green technologies** that will provide many jobs in the future. Green technologies tend to reduce, rather than increase, the flow of goods through society. The improvement in living standards that occurs with sustainable growth comes mainly from increasing the quality of items produced, not the quantity. The well-known environmental economist Herman Daly has compared the economy to a digestive tract. A sustainable economy does not increase the size of the tract, but it does improve the digestion process. Thus, the change is qualitative, not quantitative. Unfortunately, the United States may be falling behind in developing these critical technologies because of its emphasis on resource consumption compared to other modernized nations, especially in western Europe.

Measuring Sustainable Growth

The traditional economic measurement of **Gross National Product** (**GNP**) can be defined as the market value of all final goods and services during a certain period of time, such as a year, for a particular nation. Officially since 1991, the Gross Domestic Product (GDP) has replaced the GNP in the United States for most purposes. The U.S. GDP covers goods and services within the borders of the country, whereas the U.S. GNP includes all United States residents, even if some or all of their economic activity is outside of the nation's borders. Economic measurement in terms of GNP or GDP (for the purposes of the discussion below, the terms can be used somewhat interchangeably) is not very satisfactory because GNP does not subtract environmental costs that may occur when wealth is cre-

ated. For example, when trees are cut and sold for timber, they add to the GNP. In counting only immediate, short-term dollars, however, GNP ignores the greater net wealth that might have been created if the forest were managed to produce goods over a long period of time. Nigeria, for example, was once a major exporter of timber, but it overcut most of its forest resources and now imports many more forest products than it exports.

Other examples also illustrate how misleading GNP can be:

1. The Exxon Valdez oil spill in March 1989 actually caused U.S. GNP to rise because much of the $2.2 billion spent on labor and equipment for cleanup was added to income.

2. The tens of billions of dollars spent on health care for illnesses caused by air pollution in the United States generally add to GNP.

3. The timber and increased agricultural production obtained through rainforest destruction add to the GNP of those countries even though the destruction leads to the widespread extinction of many tropical species. The permanently lost genetic variations in those species might have produced lifesaving drugs, foods, and many other economic benefits that would have been much greater than the immediate income from the timber and agriculture.

4. Poverty, disease, and human misery increased globally during the 1980s and 1990s even though the Gross World Product (GWP or the global equivalent of GNP) actually rose by over 50% between 1980 and 2001.

Examples such as these have caused some economists to seek indicators of economic well-being that account for environmental costs. One of the best-known alternatives to the GNP is the **Index of Sustainable Economic Welfare** (**ISEW**), developed by the economists Herman Daly and John Cobb. The ISEW helps us measure sustainable growth that creates income and jobs without exacting a high environmental cost. It includes several environmental measures such as depletion of nonrenewable resources, loss of farmland from erosion and urbanization, wetlands loss, damage from air and water pollution, and even estimates of long-term damage from global warming and ozone depletion.

Because it includes so many variables, the ISEW is much more difficult to compute than GNP, especially for developing nations where environmental monitoring is poor. Enough U.S. data are available, however, and when the U.S. ISEW is computed, it presents a very different picture of the nation's well-being than GNP does: Instead of steadily increasing, ISEW has remained static or declined slightly since the late 1960s (**Figure 19-4**). It seems likely that once data are available for other countries, their ISEWs will show even greater declines in many cases. Developing countries with high populations and rapid environmental destruction are especially likely to show declines.

Economics of Sustainable Growth

Two basic economic systems have been used in the modern world, and neither has fostered sustainable growth. In **command economies**, the government tells individuals what to produce and sets prices for those products. The collapse of the Soviet Union demonstrated the major problem with command economies: The system develops widespread inefficiencies because product costs, which are artificially set, do not reflect real costs. These massive inefficiencies are also the reason more environmental damage has occurred in the countries of the former Soviet Union and other command economies than in countries with the other type of economy. These **market economies**, as they are called, are based on the "invisible hand," first discussed by Adam Smith over two hundred years ago. In a pure market economy, production and costs are determined by competition among individuals. Smith argued that self-interested individuals promote optimum allocation of resources in society, as if driven by an invisible hand. This system is more efficient than command economies because the laws of supply and demand set prices that generally reflect real costs to society.

Externalities: Market Failures and the Environment

Despite their general efficiency in allocating resources, market economies are not perfect. Economists have long noted that market failures occur where society is not always benefited by the individual pursuit of self-interest. For example, until the enactment of laws regulating child employment, children were used as a cheap source of labor in the sweatshops of the late 1800s (**Figure 19-5**). Child labor exacted a toll on society in the form of the psychological and physical injury to the children's development that was not included in the price of the cheap clothing or other products they made. Economists call such market failures **externalities**: Some costs of production are not included in (are external to) the price of the product. With ex-

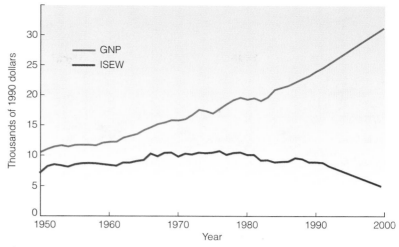

Figure 19-4 Per capita GNP and ISEW for the United States. (*Source*: Updated from *Saving the Planet: How to Shape an Environmentally Sustainable Global Economy* by Lester R. Brown, Christopher Flavin, and Sandra Postel, with the permission of W. W. Norton & Company, Inc. Copyright © 1991 by Worldwatch Institute.)

ternalities, society as a whole pays a cost that should be paid by the producer and consumer of the product. Externalities are a major reason why no pure market economy exists. Regulations and other forms of intervention that we will discuss shortly are necessary. Thus, almost all real economies in the world are mixed economies, which combine varying degrees of market and command traits. In the United States, the market economy traits are generally more prominent than in Western European countries, although both have mixed economies.

Pollution, resource depletion, and the many other ways that humans degrade their environment are environmental externalities, which society is forced to pay because they are not paid by the industry or other activity that produces the degradation. But how can markets generate such damaging externalities? Markets are generally efficient because they are based on individuals making rational decisions to maximize benefits and minimize costs. In theory, then, markets should act to reduce environmental degradation before the environmental costs exceed the benefits of production. The problem is that markets only work if the commodity is privately owned.

In Chapter 1, we examined Garrett Hardin's 1968 essay on *The Tragedy of the Commons*. Resources that are public goods, or commons, will eventually become depleted or polluted if the free market is left unchecked. Unlike a privately owned good, no one person or organization has a vested interest in taking care of public goods. If individuals

Figure 19-5   During the Industrial Revolution, (a) children often worked long hours in U.S. and European factories. (b) Child labor continues to be used in many developing nations today.

(a)

(b)

are left to pursue their own interests, they will destroy the commons, rendering it useless to everyone including themselves. This behavior is actually quite logical: Because no one owns the commons, someone else will deplete or pollute the resource if you do not.

Tragedy of the Commons: Depletion and Pollution

Depletion of a public resource is stimulated by the economic law of supply and demand: The price of goods increases as their supply diminishes. Thus, the scarcer the resource, the more expensive it will become. For example, the OPEC embargo of the early 1970s caused petroleum prices to skyrocket. A key implication for public goods is that as the resource becomes scarcer, there will be increasing pressure to exploit the resource even more. Scarcity promotes even greater scarcity because rising prices reward people who exploit what resource remains. Among the many examples that could be cited are the overexploitation of fishing grounds in many parts of today's oceans, overexploitation of game animals such as rhinos, soil exhaustion and erosion from overuse, and depletion of the highest-grade ore deposits in the United States.

Unsustainable growth is also characterized by pollution of the environmental commons for the same reason: Because no one has a vested interest in the commonly held property, people will freely pollute it. Thus, air pollution occurs because the atmosphere is a commonly held good that no one owns. The same is true for water pollution, for virtually no one owns an entire lake, much less an entire river, or ocean. Litter is usually worst in parks, along highways, and in other public places.

Solution: Internalizing Environmental Externalities

The solution to environmental externalities is to internalize the environmental costs by forcing the depleters and polluters to pay for the damage incurred to public environmental property. In the case of depletion of a commons, the greater price that accompanies scarcity must be counteracted. One way to do this is to impose a severance tax, which is levied on minerals as they are extracted. Many states levy severance taxes on coal that is mined. To offset the high price of scarce supplies, such a tax ideally would increase as the resource became depleted. Miners, drillers, loggers, fishermen, and other resource harvesters would then be discouraged from harvesting further. For renewable resources such as fishes or lobster, the tax would decrease once the supply (populations) had rebounded. For miners and other harvesters of nonrenewable resources, the tax would stay high, encouraging the industry to invest in finding alternative resources to use, conservation, and recycling. Similar taxes and other methods discussed below could be used in the case of pollution of a commons to force polluters to pay the costs of pollution.

What's the Environment Worth?
Cost Problems

A major obstacle to internalizing, or paying for, environmental costs has always been determining what those costs are. Four main costs influence and thus complicate the calculation of overall environmental costs of an activity:

- intangible costs
- hidden costs
- future costs
- unequally distributed costs.

Intangible costs include destruction of scenery and other subjective costs that are very difficult to measure because people place different values on them. To some people, a view of wilderness landscape has no value; to others, it is priceless. **Hidden costs** are environmental degradations that we are unaware of; for example, the full effects of pesticides and other pollutants with chronic toxicity may not be evident for many years. **Future costs** are costs that are passed on to subsequent generations. In theory, most environmental damage, especially to nonrenewable resources, includes at least some future costs. Some impacts, however, may have only future costs: They have no discernible cost now, but the cost will arise in the future. For example, most plants now becoming extinct have not been studied for their food and medicinal potential. A plant species that might go unnoticed now could be the best way to fight a disease that arises in the future. A major problem with all three of these costs is that society tends to ignore them. Intangible costs are not real costs to people who place little value on the environment. Hidden costs are not visible. Future costs are ignored by present generations who "discount the future" (Chapter 1) because it is uncertain.

Unequally distributed costs occur because environmental costs are almost never distributed evenly throughout society. Poorer people tend to pay relatively more of many environmental costs (**Figure 19-6**). The factory that has to pay higher costs for pollution control simply passes that cost on to the consumer by raising the price of its product. Whether the product is a car, tires, food, or clothing, the poorer person feels the impact of the higher cost more strongly. Similarly, poorer people tend to be the ones who become unemployed when a factory is closed because it pollutes too much or because the costs of controlling pollution are too high

for it to operate profitably. Unequal distribution of cost creates much of the environmental controversy you hear about in the news media. For instance, the spotted owl controversy was largely caused because loggers in the area must bear much more of the cost (their jobs) of preserving the owl's habitat than, say, fire fighters living in Oklahoma. The same is true when nuclear power plants are shut down, factories are closed, and many other activities are slowed for environmental reasons.

Pay Up! Persuasion, Regulation, Motivation

One can attempt to overcome these problems and determine true environmental costs. The Contingent Valuation Method (CVM) uses surveys and other techniques to find out how much people are willing to pay to visit national parks. This helps measure the intangible cost if a park is degraded. But the science of making such measurements is just beginning, and estimates are often controversial. Despite such problems, attempts at estimating environmental costs must be made because society can no longer afford to ignore them. Using such cost estimates, industrial societies have begun to discourage depletion and pollution by requiring resource depleters, and especially polluters, to pay costs of environmental degradation. As we will see, the current payments are generally far below the level needed for a sustainable global economy, but social pressures are growing to attain that goal. Polls regularly show

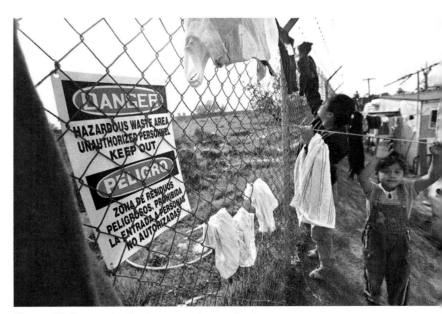

Figure 19-6   Poor families in many nations tend to bear a greater share of the burden of environmental problems. This lower-income family lives next to a waste site.

TABLE 19-3	Methods of Promoting Payment for Externalities		
	Effectiveness	**Cost of Implementing**	**Best Suited For**
Persuasion	Least	Lowest	As a supplement to other means
Regulation	Moderate	Highest	Situations with a few major violators
Economic motivation	Most	Moderate	Situations with many violators

that the majority of U.S. citizens are willing to pay extra for a cleaner environment.

Society can extract payment for formerly free environmental externalities in three basic ways: persuasion, government regulation, and economic motivation or incentives (Table 19-3). Persuasion usually only influences a few people. It is particularly ineffective for industrial activities since cheaters are rewarded by having lower costs and will soon drive out of business conscientious industries that pay more environmental costs. Persuasion's low cost, however, makes it a useful supplement in conjunction with regulation and/or economic incentives.

Government regulation restricts depletion and pollution by passing laws. Regulation is deceptively appealing. It is the most direct way to prevent environmentally damaging activities: ban plastic, require pollution controls, fine people for littering, close polluting factories, and so on (**Figure 19-7**). Many economists, however, including prominent environmental economists such as Lester Brown, agree that regulation is often more inefficient and more costly than using economic incentives. One reason is the high cost of enforcing regulations where many people are involved. Costs of enforcement are increased further because people often try to evade the laws. Thus, enforcement entails costs of monitoring plus costs of catching lawbreakers, prosecuting them, and, sometimes, keeping them in jail. For example, imagine the costs of trying to enforce a law requiring people to take public transportation to work once a week. In contrast, an economic incentive such as higher gasoline prices will encourage smaller cars, fuel economy, carpooling, and use of public transportation. Similarly, more people may recycle items if they are paid for doing so than if a law is passed requiring them to recycle.

Another criticism of regulations is that they tend to be relatively inflexible. For example, a law that requires all factories to meet the same rigid air quality standards for emissions does not take into account the age of the factories or what they produce. Such laws can lead to high prices for the consumer and force factories to close. The 1990 Clean Air Act Amendments, which allow trading and buying of pollution permits, achieve the same level of control at less social cost by recognizing that some factories can eliminate many emissions very cheaply while others cannot. Factories that would otherwise have to close can stay open and save jobs while other, newer or different factories can cut emissions to a greater degree.

Despite the drawbacks, regulation has been successfully used during the last few decades by many industrial nations, especially the United States. Regulation is most useful where there are few violators and each violator can do great harm to the environment. Nuclear waste disposal fits both of these criteria and is thus probably best controlled by regulation. Manufacturing CFCs, hunting whales, and

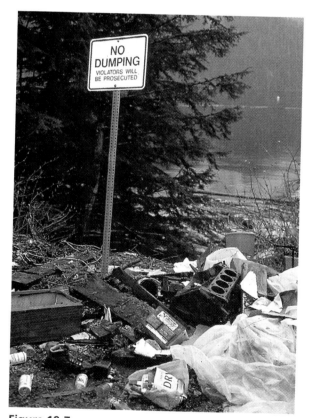

Figure 19-7   Laws against environmental damage are easy to make, but sometimes difficult and costly to enforce. This photograph of illegal dumping was taken in Alaska.

trading in endangered species are other examples of easily regulated activities. The great improvement in air and water quality in the United States has largely been due to the Clean Air Act, Safe Drinking Water Act, and other laws passed in the early 1970s that set uniform standards. In recent years, however, rising costs of regulation, changing economic conditions, increasing emphasis on input reduction, and many other factors have led to increasing interest in economic incentives as a way of sustaining the environment.

Economic Incentives to Sustain the Environment

By rewarding individuals monetarily, economic incentives help meet many goals effectively. Incentives are less expensive than regulation because they save on enforcement costs and are more flexible: Producers and consumers decide how to pay for the change to meet the goal. As each chooses the least costly way to achieve the goal, society as a whole has lower indirect costs. Environmental economists most often promote two basic types of economic incentives: government incentives and privatization.

Government incentives include taxes, subsidies, licenses, fees, vouchers, and many other ways that local, state, and national government can reward or discourage behaviors. These government incentives are often called *green fees*. The government incentives most often suggested are various types of taxes, or **green taxes**. For example, sulfur dioxide emissions can be discouraged directly by taxing emissions or indirectly by taxing high-sulfur coal, electricity consumption, or electricity generation (**Figure 19-8**). Such taxes are advocated by many economists as a relatively simple way to improve the functioning of the market to reflect an activity's true environmental cost, meaning the cost of externalities. Green taxes can do this by (1) serving as an incentive to slow depletion and pollution and (2) raising money that can go toward paying other environmental costs.

Green taxes are often an effective incentive because they promote all the major aspects of a sustainable economy: less depletion, recycling and use of renewable resources, and less pollution at the source. For example, a carbon tax on fossil fuels promotes conservation, encourages alternative energy research, and reduces greenhouse gas pollution. Similar taxes can be levied on water to promote conservation and desalination; on metals to promote conservation and recycling, and so on. Raising the price encourages people to stop discounting the future, conserve resources, and slow

pollution and solid waste disposal, as well as save money. Green taxes also raise money for environmental uses. Many billions of dollars could be raised by taxing polluting substances and scarce resources, especially from a carbon tax on fossil fuels.

Many estimates suggest that the overall burden of green taxes need not be great. The United States already has among the lowest taxes of all industrialized countries and imposes relatively few green taxes compared to Western Europe. More importantly, green taxes need not add new taxes; instead, taxes can be shifted from one activity to another. Income taxes, for instance, could be reduced by the revenue generated by the green taxes (**Table 19-4**). Taxes tend to discourage the activity being taxed. Instead of discouraging work, as income taxes do, green taxes discourage depletion and pollution.

The burden of green taxes could also be reduced by stopping subsidies on environmentally damaging activities. Many nations have a "reverse green tax," sometimes called "dirty subsidies." Instead of discouraging environmental damage with taxes, these actually encourage damage with government subsidies (payments). Examples in the United States include the following:

1. Subsidized mining of virgin ore deposits, which discourages recycling of metals (Chapter 10).

Figure 19-8   Green taxes can be indirect, such as a tax on sulfur in coal or on electricity generation or consumption, or direct, such as a tax on sulfur emitted.

TABLE 19-4	Examples of "Tax-Shifting" from Income to Environmentally Harmful Activities		
Nation, Year Implemented	**Taxes Cut On**	**Taxes Raised On**	**Revenue Shifted (%)**
Sweden, 1991	income	carbon, sulfur emissions	1.9%
Denmark, 1994	income	energy, water, landfills	2.5%
Spain, 1995	wages	gasoline	0.2%
Netherlands, 1999	income	fossil fuel, landfills, water	0.9%
Germany, 1999	wages	energy use	2.1%

Source: Modified from *Eco-Economy*, Table 11-1, by Lester Brown. New York: Norton, 2001.

2. Subsidized irrigation water for farmers, as in California (Chapter 9).
3. Subsidized grazing of animals on government land. U.S. ranchers can often graze cattle on federal land for much less than the cost of nearby private lands (Chapter 12).
4. Subsidized logging in government-owned forests (Chapter 12).

Many other countries also subsidize resource use and encourage rapid depletion. Until recently, Brazil promoted logging of rainforests by building roads, making direct payments to timber companies, and many other subsidies. According to the World Bank, many agricultural countries subsidize pesticides.

Such subsidies decrease costs below market price, resulting in massive waste and pollution. For example, the World Bank estimated that a staggering 5445 million metric tons of carbon pollution could have been eliminated between 1991 and 2000 if developing countries and the former Soviet countries had stopped subsidizing fossil fuels. Most consumption subsidies are artifacts of outdated economic policies and protectionist trade policies. For instance, the United States once wanted to encourage mining of ores to provide a growing manufacturing economy with cheap raw materials. Environmental economics argues, however, that these subsidies do not promote sustainable economies. By lowering costs well below the true cost, they discourage conservation, recycling, and pollution reduction (Chapter 1).

The second basic economic incentive is **privatization**, in which environmental resources become the property of individuals. By transferring resources from the commons to individuals or companies that have a vested interest in them, the basic cause of the tragedy of the commons is removed. Examples include commercial forests that raise trees for timber, commercial parks that raise endangered species, and individual ownership of water rights in the western United States.

Despite some success with privatization, many environmental economists believe that it is often less useful than green taxes (and other green fees) in solving environmental problems. One problem is that many environmental commons cannot practically be privatized. For instance, how can you own a part of the atmosphere? Similarly, it is impractical to have ownership of the oceans or rivers that travel through land owned by many people. Another problem is that even where private ownership is practical, there is no guarantee that market prices will always reflect true environmental costs. For example, private ownership of oil and ore deposits will still lead to depletion of those resources unless society implements regulations, taxes, or some other mechanisms that prevent market failure.

The Low Cost of Sustainability

Many environmental economists believe that the transition to a sustainable economy can be made with surprisingly little sacrifice, if it is begun soon. They say the transition can be paid for by the savings that will be realized when currently wasted resources are used more efficiently. Three types of change will help to pay for currently existing externalities:

1. Cost-free changes to eliminate policies that cause environmental damage.
2. Investments that are economically and environmentally profitable.
3. Changes that cost money but are environmentally profitable.

The first two types of changes require no economic sacrifices. First, we can simply stop subsidies and other government policies that actively promote environmental damage, such as subsidized mining and rainforest destruction. Second, public and private investments in water supplies, conservation, and other activities both pay for themselves and make a monetary profit. We saw, for example, that many investments in energy efficiency not only reduce fossil fuel pollution and conserve resources, but also increase net profits by reducing waste (Chapter 8).

In some cases, private investments in businesses that develop such green technology can benefit both the environment and the investor. **Green investing** is a type of socially responsible investing that has become increasingly popular as an alternative to investments made on the basis of profits alone (see Case Study 19-1). Ending current taxpayer subsidies that promote unsustainable businesses, such as the fossil fuel and nuclear industries, would encourage green investing by giving sustainable industries a level playing field. Businesses would have a greater incentive, for example, to develop more fuel-efficient cars and alternative fuels.

Only the third group of changes require actual monetary sacrifice. Examples would be green taxes on smokestack emissions and landfill wastes, which would raise the cost of products and garbage removal. However, even these measures have no net cost when the benefits of a cleaner environment are included in the calculation.

The basic effect of these changes would be to restructure the economy and rechannel industry and jobs into producing goods and services that the environment can sustain for many centuries. To repeat a key point, this change will not lead to unemployment, but will shift employment to sustainable jobs. For example, as internal combustion engines become less important, employment making solar batteries will increase. And, as noted above, such sustainable activities tend to provide more jobs than unsustainable ones. Of course, the transition to sustainability will not always be easy, but economic transitions have historically required social adaptations (**Figure 19-9**). The difficulties can be minimized if the transition is gradually carried out. A phase-in period of 5 to 10 years is often suggested. In his book, *Eco-Economy*, Lester Brown discusses a number of examples of fast-growing industries that are not only profitable but also more sustainable and cleaner than traditional industries: fish farming, bicycle manufacturing, wind farm construction, wind turbine and fuel cell manufacturing, hydrogen generation, light rail construction, and tree planting.

Unfortunately, the current transition to sustainability, especially in the United States, is too gradual for most who are concerned for the environment. In the United States, there has been very little support for green taxes. In 1990, Congress approved a 4 cent per gallon (3.785 liters) gasoline tax, but this is a small fraction of the tax required to significantly discourage auto use, lead to more efficient cars, and reduce emissions. Studies show that current prices must be doubled or tripled to significantly shift consumption patterns. In the case of gasoline, a tax of at least $1 per gallon will be required. Such a tax would put U.S. gasoline costs at about the level of those in Western Europe and Japan, where the high costs have greatly improved conservation and efficiency. The speed and power of green taxes to improve the environment were demonstrated in 1989, when a high tax on leaded gasoline went into effect in the United Kingdom. Just one year later, in 1990, the use of unleaded gasoline had climbed from 4% to 30%.

Figure 19-9 Economic transitions often involve the demise of some careers, forcing people to adjust by seeking new careers. The village blacksmith, as depicted in this late eighteenth century American engraving, is now a thing of the past.

Poverty and the Global Environment

Global environmental problems are tightly interwoven with poor economic conditions. Poor nations find themselves in a vicious cycle. Poverty often correlates with overpopulation, which then leads to ecological decline as greater numbers of

CASE STUDY 19-1

Would You Be a Green Investor?

Visit www.environment.jbpub.com for more information.

Economists often say that people vote with their dollars, especially in a market economy. This means that our spending reflects what we truly value. A good example is the growing popularity of socially responsible investing, also known as SRI to investment brokers.

Socially responsible investing refers to financial investments that are based on more than merely the desire to maximize your profit return. You tell your investment broker, for instance, that you want to avoid investing in certain companies because you disapprove of some of their activities and do not want to support them. Popular examples are socially screened mutual funds that seek to invest only in companies that have a clean record on the environment or other social issues. The first socially screened fund was the Pax World Fund, created in 1970 to avoid investing in suppliers to the Vietnam War.

In the early 1990s, the number of socially responsible investors grew dramatically. According to the Investor Responsibility Research Center (IRRC), which promotes SRI by supplying subscribers with information about corporate behavior on social issues, the number of socially screened mutual funds grew at four times the rate of traditional funds. Total assets of these funds are well over $9 billion.

The specific social issues promoted vary among the mutual funds. Some funds consider a wide range of issues. While these issues usually include the environment, they also include a company's hiring practices on women and minorities and its record on animal testing. Investment in tobacco and alcohol companies may also be avoided. But some funds are strictly environmental; in fact, a 1993 survey of money managers concluded that the issue that concerned most investors is the environment. Green Century Capital Management and Merrill Lynch's Ecological Trust are examples of funds that focus on a company's environmental record.

As always, one must be wary of the "greenwashing" problem. Greenwashing refers to false environmental claims that are sometimes made by companies to use the popularity of environmental issues to attract consumers or, in this case, investors. A number of waste management companies, for instance, are known as environmental sector stocks because they are in the pollution control and trash collection business. Some brokers have therefore included them as desirable green investments. But many of these companies have been fined millions of dollars for flagrant environmental violations such as illegal dumping and disposal of waste. Most environmental funds avoid them, preferring instead to invest in companies with no environmental violations, especially if the company actively promotes sustainability by making recycled

---

people overuse resources and pollute. As natural resources dwindle and are degraded, poverty increases, often leading to more population growth. Many nations, even those as different as Poland and Haiti, illustrate how economic problems are closely related to environmental problems.

In 2002, approximately one-fifth of the world's population lived in abject poverty, with incomes of a dollar a day or less. Billions more have standards of living well below that of U.S. and Canadian citizens. Increasing the standard of living in these countries involves far more than simply "developing" their economies by transferring current technologies to make their agricultural economies more industrial. Most current technologies use far more resources than the world could afford if everyone used them. The wealthiest 25% of the world's population now consume about 75% of the world's resources. The environmental economist Herman Daly has estimated that for all developing nations to attain the income levels of the leading industrial nations such as Japan and the United States, the consumption of the Earth's resources would have to increase by a factor of 36, or 3600%, if current technologies are used to industrialize.

This would require the natural resources provided by at least three more planets like Earth. For example, many geologists estimate that the world's entire oil supply would be gone in just a few years if all countries used as much per person as the United States. In addition to consuming more resources, the technologies used by the wealthiest nations also produce much more solid waste, greenhouse gases, and other pollution per person (**Figure 19-10**). The solution, as noted in many parts of this book, is to lift developing nations out of poverty by transferring sustainable, or green, technologies that produce sustainable growth (see Case Study 19-2).

products, helping rebuild wetlands on its property, and so on.

Another point to consider is how important profitability is to you. Some money managers caution investors that environmental (and all SRI) funds and stocks may, on average, produce lower rates of return. This is not always true; the Parnassus Fund is often cited as a consistently high-performing investment. But brokers often note that because SRI funds are constrained in where they can invest, they may not be as productive. On the other hand, many investors are willing to risk lower profits because they are concerned about how a company makes its money. Does this potential sacrifice have any effect? Does SRI work? Although SRI is still relatively new, it already appears to be having a significant effect. One sign of this is the rapid growth of organizations such as the IRRC that provide subscribers with information about a company's record on social issues. Perhaps even more significant is that the major pension funds, the real titans of investment, are beginning to consider SRI.

With the vast amounts of money they have at their disposal, these funds can strongly influence a company by threatening to withdraw their investments. In 1990, the world's largest pension fund, the teachers and professors retirement fund with assets of over $100 billion, began allowing members to put money into socially screened accounts. These members often send resolutions to companies requesting that they adhere to specific environmental principles. As one official noted, "companies who receive a resolution from a public pension fund take it very seriously."

So, does the investment future look green? It is already becoming standard practice before company mergers and buyouts to do an analysis on a company's environmental record. According to a veteran Wall Street broker, "the companies that are . . . getting themselves into cleaner businesses are the ones that look good for the future. We are in between a long period when it paid to be dirty and when it will be seen as profitable to be clean."

Questions

1. If you had $10,000 to invest right now, would you be a "green" investor? Why or why not? Which companies would you specifically prefer to invest in?

2. If you had $10,000 to invest right now, would socially responsible investing (of any kind) be an option for you? Why or why not? Which companies would you specifically prefer to invest in?

3. Some critics have called SRI a mere ploy to gain money from "bleeding hearts." Is this unjustified cynicism? Do they have a valid point?

Figure 19-10 (a) Municipal waste and (b) carbon dioxide are just two of many pollutants that increase rapidly with increasing wealth. (*Source:* Excerpted from *World Development Report 1992*. Copyright © 1992 by The International Bank for Reconstruction and Development/The World Bank. Reprinted by permission of Oxford University Press, Inc.)

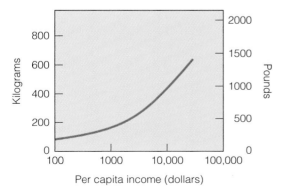

(a) Municipal wastes per capita

(b) Carbon dioxide emissions per capita

CASE STUDY 19-2

Cookstoves in Poor Nations: Small Is Beautiful (and Sustainable)

In his classic 1973 book, *Small Is Beautiful*, E. F. Schumacher argued eloquently for the importance of "appropriate technology." These are sustainable technologies that are affordable and can be produced and maintained locally. Such technologies not only have less environmental impact than traditional industrial technologies, but also promote sustainable economic growth and increase the standard of living in poor nations.

The immense power of such small-scale local technologies occurs from the cumulative effects when many people use them. Thus, as often happens, large-scale change is a product of seemingly mundane trivial changes multiplied many times. One of the most dramatic examples is the cookstove. Half of the world's population, mainly in developing nations, prepare food and heat their homes with fires that burn dung, wood, or other combustible materials.

While seemingly rather harmless, these cookstoves are major causes of massive environmental destruction. Many thousands of acres of forests and other ecosystems are denuded and degraded as people seek firewood and other biomass fuels. Besides the obvious harm to the ecosystem, deforestation and plant denudation are a major cause of soil erosion. Human health suffers too. Smoke from indoor fires often exceeds 20 times the limit of ash and other pollutant levels recommended by the World Health Organization.

Traditional cookstoves can be improved in a number of ways. The most basic is to increase the heating efficiency. Traditional open fires with a pot above are only about 10% efficient; 90% of the heat goes into the air and is wasted. Traditional metal stoves are not much better. Lining the stove with ceramic, brick, or clay increases the heating efficiency to at least 20% and often as much as 40%.

Increasing efficiency to just 20% can reduce the amount of firewood (ecological damage) and smoke (human health damage) by half, a good example of how powerful such

a simple change can be. Using less firewood or other fuel also saves money. In Kenya, switching to more efficient cookstoves typically saves about $65 per year, as much as one-fifth the annual income of many families. Since the cost of such lined cookstoves is usually less than $5, this is a good investment. China has over 120 million insulated cookstoves due to a government program promoting them. Millions of similar stoves are becoming popular throughout Asia and Africa, thanks to dozens of government and private programs.

A second, usually better, type of improvement is the solar cookstove or oven (**Figure 1**). When the Sun shines, the food cooks in a glass-covered box that traps the heat. The walls and floor of the box are lined with reflective metal or foil. On a sunny day, meat stews and rice dishes can be fully cooked in two to five hours. No biomass fuel is used, and no unhealthy smoke is produced.

The solar cookstove is still relatively new and is not yet as widely used as insulated biomass cookstoves. Cost is a factor, too, since solar cookstoves, at $20–$40 apiece, are considerably more expensive. But mass production of solar cookstoves, often by local craftspeople, is expected to greatly reduce prices, just as it has with the insulated biomass cookstoves over the last few years. Considering that solar energy greatly reduces the need to buy firewood or other fuels, and thus saves even more money, they should become quite popular.

The experience with cookstove improvements has been educational in many ways. Early efforts to introduce the new technologies were unsuccessful because many social factors, such as community needs and customs, were not considered. In many areas, people will adopt newer cookstoves only when community groups are allowed to modify the designs and local artisans become stove makers. Improved cookstoves thus not only reduce harm to the environment, they also produce jobs for the community.

Questions

1. What do you think is meant by the phrase "small is beautiful" in the context used here?
2. Think of examples of other kinds of "appropriate technology" that can benefit the environment and the economy of a country.

Figure 1 A solar oven, being used in Kenya.

Origins of Global Poverty

Economic theories of the 1940s through 1970s generally maintained that the poverty of the developing (then called Third World) countries was a temporary situation. Their agricultural economies, which produced little wealth, were largely a product of nineteenth-century colonial policies that had made little effort to modernize the native societies. Supposedly, trade with richer nations and some monetary aid would in time provide the technology and other necessities for the poor economies to reach **economic takeoff**. This was the point where poor countries could produce enough wealth to financially support themselves, export manufactured goods, and eventually attain the living standards of the United States, Canada, Europe, and other industrialized (First World) countries. As early as 1948, President Harry Truman, in his inaugural address, urged not only the rebuilding of Europe with the Marshall Plan, but the development of the world's poor countries. President John F. Kennedy declared the 1960s to be the "decade of development" and pledged 1% of the U.S. GNP toward this goal.

Unfortunately, economic takeoff has not occurred in most poor countries. Instead of becoming richer, poor nations have become poorer relative to the richer nations (**Figure 19-11**). In 1950, the average income in the rich nations was about 7 times as much as the average income in the poorest nations. By the early 2000s, the differential had risen to over 20.

Why haven't the poorer countries experienced economic takeoff? Economists disagree on the answer, but a number of factors seem important. First, much of the money acquired by developing nations has been badly mismanaged. Instead of investing in roads, education, and other necessities for building a prosperous economy, they have often spent the money on military equipment to help keep dictators and elite social classes in power. Recent trends toward democracy in Africa, Asia, and Latin America may alleviate this problem. A second, and more important, reason is that the world is much more competitive than when the United States and other developed countries were industrializing. It is very difficult for less developed countries to make products that can compete with advanced technologies.

As a result, poorer countries must sell whatever products they can, which usually means their natural resources. Often these are nonrenewable: oil, high-grade ores, native species being driven toward extinction, and so on. Rather than investing in equipment and other capital that will create more wealth, these countries are spending their resources in ways that will not provide them with long-term sustainable growth. We saw in Chapter 14 that the loss of these irreplaceable natural resources is accelerating because they are cheap. The price paid for these resources, mainly by consumers in the industrialized nations, does not reflect their true environmental costs. The value of such resources to future generations, to local peoples, and to local ecosystems is discounted (and often totally omitted) by the consumers who buy them.

Eliminating Global Poverty

The growing disparity between poor and rich countries has led to concern that a new economic war between North and South will replace the now-defunct military Cold War between East and West. Most of the wealthier nations are in the higher latitudes, mainly in the Northern Hemisphere (an exception is Australia). Poor countries are found to the south, around the equator and in most of the Southern Hemisphere (**Figure 19-12**). These poor tropical na-

Figure 19-12   This polluted slum in Brazil is an all-too-common example of the economic disparity between poor tropical countries and the rich countries to the north.

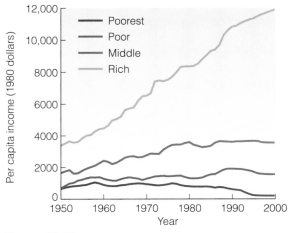

Figure 19-11   Per capita income has steadily increased in the rich nations, but has remained relatively stable or even decreased in the poorer nations. (*Source*: Updated from P. Ekins, *Green Economics* [London: Gaia Books, 1992], p. 32. Used with permission of Gaia Books.)

tions share many common interests, as is readily apparent at the United Nations, where poor countries often form voting blocs on international issues. One solution to this growing poor–rich division is political: National governments must meet and reach agreements on international issues that affect them all. International meetings on economic and military issues have been held for years, but not until 1992 did international environmental issues came to the fore. In that year, the **Earth Summit** in Rio de Janeiro, which was attended by 178 nations, addressed global concerns such as global warming, biodiversity, and deforestation. Although the summit was a great success in demonstrating global concern for these issues and some important agreements were signed, many participants fear that many of the agreements will be ineffective because they either lack means of enforcement or take steps that are too little or too late to rectify major environmental problems.

Economic Aid and Debt Cancellation. We noted above that economic aid has not led to economic takeoff because of money mismanagement and the competitive global economy, among other factors. For these and other reasons, many wealthier countries have reduced their aid to poor countries in recent years. Most industrialized countries have never come close to giving the 1% GNP pledged in the early 1960s. The United States annually gives among the lowest proportionate aid levels of any nation. Each year, the United States donates less than 0.20% of its GNP to foreign aid. Norway and the Netherlands rank among the highest in the world, donating about 1% of their GNPs. Furthermore, much U.S. aid has always gone to just a few nations, such as Israel and Egypt, for military reasons.

The lack of donated money has meant that poor countries have had to borrow from richer countries, causing a debt crisis, also called a "debt bomb" (Chapter 1). By the early 2000s, the debt of the developing nations had exceeded $1.5 trillion, amounting to about half of those countries' collective GNP. Because the interest on this debt is higher than the foreign aid provided by rich countries, since 1982, there has been a net flow of money from the poor nations to the rich nations.

As a first step in resolving this debt crisis, the United Nations estimates that the environmental and developmental programs proposed at the Earth Summit would require rich countries to contribute at least 0.7% of their GNP as aid to the poor countries. This sum is less than 10% of the annual military spending of the United States. In most other

wealthy nations, it would be less than 33% of the military budget.

Besides direct payments, the rich nations can provide many other types of economic aid to ease the debt crisis. One is cancellation of debt. By 2002, the United States, Germany, Canada, and Great Britain had canceled over $10 billion in loans to African nations, largely for humanitarian reasons. However, this is a tiny fraction compared to the more than $1.5 trillion owed by the poor countries. The Worldwatch Institute estimates that about 60% of this debt will have to be canceled if the poor countries are to save enough money (instead of making loan payments) to begin sustainable development. The sum does not seem so great when compared to the Gulf War's cost of about $40 billion, which was raised in a few months, and especially when the benefits of saving the global environment are compared to the benefits of fighting a brief war for an unsustainable resource (oil). Even so, outright cancellation of such a large debt is very unpopular among the taxpayers of richer nations, especially the United States.

To make the cancellation more appealing, some observers have suggested that the debt be traded for something valuable. Among the most notable trades are the **debt-for-nature-swaps** suggested by the biologist Thomas Lovejoy. With these, a conservation organization pays about 15%–30% of the debt to the lender, and the rest of the debt is canceled. In exchange, the poor country carries out specified environmental programs. In Chapter 12, for instance, we saw that land was set aside for preservation of tropical forests. Other examples include Costa Rica, where the Dutch government in 1989 swapped $33 million in debt for $10 million worth of programs in reforestation, water resource management, and soil conservation. Similarly, the United States has agreed to allow Poland to cancel 10% of its debt if the money was spent instead on toxic clean-up and other badly needed environmental restoration.

Private Investments and Business Loans. Debt cancellation, debt swaps, and even direct donations of money are largely short-term solutions. The fundamental problem is that the developing nations are not being paid the true costs for their resources, when such values as environmental uses are considered. If the value of reducing global warming, conserving rare species and rare genes for agriculture and pharmaceuticals, and preserving uses for future generations were included in the payments, money would flow into these poor nations. Such a flow is not aid; it is simply the proper payment for the true value of their resources.

The best way to obtain the true value of resources is to produce a healthy local economy so that profits are channeled to the local inhabitants of a poor nation. This usually requires investments and business loans from wealthy nations. An important kind of business loan are so-called **microloans**, which are small loans made to small businesses that promote sustainable activities. This is another example of the grassroots or bottom-up approach to environmental problems. In this case, such loans improve both the economy and the natural environment of a poor nation. For example, microloans have been made to family businesses in Brazil that practice sustainable harvesting of the rainforest, and to small businesses in India that build solar ovens (and thereby reduce deforestation; see Case Study 19-2). Microloans are sometimes given out by government agencies of wealthy nations, the World Bank, and other international lending agencies. They are also given out by nongovernmental organizations, often called NGOs.

Once enough money is made available to developing countries through a mixture of loans and gifts, most economists believe that further funding will be added by economic growth. Private investment from richer countries will occur as the poorer countries develop stable, growing economies. This economic growth will itself produce profits that can be reinvested and help pay off debts. However, the environmental goal is for this growth to be sustainable with minimal ecological harm.

Developing a Sustainable Economy

Most environmental economists agree that a sustainable world economy can still be built, but only if major steps are taken within the next two decades. These steps can be summarized as follows: (1) increase the flow of money into developing countries, reversing the present flow away from them; (2) channel the money flow into activities that not only produce wealth but are sustainable; and (3) establish some social order that promotes fair distribution of that wealth or access to it.

Many experts believe that these steps can be most effectively carried out if a central global financial institution oversees the money flow. For instance, if richer nations were to contribute 0.7% of their GNP as discussed above, the money would be given to this central institution, which would then distribute it. A possible candidate is the World Bank, which is already the largest international lender in the world. Despite the many advantages of having a single, central institution, political differences among countries probably render it impractical. Also, there is no guarantee that the money would be distributed in the best way. In the early 1980s, the World Bank and similar institutions came under intense criticism for disregarding the environmental impacts of their loans. They funded dam building and irrigation projects that destroyed huge tracts of land and polluted rivers; they also funded massive construction and deforestation projects in many countries (**Figure 19-13**). This criticism led to the creation of an Environment Department in the bank that investigates the environmental impacts of

Figure 19-13   Financing the building of dams in developing nations was once a very common, and very ecologically destructive, practice of the World Bank. Persistent pressures by many environmental groups have helped reduce, but not eliminate, such financing of unnecessary environmental damage. Shown here is the Itaipu Dam, a Brazilian hydroelectric project.

funded projects. However, this department has been widely criticized as being too small and ineffective; environmentally damaging loans still occur.

Aside from the World Bank, an increasingly important institution influencing developing nations is the **World Trade Organization (WTO)**. The WTO was founded in 1995 to reduce trade barriers and settle trade disputes between nations. Because of its focus on economic growth, the WTO has been increasingly criticized by environmentalists as well as advocates for social justice (see Case Study 19-3).

Besides governmental organizations, promoting sustainable economies can also be done by private groups. We discuss the importance of nongovernmental organizations in Chapter 20. Even private individuals can make a difference.

Ted Turner, founder of CNN, donated $1 billion to the United Nations in 1997 to support work on population stabilization and environmental protection. He created the UN Foundation through which other billionaires could "give back" to the global community. A similar effort has since been made by Bill Gates, founder of Microsoft, in starting a large foundation for coping with health and population issues in poor nations. Given the enormous wealth in the private sector of wealthy nations, these types of initiatives can potentially promote even greater changes than is possible with governmental initiatives.

Is a sustainable global economy just a lofty ideal that is unrealistic? Not in basic economic calculations. **Table 19-5** outlines the estimated costs and

TABLE 19-5	Estimated Costs and Long-Term Benefits of Selected Environmental Programs in Developing Countries			
Program	Billions of Dollars a Year	As a Percentage of GDP in 2000	As a Percentage of GDP Growth, 1990–2000	Long Term Benefits
Increased investment in water and sanitation	10.0	0.2	0.5	Over 2 billion more people provided with service. Major labor savings and health benefits. Child mortality reduced by more than 3 million a year.
Controlling particulate matter (PM) emissions from coal-fired power stations	2.0	0.04	0.1	PM emissions virtually eliminated. Large reductions in respiratory illnesses and acid deposition.
Reducing acid deposition from new coal-fired stations	5.0	0.1	0.25	
Changing to unleaded fuels; controls on the main pollutants from vehicles	10.0	0.2	0.5	Elimination of pollution from lead; more than 90% reductions in other pollutants.
Reducing emissions, effluents, and waste from industry	10.0–15.0	0.2–0.3	0.5–0.7	Appreciable reductions in levels of ambient pollution despite rapid industrial growth. Low-waste processes often a source of cost savings from industry.
Soil conservation and reforestation, including extension and training	15.0–20.0	0.3–0.4	0.7–1.0	Improvements in yields and productivity of agriculture and forests. Lower pressures on natural forests. All areas eventually brought under sustainable forms of cultivation.
Additional resources for agricultural and forestry research, in relation to projected levels, and for resource surveys	5.0	0.1	0.2	
Family planning (incremental costs of an expanded program)	7.0	0.1	0.3	Long-term world population stabilizes at 10 billion instead of 12.5 billion.
Increasing primary and secondary education for girls	2.5	0.05	0.1	Primary education for girls extended to 25 million more girls, and secondary education to 21 million more.

Source: *World Development Report 1992* (New York: World Bank, 1992), p. 174.

The World Trade Organization: Wealth versus Environment and Justice?

Today, global environmental issues are inexorably associated with the broader issues of the relationships between the North and the South—the industrialized nations and the developing nations—as well as the movement to liberalize world trade. Furthermore, the concept of sustainable development includes the notion of equity between different peoples and nations of the world, so it is only fair that the developed countries help the developing countries address environmental issues of global impact.

Clearly, unilateral action by a single nation will do little to correct global environmental problems that are collectively caused by many countries. For instance, curbs on greenhouse or CFC emissions by a single country will not solve the global problem of atmospheric deterioration. In addition, we now realize that environmental deterioration and the depletion of resources, particularly in developing countries, often promote cycles of poverty that lead to further environmental deterioration. The North is increasingly pressuring the South to end environmental deterioration and protect its remaining natural resources (such as tropical rainforests) for the good of the planet as a whole, yet the North has not set a very good example of resource conservation in the past. Furthermore, many developing countries are heavily in debt to developed countries and must exploit their natural resources to service that debt.

International trade agreements, such as the North American Free Trade Agreement (NAFTA), can also have direct environmental ramifications. Major issues include whether free trade in goods and products will also entail free trade in pollutants or environmentally destructive practices. A country that requires recyclable, and thus more expensive, packaging places an added burden on foreign manufacters who do not have to comply with the same requirements at home. The result can be trade disputes and international discord. NAFTA acknowledges that individual nations may enact environmental, health, and safety laws as long as they are "necessary" and based on solid scientific data. But defining "necessary" and solid science in this context can be difficult. As some commentators have noted, many U.S. environmental laws are based more on political compromise than science and thus may not meet the standards of NAFTA or other international trade agreements.

The most visible place where these issues have recently come to public attention is with the World Trade Organization (WTO). Founded in 1995 to replace GATT (General Agreement on Tariffs and Trade), the WTO has over 100 nations in its membership who attend meetings intended to promote global trade and economic growth. The WTO has gained a reputation among many critics who argue that it emphasizes economic growth at the expense of environmental regulations and social justice. For example, critics note that whenever the WTO has negotiated a case where there has been a conflict between trade versus environmental regulations or social injustice (such as the use of human labor under poor living conditions), the WTO has generally ruled in favor of trade.

As a result, WTO meetings typically attract many protestors. In late 1999, a meeting in Seattle was attended by 5000 delegates from over 150 nations, but it also attracted 50,000 protestors who used civil disobedience and other means to voice their concern over the direction that the WTO was taking (**Figure 1**).

Questions

1. Do you think civil disobedience at the WTO meetings is justified? Is it the right course of action? Would you do it?
2. Should the WTO be eliminated? Why or why not?

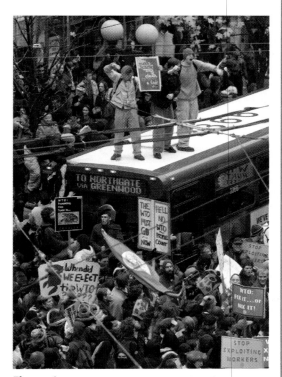

Figure 1 Protesters at the 1999 World Trade Organization meeting in Seattle.

long-term benefits of many environmental programs that can be undertaken in developing countries. The costs are estimated as a percentage of the GDP (Gross Domestic Product) of developing countries in the year 2000. GDP is the amount of wealth produced within a country, as opposed to GNP, which includes exports and wealth produced from outside the country. The key point is that the cost of all these environmental programs is only a small fraction of the wealth produced by a healthy, growing economy. The total estimated cost of the programs listed in the table is about $75 billion, or only about 1.4% of the combined GDPs of developing countries in the year 2000. Of course, Table 19-5 is not comprehensive, but the World Bank estimates that even if rainforest protection, pollution cleanup, alternative energy development, and other comprehensive items were included, the cost would still not exceed 3% of the GDP of the developing countries. These estimates are obviously very approximate, but they provide an essential view of the true economic costs of a global, sustainable society. They show that a sustainable global society is a realistic, attainable goal.

STUDY GUIDE

SUMMARY

- Environmental economics blends the short-term need for jobs and the long-term need to protect the environment.
- Environmental economics recognizes the need for sustainable growth, which uses minimal resources, controls pollution by input control, and relies on renewable resources.
- The GNP does not subtract the environmental costs incurred when wealth is created. A well-known alternative is the Index of Sustainable Economic Welfare (ISEW), which includes several environmental measures.
- Market failures occur in a market economy when society is not benefited by this individual pursuit of self-interest. These market failures, known as externalities, do not include all costs of production in the price of a product.
- When dealing with the environment, externalities include resource depletion and pollution.
- The solution to environmental externalities is to internalize the cost by forcing consumers to pay for the true environmental costs. Environmental costs are complicated by in-

tangible costs, hidden costs, future costs, and unequal distributions of cost.
- Society can extract payment for environmental externalities by persuasion, government regulation, and economic motivation, including green taxes and privatization.
- The solution to global poverty and many global environment problems is sustainable development.
- Developing countries have not experienced economic takeoff for various reasons including their mismanagement of money and the competitive world in which they are becoming industrialized.
- The developing nation's debt crisis may be at least partly resolved by direct economic aid and cancellation of debts.
- Environmental economists believe that a sustainable world economy may be built by increasing the flow of money to developing nations and channeling money into wealth-producing, sustainable activities within a social structure that fairly distributes the wealth.

KEY TERMS

Command economies	Future costs	Intangible costs
Cowboy economics	Green investing	Market economies
Debt-for-nature swaps	Greenmail	Microloan
Earth Summit	Green taxes	Privatization
Economics	Green technologies	Sustainable growth
Economic takeoff	Gross National Product (GNP)	Throwaway society
Environmental economics	Hidden costs	Unequally distributed costs
Externalities	Index of Sustainable Economic	World Trade Organization
False dichotomy	Welfare (ISEW)	(WTO)

STUDY QUESTIONS

1. What are two ways in which economics is concerned with environmental issues? Name three ways that the economic approach differs from that of environmental science.
2. What is a *false dichotomy*? Why is the issue of jobs versus environment a false dichotomy?
3. Why do environmentally friendly activities produce so many jobs? What is greenmail?
4. Define *environmental economics*, *sustainable growth*, and *cowboy economics*.
5. How does traditional economics measure the total output of goods and services of a country? What environmental measures does the ISEW include?
6. What are the two basic economic systems of the modern world? Which is more efficient? Why?
7. What are externalities? What is the tragedy of the commons?
8. Why does scarcity promote even greater scarcity? Give an example.
9. What are some criticisms of government regulation to solve pollution and other problems?
10. List and define the four main costs that complicate the calculation of the overall environmental costs of an activity.
11. What are three basic ways that society can extract payment for a formerly free environmental externality? Define them. Which is least effective?
12. What is *economic takeoff*? What are some factors that explain why it has not occurred in many poor nations?
13. In what year did the net flow of money begin to flow from poor to rich nations? What is the fundamental problem causing this net flow in the wrong direction? Give examples.
14. Explain the concept of green taxes. Provide examples and explain the advantages and disadvantages of their use.

WHAT'S THE EVIDENCE?

1. The authors state that the GNP is misleading. What evidence is presented to support this?
2. This chapter implies that green taxes are an effective way of letting the free market solve environmental problems. Is the evidence presented for this convincing? Why or why not?

CALCULATIONS

1. If the U.S. gross domestic product (GDP) was about $9.3 trillion per year (1999 figure), and the United States donated 1% of that to poor nations, how much money would be donated?
2. If a U.S. GDP of $10 trillion grew by 10% over 5 years, how much would be donated to poor nations 5 years hence if 1% were given?

ILLUSTRATION AND TABLE REVIEW

1. Based on Table 19-5, which environmental program gives the most long-term benefits, in your judgment?
2. Based on Table 19-5, which environmental program or programs will be the most expensive to implement? The least expensive?

WWW.ENVIRONMENT.JBPUB.COM

www.environment.jbpub.com

Connect to this book's web site: http://www.environment.jbpub.com/ The site features eLearning, an online review area that provides quizzes, chapter outlines, and other tools to help you study for your class. We've posted the answers to the book's odd-numbered questions. You can also follow useful links for more in-depth information. The site provides sources for the information in this chapter and recommendations for further readings.

Progress as depicted in an 1868 Currier and Ives illustration: "Across the Continent: Westward the Course of Empire Takes its Way." One of the most influential ethical traditions in the modern West is utilitarianism. The basic principle of utilitarianism is that the overall good should be maximized. Utilitarianism's emphasis on "the greatest good for the greatest number" is often closely linked with the concepts of progress and democracy. (*Source*: "Across the Continent: Westward the Course of Empire Takes its Way." Artist Frances F. Palmer; Publisher: Currier & Ives, 1868; Museum of the City of New York, 56.300.107. The Harry T. Peters Collection.)

CHAPTER 20

Historical and Cultural Aspects of Environmental Concerns

Ever since the origin of our species, humans have interacted with the natural environment. Our current relationship with the environment is the result of definite historical trends and human actions. Although we may not have intended to bring about global environmental changes, humankind has clearly and unquestionably had a global impact on natural systems. The depletion of the stratospheric ozone layer by artificial CFCs, to cite just one example, cannot be attributed to natural causes.

Humans are social creatures, and a human society's most characteristic attributes are its **social institutions**—its familial, political, educational, religious, and economic systems. The social institutions of any society reflect and express its underlying beliefs and values. Some critics argue that the values, beliefs, and institutions of modern, technologically oriented Western society (as it originated in Western Europe and then spread to America and the rest of the world) are directly responsible for fomenting much of the environmental degradation that we currently face. More optimistically, it has been suggested that even though Western society may have brought on current global environmental concerns, it also contains the necessary technology and institutions to address those concerns. Other observers disagree; they believe that the world's environmental problems will be resolved only by a radical remodeling of Western values and social institutions.

In this chapter we briefly explore social, legal, and historical aspects of current environmental concerns. As the focus of this book is environmental science and not environmental studies, we do not deal with these issues in as much depth as we have dealt with some of the more scientific aspects. Nevertheless, science cannot be divorced from society. Science does not exist in a vacuum, and of all the sciences, environmental science is perhaps the most intimately bound up with social, political, and historical factors.

Here we focus primarily on Western social institutions and values, particularly as manifested in the United States, for several reasons:

1. As we have already noted, many critics blame much of the current global environmental degradation on Western values and social institutions.
2. In many regards, these Western values and institutions find their fullest expression in the United States.
3. Much of the world is apparently striving to emulate the United States in many respects, such as in industrial development and material affluence.
4. Because the United States arguably leads the world in pollution emissions, resource exploitation, energy use, and general environmental degradation, it can provide a particularly valuable case history.
5. The environmental history of the United States is in some ways more accessible than that of other major countries, at least since the time of European contact. Since European explorers first reached the "virgin" continent about five centuries ago (of course, the Native Americans had already modified the "natural" environment to some extent), there has been a fairly continuous recorded history of the environmental changes that have occurred. In comparison, major human-induced environmental changes have been taking place in Europe, Africa, and Asia since before recorded history (although great strides have been made toward reconstructing the environmental history of Europe and, indeed the whole world).

An Historical Perspective on American Environmentalism

When Europeans first colonized America, it was certainly not an "uninhabited wilderness." The Americas had been inhabited for over fifteen thousand years by peoples who had crossed over the Bering Strait from Asia at the end of the last Ice Age. These indigenous peoples, whom the Europeans called "Indians" (and are now commonly referred to as "Native Americans" or "Native Peoples"), had spread throughout the two continents. North America may have had on the order of 5 million human inhabitants at the time of European contact (**Figure 20-1**).

It is a myth that the native North Americans did not modify the environment, for they certainly did. They cleared fields, planted crops, hunted game, built dwellings, and so on, but they did not modify the natural environment in the same way that the European settlers did. On the whole, the North American natives appear to have lived in a relatively sustainable manner for thousands of years. (However, some researchers argue that the first Native Peoples may have been at least partially responsible for massive extinctions of large mammals, such as the mastodon, that occurred at the end of the last Ice Age about 10,000 years ago.) In part, this may have been due to their small populations and relatively simple material culture, but their complex belief systems also fostered sustainable liv-

Figure 20-1   The English navigator Bartholomew Gosnold (died 1607) trading with the Native Americans, possibly in the area of Buzzards Bay, Cape Cod, Massachusetts (colored line engraving, 1634).

ing. In general, Native Americans did not believe in the ownership of land and nature. Out of necessity and circumstance, they lived their lives more as a part of nature than apart from nature.

The earliest European colonists certainly viewed nature and human society as a dichotomy. Humans were separate and above nature, not a part of it. Some scholars argue that human domination and exploitation of the land was reinforced by an interpretation of Judeo-Christian theology, which saw the world and nature as having been created by God for humankind. The Europeans considered it their right to subdue, tame, and exploit the "uninhabited wilderness" of North America for humankind's benefit.

The natural resources of the vast continent were ripe for the taking. Sustainable use was not an issue because the resources seemed limitless; there was always more to the west, on the frontier. In the 1600s, the English colonies in New England survived in large part by shipping lumber, beaver skins, and other natural resources back to Europe in exchange for manufactured goods. As forests and animal populations in coastal areas were decimated, the Europeans knew they could simply move farther inland. Now sometimes referred to as a **frontier mentality**, this attitude of the earliest colonists continued to exert a major influence on American development for the next four hundred years.

The forests in particular were seen not just as a resource to be exploited, but as a nuisance to be eliminated: They were dark and they sheltered savages and wild beasts. Accordingly, the forests were cleared to make agricultural fields and gardens. The once wild land was cultivated, as most of the land of Europe had been for many centuries (or even millennia).

This mindset of exploiting the land and its natural resources and then moving on dominated American environmental thinking for about 250 years (ca. 1620s through the 1870s) as the country continued to expand westward. Wildlife, such as the American bison (see Case Study 20-1), were decimated in many areas, forests were destroyed, and soils eroded as they were put under the plow. The U.S. government encouraged the expansion of the frontier by transferring government-owned land and natural resources to private hands. The Homestead Act of 1862 gave each qualified settler in the Great Plains area and elsewhere 160 acres (64.75 hectares) free of charge. Likewise, the Railroad Acts of the 1850s and 1860s gave away large tracts of land to railroad companies to encourage them to build railroad lines that could move people and goods across the country. Vestiges of the frontier mentality remain an integral part of the American psyche to this day.

The First Century of American Environmentalism

Even as the West was being "won," however, certain thinkers were questioning the rampant exploitation of natural resources. The New England **transcendentalists**, such as Ralph Waldo Emerson (1803–1882) and Henry David Thoreau (1817–1862), decried the human destruction of the environment. Thoreau mourned the decline and loss of numerous species from his native eastern Massachusetts, such as bear, moose, deer, porcupines, wolves, and beavers, as well as the great reduction in the size and diversity of the forests. To gain a better appreciation for nature, he built a cabin in the woods on Walden Pond near Concord, Massachusetts. From his two years living alone there came *Walden, or Life in the Woods* (1854). *Walden* has become an American classic, inspiring generations of naturalists, ecologists, and environmentalists.

In 1864, George Perkins Marsh published an exhaustive study entitled *Man and Nature; or Physical Geography as Modified by Human Action*. Marsh had grown up in Vermont, but traveled widely. His compilation documented how human intervention had resulted in the destruction of forests, soils, and waters and greatly modified flora and fauna. Marsh's work, using clear scientific evidence and case studies, challenged the notion of an inexhaustible Earth. Marsh did not argue against humankind's intervention in, and transformation of, nature, but he did suggest it must be done with knowledge and foresight, if irreparable damage and negative consequences were to be avoided. His work laid the groundwork for the late nineteenth-century conservation and preservation movements.

Two figures dominate American environmentalism in the late nineteenth and early twentieth centuries: Gifford Pinchot (1865–1946) and John Muir (1838–1914) (**Figure 20-2**). While both supported environmentalism, they took very different approaches to the problem. Pinchot, though born in Connecticut, was professionally trained in Europe as a forester and believed in using the latest scientific knowledge to manage the land. His aim was to produce the maximum sustained yield in forestry. Taking a utilitarian approach, Pinchot basically thought that the forests and other natural resources should be managed so as to obtain the most benefit for the greatest number of people. As such, he was an advocate of **conservationism**. From 1898 to 1910, Pinchot was chief forester for the U.S. Division of Forestry (reestablished as the U.S. Forest Service in 1905, with Pinchot at the helm), so he was well able to implement his philosophy.

Figure 20-2 Gifford Pinchot (above) and John Muir (right).

Muir was born in Scotland, but came to America as a child. Largely on his own, he explored much of the U.S. and Canadian wilderness, becoming an accomplished naturalist. He also visited Asia, North Africa, Australia, and New Zealand, writing about his travels for newspapers and magazines. Through his experiences, Muir came to believe that nature has an inherent value and right to exist in and of itself independent of any value it may have for humankind. Thus, Muir took a strict **preservationist** position. He had a particular interest in the Sierra Nevada Mountains of western North America and founded the environmentalist group the Sierra Club in 1892. Muir advocated the creation of national parks that would be protected from any type of human intervention. In particular, he was instrumental in convincing the government to establish Yosemite National Park in 1890 (**Figure 20-3**).

Although strong allies in the general environmental movement, Muir and Pinchot clashed over many philosophical and practical issues. Their greatest battle was over the damming of the Hetch Hetchy Valley (adjacent to Yosemite Valley) through which the Tuolumne River flows. This valley was of great scenic beauty, and Muir, with his preservationist stance, advocated preserving and protecting it. Pinchot, tak-

Figure 20-3 *Valley of the Yosemite* (1864), an oil painting by the German-American landscape painter Albert Bierstadt (1830–1902). (*Source*: Albert Bierstadt, *Valley of the Yosemite* (1864). Gift of Mrs. Maxim Karolik for the M. and M. Karolik Collection of American Paintings, 1815–1865. Courtesy of the Museum of Fine Arts, Boston.)

CASE STUDY 20-1

The North American Bison

The story of the near-extinction of the North American bison (also known as the American buffalo) is a sad tale of the exploitation and wanton destruction of the members of a species. Only a few centuries ago, some 12 to 20 million, or more, bison (*Bison bison*) roamed through North America. Two subspecies inhabited the continent. The more numerous plains bison lived mainly east of the Sierra Nevada and inhabited most of what is now the United States except the Great Lakes area, New England, and parts of the southeast coast. The plains bison extended north into Manitoba, Saskatchewan, and eastern Alberta (Canada). The woodland, wood, or mountain bison (of the same species, but generally considered a distinct subspecies) inhabited the Rocky Mountain region from Colorado to Alberta and even farther north.

Europeans and their descendants were not the only people to hunt the bison en masse. Certain Native American tribes survived almost exclusively by hunting the bison, often utilizing the entire carcass for everything from food to fuel to shelter and tools. Indeed, this reliance on the bison became a prime incentive for the white man to decimate the bison herds in the nineteenth century. It was often said that every dead buffalo (bison) meant a dead Indian (of course, this was a bit of an exaggeration).

At any rate, buffalo hunting was certainly not confined to those of European extraction. At one site in southeastern Colorado, dating from about 6500 B.C., Paleo-Indians drove a herd of close to two hundred bison to their death over the edge of a gorge. In historical times, the North American Indians sometimes hunted down buffalo in huge numbers, aided by the horse and firearms, which they adopted from the Europeans. As late as the autumn of 1883, when the buffalo was virtually extinct, Sitting Bull and his comrades took about two months to decimate a herd of about ten thousand in North Dakota. Nevertheless, despite these abuses, the major blame for the near-extinction of the bison must be placed on the heads of the white men.

As the continent was civilized, the buffalo quickly declined. By about 1800, buffalo were extinct east of the Mississippi River, yet they remained extremely abundant elsewhere in North America as late as the 1860s. On the plains, herds of bison 5 miles (8 km) in breadth and some 50 miles (80 km) long, could be seen, with the animals so closely spaced that "the whole country was covered with what appeared to be a monstrous moving brown blanket" (letter from Col. C. Goodnight in the early 1860s).

The introduction of the railroad made it easy for professional hunters to reach the herds and for bison products to be shipped back East. In the late 1860s, the first transcontinental railway divided the plains buffalo into a southern herd and a northern herd. As amateurs joined professional hunters, hundreds of thousands, then millions of bison were slaughtered. So many carcasses were left to rot on the plains that settlers complained of the stench. By 1875, so few members of the southern herd were left that both Kansas and Colorado passed laws attempting to protect the buffalo. Still, 100,000 more buffalo were killed during the winter of 1877–1878. A few hundred survivors made their way to Texas, but by 1889, they too had been shot—and thus ended the southern herd.

Between 1881 and 1883, the northern herd came under massive attack by both whites and Indians. By 1884, the northern herd was virtually wiped out in the United States, and by 1885, the wood or forest buffalo of Canada was almost extinct in the wild.

Why were the buffalo hunted so heavily? Some hunters were after the meat, for which there was a considerable market. Others wanted the hides, which could be made into good leather. And fur traders sought "buffalo robes" to be used as overcoats and wraps. Buffalo fat could be used in the manufacture of soap and candles, and buffalo horns were used in hat racks and other accessories.

ing the more utilitarian approach of the classic conservationist, supported damming the river to develop a supply of fresh water for the San Francisco region. After protracted debate, Pinchot's views won, and in 1913, the Hetch Hetchy Valley was dammed. Some have speculated that this defeat contributed to Muir's death the following year. But Muir's reputation and message have only been enhanced with time. The Sierra Club remains one of the foremost environmental groups in America today. Meanwhile, modern environmentalists continue to debate the merits of conservationism versus preservationism.

In the late nineteenth and early twentieth centuries, the most widely held environmental stance was **progressive conservationism**. Influenced by

Many buffalo were killed for pure sport, which, it was claimed, could become highly addictive. Expeditions were arranged where interested parties could test their shooting skill, and have a bit of adventure, by slaughtering wild buffalo from a railway car.

Even after the buffalo were gone, the ground in many areas was literally covered with their bones (**Figure 1**). Homesteaders collected the bones and burned them as fuel. These bones had many other uses as well. They were shipped back East by the trainload where they were ground up for fertilizer and used in sugar refining (to neutralize acids). The hooves and horns were used in the manufacture of glue. Well into the twentieth century, buffalo bones were being shipped to the East for commercial use.

The demise of the North American buffalo occurred in a flash, and only after the fact did the Canadian and U.S. governments intercede to try to halt the destruction. Canada passed legislation to protect the buffalo in 1885, but the U.S. government waited until 1899. By the 1890s, only two groups of bison remained in North America: a mixed lot of plains and forest and wood buffalo in Canada, and a small herd of plains buffalo in Yellowstone Park. At the close of the nineteenth century, William T. Hornaday, director of the New York Zoological Park, made a census of all known living North American buffalo and found just under 1100. In 1905, Hornaday founded the American Bison Society, with the primary goal of saving the American buffalo from extinction.

The American Bison Society, with support from President Theodore Roosevelt and others, quickly established reservations for the bison, where they were provided with shelter and fodder as necessary. Results were achieved quickly: In 1910, there were over 2100 bison in North America, and by the 1930s, there were over 20,000. Today, ranchers keep herds throughout the continental United States, and there are sizable herds in government reser- vations such as Yellowstone Park and the Wood Buffalo Park in Alberta, Canada.

A sad footnote to the otherwise happy resurrection of the North American buffalo is that the wood or forest bison no longer exists as a distinct subspecies. The last remaining herd of pure forest bison inhabited the Wood Buffalo Park, but unfortunately in the 1920s, plains bison were introduced to the same park. Naturally, the two subspecies mixed and hybridized, and the rarer forest bison ceased to exist as a pure-breed form.

Questions

1. Why do you think it took so long for the U.S. government to take action in an attempt to save the North American bison?

2. The near extinction of the North American bison is only one case of humans exploiting a species that was once extremely abundant. Can you name some other examples? (You might refer to previous chapters in this book.)

Figure 1 Piles of North American bison (buffalo) bones could be found throughout the West in the late nineteenth century.

Pinchot and like-minded individuals, the federal government took an increasingly active role in protecting forests and other natural resources for the public. In 1872, Yellowstone National Park was established, and in 1891, the Forest Reserve Act permitted the president to establish forest reserves (which later became national forests). The 1890 U.S. census declared that there was no longer a de- finable American frontier, conceptually closing an era in American environmental history.

Theodore Roosevelt (1858–1919), who served as president from 1901 to 1909, contributed greatly to the environmental cause. Roosevelt loved the outdoors and was an established naturalist and conservationist in his own right. Pinchot served as a key adviser to Roosevelt, and under his administration, large tracts of land were added to the national forests. Roosevelt also protected the Grand Canyon and other areas that would later become national parks, and he sponsored a White House Conference of Governors to discuss conservation issues. In 1908, the National Conservation Commission was charged with making an inventory of the country's natural resources.

The Great Depression of the late 1920s and 1930s and the administration of Franklin D. Roosevelt (1882–1945; president 1933–1945) affected attitudes toward natural resource management and conservation. The economic depression of this time can be attributed, at least in part, to the exhaustion and degradation of the land in parts of the United States. In the Appalachian region of the Southeast, for instance, the forests had been decimated and farming had exhausted the soil, causing widespread erosion and flooding. In the Midwest and Great Plains, the once fertile soils had been overcultivated, and severe soil erosion set in. With just a few dry years, the "Dust Bowl" was created. Wind storms stripped the land of the soil, creating huge dust clouds and causing havoc for the population and the ecosystem (**Figure 20-4**).

During the depression, the federal government instituted massive programs aimed at both creating employment and restoring the environment. The Civilian Conservation Corps put otherwise unemployed citizens to work planting trees, developing and maintaining park and recreation areas, restoring waterways, building flood control devices (such as levees and dams), controlling soil erosion, and protecting wildlife. During this period, the Tennessee Valley Authority was established to address economic and resource management issues in the depressed Tennessee Valley. As part of this program, forests were replanted and dams (for hydroelectric power and flood control) were built. Likewise, dams were constructed in many arid regions of the western United States. One of the most famous is Hoover Dam on the Colorado River (near the Nevada-Arizona border); built in 1935, it created the water reservoir known as Lake Mead. Dam construction promoted agriculture and provided hydroelectric energy to fuel industry. Thus, these government programs were viewed as not only conserving natural resources, but promoting economic growth as well.

Other developments included the founding of the Soil Erosion Service (later renamed the Soil Conservation Service) in 1933 and the creation of the Agricultural Adjustment Administration (later to become the Agricultural Stabilization and Conservation Service). Researchers studied the causes and prevention of soil erosion and

(a)

Figure 20-4   The Dust Bowl. (a) A photo album montage of a dust storm approaching Stratford, Texas, 1935. (b) This famous photograph, by Dorothea Lange, of a California migrant worker captured the destitution and despair caused by the Depression. The woman in the photograph was just 32 years old, with seven children.

(b)

shared their findings with farmers. In addition, the government began to pay some farmers to reduce their crop production, thereby helping to stabilize prices and reduce soil erosion. To reduce soil erosion caused by overgrazing, some limits were placed on the grazing of animals on federal lands.

The onset of World War II (America's direct involvement on the battlefield lasted from 1941 to 1945) saw conservation and resource management issues take a backseat to other concerns. The war forced the United States to mobilize; industrial and technological efforts concentrated on military production and winning the war. The experience of this war generally reinforced the American public's belief in social progress through economic growth and technological achievements. The development of the atomic bomb, for instance, led to the "atoms for peace" program and the construction of commercial nuclear power plants that would presumably generate clean and inexpensive energy for all. Science and technology could finally make the American Dream come true.

Environmentalism since World War II

Unfortunately, the technological advances of World War II and the immediate postwar period had the potential to unleash devastating effects on the environment. Production methods, using newly developed technologies, shifted from labor-intensive processes to energy-intensive processes that consumed enormous quantities of energy, often in the form of fossil fuels. Synthetic chemicals increasingly substituted for "natural" resources. This was the age of plastics and the automobile. Individualism, perhaps best expressed in the American ideal of at least one car per family and the freedom to drive on the open road, led inevitably to an unprecedented per capita consumption of natural resources. The Green Revolution, which fed the burgeoning population, was based on energy-intensive, as well as fertilizer-, herbicide-, and pesticide-intensive, "factory farming." Economic growth reached unprecedented levels.

But all was not perfect. With the new technologies and consumption habits came new environmental concerns. Greater and greater levels of toxic chemicals and other dangerous wastes were generated and, in many cases, released into the environment. Nuclear power, in some people's opinions, was a catastrophe waiting to happen. Conservationists began to suggest that we were quickly depleting our resources and degrading the quality of our air, water, and land. Concern over the quality of life, not just the quantity of material goods, became the focus of much environmentalism in the 1960s and 1970s.

In 1962, Rachel Carson's book *Silent Spring* described the adverse effects of pesticides and warned of the environmental disaster that might befall us if we continued to pollute. Carson's book made a strong impact, and some historians consider its publication the beginning of the modern era of American environmentalism. The 1960s and early 1970s were a time of change in American society. The civil rights movement was challenging long-accepted forms of discrimination (and from this, in part, arose the concept of "discrimination" against animals—see Case Study 20-2). The hippie counterculture was questioning traditional values, and as U.S. involvement in the Vietnam War escalated, more and more people were protesting the establishment. Many of the initial environmental concerns were people-oriented: We needed to protect the quality of the air, water, and land and preserve wildlife and wilderness for the benefit of humans. With the *Apollo* missions to the Moon (culminating with the first human landing on the Moon in 1969), photographs of the Earth taken from outer space were widely circulated. The concept of "Spaceship Earth" gained wide currency: We are all on one tiny planet drifting through space; Earth is all we have, so we had better take care of it.

The 1960s and 1970s also witnessed the passing of a number of federal laws that still form the basis for environmental protection in the United States. In 1970, the Environmental Protection Agency (EPA) was established, and on April 22, 1970, much of the country enthusiastically celebrated the first "Earth Day." In many ways, this was a time of idealism. Even as people finally acknowledged the existence of environmental problems, they assumed that the problems could be solved.

By the mid-1970s, however, public concern with the environment had declined. Oil shortages were occurring (though artificially created by the OPEC cartel), and the country faced economic recession. All too often, it seems, when the public has to choose between the economy and the environment, the economy wins. By the late 1970s and early 1980s, an active backlash against environmentalism was under way. Many of the dire predictions made by environmentalists in the 1960s did not seem to have come true. The Ronald Reagan administration (1981–1989) valued economic growth over environmental protection and long-term responsible resource management. Then, too, the 1980s was the decade of the "yuppies"—the young urban professionals who often seemed to care more about making money than anything else.

Nevertheless, environmentalism was not completely dead in the late 1970s and 1980s. Love Canal

CASE STUDY 20-2

Animal Rights

Visit www.environment.jbpub.com for more information.

Do animals have rights? Do animals have moral status? Or do only humans have moral status and rights? For that matter, should all humans have rights? After all, less than 150 years ago, some people in some parts of the United States still kept human slaves, and a century ago, American women were not allowed to vote (**Figure 1**). Today, there is certainly agreement that all human beings have moral status and certain basic rights, but what about nonhuman animals?

Western thought has had a long tradition of treating all nonhuman animals and plants as objects. Classical Kantian ethics treated animals and plants as merely a means for humans to achieve their goals and ends. The influential philosopher René Descartes (1596–1650) treated animals and plants as simply thoughtless machines that might respond to stimuli, but did not possess feelings or consciousness; thus, they did not rightfully deserve moral status. One could legitimately hunt, kill, eat, pen, or even "torture" animals (of course, by Cartesian standards it would not really be torture since animals have no feelings) with no moral qualms.

In recent times, an increasing number of people have begun to wonder if animals, particularly the "higher" animals such as domesticated mammals and pets, should have moral status—that is, inherent rights. In practical terms, this issue is often linked to the question of whether various types of animals can experience pleasure and pain and to what degree they experience these sensations. A related issue is how much animals understand, how (if at all) they perceive the past, and how much they can anticipate the future. In other words, how sentient are animals? These questions can be addressed scientifically and clearly demonstrate the interplay between scientific information and ethical theories.

People who are in intimate contact with animals on a regular basis, such as with dog or cat companions, have long intuitively realized that at least some animals are sentient beings. Higher mammals, for instance, seem to possess a keen awareness of pleasure and pain and can sense and express a gauntlet of emotions, as well as to some degree remember the past and anticipate the future. Evolutionary theory indicates that it is only to be expected that animals and humans should share similar experiences, given our collective common origins and kinships. Anatomical and physiological studies demonstrate that at least the higher nonhuman animals have the appropriate neural structures and networks to transmit pleasure and pain impulses. Indeed, the neural anatomy of higher mammals is remarkably similar to that of humans. It seems clear that at least the higher animals are fully sentient.

If nonhuman organisms are fully sentient, then many theorists reason that they are due some form of moral consideration. In response to this idea, a strong animal rights movement has developed over the past few decades. Animal rights activists object to the use of live animals in testing new drugs and cosmetics: The animals suffer pain and death simply for the benefit of humans. They may object to hunting animals, wearing furs and leather, exploiting animals on farms for food, locking up animals in cages or zoological parks, or even keeping animals as pets where the animal takes a position subservient to a human "master." In a larger context, animal rights activists argue that what is good for humans might not be good for animals, and human interests and concerns should not always outweigh animal interests. The benefit to humans of developing a piece of pristine forest may not be sufficient to offset the suffering of the animals that currently live in the forest.

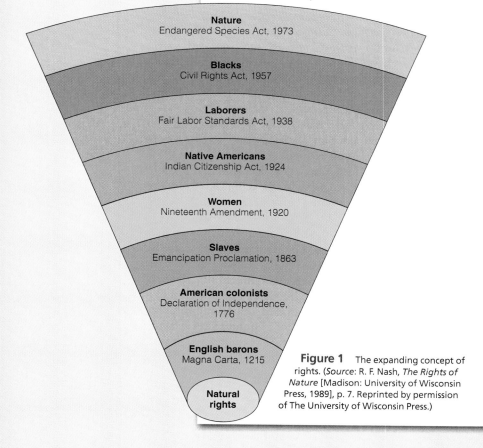

Figure 1 The expanding concept of rights. (*Source*: R. F. Nash, *The Rights of Nature* [Madison: University of Wisconsin Press, 1989], p. 7. Reprinted by permission of The University of Wisconsin Press.)

Nature
Endangered Species Act, 1973

Blacks
Civil Rights Act, 1957

Laborers
Fair Labor Standards Act, 1938

Native Americans
Indian Citizenship Act, 1924

Women
Nineteenth Amendment, 1920

Slaves
Emancipation Proclamation, 1863

American colonists
Declaration of Independence, 1776

English barons
Magna Carta, 1215

Natural rights

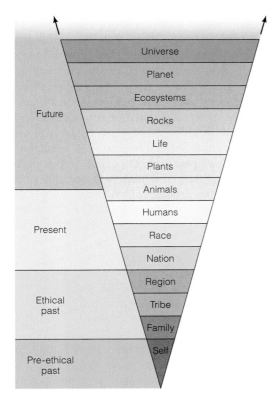

Figure 2 The evolution of ethics. (*Source*: R. F. Nash, *The Rights of Nature* [Madison: University of Wisconsin Press, 1989], p. 5. Reprinted by permission of The University of Wisconsin Press.)

Animal rights theorists view their ideas as part of the historical development of morality (**Figures 1 and 2**). Just as humans (and in particular human males) have learned to successively extend moral consideration beyond the confines of gender (acknowledging the moral status of women) and the local tribe, nation, and race, so humans must learn to extend moral consideration beyond the confines of our own species to other sentient organisms. Although some people may think it absurd to give moral consideration to animals, animal rightists point out that at one time some white American males thought it absurd to disrupt the basic economy of the southern United States simply to relieve the alleged suffering of a few black slaves. It is just as unjustifiable for members of one particular species (humans, *Homo sapiens*) to regard the feelings and concerns of other organisms as of less moral significance as it is for one race of humans to dismiss another race as of less moral significance.

Some practical criticisms of the typical animal rights stance have been advanced. If an animal rights theorist stresses the concept of sentience as the basis for moral standing, where do we draw the line? Certainly, cats, dogs, rhinos, zebras, and hippos are sentient and therefore bear moral consideration by this criterion. But what about snakes, frogs, lobsters, shrimp, clams, and bacteria? And what is the ultimate basis for sentience as a criterion for moral standing in the first place? Do animals have a degree of moral standing only insofar as they resemble humans in their abilities to feel pleasure and pain and perhaps to process cognitive information? Do life-forms that differ significantly from humans receive no moral credit? What about vegetable rights? After all, plants are living organisms too, so should not plants also be accorded some degree of moral consideration?

Some theorists also ask whether all sentient animals should be accorded equal moral status. For instance, is a stray domestic cat just as important from a moral point of view as a member of a rare, endangered species of rhino? If we had to choose between saving two hundred stray cats or a dozen rhinos, should we assume that each individual has the same moral status and therefore on utilitarian grounds maximize the good by saving the two hundred cats while letting the dozen rhinos (which might be the last members of their species) expire? To many people, such reasoning seems intuitively wrong—the rare rhinos should be saved first (although in actuality, many humans expend much time, energy, and money supporting animal shelters for stray domestic cats and dogs even as many rare and endangered species go extinct in the wild). Yet what justification would we have for favoring rhinos over stray cats? Do rhinos bear a higher moral status than stray cats, or would we save the rhinos simply so that we humans would have the pleasure of knowing that the rhinos exist as a species?

Questions

1. In your opinion, do animals have rights? If so, to what degree or extent? If not, why not? Should all animals have the same rights, or do some animals deserve more rights that others? If so, on what basis?

2. Are there laws in your local community against, for instance, torturing cats and dogs? If there are, does this mean cats and dogs have a legal status and rights?

3. Are some animals, either particular individuals or entire species, more important than others? If so, how does one judge relative importance? And who does the judging?

came to the public's attention in 1978 and led to the Superfund legislation with its efforts to clean up toxic and hazardous wastes. The Three Mile Island nuclear power plant accident in Pennsylvania occurred in 1979, effectively putting a damper on the further development of nuclear power in the United States. Then the Chernobyl nuclear power plant disaster in Russia in 1986 confirmed some people's worst fears. Even as the Reagan administration downplayed environmental issues, environmental groups such as the Sierra Club counted hundreds of thousands of new members. The radical environmental organization Earth First! was organized in 1980 and quickly grew in membership. By the end of the decade, the United States had joined other nations in signing the Montreal Protocol to save the ozone layer (1987), and the 1989 oil spill in Prince William Sound, Alaska, helped bring environmental issues to the forefront of the public's attention once again. The 1990–1991 Gulf War fought during the George H. W. Bush administration (1989–1993; Bush was vice-president under Ronald Reagan), led many people to once more reassess the wisdom of America's heavy reliance on foreign oil. Record heat waves and unexpected severe storms at the end of the decade caused many to take the notion of global warming seriously. And the global population continued to increase with no end in sight even as per capita food production declined. All in all, by the end of the 1980s, environmental issues were coming once again to center stage. Earth Day 1990 received more attention than any Earth Day since the first one in 1970. The end of the Cold War, the breakup of the former Soviet Union, and the reunification of East and West Germany allowed (indeed, required) us to focus attention on cleaning up the world and developing societies that can survive the twenty-first century and beyond.

In 1993, a Democratic administration took the reins from the Republican Reagan-Bush team that had held the White House for twelve years. President Bill Clinton (1993–2001) expressed concern and interest in environmental issues, but perhaps more significantly, Vice President Al Gore, author of the widely read book *Earth in the Balance: Ecology and the Human Spirit* (1992), was an avowed environmentalist. Before joining the Clinton ticket, Gore had launched a presidential campaign based in part on an explicit pro-environmental platform.

In these same years, however, a strong backlash against the environmental movement arose once again. Politicians who were unsympathetic to environmental concerns continued to be elected to Congress. In Washington and in state houses across the country, environmental legislation was stalled, and environmental lobbying groups lost some of their political clout. Polls showed that while the public was still generally interested in environmental issues, the economy, crime, health, and welfare were more immediate concerns. Many of the older, established environmental groups faced budget shortfalls (as generous contributions became less frequent) and declining memberships. Greenpeace and the Wilderness Society, for instance, witnessed dramatic drops in membership. The Sierra Club faced its worst financial situation in decades with multimillion-dollar deficits. A *USA Today* poll found that the 10 largest environmental groups in America collectively lost 6.5% of their membership between 1990 and 1994.

Part of the problem with the largest environmental groups and agencies may be that they grew into massive bureaucracies—and in the process lost the trust of many grassroots environmentalists. The Sierra Club, for instance, operated with an annual budget in the tens of millions of dollars, including a Washington lobby group, a book business, and a worldwide ecotourism operation. Environmental agencies in the state and federal governments became notorious for their regulations, bureaucratic hoops, and red tape. Washington pollster Celinda Lake quipped that "Today, if you ask a typical American to name a bureaucrat, he'll point to someone in some environmental agency."

But even as larger, established environmental organizations were declining, many newer, smaller, local and grassroots organizations were springing up, often in response to environmental threats in their own communities. Grassroots organizing seems to be the wave of the environmental future. Estimates are that over 6000 grassroots environmental organizations are active in the United States. On college campuses, interest in the environment is on the upswing; environmental studies programs are expanding and evolving. Environmental awareness has become part of the worldview of the generation that is now entering adulthood. And their concern is not just with local recycling programs or nuclear power plants; instead, they are focusing on the broader issues that must be faced in the twenty-first century—the achievement of long-term sustainability and the fundamental changes in society that this may entail.

In 2001, George W. Bush (son of former President George H. W. Bush) was inaugurated president of the United States after a closely contested battle with Al Gore. President Bush quickly made it clear that he was not the best friend of many staunch environmentalists. He advocated increased development of fossil fuels, and he was accused of weakening environmental and health standards. However, after the Al Qaeda attacks on the United States on September 11, 2001, and the subsequent turmoil, in many people's lives, environmental issues took a back seat to other affairs, such as security from terrorist threats.

Are Western Societal Values to Be Blamed for Current Environmental Problems?

Every human society has social institutions—based ultimately on fundamental beliefs, behaviors, and cultural norms—that govern the interactions of the vast majority of its people with each other and with the surrounding physical and biological environment. The most basic social institutions of a society are those governing (1) family organization and relationships, which provide most importantly for the raising of children; (2) economic concerns, which provide for the production and distribution of material goods and services; (3) political concerns, which provide for the distribution of power and prestige and the protection of members of society from each other and from external enemies; (4) educational concerns, which provide for the transmission of knowledge and the cultural heritage from one person and generation to another; and (5) religious concerns, which provide for the establishment of a moral and ethical code and an explanation of the meaning and purpose of human life within the society.

Fundamental values and norms, and therefore the social institutions through which these values and norms are expressed, may vary from one society to another. The social institutions of any particular society constrain the manner in which members of the society interact with each other and their environment. For instance, a particular society may value the concept of private ownership and the right of landowners to do anything they please with their land. A landowner who clear-cuts a large forest and replaces it with an open dump may cause severe problems for the other members of the society as leachate from the dump flows onto their property. Yet the norms and social institutions of the society may limit the ability of the other members to stop the landowner from pursuing such activities. The situation may effectively be unresolvable until the social institutions are modified, perhaps by incorporating the concept that private landowners can do whatever they want on their land as long as it does not affect anyone else detrimentally. Many thinkers contend that social institutions (and the underlying norms and values) must change and evolve over time as a human society expands and develops technologically. Thus, for instance, not all of the norms and values of two or three thousand years ago are necessarily appropriate to the twenty-first century.

Western Values and Social Institutions

According to many observers, the dominant social institutions in the modern world are those that incorporate the values, norms, beliefs, and ideologies found in Western European and American society. Indeed, this is arguably true despite the fact that much of the world is not, on the face of it, "Western." For instance, one-fifth of the world's population is Muslim, a "non-Western" culture and religion, yet even as some radical Muslims reject Western beliefs and ideals, many Muslims are influenced by and adapt aspects of Western thinking, institutions, and technologies to their own cultures. The same can be said for other non-Western religions and traditions, such as Hinduism and Buddhism. Around the world, indigenous cultures are going extinct, only to be replaced by the homogenizing Western culture. The whole world seems to be industrializing and adopting Western technology, and simultaneously adopting Western values, beliefs, and social institutions, either explicitly or in very subtle ways. Here we use Western beliefs and social institutions in a broad sense to include the entire spectrum from classic capitalism (with its emphasis on private property and private ownership of the means of production) to various forms of socialism and communism (which de-emphasize private property and stress varying degrees of public property and government intervention in the means of production and the distribution of goods and services). Non-Western beliefs and social institutions include those represented by traditional Islamic, Hindu, Buddhist, and Confucian societies (to give a few major examples) and those found among many indigenous peoples such as the Native Americans or Australian aborigines prior to European contact.

Classic Western institutions (whether capitalist, socialist, or communist) are based on the concept of increasing rates of economic growth, increasing development, increasing the provision of material goods and services, and increasing the standard of living for everyone. In simple terms, these institutions rely on the "pie" increasing in size so that everyone's slice of the pie gets larger. In an ideal capitalist system, for instance, the rich and elite may get richer, but the poor don't get poorer. As the pie gets bigger, even the small slices that the lower classes receive will, in absolute terms, get bigger. As long as the pie continues to grow, everyone makes material progress over time, and general standards of living rise. But it appears we will soon reach, or perhaps have already reached, the absolute limits to the size of the pie. It can be argued that we can no longer continue to expand—to pursue business as usual, because the physical environment will not support continued human expansion. The strategy that worked so successfully for hundreds of years, as long as there was a relative abundance of untapped natural resources, must now be modified.

It can be argued that current environmental concerns can ultimately be traced to a few fundamental beliefs inherent to Western culture:

1. Western society and culture are dominated by an anthropocentric and humanistic worldview. Humans are superior to all other living things, and human interests come before those of other species or inanimate objects. There is a dichotomy between nature and human society. Humans have a right to conquer, tame, and subdue nature.

2. Western culture emphasizes the individual human and often promotes individual achievements, even to the detriment of the collective good. An individual who exploits a natural resource, such as an oil field, for his or her own gain and profit is regarded as an industrious, ingenious person to be admired even though his or her actions may cause damage to society in the long run.

3. Western culture is focused on materialism, the idea that the production and consumption of material goods are necessary for a good life. To satisfy this materialistic urge, Western society must draw on the physical environment.

4. According to traditional Western thinking, the natural world is virtually unlimited, containing a wide variety of free (except for the human labor needed to procure them) and inexhaustible resources. In this view, not only can we take from the natural world with impunity, but we can dump our wastes back into nature. As a sink to absorb wastes, the atmosphere, rivers, oceans, and even the surface of the land itself are, for all practical purposes, infinite. Nature is the ultimate provider of resources and the ultimate sink for wastes.

5. Western society and culture are based on a growth ethic. Progress is measured in terms of growth; bigger and more are better. People who produce and consume large quantities of goods are better off. Social institutions depend on an expanding economy. Sustainable living, which does not involve continually exploiting new resources and new frontiers, is antithetical to this basic Western worldview.

6. Western society and cultural institutions emphasize technological and scientific knowledge and achievements over other types of knowledge, such as moral or aesthetic knowledge and achievements. This is not to say that the arts are not valued in the West, for they certainly are, but technology that can be used to manipulate the physical environment to achieve the material goals of society is even more important. Indeed, Western society seems almost to have a blind faith that technology can solve all problems and achieve all goals. We sometimes hear comments such as "If we can put a man on the Moon, we should be able to eliminate greed and social strife." The fallacy here is that putting a person on the Moon is a relatively straightforward technological problem (**Figure 20-5**). Eliminating greed or social strife, however, is a complex sociocultural problem that is not directly amenable to technological solution, although, of course, technology may contribute to the solution by providing an adequate supply of food and material goods for everyone.

Some critics, such as the protesters at the 1999 World Trade Organization meeting in Seattle, main-

Figure 20-5   Edwin E. Aldrin, Jr., stands on the Moon in a photograph taken by Neil A. Armstrong, July 1969.

tain that Western economic and political institutions in particular promote beliefs that lead to abuse of the environment. For instance, in capitalistic systems, private property owners make the decisions about the means of production of goods and services. These property owners attempt to maximize their own profits while often ignoring or discounting the damage that might ensue to the general environment that is shared with the rest of the population. Even in socialist and communist systems, where economic decisions about the means of production and the modes of distribution of goods and services are made by the government, the traditional emphasis is on increasing economic growth and supplying ever more goods and services to the populace. In practice, environmental considerations take a back seat to sociopolitical concerns. The net result is that in traditional Western systems, the environmental costs of goods and services are not factored into the price of a product. The market fails to take account of "externalities," the social and environmental costs of goods and services that are not reflected in the price that consumers pay. For instance, the $1.50 to $2.00 that a consumer may pay for a gallon of gasoline in the United States primarily reflects the cost of pumping, refining, and transporting the petroleum; it does not reflect the cost of the health and environmental damage that burning the gallon of gasoline will cause, nor does it necessarily reflect the government expenses of protecting and securing the supply of oil—think of the wars that have been fought over oil supplies. In the case of gasoline, even exploration, refining, and marketing are partially subsidized by the U.S. government.

Political decisions in traditional Western systems often promote an economic system that leads to environmental abuse. In capitalistic systems, business groups and business-funded lobbyists are extremely influential, and such groups typically wish to keep to a minimum legislation that protects the environment at the expense of business, such as legislation that would factor externalities into the cost of products. Lobbyists claim such legislation will reduce profits—reduce economic growth (upon which the entire system is based)—causing unemployment and social disruption. In socialist and communist systems, much the same line of thinking takes place except that the government and business are one and the same. Indeed, recent history has demonstrated that repressive communist systems (many of which have now collapsed) may result in even greater damage to the environment than capitalistic systems; in communist systems, any dissenters who might speak on behalf of the environment may be crushed so that the state can successfully conceal

environmental degradation from the populace and the world at large.

Alternatively, one can argue that the fundamental beliefs and social institutions of Western society are not the source of environmental problems, but rather may be our only hope for a solution. In this view, the concept of special Western guilt for global environmental destruction is a myth. Primordial innocence never existed; neither did the noble savage who lived in true harmony with nature. Aboriginal peoples in such areas as Madagascar, Hawaii, and New Zealand caused numerous species to go extinct and generally wrought ecological devastation on their lands. The extinction of many large mammal species at the end of the last Ice Age, about 10,000 years ago, may have been due in part to human overhunting. For thousands of years, the Chinese have been involved in the wholesale extermination of many wildlife species, including elephants, tigers, and rhinoceroses (**Figure 20-6**). Indigenous, traditional East Asian pharmacology advocated the use of preparations made from rhinoceros horn or tiger penis as remedies for all sorts of afflictions—Western medical science is much more effective and less damaging to the environment. Only

Figure 20-6 Traditional Chinese medicines often include ingredients derived from rare species.

through Western technology and human ingenuity, promoted by Western concepts of individualism, resource expansion, and development, can we ever satisfy the needs and cravings of all the world's population—or so some thinkers argue.

According to one school of thought, there are no resource limits as long as we do not limit human creativity to develop and exploit new resources. In the nineteenth century, for example, some pessimists believed that modern civilization would eventually run out of high-quality energy sources as coal and other fossil fuels were depleted. But in the twentieth century, Western scientists learned to harness the hitherto unimagined energy contained within the atom. We now have hundreds of fission nuclear power plants in operation around the world, and the development of fusion as a controllable energy source is just a matter of time. Likewise, aluminum was once an unknown metal. For several decades after it was first discovered (around 1825), it was extremely rare and costly. Now aluminum is very common and is put to the most mundane uses, such as packaging food and beverages. To cite just one more example, the Green Revolution after World War II allowed increased harvests that would have been unthinkable just 50 years earlier. Even the physical limits of Earth, according to extreme advocates of this point of view, represent no absolute limits. We have already proved that we are capable of space travel, and the universe is unimaginably large. Similarly, Europe may have seemed very small and crowded in 1450, but a century later, Europeans were colonizing an entire New World (North and South America) that was unimagined by the pessimists of the mid-fifteenth century.

It is not our purpose here to decide whether Western values and social institutions are the cause of, or the solution to, current environmental concerns. Ultimately, it may be that Westernization was both. Only time will tell.

Sustainable Development/ Sustainable Growth

Classically, it has been argued that a strong correlation exists between economic development and growth and environmental degradation. As humans increase their rates of growth and expand their production of goods and services, more and more resources are depleted, more wastes are spewed into the environment, and the environment suffers. But does economic growth have to lead to environmental degradation? Perhaps not. Many environmentalists believe that we must adopt a new paradigm—a new way of thinking and acting, a new ideology—that has been labeled **sustainable development** or "sus-

tainable growth." Sustainable development focuses on making social, economic, and political progress to satisfy global human needs, desires, aspirations, and potential without damaging the environment. In this context, *development* does not refer to old notions of simply increasing industrial output and increasing consumption of goods and services. Rather, "[d]evelopment involves a progressive transformation of economy and society" (World Commission on Environment and Development 1987). Sustainable development emphasizes equity in managing the world's resources—equity among different peoples and nations of the world, and equity between one generation and another. Non-sustainable growth means leaving a degraded environment to future generations simply in order to live a certain lifestyle in the present. One country should not disproportionately use global resources or cause irrevocable environmental degradation that affects other peoples or the entire Earth. The key word is sustainability, meaning to prolong indefinitely into the future the ability of Earth to support and provide for not only humans, but all life on the planet.

Addressing Environmental Concerns by Refocusing Western Political-Economic Structures

As we have already discussed, some people believe that one of the major flaws of traditional Western economics is the failure to include the environmental consequences of goods and services in market prices. The environmental costs are externalized and not paid directly by the producer or the consumer; instead, the price is absorbed by the populace as a whole. For too long, it is contended, companies and individuals have been allowed to freely exploit raw resources and dump their wastes back into the commons (such as a factory dumping waste into a public river or the global atmosphere). This allows a few to profit at the expense of the majority. By fostering economic growth, the traditional Western political-economic structure actually encourages resource depletion and pollution, at least as long as there continue to be resources to deplete and some place to dump the resultant wastes. Economic growth and the increasing profits and employment it brings have been the traditional goal of both Western business and Western governmental/political institutions. But this goal is very short-sighted if the methods used to achieve short-term economic growth discount long-term environmental degradation. In this section we briefly discuss a few of the many ways that these concerns might be addressed. We hope that you, the reader, will continue thinking about such issues.

In the United States, an environmentally conscientious company may often feel trapped. If it practices business as usual, externalizing environmental costs of production, then damage to the environment results. But if the company internalizes the environmental costs, perhaps by using a more expensive manufacturing system that eliminates harmful emissions, then the product will cost more; since the public will generally purchase the cheaper product, the company will not be able to compete and may eventually go bankrupt. Likewise, the politician who is perceived as putting the environment before jobs (although this is often a false dichotomy) may soon be voted out of office.

But this analysis is too simplistic. In at least some cases, the public does not automatically purchase the cheapest product or always vote for the politician who will lower taxes and increase income. Education has already gone a long way toward changing public perceptions and actions. The fact that a product is environmentally sound can be used to help market it even if the product does cost a bit more than its competitors. Likewise, throughout the world, "green" politicians are gaining influence and power.

Playing on the environmental sympathies of the public, however, can carry some businesses only so far—governments must also become involved. Environmentally progressive businesspersons may favor government intervention that allows for the internalization of externalities across an industry. That way, different businesses and corporations can compete fairly while also reducing or eliminating the environmental damage they cause. Government intervention can take the form of regulations that protect the environment. Government subsidies that promote the exploitation of virgin resources can be eliminated. Government-imposed limitations on corporate liability for environmental damage could be eliminated; likewise, government-subsidized insurance programs that cover environmental hazards and damages could be curtailed. Why, for instance, do we need the Price-Anderson Act of 1957, which limits the liability of the nuclear industry in case of an accident, other than to promote the development of nuclear power? Laws can be (and are) passed that set limits on permissible amounts of pollution that can be pumped into the environment. Certain types of toxic or hazardous materials can be banned. Particular resources, such as a forest or wilderness area, can be preserved through government intervention. Government can also help internalize environmental costs by instituting stiff "green" taxes on waste emissions, resource depletion, energy use (such as carbon taxes or taxes on nuclear energy production to reduce the accumulation of radioactive wastes), and so on. Such environmental taxes can go a long way toward correcting the market by internalizing the indirect costs of products.

The government could also actively subsidize environmentally compatible technologies and activities. For instance, it could give tax breaks to companies that install environmentally friendly technologies (such as solar heating systems or pollution-reducing equipment) or pay bonuses for devising more efficient means of production.

Some analysts argue that we should also create markets for the "right" or "privilege" to pollute (as was done to a certain extent with the 1990 Clean Air Act Amendments). For instance, the government could sell permits that allow the holder to dump wastes, up to a certain level, into a particular river or the air. The sum total of the pollution permitted would not be more than can be safely and sustainably handled by the environment. Pollution permits could be bought, sold, and traded on the open market. A manufacturer that wanted to discharge pollutants into a river would have to pay for the appropriate permits and thus would be forced to internalize the cost of the pollution. A competing manufacturer might be able to produce the same product by a different (perhaps more expensive), nonpolluting technology and therefore would save the expense of purchasing pollution permits. Concerned citizens, environmental organizations, or even the government might buy and hold pollution permits in order to curb waste emissions. Savvy investors might put their money into pollution permits just as today they invest in real estate.

A conceptual problem with pollution permits concerns who has the right to establish and sell them in the first place. Perhaps a global governmental authority would be needed since some types of pollution are global in nature. But should the "privilege" to pollute be legitimized and privatized in such a manner? On the face of it, one can argue that such a system will increase inequity between peoples—those who hold pollution permits will be a rung above those who do not. Those who are allowed to pollute can cause damage and injury to those who cannot, while the latter have no way to reciprocate. As the global population expands (and every indication is that it will continue to grow for many years), pollution will inevitably increase, and those holding the limited number of pollution permits will gain in power and wealth at the expense of everyone else. Thus, pollution permits may be at odds with the concept of sustainable development and the goal of global equity among all peoples.

Alternatively, we could simply insist that individuals and corporations not pollute and instead carry on their businesses in a sustainable manner that does not threaten the environment or human health.

In the United States and many other countries, corporations can be forced to internalize external costs to a certain extent through the court system. An individual or community that is hurt or damaged by a company's negligence or illegal acts can file a lawsuit against the corporation. The courts may force the company to comply with existing regulations, change its activities, or pay a large monetary settlement. In this manner, the company is made to internalize certain factors that were previously mere externalities. For example, a company may be compelled to take responsibility for the injuries that ensue when it dumps toxic waste into a river. Using the judiciary system in this way leads to numerous problems, however. On a practical level, considerable knowledge and a substantial initial outlay of money are generally needed to attack a large corporation that can afford to hire good lawyers. In most areas, the court systems are already crowded, and a case may take years to come to trial or arbitration. In many situations, such as damage due to air pollu-

tion, it may be difficult or impossible to identify the particular corporation that is directly responsible for the damage in order to take it to court. Lawsuits are perhaps most effective when used by individuals or small groups of victims to remedy damages that can be clearly attributed to the actions of a single or possibly a few corporations.

Political solutions to environmental problems are becoming increasingly prominent as more and more citizens are becoming environmentally aware. Citizens can form environmental interest groups that lobby the government at all levels, or an individual can write directly to governmental officials about environmental concerns. Citizens can support (with donations, time, and votes) candidates who are pro-environment. Individuals and groups can also influence public opinion and governmental decisions through the educational system and the media. Sitting on a local city council or writing letters to the local newspaper can be very effective.

Beyond working within the confines of the current political system by voting, lobbying, testifying before committees, and so forth, one can also engage in political action outside the established route. Some environmentalists contend that because the established political system encourages the continued degradation of the environment, we have a moral obligation to challenge the legitimacy of the current system and the status quo. Increasingly, some people are calling for an environmental revolution that will radically change our values and the way we do business. Grassroots movements of concerned individuals on a local, national, and ultimately international level may be necessary to challenge and overturn the status quo. Tactics that can be used include marches, boycotts, protests, and mass meetings demanding change (**Figure 20-7**). If nothing else, such activities can focus awareness and concern, forcing politicians to make what were once peripheral issues central concerns. Just as the U.S. civil rights movement used such tactics with great success, so too might the current environmental movement.

Environmental Law

Law and the Environment— A General Overview

Over the past century, and especially over the last 60 years, society has turned increasingly to legal avenues to protect the environment. A number of laws dealing with environmental issues have been enacted on the local, national, and international levels. Yet there are many gaps in these laws, for this field is still developing. Initially, the prime motivating force behind most environmental legislation was

Figure 20-7   A mass rally protesting the importation of French plutonium into Japan.

to protect human safety and welfare, but now certain aspects of the environment (for instance, endangered organisms) are coming to be viewed as having legal standing and legal rights unto themselves. Increasingly, environmental advocates and their attorneys are defending the environment from assault and injury by human beings.

In the broadest sense, the field of **environmental law** encompasses all of the laws, statutes, regulations, agreements, treaties, declarations, resolutions, and the like that have bearing on environmental issues. Environmental laws range in scale from local community ordinances prohibiting litter on the streets to international treaties regulating trade in endangered species or the release of stratospheric ozone-destroying substances. Many authorities divide environmental **legal instruments** (laws, treaties, regulations, conventions, and so on) into two categories: **hard laws**, which are legally binding and mandatory, and **soft laws**, which are not legally binding, but act more as a guide to policy. This distinction is particularly useful in analyzing international treaties, conventions, and agreements. Many international legal instruments are not binding on the signing nations, but tremendous moral and public pressure may be brought to ensure that the signatories more or less conform to the agreement. In contrast, hard laws force the parties to comply or else. But or else what? Enforcing hard laws at the international level may be difficult. Countries wishing to see an environmentally friendly treaty honored may impose trade or other sanctions against an offender, but to this date, nations rarely go to war over environmental protection per se, although certainly countries have gone to war over the control of scarce natural resources (such as metals, oil, or water). Many international environmental treaties are initiated by, or fall under the aegis of, the United Nations (UN), but the UN has only limited powers of enforcement.

The legal and philosophical basis for environmental laws can be approached from several different directions. In a narrow sense, one can argue that pollution is a nuisance or negligent behavior that detracts from others' property and personal rights. Thus, pollution in particular and environmental degradation in general, can be viewed as **common law torts** (wrongs) that are within the jurisdiction of the law. More broadly, one can contend that fundamental human rights include the right to a safe, healthy environment. Any government should use its authority and power to promote the general welfare of its citizens and subjects, and this includes enacting laws that address environmental concerns. Using the U.S. legal system as an example, some

scholars have argued that the Ninth Amendment to the U.S. Constitution (which reads "The enumeration in the Constitution, of certain rights, shall not be construed to deny or disparage others retained by the people.") can be interpreted as guaranteeing the people's right to a healthy, clean, and safe environment. Nowhere in the Constitution are environmental issues mentioned explicitly (of course, such topics were perhaps not a major concern in the late eighteenth century), but one can argue that the right to a healthy and healthful environment is so fundamental that it did not need to be explicitly enumerated in the Constitution.

In the broadest context, one can justify environmental legislation as simply protecting the fundamental rights of not only humans, but all organisms and even inanimate objects and Earth itself (of course, some people deny that nonhumans have fundamental rights, but then at one time human slaves were also denied fundamental rights). Human-mandated environmental laws are necessary only to the degree that humans fail to be considerate of the larger nature of which they are necessarily a part. If people did not exploit and abuse the environment, there would be no need for environmental legislation.

International Environmental Law

At the international level, a number of environmental treaties, conventions, agreements, and protocols (approximately 200 at last count) have been established. These legal instruments are binding and enforceable to varying degrees, and all are dependent on independent and autonomous governments becoming signatories and pledging their support. Unfortunately, many "global" treaties and agreements have received less than global support, and even when technically binding on participating countries, they are very difficult to enforce. Nevertheless, the latter half of the twentieth century and early twenty-first century has witnessed great strides in international cooperation to protect and preserve the environment. Conventions and agreements have been established that provide for, among other things, the protection of areas of outstanding natural or cultural value; restrictions on trade in endangered species; protection of migratory species; a legal framework for the use of the oceans and their resources, including which areas fall under national jurisdictions; and the prevention of marine pollution.

Some of the most important soft legal instruments of recent decades trace their origins to the United Nations Conference on Environment and Development (UNCED, popularly known at the time as the "Earth Summit"), which was held in

Figure 20-8   Delegates meet during the "Earth Summit" held in Rio de Janeiro, June 1992.

Rio de Janeiro in June 1992 (**Figure 20-8**). It is generally considered to be one of the largest summits ever held on any topic—delegates from more than 175 countries, including more than 100 heads of state, attended. This conference can be viewed as the final ending of the Cold War, marking a new beginning where money that once would have been spent on military expenditures could now be funneled into environmental protection and sustainable global development (although a decade later, some were arguing that now money must be spent fighting terrorism, detracting from environmental spending). As a result of the Earth Summit, a number of soft treaties and declarations were formulated that lay the groundwork for future discussions and agreements.

- *Climate Treaty (United Nations Framework Convention on Climate Change)*. This treaty, formulated at the Rio Conference, established broad principles and outlined the potential moral and legal obligations of nations to curb the release of greenhouse gases that could cause global warming. Overall, the treaty is very weak and effectively nonbinding, but it set a precedent and lay the groundwork for further international discussions on the topic (see Chapter 17 for more discussion of global climate change).
- *Biodiversity Treaty (United Nations Convention on Biological Diversity)*. This treaty is aimed at promoting the preservation and

careful management (including sustainable use) of biological diversity and begins to address the issue of genetic engineering using genes found in rare species. It imposes no binding legal obligations on the signing nations, however, and it does not totally clarify the relationship between nations that exploit the genes of species and the nations that are the repositories of the species bearing the genes. For instance, if a representative of a developed country (such as a scientist employed by a large pharmaceutical corporation) finds that a gene of a newly discovered indigenous species in a developing country has great medicinal, and therefore economic, value, who shares in the fruits of this discovery? Does the developing nation have a right to profit from the discovery? It is generally hoped that the preliminary Biodiversity Treaty formulated at the Earth Summit in 1992 will lay the groundwork for hard agreements addressing these types of issues.

- *Forest Agreement (Statement of Agreement on Forest Principles)*. Environmentalists hoped that the Earth Summit would provide a forum to develop an international treaty aimed at protecting the world's remaining forests. However, the developing nations (especially India and Malaysia) generally took the stance that forests within their boundaries were subject only to their authority and thus could not be the subject of an international agreement. Part of the problem is that the developed nations (the North) long ago cut down much of their forest areas en route to achieving their wealthy, developed status. The developing nations (the South) demand the right to follow a similar path in promoting their own internal development. Essentially, the countries of the South do not want to remain in relative poverty while protecting their forests for, as they see it, the relative benefit of the North. The final general statement on world forests produced by the Earth Summit contained few new ideas and accomplished little in the way of promoting the preservation of the Earth's forests.
- *Rio Declaration*. This statement, which came out of the Earth Summit, was signed by many nations, generally committing participants to pursue sustainable development and work to rid the world of poverty. The statement declares that developed countries have a special responsibility to help in global restoration efforts since they have the financial resources to

afford such efforts, and historically, the developed countries have been the largest contributors to global environmental degradation. The Rio Declaration, like other Earth Summit documents, is nonbinding but it does carry some moral and political obligations.

- *Agenda 21*. This 800-page document from the Earth Summit contains recommendations as to how countries can pursue sustainable development and protect the environment. It covers such topics as decentralization of decision making relative to the management of local natural resources, land reform to increase the land rights of rural and indigenous peoples, ways to improve the status and participation of women in the context of sustainable development, and the development and adoption of taxes and other economic incentives that promote sustainability with the concomitant removal of subsidies that are at odds with the preservation of natural resources.

The negotiation of international environmental treaties and other legal instruments, such as those just outlined, is an important step. Nevertheless, it is generally extremely difficult to monitor and enforce international treaties. Indeed, some critics argue that many environmental treaties currently in existence are virtually meaningless because they are not enforced. Many treaties require signatory countries to self-report on their progress in fulfilling the requirements of the treaty, but this self-monitoring is generally not sufficient. In some cases, only a minority of nations submit the required reports, and even then the data may be incomplete.

Some treaties, such as the Montreal Protocol, include voting mechanisms that allow a majority of participating nations to vote stricter measures into effect after the fact; these measures then become binding on all original signatories. Thus, unanimous consent is not required to modify the agreement (some argue that consensus may actually be reached more quickly because dissenting countries know they may be outvoted). Still, once the agreement is in place, ensuring that all participating countries live up to it can be difficult.

Short of military action, which very few people would advocate, peer pressure, moral persuasion, and public embarrassment have been found to be surprisingly effective in forcing nations to comply with the provisions of environmental treaties. For most nations and their leaders, "public face," or how the nation and its government is viewed by the world community, is very important. To maintain or improve their image, governments may comply with environmental treaties rather than risk being publicly embarrassed. In recent years, nongovernmental organizations (NGOs), such as citizens' groups, business organizations, and environmental coalitions, have been extremely helpful in forcing compliance of international environmental treaties. NGOs are often quite willing to gather and publicize incriminating information on nations that are unwilling to live up to the letter, or even spirit, of an environmental treaty. Acknowledging the value and contributions of NGOs, the UN and its affiliated agencies have been encouraging the increased involvement of NGOs not just in monitoring environmental treaties but also in helping to initiate and formulate them.

Money and trade can also be useful in implementing and enforcing international environmental agreements. Understandably, many developing nations are much more willing to implement a treaty if they receive financial assistance. Thus, under the 1990 amendments to the Montreal Protocol, a fund containing several hundred million dollars was established to help developing countries finance the phaseout of ozone-depleting chemicals.

Another way to force compliance with environmental treaties is by trade incentives and disincentives. For instance, Montreal Protocol members are forbidden to purchase CFCs from countries that have not signed the convention. A country or group of countries can impose trade embargoes on nations that do not fulfill the requirements of an environmental treaty. Some environmentalists fear, however, that international trade agreements may interfere with the use of trade sanctions to force nations to behave in an environmentally sound manner. Depending on the interpretation, certain provisions of the World Trade Organization (WTO), which sets the basic rules for global trade, may make it illegal to restrict trade based on certain environmental concerns.

Environmental Law and Regulation in the United States

In the United States, the national government is composed of three branches: the legislative, the executive, and the judicial. The **legislative branch** makes the basic laws and also allocates funds; the federal legislative body is the Congress, composed of the House of Representatives and the Senate. The **executive branch**, consisting of the president, vice president, and the president's appointees, is charged with administering and enforcing the laws of the country. The **judicial branch**, or the court system (at the highest level, the U.S. Supreme Court), interprets the laws of the land, including the U.S. Constitution, and renders judgments in trials and disputes. Each state has a somewhat similar government. At the local level, a city may have a mayor (the executive) and a city council (the legislative branch).

At all levels, all three branches of government have a role to play in environmental issues. Legislatures may pass laws that protect, or harm, the environment—or at least regulate human affairs that affect the environment. The actual influence that a particular law has depends on how it is administered and enforced. Finally, disputes over the implementation of laws are often decided in the courts. In some cases, the judiciary may define and clarify what originally may have been a fairly vague law.

Extremely important on a national level are the **regulatory agencies** that deal with environmental concerns. Once a law is enacted by Congress, it is typically administered by a regulatory agency in the executive branch of government. From an environmental perspective, some of the more important national regulatory agencies are the Environmental Protection Agency, the Department of Energy, the Department of the Interior, the Department of Agriculture, and the Bureau of Land Management. The effectiveness of a particular law often depends on the agency administering it. Furthermore, the regulatory agencies often have considerable latitude in interpreting and implementing a law. For instance, Congress may pass a law requiring certain industries to maintain emissions of specific pollutants at acceptable levels; the appropriate regulatory agency must then determine exactly what constitutes "acceptable" levels. The influence that regulatory agencies can have was demonstrated during the Reagan administration. The Reagan era was marked by the weakening and dismantling of much of the federal bureaucracy charged with administering environmental laws. As a result, between 1981 and 1988, enforcement of laws governing pollution and hazardous waste was extremely lax, and many government-owned natural resources were made available to individuals and companies for private profit. For instance, the Reagan administration increased the leasing of continental shelf areas and national forests for oil and mineral exploration, exploitation, and timbering. These actions were not simply the result of negligence when it came to environmental concerns, but were pursued as an integral part of the administration's pro-business, free-market approach to government and economic affairs.

Regulatory agencies are subject to the influences of political opinion and professional lobbyists. As the agencies set standards and guidelines for the enforcement of environmental laws, they often expend considerable energy gathering information and conducting studies. Advisory committees may make recommendations, and interested parties may testify at public and private hearings. The levels set for "acceptable" emissions of pollutants, for instance, may be the result of negotiation and compromise among different interests. Ultimately, the existence of any regulatory agency is dependent upon the legislature, for it supplies the agency's funds. If members of Congress perceive strong public opposition to the policies of a certain agency, they may simply cut the agency's funding.

Decision Making in the Public Arena

As administrators implement environmental laws by developing specific rules, regulations, and guidelines, many decisions must be made at many different levels. All decisions must include some political, social, and ethical values; there is no such thing as a politically neutral decision, although certainly some decisions come closer than others. Many "rational" decisions are based on evidence and "proof." At a fundamental level, the standards of proof that are accepted can vary greatly depending on a particular person's outlook and inclinations. For instance, given the question of whether small concentrations of a certain chemical are harmful to human or environmental health, an identical study can be interpreted in more than one way. The chemical's manufacturer may presume the chemical to be harmless until "proven" otherwise, so a scientific study that simply suggests (perhaps based on a weak statistical correlation) that the chemical has ill effects will be rejected. On the other hand, an environmentalist or health advocate may, in the name of safety, presume the chemical is dangerous if there is any evidence, no matter how weak, to that effect. Thus, both the manufacturer and the environmentalist will use the same study to bolster their arguments. In this case, the disagreement boils down to the use of different standards of proof for the chemical's harmfulness or harmlessness. How stringent a standard of proof should be is not a scientific question, but a political, moral, and philosophical question. In such a case, there is no neutral middle ground. Accepting a middle-of-the-road standard of proof for our hypothetical chemical is not a neutral decision, but a decision that essentially weighs environmental and health issues equally with the economic concerns of the manufacturer and thus arrives at a practical compromise between the two camps.

In making policy decisions concerning environmental affairs, three basic techniques are commonly relied upon: (1) soliciting the best professional judgments of experts in the relevant field or fields; (2) basing new policies on the precedents of previous ones (sometimes known as **bootstrapping**); and (3) using **benefit-cost analyses** (also referred to as risk-benefit analyses). These techniques are not mutually exclusive; in practice, they are often used in conjunction with one another. Here we briefly discuss each of these techniques.

Judgments of Experts. Relying on the judgments of experts would seem to make good sense, for who is more knowledgeable in a particular field than a trained expert? But such a strategy quickly runs into problems. Equally competent and well-credentialed experts within even a narrowly circumscribed field may disagree. On the other hand, if the professionals agree, it may simply be because they have all been subjected to similar training and backgrounds and therefore carry similar underlying assumptions and biases—they represent a single point of view, but not necessarily the only point of view. And for any given situation, which experts should be consulted? If a decision has to be made as to whether, how, and to what extent to log an old-growth forest, who should be consulted? Professional foresters, zoologists, logging company executives, professional employees of environmental organizations, economists, and others might all be considered relevant experts, but they could have very different views on the situation. Ultimately, the relevant knowledge that professionals can offer should be taken into consideration, but as we have already pointed out, the ultimate decision-making process must include more than simply objective knowledge. A decision can only be reached by combining the knowledge and judgments of the experts with political, social, and ethical values.

Precedents. In many situations, new or revised policies and decisions are based on the precedents set by older decisions and policies. Previously established standards are applied to new situations. As an example, in attempting to establish health standards for the use of a new synthetic chemical, one might allow risks to the health of the public and the environment that are similar to the risks posed by other chemicals that are already in use. This method of decision making and policy setting puts great emphasis on the status quo; essentially, the decision maker assumes that the current standard is the best until trouble ensues. Then the policy or standard is revised just enough to solve the immediate problem.

This mode of policy making is sometimes referred to as "bootstrapping" (after the old adage of pulling oneself up by one's bootstraps, that is, without anyone else's help) because present policies and decisions are the result of the slow accumulation and refinement of past policies and decisions. This method of policy making is also referred to as **muddling along** or following the political path of least resistance. Since the decision makers are tied to precedent, they have little opportunity to make major policy revisions or consider new types of data and radical alternatives to current policies. Such muddling along tends to promote business as usual and can be an effective way of dealing with slight variations on old themes. It may fail completely, however, when novel and unprecedented situations arise. For instance, scientific and technological innovations such as bioengineered organisms, new medical techniques, or sophisticated electronic wizardry may initially fall outside the boundaries of the issues addressed by established policies. New policies, not just refinements of the old, may be needed.

Benefit-Cost Analysis. Another popular approach to decision making and policy formulation is benefit-cost (or risk-benefit) analysis. The task of a benefit-cost analysis is to establish what potential decisions could be made and then determine for each the costs, or risks, versus the benefits. For any particular decision, there may be a statistical probability of good or bad effects, and these probabilities must be taken into account. Benefit-cost analyses usually quantify in one way or another the benefits and costs of each potential decision, and the rational course to pursue is considered to be the decision that bears the least cost but most benefit. Benefits and costs are often quantified in monetary terms, but other measures can also be used such as human lives lost or saved, infant mortality rates, loss of productive work time at a factory, and so on. Classic Western benefit-cost analysis is often closely associated with utilitarianism and its efforts to maximize "the greatest good for the greatest number." Of course, it is often difficult, if not impossible, to simultaneously maximize two different variables.

Benefit-cost analyses often have a broad appeal because, at least superficially, they use numbers and appear to be scientific and rigorous. Naively, people may believe that such analyses can lead to purely rational, objective decisions and policies. This is not to deny the importance of benefit-cost analysis, which can lead to refined and critical thinking about a problem. But, as we have already stressed, no decision can be made in a political, social, or ethical vacuum. Unfortunately, benefit-cost analyses may contain built-in, and perhaps unacknowledged, political, social, and ethical variables that become incorporated into the ultimate decision.

Benefit-cost analyses often consider very different types of entities and need to equate them in order to quantify them. For instance, in designing a new car, an automobile manufacturer may do a benefit-cost analysis. The more the company must spend to make a better car, the higher the price it has

CASE STUDY 20-3

Global Population and Ethical Considerations

With over six billion people on Earth, it can be argued that virtually every major environmental problem is exacerbated by overpopulation. Many environmental activists regard overpopulation as the root cause of most current global environmental problems, yet in many circles, discussion of population control is taboo. Out of religious or ethical considerations, many people object to any control on human reproduction. To have children is a fundamental desire of many persons. Some have suggested that the family is the fundamental and natural unit of society, and that it is a fundamental human right for the family itself to make all choices and decisions about family size, that is, how many children a couple chooses to have. Those who advocate population control have been accused of racism, because they tend to be wealthy whites from the industrialized nations urging the use of birth control in nonwhite developing nations. Abortion, which is often used where contraception fails, is viewed as murder of the unborn in some quarters.

Out of equally strong ethical convictions, many people regard population control as a primary duty. Given the current state of affairs, limiting human population is the only way that the globe can be saved. Neglecting population control in the developing countries results in increased poverty, starvation, and death among the already poor peoples of the world, and increasing the wealthy population of the developed world places further burdens on the already overstressed global environmental support system. In the end, if the global population continues to grow, we may all be doomed. The "right" of a couple to have as many children as they please when they please ends when this "right" affects the "rights" of other humans (and possibly nonhuman organisms also) to live in a healthy, safe, functioning environment.

Taking a somewhat utilitarian view, we might wish to maximize the "good" in the world. If we focus only on human good, then we might maximize the total amount of human good by increasing the number of people in the world. As long as a human life, on average, contains more good than bad, then the more people there are, the more good there is. According to this way of thinking, population control will only be called for when adding more lives begins to bring more bad into the world than good. But how do we determine when this point is reached? Who can judge for the world as a whole? Even in a region plagued by starvation and disease, not everyone will necessarily agree that on average another human life will create more bad than good. And in an affluent region, not everyone will necessarily agree that an additional human life will bring more good than bad.

Even within this framework, other considerations must be taken into account. What are the future implications of increasing (or decreasing) the human population now? Perhaps increasing the human population will increase the amount of human good in this generation, but bring about untold misery in the long run. Do we factor potential future generations into our analysis?

Maximizing the total human good in the world may not even be an appropriate strategy. If we maximize the total human good by increasing the population to the point where

to charge for the car. All other things being equal, higher-priced cars tend to sell more slowly, but better-quality cars tend to sell more quickly. A benefit-cost analysis could help the manufacturer optimize profits by determining how well a car should be built; the benefits of better sales are weighed against the costs of building the better car, and the increase in sales of cheaper cars is weighed against the decline in sales if the car is more expensive. But what if the more expensive, but less profitable, car is safer and better for the environment? The typical capitalistic business perspective does not take such considerations into account as long as they do not affect the bottom line—the company's profit. From an ethical perspective, however, such considerations might be more important than the bottom line.

If the car manufacturer is forced to internalize the costs/risks the cars pose to human health and the environment, these considerations will enter the analysis. Assume, for instance, that if the car crashes, passengers are likely to die because the car was poorly designed and inexpensively built. The manufacturer is sued and has to pay, on average, $1 million per life lost in such accidents. In terms of the manufacturer's benefit-cost analysis, a human life is now worth $1 million. Assume further that such a death occurs in 1 in every 100,000 cars, and the manufacturer would have to pay $100 per car to adding more people will start to decrease the total amount of good, then we will end up with many people whose lives are only marginally good. This consideration has led some people to argue that instead of maximizing the total human good in the world, we should maximize the average amount of good per human life. Essentially, this approach emphasizes quality over quantity. By this way of thinking, a small population where individual lives are characterized by a great deal of "good" (however good is defined) would be preferable to a larger population where the average human life contains even slightly less good. In extreme cases, one could even contend that it is ultimately better to allow an overly large population to reduce its size by such natural (if somewhat cruel) means as starvation and disease so that those who survive can have a higher average amount of good in their lives. The alternative might be to barely maintain, perhaps through relief efforts, those on the brink of starvation—but such actions can result in a general lowering of the average amount of good per individual. Certainly, many people find it morally more acceptable to control population growth by the use of birth control so that more lives are not conceived in the first place. Actual population reductions will come about as people die of old age.

Those who espouse the maximization of the average human good generally believe the present world needs a strong policy of population reduction. In both the developed and the developing nations, most people would probably be better off, on average, if there were fewer people around. Fewer people would mean more resources per person and less poverty, starvation, and crowding throughout the world.

So far in this discussion, we have focused only on human good, but what about the rights, interests, and therefore the good of the nonhuman organisms, species, and ecosystems with whom we humans share the planet? If we take these other creatures into account, as the ecocentrists (those taking a holistic and non-anthropocentric, nonhuman-oriented perspective that focuses on larger entities, such as entire ecosystems) insist we must, then clearly the humans already inhabiting the globe are having detrimental effects on other life-forms and the Earth as a whole (as is discussed at length elsewhere in this book). From the ecocentric perspective, it is clear that drastic reductions in the human population are needed.

Questions

1. Do you believe that each person has a fundamental human right to decide how many children to have? Or do people have a fundamental duty to control their own breeding? Justify your answer.

2. Do you think that it is appropriate to attempt to maximize the "good" in the world? If so, what exactly is "good" and how should it be maximized? If you don't believe that the "good" should be maximized, explain why not.

3. Do nonhuman species have "rights" or "interests"? Justify your answer. If they do have rights and interests, how do those rights and interests affect the issue of global human population?

avoid the problem that leads to the deaths. A simple benefit-cost analysis indicates the manufacturer is much better off paying $1 million to the family of a dead victim once in every 100,000 cars than paying $10 million ($100 × 100,000) to make 100,000 cars safer. Of course, this overly simplistic analysis ignores such issues as the loss in sales that might ensue from bad publicity. Many people, however, would find the decision to build a cheaper car at a greater profit, even though an occasional life is lost, to be morally reprehensible. This example shows one of the major limitations of benefit-cost analysis: All factors must be quantified in terms of a common unit (in this case, dollars).

Many benefit-cost analyses do not take all considerations into account or at least do not weight them equally. Typically, considerations that directly affect the decision or policy maker are given the most weight. Returning to our hypothetical car manufacturer, perhaps a more expensive car that yields less profit also produces less pollution and waste and therefore is considerably less costly (that is, less damaging) to the global environment. This will be of no direct concern to the car manufacturer who is only trying to maximize profits, as long as the company does not need to pay the costs of environmental damage caused by the cheaper, and therefore more profitable, cars. A strong governmental authority charged with regulating the automobile industry, however, may factor environmental costs into its benefit-cost analysis when determining what policies and regulations to set for the industry. But even if we take environmental damages into account, how do we place a dollar figure on them so they can be incorporated into our quantitative analysis? What is the value of a forest undamaged by acid rain? Is the value simply the value of the lumber that resides in the trees, or is it something more? How much is it worth, in dollars, to have a city free of photochemical smog? Do we simply add the medical costs of pollution-induced diseases and the costs of pollution-induced damage to buildings and other structures? Or is there a value in having a clean, healthy environment beyond the immediate expenses avoided by not living in a polluted environment? What is the value of a scenic vista, a wildlife preserve, or a healthy forest with no immediate utilitarian use?

These examples are only a few of the issues that can be very difficult to factor into a benefit-cost analysis. Another criticism is that very often benefit-cost analyses consider only a subset of contemporary humans (in our hypothetical example, perhaps only the owners of the automobile manufacturing company). However, it can be cogently argued that the environment belongs not just to living humans, but to future generations and to non-human beings as well. In a benefit-cost analysis of, say, open-pit coal mining or a major hydroelectric power dam, to what extent should the benefits and costs for nonhuman organisms and future generations be taken into account?

These criticisms of benefit-cost analyses do not mean that this technique is without legitimate uses. Such analyses can often be extremely useful, even if they are not used as the sole criterion for arriving at a decision. Especially when opposing groups apply independent benefit-cost analyses to the same problem, these techniques can help clarify differences in underlying assumptions and values that lead to disagreements. Perhaps the greatest benefit of the technique is that when used correctly, it can promote thoughtful, rational, critical thinking on important issues.

Ethical Considerations.   Another way to arrive at decisions and establish policy is to formulate opinions solely on the basis of an underlying, well thought out philosophy and ethics (see, for example, Case Study 20-3). This means making decisions and establishing policies on the basis of what is considered to be inherently right and good, not by simply balancing immediate, or even long-term, costs and benefits. Of course, not everyone agrees on what is ultimately "right" and "good" or shares the same values, and some persons consider making decisions on the basis of a benefit-cost analysis to be inherently "right." Others think decisions based on ethical considerations are non-rational and simply political or ideological. But, as we have tried to show, any particular benefit-cost analysis is inherently value-laden simply by virtue of the way it is formulated and the potential alternatives, costs, and benefits it incorporates. As long as humans must make decisions, there will be room for discussion and argument as to which paths should be followed.

STUDY GUIDE

SUMMARY

- Social institutions reflect the beliefs and values of human societies.
- Early Europeans coming to North America initially tended to tame, use, and "exploit" the "uninhabited wilderness."
- By the middle 1800s, writers and thinkers, such as George Perkins Marsh, documented how human intervention was modifying the natural environment.
- In the late nineteenth century, the beginning of modern environmentalism took root with the conservationism of Gifford Pinchot and the preservationism of John Muir.
- "Modern" U.S. environmentalism came to the fore in the 1960s and 1970s with such events as the publication of Rachel Carson's *Silent Spring* (1962), the establishment of the Environmental Protection Agency (1970), and the first "Earth Day" (1970).

- In some people's opinions, Western societal values have promoted current environmental problems, but others argue that it is Western values and technologies that are best suited to address current environmental concerns.
- Sustainable development focuses on social, economic, and political progress without damaging the environment.
- Environmental law, at both national and international levels, can be viewed as protecting the fundamental right of all humans (and perhaps all organisms) to live in a healthy environment.
- At a public level, decisions affecting the environment may be made using any of the following techniques/bases: (1) professional judgments of experts in relevant fields, (2) basing new policies on precedents, (3) benefit-costs analyses, and (4) philosophical, ethical, and moral considerations.

KEY TERMS

Benefit-cost analysis
Bootstrapping
Common law torts
Conservationism
Environmental law
Executive branch
Frontier mentality

Hard laws
Judicial branch
Legal instrument
Legislative branch
Muddling along
Preservationism

Progressive conservationism
Regulatory agency
Social institution
Soft laws
Sustainable development
Transcendentalist

STUDY QUESTIONS

1. Give a brief historical overview of American environmentalism. What are the major chapters or periods in this history? What was accomplished in each?

2. Discuss the historical trends in federal legislation and interest in and public sympathy toward environmental issues. What is the mood of the country relative to these matters today?

3. List some of the basic social institutions of Western society. In your opinion, are they to be blamed for the current environmental concerns?

4. How might it be argued that Western social institutions and technology are the keys to solving our environmental problems?

5. Compare and contrast the approaches of conservationists and preservationists to environmental issues. Who was the outstanding early advocate for each?

6. Describe the environmental stance that was known as progressive conservationism.

7. How did the Great Depression and World War II affect environmental attitudes in the United States?

8. Describe and discuss the concept of sustainable development. Do you believe we should

be striving toward this goal? If your answer is yes, how might it be accomplished? If your answer is no, what alternatives do you propose?

9. What are the philosophical and historical bases for environmental laws?

10. Distinguish between hard and soft laws.

11. Describe some of the major international environmental laws, treaties, conventions, declarations, and protocols that are currently on the books.

12. What was accomplished at the 1992 Earth Summit?

13. Why is it difficult to enforce international environmental laws? How can compliance be encouraged?

14. Explain how environmental laws and regulations are arrived at and instituted in the United States. Who makes the laws? Who makes the regulations? Who enforces them?

15. List the various ways that environmental decisions, especially public policy decisions, may be arrived at.

16. What are the pros and cons of benefit-cost analyses? Why is this method of decision making often strongly favored as well as sometimes strongly criticized?

WHAT'S THE EVIDENCE?

1. Cite evidence for and against the thesis that "Western societal values are to be blamed for current environmental problems." Is this an overstatement? Is it simply true? Or is it false?

2. When it comes to decision making, some people put most of their faith in benefit-cost analyses. Others, in contrast, believe that such

analyses can be dehumanizing and may overlook important factors that are not readily quantified (perhaps the value of a human life, the concept of "quality of life," or aesthetic considerations). Develop a compelling argument for or against the use of benefit-cost analyses, citing evidence that supports your position.

WWW.ENVIRONMENT.JBPUB.COM

www.environment.jbpub.com

Connect to this book's web site: http://www.environment.jbpub.com/ The site features eLearning, an online review area that provides quizzes, chapter outlines, and other tools to help you study for

your class. We've posted the answers to the book's odd-numbered questions. You can also follow useful links for more in-depth information. The site provides sources for the information in this chapter and recommendations for further readings.

The technology of the Information Age can now reach all over the globe. Here, people watch a soccer match on a solar-powered television in the Republic of Niger.

EPILOGUE Environmental Literacy

Experts can explain anything in the objective world to us, yet we understand our own lives less and less . . . We live in the postmodern world, where everything is possible and almost nothing is certain. Something is being born . . . one age is succeeding another.

Václav Havel, *playwright, human rights campaigner, and president of the Czech Republic*

Our global society and the global environment continue to change at a quickening pace. With the beginning of the third millennium, we are, indeed, entering a new age of increased environmental awareness among all peoples. Humankind can no longer be insensitive to environmental issues—for if we do not take concerted action to address environmental problems, they will quickly overwhelm us. Environmental issues are no longer a "special interest," but are taking center stage in the political, economic, and social arenas. Every citizen of the planet must have a working knowledge of environmental issues; everyone must be environmentally literate.

In the past, environmental science has often been considered a negative discipline that held out little hope for the future. Throughout this book, we have tried to demonstrate the various ways that environmental science can, and indeed must, become the "science of reality." This means that environmentalists must do more than criticize the status quo. It requires that we move beyond simply reacting to problems and relying on emotional persuasion and abstract formulations. We must proactively seek long-term solutions that are practical, sustainable, and realistic—solutions grounded in factual knowledge. Doing so will not always be easy, for in some cases it will require more than simple technological fixes: The very fabric and mind-set of modern society may have to be modified. This can only come about through increasing environmental literacy.

In his book, *Ecological Literacy*, the philosopher David Orr discusses the importance of moving beyond heated dialogues toward cooperation in building a sustainable society for all living things. This means that we should approach environmental problems by first becoming environmentally literate—by taking the time learn the basic facts. Without this step, political, social, and personal biases will determine the outcome of environmental problem solving, often to the long-term detriment of both the environment and society. Admittedly, environmental literacy is not easy in today's "information age," which immerses all of us in a sea of facts. Sorting out the relevant from the irrelevant, the important from the trivial, can be extremely difficult. But this is what textbooks are for—to provide a distillation and overview of the fundamentals. Here we have provided a foundation upon which you, the reader of this book, can build as you learn more about environmental issues.

But we hope you will not be satisfied to simply learn about environmental issues. Become actively involved! You should never think that you, as an environmentally literate individual, cannot make a difference. You can have an impact—through the way you think, the way you vote, and the way you live your life. Nearly all so-called global environmental problems are aggregations of local problems. Global warming, species extinctions, ozone depletion, and certainly many regional problems result from the nearly countless actions of numerous individuals. Since the problems arise from individuals, the solution must also arise from each of us. Global change will occur; the global society of the future will not be the society we know today. The question is: How will society change? Will the change be environmentally sustainable? The collective actions of all of us will determine the future. When global society consists entirely of people who care enough about the environment and future generations to act accordingly, we will finally be able to solve our environmental problems. The very notion of global environmental problems will become an historical relic. We look forward to a society that is at peace with the environment.

APPENDIX

English/Metric Conversion Table

This table is meant as a handy reference guide to some of the basic English and metric units involving length, mass (weight), area, volume, and temperature. It is not intended to be comprehensive. For common measures of energy and power, see Fundamentals 7-2, page 190.

  Conversion values (from English to metric, or vice versa) are rounded unless designated as exact equivalents.

English to Metric

	English	Metric
Length:	1 inch (in)	2.54 centimeters (cm) (exactly)
	1 foot (ft)	30.48 cm (exactly)
		0.3048 meters (m) (exactly)
	1 yard (yd)	91.44 cm (exactly)
		0.9144 m (exactly)
	1 mile (5280 ft)	1609.34 m (exactly)
		1.60934 kilometers (km) (exactly)
Mass (weight):	1 ounce, avoirdupois	23.3495 grams (gr)
	1 pound (lb), avoirdupois	453.59237 gr (exactly)
		0.45359237 kilograms (kg) (exactly)
	1 ton, net or short (2000 lb)	907.18474 kg
		0.90718474 metric ton
Area:	1 square inch (in²)	6.4516 cm² (exactly)
	1 square foot (ft²)	0.092903 m² (exactly)
	1 square yard (yd²)	0.836127 m²
	1 acre (43,560 ft²)	4046.856 m²
		0.4046856 hectares (ha)
	1 square mile	258.9975 ha
		2.589975 km²
Volume:	1 cubic inch (in³)	16.38706 cm³
	1 cubic foot (ft³)	0.028317 m³
	1 cubic yard (yd³)	0.764555 m³
	1 gallon (U.S.)	3.785 liters (L)
	1 liquid quart (U.S.)	0.946 L
	1 liquid ounce (U.S.)	29.573 milliliters (ml)

Temperature: 1 degree Fahrenheit equals 5/9 (.555555) degree Celsius (centigrade). To convert a temperature in degrees Fahrenheit to the equivalent temperature in Celsius, subtract 32 and divide by 1.8:
degrees C = (degrees F − 32)/1.8

Metric to English

	Metric	English
Length:	1 centimeter (cm)	0.393701 in
	1 meter (m)	39.3701 in
		3.28084 ft
		1.09361 yd
	1 kilometer (km)	3280.84 ft
		0.62137 mile
Mass (weight):	1 gram (gr)	0.042827 ounce, avoirdupois
	1 kilogram (kg)	2.204623 lb
	1 metric ton	1.10231 short tons

English/Metric Conversion Table—Continued

Metric to English

	Metric	English
Area:	1 cm^2	0.155 in^2
	1 m^2	10.7639 ft^2
		1.19599 yd^2
	1 hectare	2.47105 acres
	1 km^2	247.105 acres
		0.3861 mile^2
Volume:	1 cm^3	0.06102 in^3
	1 m^3	35.3145 ft^3
		1.3079 yd^3
	1 liter (L)	1.057 quarts
		0.2642 gallon (U.S.)
	1 milliliter (ml)	0.0338 liquid ounce (U.S.)

Temperature: 1 degree Celsius (centigrade) equals 9/5 (1.8 exactly) degrees Fahrenheit. To convert a temperature in degrees Celsius to the equivalent temperature in Fahrenheit, multiply by 1.8 and then add 32:

degrees F = (degrees C $\times$ 1.8) + 32

abatement devices "End of the pipe" pollution control equipment.

abortion The termination of a pregnancy before the term is up (before the child is scheduled to be born).

absorptive ability The ability of a natural sink (such as a river, lake, or the atmosphere) to absorb pollutants or potential pollutants.

acid rain Precipitation (rain, snow, sleet, and so forth) that is more acidic than normal (generally due to human-produced air pollutants); also known as acid precipitation.

acoustical materials Sound-absorbing materials that can be used to reduce noise.

active solar techniques Mechanisms, such as flat-plate collectors, that are designed to actively collect the energy of sunlight and use it; for example, to heat a building or to heat water.

acute toxicity When toxic substances are harmful shortly after exposure.

additivity With reference to toxicology, when the toxic effects of two substances are added together.

ADMSE Acronym for absorption, distribution, metabolism, storage, and excretion—the five basic steps in the path of a toxic substance through the body.

aeration Exposing water to the air; often results in the release into the atmosphere of gaseous impurities found in polluted water.

aeration, zone of The zone above the water table where the voids in the soil or rock may contain water but are not fully saturated.

aerosol spray Products that are sprayed as a fine mist during use, such as canned spray paints, deodorants, and so forth. For many years such products used ozone-depleting CFCs as propellants.

age profile The population age structure profile is a graphic representation of the age structure of a population at a given time.

age structure The relative proportion of individuals in each age group in a population.

agricultural (Neolithic) revolution The advent of domestication and agriculture beginning around 10,000 years ago.

agriculture The cultivation or raising of plant crops and livestock.

alkaline The opposite of acidic; basic. Alkaline soil or rock may neutralize acid rain.

alpha particles A type of radiation essentially composed of energetic helium nuclei.

alternative energy sources Energy sources, such as solar power, wind power, and so forth, that are alternatives to the fossil fuels, nuclear power, and large-scale hydroelectric power.

amensalism A form of symbiosis where one species inhibits another while being relatively unaffected itself.

amino acids The complex molecules that are the "building blocks" of proteins.

animal rights The concept that nonhuman animals have inherent ethical moral status and rights.

annual growth rate How much a population grows in one year. *See also* percent annual growth.

antagonism With reference to toxicology, when substances work against each other and cancel each other's effects.

anthracite Hard coal.

anthropocentric Human-centered.

anthropogenic Formed, produced, or caused by humans.

appropriation law A law under which owners of land may be denied the right to withdraw water from a lake or stream if a more beneficial use for the water is found (government can appropriate the use of the water).

aquaculture Refers to aquatic organism farming in general and sometimes to freshwater organism farming in particular.

aquiclude A relatively impermeable rock layer that obstructs the flow of water.

aquifer A relatively permeable rock layer below the water table that contains a significant amount of water.

artesian well A well from which water flows freely without pumping due to water pressure built up in the recharge area.

asthenosphere A layer in the mantle that is relatively weak and viscous; lies directly underneath the solid lithosphere.

atmosphere The sphere or "layer" of gases that surrounds the Earth.

atmospheric cycles The large-scale movements in the Earth's atmosphere that give rise to the major wind belts and so forth.

atom The smallest particle of an element.

baghouses A series of fabric bags that act as filters to remove particulate matter from polluted air.

battery A device that when charged with electricity stores the energy in the form of chemical energy. When the battery is discharged, the energy is converted back into electrical energy.

behavioral toxicology The study of chemicals that cause disorders of behavior or learning.

benefit-cost analysis (BCA) A method of comparing the benefits of an activity to its cost; also known as risk-benefit analysis.

benthic Refers to the bottom-dwelling zone of the aquatic (specifically marine) biome.

beta particles A type of radiation, essentially high-speed electrons.

Bhopal The city in India where, in December 1984, a major leak occurred at the Union Carbide pesticide-producing plant, ultimately killing thousands and injuring tens of thousands more. The Bhopal incident helped inspire Congress to pass the Emergency Planning and Community Right-to-Know Act that established the Toxics Release Inventory. *See also* Toxics Release Inventory.

big five energy sources Coal, oil, natural gas, large-scale hydroelectric, and nuclear power.

bioassay An hierarchical sequence of procedures used to test the toxicity of a chemical substance. Initially, a relatively short, inexpensive test is administered; if harmful effects are indicated, then progressively more thorough tests are carried out.

biocentric Life-centered.

biochemical conversion The harnessing of microorganisms to convert biomass into certain fuels.

bioconcentration Bioaccumulation or biomagnification; the tendency for a substance to accumulate in living tissue. An organism may take in a substance, but not excrete or metabolize it; then, a predator eats the organism and ingests the substance already accumulated in the prey.

bioconcentration factor (BCF) A measure of the tendency for bioconcentration by a given substance.

biodegradable plastics Biodegradable generally refers to a substance that can be degraded, decomposed, or broken down by microorganisms into simple compounds such as water and carbon dioxide. A biodegradable plastic is a plastic that can be broken down in such a manner; traditionally, most plastics have been synthetic compounds that are not biodegradable.

biodiversity Biological diversity, or the variety of living things in a given area. Biodiversity can be defined in many different ways, such as in terms of genetic diversity or ecosystem diversity, but it is most often defined in terms of species diversity.

bioengineering Genetic manipulations and engineering to produce new varieties and types of organisms.

biogas digester A special chamber or reactor used to promote biochemical conversion of biomass.

biogeochemical cycles The cycles of elements and compounds through the atmosphere, lithosphere, hydrosphere, and biosphere.

biological control *See* Integrated Pest Management.

biological environment The living world or biosphere.

biological impoverishment The loss of variety in the biosphere (even when species have not gone completely extinct).

biological oxygen demand (BOD) The amount of oxygen used by organisms and chemical processes in a particular stream, lake, or other body of water to carry out decomposition.

biomass The total weight of living tissue in a community.

biomass energy Energy produced by the burning of such biomass as organic wastes, standing forests, and energy crops.

biomass pyramid A graphic depiction of the amount of biomass that occurs at each trophic level in a particular community or ecosystem.

biome A large-scale category that includes many communities of a similar nature.

bioprospecting The developing field wherein biologists, chemists, and other researchers are compiling a database of the commercial potential of many species; bioprospecting can, for example, draw from the knowledge of native healers or shamans in tropical countries about the medicinal properties of local plants and animals, but it ultimately seeks information about all species, including even deepwater marine organisms.

bioremediation The use of bacteria and other small organisms (such as single-celled and multicellular microbes and fungi) to clean up or reduce unwanted concentrations of certain substances; also known as biotreatment.

biosphere The sphere or "layer" of living organisms on Earth.

biotechnology The artificial use and manipulation of organisms toward human ends.

biotoxin A poisonous substance produced by an organism, often of high potency such that a small amount can kill an adult human.

birth control The artificial control, or prevention, of unwanted births.

birth rate How often members of a population give birth, or number of births per unit time per unit number of members of a population. *See* crude birth rate.

bituminous coal Soft coal.

blackout When a region served by an electric power plant is left without power for an extended period of time, perhaps due to a major breakdown.

bootstrapping Basing new policies on the precedents of previous ones; sometimes referred to as "muddling along" or following the path of least resistance.

bottom-up approach The development and encouragement of sustainable uses of biodiversity that provides incentives to save species while also respecting the right of all people to support their families and have a decent quality of life.

breeder reactor A nuclear rector that is especially designed to actively convert nonfissionable isotopes into fissionable isotopes that can then be used as fuel.

brownfield development Improving old, run-down, and abandoned parts of a city or town by renovating them for new industries, stores, businesses, or multifamily housing.

brownout When the capacity of a power plant is exceeded by a few percent and the voltage to consumers is inadequate such that lights often dim.

BTU British thermal unit; the amount of energy that when converted completely to heat will warm one pound of water by one degree Fahrenheit.

buffer zone In a preserve, an area of moderately utilized land that provides a transition into the unmodified natural habitat in the core preserve where no human disturbance is allowed.

calorie The amount of energy that when converted completely to heat will warm one gram of water by one degree Celsius.

carbamates A group of synthetic organic pesticides that are less persistent, but more toxic to humans, than chlorinated hydrocarbons; includes sevin.

carbon cycle The biogeochemical cycle of carbon.

carbon dioxide CO_2, the primary greenhouse gas.

carbon efficiency The amount of economic output per unit of carbon released.

carbon emission The emission of carbon, primarily as carbon dioxide, into the atmosphere during the burning of fossil fuels and other organic matter and similar activities.

carbon tax A tax levied on fossil fuels (or any fuels) in proportion to the amount of carbon emitted during combustion.

carcinogen A cancer-producing substance.

carrying capacity The maximum population size that can be sustained by a certain environment for a long period of time (potentially indefinitely); often represented by the symbol K.

carryover grain stocks Stocks of grain that are saved from one harvest year and remain at the beginning of the next harvest year.

cassandras People who argue that humans have always altered the environment and managed things poorly; they maintain that ultimately exponential growth of populations and technologies will totally degrade the environment.

catalytic converter A device that carries out a number of chemical reactions that convert air pollutants to less harmful substances.

cataract A condition where the lens of the eye becomes opaque.

chain reaction In a nuclear reactor, when the fissioning of one atom releases neutrons that induce the fissioning of other atoms, and so forth.

channelization The artificial straightening of a river or stream.

chaparral Savanna-like environments with dense, scrubby vegetation, found, for instance, in the southwestern United States and Mexico.

charismatic species A high-profile endangered or threatened species that attracts broad public concern.

chemical prospecting The search for data on the chemical properties and potential of various species (such as for use in medicines, cosmetics, genetic engineering, and so forth).

chlorinated hydrocarbons A group of synthetic organic pesticides that includes chlordane and DDT.

chlorine A reactive element, found in chlorofluorocarbons and other substances, that has been implicated as a prime factor in the deterioration of the ozone layer.

chlorofluorocarbons (CFCs) Artificially produced compounds composed primarily of carbon, fluorine, and chlorine. CFCs have been implicated in the deterioration of the ozone layer.

chronic toxicity Long-term effects of toxic substances, especially relative to long exposures of low doses.

circle of poison Due to the fact that much produce consumed in the United States is grown in developing countries, U.S. citizens continue to be exposed to various pesticides that have been banned in the United States but are still used in various developing countries; this is sometimes referred to as the circle of poison.

Clean Air Act A federal statute enacted in 1963 that was the first of a series of acts and amendments that exerted increasing federal pressure on air polluters to clean up their emissions.

Clean Water Act (CWA) A federal statute enacted in 1972 that has been very successful in improving the water quality of lakes and rivers.

clear-cutting The harvesting of trees such that an entire stand of trees is completely removed; also known as even-aged management.

climate The average weather in a certain area over time ranges of decades to millennia to hundreds of millions of years.

climax community The last, relatively stable and diverse, community in the sequence of community succession.

closed system An isolated system that exchanges nothing with other systems.

closed-loop reclamation (water) Treating wastewater to the level needed before direct reuse.

closed-loop recycling The indefinite recycling of a material or substance without degradation or deterioration, such as the recycling of many metals and glasses.

clustered housing A way to increase the efficiency of land use by clustering houses fairly close together while leaving relatively large amounts of open or undeveloped land that can be used for social commons, parks, wildlife habitats, or other non-residential use.

coal A general term used to refer to various solid fossil fuels.

cogeneration A power plant produces several types of energy simultaneously, such as electricity and heat, that can be used locally.

cold fusion Promotion of fusion reactions at relatively low ("room temperatures") temperatures.

coliform bacteria Bacteria, such as *Escherichia coli*, that live in the human intestine in huge numbers. The abundance of coliform bacteria in water is a good indicator of how much human feces had recently entered the water.

combined cycle turbine plant A power plant that not only uses a gas turbine, but utilizes excess heat to heat water and power a steam turbine.

command economy A national economy where the central government tells individuals and companies what to produce and sets prices for the resultant products (for example, the economy of the former Soviet Union before its collapse). *See also* market economy for comparison.

commensalism A form of symbiosis where one species benefits and the other is not affected.

commercial extinction When a species becomes so rare that harvesting it is no longer economically viable.

common law A body of law based primarily on judicial rulings, custom, and precedent.

common law torts Basically "wrongs" or wrongful acts against a person, people, or institution.

community All of the populations of different species that inhabit a certain area.

community succession The sequential replacement of species in a community by immigration of new species and the local extinction of old species. The first stage is the pioneer community, and the last stage is the climax community.

compact development A way to minimize the amount of geographic area occupied by a city, for instance, by setting boundaries on the city's growth; compact development can encourage development that builds "up" rather than "out."

competition Organisms competing ("fighting") for the same limited resource.

competitive exclusion A situation where niche overlap is very great and competition is so intense that one species eliminates another from a particular area.

complexity How many kinds of parts a system has.

compost The product of composting. *See* composting.

composting The decomposition of organic materials by microorganisms; produces various forms of "soils."

compound Two or more atoms chemically bonded together. For instance, water is a compound of hydrogen and oxygen.

compressed air energy storage (CAES) Electricity is used to pump air under pressure into a storage reservoir; when the energy is needed, the pressurized air is released.

cone of depression The localized lowering of the water table around a well from which water is being withdrawn faster than it is replenished.

confined aquifer An aquifer bounded above and below by aquicludes.

conservation Refers to attempts to minimize the use of a natural resource.

conservation biology A subdiscipline of biology that draws on genetics, ecology, and other fields to find practical ways to save species from extinction and preserve natural habitats.

conservation easements A preservation tool that may be used by a land trust or conservation group to limit development; under a conservation easement a landowner may give, lease, or sell the development rights to the land to a preservation group.

conservation of matter and energy, law of The concept that matter and energy cannot be created or destroyed; they can only be transformed.

conservationism The classical view that forests and other natural resources should be managed to provide the most benefit for the greatest number of people.

consumed water Water that is withdrawn and not returned to its original source.

consumption The use of goods and services, materials and energy, by humans.

containment structure In a nuclear reactor, a housing around the reactor vessel designed to protect the outside environment from major radioactive contamination if an accident should occur.

contaminated water Water that is rendered unusable for drinking.

continental crust The layer of rock that underlies the continents, generally containing a higher percentage of silicon dioxide than oceanic crust. *See* crust.

Contingent Valuation Method (CVM) A method that attempts to "objectively" measure the dollar values of changes in environmental quality; often uses questionnaires and other surveys that ask people what they would pay for various environmental improvements.

contraception The prevention of conception or impregnation.

control rod Made of some substance with the ability to absorb neutrons, such as cadmium or boron, the control rods are used in a nuclear reactor to control or even halt the nuclear chain reaction.

convection cell A circulation pattern set up in a hot fluid, such as in the Earth's mantle or in the atmosphere. A hot liquid rises, flows laterally, and sinks as it cools.

Convention on Biological Diversity (CBD) A document signed at the 1992 Earth Summit that acknowledges that developing nations deserve a share of the profits generated from their genetic resources by agricultural activities and genetic technology.

Convention on International Trade in Endangered Species (CITES) An international convention that outlaws trade in endangered species and products made from endangered species.

core (a) The innermost portion of the Earth; thought to be composed primarily of an iron-nickel alloy. (b) The interior of a nuclear reactor containing the fuel, moderator, and control rods.

cornucopians People who argue that human ingenuity always has, and always will, overcome environmental limitations.

corrosive When used in reference to hazardous wastes, generally referring to liquids that are highly acidic, very alkaline, or otherwise chemically very reactive.

cowboy economics A way of thinking that views the environment as a source of endless space and riches (as the frontier in the Old West was once viewed).

criteria pollutants With reference to air pollution, the six basic criteria pollutants are particulate matter, sulfur dioxide, nitrogen dioxide, ozone, lead, and carbon monoxide.

critical mineral A mineral that is necessary for the production of essential goods.

crop rotation Planting different crops on a particular field in different years—the same crop is not planted on the same field year after year.

crude birth rate The number of births per year per 1000 members of a population.

crude death rate The number of deaths per year per 1000 members of a population.

crust The outermost layer of rock that forms the solid surface of our planet Earth; divided into continental crust and oceanic crust.

cultivable land Land that can be successfully cultivated to grow crops.

cultural extinction The extinction of a tribal people or indigenous culture.

cyclone An intense storm that typically develops over a warm tropical sea.

dams Structures that obstruct river or stream flow to form artificial lakes or reservoirs.

daughter products The atoms resulting from the splitting, or fission, of a large atom such as uranium or plutonium.

dead zone An area around a mine in which no vegetation or animal life can survive.

debt bomb The more than $1.5 trillion that the poor, developing countries (generally found in the Southern Hemisphere) owe the developed countries; sometimes referred to as the Third World debt crisis.

debt-for-nature swaps In a typical debt-for-nature swap, an environmental organization pays a percentage of a debt to a lender on behalf of a poor country and in exchange, the country carries out a specified environmental program, such as a reforestation program or setting aside land for a preserve.

decentralization The movement away from large, centralized sources of power and production.

decibel (dB) scale A scale used to measure the loudness of sounds.

decommission To take out of service, dismantle, and dispose of a nuclear power plant.

deep ecology A philosophical outlook that takes a holistic and nonanthropocentric perspective, and rejects the human versus environment/nature dichotomy.

deep-well injection A method of disposing of liquid wastes, such as industrial or hazardous wastes, wherein they are pumped or injected down wells deep below the Earth's surface.

deficit areas (water) Areas that receive less precipitation than is needed by well-established vegetation.

deforestation The removal of forest cover from an area.

Delaney Clause A controversial provision, passed in 1958, of the Food, Drug and Cosmetics Act that explicitly prohibits even minuscule amounts of pesticides and other chemicals in processed foods if the chemicals are found to harm laboratory animals. Critics of the clause argue that the costs of regulating and removing such minuscule amounts of chemicals (which may cause little, if any, harm in humans) far outweigh any potential benefits.

delisted species Species that are removed from the endangered species list; a species can be delisted either because it is no longer endangered or threatened, or because it has gone extinct.

dematerialization The reduction of the size of products, particularly as a way to conserve mineral resources.

demographic transition Essentially, the concept or theory that as a nation undergoes technological and economic development, its population growth rate will decrease.

demography The study of the size, growth, density, distribution, and other characteristics of human populations.

density-dependent regulation Biological interactions, such as competition and predation, acting to control the abundance of a population of organisms.

density-independent regulation Physical processes, such as droughts or volcanic eruptions, acting to control the abundance of a population of organisms.

deontology The study of moral obligations and duties.

depth diversity gradient The concept that among aquatic communities species richness generally increases with water depth down to about 6560 feet (2000 m) and then declines with further depth.

descriptive ethics The cataloging and description of the various ethical approaches that have been used by humans.

desertification The spread of desertlike conditions due to human exploitation and misuse of the land.

developer pays An approach where a city requires that a developer of a residential subdivision pay for the required infrastructure of roads, sewer lines, utility lines, and connections to other services that are often covered by the local government and thus ultimately the taxpayers; this is a method to discourage the growth of urban and suburban sprawl and instead use land, such as brownfields, that is already available more efficiently.

dilution The reduction in concentration of a pollutant when it is discharged into water.

diminishing returns, law of As the supply of a resource declines, increasing efforts to extract the resource produce progressively smaller relative amounts of the resource.

dioxin A group of more than seventy-five related compounds that are extremely toxic, artificially produced chemicals. Dioxins can be inadvertently synthesized in incinerators when trash and garbage are burned.

direct value The value of utilizing a particular resource in such a way that it may be depleted or destroyed—for instance, the value of logging a forest.

discharge The volume of water carried by a channel.

discounting by distance Ignoring or not fully paying for the environmental costs of our actions on people living in another area.

discounting the future Focusing on the present and ignoring or discounting future costs of resource depletion and environmental degradation.

distillation A method of desalination whereby salt water is evaporated so as to remove the dissolved salts.

DNA Deoxyribonucleic acid, the material of which genes are composed.

dosage The amount of a substance, such as a poison, medicine, or vitamin, administered to an individual organism. Dosage is generally measured as the amount of a substance administered to an organism divided by the body weight of the organism.

doubling time The amount of time that a population of a given size at time zero, increasing at a fixed rate, will take to double in size.

drainage basin The region drained by a particular network of rivers and streams.

drought A period of abnormally low rainfall for an extended period of time in a particular area.

dump A place where trash, garbage, and other waste are piled—often in a relatively out-of-the-way area such as on the outskirts of a town.

durable goods Material goods or products that are designed to last a relatively long period of time.

duty A moral or required action or obligation.

Earth Summit The United Nations Conference on Environment and Development (UNCED), an international meeting held in Rio de Janeiro in June 1992 to discuss environmental and development issues; delegations from over 175 countries, including more than a hundred heads of state, attended.

earthquake Shock waves that originate when large masses of rocks, generally located below the surface of the Earth, suddenly move relative to each other.

ecocentric Ecological-centered.

ecofeminism Analyzes societal and cultural traits that may have led to the degradation of the environment; specifically concentrates on the oppression of women by a traditional patriarchal society as a major form of social domination that has precipitated other forms of exploitation and domination, including exploitation of the environment.

ecological extinction Occurs when a species, although not totally extinct in an area, has become so rare that it has essentially no role or impact on its ecosystem.

ecological release The population of a particular species increases greatly in size when a competitor is removed.

ecological succession The successive groups of plants and animals that will colonize a newly cleared patch of land or uncolonized body of water.

ecology The study of how organisms interact with each other and their environment.

ecology, first law of Garrett Hardin's concept that "we can never do merely one thing"; sometimes referred to as the "law of unintended consequences."

economic take-off The point at which a poor country can presumably produce enough wealth to financially support itself, export manufactured goods, and eventually attain the living standards of a wealthy, industrialized country.

economic value Directly tangible and monetary value of a resource, such as when a resource is bought or sold, perhaps for food, materials, or energy.

economics The study of the production, distribution, and consumption of goods and services; in an environmental context, economics is often defined as the study of scarcity and how humans can cope with scarcity.

ecoscam *See* greenwashing.

ecosystem A biological community plus the surrounding physical environment.

ecosystem simplification Occurs when the number of species in an ecosystem declines.

ecoterrorism Committing terrorist-type acts, such as sabotage against property or people, in the defense of the environment and environmental issues and ideals.

ecotone A sharp boundary between adjacent biological communities.

ecotourism Responsible travel to natural areas, often to see wild flora and fauna, that conserves the local environment and supports the local people.

ecotoxicology The study of how chemicals affect entire ecosystems.

edge effects Disturbances from the surrounding area (perhaps dogs, housecats, humans, wind, temperature changes, pollution, and so forth) penetrate along the edges of a preserved area, resulting in habitat loss.

efficiency The useful work that is performed relative to the total energy input to a system.

efficiency improvements Innovations that reduce the flow of throughput by decreasing the per capita resource use.

effluent charges (water) Taxes or fees levied on the discharge of industrial wastewater.

effluent taxes Taxes charged on the basis of how much pollution (effluent) is discharged by a company, industry, or operation.

electricity An electric current or flow of electrons through a conductor.

electrostatic precipitator A device that uses a high voltage to remove charged particulates from polluted air (such as the smoke coming out of a smokestack).

element A fundamental substance that cannot be broken down further into other elements by standard chemical means. Gold, iron, hydrogen, and oxygen are examples of elements.

embodied energy The energy used in producing a product.

emissions offsets The concept that in order to receive a construction permit to build a factory or other industrial operation that may be a new source of air pollution in a particular area, ways must be found to reduce emissions from existing sources in that area.

emotional value Emotional bonds by a human or humans to a resource; the value of a resource beyond "practical value" or simple sensory enjoyment.

Endangered Species Act of 1973 An act that directs the U.S. Fish and Wildlife Service to maintain a list of species that are endangered (in immediate danger of extinction) or threatened (likely to be endangered soon).

endemic species A very localized species that inhabits only a relatively small area.

energy The ability to do work.

energy budget The Earth's energy budget is, collectively, all of the various flow pathways of all energy on Earth.

energy conservation Decreasing the demand for energy.

energy efficiency The usable output per unit of energy.

energy farm A farm that produces biomass to be used as an energy source.

energy intensity index The energy consumption per gross national product of a nation.

energy minerals The fossil fuels (oil, coal, and natural gas) and uranium ore.

energy storage Storing energy in a form that is readily accessible to humans.

enriched Referring to uranium, *see* enrichment.

enrichment Increasing the percentage of fissionable uranium above that found in natural uranium as it is mined and processed.

entropy The amount of low-quality energy, or the amount of disorder and randomness, in a system.

environment In the broadest sense, all aspects of the natural environment plus human manipulations and additions to the natural environment.

Environmental Impact Statement (EIS) A key part of the National Environmental Policy Act, anyone proposing action on public lands, such as logging or construction of roads or buildings, must first perform a study that documents the project, reasonable alternatives to the proposed project, an overview of the natural environment that will be affected, and the environmental consequences of the project. Most states also require environmental impact statements at the state level.

environmental economics Economics with an emphasis on the study of the integration of jobs and environmental preservation.

environmental equity Treating all persons, regardless of color, creed, or social status, equally when developing environmental policies and enforcing environmental laws and regulations.

environmental hormones Pollutants that send false signals to the complex hormonal systems that regulate reproduction, immunity, behavior, and growth of organisms.

environmental justice The concept of implementing environmental equity and also reducing the environmental risks to all people.

environmental law Laws, statutes, regulations, treaties, agreements, declarations, resolutions, and the like that have bearing on environmental issues.

environmental racism Racial discrimination in environmental policy making and the racially unequal enforcement of environmental laws and regulations.

environmental science The systematic study of all aspects of the environment and their interactions.

environmental service Values of a resource in providing services that allow humans to exist on Earth, such as the production of atmospheric oxygen by photosynthetic plants.

environmental wisdom The ability to sort through facts and information about the environment and make correct decisions and plan long-term strategies.

epidemiological statistics Statistics used to examine correlations between environmental factors and the health of humans and other organisms.

erosion The deterioration and weathering away of soil or rock.

esthetic (aesthetic) value Value of a resource in making the world more beautiful, more appealing to the senses, and generally more pleasant.

estuaries Transitional ecosystems between ocean and freshwater biomes.

ethanol Ethyl alcohol, commonly used biomass-derived fuel.

ethical value Value of a resource unto itself, regardless of its value to humans.

ethics The philosophical study of moral values.

eukaryote A cell with a true nucleus, true chromosomes, and various specialized cellular organelles.

eutrophication A rapid increase in algae or plant growth in an aquatic system due to the influx of a limiting nutrient that was in short supply previously.

evaporite A mineral deposit formed when salt ions precipitate out of a natural body of water.

evapotranspiration The transfer of water into the atmosphere by evaporation and transpiration (the release of water vapor by plants).

executive branch The branch of the government charged with administering and enforcing the laws; the federal executive branch consists of the president, the vice president, and the president's appointees.

exotic species A nonnative species that is artificially introduced to an area.

explosion of life A phrase sometimes used to refer to the rapid diversification of life on Earth about 570 million years ago.

exponential growth The relatively rapid ("geometric") growth phase experienced by many populations during their history. Such growth can be expressed, at least approximately, using an exponential equation or curve.

exponential phase Period or phase of exponential growth in a population. *See* exponential growth.

exponential reserve How long a resource reserve, such as a mineral reserve, will last if demand changes (specifically, if it increases) over time.

externality Environmental, social, and other costs of production that are not included in the price of the product that causes the costs. *See also* market failure.

extinction The loss or death of a group of organisms.

extinction vortex Even if some individuals of a species or population survive disturbances, the population may never fully recover if it becomes too small. The species is said to fall into an extinction vortex and is doomed to eventual extinction.

extirpation The extinction of a species or other group of organisms in a particular local area.

extrinsic values Values that are external to a resource's own right to exist; values based on the ability of a resource to provide something valued by humans.

fallacy of enlightenment The idea or notion that education of people alone, without any other action, will solve overpopulation and other associated environmental problems of the world.

false dichotomy When a complex question inappropriately becomes a polarized into an "either-or" issue (for instance, jobs or the environment).

Federal Insecticide, Fungicide, and Rodenticide Act (FIFRA) A law that requires anyone wishing to manufacture a pesticide to register that product with the EPA.

ferrous Referring to iron and related metals that are commonly alloyed with iron, such as nickel and chromium.

fertile isotope An isotope, such as U-238, that can absorb a neutron and form a new element/isotope (such as Pu-239) that is readily fissionable.

fertility rate The number of births in a population. The general fertility rate is the total number of births in a population in any given year as a function of the number of women in their reproductive years. The total fertility rate is the number of children a woman in a given population will have, on average, during her childbearing years.

fertilizer A substance, often an artificial chemical mixture, that is spread on or through the soil to make it more fertile.

filtration The percolation of water through sand and other settled sediment to remove suspended particles.

fission The splitting of an atom, such as uranium or plutonium, to release energy.

fissionable atom An atom that is easily split by neutron penetration, such as U-235.

five e's The five potential values of environmental resources: esthetic (aesthetic), emotional, economic, environmental services, and ethical.

flat-plate collector A device that usually consists of a black metal plate that absorbs heat from the Sun; the heat can be transferred to a liquid and then used as desired (for instance, to heat a building).

flood A high flow of water that overruns its normal confinement area and covers land that is usually dry.

fluidized bed combustion A way to reduce air pollution by burning very small coal particles at very high temperatures in the presence of limestone particles (the limestone helps to capture sulfur and other pollutants).

flywheel energy storage system Uses a rapidly rotating flywheel to store energy.

food web A graphic depiction of the interrelationships by which organisms consume other organisms.

Food, Drug and Cosmetics Act A U.S. federal act (1938) that authorizes the Food and Drug Administration (FDA) to test foods, drugs, and cosmetics to see if they are safe to market for human use.

fossil fuels Coal, oil, natural gas, and related organic materials that have formed over geologic time.

Framework Convention on Climate Change A convention agreed to by many nations at the 1992 Earth Summit. Although the convention does not establish legal obligations or specific target dates, it requires signing countries to use their best efforts to control emissions of greenhouse gases.

frontier mentality The general concept that humans are separate from and above nature, that humans need to subdue the uninhabited wilderness, and that there are always plenty of fresh natural resources to exploit on the edge of the frontier.

fuel cell In a fuel cell electrons are removed from hydrogen atoms to form an electric current; the hydrogen ions combine with oxygen to form water.

fusion The combining or fusing of isotopes of light elements to form a heavier element—in the process, energy may be released.

future costs Environmental costs of a product or service that are not paid now, but rather are passed on to future generations.

Gaia hypothesis The hypothesis that the Earth is similar to an organism and its component parts are integrated analogously to the cells and organs in a living body.

gamma rays A type of radiation consisting of short-wavelength, high-energy electromagnetic radiation.

garbage "Wet" and generally edible (perhaps by pigs or other animals) discarded matter, such as old food remains, yard clippings, dead animals, leftovers from meat packing operations and butcher shops, and so forth.

gene The basic unit of heredity.

General Circulation Models (GCMs) Complex mathematical models that, with the help of supercomputers, simulate the Earth's climatic patterns.

generator A machine that converts mechanical energy (rotational energy) into electrical energy.

genetic patent rights The patenting of genetic resources, often advocated as a way to allow native peoples to receive profits derived from pharmaceuticals, agriculture, and other uses of genetic resources found in their territories.

geothermal energy Energy (heat) originating from deep within the Earth.

grain Cereals and similar plants that are used for human or animal consumption, such as wheat, corn (maize), rice, barley, and so forth.

grain production The growing, or production, of grains such as rice, wheat, maize (corn), sorghum, barley, oats, rye, millet, and so on.

grassroots activism Local participation in environmental issues.

gray water Untreated or partially treated wastewater that is used for such purposes as watering golf courses and lawns or flushing toilets (rather than using cleaner water of drinkable quality).

Green movement A political and social movement of the late twentieth century and twenty-first century that addresses the root causes of environmental degradation and generally advocates a revolutionary break from the past and the establishment of new, Earth-amenable social institutions.

Green Revolution Modern, chemically based, usually mechanized agriculture that was first used on a large scale in the industrialized countries after World War II.

green fees Fees or taxes (such as fees that increase the price of a resource) that are applied to generally reduce throughput (reduce the use of a resource and/or to reduce the production of pollutants).

green investing Investing in environmentally friendly and sustainable businesses and industries.

green taxes *See* green fees.

green technologies Environmentally friendly technologies, including technologies that promote sustainability via efficiency improvements, reuse/recycling, and substitution.

greenbelt zones Zones or areas in or around a city where the removal of native vegetation is prohibited and/or parks and other open, undeveloped, and vegetated space is protected.

greenhouse effect The warming up of the lower atmosphere due to the accumulation of greenhouse gases that trap heat near the surface of the Earth.

greenhouse gases Gases, such as carbon dioxide, methane, and CFCs, that are relatively transparent to the higher-energy sunlight, but trap lower-energy infrared radiation. Greenhouse gases that accumulate in the atmosphere promote global warming.

greenmail When companies threaten to close or relocate (often to another country) if they are forced to comply with environmental laws; also known as job blackmail.

greenwashing Marketing that unscrupulously seeks to profit from environmental concerns; also known as ecoscams.

Gross National Product (GNP) A standard index that measures a nation's output of goods and services.

groundwater A general term for the water beneath the Earth's surface.

growth rate The rate at which a population is increasing (or decreasing, in the case of a negative growth rate) in size. *See also* percent annual growth.

habitat The general place or physical environment in which a population lives.

habitat conservation plan A plan where habitat of an endangered species is preserved (in some cases, a portion of the habitat may be destroyed but the remainder is protected).

habitat fragmentation Habitat disruption where natural habitat is broken into small, relatively isolated fragments.

hard laws Statutes or legal instruments that are legally binding and mandatory.

hard technologies Energy technologies that depend on large scale, centralized, complex, and expensive plants and infrastructures.

hardness The amount of calcium, magnesium, and certain other ions in water; "hard water" contains high amounts of these ions.

hazardous waste Wastes that are particularly dangerous or destructive; specifically characterized by one or more of the following properties: ignitable, corrosive, reactive, or toxic.

heat Infrared radiation. *See* infrared radiation.

heavy metals Elements, such as lead, mercury, zinc, copper, cadmium, and so forth, that may be required in trace amounts by organisms, but can cause damage when ingested in larger quantities (such as binding with enzymes and thus impairing their functions).

herbicide A chemical substance used to kill plant weeds.

hidden costs Environmental costs of a product or service that we are initially unaware of, such as future environmental degradation from the full effects of pesticides or other pollutants.

holistic Seeking connections among all aspects of a problem, concern, or issue.

hot spot An area of exceptionally high species richness, especially of concentrations of localized rare species that occur nowhere else.

Hubbert's bubble In the 1950s the geologist M. King Hubbert accurately predicted that U.S. oil production would peak around 1970 and subsequently decline.

Human Suffering Index (HSI) An index developed by the Population Crisis Committee in the 1980s. This index is a summation of a number of different ratings of a country, such as GNP per capita, food sufficiency, inflation, accessibility of clean drinking water, literacy, energy consumption, growth of the labor force, urbanization, and political freedom.

hunger Not having enough food to sustain oneself adequately; a weakened condition that is brought on by the lack of food.

hydroelectric power *See* hydropower.

hydrologic cycle Movement of water about the surface of the Earth, driven by energy from the Sun.

hydropower The use of artificial or natural waterfalls to generate electricity.

hydrosphere The liquid water sphere or "layer" on Earth; it includes the oceans, rivers, lakes, and so on.

hydrothermal fluid reservoir An area where hot rock occurs at relatively shallow depth and natural groundwater is heated, sometimes to extremely high temperatures.

hydrothermal processes Hot water dissolves, transports, and subsequently re-precipitates and concentrates elements and minerals into deposits.

ice ages Intervals in the history of the Earth, especially during the last two million years, when average global surface temperatures were lower than they are currently and continental ice sheets were much more extensive than they are today.

igneous rock A rock formed or crystallized from molten rock (magma).

ignitable A substance that will easily ignite and burn rapidly; a characteristic of some types of hazardous waste.

immigration The movement of people into a country or, more generally, the movement of individuals into a particular population.

immunotoxicology The examination of how chemicals affect the immune system.

incineration The burning of trash and garbage at high temperatures in a large furnace so as to get rid of as much as possible.

independent power producers (IPPs) IPPs construct electricity-generating plants and then sell the electricity to the large utilities.

Index of Sustainable Economic Welfare (ISEW) An alternative to the traditional Gross National Product, the ISEW measures sustainable growth. The ISEW includes environmental measures such as depletion of nonrenewable resources, loss of farmland from erosion and urbanization, loss of wetlands, damage from air and water pollution, and so forth.

indicator species A species in a community or ecosystem that is more susceptible to disturbances than most other species.

indirect value The way that a resource may be valued other than its direct value, such as the emotional and aesthetic values of a forest.

Industrial Revolution A series of industrial and technological inventions, beginning in England in the late eighteenth and early nineteenth centuries, that resulted in the cheap and efficient mass production of many commodities.

industrial ecosystem An industrial situation that mimics the principles of ecosystems in nature. No "waste" is produced; rather, the effluents of one industrial process form the raw materials for another industrial process.

industrial materials Nonmetallic materials/minerals necessary to industry, such as salts, fertilizer components, sulfur, asbestos, abrasive minerals, and so forth.

industrial solid waste Solid wastes produced by industries, including, for instance, wastes from large-scale manufacturing, mining, resource processing, and so forth.

infant mortality The number of babies that die before their first birthday, given that they are born alive.

infrared radiation Low-energy, long-wave-length electromagnetic radiation that humans perceive as heat.

infrastructure Roads, bridges, buildings, electric power lines, water and sewer pipes, and other relatively durable material objects upon which modern societies depend.

input The materials and energy that enter the societal system and flow through society as throughput and exit the societal system as output (deposited into sinks or possibly reused or recycled in the cases of some materials, in which case the output becomes input and throughput once again); such input ultimately comes from natural resources (also known as sources). *See* output and throughput.

input reduction Reducing the flow of materials and energy through society.

intangible costs Environmental costs of a product or service that include destruction of scenery or other subjective costs that are very difficult to measure because different people place different values on them.

Integrated Pest Management (IPM) Integrated Pest Management and the concept of biological control follow the basic philosophy that the farmer should not try to totally eliminate pests, but should simply attempt to control them so that they do not cause serious damage. IPM and biological control often use "natural" controls, such as the pests' natural biological predators. IPM seeks to reduce the use of artificial chemical pesticides.

integration The strength of the interactions among the parts of a system.

Intergovernmental Panel on Climate Change (IPCC) A large, international group of officials, scientists, and other researchers who, under the auspices of the United Nations, have been investigating the issue of global climate change, particularly potential future global warming.

intermittent power source A power source, such as the Sun, that can only be used periodically (the Sun can only be used directly during daylight hours).

intrinsic factors The subtle ways that the complex biochemistries of even closely related species can differ such that a chemical that has no effect on one species may have acute or long-term effects on another species.

intrinsic rate of increase The potential for increase in a given population, often symbolized by r; in large part it is determined by the birth rate and death rate of the population.

intrinsic value The value of a resource unto itself, regardless of its value to humans; often considered the ethical value of a resource, or the right of the resource to exist.

introduction The release of species into new areas that they did not formerly (naturally) occupy, such as the introduction of African savanna species onto game preserves in Florida and Texas.

ionizing radiation Radiation that upon hitting electrically neutral atoms may cause the atoms to lose electrons and thus gain an electrical charge—that is, become ionized.

irrigation The artificial watering of land.

isotopes Atoms of the same element that differ from each other in weight because they have differing numbers of neutrons.

joule A basic unit of energy and work, equal to a force of one newton times one meter.

judicial branch The branch of a government that interprets the laws of the land; in the United States, this is the court system, and the highest federal level is the U.S. Supreme Court.

keystone species A certain species that one or more other species are dependent upon for food, reproduction, or some other basic need.

lag phase The relatively slow growth exhibited by many populations at the beginning of their history.

Land Trust Alliance An association of hundreds of local land trusts, dedicated to preserving open space and natural habitat, in the United States.

landfill In the simplest sense, a hole in the ground where solid waste is deposited. In a modern sanitary landfill, the hole is lined so that materials will not escape, and it is covered with layers of dirt as it is progressively filled. When completely filled, it is capped and sealed with more dirt and topsoil.

latent heat storage Heat storage based on phase changes in material objects, such as the melting of ice.

latitudinal diversity gradient The steady decrease in species richness in most groups as one moves away from the equator.

law of the minimum Holds that the growth of a population is limited by the resource in shortest supply.

leachate A liquid solution that forms as water percolates through waste, such as refuse in a landfill or old mining tailings. Leachate may contain any chemicals that can be dissolved, particles, and even live microorganisms.

legal instruments Laws, treaties, regulations, conventions, and the like.

legislative branch The branch of a government that is responsible for making basic laws; in the United States, the federal legislative body is the Congress, composed of the House of Representatives and the Senate.

less developed countries (LDCs) The less rich and less industrialized countries of the world; basically, the countries that are not considered to be more developed countries. *See also* more developed countries.

Lethal Dose-50 (LD-50) The dose of a toxic substance that will kill 50% of a certain population. LD-50 has become the standard reference for summarizing the toxicity of substances.

Life Cycle Analysis (LCA) Sometimes known as cradle to grave analysis, LCA analyzes the entire "life cycle" of a product, from procurement of the raw materials, through its use, to its eventual disposal and the possible reuse or recycling of its components.

life expectancy The average number of years that a typical person at a certain age can expect to live.

life history Age of sexual maturation, age of reproduction, age of death, and other important events in an individual's lifetime, particularly as they influence reproductive traits.

light pollution Excess "waste" light given off by outside sources (or sources visible from the outside) at night.

light water reactor (LWR) A common type of commercial reactor that uses ordinary (light) water as the moderator.

lignite A soft, dark brown, coal-like material.

limiting nutrient The nutrient in shortest supply in a particular ecosystem.

lithosphere The rock sphere or layer that forms the surface of the Earth; composed of the crust and uppermost portion of the mantle.

Little Ice Age A period in Earth history, beginning in late medieval/early renaissance times (ca. A.D. 1300) and ending only two or three hundred years ago, during which average global temperatures were slightly lower than before or after.

loss of life expectancy (LLE) The average amount that a life will be shortened by a particular risk under consideration.

Love Canal A site in the town of Niagara Falls, New York, that gained national attention in the late 1970s when hazardous chemicals that were buried in the area began to adversely affect the residents. The disaster of Love Canal helped to spur Congress to pass the Comprehensive Environmental Response, Compensation, and Liability Act of 1980.

magmatic processes Processes during which certain minerals may selectively crystallize out of a hot, molten body of rock.

mantle The thick, at least partially molten layer of rock found between the Earth's core and crust.

mariculture Saltwater seafood farming. *See also* aquaculture.

marine protected areas Areas of ocean that are set aside as preserves for marine life.

market economy A national economy in which production and costs are determined by competition among individuals. *See also* command economy for comparison.

market failure When market prices do not reflect all the true costs of a product or service (often referred to as an "externality").

mass extinction A catastrophic event in Earth history that kills large numbers of species. Some mass extinctions killed more than 60% of all living species on Earth at the time of the extinction.

maximum contaminant levels (MCLs) The highest concentrations allowed by the EPA in water designated for certain uses.

maximum sustainable yield (MSY) The concept that the optimum way to exploit a renewable resource is to harvest as much as possible up to the point where the harvest rate equals the renewal rate.

megacity A large city, generally with a population of more than 10 million.

megalopolis A single vast urban area formed by the expansion and merging of adjacent cities and their suburbs.

megawatt (MW) A million watts.

melanoma A condition of malignant skin cancer.

meltdown A major accident at a nuclear power plant where the core is damaged and begins to "melt."

membrane method A way of producing freshwater from saltwater by forcing the saltwater through a fine membrane; also known as the filter or reverse-osmosis method.

metallic minerals Minerals containing significant (and often minable) amounts of iron, aluminum, copper, zinc, lead, gold, silver, and so forth.

metamorphic rock A rock formed by modifying a preexisting rock, usually through high temperatures and/or pressures.

methane Natural gas, CH_4, a fossil fuel and potent greenhouse gas.

methanol Methyl alcohol; a fuel that can replace gasoline in many situations and can be produced from natural gas, coal, or biomass.

microirrigation Method of irrigation in which water is transported to crops through pipes and then dripped onto the plants through tiny holes in the pipes, which are installed on or below the surface of the soil; sometimes called drip irrigation.

microloan A small loan made to a small business that promotes sustainable activities.

milling process The process, including crushing, grinding, and leaching, during which a mineral, such as copper or uranium oxide (yellowcake), is removed from the raw ore and concentrated.

mineral A naturally occurring inorganic solid that has a regular crystalline internal structure and composition—for instance, the mineral quartz.

mineral deposit An area where a certain mineral has been concentrated by natural processes.

mineral resources Minerals and earth materials (sometimes including the energy minerals, such as fossil fuels and uranium ore) that form natural resources from which humans draw.

minimum viable population The smallest population size (for a certain population or species) that can stay above the extinction vortex.

mitigation When humans replace a natural ecosystem that was destroyed with a new ecosystem.

moderator A substance, such as water, graphite, or beryllium, used in a nuclear reactor to slow down fast neutrons.

molecular toxicology The examination of the interaction of toxic chemicals with cellular enzymes.

monoculture A form of agriculture where only a single species is grown in a particular field, such as a field devoted entirely to wheat.

Montreal Protocol An agreement reached in 1987 at a meeting in Montreal, Canada, whereby a number of industrialized countries pledged to freeze CFC production at 1986 levels and then gradually decrease CFC production to 50% of 1986 levels by 1999.

moral antirealists Persons who maintain that moral values are created by humans, either consciously or unconsciously.

moral realists Persons who believe that moral values are objective and real and can be discovered by humans.

moral status The question of who, or what, is to be considered worthy of having rights (or what level of rights) in an ethical system.

moral values *See* moral.

moral/moral values Relating to the principles of right or wrong, good or bad behavior. Moral values involve distinguishing or discovering the principles of right and wrong and determining their relative importance and relations.

more developed countries (MDCs) The richer, industrialized countries of the world; basically, the countries of North America, Europe, Japan, Australia, New Zealand, and the former Soviet Union.

Mount Pinatubo The site of a 1991 volcanic eruption in the Philippines that spewed so much smoke and ash into the atmosphere that it temporarily depressed global temperatures (had a minor cooling effect that was experienced worldwide for several years).

muddling along *See* bootstrapping.

multiple-use development The use of a piece of land for different purposes simultaneously, such as the use of riverside land for water filtration by plants, recreation, flood buffers, and wildlife habitat.

multiple-use principle When land is put to many uses at the same time, such as logging, mining, grazing, farming, oil exploration, hunting, fishing, and so forth.

multistage flash distillation (MSF) A method of distillation used to desalinate seawater; cold seawater is run through a series of coils in chambers that become progressively hotter.

municipal solid waste The solid waste produced by the residents and businesses of a city, town, or other municipality; includes old newspapers, packaging materials, empty bottles, leftover foods, leaves and grass clippings, and so forth.

mutagen A substance that causes genetic mutations in sperm or egg cells.

mutation A change in a gene, ultimately caused by a change in the DNA sequence.

mutualism A form of symbiosis that benefits both species.

National Environmental Policy Act (NEPA) A federal act (1969) requiring that all major federal actions, including activities on federal public lands, be reviewed for their environmental impacts before such actions take place. *See* Environmental Impact Statement.

National Marine Sanctuary A marine area that is protected under the auspices of the U.S. government.

natural environment The physical and biological environments independent of human technological intervention.

natural gas A term used for fossil fuels in the gaseous state, particularly methane.

natural hazard An "unpredictable" natural event, such as an earthquake, volcanic eruption, flood, avalanche, drought, fire, tornado, hurricane, and so forth.

natural selection The basic mechanism of evolutionary change, first formulated by Charles Darwin. Biological populations of organisms exhibit variations among the individuals, and the individuals with more advantageous traits tend to contribute more offspring (bearing the advantageous traits) to the next generation. Thus, this natural selection determines which individuals will contribute most to the next generation.

natural siting criteria An approach where the locations (sites or sitings) of roads, buildings, and other structures are selected to be where the geological and biological factors are most favorable; essentially, "working with nature" when selecting locations for human-made structures.

Nature Conservancy Founded in 1951, the Nature Conservancy is an organization that preserves natural areas and wildlife sanctuaries around the world.

neo-Malthusian A term often used to refer to persons who believe that the modern rapid increase in the human population is extremely detrimental.

net primary productivity (NPP) The rate at which producer or primary, usually plant, biomass is created.

net secondary productivity (NSP) The rate at which consumer and decomposer, or secondary, biomass is created.

net yield The concept of net yield for nonrenewable resources holds that a resource can continue to be extracted as long as the resources used in extraction do not exceed the resources gained.

neutron A subatomic particle that has approximately the same mass as a proton, but does not bear an electric charge.

niche An organism's "occupation," or how it lives.

Nimbus-7 A NASA satellite that confirmed the depletion of the ozone layer over the South Pole in the 1980s.

nitrogen cycle The biogeochemical cycle of nitrogen.

nitrogen fixing The conversion of atmospheric nitrogen, usually by nitrogen-fixing bacteria, into forms such as ammonia that are more chemically reactive than atmospheric nitrogen and can also take a nongaseous form under various conditions found on the surface of Earth.

nitrogen oxides NO_x, important components of both lower atmospheric pollution and the upper atmospheric greenhouse gases that promote global warming.

noise Unwanted sound.

nonferrous Metals such as gold, copper, silver, and so forth that are not commonly alloyed with iron. *See also* ferrous.

nonhazardous waste Waste that is not classified as hazardous waste. *See* hazardous waste.

nonmetallic minerals Structural materials, such as sand, gravel, and building stone, and nonmetallic industrial minerals, such as salts, sulfur, fertilizer components, abrasives, gemstones, and so forth.

nonrenewable resource A resource, such as fossil fuels, that does not significantly regenerate itself on a human time scale.

normative ethics Drawing upon a certain ethical system, normative ethics renders ethical opinions and judgments of actions and prescribes appropriate behavior.

nuclear power The use of nuclear fission reactions to generate electricity.

nuclear proliferation The spread of atomic weapons.

ocean A large body of saltwater on the Earth's surface, such as the Atlantic or Pacific Ocean; also refers to the entire body of saltwater that covers much of the surface of our planet.

ocean energy Waves, tides, differential heat layers, and other sources of energy directly related to the world's oceans.

ocean thermal energy conversion (OTEC) A proposed way of extracting usable energy from the oceans based on the fact that in tropical areas the surface waters are warmer than the waters at depth.

oceanic crust The layer of rock that underlies the ocean basins, generally containing a lower percentage of silicon dioxide than continental crust. *See* crust.

oil Petroleum and related liquid fossil fuels.

oil shale A rock that contains the organic precursors of oil known as kerogen. Kerogen can be converted to oil by heating.

open system A system that is not isolated in that it exchanges matter and/or energy with other systems.

open-loop recycling A situation where a material or substance can be recycled once or a few times, but not indefinitely because the material is damaged or degraded each time it is recycled (for instance, paper fibers or some plastics).

openness Refers to whether a system is isolated from other systems.

optimum sustainable yield (OSY) The concept that the optimum harvestable rate for a renewable resource, such as a wild population of fish, must consider factors beyond simple maximum yield, for instance, effects on other species in the ecosystem and various human uses of the ecosystem. *See* maximum sustainable yield.

ore A mineral deposit that can be economically mined at a certain time and place with a certain technology.

organic farming Avoiding the use of synthetic chemicals, such as synthetic or artificial fertilizers, pesticides, and herbicides, when farming; many organic farmers also avoid the use of genetically modified or bioengineered organisms.

organophosphates A group of synthetic organic pesticides that are less persistent, but more toxic to humans, than chlorinated hydrocarbons; includes malathion and parathion.

output Materials and energy that flow out of the societal system. *See* input and throughput.

overshoot When a population exceeds its carrying capacity.

oxygenation A method of reformulating gasoline by blending oxygen-rich liquids into it such that carbon monoxide will be converted to carbon dioxide and volatile organic compounds will be better oxidized.

ozonation The use of the reactive chemical ozone (O_3) to disinfect water.

ozone An O_3 molecule. Ozone contributes to air pollution in the troposphere, but is an important natural component of the stratosphere. The stratospheric ozone layer protects the Earth's surface from excessive levels of ultaviolet radiation.

ozone layer A layer of ozone in the stratosphere, most concentrated at an altitude between about 12 and 16 miles (20-25 km).

parasitism Occurs where one species (the parasite) lives off another species (the host) and may actively harm the host; often considered a form of symbiosis.

particulates Small particles of dispersed matter, in either a solid or a liquid state, that are larger than individual molecules and are one of the categories of air pollution.

passive solar design A type of architecture that uses the inherent characteristics of a building to capture heat and light from the Sun.

peak load The amount of electricity needed at the time of highest demand.

peat A thick accumulation of partially decayed plant material.

pegmatite A rock type formed from residual magma that is often characterized by very large crystals that may contain high concentrations of otherwise relatively rare elements.

pelagic Refers to the water column zone of the aquatic (specifically, marine) biome.

percent annual growth The rate of natural increase expressed as a percentage of the given population.

persistence How long a pollutant stays in the environment in unmodified form.

pesticide A chemical substance used to destroy animal pests, such as insects, that might attack a crop. More generally, a pesticide can be any chemical manufactured to kill any organisms that humans consider undesirable.

pesticide treadmill The concept that to keep up with the rapid pace of evolving pesticide immunity, new pesticides must constantly be developed.

petajoule 10^{15} joules.

Petkau effect The general concept that sustained low doses of radiation may be more damaging to organisms than single large doses. More specifically, the Petkau effect documents how cell membranes are damaged by low but sustained doses of radiation.

petroleum Essentially, oil or liquid fossil fuel.

pH scale A scale that is used to measure acidity; 1 is very acidic, 7 is neutral, and 14 is very basic (alkaline).

pheromones Biochemical scents used by males or females to attract the opposite sex.

philosophical ethics The consideration of ethical controversies and underlying values, principles, and rules at an abstract level.

photic zone The upper part of the aquatic biome where light can penetrate the water.

photochemical pollutant A pollutant produced when sunlight initiates chemical reactions among NO_x, volatile organic compounds, and other substances found in the air.

photosynthesis The process by which organisms such as green plants convert light energy to chemical energy and synthesize organic compounds from water and carbon dioxide.

photovoltaics The use of semiconductor technology to generate electricity directly from sunlight.

Physical Quality of Life Index (PQLI) An index developed by the Overseas Development Council in the late 1970s that rates a country on the basis of the average life expectancy, infant mortality, and literacy rates of its citizens.

physical environment The natural physical nonliving world, including the lithosphere, hydrosphere, and atmosphere.

pioneer community The initial community of colonizing species in a particular area.

pioneer stage *See* pioneer community.

placer deposit Heavy mineral grains that have weathered from rocks, been transported, sorted, and finally settled in an ore deposit, such as gold nuggets in the bottom of a stream.

plate tectonics The concept or theory that the Earth's lithosphere is divided into numerous plates that are in motion relative to each other; the continents ride on the backs of the plates and thus move ("drift") over geologic time. Plate tectonics accounts for the distribution of most earthquakes, volcanoes, and other important geological phenomena.

plutonium A heavy element that contains 94 protons. Fissionable isotopes of plutonium can be used as fuel in nuclear reactors and can also be manufactured into bombs.

Pollutant Standards Index (PSI) An index that measures air quality, ranging from 0 (best air quality) to over 400 (worst air quality).

polluted water Water that is rendered unusable for its intended purpose.

pollution A term often used to refer to excess outputs by society into the environment. Depending on how the term is used, pollution can also refer to excess outputs by natural processes (such as "pollution" caused by a volcanic eruption).

population A group of individuals of the same species living in the same area.

population bomb A term used to refer to the fact that the vast majority of the world's population growth is currently taking place in the Southern Hemisphere.

positive feedback The process in which part of a system responds to change in a way that magnifies the initial change.

potable water Water that is safe to drink.

power Work (requiring energy) divided by the time period over which the work is done.

power plant A plant that converts some form of energy, for instance, chemical energy found in coal, into electrical energy.

precautionary principle The principle that advises that, in the face of uncertainty, the best course of action is to assume that a potential problem is real and should be addressed ("better safe than sorry").

precipitation Removal of water from the atmosphere as rain, snow, sleet, or hail.

precycling The reduction of packaging materials by manufacturers.

predation The process in which certain organisms kill and consume other organisms.

preservation Refers to nonuse, such as a "preserve" that is set aside and protected in its pristine natural state.

preservationism The classical view that nature should be preserved for its own sake.

primary pollutants Pollutants that are directly emitted, such as SO_x by coal-burning power plants and other industries.

primary standards Under the drinking water standards of the Safe Drinking Water Act, primary standards specify contaminant levels based on health-related criteria.

primary treatment The use of physical processes, especially screening and settling, to remove materials from water.

privatization When a resource, such as an environmental resource, is transferred from the commons to an individual or company that has a vested interest in the particular resource.

progressive conservationism The view that the federal government in particular should take an active role in protecting forests and other natural resources for the public; popular during the late nineteenth and early twentieth centuries.

property rights Rights associated with ownership or use of material property, including land.

protocells Cell-like structures artificially produced by heating amino acids and other organic molecules.

pumped hydroelectric storage (PHS) In PHS, electricity is used to drive pumps that transfer water from a lower reservoir to a higher one. Electricity can then be regenerated by allowing water to flow through a turbine on its way back to the lower reservoir.

quad A quadrillion BTUs.

radiation Electromagnetic radiation; includes visible light, heat, ultraviolet radiation, gamma rays, X rays, and so forth. The term radiation is also sometimes used to refer to the emission of particles (such as alpha and beta particles) from a radioactive atom.

radioactivity The emission of particles (such as alpha and beta particles) and rays (energy, such as gamma rays) from a nucleus as it disintegrates.

radionuclides Unstable atoms that undergo spontaneous disintegration and give off radiation in the process.

rate of natural increase The crude birth rate minus the crude death rate of a population.

reactive With reference to hazardous waste, materials that are very active chemically and can easily cause explosions and/or release harmful fumes.

reactor vessel In a nuclear power plant, a thick steel tank that usually contains the reactor core and primary water loop.

recharge area An area where rainfall can infiltrate into an aquifer.

recovery rate The amount of the original material that can actually be recovered and recycled.

recovery ratio The recovery rate expressed as a ratio. *See* recovery rate.

recycling Using the same resource over and over, but in modified form.

recycling loop The use of a resource, followed by the discarding and reprocessing of the resource, and then the reuse of the resource.

refuse Refers to both trash and garbage. *See also* garbage; trash.

regulatory agencies Agencies that are part of the executive branch of the government and are responsible for administering and enforcing law enacted by Congress.

reintroduction Release of plants and animals back into habitat that they formerly occupied, but in which they may have become locally scarce or extinct.

relativists Persons who believe that there is no objective or rational way to determine right or wrong, good or bad, in an ultimate (versus relative) sense.

remediation Efforts to counteract some or all of the effects of pollution after it has been released into the environment.

renewable energy An energy source that, from an Earth perspective, is continually renewed (for example, solar energy).

renewable resource A resource that will regenerate within a human time scale; for example, crops and energy received from the Sun.

replacement level fertility The number of children needed to keep a population at a stable size; generally considered to be a total fertility rate of about 2.1 children per woman.

reprocessing facility A facility designed to reprocess spent nuclear fuel in order to recover fissionable materials.

reproductive rights Essentially, a woman's right to determine for herself if, and when, she will bear children.

reserve A resource that has been located and can be profitably extracted at the current market price. For minerals in particular, a reserve is an identified ore deposit that has yet to be exploited.

residence time (a) The amount of time that a certain atom or molecule spends, on average, in a certain portion of its biogeochemical cycle. (b) The time it takes for a pollutant to move through the environment.

resource A source of raw materials used by society.

resource recovery facility A plant where trash and garbage are burned and the heat generated is recovered and applied to some useful end, such as generating electricity.

respiration Biological combustion or the "burning" of food molecules in an organism. During respiration large organic molecules are broken down into simpler organic molecules, and energy is released.

restoration The process of returning a degraded resource to its natural state.

reuse Using the same resource over and over in the same form.

right Something to which one has a just, inherent, or natural claim—such as the right to life.

riparian law A law under which the owner of land has the right to withdraw water that is adjacent to the land, such as from a river or lake.

riparian zone Area along a river bank or stream bank.

risk A dangerous situation, or the chance of loss or peril to human life or other valuable entities; generally refers to any potential danger.

rock cycle The cycling of solid earth materials—rocks—from one form to another. Sedimentary rocks may be subjected to great pressures and/or temperatures and be transformed into metamorphic rocks. Metamorphic rocks may be heated to the point where they melt, forming a magma that later crystallizes as igneous rocks. Igneous rocks may be weathered and broken down into grains that form the basis of a sedimentary rock.

rubbish A very general term that includes trash, garbage, and other items such as construction and demolition debris.

Safe Drinking Water Act (SDWA) A federal statute enacted in 1974 that aimed explicitly at improving the quality of drinking water by establishing primary and secondary standards for contaminant levels in water.

salinization An increase in soil salt content that sometimes occurs due to prolonged irrigation, especially in poorly drained arid regions.

sanitary landfill *See* landfill.

saturation, zone of The region below the water table where all voids in the soil and rock are fully filled with water.

scaling factors In the analysis of dose-response assessment and toxic risk assessment of substances, a way of extending animal results to humans by taking differences in body sizes (masses) into account.

scarcity With reference to natural resources, such as minerals, a lack of such resources as compared to demand for the resources.

scoping When local government agencies and the general public are involved in producing an Environmental Impact Statement for a proposed project.

scrubbers Devices that cleanse emissions, usually with water, before they are released into the air.

sea level rise Worldwide rises in sea level (sea level height increasing relative to the elevations of the continents), such as has been predicted as a result of global warming.

secondary pollutants Pollutants produced by reactions among other air pollutants, such as photochemical pollutants.

secondary standards Under the drinking water standards of the Safe Drinking Water Act, secondary standards specify contaminant levels based on nonhealth-related criteria (such as color or odor, which are not strictly health issues).

secondary treatment The use of biological (microbial) processes to remove materials from polluted water.

sedimentary processes Natural processes that may concentrate minerals through precipitation from a solution or by differential settling of grains in moving or still water.

sedimentary rock A rock formed on or near the Earth's surface by the settling or precipitation of materials. A sedimentary rock may be composed of grains or clasts of particulate matter (usually older rocks or mineral grains), or it may be formed from the chemical or organic precipitation of minerals from an aqueous solution (such as certain limestones formed in an ocean).

sedimentation The settling out of suspended particles from a body of water (or in some cases, very fine particles settled from the air or blown by the wind).

selective cutting The harvesting of trees such that only certain trees are cut down and the land is not stripped bare; also known as uneven-aged management.

sensible heat storage Allowing a material substance to heat up and then release its heat again.

septic tank A large concrete tank buried underground to receive household sewage. Septic tanks are used in areas without a sewer system.

sequential use development When land that is no longer needed in one capacity is used in another way, as for instance when a capped landfill is converted to a park.

severance tax A tax levied on minerals or other resources as they are extracted.

shore zone Transitional zone of a body of water with land.

sick building syndrome Refers to ailments such as headaches, nausea, allergic reactions, and other symptoms that are caused by indoor air pollutants in a building where people work or live.

sinkhole A type of land subsidence, taking the form of a large depression in the ground, caused by water withdrawal. A sinkhole occurs when a thin layer of rock overlying an underground cavern collapses.

sinks Environmental reservoirs that receive the throughput of society.

six e's The six potential values of wild biological resources: esthetic (aesthetic), emotional, economic, environmental services, ethical, and evolutionary.

sky glow The phenomenon when the sky at night in a certain area is not dark as it would be naturally, but glows (perhaps in shades of pink, orange, white, yellow, or gray) due to excess light in the local atmosphere. *See* light pollution.

slash and burn agriculture A form of agriculture where trees and other vegetation are cut down and burned in order to clear the land and release nutrients into the soil.

smelting A process in which concentrated ore, such as copper one, is roasted and subjected to high temperatures in a smelting furnace to produce crude metal (for instance, crude copper).

smog Originally referred to a mixture of smoke and fog, but now generally refers to the brown haze of photochemical pollutants found in some urban areas.

social ecology An approach to environmental concerns that attempts to identify social factors that are the underlying causes of current environmental degradation.

social institutions Generally, the institutions or constructs that are important in regulating society, such as familial, political, educational, religious, and economic systems and institutions.

soft laws Legal instruments that are not legally binding, but act more as guides to policy.

soft technology Energy technologies that are generally small scale, relatively inexpensive, and localized.

soil A mixture of weathered rocks and minerals, decayed organic matter, living organisms, air, and water.

soil degradation The damaging or destruction of natural soils; often due to overuse, abuse, and neglect by humans.

soil fertility The ability of the soil to support plant life and associated fauna.

soil horizons Layers of soil that form approximately parallel to the surface of the land; may include the topsoil, subsoil, a layer of partially distintegrated rock, and finally the underlying bedrock.

solar energy Energy derived from the Sun.

solar thermal technology The use of the Sun's energy to heat substances such as water to produce steam that drives a turbine and generates electricity.

solar-hydrogen economy An economy based primarily on solar power (in all its forms, including wind power and hydroelectric power) and using hydrogen as a convenient way to store and transport energy.

solid waste Broadly defined, soil waste includes such items as household garbage, trash, refuse, and rubbish, as well as various solids, semisolids, liquids (such as sludge or liquids in solid containers), and gases (often contained in solid containers, such as gas canisters), that result from mining, agricultural, commercial, and industrial activities.

source reduction As applied to solid waste, reducing the generation of waste in the first place (as opposed to later reusing or recycling waste).

sources The environmental resources (matter and energy) that are taken from nature and used by society.

species Often defined, at least for sexual organisms, as all of the organisms that can interbreed (or potentially interbreed) to produce fertile offspring. Different species are reproductively isolated from one another.

species recovery Increasing the population size of a threatened or endangered species so as to ensure that it will not go extinct.

species richness The number of different species that occur in at given area.

species triage The concept that conservation efforts should be directed at those species that are at risk of extinction and can still be saved.

species-area curve A curve or graph that shows the number of different species found in a gradually enlarged area of sampling.

static reserve How long a resource reserve, such as a minimal reserve, will last if demand does not change with time.

stock pollutant Refers to synthetic materials that are not found in nature and tend to be among the most persistent pollutants.

strategic mineral A critical mineral that a particular country must import from areas that are potentially unstable politically, militarily, or socially.

stratosphere The thermal layer of the atmosphere above the troposphere in which temperature increases with altitude. The ozone layer occurs within the stratosphere.

strip-mining A form of surface mining, especially for coal, that is very destructive to the landscape.

structural materials Nonmetallic materials/minerals used in building, such as building stone, sand, gravel, and other components of cement and concrete.

substitutability Finding substitutes for various materials. *See* substitution.

substitution The use of one resource in place of another.

suburban sprawl The spreading of a city's population out into the surrounding countryside, forming suburbs.

superconducting magnetic energy storage (SMES) A system in which superconducting loops or coils would be used to store electrical energy by allowing the current to circulate around a closed loop of nearly zero resistance.

superconductivity Superconductors are substances through which electrons can pass with virtually no friction or resistance, thus allowing almost 100% energy transmission.

Superfund The common name for the federal Hazardous Substance Response Fund that is used for cleanup and related expenses associated with hazardous waste sites on the EPA's National Priorities List.

supertoxin A poisonous substance of such high potency that less than a drop can kill an adult human.

surplus areas (water) Areas that receive more precipitation than is needed by well-established vegetation, including crops.

sustainability Meeting the needs of today without reducing the quality of life for future generations.

sustainable development Development that focuses on making social, economic, and political progress to satisfy global human needs, desires, aspirations, and potential without damaging the environment; sometimes known as sustainable growth.

sustainable economy An economy that produces wealth and provides jobs for many human generations without degrading the environment.

sustainable growth *See* sustainable development.

sustainable harvesting The sustainable use/harvesting of nuts, fruits, and other products that can be extracted from an ecosystem without causing damage; sometimes known as extractive forestry.

sustainable technology Technology that permits humans to meet their needs with minimum impact on the environment.

symbiosis Organisms of different species living together; includes such relationships as amensalism, commensalism, mutualism, and parasitism.

synergism With reference to toxicology, when the toxic effects of different substances are multiplied through interaction.

syngas A mixture of hydrogen gas and carbon monoxide; also known as coal gas or town gas.

system A set of components functioning together as a whole.

systems approach The study of phenomena as systems and subsystems, allowing the isolation of various parts or portions of the world so as to focus on those aspects that interact with one another more closely than others. *See* system.

tailings In mining, the residue after high-grade ore is extracted.

taxonomists Practitioners of taxonomy. *See* taxonomy.

taxonomy In biology, the description, classification, and naming of groups of organisms.

teleological tradition According to this way of thinking, every being and object in nature (as well as among human artifacts) has a purpose, function, end, final cause, or utility in the overall natural design and order of the world and universe.

teratogen A substance that adversely affects fetal development, such as when a pregnant woman ingests harmful chemicals.

terawatt (TW) 10^{12} watts.

thermal inversion Occurs when a layer of warm air overlies cooler air in the trophosphere (lower atmosphere), thus inverting the usual condition in which air becomes cooler as altitude increases.

thermochemical conversion The heating of biomass in an oxygen-deficient environment in order to produce substances that can be used as fuels.

thermochemical heat storage The use of reversible chemical reactions to store heat.

thermocline The boundary between warmer surface waters and colder deeper waters in a body of water.

thermodynamics The study of energy and energy conversions.

thermodynamics, first law of States that energy can be neither created nor destroyed, but only transformed.

thermodynamics, second law of States that when energy is transformed from one form to another, it is degraded. This is sometimes known as the law of entropy because energy transformations increase the entropy of a closed system.

Third World debt crisis *See* debt bomb.

thorium An element that contains 90 protons; thorium-232 can be used in a breeder reactor to produce fissionable U-233.

three R's Return, repurify, and reuse (with reference to water).

threshold of exposure The smallest amount of a poison or other toxic substance that is necessary to cause harm.

throughput The movement of materials and energy through society, or the materials and energy so moved. There can be continuous flow from input to throughput to output. *See* input and output.

throwaway society A society based on the one-way flow of matter; products are simply discarded after being used.

tidal power The harnessing of the tides to produce energy in a form that humans can readily utilize.

tokamak A large machine that uses magnetic fields to confine and promote controlled fusion reactions.

topsoil An upper layer of the soil, composed primarily of a mixture of organic matter and mineral matter; it is alive with microscopic and small macroscopic organisms.

tornado Typically consists of a rapidly rotating vortex of air that forms a funnel.

torts *See* common law torts.

total fertility rate *See* fertility rate.

Toxic Substances Control Act (TSCA) A federal statute enacted in 1976 to regulate toxic substances not covered under other laws. The basic goals of the TSCA are to require industry to produce data on environmental effects of chemicals and to prevent harm to humans and the environment, while not creating unnecessary barriers to technology.

toxic Toxic materials or substances are harmful or fatal when consumed by organisms in relatively small amounts.

toxic threshold *See* threshold of exposure.

toxic tort A lawsuit against a manufacturer of a toxic substance for harm caused by that substance.

toxicology The study of the effects of chemicals that are harmful or fatal when consumed by organisms in relatively small amounts.

Toxics Release Inventory (TRI) A report compiled annually by the Environmental Protection Agency on toxics released by U.S. industries based on data reported to the EPA by those industries under the Emergency Planning and Community Right-to-Know Act of 1986.

toxin A poisonous substance produced by an organism; the term is sometimes used to refer to any toxic substance in general.

Tragedy of the Commons The concept that property held in common by many people will generally be overused until it deteriorates or is even destroyed.

transcendentalism A general philosophy espoused in the nineteenth century by such writers as Ralph Waldo Emerson and Henry David Thoreau. In many ways, transcendentalism emphasized the spiritual over the material, especially the materialism of society. This resulted in a concern for the natural environment and a mourning and decrying of the destruction of the natural environment by humans.

transcendentalist An adherent of the philosophy of transcendentalism. *See* transcendentalism.

transgenic crops Genetically transformed crops; crops that have been artificially engineered using bioengineering.

trash Waste that is "dry" (as opposed to liquid or gas) and nonedible, such as newspapers, boxes, cans, containers, and so forth.

troposphere The lowermost thermal layer of the atmosphere, wherein temperatures normally decline with increasing altitude; the layer of the atmosphere in which most weather phenomena take place.

true environmental costs The concept that the cost or value of a resource should include all indirect, as well as direct, values and costs of the resource.

turbine A machine that converts the lateral motion of a liquid or gas into rotational motion.

ultraviolet (UV) radiation Relatively high-energy, short-wavelength, electromagnetic radiation (light). UV has wavelengths in the range of 200 to 400 nanometers (one nanometer equals a billionth of a meter).

umbrella species A large, charismatic species (e.g., the Florida panther). When the habitat for such a species is protected, many other species will be protected as well.

unequally distributed costs Environmental costs of a product or service that are not distributed evenly among socio-economic groups of people, such as when poorer people either pay relatively more or absorb relatively more of the environmental costs (for instance, being forced by economics to live in more polluted areas).

unique species Species that are not closely related to any other living species.

uranium A heavy element that contains 92 protons. Fissionable isotopes of uranium can be used as fuel in nuclear reactors and can also be manufactured into bombs.

urban ore The concept that wastes of today, such as materials disposed of in landfills, may serve as sources of valuable raw materials, such as metals, in the future.

urbanization The trend toward increasing numbers of people living in cities.

utilitarianism A philosophy whose basic principle is that the overall good should be maximized ("the greatest good for the greatest number").

value The relative worth, utility, or importance of an object or idea. *See also* moral.

virgin ores or resources Original natural ores or resources that are extracted from the Earth or from "Nature."

virgin supplies of ore *See* virgin ores.

volcano A place on the Earth's crust where hot, molten rock (magma) wells up to the surface.

waste-to-energy The burning of municipal solid waste to produce energy.

water table The boundary between the zones of aeration and saturation.

waterlogging The rising of the water table over time, and the soaking of soils, in areas where irrigated land is poorly drained. Water-logging is often associated with salinization.

watt A common unit of power defined as one joule of work or energy per second.

weather Short-term, daily perturbations in the atmospheric/hydrologic cycles.

Wilderness Act of 1964 A federal statute that allowed for the designation of wilderness areas on federal lands that are to be managed so as to retain their primeval character.

Wildlands Project An ambitious plan to designate a network of interconnected wildlife preserves and habitat corridors throughout the United States.

wildscaping Retention of native soil, vegetation, and other natural features when building on the land, rather than the removal of soil, vegetation, and natural features followed by artificial landscaping once building is completed.

wind farm A vast trace of land covered with wind-powered turbines that are used to drive generators that produce electricity.

wind power The harnessing of the wind's energy for human applications.

Wise Use movement A movement with the avowed goal of nullifying or eliminating most environmentally based laws and regulations. Wise Users generally see such laws as overly expensive, disruptive of free enterprise, and infringing on property rights.

withdrawn water Water that is taken from its source (such as a river, lake, or aquifer); it may be returned to its source after use.

work A force applied to a material object times the distance that the object is moved.

World Trade Organization An international organization founded to reduce trade barriers and settle trade disputes between nations.

xeriscaping Landscaping designed to save water.

xerophytes Plants, such a plants found in deserts, with adaptations that allow them to survive where there is a scarce and unpredictable water supply.

yellowcake Uranium oxide (U_3O_8) or "natural uranium."

Photograph Acknowledgments

Section 1. Section opener: Lee Hannah/Conservation International

Chapter 1. Chapter opener: NOAA Restoration Center, Mike Devany & Marine Resource Council. 1.2: William Larose/Greenpeace. 1.4: Paul Bierman/Visuals Unlimited. 1.5: Inserts ©1999 PhotoDisc, Inc. 1.8: Kit King/Getty Images. 1.10: NOAA Restoration Center and Damage Assessment and Restoration Program. 1.12: ©Dennis Capolongo/Greenpeace. 1.13: UN Photo Library. 1.14: Theodore Roosevelt Collection/Harvard College Library. 1.15: AP Photo/Rich Pedroncelli. 1.16: Courtesy of Gene Carl Feldman/NASA Goddard Space Flight Center. Case Study 1: ©1999 PhotoLink/PhotoDisc, Inc. Page 29: ©1999 Karl Weatherly/PhotoDisc, Inc.

Chapter 2. Chapter opener: UN Photo Library. 1.10b: Robert Nickelsberg/Getty Images. 2.11: Mark Richards/PhotoEdit. 2.13b: Robert Nickelsberg/Getty Images. 2.14: Owen Franken/Corbis. Case Study 1: Corbis. Case Studies 2.1 and 2.2: The Granger Collection, New York.

Chapter 3. Chapter opener: ©1997 PhotoLink/PhotoDisc, Inc. 3.6: L. Lee Grisneer/US Fish and Wildlife Service. 3.7: Glenn Oiver/Visuals Unlimited. 3.8: Neville Coleman/Visuals Unlimited. 3.10: ©1996, PhotoDisc, Inc. 3.11b: Harold Hoffman/Photo Researchers, Inc. 3.13: ©1999 PhotoLink/PhotoDisc, Inc. 3.18: Tracy Brooks–Mission Wolf/U.S. Fish and Wildlife Service. 3.22b: ©1996 D. Normark/PhotoLink/PhotoDisc, Inc. 3.23b: Laguna Photo/Gamma Liaison. 3.26b: ©1996 T. O'Keefe/PhotoLink/PhotoDisc, Inc. Page 91: ©1999 Jeremy Woodhouse/PhotoDisc, Inc.

Chapter 4. Chapter opener: Reto Stöckli/NASA Goddard Space Flight/MODIS. 4.2d and e: Jack Jeffrey Photography. 4.5: ©1996 L. Hobbs/PhotoLink/PhotoDisc, Inc. 4.10: ©PhotoDisc. 4.11: ©1999 PhotoLink/PhotoDisc, Inc. 4.12A: ©1999 PhotoLink/PhotoDisc, Inc. 4.13: E. R. Degginger/Color-Pic, Inc. 4.14: Tim McCabe, USDA Natural Resources Conservation Service. 4.15a: ©1996 PhotoLink/PhotoDisc, Inc. 4.15b: Robin West/U.S. Fish and Wildlife Service. 4.16: Charlie Ott/Photo Researchers, Inc. 4.17a: Thomas Kitchin/Tom Stack & Associates, Inc. 4.17b: ©1999 PhotoLink/PhotoDisc, Inc. 4.19a: NOAA National Estuarine Research Reserve Collection. 4.19b: NOAA National Undersea Research Program (NURP) Collection. 4.20a: Dr. Neil Sullivan/NOAA Corps Collection. 4.20b: Reese/OAR/National Undersea Research Program (NURP) Collection. 4.21: Jacques Descloitres, MODIS Land Science Team/NASA. 4.22: ©1999/PhotoLink/PhotoDisc, Inc. Case Study 2.1: ©Brad Moon.

Chapter 5. Chapter opener: Cotopaxi, 1862. Frederic Edwin Church. Founders Society Purchase, Robert H. Tannahill Foundation Fund, Gibbs-Williams Fund, Dexter M. Ferry, Jr. Fund, Merrill Fund, Beatrice W. Rogers Fund, and Richard A. Monoogian Fund. Photograph ©1985 The Detroit Institute of Arts. 5.1: NASA. Fig. 5.7: J. Frederick Grassle/Woods Hole Oceanographic Institution. 5.11b. NOAA. 5.13: NOAA National Geophysical Data Center. 5.14: ©1996 Sherburne/PhotoLink/PhotoDisc, Inc. 5.15: USGS. 5.16: FEMA News Photo. 5.17: NOAA Photo Library, NOAA Central Library; OAR/ERL/National Severe Storms Laboratory (NSSL). 5.18: NOAA Historic National Weather Service Collection. 5.19: Jacques Descloitres, MODIS Land Rapid Response Team, NASA Goddard Space Flight Center. 5.20: FEMA News Photo. 5.21: Alain Nogues/Corbis Sygma. 5.22: Alaskan Type I Incident Management Team, Photo by John McColgan, Bureau of Land Management, Alaska Fire Service. 5.23: Richard Mieremet/NOAA, OSDIA. Case Study 2.1: Katsuhiko Ishida.

Chapter 6. Chapter opener: ©1996 C. Borland/PhotoLink/PhotoDisc, Inc. 6.1: Tim McCabe/National Resources Conservation Service. 6.8: Photographed by Pat Toth-Smith, Aesthetic Images, for the catalog of Green Earth Office Supply, Los Gatos, California. 6.11: ©2000, Eyewire. Case Study 1.1: Tom Stewart/Corbis.

Chapter 7. Chapter opener: ©1999 GeoStock/PhotoDisc, Inc. 7.2: Michaud Grapes/Photo Researchers, Inc. 7.3: NOAA/EXXON VALDEZ Oil Spill Trustee Council. 7.5: U.S. Department of Energy. 7.7: Mark Marten/U.S. Department of Energy. 7.10: U.S. Department of Energy. 7.11: Sovfoto/Eastfoto. 7.12: Sovfoto/Eastfoto. 7.13: F. Edwards/Visuals Unlimited. 7.15: Science VU/Visuals Unlimited. 7.16: Eric Sander/Gamma. Case

Study 2.1: NOAA/Office of Response and Restoration, National Ocean Service. Case Study 3.1a: John and Karen Hollingsworth/U.S. Fish and Wildlife Service.

Chapter 8. Chapter opener: Tim Ellison/National Renewable Energy Laboratory. 8.1a and b: Courtesy of U.S. Army Corps of Engineers. 8.2: Warren Gretz/Courtesy of National Renewable Energy Laboratory. 8.3: Andrew Carlin–Tracy Operators/Courtesy of National Renewable Energy Laboratory. 8.4: Courtesy of National Renewable Energy Laboratory/West Bengal Renewable Energy Development Agency. 8.6: Kramer Junction Company/Courtesy of National Renewable Energy Laboratory. 8.7: Courtesy of Southern California Edison Company. 8.9: ©2000 Eyewire. 8.10: Pacific Gas & Electric/Courtesy of National Renewable Energy Laboratory. 8.12a and b: Courtesy of the French Technology Office. Case Study 1.1b: Courtesy of Ballard Power Systems Inc. Case Study 2.1: Toyota Motor Corporation. Case Study 2.2: Jeff Greenberg/Visuals Unlimited.

Chapter 9. Chapter opener: Edwin G. Huffman/World Bank. 9.3f: ©1996 K. Knudson/PhotoLink/PhotoDisc, Inc. 9.6: Peter Brown/Rocky Mountain Tree-Ring Research, Ft. Collins, CO/NOAA Paleoclimatology Program. 9.11: U.S. Geological Survey. 9.14: A.S Navoy/U.S. Geological Survey. 9.18: Margaret Luzier/U.S. Army Corps of Engineers. 9.21: ©1999 Ken Samuelsen/PhotoDisc, Inc. Case Study 1.1: George Chan/Photo Researchers, Inc.

Chapter 10. Chapter opener: Courtesy of the Kennecott Utah Copper Corporation. 10.4: Carlyn Galati/Visuals Unlimited. 10.5: ©1996 K. Knudson/PhotoLink/PhotoDisc, Inc. Case Study 1.1: Dr. David Hyndman, The University of Queensland. Page 259: ©1999 Geostock/PhotoDisc, Inc.

Chapter 11. Chapter opener: Courtesy of South Florida Water Management District. 11.1: National Ocean Service Photo Gallery. 11.5: Peggy Greb/USDA Agricultural Research Service. 11.6: Rob Suisted/www.naturespic.com. 11.7: Courtesy of the John Hay Library, Brown University. 11.8: Mary DeDecker/U.S. Fish and Wildlife Service. 11.9: Lynn Betts/USDA Natural Resources Conservation Service. 11.11: Robert K. Boone/U.S. Fish and Wildlife Service. 11.14a: John and Karen Hollingsworth, U.S. Fish and Wildlife Service. 11.14b: Steve Hillebrand, U.S. Fish and Wildlife Service. 11.15: Guillermo Gonzalez/Visuals Unlimited. Case Study 1.1: Peter Yates/Science Photo Library/Photo Researchers, Inc.

Case Study 2.1: Claire Dobert, U.S. Fish and Wildlife Service. Case Study 2.2: Anthony Mercieca/Photo Researchers, Inc. Page 281: Craig Koppie/U.S. Fish and Wildlife Service.

Chapter 12. Chapter opener: NASA/Goddard Space Flight Center Scientific Visualization Studio. 12.2: NPS photo. 12.3: ©1997 PhotoLink/PhotoDisc, Inc. 12.4: ©1996 PhotoLink/PhotoDisc, Inc. 12.5: Jeff Greenberg/Visuals Unlimited. 12.8: NPS Photo. 12.9a: Terranova International/Science Source/Photo Researchers, Inc. 12.10: ©1997 PhotoLink/PhotoDisc, Inc. 12.11: ©1999 Alan and Sandy Carey/PhotoDisc, Inc. 12.12a: NOAA/Florida Keys National Marine Sanctuary. 12.14: Lynn Betts, USDA Natural Resources Conservation Service. 12.16: ©Phil Schermeister/Corbis. Case Study 1.1: NPS Photo. Case Study 1.2: NPS Photo. Case Study 2.1: Courtesy of Ijams Nature Center. Case Study 3.1: ©Dean DeChambeau.

Chapter 13. Chapter opener: Tim McCabe/USDA. 13.1b: UN/DPI Photo. 13.2: Art Zamur/Getty Images. 13.4: Agricultural Research Service/USDA. 13.5: Bill Bachman/Photo Researchers, Inc. 13.6: Tim McCabe/USDA Natural Resources Conservation Service. 13.7: Jeff Vanuga/USDA Natural Resources Conservation Service. 13.10: Gene Alexander/USDA Natural Resources Conservation Service. 13.11: S. Berry/Visuals Unlimited. 13.12: Tim McCabe/USDA Natural Resources Conservation Service. Case Study 2.2: Scott Bauer/Agricultural Research Service/USDA. Page 333: NASA.

Chapter 14. Chapter opener: David Parsons/National Renewable Energy Laboratory. 14.2: John D. Cunningham/Visuals Unlimited. 14.5L: ©Charles Feil/Stock, Boston, Inc./PictureQuest. 14.5R: ©1999 PhotoLink/PhotoDisc, Inc. 14.8: Yves Forestier/Corbis Sygma. 14.12: Roy David Farris/Visuals Unlimited. 14.13aF: UN/DPI Photo. 14.13b: Tim McCabe/USDA. 14.15a and b: Scott Bauer/USDA Agricultural Research Service. 14.16: Baldev/Corbis Sygma.

Chapter 15. Chapter opener: David R. Frazier Photolibrary. 15.1: David S. Addison/Visuals Unlimited. 15.4: ©1996 PhotoLink/PhotoDisc, Inc. 15.5: Tim McCabe/USDA. 15.6: Tim McCabe/USDA Natural Resources Conservation Service. 15.8: ©Dean DeChambeau. 15.9: Ken Winters/U.S. Army Corps of Engineers. 15.12: ©PhotoDisc, Inc. 15.15: ©Dean DeChambeau. 15.16: Elizabeth H. White/CDC. 15.17: Jeff Vanuga/USDA Natural Resources

Conservation Service. Case Study 2.1: NASA/GSFC/JPL, MISR Team.

Chapter 16. Chapter opener: Jeff Greenberg/Visuals Unlimited. 16.2a: Edwin P. Ewing, Jr./CDC. 16.2b: Courtesy Jacques Descloitres/MODIS Land Rapid Response Team at NASA/GSFC. 16.7: Hans Reinhard/Bruce Coleman, Inc. 16.8a: © Dean DeChambeau. 16.11: ©2000 Corbis. 16.12: ©1999 Scott T. Baxter/PhotoDisc, Inc. Case Study 2.1: Adrien Lamarre/U.S. Army Corps of Engineers.

Chapter 17. Chapter opener: Image courtesy Greg Shirah GSFC Scientific Visualization Studio/NASA. 17.3: K. Jackson/U.S. Air Force. 17.5: NASA. 17.6: ©1999 Emma Lee/Life File/PhotoDisc, Inc. 17.9: Guiseppe Zibordi/Michael Van Woert/NOAA NESDIS, ORA. 17.10: Courtesy of Florida Keys National Marine Sanctuary. 17.11: Jacques Descloitres/MODIS Land Rapid Response Team, NASA/GSFC. 17.12: C. Mayhew & R. Simmon (NASA/GSFC), NOAA/NGDC, DMSP Digital Archive. Case Study 2.1: Jack Dykinga/USDA Argricultural Research Service. Case Study: 3.1: James D. Gathany/CDC.

Chapter 18. Chapter opener: NOAA Restoration Center, NE Region. 18.4: Bill Eager/NREL. 18.5: Hank Morgan/Science Source/Photo Researchers, Inc. 18.6: Robert Etzel/U.S. Army Corp of Engineers. 18.7: Mark Warford/Greenpeace. 18.9: Bernd Wittich/Visuals Unlimited. 18.10: Lisa Bunin/Greenpeace. 18.11: EXXON VALDEZ Oil Spill Trustee Council/NOAA. Case Study 1.1: R. Calentine/Visuals Unlimited. Case Study 1.2: Mitsuhiro Wada/Gamma. Page 473: Harry Weddington/U.S. Army Corp of Engineers.

Chapter 19. Chapter opener: NREL. 19.2: Jeffrey Howe/Visuals Unlimited. 19.5a: Lewis W. Hine/Courtesy of the U.S. National Archive and Record Administration. 19.5b: J.P. Laffont/Corbis Sygma. 19.6: ©Gary Kazzanjian/AP Photo. 19.7: Ken Graham/Greenpeace. 19.9: The Granger Collection, New York. 19.12: Francis Dobbs/The World Bank. 19.13: Andrew Davis/Greenpeace Case Study 1.1: Daniel M. Kammen, University of California, Berkeley. Case Study 3.1: Barry Wong/The Seattle Times.

Chapter 20. Chapter opener: Source: "Across the Continent: Westward the Course of Empire Takes its Way." Artist: Frances F. Palmer; Publisher: Currier & Ives, 1868; Museum of the City of New York, 56.300.107. The Harry T. Peters Collection. 20.1: The Granger Collection, New York. 20.2a: The Granger Collection, New York. 20.2b: Bettmann/Corbis. 20.3: Courtesy of the Museum of Fine Arts, Boston. 20.4a: NOAA George E. Marsh Album. 20.4b: Photo by Dorothea Lange from the collection of the USDA Natural Resources Conservation Service. 20.5: NASA. 20.6: Will and Deni McIntyre/Photo Researchers, Inc. 20.7: Satoru Ohmori/Gamma. 20.8: Allan Tannenbaum. Case Study 1.1: Science VU/Visuals Unlimited.

Political Map of the World